OM NAMA SHIVA YA.
OM SAI RAM
SHREE JATSIMHA

A Course in Digital

Signal Processing

A Course in Digital Signal Processing

BOAZ PORAT

Technion, Israel Institute of Technology
Department of Electrical Engineering

JOHN WILEY & SONS, INC.

NEW YORK   CHICHESTER   BRISBANE   TORONTO   SINGAPORE

Acquisitions Editor	*Charity Robey*
Marketing Manager	*Harper Mooy*
Senior Production Editor	*Cathy Ronda*
Designer	*Karin Kincheloe*
Manufacturing Manager	*Mark Cirillo*

This book was set in Lucida Bright, and printed and bound by Hamilton Printing. The cover was printed by The Lehigh Press, Inc.

Recognizing the importance of preserving what has been written, it is a policy of John Wiley & Sons, Inc. to have books of enduring value published in the United States printed on acid-free paper, and we exert our best efforts to that end.

The paper in this book was manufactured by a mill whose forest management programs include sustained yield harvesting of its timberlands. Sustained yield harvesting principles ensure that the number of trees cut each year does not exceed the amount of new growth.

Library of Congress Cataloging-in-Publication Data

Porat, Boaz.
A course in digital signal processing / Boaz Porat.
p. cm.
Includes bibliographical references.
ISBN: 0-471-14961-6 (alk. paper)
1. Signal processing--Digital techniques.   I. Title.
TK5102.9.P66 1997
621.382'2--dc20   96-38470

Printed in the United States of America

10 9 8 7 6 5 4 3 2

To Aliza
"The first time ever ... "

To Ofer and Noga

and

In Memory of David, Tova, and Ruth Freud

The Author

**Boaz Porat** was born in Haifa, Israel, in 1945. He received the B.S. and M.S. degrees in electrical engineering from the Technion, in Haifa, Israel, in 1967 and 1975, respectively, and the M.S. degree in statistics and Ph.D. in electrical engineering from Stanford University in 1982. Since 1983, he has been with the Department of Electrical Engineering at the Technion, Haifa, where he is now a professor. He has held visiting positions at University of California at Davis, California; Yale University, New Haven, Connecticut; and Ben-Gurion University, Beer-Sheba, Israel. He also spent various periods with Signal Processing Technology, California, and served as a consultant to electronics industries in Israel on numerous occasions. He is a Fellow of the Institute of Electrical and Electronics Engineers.

Dr. Porat received the European Association for Signal Processing Award for the Best Paper of the Year in 1985; the Ray and Miriam Klein Award for Excellence in Research in 1986; the Technion's Distinguished Lecturer Award in 1989 and 1990; and the Jacknow Award for Excellence in Teaching in 1994. He was an Associate Editor of the IEEE TRANSACTIONS ON INFORMATION THEORY from 1990 to 1992, in the area of estimation. He is author of the book *Digital Processing of Random Signals: Theory and Methods,* published by Prentice Hall, and of 120 scientific papers. His research interests are in statistical signal processing, estimation, detection, and applications of digital signal processing in communications, biomedicine, and music.

The Software

The MATLAB software and the data files for this book are available by anonymous file transfer protocol (ftp) from:
`ftp.wiley.com /public/college/math/matlab/bporat`
or from:
`ftp.technion.ac.il /pub/supported/ee/Signal_processing/B_Porat`
See the file `readme.txt` for instructions.

Additional information on the book can be found on the World-Wide Web at:
`http://www-ee.technion.ac.il/~boaz`

The author welcomes comments, corrections, suggestions, questions, and any other feedback on the book; send e-mail to `boaz@ee.technion.ac.il`.

Preface

The last thing one discovers in composing a work is what to put first.

Blaise Pascal (1623–62)

This book is a text on digital signal processing, at a senior or first-year-graduate level. My purpose in writing it was to provide the reader with a precise, broad, practical, up-to-date exposition of digital signal processing. Accordingly, this book presents DSP theory in a rigorous fashion, contains a wealth of material—some not commonly found in general DSP texts, makes extensive use of MATLAB[†] software, and describes numerous state-of-the-art applications of DSP.

My students often ask me, at the first session of an undergraduate DSP course that I teach: "Is the course mathematical, or is it useful?" to which I answer: "It is both." To convince yourself that DSP is mathematical, take a moment to flip through the pages of this book. See? To convince yourself that DSP is useful, consider your favorite CD music recordings; your cellular phone; the pictures and sound you get on your computer when you connect to the Internet or use your multimedia CD-ROM software; the electronic medical instruments you might see in hospitals; radar systems used for air traffic control and for meteorology; the digital television you may have in the near future. All these rely to some extent on digital signal processing.

What does this book have to offer that other DSP texts don't? There is only one honest answer: my personal perspective on the subject and on the way it should be taught. So, here is my personal perspective, as it is reflected in the book.

1. Theory and practice should be balanced, with a slight tilt toward theory. Without theory, there is no practice. Accordingly, I always explain *why* things work before explaining *how* they work.

2. In explaining theories, accuracy is crucial. I therefore avoid cutting corners but spend the necessary time and effort to supply accurate and detailed derivations. Occasionally, there are results whose derivation is too advanced for the level of this book. In such cases, I only state the result, and alert the reader to the missing derivation.

3. Consistent notation is an indispensable part of accuracy; ambiguous notation leads to confusion. The theory of signals and systems is replete with mathematical objects that are similar, but not identical: signals in continuous and discrete time, convolutions of various kinds, Fourier transforms, Laplace transforms, z-transforms, and a host of discrete transforms. I have invested effort in developing a consistent notation for this book. Chapter 1 explains this notation in detail.

4. Examples should reflect real-life applications. Drill-type examples should not be ignored, but space should also be allocated to engineering examples. This is

---

[†]MATLAB is a registered trademark of The MathWorks, Inc., Natick, MA, U.S.A.

not easy, since the beginning student often has not been exposed to engineering reality. In constructing such examples, I have tried to be faithful to this reality, while keeping the discussion as elementary as possible.

5. The understanding of DSP algorithms can be greatly enhanced by reading a piece of software code that implements the algorithm. A software code must be accurate, otherwise it will not work. Illustrating algorithms through software codes used to be a nightmare in the old days of FORTRAN and even during the present days of the C language. Not any more! Now we have MATLAB, which is as easy to read as plain English. I therefore have made the effort to illustrate every computational procedure described in the book by a MATLAB code. The MATLAB programs are also available via the Internet from the publisher or the author, see instructions preceding this preface. Needless to say, I expect every student to be MATLAB literate.

A problem in writing a textbook for a course on DSP is that the placement of such courses in the curriculum may vary, as also the level and background assumed of the students. In certain institutes (such as the one I am in), the first DSP course is taken at a junior or senior undergraduate level, right after a signals and systems course; therefore, mostly basic material should be taught. In other institutes, DSP courses are given at a first-year graduate level. Graduate students typically have better backgrounds and wider experiences, so they can be exposed to more advanced material. In trying to satisfy both needs, I have included much more material than can be covered in a single course. A typical course should cover about two thirds of the material, but undergraduate and graduate courses should not cover the same two thirds.

I tried to make the book suitable for the practicing engineer as well. A common misconception is that "the engineer needs practice, not theory." An engineer, after a few years out of college, needs updating of the theory, whether it be basic concepts or advanced material. The choice of topics, the detail of presentation, the abundance of examples and problems, and the MATLAB programs make this book well suited to self study by engineers.

The main prerequisite for this book is a solid course on signals and systems at an undergraduate level. Modern signals and systems curricula put equal (or nearly so) emphases on continuous-time and discrete-time signals. The reader of this book is expected to know the basic mathematical theory of signals and their relationships to linear time-invariant systems: convolutions, transforms, frequency responses, transfer functions, concepts of stability, simple block-diagram manipulations, and some applications of signals and systems theory.

I use the following conventions in the book:

1. Sections not marked include basic-level material. I regard them as a must for all students taking a first DSP course. I am aware, however, that many instructors disagree with me on at least two subjects in this class: IIR filters and the FFT. Instructors who do not teach one of these two subjects (or both) can skip the corresponding chapters (10 and 5, respectively).

2. Sections marked by an asterisk include material that is either optional (being of secondary importance) or more advanced (and therefore, perhaps, more suitable for a graduate course). Advanced problems are also marked by asterisks.

3. Superscript numerals denote end notes. End notes appear at the end of the chapter, in a section named "Complements." Each end note contains, in square brackets, backward reference to the page referring to it. Most end notes are of a more advanced nature.

4. Occasionally I put short paragraphs in boxes, to emphasize their importance. For example:

> Read the Preface before you start Chapter 1!

5. Practical design procedures are highlighted; see page 284 for an example.

6. The symbol □ denotes the end of a proof (*QED*), as well as the end of an example.

7. The MATLAB programs are mentioned and explained at the points where they are needed to illustrate the material. However, the program listings are collected together in a separate section at the end of each chapter. Each program starts with a description of its function and its input–output parameters.

Here is how the material in the book is organized, and my recommendations for its usage.

1. Chapter 1, beside serving as a general introduction to the book, has two goals:

    (a) To introduce the system of notations used in the book.

    (b) To provide helpful hints concerning the use of summations.

    The first of these is a must for all readers. The second is mainly for the relatively inexperienced student.

2. Chapter 2 summarizes the prerequisites for the remainder of the book. It can be used selectively, depending on the background and level of preparation of the students. When I teach an introductory DSP course, I normally go over the material in one session, and assign part of it for self reading. The sections on random signals may be skipped if the instructor does not intend to teach anything related to random signals in the course. The section on real Fourier series can be skipped if the instructor does not intend to teach the discrete cosine transform.

3. Chapter 3 concerns sampling and reconstruction. These are the most fundamental operations of digital signal processing, and I always teach them as the first subject. Beside the basic-level material, the chapter contains a rather detailed discussion of physical sampling and reconstruction, which the instructor may skip or defer until later.

4. Chapter 4 is the first of three chapters devoted to frequency-domain analysis of discrete-time signals. It introduces the discrete Fourier transform (DFT), as well as certain related concepts (circular convolution, zero padding). It also introduces the geometric viewpoint on the DFT (orthonormal basis decomposition). Also introduced in this chapter is the discrete cosine transform (DCT). Because of its importance in DSP today, I have decided to include this material, although it is not traditionally taught in introductory courses. I included all four DCT types for completeness, but the instructor may choose to teach only type II, which is the most commonly used, and type III, its inverse.

5. Chapter 5 is devoted to the fast Fourier transform (FFT). Different instructors feel differently about this material. Some pay tribute to its practical importance by teaching it in considerable detail, whereas some treat it as a black box, whose details should be of interest only to specialists. I decided to present the Cooley–Tukey algorithms in detail, but omit other approaches to the FFT. The way I teach FFT is unconventional: Instead of starting with the binary case, I start with the general Cooley–Tukey decomposition, and later specialize to the binary case. I regard this as a fine example of a general concept being simpler than its special cases, and I submit to the instructor who challenges this approach to try it once.

This chapter also includes a few specialized topics: the overlap-add method of linear convolution, the chirp Fourier transform, and the zoom FFT. These are, perhaps, more suitable for a graduate course.

6. Chapter 6 is concerned with practical aspects of spectral analysis, in particular with short-time spectral analysis. It starts by introducing windows, the working tool of spectral analysis. It then discusses in detail the special, but highly important, problem of the measurement of sinusoidal signals. I regard these two topics, windows and sinusoid measurements, as a must for every DSP student. The last topic in this chapter is estimation of sinusoids in white noise. I have included here some material rarely found in DSP textbooks, such as detection threshold and the variance of frequency estimates based on a windowed DFT.

7. Chapter 7 provides the preliminary background material for the second part of the book, the part dealing with digital filters. It introduces the z-transform and its relationship to discrete-time, linear time-invariant (LTI) systems. The z-transform is usually taught in signals and systems courses. However, my experience has shown that students often lack this background. The placement of this material in this book is unconventional: in most books it appears in one of the first chapters. I have found that, on the one hand, the material on z-transforms is not needed until one begins to study digital filters; on the other hand, this material is not elementary, due to its heavy dependence on complex function theory. Teaching it within the middle of an introductory course, exactly at the point where it is needed, and after the student has developed confidence and maturity in frequency-domain analysis, has many pedagogical advantages. As in other books, the emphasis is on the two-sided transform, whereas the one-sided z-transform is mentioned only briefly.

8. Chapter 8 serves as an introduction to the subject of digital filters. It contains a mixture of topics, not tightly interrelated. First, it discusses the topic of filter types (low pass, high pass, etc.) and specifications. Next, it discusses in considerable detail, the phase response of digital filters. I decided to include this discussion, since it is missing (at least at this level of detail) from many textbooks. It represents, perhaps, more than the beginning student needs to know, but is suitable for the advanced student. However, the concept of linear phase and the distinction between constant phase delay and constant group delay should be taught to all students. The final topic in this chapter is an introductory discussion of digital filter design, concentrating on the differences between IIR and FIR filters.

9. Chapters 9 and 10 are devoted to FIR and IIR filters, respectively. I spent time trying to decide whether to put FIR before IIR or vice versa. Each of the two choices has its advantages and drawbacks. I finally opted for FIR first, for the following reasons: (1) this way there is better continuity between the discussion on linear phase in Chapter 8 and the extended discussion on linear phase in FIR filters at the beginning of Chapter 9; (2) there is also better continuity between Chapters 10 and 11; (3) since FIR filters appear more commonly than IIR filters in DSP applications, some instructors may choose to teach only FIR filters, or mention IIR filters only briefly. An introductory course that omits IIR is most likely to omit Chapters 11 and 12 as well. This enables the instructor to conveniently end the course syllabus with Chapter 9.

The chapter on FIR filters contains most of what is normally taught on this subject, except perhaps design by frequency sampling. Design by windows is explained in detail, as well as least-squares design. Equiripple design is covered, but in less detail than in some books, since most engineers in need of equiripple filters would have to rely on canned software anyway.

The chapter on IIR filters starts with low-pass analog filter design. Butterworth and Chebyshev filters are suitable for a basic course, whereas elliptic filters should be left to an advanced course. Analog filters, other than low pass, are constructed through frequency transformations. The second half of the chapter discusses methods for transforming an analog filter to the digital domain. Impulse invariant and backward difference methods are included for completeness. The bilinear transform, on the other hand, is a must.

10. Chapter 11 represents the next logical step in digital filter design: constructing a realization from the designed transfer function and understanding the properties of different realizations. Certain books treat digital system realizations before they teach digital filters. It is true that realizations have uses other than for digital filters, but for the DSP student it is the main motivation for studying them.

I decided to include a brief discussion of state space, following the material on realizations. Beside being a natural continuation of the realization subject, state space has important uses for the DSP engineer: impulse response and transfer function computations, block interconnections, simulation, and the like. I realize, however, that many instructors will decide not to teach this material in a DSP course.

The bulk of Chapter 11 is devoted to finite word length effects: coefficient quantization, scaling, computation noise, and limit cycles. Much of the material here is more for reference than for teaching. In a basic course, this material may be skipped. In an advanced course, selected parts can be taught according to the instructor's preferences.

11. Chapter 12 concerns multirate signal processing. This topic is usually regarded as specialized and is seldom given a chapter by itself in general DSP textbooks (although there are several books completely devoted to it). I believe that it should be included in general DSP courses. The chapter starts with elementary material, in particular: decimation, interpolation, and sampling-rate conversion. It then moves on to polyphase filters and filter banks, subjects better suited to a graduate course.

12. Chapter 13 is devoted to the analysis and modeling of random signals. It first discusses nonparametric spectrum estimation techniques: the periodogram, the averaged (Welch) periodogram, and the smoothed (Blackman–Tukey) periodogram. It then introduces parametric models for random signals and treats the autoregressive model in detail. Finally, it provides a brief introduction to Wiener filtering by formulating and solving the simple FIR case. The extent of the material here should be sufficient for a general graduate DSP course, but not for a specialized course on statistical signal processing.

13. Chapter 14 represents an attempt to share my excitement about the field with my readers. It includes real-life applications of DSP in different areas. Each application contains a brief introduction to the subject, presentation of a problem to be solved, and its solution. The chapter is far from being elementary; most beginning students and a few advanced ones may find it challenging on first reading. However, those who persist will gain (I hope) better understanding of what DSP is all about.

Many people helped me to make this book a better one—Guy Cohen, Orli Gan, Isak Gath, David Malah, Nimrod Peleg, Leonid Sandomirsky, Adam Shwartz, David Stanhill, Virgil Stokes, Meir Zibulsky—read, found errors, offered corrections, criticized, enlightened. Benjamin Friedlander took upon himself the tedious and unrewarding task of teaching from a draft version of the book, struggling with the rough edges and helping smooth them, offering numerous suggestions and advice. Shimon Peleg read the book with the greatest attention imaginable; his detailed feedback on almost every page greatly improved the book. Simon Haykin was instrumental in having this book accepted for publication, and gave detailed feedback both on the early draft and later. William J. Williams and John F. Doherty reviewed the book and made many helpful suggestions. Irwin Keyson, Marge Herman, and Lyn Dupré, through her excellent book [Dupré, 1995], helped me improve my English writing. Brenda Griffing meticulously copyedited the book. Aliza Porat checked the final manuscript. Ezra Zeheb provided me with Eliahu Jury's survey on the development of the z-transform. James Kaiser helped me trace the original reference to the Dolph window. Thomas Barnwell kindly permitted me to quote his definition of digital signal processing; see page 1. Steven Elliot, the former acquisition editor at Wiley, and Charity Robey, who took over later, gave me a lot of useful advice. Jennifer Yee, Susanne Dwyer, and Paul Constantine at Wiley provided invaluable technical assistance. Yehoshua Zeevi, chairman of the Department of Electrical Engineering at the Technion, allowed me to devote a large part of my time to writing during 1996. Yoram Or-Chen provided moral support. Toshiba manufactured the T4800CT notebook computer, Y&Y, Inc. provided the LATEX software, and Adobe Systems, Inc. created PostScript. Ewan MacColl wrote the song and Gordon Lightfoot and the Kingston Trio (among many others) sang it. I thank you all.

I try never to miss an opportunity to thank my mentors, and this is such an opportunity: Thank you, Tom Kailath and Martin Morf, for changing my course from control systems to signal processing and, indirectly, from industry to academia. If not for you, I might still be closing loops today! And thank you, Ben, for expanding my horizons in so many ways and for so many years.

And finally, to Aliza: The only regret I may have for writing this book is that the hours I spent on it, I could have spent with you!

Boaz Porat

Haifa, August 1996

Contents

Symbols and Abbreviations

Symbols

1. The symbols are given in an alphabetical order. Roman symbols are given first, followed by Greek symbols, and finally special symbols.

2. Page numbers are the ones in which the symbol is either defined or first mentioned. Symbols for which there are no page numbers are used throughout the book.

3. Section 1.2 explains the system of notation in detail.

Symbol	Meaning	Page
$a_1, \ldots, a_p$	denominator coefficients of a difference equation	214
$\boldsymbol{a}_i$	solution of ith-order Yule–Walker equation	527
$\boldsymbol{b}_i$	solution of the ith-order Wiener equation	538
$\tilde{\boldsymbol{a}}_i$	the vector $\boldsymbol{a}_i$ in reversed order	527
$A(\theta)$	amplitude function of a digital filter	256
A_p	pass-band ripple	246
A_s	stop-band attenuation	247
$\mathcal{A}_\mathrm{c}(N)$	number of complex additions in FFT	138
$\mathcal{A}_\mathrm{r}(N)$	number of real additions in FFT	141
$\boldsymbol{A}, \boldsymbol{B}, \boldsymbol{C}, \boldsymbol{D}$	state-space matrices	403
$a(z), b(z)$	denominator and numerator polynomials of a rational transfer function	215
adj	adjugate of a matrix	407
$b_0, \ldots, b_q$	numerator coefficients of a difference equation	214
B	number of bits (Chapter 11)	
$\mathbb{C}$	the complex plane	
$\boldsymbol{C}_N$	the discrete cosine transform matrix	114
CG	coherent gain of a window	187
d	discrimination factor	329
D	duration of a signal	189
$\mathcal{D}$	discrete Fourier transform (DFT) operator	94
$D(\theta, N)$	Dirichlet kernel	167
det	determinant of a matrix	407
e	$\sum_{n=0}^{\infty} \frac{1}{n!}$	
$e[n]$	quantization noise in a digital filter	429
e_i, f_i	coefficients in parallel realization	397
$E(\cdot)$	expectation	
$\exp\{a\}$	e^a	
f	continuous-time frequency	

$\text{Si}(x)$	sine integral	292
$\text{sign}(t)$	the sign function	37
$\text{sinc}(t)$	the sinc function	16
$\text{sn}(u, m)$	Jacobi elliptic sine function	343
SNR_i	input signal-to-noise ratio	188
SNR_o	output signal-to-noise ratio	188
t	time	
T	sampling interval	
$\boldsymbol{T}$	state-space transformation	405
$T_N(x)$	Chebyshev polynomial	333
$u(\phi, m)$	elliptic integral of the first kind	343
$w[n]$	a window	168
$w_\text{b}[n]$	Blackman window	174
$w_\text{d}[n]$	Dolph window	175
$w_\text{hm}[n]$	Hamming window	172
$w_\text{hn}[n]$	Hann window	171
$w_\text{k}[n]$	Kaiser window	175
$w_\text{r}[n]$	rectangular window	165
$w_\text{t}[n]$	triangular (Bartlett) window	169
W_N	$\exp\{j2\pi/N\}$	94
$x[n], y[n]$, etc.	discrete-time signals	
$x_{(\downarrow M)}[n]$	M-fold decimation	464
$x_{(\uparrow L)}[n]$	L-fold expansion	464
$x(t), y(t)$, etc.	continuous-time signals	
$\tilde{x}(t), \tilde{x}[n]$	periodic extension of a signal	6
$x(t_0^+)$	limit from the right of $x(t)$ at t_0	
$x(t_0^-)$	limit from the left of $x(t)$ at t_0	
$x_\text{p}(t)$	impulse sampling of $x(t)$	46
$X^\text{c1}[k]$	type-I discrete cosine transform (DCT-I)	116
$X^\text{c2}[k]$	type-II discrete cosine transform (DCT-II)	118
$X^\text{c3}[k]$	type-III discrete cosine transform (DCT-III)	118
$X^\text{c4}[k]$	type-IV discrete cosine transform (DCT-IV)	119
$X^\text{d}[k]$	discrete Fourier transform (DFT)	94
$X^\text{f}(\theta)$	discrete-time Fourier transform	27
$X^\text{h}[k]$	discrete Hartley transform	131
$X^\text{s1}[k]$	type-I discrete sine transform (DST-I)	120
$X^\text{s2}[k]$	type-II discrete sine transform (DST-II)	120
$X^\text{s3}[k]$	type-III discrete sine transform (DST-III)	120
$X^\text{s4}[k]$	type-IV discrete sine transform (DST-IV)	120
$X^\text{z}(z)$	z-transform	206
$X_+^\text{z}(z)$	unilateral z-transform	226
$X^\text{F}(\omega)$	continuous-time Fourier transform	11
$X^\text{L}(s)$	Laplace transform	7
$X^\text{S}[k]$	Fourier series coefficients	17
$y_\text{zir}[n]$	zero-input response of a difference equation	228
$y_\text{zsr}[n]$	zero-state response of a difference equation	228
z	z-transform variable	
$\mathbb{Z}$	the set of integers	
$\mathcal{Z}$	z-transform operator	206

$\alpha_1, \ldots, \alpha_p$	poles of a rational transfer function	216
α_r, α_i	real and imaginary parts of the pole α	401
$\beta_1, \ldots, \beta_r$	zeros of a rational transfer function	216
γ_x	variance of the random variable x	21
$\gamma_{x,y}$	covariance of the random variables x, y	22
$\Gamma_w(x)$	integral of an amplitude function of a window	294
$\delta^+, \delta^-, \delta_p$	pass-band tolerances	246
δ_s	stop-band tolerance	246
$\delta(t)$	the Dirac delta (impulse) function	12
$\delta[n]$	the unit sample (discrete-time impulse)	29
ε	parameter for Chebyshev and elliptic filters	335
ζ	angle of a complex zero	282
θ	discrete-time angular frequency	
θ_p	pass-band edge frequency	246
θ_s	stop-band edge frequency	246
$\kappa_x(\tau)$	covariance function of a continuous-time random signal $x(t)$	22
$\kappa_x[m]$	covariance sequence of a discrete-time random signal $x[n]$	30
$\hat{\kappa}_x[m]$	estimated covariance	521
$\kappa_{yx}[m]$	cross-covariance sequence of two discrete-time random signals $x[n], y[n]$	537
λ_k	scaling parameters in filter realizations	424
$\boldsymbol{\lambda}_i$	right side of the ith-order Wiener equation	538
$\Lambda(\cdot)$	attenuation function	329
μ_x	mean of the random variable x	21
$\hat{\mu}_x$	estimated mean	521
$\boldsymbol{\xi}_p$	vector used in least-squares AR modeling	536
Ξ_p	matrix used in least-squares AR modeling	536
π	the ratio between the circumference and the diameter of a circle	
ρ_i	partial correlations, or reflection coefficients	528
σ_x	standard deviation of the random variable x	21
τ_g	group delay of a digital filter	259
τ_p	phase delay of a digital filter	258
$\phi(\theta)$	continuous phase of a digital filter	256
$\psi(\theta)$	phase response of a digital filter	253
$\upsilon(t)$	the continuous-time unit-step function	48
$\upsilon[n]$	the discrete-time unit-step function	48
ω	continuous-time angular frequency	
ω_{3db}	$-3\,dB$ angular frequency	329
$\omega_c, \omega_l, \omega_h$	parameters in frequency transformations	347
(a,b)	the set $\{x : a < x < b\}$ (open interval)	
$(a,b]$	the set $\{x : a < x \le b\}$ (left semiopen interval)	
$[a,b)$	the set $\{x : a \le x < b\}$ (right semiopen interval)	
$[a,b]$	the set $\{x : a \le x \le b\}$ (closed interval)	
$\bar{x}$	complex conjugate of x	
$\Im\{x\}$	imaginary part of the complex number x	
$\Re\{x\}$	real part of the complex number x	

$\sphericalangle x$	angle of the complex number x	
$\lvert x \rvert$	absolute value of x	
$\lVert x \rVert$	norm of x (Chapter 11)	
$\boldsymbol{x}'$	transpose of the vector (or the matrix) $\boldsymbol{x}$	
$\lfloor x \rfloor$	the floor of x (rounding downward)	
$\lceil x \rceil$	the ceiling of x (rounding upward)	
$\langle x \rangle$	empirical mean of x	
$1(t)$	the DC function	14
$\stackrel{\triangle}{=}$	equality by definition	
$*$	convolution (in continuous or discrete time)	6
$\circledast$	circular convolution (in discrete time)	108
$\star$	correlation (in continuous or discrete time)	41
$\cup$	set union	
$\cap$	set intersection	

Abbreviations

Abbreviation	Meaning	Page
A/D	analog-to-digital (converter)	65
AM	amplitude modulation	42
AR	autoregressive (model or signal)	523
ARMA	autoregressive moving-average (model or signal)	523
BIBO	bounded-input, bounded-output	213
BP	band-pass (filter)	249
BPSK	binary phase-shift keying	75
BS	band-stop (filter)	250
CQF	conjugate quadrature filter (bank)	490
D/A	digital-to-analog (converter)	63
DC	direct current (function)	14
DCT	discrete cosine transform	114
DFT	discrete Fourier transform	93
DHT	discrete Hartley transform	131
DSB	double-side-band (modulation)	38
DSP	digital signal processing	2
DST	discrete sine transform	120
ECG	electrocardiogram	580
EEG	electroencephalogram	21
FFT	fast Fourier transform	133
FIR	finite impulse response (filter)	244
GLP	generalized linear phase	259
GSM	Groupe Special Mobile	550
HP	high-pass (filter)	249
IDFT	inverse discrete Fourier transform	97
IF	intermediate frequency	73
IIR	infinite impulse response (filter)	244
IRT	impulse response truncation	284
ISI	intersymbol interference	539
ISO	International Standards Organization	503

LAR	log–area ratio	562
LP	low-pass (filter)	246
LPC	linear predictive coding	556
LSB	least significant bit	412
LTI	linear time-invariant (system)	13
MA	moving-average (model or signal)	523
MAC	multiplier–accumulator	583
MMSE	minimum mean-square error	525
MOS	mean opinion score	493
MPEG	Moving Picture Expert Group	495
MSB	most significant bit	67
NRZ	non–return to zero	75
OLA	overlap-add (convolution)	149
OQPSK	offset quadrature phase-shift keying	517
PAM	pulse amplitude modulation	539
PSD	power spectral density	24
QMF	quadrature mirror filter (bank)	489
RAM	random-access memory	585
RCSR	real, causal, stable, rational (filter)	253
RMS	root mean square	191
ROC	region of convergence	206
ROM	read-only memory	585
S/H	sample and hold	65
SISD	single-instruction, single-data	582
SISO	single-input, single-output (system)	13
SLL	side-lobe level	189
SNR	signal-to-noise ratio	188
SSB	single-side-band (modulation)	274
WSS	wide-sense stationary (signal)	13
ZOH	zero-order hold	59

Chapter 1

Introduction

Digital Signal Processing:

That discipline which has allowed us to replace a circuit previously composed of a capacitor and a resistor with two antialiasing filters, an A-to-D and a D-to-A converter, and a general purpose computer (or array processor) so long as the signal we are interested in does not vary too quickly.

Thomas P. Barnwell, 1974

Signals encountered in real life are often in continuous time, that is, they are waveforms (or functions) on the real line. Their amplitude is usually continuous as well, meaning that it can take any real value in a certain range. Signals continuous in time and amplitude are called *analog signals*. There are many kinds of analog signals appearing in various applications. Examples include:

1. Electrical signals: voltages, currents, electric fields, magnetic fields.

2. Mechanical signals: linear displacements, angles, velocities, angular velocities, forces, moments.

3. Acoustic signals: vibrations, sound waves.

4. Signals related to physical sciences: pressures, temperatures, concentrations.

Analog signals are converted to voltages or currents by *sensors*, or *transducers*, in order to be processed electrically. Analog signal processing involves operations such as amplification, filtering, integration, and differentiation, as well as various forms of nonlinear processing (squaring, rectification). Analog processing of electrical signals is typically based on electronic amplifiers, resistors, capacitors, inductors, and so on. Limitations and drawbacks of analog processing include:

1. Accuracy limitations, due to component tolerances, amplifier nonlinearity, biases, and so on.

2. Limited repeatability, due to tolerances and variations resulting from environmental conditions, such as temperature, vibrations, and mechanical shocks.

3. Sensitivity to electrical noise, for example, internal amplifier noise.

4. Limited dynamic range of voltages and currents.

5. Limited processing speeds due to physical delays.

6. Lack of flexibility to specification changes in the processing functions.

7. Difficulty in implementing nonlinear and time-varying operations.

8. High cost and accuracy limitations of storage and retrieval of analog information.

Digital signal processing (DSP) is based on representing signals by numbers in a computer (or in specialized digital hardware), and performing various numerical operations on these signals. Operations in digital signal processing systems include, but are not limited to, additions, multiplications, data transfers, and logical operations. To implement a DSP system, we must be able:

1. To convert analog signals into digital information, in the form of a sequence of binary numbers. This involves two operations: sampling and analog-to-digital (A/D) conversion.

2. To perform numerical operations on the digital information, either by a computer or special-purpose digital hardware.

3. To convert the digital information, after being processed, back to an analog signal. This again involves two operations: digital-to-analog (D/A) conversion and reconstruction.

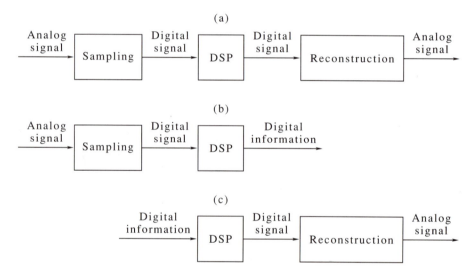

Figure 1.1 Basic DSP schemes: (a) general signal processing system; (b) signal analysis system; (c) signal synthesis system.

There are four basic schemes of digital signal processing, as shown in Figure 1.1:

1. A general DSP system is shown in part a. This system accepts an analog input signal, converts it to a digital signal, processes it digitally, and converts it back to analog. An example of such a system is digital recording and playback of music. The music signal is sensed by microphones, amplified, and converted to digital. The digital processor performs such tasks as filtering, mixing, and reverberation control. Finally, the digital music signal is converted back to analog, in order to be played back by a sound system.

2. A signal analysis system is shown in part b. Such systems are used for applications that require us to extract only certain information from the analog signal. As an example, consider the Touch-Tone system of telephone dialing. A Touch-Tone

dial includes 12 buttons arranged in a 4×3 matrix. When we push a button, two sinusoidal signals (tones) are generated, determined by the row and column numbers of the button. These two tones are added together and transmitted through the telephone lines. A digital system can identify which button was pressed by determining the frequencies of the two tones, since these frequencies uniquely identify the button. In this case, the output information is a number between 1 and 12.

3. A signal synthesis system is shown in part c. Such systems are used when we need to generate an analog signal from digital information. As an example, consider a text-to-speech system. Such a system receives text information character by character, where each character is represented by a numerical code. The characters are used for constructing syllables; these are used for generating artificial digital sound waveforms, which are converted to analog in order to be played back by a sound system.

4. A fourth type of DSP system is purely digital, accepting and yielding digital information. Such a system can be regarded as a degenerate version of any of the three aforementioned types.

As we see, Thomas Barnwell's definition of DSP (quoted in the beginning of the chapter), although originally meant as ironic, is essentially correct today as it was when first expressed, in the early days of DSP. However, despite the relative complexity of DSP systems, there is much to gain for this complexity. Digital signal processing has the potential of freeing us from many limitations of analog signal processing. In particular:

1. Computers can be made accurate to any desired degree (at least theoretically), by choosing their word length according to the required accuracy. Double precision can be used when single precision is not sufficient, or even quadruple precision, etc.

2. Computers are perfectly repeatable, as long as they do not malfunction (due to either hardware or software failure).

3. The sensitivity of computers to electrical noise is extremely low (but not nil, as is commonly believed; electrical noise can give rise to bit errors, although rarely).

4. Use of floating point makes it possible, by choosing the word length, to have a practically infinite dynamic range.

5. Speed is a limiting factor in computers as well as in analog devices. However, advances in technology (greater CPU and memory speeds, parallel processing) push this limit forward continually.

6. Changes in processing functions can be made through programming. Although programming (or software development in general) is usually a difficult task, its implementation (by loading the new software into the computer storage devices) is relatively easy.

7. Implementing nonlinear and time-varying operations (e.g., in adaptive filtering) is conceptually easy, since it can be accomplished via programming, and there is usually no need to build special hardware.

8. Digital storage is cheap and flexible.

9. Digital information can be encrypted for security, coded against errors, and compressed to reduce storage and transmission costs.

Digital signal processing is not free of drawbacks and limitations of its own:

1. Sampling inevitably leads to loss of information. Although this loss can be minimized by careful sampling, it cannot be completely avoided.

2. A/D and D/A conversion hardware may be expensive, especially if great accuracy and speed are required. It is also never completely free of noise and distortions.

3. Although hardware becomes cheaper and more sophisticated every year, this is not necessarily true for software. On the contrary, software development and testing appear more and more often to be the main bottleneck in developing digital signal processing applications (and in the digital world in general).

4. In certain applications, notably processing of RF signals, digital processing still cannot meet speed requirements.

The theoretical foundations of digital signal processing were laid by Jean Baptiste Joseph Fourier who, in 1807, presented to the Institut de France a paper on what we call today *Fourier series*.[1] Major theoretical developments in digital signal processing theory were made in the 1930s and 1940s by Nyquist and Shannon, among others (in the context of digital communication), and by the developers of the z-transform (notably Zadeh and Ragazzini in the West, and Tsypkin in the East). The history of applied digital signal processing (at least in the electrical engineering world) began around the mid-1960s with the invention of the fast Fourier transform (FFT). However, its rapid development started with the advent of microprocessors in the 1970s. Early DSP systems were designed mainly to replace existing analog circuitry, and did little more than mimicking the operation of analog signal processing systems. It was gradually realized that DSP has the potential for performing tasks impractical or even inconceivable to perform by analog means. Today, digital signal processing is a clear winner over analog processing. Whereas analog processing is—and will continue to be—limited by technology, digital processing appears to be limited only by our imagination.[2]

We cannot do justice to all applications of DSP in this short introduction, but we name a few of them without details:

Biomedical applications: analysis of biomedical signals, diagnosis, patient monitoring, preventive health care, artificial organs.

Communication: encoding and decoding of digital communication signals, detection, equalization, filtering, direction finding.

Digital control: servomechanism, automatic pilots, chemical plants.

General signal analysis: spectrum estimation, parameter estimation, signal modeling, signal classification, signal compression.

Image processing: filtering, enhancement, coding, compression, pattern recognition.

Instrumentation: signal generation, filtering.

Multimedia: generation, storage, and transmission of sound, still images, motion pictures, digital TV, video conferencing.

Music applications: recording, playback and manipulation (mixing, special effects), synthesis of digital music.

Radar: radar signal filtering, target detection, position and velocity estimation, tracking, radar imaging.

Sonar: similar to radar.

Speech applications: noise filtering, coding, compression, recognition, synthesis of artificial speech.

Telephony: transmission of information in digital form via telephone lines, modem technology, cellular phones.

Implementation of digital signal processing varies according to the application. Off-line or laboratory-oriented processing is usually done on general purpose computers using high-level software (such as C, or more recently MATLAB). On-line or field-oriented processing is usually performed with microprocessors tailored to DSP applications. Applications requiring very high processing speeds often use special-purpose very-large-scale integration (VLSI) hardware.

1.1 Contents of the Book

Teaching of digital signal processing begins at the point where a typical signals and systems course ends. A student who has learned signals and systems knows the basic mathematical theory of signals and their relationships to linear time-invariant systems: convolutions, transforms, frequency responses, transfer functions, concepts of stability, simple block-diagram manipulations, and more. Modern signals and systems curricula put equal emphases on continuous-time and discrete-time signals. Chapter 2 reviews this material to the extent needed as a prerequisite for the remainder of the book. This chapter also contains two topics less likely to be included in a signals and systems course: real Fourier series (also called Fourier cosine and sine series) and basic theory of random signals.

Sampling and reconstruction are introduced in Chapter 3. Sampling converts a continuous-time signal to a discrete-time signal, reconstruction performs the opposite conversion. When a signal is sampled it is irreversibly distorted, in general, preventing its exact restoration in the reconstruction process. Distortion due to sampling is called *aliasing*. Aliasing can be practically eliminated under certain conditions, or at least minimized. Reconstruction also leads to distortions due to physical limitations on realizable (as opposed to ideal) reconstructors. These subjects occupy the main part of the chapter.

Chapter 3 also includes a section on physical aspects of sampling: digital-to-analog and analog-to-digital converters, their operation, implementation, and limitations.

Three chapters are devoted to frequency-domain digital signal processing. Chapter 4 introduces the discrete Fourier transform (DFT) and discusses in detail its properties and a few of its uses. This chapter also teaches the discrete cosine transform (DCT), a tool of great importance in signal compression. Chapter 5 concerns the fast Fourier transform (FFT). Chapter 6 is devoted to practical aspects of frequency-domain analysis. It explains the main problems in frequency-domain analysis, and teaches how to use the DFT and FFT for solving these problems. Part of this chapter assumes knowledge of random signals, to the extent reviewed in Chapter 2.

Chapter 7 reviews the z-transform, difference equations, and transfer functions. Like Chapter 2, it contains only material needed as a prerequisite for later chapters.

Three chapters are devoted to digital filtering. Chapter 8 introduces the concept of filtering, filter specifications, magnitude and phase properties of digital filters, and review of digital filter design. Chapters 9 and 10 discuss the two classes of digital filters: finite impulse response (FIR) and infinite impulse response (IIR), respectively. The focus is on filter design techniques and on properties of filters designed by different techniques.

The last four chapters contain relatively advanced material. Chapter 11 discusses filter realizations, introduces state-space representations, and analyses finite word

length effects. Chapter 12 deals with multirate signal processing, including an introduction to filter banks. Chapter 13 concerns the analysis and modeling of random
signals. Finally, Chapter 14 describes selected applications of digital signal processing: compression, speech modeling, analysis of music signals, digital communication,
analysis of biomedical signals, and special DSP hardware.

1.2 Notational Conventions

In this section we introduce the notational conventions used throughout this book.
Some of the concepts should be known, whereas others are likely to be new. All concepts will be explained in detail later; the purpose of this section is to serve as a
convenient reference and reminder.

1. **Signals**

    (a) We denote the real line by $\mathbb{R}$, the complex plane by $\mathbb{C}$, and the set of integers
        by $\mathbb{Z}$.

    (b) In general, we denote temporal signals by lowercase letters.

    (c) Continuous-time signals (i.e., functions on the real line) are denoted with
        their arguments in round parentheses; for example: $x(t)$, $y(t)$, and so on.

    (d) Discrete-time signals (i.e., sequences, or functions on the integers) are de-
        noted with their arguments in square brackets; for example: $x[n]$, $y[n]$, and
        so on.

    (e) Let the continuous-time signal $x(t)$ be defined on the interval $[0, T]$. Its
        periodic extension on the real line is denoted as

        $$\tilde{x}(t) = x(t \bmod T).$$

    (f) Let the discrete-time signal $x[n]$ be defined for $0 \leq n \leq N - 1$. Its periodic
        extension on the integers is denoted by

        $$\tilde{x}[n] = x[n \bmod N].$$

2. **Convolutions**

    (a) The *convolution* of continuous-time signals is (whenever the right side exists)

        $$\{x * y\}(t) = \int_{-\infty}^{\infty} x(\tau)y(t - \tau)d\tau, \quad t \in \mathbb{R}. \tag{1.1}$$

        Convolution is an operator acting on a pair of continuous-time signals x and
        y, and producing a continuous-time signal $x * y$, whose value at time t is
        $\{x * y\}(t)$.

    (b) The *discrete-time convolution* of discrete-time signals is (whenever the right
        side exists)

        $$\{x * y\}[n] = \sum_{m=-\infty}^{\infty} x[m]y[n - m], \quad n \in \mathbb{Z}. \tag{1.2}$$

        Discrete-time convolution is an operator acting on a pair of discrete-time
        signals x and y, and producing a discrete-time signal $x * y$, whose value at
        time n is $\{x * y\}[n]$.

    (c) The *circular convolution* of finite-duration, discrete-time signals
        $\{x[n], y[n], 0 \leq n \leq N - 1\}$ is

        $$\{x \circledast y\}[n] = \sum_{m=0}^{N-1} x[m]y[(n - m) \bmod N], \quad 0 \leq n \leq N - 1. \tag{1.3}$$

Circular convolution can be regarded as an operator acting on a pair of N-dimensional vectors x and y, and producing an N-dimensional vector $x \circledcirc y$, whose nth component is $\{x \circledcirc y\}[n]$.

3. **Transforms**

For transforms we use two types of notation: a script letter to denote the *transform operator*, and an uppercase letter modified by a superscript to denote the resulting function. The superscript specifies the transform in question. An uppercase superscript indicates that the transformed signal is in continuous time, whereas a lowercase superscript indicates that the transformed signal is in discrete time.

(a) The *two-sided Laplace transform* of the continuous-time signal $x(t)$ is[3]

$$X^{\mathrm{L}}(s) = \{\mathcal{L}x\}(s) = \int_{-\infty}^{\infty} x(t)e^{-st}dt, \quad s \in \mathbb{C}, \tag{1.4}$$

whenever the right side exists.

(b) The *Fourier transform* of the continuous-time signal $x(t)$ is

$$X^{\mathrm{F}}(\omega) = \{\mathcal{F}x\}(\omega) = \int_{-\infty}^{\infty} x(t)e^{-j\omega t}dt, \quad \omega \in \mathbb{R}, \tag{1.5}$$

whenever the right side exists. The inverse relationship is

$$x(t) = \{\mathcal{F}^{-1}X^{\mathrm{F}}\}(t) = \frac{1}{2\pi} \int_{-\infty}^{\infty} X^{\mathrm{F}}(\omega)e^{j\omega t}d\omega, \quad t \in \mathbb{R}. \tag{1.6}$$

We have, when both Laplace and Fourier transforms exist,

$$X^{\mathrm{F}}(\omega) = X^{\mathrm{L}}(j\omega). \tag{1.7}$$

(c) The *Fourier series* of the continuous-time signal $x(t)$ on the interval $[0, T]$ (or of its periodic extension) is

$$X^{\mathrm{S}}[k] = \{Sx\}[k] = \frac{1}{T} \int_0^T x(t) \exp\left(-\frac{j2\pi kt}{T}\right) dt, \quad k \in \mathbb{Z}. \tag{1.8}$$

The inverse relationship is

$$x(t) = \{S^{-1}X^{\mathrm{S}}\}(t) = \sum_{k=-\infty}^{\infty} X^{\mathrm{S}}[k] \exp\left(\frac{j2\pi kt}{T}\right), \quad 0 \le t \le T. \tag{1.9}$$

(d) The *two-sided z-transform* of the discrete-time signal $x[n]$ is

$$X^{\mathrm{z}}(z) = \{\mathcal{Z}x\}(z) = \sum_{n=-\infty}^{\infty} x[n]z^{-n}, \quad z \in \mathbb{C}, \tag{1.10}$$

whenever the right side exists.

(e) The *Fourier transform* of the discrete-time signal $x[n]$ is

$$X^{\mathrm{f}}(\theta) = \{\mathcal{F}x\}(\theta) = \sum_{n=-\infty}^{\infty} x[n]e^{-j\theta n}, \quad \theta \in \mathbb{R}, \tag{1.11}$$

whenever the right side exists. The inverse relationship is

$$x[n] = \{\mathcal{F}^{-1}X^{\mathrm{f}}\}[n] = \frac{1}{2\pi} \int_{-\pi}^{\pi} X^{\mathrm{f}}(\theta)e^{j\theta n}d\theta, \quad n \in \mathbb{Z}. \tag{1.12}$$

We have, when both z- and Fourier transforms exist,

$$X^{\mathrm{f}}(\theta) = X^{\mathrm{z}}(e^{j\theta}). \tag{1.13}$$

(f) The *discrete Fourier transform* of the finite-duration, discrete-time signal $\{x[n], \ 0 \le n \le N - 1\}$ is

$$X^{\mathrm{d}}[k] = \{\mathcal{D}x\}[k] = \sum_{n=0}^{N-1} x[n] \exp\left(-\frac{j2\pi kn}{N}\right), \quad 0 \le k \le N - 1. \tag{1.14}$$

The inverse relationship is

$$x[n] = \{\mathcal{D}^{-1}X^{\mathrm{d}}\}[n] = \frac{1}{N}\sum_{k=0}^{N-1} X^{\mathrm{d}}[k]\exp\left(\frac{j2\pi kn}{N}\right), \quad 0 \le n \le N - 1. \quad (1.15)$$

1.3 Summation Rules

The notation

$$\sum_{n=1}^{N} a[n] = a[1] + a[2] + \cdots + a[N]$$

should be familiar to readers of this book. Digital signal processing uses the summation notation extensively. We therefore present here a few rules for proper handling of summations.

1. In the sum

$$\sum_{n=n_1}^{n_2} a[n,k],$$

the variables n and k play completely different roles. The variable n is *dummy*: It does not exist outside the sum, and it can be renamed arbitrarily without affecting the value of the expression. In mathematical logic, it is known as a *bound variable*. The variable k is *free*: The result depends on it and it cannot be renamed, lest it change the value of the expression. The first rule of summation is therefore: Always be sure to distinguish between free variables and bound variables in the expression you are reading or writing. Never use the same symbol for a free variable and a dummy variable in the same expression. For example, the expression

$$a[n]\sum_{n=n_1}^{n_2} b[n],$$

which is perfectly valid mathematically, will create havoc when the product is expanded. Use

$$a[n]\sum_{m=n_1}^{n_2} b[m]$$

instead, and now you can safely expand it to

$$\sum_{m=n_1}^{n_2} a[n]b[m].$$

2. A sum can be broken up into several sums, and several sums can be joined into a single sum as in

$$\sum_{n=n_1}^{n_3} a[n] = \sum_{n=n_1}^{n_2} a[n] + \sum_{n=n_2+1}^{n_3} a[n],$$

provided n_2 is between n_1 and n_3.

3. If the upper limit is smaller than the lower limit, the sum is zero by definition, that is

$$\sum_{n=n_1}^{n_2} a[n] = 0, \quad \text{if} \quad n_2 < n_1.$$

We call this an *empty sum*. Note that this convention is *not* shared by integrals, where $\int_a^b x(t)dt = -\int_b^a x(t)dt$.

4. Zero terms do not affect the sum. Thus, if $a[n] = 0$ for $n_2 + 1 \le n \le n_3$, then

$$\sum_{n=n_1}^{n_2} a[n] = \sum_{n=n_1}^{n_3} a[n].$$

5. Here is how a change of variables appears:

$$\sum_{n=n_1}^{n_2} a[n+k] = \sum_{m=n_1+k}^{n_2+k} a[m] = \sum_{n=n_1+k}^{n_2+k} a[n].$$

In passing from the first form to the second, we made the substitution $m = n + k$. Note how this affects the summation limits. Since m is a dummy variable, we can revert from m to n and use the third form.

6. Here is how a change of variables with reversal of direction appears:

$$\sum_{n=n_1}^{n_2} a[k-n] = \sum_{m=k-n_2}^{k-n_1} a[m] = \sum_{n=k-n_2}^{k-n_1} a[n].$$

Carefully observe the change of summation limits in this case.

7. Double sums typically appear in two situations:

    (a) When expanding a product of sums as in

$$\left[\sum_{n=n_1}^{n_2} a[n]\right]\left[\sum_{m=m_1}^{m_2} b[m]\right] = \sum_{n=n_1}^{n_2}\sum_{m=m_1}^{m_2} a[n]b[m].$$

    (b) When substituting an entity involving one sum in another sum as in

$$\sum_{n=n_1}^{n_2} a[n]\left[\sum_{m=m_1}^{m_2} b[m,n]\right] = \sum_{n=n_1}^{n_2}\sum_{m=m_1}^{m_2} a[n]b[m,n].$$

Be sure to use different dummy variables in the two sums. Never use the same dummy variable in two different sums appearing in the same expression. For example, never try to expand the left side of 7a in the form $\left[\sum_{n=n_1}^{n_2} a[n]\right]\left[\sum_{n=m_1}^{m_2} b[n]\right]$. Change the dummy variable in one of the sums, as shown above.

8. Convolution of finite sequences is often confusing to beginners, due to the need to set the summation limits properly. Consider the convolution of the sequences $\{x[n],\ 0 \le n \le N_1 - 1\}$ and $\{y[n],\ 0 \le n \le N_2 - 1\}$. In theory, the convolution is

$$z[n] = \sum_{m=-\infty}^{\infty} x[m]y[n-m].$$

However, due to the finite length of the two operands, the terms in the sum are nonzero only when

$$0 \le m \le N_1 - 1, \quad 0 \le n - m \le N_2 - 1.$$

This is the same as

$$0 \le m \le N_1 - 1, \quad n - N_2 + 1 \le m \le n,$$

which can also be written as

$$\max\{0, n - N_2 + 1\} \le m \le \min\{N_1 - 1, n\}.$$

In summary, the convolution of the two sequences is given by

$$z[n] = \sum_{m=\max\{0,n-N_2+1\}}^{\min\{N_1-1,n\}} x[m]y[n-m]. \qquad (1.16)$$

1.4   Summary and Complements

1.4.1   Summary

In this chapter we introduced the concept of digital signal processing, compared digital and analog processing methodologies, mentioned a few applications of DSP, and presented a brief summary of the contents of the book. We also introduced the system of notation used in this book and included some guidelines for the use of summation.

Of the many textbooks on digital signal processing, some of the better known are Oppenheim and Schafer [1975, 1989], Parks and Burrus [1987], Roberts and Mullis [1987], Proakis and Manolakis [1992], Antoniou [1993], Kuc [1988], Strum and Kirk [1989], Haddad and Parsons [1991], and Jackson [1996].

1.4.2   Complements

1. [p. 4] This paper was not published at that time, due to the unyielding opposition of J. L. Lagrange.

2. [p. 4] We do not wish to form the impression that analog signal processing has become obsolete. Analog signal processing is used, for example (1) for relatively simple tasks, in which analog implementation is the most economical, (2) when speed requirements imposed by the frequency of the signal render digital processing impractical, (3) when interfacing analog signals to digital processors (more on this will be said in Chapter 3). Also, some recent electronic devices operate on analog signals (such as voltages and charges), but in discrete time; examples include charge-coupled devices and switched-capacitor circuits.

3. [p. 7] The *one-side Laplace transform* is defined as

$$X_+^L(s) = \{\mathcal{L}x\}(s) = \int_0^\infty x(t)e^{-st}dt, \quad s \in \mathbb{C}. \tag{1.17}$$

Note that the two Laplace transforms of $x(t)$ coincide if and only if $x(t) = 0$ for $t < 0$. Also, (1.7) does not hold in general for $X_+^L(s)$.

Chapter 2

Review of Frequency-Domain Analysis

In this chapter we present a brief review of signal and system analysis in the frequency domain. This material is assumed to be known from previous study, so we shall keep the details to a minimum and skip almost all proofs. We consider both continuous-time and discrete-time signals. We pay attention to periodic signals, and present both complex (conventional) and real (cosine and sine) Fourier series. We include a brief summary of stationary random signals, since background on random signals is necessary for certain sections in this book.

2.1 Continuous-Time Signals and Systems

Let $x(t)$ be a continuous-time signal whose values can be real or complex. The Fourier transform of the signal is[1]

$$X^{\mathrm{F}}(\omega) = \{\mathcal{F}x\}(\omega) = \int_{-\infty}^{\infty} x(t)e^{-j\omega t}dt, \quad \omega \in \mathbb{R}, \tag{2.1}$$

subject to the existence of the right side in a well-defined sense. A sufficient condition for the existence of the Fourier transform (2.1) is

$$\int_{-\infty}^{\infty} |x(t)|dt < \infty. \tag{2.2}$$

Subject to further conditions, the inverse Fourier transform exists and is given by

$$x(t) = \{\mathcal{F}^{-1}X^{\mathrm{F}}\}(t) = \frac{1}{2\pi}\int_{-\infty}^{\infty} X^{\mathrm{F}}(\omega)e^{j\omega t}d\omega, \quad t \in \mathbb{R}, \tag{2.3}$$

for every point t at which $x(t)$ is continuous. The following three conditions together, known as *Dirichlet conditions*, are sufficient for (2.3) to hold at every continuity point of $x(t)$:

1. $x(t)$ satisfies (2.2).

2. $x(t)$ is continuous, except for discontinuity points whose number on any finite interval is finite; the limits at both sides of each discontinuity point exist.

3. $x(t)$ has a finite number of minima and maxima on any finite interval.

The variable ω is called the *angular frequency,*[2] and is measured in radians per second (rad/s). The variable $f = \omega/2\pi$ is called the *frequency*, and is measured in hertz (Hz).

11

The main properties of the Fourier transform are as follows.

1. **Linearity**

$$z(t) = ax(t) + by(t) \iff Z^F(\omega) = aX^F(\omega) + bY^F(\omega), \quad a, b \in \mathbb{C}. \tag{2.4}$$

2. **Time shift**

$$y(t) = x(t - \tau) \iff Y^F(\omega) = e^{-j\omega\tau} X^F(\omega), \quad \tau \in \mathbb{R}. \tag{2.5}$$

3. **Frequency shift (modulation)**

$$y(t) = e^{j\lambda t} x(t) \iff Y^F(\omega) = X^F(\omega - \lambda), \quad \lambda \in \mathbb{R}. \tag{2.6}$$

4. **Time and frequency scale**

$$y(t) = x(at) \iff Y^F(\omega) = \frac{1}{|a|} X^F\left(\frac{\omega}{a}\right), \quad a \in \mathbb{R}, \, a \neq 0. \tag{2.7}$$

5. **Time-domain convolution**

$$z(t) = \{x * y\}(t) \iff Z^F(\omega) = X^F(\omega) Y^F(\omega). \tag{2.8}$$

6. **Time-domain multiplication**

$$z(t) = x(t)y(t) \iff Z^F(\omega) = \frac{1}{2\pi} \{X^F * Y^F\}(\omega). \tag{2.9}$$

7. **Parseval's theorem**

$$\int_{-\infty}^{\infty} x(t)\bar{y}(t)dt = \frac{1}{2\pi} \int_{-\infty}^{\infty} X^F(\omega) \bar{Y}^F(\omega)d\omega, \tag{2.10}$$

(where the bar denotes complex conjugation) and its special case

$$\int_{-\infty}^{\infty} |x(t)|^2 dt = \frac{1}{2\pi} \int_{-\infty}^{\infty} |X^F(\omega)|^2 d\omega. \tag{2.11}$$

8. **Conjugation**

$$y(t) = \bar{x}(t) \iff Y^F(\omega) = \bar{X}^F(-\omega). \tag{2.12}$$

9. **Symmetry** If $x(t)$ is real valued then

$$X^F(-\omega) = \bar{X}^F(\omega), \tag{2.13a}$$

$$\mathfrak{R}\{X^F(-\omega)\} = \mathfrak{R}\{X^F(\omega)\}, \tag{2.13b}$$

$$\mathfrak{I}\{X^F(-\omega)\} = -\mathfrak{I}\{X^F(\omega)\}, \tag{2.13c}$$

$$|X^F(-\omega)| = |X^F(\omega)|, \tag{2.13d}$$

$$\angle X^F(-\omega) = -\angle X^F(\omega). \tag{2.13e}$$

(where $\mathfrak{R}$ denotes the real part, $\mathfrak{I}$ denotes the imaginary part, and $\angle$ denotes the angle of the corresponding complex number).

10. **Realness**

$$x(t) = \bar{x}(-t) \iff X^F(\omega) \text{ is real.} \tag{2.14}$$

A continuous-time signal of great importance is the *impulse* (or *delta*) *function* $\delta(t)$, formally defined by the *sifting property*[3]

$$\int_{-\infty}^{\infty} f(t)\delta(t)dt = f(0) \tag{2.15}$$

for any function $f(t)$ that is continuous at $t = 0$.

A *dynamic system* (or simply a *system*) is an object that accepts signals, operates on them, and yields other signals. The eventual interest of the engineer is in physical systems: electrical, mechanical, thermal, physiological, and so on. However, here we regard a system as a mathematical operator. In particular, a continuous-time, *single-input, single-output* (SISO) system is an operator that assigns to a given input signal $x(t)$ a unique output signal $y(t)$. A SISO system is thus characterized by the family of signals $x(t)$ it is permitted to accept (the input family), and by the mathematical relationship between signals in the input family and their corresponding outputs $y(t)$ (the output family). The input family almost never contains all possible continuous-time signals. For example, consider a system whose output signal is the time derivative of the input signal (such a system is called a differentiator). The input family of this system consists of all differentiable signals, and only such signals.

When representing a physical system by a mathematical one, we must remember that the representation is only approximate in general. For example, consider a parallel connection of a resistor R and a capacitor C, fed from a current source $i(t)$. The common mathematical description of such a system is by a differential equation relating the voltage across the capacitor $v(t)$ to the input current:

$$\frac{v(t)}{R} + C\frac{dv(t)}{dt} = i(t). \tag{2.16}$$

However, this relationship is only approximate. It neglects effects such as nonlinearity of the resistor, leakage in the capacitor, temperature induced variations of the resistance and the capacitance, and energy dissipation resulting from electromagnetic radiation. Approximations of this kind are made in all areas of science and engineering; they are not to be avoided, only used with care.

Of special importance to us here (and to system theory in general) is the class of linear systems. A SISO system is said to be *linear* if it satisfies the following two properties:

1. **Additivity:** The response to a sum of two input signals is the sum of the responses to the individual signals. If $y_i(t)$ is the response to $x_i(t)$, $i = 1, 2$, then the response to $x_1(t) + x_2(t)$ is $y_1(t) + y_2(t)$.

2. **Homogeneity:** The response to a signal multiplied by a scalar is the response to the given signal, multiplied by the same scalar. If $y(t)$ is the response to $x(t)$, then the response to $ax(t)$ is $ay(t)$ for all a.

Another important property that a system may possess is time invariance. A system is said to be *time invariant* if shifting the input signal in time by a fixed amount causes the same shift in time of the output signal, but no other change. If $y(t)$ is the response to $x(t)$, then the response to $x(t - t_0)$ is $y(t - t_0)$ for every fixed t_0 (positive or negative).

The resistor–capacitor system described by (2.16) is linear, provided the capacitor has zero charge in the absence of input current. This follows from linearity of the differential equation. The system is time invariant as well; however, if the resistance R or the capacitance C vary in time, the system is not time invariant.

A system that is both linear and time invariant is called *linear time invariant*, or LTI. All systems treated in this book are linear time invariant.

The Dirac delta function $\delta(t)$ may or may not be in the input family of a given LTI system. If it is, we denote by $h(t)$ the response to $\delta(t)$, and call it the *impulse response* of the system.[4] For example, the impulse response of the resistor–capacitor circuit described by (2.16) is

$$h(t) = \begin{cases} e^{-t/RC}, & t \geq 0, \\ 0, & t < 0. \end{cases}$$

If the impulse response of an LTI system exists, it completely characterizes the input–output relationship of the system. Given an input signal $x(t)$, the (so-called *zero-state*) output signal $y(t)$ is the convolution[5]

$$y(t) = \{h * x\}(t). \tag{2.17}$$

The signal is in the input family of the system if and only if the right side of (2.17) exists.

The *frequency response* of an LTI system is the Fourier transform $H^F(\omega)$ of the impulse response. The frequency-domain counterpart of (2.17) is

$$Y^F(\omega) = H^F(\omega)X^F(\omega). \tag{2.18}$$

Not every LTI system possessing an impulse response necessarily has frequency response, since $H^F(\omega)$ may not exist. For example, the frequency response of a system whose impulse response is $h(t) = e^t$ is not defined.

2.2 Specific Signals and Their Transforms

In this section we list a few continuous-time signals that will be used repeatedly in this book, along with their Fourier transforms.

2.2.1 The Delta Function and the DC Function

We have already introduced the Dirac delta function $\delta(t)$ in (2.15). The Fourier transform of the delta function is

$$\{\mathcal{F}\delta\}(\omega) = \int_{-\infty}^{\infty} \delta(t)e^{-j\omega t}dt = 1. \tag{2.19}$$

The function whose value is 1 for all t is called the DC function, and will be denoted by $1(t)$ (DC stands for *direct current*, a traditional electrical engineering term). By an argument dual to (2.19), the Fourier transform of the DC function is

$$\{\mathcal{F}1\}(\omega) = 2\pi\delta(\omega). \tag{2.20}$$

Figure 2.1 illustrates the delta and the DC functions.

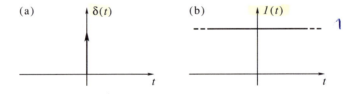

Figure 2.1 The Dirac delta function (a) and the DC function (b).

2.2.2 Complex Exponentials and Sinusoids

The *complex exponential function* is

$$x(t) = e^{j\omega_0 t}. \tag{2.21}$$

The parameter ω_0 is the *angular frequency* of the complex exponential; it is a real number, either positive or negative. In the special case $\omega_0 = 0$ we get the DC function.

The Fourier transform of the complex exponential function is obtained from (2.20) and the modulation property (2.6) as

$$X^F(\omega) = 2\pi\delta(\omega - \omega_0). \tag{2.22}$$

A *sinusoidal* function has the general form

$$x(t) = A\cos(\omega_0 t + \phi_0). \tag{2.23}$$

The parameters A, ω_0, and ϕ are, respectively, the amplitude, angular frequency, and initial phase. The amplitude and the angular frequency are real and positive. The initial phase is assumed to be in the range $[-\pi, \pi)$ for uniqueness. In the special case $A = 1$, $\phi = 0$ we get the cosine function. In the special case $A = 1$, $\phi = -0.5\pi$ we get the sine function.

From (2.22) and the two Euler formulas

$$e^{j\omega_0 t} \rightleftharpoons \delta(\omega - \omega_0)$$

$$\cos(\omega_0 t) = 0.5(e^{j\omega_0 t} + e^{-j\omega_0 t}), \quad \sin(\omega_0 t) = 0.5j(e^{-j\omega_0 t} - e^{j\omega_0 t}), \tag{2.24}$$

we get the Fourier transforms of the cosine and sine functions:

$$1 \rightleftharpoons \delta(\omega).$$

$$\{\mathcal{F}\cos\}(\omega) = \pi[\delta(\omega - \omega_0) + \delta(\omega + \omega_0)], \tag{2.25a}$$

$$\{\mathcal{F}\sin\}(\omega) = j\pi[\delta(\omega + \omega_0) - \delta(\omega - \omega_0)]. \tag{2.25b}$$

Using the trigonometric formula

$$\cos(\omega_0 t + \phi_0) = \cos\phi_0\cos(\omega_0 t) - \sin\phi_0\sin(\omega_0 t),$$

we get the following for the general sinusoidal function $x(t)$ in (2.23):

$$X^F(\omega) = A\pi(\cos\phi_0 + j\sin\phi_0)\delta(\omega - \omega_0) + A\pi(\cos\phi_0 - j\sin\phi_0)\delta(\omega + \omega_0)$$

$$= A\pi e^{j\phi_0}\delta(\omega - \omega_0) + A\pi e^{-j\phi_0}\delta(\omega + \omega_0). \tag{2.26}$$

Suppose we feed the sinusoidal signal $x(t)$ defined in (2.23) to the input of an LTI system whose frequency response is $H^F(\omega)$. Then, according to (2.18) and (2.26), the Fourier transform of the output signal will be

$$Y^F(\omega) = H^F(\omega_0)\pi e^{j\phi_0}\delta(\omega - \omega_0) + H^F(-\omega_0)\pi e^{-j\phi_0}\delta(\omega + \omega_0). \tag{2.27}$$

↳ bcos $\omega = \omega_0$ a   ↳ bcos $\omega = -\omega_0$ a

Let us write the frequency response as

$$H^F(\omega) = |H^F(\omega)|e^{j\psi(\omega)}, \tag{2.28}$$

where $\psi(\omega)$ is the phase response of the system. Assuming that the system's impulse response is real, we have   if real, Then phase $\phi(\omega) = -\phi(-\omega)$

$$|H^F(\omega)| = |H^F(-\omega)|, \quad \psi(\omega) = -\psi(-\omega). \tag{2.29}$$

magnitude $|H(\omega)| = |H(-\omega)|$

Therefore,

$$Y^F(\omega) = |H^F(\omega_0)|[\pi e^{j[\phi_0 + \psi(\omega_0)]}\delta(\omega - \omega_0) + \pi e^{-j[\phi_0 + \psi(\omega_0)]}\delta(\omega + \omega_0)]. \tag{2.30}$$

It follows that   ↳ system response inserts an addition phase shift.

$$y(t) = |H^F(\omega_0)|\cos[\omega_0 t + \phi_0 + \psi(\omega_0)]. \tag{2.31}$$

This result is a fundamental relationship between sinusoidal signals and LTI systems; it can be expressed as follows:

> When a sinusoidal signal at frequency ω_0 is fed to a real LTI system, the output is a sinusoidal signal at the same frequency. The ratio of magnitudes of the output and input signals is the magnitude response of the system at ω_0, and the difference in phases is the phase response of the system at ω_0.

Because of this property, sinusoidal signals are said to be *eigenfunctions* of LTI systems. For comparison, recall that multiplication of a matrix by an eigenvector of the same matrix yields the eigenvector again, multiplied by the corresponding eigenvalue. Thus eigenfunctions of an LTI system fulfill a role similar to that of eigenvectors of a matrix, and the system's frequency response fulfills a role similar to that of eigenvalues.

2.2.3   The rect and the sinc

The *rectangular* function, or *rect* (also known as *box, boxcar,* or *pulse*), is defined by

$$\text{rect}(t) = \begin{cases} 1, & |t| < 0.5, \\ 0.5, & |t| = 0.5, \\ 0, & |t| > 0.5. \end{cases} \tag{2.32}$$

The *sinc* function is defined by

$$\text{sinc}(t) = \begin{cases} \frac{\sin(\pi t)}{\pi t}, & t \neq 0, \\ 1, & t = 0. \end{cases} \tag{2.33}$$

The Fourier transform of the rect function is

$$\{\mathcal{F}\text{rect}\}(\omega) = \int_{-0.5}^{0.5} e^{-j\omega t}\,dt = -\frac{1}{j\omega}(e^{-j0.5\omega} - e^{j0.5\omega}) = \text{sinc}\left(\frac{\omega}{2\pi}\right). \tag{2.34}$$

By a dual argument, the Fourier transform of the sinc function is

$$\{\mathcal{F}\text{sinc}\}(\omega) = \text{rect}\left(\frac{\omega}{2\pi}\right). \tag{2.35}$$

Figure 2.2 illustrates the rect and the sinc functions.

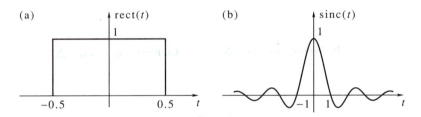

Figure 2.2 The rect function (a) and the sinc function (b).

2.2.4   The Gaussian Function

The Gaussian function is defined by

$$g(t) = e^{-0.5t^2}, \quad t \in \mathbb{R}. \tag{2.36}$$

The Fourier transform of the Gaussian function is derived as follows:

$$G^{\text{F}}(\omega) = \int_{-\infty}^{\infty} e^{-0.5t^2} e^{-j\omega t}\,dt = e^{-0.5\omega^2} \int_{-\infty}^{\infty} e^{-0.5(t+j\omega)^2}\,dt = \sqrt{2\pi}e^{-0.5\omega^2}. \tag{2.37}$$

As we see, the Fourier transform of a Gaussian function is a Gaussian function of ω (up to a scale factor). The pair $g(t)$ and $G^{\text{F}}(\omega)$ are shown in Figure 2.3. The Gaussian is a rare example of a signal and its Fourier transform being both real and positive valued for all t and ω.

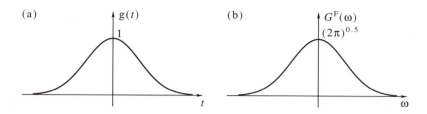

Figure 2.3 The Gaussian function (a) and its Fourier transform (b).

2.3 Continuous-Time Periodic Signals

A continuous-time signal $x(t)$ is said to be periodic if there exists $T > 0$ such that

$$x(t) = x(t + T), \quad \text{for all } t \in \mathbb{R}. \tag{2.38}$$

The smallest T for which (2.38) holds is called the *period*. A periodic signal satisfying certain smoothness conditions admits a Fourier series representation:

$$x(t) = \{S^{-1}X^S\}(t) = \sum_{k=-\infty}^{\infty} X^S[k] \exp\left(\frac{j2\pi kt}{T}\right), \tag{2.39}$$

where the sequence of coefficients $\{X^S[k]\}$ is given by

$$X^S[k] = \{Sx\}[k] = \frac{1}{T}\int_0^T x(t) \exp\left(-\frac{j2\pi kt}{T}\right) dt, \quad k \in \mathbb{Z}. \tag{2.40}$$

Note that (2.39) applies equally well to $x(t)$ and its restriction on any single period, say on $[0, T]$, or on $[-0.5T, 0.5T]$. Therefore, we shall use Fourier series interchangeably for signals defined on intervals and for their periodic extensions.

Parseval's theorem for Fourier series is

$$\sum_{k=-\infty}^{\infty} |X^S[k]|^2 = \frac{1}{T}\int_0^T |x(t)|^2 dt. \tag{2.41}$$

When $x(t)$ is a real signal, its Fourier coefficients satisfy $\bar{X}^S[k] = X^S[-k]$. Then we get from (2.39)

$$x(t) = X^S[0] + \sum_{k=1}^{\infty} 2|X^S[k]| \cos\left(\frac{2\pi kt}{T} + \phi[k]\right), \quad \phi[k] = \angle X^S[k]. \tag{2.42}$$

The individual terms $2|X^S[k]| \cos(2\pi kt/T + \phi[k])$ are called the *harmonics* of $x(t)$.

Periodic signals do not satisfy the condition (2.2), so they do not possess a Fourier transform in the usual sense. However, they can be formally represented as

$$x(t) = \frac{1}{2\pi}\int_{-\infty}^{\infty} X^F(\omega)e^{j\omega t}d\omega, \tag{2.43}$$

where

$$X^F(\omega) = 2\pi \sum_{k=-\infty}^{\infty} X^S[k]\delta\left(\omega - \frac{2\pi k}{T}\right). \tag{2.44}$$

The right side of (2.44) is the accepted definition of the Fourier transform of the periodic signal (2.39). As we see, the Fourier transform of a periodic signal is supported on a discrete set of frequencies—the integer multiples of $2\pi/T$. It is common to refer to these frequencies as *spectral lines*. More generally, the Fourier transform of any sum (finite or convergent infinite) of complex exponential signals consists of spectral lines at the frequencies of the terms of the sum.

The following relation between periodic signals and linear systems is of special interest:

Theorem 2.1 When the input signal of an LTI system is periodic, the output (if exists) is also a periodic signal.

Proof Let $x(t)$ be periodic with period T, and let $h(t)$ be the impulse response of the LTI system. Then the output $y(t)$ is given by

$$y(t) = \int_{-\infty}^{\infty} h(\tau)x(t - \tau)d\tau. \tag{2.45}$$

Therefore,

$$y(t + T) = \int_{-\infty}^{\infty} h(\tau)x(t + T - \tau)d\tau = \int_{-\infty}^{\infty} h(\tau)x(t - \tau)d\tau = y(t), \tag{2.46}$$

so $y(t)$ is periodic.   □

2.4   The Impulse Train

The *impulse train* $p_T(t)$ is defined by

$$p_T(t) = \sum_{n=-\infty}^{\infty} \delta(t - nT), \tag{2.47}$$

where T is a positive constant, called the *period*. Figure 2.4 depicts the impulse train.

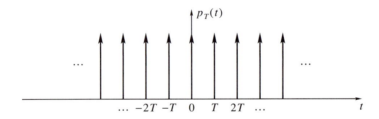

Figure 2.4 The impulse train $p_T(t)$.

An alternative expression for the impulse train is given by the *Poisson sum formula*:

Theorem 2.2

$$p_T(t) = \sum_{n=-\infty}^{\infty} \delta(t - nT) = \frac{1}{T} \sum_{k=-\infty}^{\infty} \exp\left(\frac{j2\pi kt}{T}\right). \tag{2.48}$$

An informal proof Since $p_T(t)$ is periodic, it admits the Fourier series

$$p_T(t) = \sum_{k=-\infty}^{\infty} P_T^S[k] \exp\left(\frac{j2\pi kt}{T}\right), \tag{2.49}$$

where

$$P_T^S[k] = \frac{1}{T} \int_{-T/2}^{T/2} p_T(t) \exp\left(-\frac{j2\pi kt}{T}\right) dt = \frac{1}{T}. \tag{2.50}$$

Substitution of (2.50) in (2.49) yields (2.48). This proof is informal, since we have not proved the existence of the Fourier series (2.49) in this case. A formal proof can be found in the mathematical literature.   □

A dual form of the Poisson formula is

$$\sum_{k=-\infty}^{\infty} \delta\left(\omega - \frac{2\pi k}{T}\right) = \frac{T}{2\pi} \sum_{n=-\infty}^{\infty} e^{jn\omega T}. \tag{2.51}$$

Freq. Domain — *Time Domain*

The dual form is obtained by substituting

$$t \leftarrow \omega, \quad T \leftarrow \frac{2\pi}{T}$$

in (2.48). In both (2.48) and (2.51), n (or k) can be replaced by $-n$ ($-k$, respectively), since the summations are from $-\infty$ to ∞.

Using the Poisson formula, we can compute the Fourier transform of $p_T(t)$ as follows (this is again an informal derivation):

$$P_T^F(\omega) = \int_{-\infty}^{\infty} p_T(t)e^{-j\omega t}dt = \sum_{n=-\infty}^{\infty} \int_{-\infty}^{\infty} \delta(t - nT)e^{-j\omega t}dt$$

$\Rightarrow t = nT$

$$= \sum_{n=-\infty}^{\infty} e^{-j\omega n T} = \frac{2\pi}{T} \sum_{k=-\infty}^{\infty} \delta\left(\omega - \frac{2\pi k}{T}\right). \tag{2.52}$$

The last equality in (2.52) follows from (2.51). We have thus obtained the following interesting result: The Fourier transform of an impulse train in t is an impulse train in ω, up to a scale factor. If the period of the given function is T, the period of the transform is $2\pi/T$.

2.5   Real Fourier Series*

Consider again the Fourier series pair (2.39), (2.40). As we know, the Fourier coefficients $X^S[k]$ are complex even when the signal $x(t)$ is real. The Fourier series of a real signal can also be written in the real form (2.42), which is equivalent to

$$x(t) = X^S[0] + 2 \sum_{k=1}^{\infty} \left[\Re\{X^S[k]\} \cos\left(\frac{2\pi k t}{T}\right) - \Im\{X^S[k]\} \sin\left(\frac{2\pi k t}{T}\right) \right]. \tag{2.53}$$

This representation requires two sets of basis functions—sines and cosines. There exist, however, modified forms of Fourier series for which the coefficients of a real signal are real and the basis functions are either sines or cosines (but not both simultaneously). The derivation of the modified series relies on the following idea: Instead of regarding $x(t)$ as a signal supported on the interval $[0, T]$, we regard it as half of another signal, supported on the interval $[-T, T]$. The extended signal can be arbitrary on $[-T, 0)$, but two natural choices suggest themselves:

1. Force the extended signal to be a real even function of t, that is, define

$$x_e(t) = \begin{cases} x(t), & 0 \le t \le T, \\ x(-t), & -T \le t < 0. \end{cases} \tag{2.54}$$

Note that $x_e(t)$ has no jump at $t = 0$. Moreover, if we extend $x_e(t)$ periodically with period $2T$, there will be no jumps at $t = \pm T$. However, there is a possible discontinuity of the derivative of $x_e(t)$ [assuming that $x(t)$ is differentiable] at $t = 0$, and at $t = \pm T$ if $x_e(t)$ is extended periodically.

2. Force the extended signal to be an imaginary odd function of t, that is, define

$$x_o(t) = \begin{cases} jx(t), & 0 \le t \le T, \\ -jx(-t), & -T \le t < 0. \end{cases} \tag{2.55}$$

In this case there is a possible discontinuity at $t = 0$, and at $t = \pm T$ if $x_o(t)$ is extended periodically.

Let us write the Fourier series pair for $x_e(t)$. We have from (2.40),

$$
\begin{aligned}
X_e^S[k] &= \frac{1}{2T} \int_{-T}^{T} x_e(t) \exp\left(-\frac{j2\pi kt}{2T}\right) dt \\
&= \frac{1}{2T} \int_{-T}^{0} x(-t) \exp\left(-\frac{j\pi kt}{T}\right) dt + \frac{1}{2T} \int_{0}^{T} x(t) \exp\left(-\frac{j\pi kt}{T}\right) dt \\
&= \frac{1}{T} \int_{0}^{T} x(t) \cos\left(\frac{\pi kt}{T}\right) dt.
\end{aligned}
\tag{2.56}
$$

The coefficients $X_e^S[k]$ are real and symmetric. Furthermore, $x(t)$, which is equal to $x_e(t)$ on $[0, T]$, is given by (2.39) as

$$
\begin{aligned}
x(t) &= \sum_{k=-\infty}^{\infty} X_e^S[k] \exp\left(j\frac{2\pi kt}{2T}\right) \\
&= X_e^S[0] + \sum_{k=-\infty}^{-1} X_e^S[k] \exp\left(\frac{j\pi kt}{T}\right) + \sum_{k=1}^{\infty} X_e^S[k] \exp\left(\frac{j\pi kt}{T}\right) \\
&= X_e^S[0] + 2 \sum_{k=1}^{\infty} X_e^S[k] \cos\left(\frac{\pi kt}{T}\right).
\end{aligned}
\tag{2.57}
$$

The representation (2.57) is known as the *cosine Fourier series*. Both the coefficients and the basis functions of this series are real, and only cosine basis functions are used. The cosine series is especially useful when the derivative of $x(t)$ is zero at $t = 0$ and $t = T$, because then the derivative of $x_e(t)$ has no jumps at these points.

Next let us write the Fourier series pair for $x_o(t)$. We have from (2.40),

$$
\begin{aligned}
X_o^S[k] &= \frac{1}{2T} \int_{-T}^{T} x_o(t) \exp\left(-\frac{j2\pi kt}{2T}\right) dt \\
&= \frac{1}{2jT} \int_{-T}^{0} x(-t) \exp\left(-\frac{j\pi kt}{T}\right) dt - \frac{1}{2jT} \int_{0}^{T} x(t) \exp\left(-\frac{j\pi kt}{T}\right) dt \\
&= \frac{1}{T} \int_{0}^{T} x(t) \sin\left(\frac{\pi kt}{T}\right) dt.
\end{aligned}
\tag{2.58}
$$

The coefficients $X_o^S[k]$ are real and antisymmetric. In particular, $X_o^S[0] = 0$. The signal $x(t)$, which is equal to $x_o(t)/j$ on $[0, T]$, is given by (2.39) as

$$
\begin{aligned}
x(t) &= \frac{1}{j} \sum_{k=-\infty}^{\infty} X_o^S[k] \exp\left(j\frac{2\pi kt}{2T}\right) \\
&= \frac{1}{j} \sum_{k=-\infty}^{-1} X_o^S[k] \exp\left(\frac{j\pi kt}{T}\right) + \frac{1}{j} \sum_{k=1}^{\infty} X_o^S[k] \exp\left(\frac{j\pi kt}{T}\right) \\
&= 2 \sum_{k=1}^{\infty} X_o^S[k] \sin\left(\frac{\pi kt}{T}\right).
\end{aligned}
\tag{2.59}
$$

The representation (2.59) is known as the *sine Fourier series*. As in the case of the cosine Fourier series, both the coefficients and the basis functions of the sine series are real. This series is especially useful in cases where $x(t)$ is zero at $t = 0$ and $t = T$, because then $x_o(t)$ has no jumps at these points.

Fourier cosine and sine series are useful for solving partial differential equations with boundary conditions. Such equations are common in physics. When physical laws impose a zero value of the function at the boundaries, the sine series is useful. When they impose a zero value of the derivative at the boundaries, the cosine series is useful.

2.6 Continuous-Time Random Signals*

A random signal cannot be described by a unique, well-defined mathematical formula. Instead, it can be described by probabilistic laws. In this book we shall use random signals only occasionally, so detailed knowledge of them is not required. We do assume, however, familiarity with the notions of probability, random variables, expectation, variance and covariance. We give here the basic definitions pertaining to random signals and a few of their properties. We shall limit ourselves to real-valued random signals.

A continuous-time *random signal* (or *random process*) is a signal $x(t)$ whose value at each time point is a random variable. Random signals appear often in real life. Examples include:

1. The noise heard from a radio receiver that is not tuned to an operating channel.
2. The noise heard from a helicopter rotor.
3. Electrical signals recorded from a human brain through electrodes put in contact with the skull (these are called *electroencephalograms*, or EEGs).
4. Mechanical vibrations sensed in a vehicle moving on a rough terrain.
5. Angular motion of a boat in the sea caused by waves and wind.

Common to all these examples is the irregular appearance of the signal—see Figure 2.5.

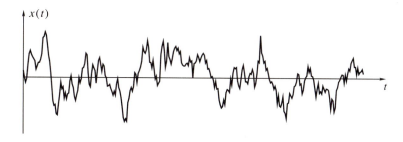

Figure 2.5 A continuous-time random signal.

A signal of interest may be accompanied by an undesirable random signal, which interferes with the signal of interest and limits its usefulness. For example, the typical hiss of audiocassettes limits the usefulness of such cassettes in playing high-fidelity music. In such cases, the undesirable random signal is usually called *noise*. Occasionally "noise" is understood as a synonym for a random signal, but more often it is used only when the random signal is considered harmful or undesirable.

2.6.1 Mean, Variance, and Covariance

The value of a random variable x is governed by its probability distribution. This distribution has, in general, a *mean*, which we denote by

$$\mu_x = E(x)$$

[where $E(\cdot)$ denotes expectation[6]], and a *variance*, which we denote by

$$y_x = E(x - \mu_x)^2.$$

The square root of the variance, $y_x^{1/2}$, is called the *standard deviation* of the random variable and is denoted by σ_x. The mean and the standard deviation are measured in the same physical units as the random variable itself. For example, if the random variable is a voltage across the terminals of a battery, the mean and the standard deviation are measured in volts, whereas the variance is measured in volts squared.

Two random variables x and y governed by a joint probability distribution have a *covariance*, defined by

$$\gamma_{x,y} = E[(x - \mu_x)(y - \mu_y)].$$

The covariance can be positive, negative, or zero; it obeys the *Cauchy–Schwarz inequality*[7]

$$|\gamma_{x,y}| \leq \sigma_x \sigma_y. \tag{2.60}$$

Two random variables are said to be *uncorrelated* if their covariance is zero. If x and y are the same random variable, their covariance is equal to the variance of x.

A random signal $x(t)$ has mean and variance at every time point. The mean and the variance depend on t in general, so we denote them as functions of time, $\mu_x(t)$ and $\gamma_x(t)$, respectively. Thus, the mean and the variance of a random signal are

$$\mu_x(t) = E[x(t)], \quad \gamma_x(t) = E[x(t) - \mu_x(t)]^2. \tag{2.61}$$

The covariance of a random signal at two different time points t_1, t_2 is denoted by

$$\gamma_x(t_1, t_2) = E\{[x(t_1) - \mu_x(t_1)][x(t_2) - \mu_x(t_2)]\}. \tag{2.62}$$

Note that, whereas $\mu_x(t)$ and $\gamma_x(t)$ are functions of a single variable (the time t), the covariance is a function of two time variables.

2.6.2 Wide-Sense Stationary Signals

A random signal $x(t)$ is called *wide-sense stationary* (WSS) if it satisfies the following two properties:[8]

1. The mean $\mu_x(t)$ is the same at all time points, that is,

$$\mu_x(t) = \mu_x = \text{const.} \tag{2.63}$$

2. The covariance $\gamma_x(t_1, t_2)$ depends only on the difference between t_1 and t_2, that is,

$$\gamma_x(t_1, t_2) = \kappa_x(t_1 - t_2). \tag{2.64}$$

For a WSS signal, we denote the difference $t_1 - t_2$ by τ, and call it the *lag variable* of the function $\kappa_x(\tau)$. The function $\kappa_x(\tau)$ is called the *covariance function* of $x(t)$. Thus, the covariance function of a WSS random signal is

$$\kappa_x(\tau) = E\{[x(t + \tau) - \mu_x][x(t) - \mu_x]\}. \tag{2.65}$$

The right side of (2.65) is independent of t by definition of wide-sense stationarity. Note that (2.65) can also be expressed as (see Problem 2.37)

$$\kappa_x(\tau) = E[x(t + \tau)x(t)] - \mu_x^2. \tag{2.66}$$

The main properties of the covariance function are as follows:

1. $\kappa_x(0)$ is the variance of $x(t)$, that is,

$$\kappa_x(0) = E[x(t) - \mu_x]^2 = \gamma_x. \tag{2.67}$$

2. $\kappa_x(\tau)$ is a symmetric function of τ, since

$$\kappa_x(\tau) = E\{[x(t + \tau) - \mu_x][x(t) - \mu_x]\}$$
$$= E\{[x(t) - \mu_x][x(t + \tau) - \mu_x]\} = \kappa_x(-\tau). \tag{2.68}$$

3. By the Cauchy–Schwarz inequality we have

$$|\kappa_x(\tau)| \le \sigma_x \sigma_x = \gamma_x, \quad \text{for all } \tau. \tag{2.69}$$

The random signal shown in Figure 2.5 is wide-sense stationary. As we see, a WSS signal looks more or less the same at different time intervals. Although its detailed form varies, its overall (or macroscopic) shape does not. An example of a random signal that is not stationary is a seismic wave during an earthquake. Figure 2.6 depicts such a wave. As we see, the amplitude of the wave shortly before the beginning of the earthquake is small. At the start of the earthquake the amplitude grows suddenly, sustains its amplitude for a certain time, then decays. Another example of a nonstationary signal is human speech. Although whether a speech signal is essentially random can be argued, it definitely has certain random features. Speech is not stationary, since different phonemes have different characteristic waveforms. Therefore, as the spoken sound moves from phoneme to phoneme (for example, from "f" to "i" to "sh" in "fish"), the macroscopic shape of the signal varies.

Figure 2.6 A seismic wave during an earthquake.

2.6.3 The Power Spectral Density

The Fourier transform of a WSS random signal requires a special definition, because (2.1) does not exist as a standard integral when $x(t)$ is a WSS random signal. However, the restriction of such a signal to a finite interval, say $[-0.5T, 0.5T]$, does possess a standard Fourier transform. The Fourier transform of a finite segment of a random signal appears random as well. For example, Figure 2.7 shows the magnitude of the Fourier transform of a finite interval of the random signal shown in Figure 2.5. As we see, this figure is difficult to interpret and its usefulness is limited.

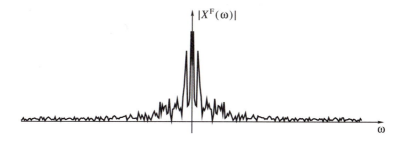

Figure 2.7 Magnitude of the Fourier transform of a finite segment of the signal in Figure 2.5.

A more meaningful way of representing random signals in the frequency domain is by their power spectra. The *power spectral density* (PSD), or *power spectrum*, of a WSS signal is defined as the Fourier transform of its covariance function,

$$K_x^F(\omega) = \int_{-\infty}^{\infty} \kappa_x(\tau) e^{-j\omega\tau} d\tau, \tag{2.70}$$

provided the right side exists. A sufficient condition for existence of the PSD is

$$\int_{-\infty}^{\infty} |\kappa_x(\tau)| d\tau < \infty. \tag{2.71}$$

The inverse relationship is

$$\kappa_x(\tau) = \frac{1}{2\pi} \int_{-\infty}^{\infty} K_x^F(\omega) e^{j\omega\tau} d\omega. \tag{2.72}$$

The PSD is not a random entity, since $\kappa_x(\tau)$ is not a random function. For example, the PSD of the random signal shown in Figure 2.5 is shown in Figure 2.8. Comparing this figure with Figure 2.7, we see that it looks smooth and well behaved.

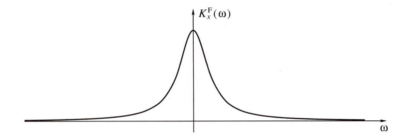

Figure 2.8 Power spectral density of the signal in Figure 2.5.

The PSD is expressed in physical units of power per hertz. For example, if $x(t)$ is measured in volts, then $K_x^F(\omega)$ is measured in V^2/Hz. The value of $K_x^F(\omega)$ can be interpreted as the average power density of the signal at frequency ω; stated in different words, the total average power of the signal in the frequency range $[\omega - 0.5d\omega, \omega + 0.5d\omega]$ is approximately $(2\pi)^{-1}K_x^F(\omega)d\omega$ when $d\omega$ is small. This shows why the units of $K_x^F(\omega)$ are power per hertz: It is because $(2\pi)^{-1}d\omega = df$, and df is measured in hertz.

The main properties of the power spectral density of a WSS signal are as follows:

1. It is real and symmetric in ω, because $\kappa_x(\tau)$ is real and symmetric in τ.

2. It is nonnegative, that is,

$$K_x^F(\omega) \geq 0, \quad \text{for all } \omega. \tag{2.73}$$

   This property will follow from Theorem 2.3; see page 26.

3. The integral of the PSD over all frequencies is proportional to the variance of the signal, that is,

$$\frac{1}{2\pi} \int_{-\infty}^{\infty} K_x^F(\omega) d\omega = \kappa_x(0) = y_x. \tag{2.74}$$

   This follows from (2.72) when substituting $\tau = 0$.

Example 2.1 Here are a few examples of covariance functions and their PSDs.

1. A random signal $v(t)$ whose power spectral density is constant at all frequencies is called *white noise*. Thus, for white noise we have

$$K_v^F(\omega) = N_0 1(\omega). \tag{2.75}$$

Correspondingly,

$$\kappa_v(\tau) = N_0 \delta(\tau). \tag{2.76}$$

As we see from (2.76) [or from (2.75) and (2.74)], white noise has infinite variance, so it exists only as a mathematical object, rather than as a physical signal. Nevertheless, it is a useful approximation of certain physical signals, as well as an important mathematical tool in the theory of random signals. The name *white* is borrowed from optics: The spectrum of white light is approximately constant at all frequencies within the range of visible light. Figure 2.9 illustrates (rather crudely) white noise.

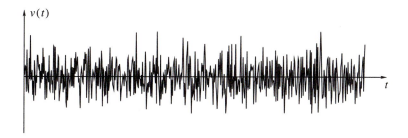

Figure 2.9 White noise.

2. A random signal $v(t)$ whose power spectral density is constant at all frequencies on a finite frequency interval and zero elsewhere is called *band-limited white noise*. Thus, for band-limited white noise we have

$$K_v^{\mathrm{F}}(\omega) = N_0 \mathrm{rect}\left(\frac{\omega}{2\omega_{\mathrm{m}}}\right) \tag{2.77}$$

for some constant ω_{m}. Correspondingly,

$$\kappa_v(\tau) = \frac{N_0 \omega_{\mathrm{m}}}{\pi} \mathrm{sinc}\left(\frac{\omega_{\mathrm{m}}\tau}{\pi}\right). \tag{2.78}$$

Band-limited white noise has a finite variance, equal to $N_0 \omega_{\mathrm{m}}/\pi$.

3. A WSS signal $x(t)$ whose covariance function is

$$\kappa_x(\tau) = N_0 e^{-\omega_0 |\tau|} \tag{2.79}$$

for some positive constant ω_0 is said to be *exponentially correlated*. The parameter $1/\omega_0$ is called the *correlation time constant* of the signal. The PSD of (2.79) is computed as follows:

$$K_x^{\mathrm{F}}(\omega) = N_0 \int_{-\infty}^{0} e^{(\omega_0 - j\omega)\tau} d\tau + N_0 \int_{0}^{\infty} e^{-(\omega_0 + j\omega)\tau} d\tau$$
$$= \frac{N_0}{\omega_0 - j\omega} + \frac{N_0}{\omega_0 + j\omega} = \frac{2N_0 \omega_0}{\omega^2 + \omega_0^2}. \tag{2.80}$$

The random signal shown in Figure 2.5 is exponentially correlated and the PSD shown in Figure 2.8 corresponds to (2.80).   □

An important relationship exists between the PSD of a WSS random signal and the Fourier transform of a finite segment of the signal. This relationship is known as the *Wiener–Khintchine theorem*, and is given as follows:

Theorem 2.3 (Wiener–Khintchine) Let $x(t)$ be a WSS random signal whose mean is zero and whose covariance function $\kappa_x(\tau)$ satisfies (2.71). Let

$$y(t) = \begin{cases} x(t), & -0.5T \le t \le 0.5T, \\ 0, & \text{otherwise,} \end{cases} \tag{2.81}$$

that is, $y(t)$ is the restriction of $x(t)$ to the interval $[-0.5T, 0.5T]$. Then

$$\lim_{T \to \infty} \frac{1}{T} E(|Y^F(\omega)|^2) = K_x^F(\omega). \tag{2.82}$$

Partial proof We have

$$\frac{1}{T} E(|Y^F(\omega)|^2) = \frac{1}{T} E\left\{ \left[\int_{-0.5T}^{0.5T} x(t) e^{-j\omega t} dt \right] \left[\int_{-0.5T}^{0.5T} x(s) e^{j\omega s} ds \right] \right\}$$

$$= \frac{1}{T} \int_{-0.5T}^{0.5T} \int_{-0.5T}^{0.5T} E[x(t)x(s)] e^{-j\omega(t-s)} dt ds = \frac{1}{T} \int_{-0.5T}^{0.5T} \int_{-0.5T}^{0.5T} \kappa_x(t-s) e^{-j\omega(t-s)} dt ds$$

$$= \frac{1}{T} \int_{-T}^{T} \kappa_x(u) e^{-j\omega u} \left[\int_0^{T-|u|} 1 \cdot dv \right] du = \int_{-T}^{T} \left(1 - \frac{|u|}{T} \right) \kappa_x(u) e^{-j\omega u} du. \tag{2.83}$$

The part of the proof that is skipped, since it is beyond the mathematical level of this book, is the equality of the following two limits:

$$\int_{-\infty}^{\infty} |a(t)| dt < \infty \implies \lim_{T \to \infty} \int_{-T}^{T} \left(1 - \frac{|t|}{T} \right) a(t) dt = \lim_{T \to \infty} \int_{-T}^{T} a(t) dt = \int_{-\infty}^{\infty} a(t) dt. \tag{2.84}$$

Taking the limit of (2.83) and using (2.84) finally gives

$$\lim_{T \to \infty} \frac{1}{T} E(|Y^F(\omega)|^2) = \int_{-\infty}^{\infty} \kappa_x(u) e^{-j\omega u} du = K_x^F(\omega). \tag{2.85}$$

Corollary The PSD $K_x^F(\omega)$ is a nonnegative function of ω, since it is the limit of nonnegative functions. □

2.6.4   WSS Signals and LTI Systems

Consider an LTI system whose impulse response $h(t)$ satisfies

$$\int_{-\infty}^{\infty} |h(t)| dt < \infty. \tag{2.86}$$

When a WSS signal is given at the input of such a system, the output signal is also WSS. This is formally stated and proved in the following theorem.

Theorem 2.4 Let $x(t)$ be a WSS signal with mean μ_x and covariance function $\kappa_x(\tau)$. Let

$$y(t) = \{h * x\}(t),$$

where $h(t)$ satisfies (2.86); then $y(t)$ is WSS. The mean and covariance function of $y(t)$ are given by

$$\mu_y = \mu_x \int_{-\infty}^{\infty} h(s) ds, \tag{2.87a}$$

$$\kappa_y(\tau) = \int_{-\infty}^{\infty} \int_{-\infty}^{\infty} h(s) h(u) \kappa_x(\tau - s + u) ds du. \tag{2.87b}$$

The PSD of $y(t)$ is given by

$$K_y^F(\omega) = K_x^F(\omega) |H^F(\omega)|^2. \tag{2.88}$$

Proof

1.

$$
\mu_y = E[y(t)] = E\left\{\int_{-\infty}^{\infty} h(s)x(t-s)ds\right\} = \int_{-\infty}^{\infty} h(s)E[x(t-s)]ds
$$

$$
= \int_{-\infty}^{\infty} h(s)\mu_x ds = \mu_x \int_{-\infty}^{\infty} h(s)ds. \tag{2.89}
$$

2.

$$
\kappa_y(\tau) = E\left\{\left[\int_{-\infty}^{\infty} h(s)x(t+\tau-s)ds\right]\left[\int_{-\infty}^{\infty} h(u)x(t-u)du\right]\right\} - \mu_y^2
$$

$$
= \int_{-\infty}^{\infty}\int_{-\infty}^{\infty} h(s)h(u)E[x(t+\tau-s)x(t-u)]dsdu - \mu_y^2
$$

$$
= \int_{-\infty}^{\infty}\int_{-\infty}^{\infty} h(s)h(u)[\kappa_x(\tau-s+u) + \mu_x^2]dsdu - \mu_y^2
$$

$$
= \int_{-\infty}^{\infty}\int_{-\infty}^{\infty} h(s)h(u)\kappa_x(\tau-s+u)dsdu. \tag{2.90}
$$

Since both μ_y and κ_y are found to be independent of t, the output signal $y(t)$ is indeed WSS.

3. Using (2.87b) and the definition of PSD, we can write

$$
K_y^{\mathrm{F}}(\omega) = \int_{-\infty}^{\infty} \kappa_y(\tau)e^{-j\omega\tau}d\tau = \int_{-\infty}^{\infty}\int_{-\infty}^{\infty}\int_{-\infty}^{\infty} h(s)h(u)\kappa_x(\tau-s+u)e^{-j\omega\tau}dsdud\tau
$$

$$
= \int_{-\infty}^{\infty}\int_{-\infty}^{\infty} h(s)h(u)e^{-j\omega(s-u)}K_x^{\mathrm{F}}(\omega)dsdu
$$

$$
= \left[\int_{-\infty}^{\infty} h(s)e^{-j\omega s}ds\right]\left[\int_{-\infty}^{\infty} h(u)e^{j\omega u}du\right]K_x^{\mathrm{F}}(\omega)
$$

$$
= H^{\mathrm{F}}(\omega)\bar{H}^{\mathrm{F}}(\omega)K_x^{\mathrm{F}}(\omega) = K_x^{\mathrm{F}}(\omega)|H^{\mathrm{F}}(\omega)|^2. \tag{2.91}
$$

In passing from the first to the second line, we integrated over τ and used the time-shift property of the Fourier transform (2.5). □

Formula (2.88) is a major result in random signal theory. It gives a new interpretation to the frequency response of an LTI system. In Section 2.2.2 we saw that an LTI system operates on a sinusoidal signal to multiply its magnitude by the magnitude response of the system and shift its phase by the phase response. Now we see that the square-magnitude response $|H^{\mathrm{F}}(\omega)|^2$ multiplies the power spectral density of a WSS random signal at each frequency.

An important special case of (2.88) is applicable when the input signal is white noise. In this case we get

$$
K_y^{\mathrm{F}}(\omega) = N_0|H^{\mathrm{F}}(\omega)|^2. \tag{2.92}
$$

Therefore, the PSD of the response to white noise is proportional to the square magnitude of the frequency response of the system.

2.7 Discrete-Time Signals and Systems

Let $x[n]$ be a discrete-time signal whose values can be real or complex. The formal definition of the Fourier transform of the signal is[9]

$$
X^{\mathrm{f}}(\theta) = \{\mathcal{F}x\}(\theta) = \sum_{n=-\infty}^{\infty} x[n]e^{-j\theta n}, \quad \theta \in \mathbb{R}, \tag{2.93}
$$

subject to the existence of the right side in a well-defined sense. A sufficient condition for the existence of the right side is

$$\sum_{n=-\infty}^{\infty} |x[n]| < \infty. \tag{2.94}$$

The inverse Fourier transform is then given by

$$x[n] = \{\mathcal{F}^{-1}X^{\mathrm{f}}\}[n] = \frac{1}{2\pi} \int_{-\pi}^{\pi} X^{\mathrm{f}}(\theta)e^{j\theta n}d\theta, \quad n \in \mathbb{Z}. \tag{2.95}$$

We remark that the Fourier transform of $x[n]$ can also be defined if (2.94) is violated, but the weaker condition

$$\sum_{n=-\infty}^{\infty} |x[n]|^2 < \infty \tag{2.96}$$

holds. In this case, the sum (2.93) exists in the sense that[10]

$$\lim_{N \to \infty} \int_{-\pi}^{\pi} \left| X^{\mathrm{f}}(\theta) - \sum_{n=-N}^{N} x[n]e^{-j\theta n} \right|^2 d\theta = 0. \tag{2.97}$$

The variable θ is called the *angular frequency* and is measured in radians per sample.

The main properties of the Fourier transform of a sequence are as follows.

1. **Linearity**

$$z[n] = ax[n] + by[n] \iff Z^{\mathrm{f}}(\theta) = aX^{\mathrm{f}}(\theta) + bY^{\mathrm{f}}(\theta), \quad a, b \in \mathbb{C}. \tag{2.98}$$

2. **Periodicity**

$$X^{\mathrm{f}}(\theta) = X^{\mathrm{f}}(\theta + 2\pi k), \quad \theta \in \mathbb{R}, \ k \in \mathbb{Z}. \tag{2.99}$$

3. **Time shift**

$$y[n] = x[n-m] \iff Y^{\mathrm{f}}(\theta) = e^{-j\theta m}X^{\mathrm{f}}(\theta), \quad m \in \mathbb{Z}. \tag{2.100}$$

4. **Frequency shift (modulation)**

$$y[n] = e^{j\lambda n}x[n] \iff Y^{\mathrm{f}}(\theta) = X^{\mathrm{f}}(\theta - \lambda), \quad \lambda \in \mathbb{R}. \tag{2.101}$$

5. **Time-domain convolution**

$$z[n] = \{x * y\}[n] \iff Z^{\mathrm{f}}(\theta) = X^{\mathrm{f}}(\theta)Y^{\mathrm{f}}(\theta). \tag{2.102}$$

6. **Time-domain multiplication**

$$z[n] = x[n]y[n] \iff Z^{\mathrm{f}}(\theta) = \frac{1}{2\pi}\{X^{\mathrm{f}} * Y^{\mathrm{f}}\}(\theta). \tag{2.103}$$

7. **Parseval's theorem**

$$\sum_{n=-\infty}^{\infty} x[n]\bar{y}[n] = \frac{1}{2\pi} \int_{-\pi}^{\pi} X^{\mathrm{f}}(\theta)\bar{Y}^{\mathrm{f}}(\theta)d\theta, \tag{2.104}$$

and its special case

$$\sum_{n=-\infty}^{\infty} |x[n]|^2 = \frac{1}{2\pi} \int_{-\pi}^{\pi} |X^{\mathrm{f}}(\theta)|^2 d\theta. \tag{2.105}$$

8. **Conjugation**

$$y[n] = \bar{x}[n] \iff Y^{\mathrm{f}}(\theta) = \bar{X}^{\mathrm{f}}(-\theta). \tag{2.106}$$

9. **Symmetry** if $x[n]$ is real valued then

$$X^f(-\theta) = \bar{X}^f(\theta), \tag{2.107a}$$

$$\mathfrak{R}\{X^f(-\theta)\} = \mathfrak{R}\{X^f(\theta)\}, \tag{2.107b}$$

$$\mathfrak{I}\{X^f(-\theta)\} = -\mathfrak{I}\{X^f(\theta)\}, \tag{2.107c}$$

$$|X^f(-\theta)| = |X^f(\theta)|, \tag{2.107d}$$

$$\measuredangle X^f(-\theta) = -\measuredangle X^f(\theta). \tag{2.107e}$$

10. **Realness**

$$x[n] = \bar{x}[-n] \iff X^f(\theta) \text{ is real.} \tag{2.108}$$

A discrete-time signal of great importance is the *unit sample*, or *unit impulse*, denoted by $\delta[n]$ and defined by

$$\delta[n] = \begin{cases} 1, & n = 0, \\ 0, & \text{otherwise.} \end{cases} \tag{2.109}$$

Note that we distinguish between the continuous-time delta and the discrete-time delta by the typography of the argument (in parentheses for the former, in brackets for the latter).

A *discrete-time SISO system* is a system whose input and output signals, $x[n]$ and $y[n]$, are in discrete time. The notions of linearity and time invariance apply to discrete-time systems in the same way as to continuous-time systems. The response of a discrete-time LTI system to the unit-sample signal $\delta[n]$ is called the *unit-sample response*, or the *impulse response* of the system, and is denoted by $h[n]$. The response of the system to an input signal $x[n]$ is given by the discrete-time convolution

$$y[n] = \{h * x\}[n], \tag{2.110}$$

provided the right side exists. The *frequency response* of the system is $H^f(\theta)$, the Fourier transform of the impulse response (if the Fourier transform exists). The frequency-domain relationship between input and output is

$$Y^f(\theta) = H^f(\theta)X^f(\theta). \tag{2.111}$$

2.8   Discrete-Time Periodic Signals

A discrete-time signal $x[n]$ is called *periodic* if there exists an integer $N > 0$ such that

$$x[n] = x[n + N], \quad \text{for all } n \in \mathbb{Z}. \tag{2.112}$$

The smallest N for which (2.112) holds is called the *period*.

Unlike continuous-time signals, a discrete-time sinusoidal signal

$$x[n] = \cos(\theta_0 n + \phi_0) \tag{2.113}$$

is not necessarily periodic. It is periodic if and only if $\theta_0/2\pi$ is a rational number (Problem 2.35). The formal Fourier transform of this signal (regardless of whether it is periodic) is

$$X^f(\theta) = \pi e^{j\phi_0} \delta(\theta - \theta_0) + \pi e^{-j\phi_0} \delta(\theta + \theta_0). \tag{2.114}$$

The Fourier series of a discrete-time periodic signal is called the *discrete Fourier transform* (DFT), and we shall discuss it in detail in Chapter 4.

As in the case of continuous-time signals, we have the following relation between discrete-time periodic signals and linear systems:

Theorem 2.5 When the input signal of a discrete-time LTI system is periodic, the output is also a periodic signal.

The easy proof is left as an exercise to the reader; see Problem 2.33.   □

2.9 Discrete-Time Random Signals*

Most of the basic definitions and properties of continuous-time random signals carry over to discrete-time random signals. Therefore, we shall briefly summarize the concepts and results pertaining to discrete-time random signals, rather than repeating the material of Section 2.6 in full detail.

A discrete-time random signal is a sequence $x[n]$ whose value at each time point is a random variable. The mean and the variance of a discrete-time random signal are

$$\mu_x[n] = E(x[n]), \quad \gamma_x[n] = E(x[n] - \mu_x[n])^2. \tag{2.115}$$

The covariance at two different time points n_1, n_2 is denoted by

$$\gamma_x[n_1, n_2] = E\{(x[n_1] - \mu_x[n_1])(x[n_2] - \mu_x[n_2])\}. \tag{2.116}$$

Note that the covariance is a function of two integer variables.

A discrete-time random signal $x[n]$ is called *wide-sense stationary* (WSS) if it satisfies the following two properties:

1. The mean $\mu_x[n]$ is the same at all time points, that is,

$$\mu_x[n] = \mu_x = \text{const.} \tag{2.117}$$

2. The covariance $\gamma_x[n_1, n_2]$ depends only on the difference between n_1 and n_2, that is,

$$\gamma_x[n_1, n_2] = \kappa_x[n_1 - n_2]. \tag{2.118}$$

For a WSS signal, we denote the difference $n_1 - n_2$ by m and call it the *lag variable* of the sequence $\kappa_x[m]$. The sequence $\kappa_x[m]$ is called the *covariance sequence* of $x[n]$. Thus, the covariance sequence of a WSS random signal is

$$\kappa_x[m] = E\{(x[n + m] - \mu_x)(x[n] - \mu_x)\} = E(x[n + m]x[n]) - \mu_x^2. \tag{2.119}$$

The main properties of the covariance sequence are as follows:

1. $\kappa_x[0]$ is the variance of $x[n]$, that is,

$$\kappa_x[0] = E(x[n] - \mu_x)^2 = \gamma_x. \tag{2.120}$$

2. $\kappa_x[m]$ is a symmetric function of m.

3.

$$|\kappa_x[m]| \le \gamma_x, \quad \text{for all } m. \tag{2.121}$$

The *power spectral density* (PSD) of a WSS signal is the Fourier transform of its covariance sequence:

$$K_x^f(\theta) = \sum_{m=-\infty}^{\infty} \kappa_x[m]e^{-j\theta m}, \tag{2.122}$$

provided the right side exists. A sufficient condition for existence of the PSD is

$$\sum_{m=-\infty}^{\infty} |\kappa_x[m]| < \infty. \tag{2.123}$$

The inverse relationship is

$$\kappa_x[m] = \frac{1}{2\pi} \int_{-\pi}^{\pi} K_x^f(\theta) e^{j\theta m} d\theta. \tag{2.124}$$

The main properties of the power spectral density of a WSS signal are as follows:

1. It is real and symmetric in θ.

2. It is nonnegative, that is,

$$K_x^f(\theta) \geq 0, \quad \text{for all } \theta. \tag{2.125}$$

3. The integral of the PSD over $[-\pi, \pi]$ is proportional to the variance of the signal, that is,

$$\frac{1}{2\pi} \int_{-\pi}^{\pi} K_x^f(\theta) d\theta = \kappa_x[0] = \gamma_x. \tag{2.126}$$

Example 2.2 Here are two examples of covariance sequences of discrete-time signals and their PSDs.

1. A discrete-time random signal whose power spectral density is constant at all frequencies is called *discrete-time white noise*. Thus, for a discrete-time white noise $v[n]$ we have

$$K_v^f(\theta) = \gamma_v, \quad \kappa_v[m] = \gamma_v \delta[m]. \tag{2.127}$$

Unlike continuous-time white noise, there is nothing nonphysical in a discrete-time white noise. Its variance is finite and the covariance of any two different samples is zero. The MATLAB command `randn(1,N)` can be used to generate N samples of discrete-time, zero mean, unit variance white noise.

2. A WSS signal $x[n]$ whose covariance sequence is

$$\kappa_x[m] = \gamma_x e^{-\theta_0|m|} \tag{2.128}$$

for some positive constant θ_0 is said to be *exponentially correlated*. The PSD of (2.128) is computed as follows:

$$K_x^f(\theta) = \gamma_x \sum_{m=-\infty}^{-1} e^{(\theta_0 - j\theta)m} + \gamma_x \sum_{m=0}^{\infty} e^{-(\theta_0 + j\theta)m} = \frac{\gamma_x e^{-\theta_0 + j\theta}}{1 - e^{-\theta_0 + j\theta}} + \frac{\gamma_x}{1 - e^{-\theta_0 - j\theta}}$$

$$= \frac{\gamma_x(1 - e^{-2\theta_0})}{1 - 2e^{-\theta_0} \cos\theta + e^{-2\theta_0}}. \tag{2.129}$$

□

The Wiener–Khintchine theorem for discrete-time signals is:

Theorem 2.6 Let $x[n]$ be a WSS discrete-time random signal whose mean is zero and whose covariance sequence $\kappa_x[m]$ satisfies (2.123). Let N be odd and define

$$y[n] = \begin{cases} x[n], & -0.5(N-1) \leq n \leq 0.5(N-1), \\ 0, & \text{otherwise,} \end{cases} \tag{2.130}$$

that is, $y[n]$ is the restriction of $x[n]$ to the interval $[-0.5(N-1), 0.5(N-1)]$. Then

$$\lim_{N \to \infty} \frac{1}{N} E(|Y^f(\theta)|^2) = K_x^f(\theta). \tag{2.131}$$

The proof is similar to that of Theorem 2.3 and will be skipped. □

Consider a discrete-time LTI system whose impulse response $h[n]$ satisfies

$$\sum_{n=-\infty}^{\infty} |h[n]| < \infty. \tag{2.132}$$

Then,

Theorem 2.7 Let $x[n]$ be a discrete-time WSS signal with mean μ_x and covariance sequence $\kappa_x[m]$. Let

$$y[n] = \{h * x\}[n],$$

where $h[n]$ satisfies (2.132). Then $y[n]$ is WSS. The mean and the covariance sequence of $y[n]$ are given by

$$\mu_y = \mu_x \sum_{n=-\infty}^{\infty} h[n], \tag{2.133a}$$

$$\kappa_y[m] = \sum_{n=-\infty}^{\infty} \sum_{k=-\infty}^{\infty} h[n]h[k]\kappa_x[m - n + k]. \tag{2.133b}$$

The PSD of $y[n]$ is given by

$$K_y^f(\theta) = K_x^f(\theta)|H^f(\theta)|^2. \tag{2.134}$$

The proof is similar to that of Theorem 2.4 and will be skipped. $\square$

An important special case of (2.134) is applicable when the input signal is white noise $v[n]$. In this case we get

$$K_y^f(\theta) = \gamma_v |H^f(\theta)|^2. \tag{2.135}$$

As we see, the PSD of the response to white noise is proportional to the square magnitude of the frequency response of the system. Using (2.126), we get from (2.135)

$$\gamma_y = \frac{\gamma_v}{2\pi} \int_{-\pi}^{\pi} |H^f(\theta)|^2 d\theta. \tag{2.136}$$

This relationship provides a convenient means of computing the variance of the response of an LTI system to white noise input. The quantity

$$\text{NG} = \frac{1}{2\pi} \int_{-\pi}^{\pi} |H^f(\theta)|^2 d\theta \tag{2.137}$$

is called the *noise gain* of the system. The noise gain of an LTI system is a measure of the ratio between the output and input variances when the input is white noise.

2.10 Summary and Complements

2.10.1 Summary

In this chapter we reviewed frequency-domain analysis and its relationships to linear system theory. The fundamental operation is the Fourier transform of a continuous-time signal, defined in (2.1), and the inverse transform, given in (2.3). The Fourier transform is a mathematical operation that (1) detects sinusoidal components in a signal and enables the computation of their amplitudes and phases and (2) provides the amplitude and phase density of nonperiodic signals as a function of the frequency. Among the properties of the Fourier transform, the most important is perhaps the convolution property (2.8). The reason is that the response of a linear time-invariant (LTI) system to an arbitrary input is the convolution between the input signal and the impulse response of the system (2.17). From this it follows that the Fourier transform of the output signal is the Fourier transform of the input, multiplied by the frequency response of the system (2.18).

We introduced a few common signals and their Fourier transforms; in particular: the delta function $\delta(t)$, the DC function $1(t)$, the complex exponential, sinusoidal signals, the rectangular function rect(t), the sinc function sinc(t), and the Gaussian function. Complex exponentials and sinusoids are eigenfunctions of LTI systems: They undergo change of amplitude and phase, but their functional form is preserved when passed through an LTI system.

Continuous-time periodic signals were introduced next. Such signals admit a Fourier series expansion (2.39). A periodic signal of particular interest is the impulse train $p_T(t)$. The Fourier transform of an impulse train is an impulse train in the frequency domain (2.52). The impulse train satisfies the Poisson sum formula (2.48). A continuous-time signal on a finite interval can be represented by a Fourier series (2.39). A continuous-time, real-valued signal on a finite interval can also be represented by either a cosine Fourier series (2.57) or a sine Fourier series (2.59).

We reviewed continuous-time random signals, in particular wide-sense stationary (WSS) signals. A WSS signal is characterized by a constant mean and a covariance function depending only on the lag variable. The Fourier transform of the covariance function is called the power spectral density (PSD) of the signal. The PSD of a real-valued WSS signal is real, symmetric, and nonnegative. Two examples of WSS signals are white noise and band-limited white noise. The PSD of the former is constant for all frequencies, whereas that of the latter is constant and nonzero on a finite frequency interval. The PSD of a WSS signal satisfies the Wiener–Khintchine theorem 2.3.

When a WSS signal passes through an LTI system, the output is WSS as well. The PSD of the output is the product of the PSD of the input and the square magnitude of the frequency response of the system (2.88). In particular, when the input signal is white noise, the PSD of the output signal is proportional to the square magnitude of the frequency response (2.92).

Frequency-domain analysis of discrete-time signals parallels that of continuous-time signals in many respects. The Fourier transform of a discrete-time signal is defined in (2.93) and its inverse is given in (2.95). The Fourier transform of a discrete-time signal is periodic, with period 2π. Discrete-time periodic signals and discrete-time random signals are defined in a manner similar to the corresponding continuous-time signals and share similar properties.

The material in this chapter is covered in many books. For general signals and system theory, see Oppenheim and Willsky [1983], Gabel and Roberts [1987], Kwakernaak and Sivan [1991], or Haykin and Van Veen [1997]. For random signals and their relation to linear systems, see Papoulis [1991] or Gardner [1986].

2.10.2 Complements

1. [p. 11] The definition of the Fourier transform is not completely standard. Variations of the definition (2.1) include:

   (a) Normalization:

   $$X^F(\omega) = \frac{1}{\sqrt{2\pi}} \int_{-\infty}^{\infty} x(t) e^{-j\omega t} dt. \tag{2.138}$$

   With this definition, the inverse transform (2.3) has a factor $1/\sqrt{2\pi}$ (instead of $1/2\pi$).

(b) Sign reversal of the frequency variable:

$$X^F(\omega) = \int_{-\infty}^{\infty} x(t)e^{j\omega t}dt. \tag{2.139}$$

In this case, the inverse transform has a negative sign in the exponent.

(c) Use of frequency (instead of angular frequency) as the transform variable:

$$X^F(f) = \int_{-\infty}^{\infty} x(t)e^{-j2\pi ft}dt. \tag{2.140}$$

In this case, the inverse transform is

$$x(t) = \int_{-\infty}^{\infty} X^F(f)e^{j2\pi ft}df. \tag{2.141}$$

(d) Various combinations of the above.

2. [p. 11] Beginners sometimes find the notion of negative frequencies difficult to comprehend. Often a student would say: "We know that only positive frequencies exist physically!" A possible answer is: "Think of a rotating wheel; a wheel rotating clockwise is certainly different from one rotating counterclockwise. So, if you define the angular velocity of a clockwise-rotating wheel as positive, you must define that of a counterclockwise-rotating wheel as negative. The angular frequency of a signal fulfills the same role as the angular velocity of a rotating wheel, so it can have either polarity." For real-valued signals, the conjugate symmetry property of the Fourier transform (2.13a) disguises the existence of negative frequencies. However, for complex signals, positive and negative frequencies are fundamentally distinct. For example, the complex signal $e^{j\omega_0 t}$ with $\omega_0 > 0$ is different from a corresponding signal with $\omega_0 < 0$.

   Complex signals are similarly difficult for beginners to comprehend, since such signals are not commonly encountered as physical entities. Rather, they usually serve as convenient mathematical representations for real signals of certain types. An example familiar to electrical engineering students is the phasor representation of AC voltages and currents (e.g., in sinusoidal steady-state analysis of electrical circuits). A real voltage $v_m\cos(\omega_0 t + \phi_0)$ is represented by the phasor $V = v_m e^{j\phi_0}$. The phasor represents the real AC signal by a complex DC signal V.

3. [p. 12] The Dirac delta function is not a function in the usual sense, since it is infinite at $t = 0$. There is a mathematical framework that treats the delta function (and its derivatives) in a rigorous manner. This framework, called the *theory of distributions*, is beyond the scope of this book; see Kwakernaak and Sivan [1991, Supplement C] for a relatively elementary treatment of this subject. We shall continue to use the delta function but not in rigor, as common in engineering books.

4. [p. 13] A common misconception is that every LTI system has an impulse response. The following example shows that this is not true. Let $x(t)$ be in the input family if and only if (1) $x(t)$ is continuous, except at a countable number of points t; (2) the discontinuity at each such point is a finite jump, that is, the limits at both sides of the discontinuity exist; (3) the sum of absolute values of all discontinuity jumps is finite. Let $y(t)$ be the sum of all jumps of $x(\tau)$ at discontinuity points $\tau < t$. This system is linear and time invariant, but it has no impulse response because $\delta(t)$ is not in the input family. Consequently, its response to $x(t)$ cannot be described by a convolution. This example is by Kailath [1980].

5. [p. 14] The term *zero state* is related to the assumption that the system has no memory, or that its response to the input $x(t) = 0$, $-\infty < t < \infty$ is identically

zero. For example, an electrical circuit containing capacitors is in zero state if its capacitors are completely discharged before the input signal is applied. If some capacitors are charged, the circuit may have nonzero output even in the absence of any input. In this case, the system is not strictly linear since a linear system must, by definition, give zero response to zero input.

6. [p. 21] If x is a random variable possessing a probability density function $p(x)$ and $g(x)$ is any function of x, the expectation of $g(x)$ is

$$E[g(x)] = \int_{-\infty}^{\infty} g(x)p(x)dx, \tag{2.142}$$

provided the right side exists. If x is a vector random variable, the density $p(x)$ is multidimensional and then the integral should be understood as a multiple integral.

7. [p. 22] The Cauchy–Schwarz inequality applies to any pair of mathematical objects for which an *inner product* is defined. If we denote the inner product of two objects x and y by $\langle x, y \rangle$, then the Cauchy–Schwarz inequality is

$$|\langle x, y \rangle| \leq \langle x, x \rangle^{1/2} \langle y, y \rangle^{1/2}, \tag{2.143}$$

with equality if and only if x and y are equal up to a proportionality constant. For example, if x and y are vectors in the three-dimensional Euclidean space, then

$$\langle x, y \rangle = \langle x, x \rangle^{1/2} \langle y, y \rangle^{1/2} \cos \alpha, \tag{2.144}$$

where α is the angle between the two vectors. In this case, the Cauchy–Schwarz inequality simply states that the cosine of any angle is not larger than 1 in magnitude. A few other objects to which the Cauchy–Schwarz inequality applies are:

(a) Vectors in the complex N-dimensional Euclidean space; for those

$$\left| \sum_{n=1}^{N} x_n \bar{y}_n \right| \leq \left[\sum_{n=1}^{N} |x_n|^2 \right]^{1/2} \left[\sum_{n=1}^{N} |y_n|^2 \right]^{1/2}. \tag{2.145}$$

(b) Complex sequences having finite energy; for those

$$\left| \sum_{n=-\infty}^{\infty} x_n \bar{y}_n \right| \leq \left[\sum_{n=-\infty}^{\infty} |x_n|^2 \right]^{1/2} \left[\sum_{n=-\infty}^{\infty} |y_n|^2 \right]^{1/2}. \tag{2.146}$$

(c) Complex-valued functions, square integrable on a certain domain; for those

$$\left| \int x(t)\bar{y}(t)dt \right| \leq \left[\int |x(t)|^2 dt \right]^{1/2} \left[\int |y(t)|^2 dt \right]^{1/2}. \tag{2.147}$$

(d) Random variables having finite second moments; for those

$$E(xy) \leq [E(x^2)]^{1/2}[E(y^2)]^{1/2}. \tag{2.148}$$

8. [p. 22] Wide-sense stationarity should be distinguished from *strict-sense stationarity*, which is a much stronger property; see Papoulis [1991]. Strict-sense stationarity is outside the scope of our discussion.

9. [p. 27] Many books refer to the Fourier transform of a sequence as the *discrete-time Fourier transform*. We avoid this terminology for the following reasons: (1) to avoid confusion with the discrete Fourier transform, which we shall study in Chapter 4; (2) because the transform itself is not in discrete time, but in the continuous variable θ; (3) because it is redundant—the transformed object being a sequence uniquely defines the transform.

10. [p. 28] The integral in (2.97) is *not* a standard (Riemann) integral, but a *Lebesgue* integral. The discussion of such integrals is outside the scope of this book; see Rudin [1964].

2.11   Problems

2.1 Compute the Fourier transform of the continuous-time signal

$$x(t) = \begin{cases} C, & a \le t \le b, \\ 0, & \text{otherwise}, \end{cases}$$

where a, b, C are real constants. Use the Fourier transform of the rect function, and the shift and scale properties (2.5), (2.7) of the Fourier transform. Then verify the result by a direct computation.

2.2 Compute the Fourier transform of the continuous-time signal

$$x(t) = \begin{cases} 1 - |t|, & -1 \le t \le 1, \\ 0, & \text{otherwise}, \end{cases}$$

in two different ways.

2.3 Compute the Fourier transforms of the continuous-time signals

$$x_1(t) = \begin{cases} e^{-\alpha t}, & t \ge 0, \\ 0, & t < 0, \end{cases} \qquad x_2(t) = \begin{cases} te^{-\alpha t}, & t \ge 0, \\ 0, & t < 0, \end{cases}$$

where $\alpha < 0$.

2.4 Compute the Fourier transform of the continuous-time signal

$$x(t) = \text{rect}(t) \cos(\omega_0 t),$$

where ω_0 is a real positive constant.

2.5 Does the signal $x(t) = \cosh(t)$ have a Fourier transform? If so, compute it. If not, explain the reason (cosh denotes a hyperbolic cosine).

2.6 Repeat Problem 2.5 for the signal $x(t) = \cosh(t)\text{rect}(t)$.

2.7 Let $g(t)$ be the Gaussian function, defined in (2.36), and let $h(t) = \{g * g\}(t)$. Compute $h(t)$, using the Fourier transform of $g(t)$.

2.8 The signal $x(t)$ is passed through the LTI system whose impulse response is $h(t)$, where

$$x(t) = \text{sinc}^2 \left(\frac{\omega_0 t}{\pi} \right), \quad h(t) = \text{sinc} \left(\frac{\omega_0 t}{\pi} \right).$$

Compute the output $y(t)$ of the system.

2.9 Let $X^F(\omega)$ be as shown in Figure 2.10. Compute the inverse Fourier transform $x(t)$. Hint: You can use your solution to Problem 2.8.

2.10 We saw in (2.36), (2.37) that the inverse Fourier transform of $G^F(\omega) = e^{-0.5\omega^2}$ is a positive function. Is the same true for

$$X^F(\omega) = e^{-0.5(\omega - \omega_0)^2} + e^{-0.5(\omega + \omega_0)^2},$$

where ω_0 is a constant real number?

2.11 Let $x(t)$ be a continuous-time signal possessing a derivative $dx(t)/dt$ and a Fourier transform $X^F(\omega)$.

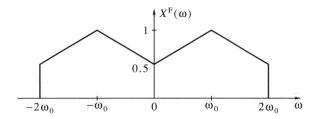

Figure 2.10 Pertaining to Problem 2.9.

(a) Prove the derivative property of the Fourier transform

$$y(t) = \frac{dx(t)}{dt} \implies Y^F(\omega) = j\omega X^F(\omega). \tag{2.149}$$

(b) Use part a and Parseval's theorem to express the quantity

$$\int_{-\infty}^{\infty} \left| \frac{dx(t)}{dt} \right|^2 dt$$

in terms of $X^F(\omega)$.

2.12 In this problem we derive the Fourier transforms of the sign function and the unit-step function.

(a) Find the Fourier transform of the signal

$$x(t) = \begin{cases} e^{-\alpha t}, & t > 0, \\ 0, & t = 0, \\ -e^{\alpha t}, & t < 0, \end{cases} \quad \text{where} \quad \alpha > 0.$$

(b) The *sign function* sign(t) is defined by

$$\text{sign}(t) = \begin{cases} 1, & t > 0, \\ 0, & t = 0, \\ -1, & t < 0. \end{cases}$$

Find the Fourier transform of the sign function.

(c) The *unit-step function* $v(t)$ is defined by

$$v(t) = \begin{cases} 1, & t > 0, \\ 0.5, & t = 0, \\ 0, & t < 0. \end{cases}$$

Find the Fourier transform of the unit-step function. Hint: Express $v(t)$ in terms of sign(t) and another signal.

2.13 Find the inverse Fourier transform of

$$H^F(\omega) = -j \,\text{sign}(\omega). \tag{2.150}$$

Hint: Use the solution to Problem 2.12. The LTI system whose frequency response is $H^F(\omega)$ is called a *Hilbert transformer*. The signal

$$y(t) = \{x * h\}(t)$$

is called the *Hilbert transform* of $x(t)$.

2.14 Let
$$y(t) = x(t) + j\{x * h\}(t),$$
where $h(t)$ is the impulse response of $H^F(\omega)$ given in (2.150). Find $Y^F(\omega)$ in terms of $X^F(\omega)$. The signal $y(t)$ is called the *analytic signal* of $x(t)$.

2.15 This problem discusses the spectra of certain modulated signals.

(a) Let
$$y(t) = x(t)\cos(\omega_0 t), \quad \omega_0 > 0.$$
Express $Y^F(\omega)$ in terms of $X^F(\omega)$.

(b) Repeat part a for
$$y(t) = x(t)\sin(\omega_0 t), \quad \omega_0 > 0.$$
The operation in either part a or part b is called *double-side-band modulation* (DSB).

(c) Let
$$y(t) = x_1(t)\cos(\omega_0 t) - x_2(t)\sin(\omega_0 t), \quad \omega_0 > 0.$$
Express $Y^F(\omega)$ in terms of $X_1^F(\omega)$ and $X_2^F(\omega)$.
This operation is called *quadrature amplitude modulation* (QAM); it is widely used in digital communication.

2.16 Consider the modulation operations described in Problem 2.15. Assume that the signal $x(t)$ is such that $X^F(\omega) = 0$ for $|\omega| > \omega_m$, where ω_m is fixed [for part c assume the same for both $x_1(t)$ and $x_2(t)$]. Assume further that $\omega_0 > \omega_m$. For each of the three modulation types, let $z(t)$ be the corresponding analytic signal of $y(t)$, as defined in Problem 2.14. Find $Z^F(\omega)$ in all three cases.

2.17 Let $x(t)$ be a real signal. Prove that:

(a)
$$x(t) = x(-t) \implies X^F(\omega) \text{ is real}, \tag{2.151}$$

(b)
$$x(t) = -x(-t) \implies X^F(\omega) \text{ is imaginary}. \tag{2.152}$$

2.18 For a real signal $x(t)$ we define
$$x_e(t) = 0.5x(t) + 0.5x(-t), \tag{2.153a}$$
$$x_o(t) = 0.5x(t) - 0.5x(-t). \tag{2.153b}$$

The signals $x_e(t)$, $x_o(t)$ are called the *even* and *odd* parts of $x(t)$, respectively. Prove that
$$X_e^F(\omega) = \Re\{X^F(\omega)\}, \quad X_o^F(\omega) = j\Im\{X^F(\omega)\}. \tag{2.154}$$

2.19 Let $x(t)$ be a real signal satisfying
$$x(t) = 0, \quad t < 0,$$
and $x(t)$ has no delta component at $t = 0$.

(a) Show that
$$x_o(t) = \text{sign}(t)x_e(t).$$

(b) Show that $\Im\{X^F(\omega)\}$ is the Hilbert transform of $\Re\{X^F(\omega)\}$, as defined in Problem 2.13.

2.20 For a signal $x(t)$, let

$$P_x = \int_{-\infty}^{\infty} |x(t)|^2 dt = \frac{1}{2\pi} \int_{-\infty}^{\infty} |X^F(\omega)|^2 d\omega.$$

The *centroid* of a signal $x(t)$ and that of its Fourier transform are defined as

$$C_x = P_x^{-1} \int_{-\infty}^{\infty} t|x(t)|^2 dt, \qquad (2.155a)$$

$$C_X = P_x^{-1} \frac{1}{2\pi} \int_{-\infty}^{\infty} \omega|X^F(\omega)|^2 d\omega. \qquad (2.155b)$$

The *effective width* of a signal $x(t)$ and that of its Fourier transform are defined as

$$W_x = \left[P_x^{-1} \int_{-\infty}^{\infty} (t - C_x)^2 |x(t)|^2 dt \right]^{1/2}, \qquad (2.156a)$$

$$W_X = \left[P_x^{-1} \frac{1}{2\pi} \int_{-\infty}^{\infty} (\omega - C_X)^2 |X^F(\omega)|^2 d\omega \right]^{1/2}. \qquad (2.156b)$$

(a) Compute the centroid and effective width of the signal

$$x(t) = e^{-t}, \quad t \geq 0.$$

(b) Compute the effective width of the signal $\text{rect}(t)$.

(c) Explain why, for any real signal $x(t)$, the centroid of $X^F(\omega)$ is zero.

2.21 Let $x(t)$ be a signal whose Fourier transform $X^F(\omega)$ exists. Prove that

$$\sum_{n=-\infty}^{\infty} x(t - nT) = \frac{1}{T} \sum_{k=-\infty}^{\infty} X^F\left(\frac{2\pi k}{T}\right) \exp\left(j\frac{2\pi kt}{T}\right). \qquad (2.157)$$

Hint: Derive the Fourier series representation of the left side of (2.157).

2.22 Let

$$x(t) = \begin{cases} e^{-\alpha t}, & t \geq 0, \\ 0, & t < 0, \end{cases} \qquad y(t) = \sum_{n=-\infty}^{\infty} x(t - nT).$$

Find the Fourier series coefficients of the periodic signal $y(t)$.

2.23 Define a *pulse train* signal by

$$r_{T,\Delta}(t) = \frac{1}{\Delta} \sum_{n=-\infty}^{\infty} \text{rect}\left(\frac{t - nT}{\Delta}\right), \quad \text{where} \quad 0 < \Delta < T. \qquad (2.158)$$

Compute the Fourier transform of $r_{T,\Delta}(t)$.

2.24 Define an *alternating impulse train*

$$q_T(t) = \sum_{n=-\infty}^{\infty} (-1)^n \delta(t - nT),$$

see Figure 2.11. Compute $Q_T^F(\omega)$, the Fourier transform of $q_T(t)$.

2.25 The impulse train $p_T(t)$ is passed through an LTI filter whose impulse response is

$$h(t) = \text{sinc}\left(\frac{0.5t}{T}\right) \cos\left(\frac{\pi t}{T}\right).$$

Find the output $y(t)$ of the filter. Hint: Solve in the frequency domain.

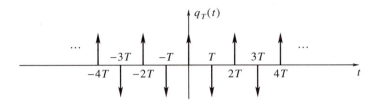

Figure 2.11 Pertaining to Problem 2.24.

2.26 The form of the Poisson sum formula given in (2.48) is

$$\sum_{n=-\infty}^{\infty} \delta(t - nT) = \frac{1}{T} \sum_{k=-\infty}^{\infty} \exp\left(j\frac{2\pi kt}{T}\right). \tag{2.159}$$

In mathematical texts, the formula is usually stated as

$$\sum_{n=-\infty}^{\infty} x(nT) = \frac{1}{T} \sum_{k=-\infty}^{\infty} X^{\mathrm{F}}\left(\frac{2\pi k}{T}\right), \tag{2.160}$$

provided $x(t)$ is continuous at the points nT and $X^{\mathrm{F}}(\omega)$ is continuous at the points $2\pi k/T$. Show that (2.159) implies (2.160).

2.27 Compute the Fourier transform of the discrete-time signal

$$x[n] = \begin{cases} C, & n_1 \le n \le n_2, \\ 0, & \text{otherwise,} \end{cases}$$

where n_1, n_2 are integer constants, and C is a real constant.

2.28 The Fourier transform of a discrete-time signal $x[n]$ is

$$X^{\mathrm{f}}(\theta) = \cos\theta + \sin(2\theta).$$

Compute $x[n]$.

2.29 Let $x[n]$ be the signal

$$x[n] = \begin{cases} \{1, -1, -2, 4, -2, -1, 1\}, & -3 \le n \le 3, \\ 0, & \text{otherwise.} \end{cases}$$

Compute the following quantities without finding $X^{\mathrm{f}}(\theta)$ first.

(a) $X^{\mathrm{f}}(0)$.

(b) $\measuredangle X^{\mathrm{f}}(\theta)$.

(c) $\int_{-\pi}^{\pi} X^{\mathrm{f}}(\theta)d\theta$.

(d) $X^{\mathrm{f}}(\pi)$.

(e) $\int_{-\pi}^{\pi} |X^{\mathrm{f}}(\theta)|^2 d\theta$.

(f) $\frac{dX^{\mathrm{f}}(\theta)}{d\theta}\big|_{\theta=0}$.

2.30 Can the function

$$X^{\mathrm{f}}(\theta) = \cos(0.5\theta), \quad \theta \in \mathbb{R}$$

be the Fourier transform of a discrete-time signal $x[n]$? If so, find $x[n]$. If not, explain the reason.

2.31 This problem discusses the Fourier transform of a sequence modulated by sign alternations.

(a) If $X^f(\theta)$ is the Fourier transform of $x[n]$, what is the Fourier transform of $(-1)^n x[n]$?

(b) Express $\sum_{n=-\infty}^{\infty} (-1)^n x[n]$ in terms of the Fourier transform of $x[n]$.

2.32 We are given two linear discrete-time systems. The response of the first to a unit impulse $\delta[n-k]$ is $h_1[n] = \sin[3(n-k)]$ and that of the second is $h_2[n] = \sin[3(n+k)]$. Is either system time invariant?

2.33 Prove Theorem 2.5.

2.34 Let $x[n]$ be a discrete-time signal with Fourier transform $X^f(\theta)$, and let

$$Y^f(\theta) = X^f(\theta) + X^f(\theta - \pi).$$

Prove that $Y^f(\theta)$ depends only on the even-indexed signal values $x[2m]$, and is independent of the odd-indexed values $x[2m + 1]$.

2.35 Prove that

$$x[n] = \cos(\theta_0 n + \phi_0)$$

is periodic if and only if $\theta_0 = 2\pi p/q$, where p and q are positive integers. Find the period N in case the condition is satisfied.

2.36 The *correlation* (or *cross-correlation*) of two continuous-time signals is defined as

$$z(t) = \{x \star y\}(t) = \int_{-\infty}^{\infty} x(t + \tau)\bar{y}(\tau)d\tau. \qquad (2.161)$$

Similarly, the correlation of two discrete-time signals is

$$z[n] = \{x \star y\}[n] = \sum_{m=-\infty}^{\infty} x[n + m]\bar{y}[m]. \qquad (2.162)$$

Express the correlation operation in the frequency domain, for both continuous-time and discrete-time signals.

2.37* Prove that (2.65) implies (2.66).

2.38* Suppose we have two discrete-time LTI systems connected in series, so the frequency response of the series connection is

$$H^f(\theta) = H_1^f(\theta)H_2^f(\theta).$$

Recall the definition of noise gain (2.137). Let NG_1, NG_2, NG be the noise gains of the corresponding frequency response functions.

(a) Show that, in general,

$$NG \neq NG_1 NG_2.$$

(b) If $H_1^f(\theta) = C$, where C is a real positive constant, what can you say about NG in this special case?

2.39* This problem discusses the effect of an LTI system on white noise in the covariance domain.

(a) White noise $x(t)$, whose covariance function $\kappa_x(\tau)$ is as in (2.76), is given at the input of an LTI system with impulse response $h(t)$. Use (2.87) to compute $\kappa_y(\tau)$, the covariance function of the output, in this special case.

(b) Repeat part a for the discrete-time case. The white noise at the input has covariance sequence $\kappa_x[m]$, as given in (2.127), and the LTI system has impulse response $h[n]$. Find $\kappa_y[m]$.

2.40* A discrete-time white noise $x[n]$ with zero mean and variance γ_x is fed to an ideal low-pass filter whose frequency response is

$$H^f(\theta) = \text{rect}\left(\frac{\theta}{2\theta_c}\right), \quad 0 < \theta_c < \pi.$$

Let $y[n]$ be the response of the filter to $x[n]$. Compute the covariance sequence $\kappa_y[m]$ of $y[n]$.

2.41* Show that the Cauchy–Schwarz inequality for random variables (2.148) follows from the inequality for functions (2.147). Hint: Use (2.142) for the vector random variable consisting of $\{x, y\}$.

2.42* Let $x(t)$ be a continuous-time signal on $[-0.5T, 0.5T]$. The signal has both a Fourier series (2.39) and a Fourier transform (2.1).

(a) Express the Fourier coefficients $X^S[k]$ in terms of the Fourier transform $X^F(\omega)$.

(b) Express the Fourier transform $X^F(\omega)$ in terms of the Fourier coefficients $X^S[k]$.

2.43* This problem illustrates the phenomenon of *harmonic distortions* caused by nonlinear operations on sinusoidal signals, in this case by hard limiting. Let $x(t)$ be the periodic signal

$$x(t) = \begin{cases} \cos(2\pi t), & -A \le \cos(2\pi t) \le A, \\ -A, & \cos(2\pi t) < -A, \\ A, & \cos(2\pi t) > A, \end{cases} \quad \text{where} \quad 0 < A < 1.$$

(a) Compute the Fourier series coefficients of $x(t)$ as a function of A. Hint: Compute separately for even k, for $k = \pm 1$, and for all other odd k.

(b) Compute the numerical values of the ratios $\{X^S[k]/X^S[1], 2 \le k \le 10\}$ for $A = 0.9, 0.5, 0.1, 0.01$.

(c) State your conclusions from these results.

2.44* Let

$$x(t) = \cos(\omega_m t), \quad y(t) = [1 + ax(t)]\sin(\omega_c t),$$

where $0 < a < 1$ and $\omega_m \ll \omega_c$. The operation of constructing $y(t)$ from $x(t)$ is called *amplitude modulation* (AM). In this problem we assume, for simplicity, that $\omega_c = M\omega_m$, where M is an integer much larger than 1.

(a) Show that $y(t)$ is periodic with period $2\pi/\omega_m$.

(b) Find the Fourier series coefficients of $y(t)$.

(c) Define

$$z(t) = \begin{cases} y(t), & y(t) \ge 0, \\ 0, & y(t) < 0. \end{cases}$$

The signal $z(t)$ is called *half-wave rectification* of $y(t)$. Since $a < 1$, we have

$$z(t) = \begin{cases} y(t), & \sin(\omega_c t) \geq 0, \\ 0, & \sin(\omega_c t) < 0. \end{cases}$$

Show that $z(t)$ is periodic with period $2\pi/\omega_m$.

(d) Find the Fourier series coefficients $Z^S[k]$ of $z(t)$ in the range

$$-(M+1) \leq k \leq M+1.$$

(e) Suggest a way to extract the signal $x(t)$ from $z(t)$.

2.45* In this problem we seek to prove the *uncertainty principle*:

> The product of the effective width of a signal and the effective width of its Fourier transform is not smaller than 0.5.

We use the definitions of centroids C_x, C_X and effective widths W_x, W_X, as given in Problem 2.20. We assume, for simplicity, that the signal is real and that the centroids of both the signal and its transforms are zero. However, the proof can be extended to signals not thus restricted.

(a) Prove that, if

$$\lim_{t \to \infty} t|x(t)|^2 = \lim_{t \to -\infty} t|x(t)|^2 = 0,$$

then

$$-\int_{-\infty}^{\infty} t \frac{d|x(t)|^2}{dt} dt = \int_{-\infty}^{\infty} |x(t)|^2 dt = P_x.$$

Hint: Use integration by parts.

(b) Use Problem 2.11 to express W_X in terms of $dx(t)/dt$.

(c) Use the Cauchy–Schwarz inequality (2.147) to find a lower bound on $W_x W_X$.

(d) Finally, use part b to show that

$$W_x W_X \geq 0.5. \tag{2.163}$$

The uncertainty principle is of fundamental importance in Fourier theory. It shows that a signal cannot be arbitrarily narrow in both time and frequency. The celebrated *Heisenberg uncertainty principle* in quantum mechanics is a consequence of (2.163) as well.

2.46* Let $x[n]$ be a discrete-time signal that is identically zero for $n \geq N$ and $n < 0$, for some integer N. Prove that the set

$$\{\theta : -\pi \leq \theta < \pi, \ X^f(\theta) = 0\}$$

contains a finite number of points at most. Hint: A polynomial equation has only a finite number of solutions.

2.47* Consider the claim that for two sequences $x[n]$, $y[n]$, $\{x * y\} = 0$ implies that at least one of the two sequences is identically zero. Is this claim true? If so, prove it; if not, give a counterexample.

2.48* Repeat Problem 2.47 if the two sequences $x[n]$, $y[n]$ have finite durations.

2.49* This problem illustrates an application of Fourier transform for evaluation of infinite sums.

(a) Evaluate the infinite sum

$$\sum_{m=0}^{\infty} \frac{1}{(2m+1)^2}.$$

Hint: Find a sequence $x[n]$ and its Fourier transform $X^f(\theta)$ such that Parseval's theorem (2.105) will yield the required result.

(b) Use part a to show that

$$\sum_{k=1}^{\infty} \frac{1}{k^2} = \frac{\pi^2}{6}.$$

Chapter 3

Sampling and Reconstruction

Signals encountered in real-life applications are usually in continuous time. To facilitate digital processing, a continuous-time signal must be converted to a sequence of numbers. The process of converting a continuous-time signal to a sequence of numbers is called *sampling*. A motion picture is a familiar example of sampling. In a motion picture, a continuously varying scene is converted by the camera to a sequence of frames. The frames are taken at regular time intervals, typically 24 per second. We then say that the scene is sampled at 24 frames per second. Sampling is essentially a selection of a finite number of data at any finite time interval as representatives of the infinite amount of data contained in the continuous-time signal in that interval. In the motion picture example, the frames taken at each second are representatives of the continuously varying scene during that second.

When we watch a motion picture, our eyes and brain fill the gaps between the frames and give the illusion of a continuous motion. The operation of filling the gaps in the sampled data is called *reconstruction*. In general, reconstruction is the operation of converting a sampled signal back to a continuous-time signal. Reconstruction provides an infinite (and continuously varying) number of data at any given time interval out of the finite number of data in the sampled signal. In the motion picture example, the reconstructed continuous-time scene exists only in our brain.

Naturally, one would not expect the reconstructed signal to be absolutely faithful to the original signal. Indeed, sampling leads to distortions in general. The fundamental distortion introduced by sampling is called *aliasing*. Aliasing in motion pictures is a familiar phenomenon. Suppose the scene contains a clockwise-rotating wheel. As long as the speed of rotation is lower than half the number of frames per second, our brain perceives the correct speed of rotation. When the speed increases beyond this value, the wheel appears to rotate *counterclockwise* at a reduced speed. Its apparent speed is now the number of frames per second minus its true speed. When the speed is equal to the number of frames per second, it appears to stop rotating. This happens because all frames now sample the wheel in an identical position. When the speed increases further, the wheel appears to rotate clockwise again, but at a reduced speed. In general, the wheel always appears to rotate at a speed not higher than half the number of frames per second, either clockwise or counterclockwise.

Sampling is *the* fundamental operation of digital signal processing, and avoiding (or at least minimizing) aliasing is the most important aspect of sampling. Thorough understanding of sampling is necessary for any practical application of digital signal processing. In most engineering applications, the continuous-time signal is given in an electrical form (i.e., as a voltage waveform), so sampling is performed by an electronic circuit (and the same is true for reconstruction). Physical properties (and limitations) of electronic circuitry lead to further distortions of the sampled signal, and these need to be thoroughly understood as well.

In this chapter we study the mathematical theory of sampling and its practical aspects. We first define sampling in mathematical terms and derive the fundamental result of sampling theory: the sampling theorem of Nyquist, Whittaker, and Shannon. We examine the consequences of the sampling theorem for signals with finite and infinite bandwidths. We then deal with reconstruction of signals from their sampled values. Next we consider physical implementation of sampling and reconstruction, and explain the deviations from ideal behavior due to hardware limitations. Finally, we discuss several special topics related to sampling and reconstruction.

3.1   Two Points of View on Sampling

Let $x(t)$ be a continuous function on the real line. Sampling of the function amounts to picking its values at certain time points. In particular, if the sampling points are nT, $n \in \mathbb{Z}$, it is called *uniform sampling*, and T is called the *sampling interval*. The numbers

$$f_{\text{sam}} = \frac{1}{T}, \quad \omega_{\text{sam}} = \frac{2\pi}{T}$$

are called the *sampling frequency* (or *sampling rate*) and *angular sampling frequency*, respectively. The sampled function is then the sequence

$$x[n] = x(nT), \quad n \in \mathbb{Z}. \tag{3.1}$$

This definition can be extended to discontinuous functions, provided the discontinuities are isolated (no more than a finite number of them on any finite interval), and the limits at each discontinuity point exist from both left and right. In this case it is common to define

$$x[n] = 0.5[x(nT^-) + x(nT^+)] \tag{3.2}$$

if nT is a discontinuity point.

An alternative description of the sampling operation is as follows. Recall the impulse train $p_T(t)$ defined in (2.47) and let

$$x_{\text{p}}(t) = x(t)p_T(t) = \sum_{n=-\infty}^{\infty} x(t)\delta(t - nT) = \sum_{n=-\infty}^{\infty} x(nT)\delta(t - nT). \tag{3.3}$$

Then $x_{\text{p}}(t)$ is, formally, a continuous-time signal, since it is defined for all t. However, it is clear that the information it conveys about $x(t)$ is limited to the values $x(nT)$, $n \in \mathbb{Z}$, since $x_{\text{p}}(t)$ is identically zero at all other points.

We thus have two ways to look at the sampled signal:

1. To consider it as a sequence of numbers $x[n] = x(nT), n \in \mathbb{Z}$, or as a discrete-time signal. We refer to $x[n]$ as *point sampling* of $x(t)$.

2. To consider it as a continuous-time signal $x_{\text{p}}(t)$. We refer to $x_{\text{p}}(t)$ as *impulse sampling* of $x(t)$.

Figure 3.1 illustrates the sampling operation from the two points of view. Part a shows a continuous-time signal, with the sampling points emphasized. Part b shows the discrete-time signal $x[n]$ obtained by point sampling. Part c shows the continuous-time signal $x_{\text{p}}(t)$ obtained by impulse sampling.

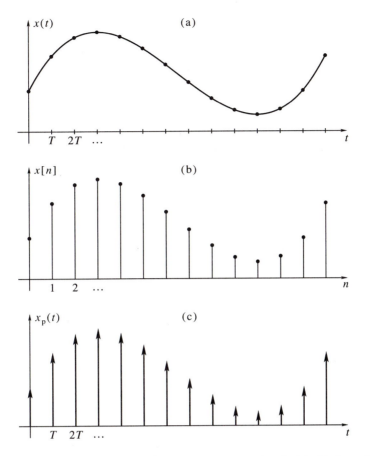

Figure 3.1 The sampling operation: (a) the continuous-time signal $x(t)$; (b) the point-sampled sequence $x[n]$; (c) the impulse-sampled signal $x_p(t)$.

The two points of view on sampling are equivalent. Physically, only point sampling is possible (and even this is an idealization of real-life sampling). Impulse sampling is convenient for mathematical derivations, since known results from continuous-time signal analysis can be used.

Example 3.1 Here are a few examples of sampled signals.

1. Let $x(t)$ be the sinusoidal signal

$$x(t) = A\cos(\omega_0 t + \phi_0).$$

Then,

$$x[n] = A\cos(\omega_0 Tn + \phi_0).$$

The sequence $x[n]$ is a *discrete-time sinusoid*, with a discrete angular frequency $\omega_0 T$. The discrete frequency is measured in radians (or radians per sample, but not radians per second).

2. Let $x(t)$ be the one-sided exponential function

$$x(t) = \begin{cases} 0, & t < 0, \\ Ae^{-\alpha t}, & t \geq 0, \end{cases} \tag{3.4}$$

where $\alpha \geq 0$. Then,

$$x[n] = \begin{cases} 0, & n < 0, \\ Ae^{-\alpha Tn}, & n \geq 0. \end{cases} \tag{3.5}$$

We emphasize that, in this case, it is common to define $x[0]$ as $x(0^+)$, rather than as the midpoint between the two discontinuity limits. The signal $x[n]$ is a geometric series, with parameter $e^{-\alpha T}$.

3. A special case of a one-sided exponential occurs when $\alpha = 0$, $A = 1$. The continuous-time and discrete-time signals are then

$$v(t) = \begin{cases} 0, & t < 0, \\ 1, & t \geq 0, \end{cases} \quad \text{and} \quad v[n] = \begin{cases} 0, & n < 0, \\ 1, & n \geq 0. \end{cases} \tag{3.6}$$

These are the continuous-time and discrete-time *unit-step* signals, respectively.

4. The continuous-time delta function $\delta(t)$ *cannot be sampled*, since it does not have a finite value at $t = 0$. It is a gross mistake to assume that its sampled version is the discrete-time delta $\delta[n]$.

5. Let $x(t) = \text{rect}(t/T_0)$. Then the sampled signal is

$$x[n] = \begin{cases} 1, & |n| \leq \left\lfloor \dfrac{0.5T_0}{T} \right\rfloor, \\ 0, & \text{otherwise.} \end{cases}$$

In particular, if $T_0 < 2T$, then the sampled signal is $\delta[n]$. $\square$

3.2 The Sampling Theorem

Suppose we are given a continuous-time signal $x(t)$ whose Fourier transform is $X^F(\omega)$. Then the impulse-sampled signal $x_p(t)$ has Fourier transform $X_p^F(\omega)$, and the point-sampled signal $x[n]$ has Fourier transform $X^f(\theta)$, defined as

$$X_p^F(\omega) = \int_{-\infty}^{\infty} x_p(t)e^{-j\omega t}dt, \quad X^f(\theta) = \sum_{n=-\infty}^{\infty} x(nT)e^{-j\theta n}. \tag{3.7}$$

The questions we address in this section are (1) How are $X_p^F(\omega)$ and $X^f(\theta)$ related to each other, and (2) how are they related to $X^F(\omega)$? The answers would enable us to understand the consequences of sampling, to assess the nature and amount of distortion introduced by sampling, and to tell under what conditions it is possible to completely avoid distortions due to sampling.

The relationships among the three Fourier transforms are given by the famous *sampling theorem*.[1]

Theorem 3.1 The Fourier transforms $X^F(\omega)$, $X_p^F(\omega)$, and $X^f(\theta)$ satisfy the following relationships:

$$X_p^F(\omega) = \sum_{n=-\infty}^{\infty} x(nT)e^{-jn\omega T} \tag{3.8}$$

$$= \frac{1}{T} \sum_{k=-\infty}^{\infty} X^F\left(\omega - \frac{2\pi k}{T}\right), \tag{3.9}$$

$$X^f(\theta) = X_p^F\left(\frac{\theta}{T}\right) = \frac{1}{T} \sum_{k=-\infty}^{\infty} X^F\left(\frac{\theta - 2\pi k}{T}\right). \tag{3.10}$$

Proof To prove (3.8), substitute the definition (3.3) of $x_p(t)$ in the definition of Fourier transform:

$$X_p^F(\omega) = \int_{-\infty}^{\infty} \left[\sum_{n=-\infty}^{\infty} x(nT)\delta(t - nT) \right] e^{-j\omega t} dt$$

$$= \sum_{n=-\infty}^{\infty} x(nT) \int_{-\infty}^{\infty} \delta(t - nT)e^{-j\omega t} dt = \sum_{n=-\infty}^{\infty} x(nT)e^{-jn\omega T}. \qquad (3.11)$$

To prove (3.9), recall that (1) a product in the time domain translates to a convolution in the frequency domain; (2) the Fourier transform of an impulse train is an impulse train in the frequency domain; see (2.52). Therefore,

$$X_p^F(\omega) = \frac{1}{2\pi} \{P_T^F * X^F\}(\omega) = \frac{1}{2\pi} \int_{-\infty}^{\infty} X^F(\lambda) \left[\frac{2\pi}{T} \sum_{k=-\infty}^{\infty} \delta\left(\omega - \lambda - \frac{2\pi k}{T} \right) \right] d\lambda$$

$$= \frac{1}{T} \sum_{k=-\infty}^{\infty} X^F\left(\omega - \frac{2\pi k}{T} \right). \qquad (3.12)$$

Finally, (3.10) follows when substituting $\theta = \omega T$ in (3.8), (3.9).

An alternative proof of (3.9) The following proof does not rely on the properties of the impulse train:

$$x(nT) = \frac{1}{2\pi} \int_{-\infty}^{\infty} X^F(\omega)e^{j\omega nT} d\omega = \sum_{k=-\infty}^{\infty} \left[\frac{1}{2\pi} \int_{-(2k+1)\pi/T}^{-(2k-1)\pi/T} X^F(\omega)e^{j\omega nT} d\omega \right]$$

$$= \sum_{k=-\infty}^{\infty} \left[\frac{1}{2\pi} \int_{-\pi/T}^{\pi/T} X^F\left(\omega - \frac{2\pi k}{T} \right) e^{j\omega nT} d\omega \right]$$

$$= \frac{1}{2\pi} \int_{-\pi}^{\pi} \left[\frac{1}{T} \sum_{k=-\infty}^{\infty} X^F\left(\frac{\theta - 2\pi k}{T} \right) \right] e^{j\theta n} d\theta. \qquad (3.13)$$

In passing from the first line of (3.13) to the second, we used the property

$$e^{-j2\pi nk} = 1, \quad n, k \in \mathbb{Z}.$$

Equation (3.13) gives, by the definition of the Fourier transform (2.93) and its inverse formula (2.95),

$$\sum_{n=-\infty}^{\infty} x(nT)e^{-j\omega nT} = \frac{1}{T} \sum_{k=-\infty}^{\infty} X^F\left(\omega - \frac{2\pi k}{T} \right). \qquad (3.14)$$

This is the same as (3.9). $\qquad\qquad\qquad\qquad\qquad\qquad\qquad\qquad\qquad\qquad\qquad\qquad$ □

The sampling theorem tells us that the Fourier transform of a discrete-time signal obtained from a continuous-time signal by sampling is related to the Fourier transform of the continuous-time signal by three operations:

1. Transformation of the frequency axis according to the relation $\theta = \omega T$.

2. Multiplication of the amplitude axis by a factor $1/T$.

3. Summation of an infinite number of replicas of the given spectrum, shifted horizontally by integer multiples of the angular sampling frequency ω_{sam}.

As a result of the infinite summation, the Fourier transform of the sampled signal is periodic in θ with period 2π. We therefore say that *sampling in the time domain gives rise to periodicity in the frequency domain.*

Example 3.2 Consider the signal $x(t) = te^{-\alpha t}v(t)$, where $\alpha > 0$. Its Fourier transform is

$$X^{\mathrm{F}}(\omega) = \int_0^\infty te^{-(\alpha+j\omega)t}dt = \frac{1}{(\alpha + j\omega)^2}. \tag{3.15}$$

The sampled signal is $x[n] = nTe^{-\alpha Tn}v[n]$, and its Fourier transform is

$$X^{\mathrm{f}}(\theta) = \sum_{n=0}^\infty nTe^{-(\alpha T+j\theta)n} = \frac{Te^{-(\alpha T+j\theta)}}{[1 - e^{-(\alpha T+j\theta)}]^2}. \tag{3.16}$$

The sampling theorem then leads us to conclude that

$$\sum_{k=-\infty}^\infty \frac{1}{[\alpha T + j(\theta - 2\pi k)]^2} = \frac{e^{-(\alpha T+j\theta)}}{[1 - e^{-(\alpha T+j\theta)}]^2}. \tag{3.17}$$

This infinite sum is not straightforward to prove directly, but it follows easily from sampling theory; see Problem 3.6 for another example.   □

3.3   The Three Cases of Sampling

A continuous-time signal $x(t)$ is called *band limited* if its Fourier transform vanishes outside a certain frequency range, that is, if there exists $\omega_{\mathrm{m}} > 0$ such that

$$X^{\mathrm{F}}(\omega) = 0 \quad \text{for} \quad |\omega| \geq \omega_{\mathrm{m}}. \tag{3.18}$$

The angular frequency ω_{m} is called the *bandwidth* of the signal. (When we learn about band-pass signals, in Section 3.6, we shall modify this definition.) Sometimes we shall say that the bandwidth is $\pm\omega_{\mathrm{m}}$, to emphasize that the Fourier transform is nonzero for both positive and negative frequencies. The definition can be extended to allow $X^{\mathrm{F}}(\pm\omega_{\mathrm{m}}) \neq 0$, provided there are no delta functions at $\omega = \pm\omega_{\mathrm{m}}$. The interval $[-\omega_{\mathrm{m}}, \omega_{\mathrm{m}}]$ is called the *frequency support* of the signal. For a signal that is not band limited, the frequency support is $(-\infty, \infty)$. Bandwidth thus defined is measured in radians per second. Expressed in hertz, the bandwidth is

$$f_{\mathrm{m}} = \frac{\omega_{\mathrm{m}}}{2\pi}.$$

Figure 3.2(a) illustrates the Fourier transform of a band-limited signal. Figure 3.2(b) shows, for comparison, the Fourier transform of a signal that is not band limited.

For now, we regard the property of being band limited as purely mathematical and defer the question of physical existence of such signals until later.

Example 3.3 The Fourier transform of the function $\mathrm{sinc}(t)$ is $\mathrm{rect}(\omega/2\pi)$, compare (2.35). The Fourier transform is nonzero only on the interval $|\omega| \leq \pi$, hence $\mathrm{sinc}(t)$ is a band-limited signal, with $\omega_{\mathrm{m}} = \pi$. On the other hand, the signal $x(t)$ in Example 3.2 is not band limited, because, as we see from (3.15), its Fourier transform does not vanish for any ω.   □

Suppose we sample a band-limited signal $x(t)$ and we choose the sampling frequency such that $f_{\mathrm{sam}} \geq 2f_{\mathrm{m}}$. The spectra of the continuous-time signal and the sampled signal are shown in Figure 3.3. *The replicas do not overlap in this case.* In particular, the Fourier transform of the sampled signal in the range $\theta \in [-\pi, \pi]$ is given by [cf. (3.10)]

$$X^{\mathrm{f}}(\theta) = \frac{1}{T}X^{\mathrm{F}}\left(\frac{\theta}{T}\right), \quad -\pi \leq \theta \leq \pi. \tag{3.19}$$

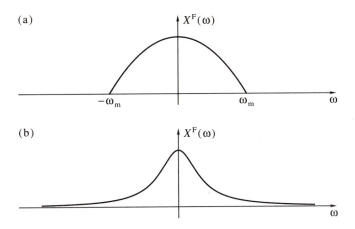

Figure 3.2 Fourier transform: (a) of a band-limited signal; (b) of a signal that is not band limited.

The conclusion is that the Fourier transform of the sampled signal in the principal frequency range $[-\pi, \pi]$ preserves the shape of the Fourier transform of the given signal, except for multiplication of the frequency and amplitude axes by constant factors. The shape at frequencies outside the interval $[-\pi, \pi]$ is obtained by periodic extension, as is always true for Fourier transforms of discrete-time signals.

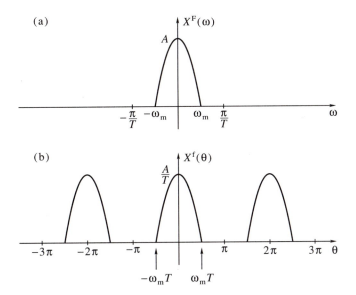

Figure 3.3 Sampling of a band-limited signal above the Nyquist rate: (a) Fourier transform of the continuous-time signal; (b) Fourier transform of the sampled signal.

The lowest sampling rate for which (3.19) holds is

$$f_{sam} = \frac{1}{T} = 2f_m.$$

This is called the *Nyquist rate*, or the *critical rate* for sampling of a band-limited signal. The first case of sampling can therefore be summarized as follows:

> If a band-limited signal is sampled at a rate equal to or greater than the Nyquist rate $2f_m$, the shape of the Fourier transform of the sampled signal in the range $\theta \in [-\pi, \pi]$ is identical to the shape of the Fourier transform of the given signal, except for multiplication of the frequency axis by a factor T, and multiplication of the amplitude axis by a factor $1/T$.

The fundamental relationship between continuous-time and discrete-time frequencies

$$\theta = \omega T = 2\pi f T$$

is of extreme importance, and should be borne in mind at all times. The quantities ω, f measure the number of radians or cycles, respectively, per second. The quantity θ measures the number of radians per sample. Since each sample extends over T seconds, we get the relationship shown in the box.

Since the Fourier transform of a band-limited signal is not distorted if the sampling rate meets the Nyquist condition, we expect to be able to reconstruct the continuous-time signal $x(t)$ from it samples $x(nT)$ in this case. Reconstruction is discussed in Section 3.4.

Example 3.4 Consider the signal

$$x(t) = \text{sinc}(f_0 t).$$

It follows from (2.35) and (2.7) that

$$X^{\text{F}}(\omega) = \frac{1}{f_0} \text{rect}\left(\frac{\omega}{2\pi f_0}\right).$$

Therefore, the Nyquist rate is $\omega_{\text{sam}} = 2\pi f_0$, or $f_{\text{sam}} = f_0$. The sampled signal is then

$$x[n] = \text{sinc}(n) = \delta[n].$$

The Fourier transform of the sampled signal is

$$X^{\text{f}}(\theta) = 1.$$

This result is perhaps intriguing: Surely, there are many continuous-time signals whose sampling at $T = 1/f_0$ would give the unit-sample signal. It follows from this result that, of all such signals, the sinc is the only one whose bandwidth is limited to $0.5f_0$ Hz, and no such signal can have a smaller bandwidth. This result has profound consequences in digital communication theory. A signal that, when sampled at an interval T, gives a unit impulse, is called a *Nyquist-T signal*.[2] We conclude that a Nyquist-T signal must have a bandwidth not smaller than $(0.5/T)$ Hz, and only the sinc signal achieves this lower bound.   □

Example 3.5 Consider the signal

$$y(t) = \text{sinc}^2(f_0 t).$$

We recall that multiplication in the time domain translates to convolution in the frequency domain. The convolution of two identical rect functions is a triangular function; therefore,

$$Y^{\text{F}}(\omega) = \frac{1}{2\pi}\{X^{\text{F}} * X^{\text{F}}\}(\omega) = \begin{cases} \frac{1}{f_0}\left(1 - \frac{|\omega|}{2\pi f_0}\right), & |\omega| \leq 2\pi f_0, \\ 0, & |\omega| > 2\pi f_0. \end{cases}$$

The Nyquist rate for $y(t)$ is $\omega_{sam} = 4\pi f_0$, or $f_{sam} = 2f_0$. The sampled signal is

$$y[n] = \text{sinc}^2(0.5n).$$

The Fourier transform of the sampled signal is

$$Y^f(\theta) = 2\left(1 - \frac{|\theta|}{\pi}\right), \quad -\pi \le \theta \le \pi.$$

□

We now consider the case of a band-limited signal sampled at a rate *lower* than the Nyquist rate. Figure 3.4 shows what happens in this case, using a sampling rate $f_{sam} = 3f_m/2$ as an example. Now the shape of the Fourier transform in the range $\theta \in [-2\pi/3, 2\pi/3]$ is preserved, but the shape in the range $|\theta| \in (2\pi/3, \pi]$ is distorted. Distortion occurs because, in this frequency range, two adjacent replicas overlap and their superposition [as expressed in (3.10)] gives rise to the shape shown in the figure. An alternative way to describe this phenomenon is this: The high-frequency contents of the continuous-time signal (in the range $|\omega| \in (3\omega_m/4, \omega_m]$) disguises itself as a low-frequency contents (in the range $|\omega| \in [\omega_m/2, 3\omega_m/4)$) and is added to the original contents in this range.

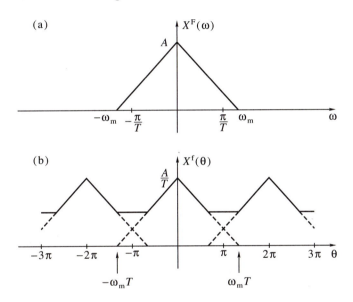

Figure 3.4 Sampling of a band-limited signal below the Nyquist rate: (a) Fourier transform of the continuous-time signal; (b) Fourier transform of the sampled signal.

The phenomenon we have described is called *aliasing.* Aliasing is caused by sampling at a rate lower than the Nyquist rate for the given signal. Since the shape of the Fourier transform in the range $\theta \in [-\pi, \pi]$ is not preserved, we expect to be unable to reconstruct the continuous-time signal from its samples. In conclusion:

> If a band-limited signal is sampled at a rate lower than the Nyquist rate $2f_m$, the shape of the Fourier transform of the sampled signal in the range $\theta \in [-\pi, \pi]$ is distorted relative to the Fourier transform of the given signal. This distortion, which is called aliasing, results from overlapping of the replicas in the sampling formula (3.10).

Example 3.6 Let

$$x_1(t) = \cos(1.2\pi t), \quad x_2(t) = \cos(0.8\pi t).$$

Suppose the two signals are sampled at interval $T = 1$ second. Then

$$x_1(nT) = \cos(1.2\pi n) = \cos(0.8\pi n) = x_2(nT).$$

Therefore, the signals $x_1(t)$ and $x_2(t)$ become indistinguishable when sampled at interval $T = 1$ second. This is illustrated in Figure 3.5. We say that $x_1(t)$ is *aliased* as $x_2(t)$. □

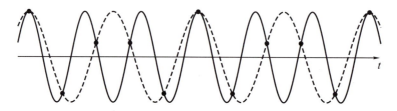

Figure 3.5 Aliasing of a sinusoidal signal. Solid line: signal at 0.6 Hz; dashed line: signal at 0.4 Hz; circles: samples at $T = 1$ second.

Example 3.7 Consider the signal

$$x(t) = a_1 \cos(0.5\pi t) + a_2 \cos(1.75\pi t).$$

Let the sampling interval be $T = 1$ second. The sampled signal is then

$$x[n] = a_1 \cos(0.5\pi n) + a_2 \cos(1.75\pi n) = a_1 \cos(0.5\pi n) + a_2 \cos(0.25\pi n),$$

since $\cos(1.75\pi n) = \cos(0.25\pi n)$ for all n. As we see, the frequency of the first component of the sampled signal is $\theta = 0.5\pi$, which is related to the corresponding frequency of $x(t)$ by the correct relationship $\theta = \omega T$. Since, however, the second component becomes a low frequency signal, it is aliased. □

Example 3.8 Consider the borderline case of sampling the signal

$$x(t) = \cos(2\pi f_0 t - \phi_0)$$

at a rate $f_{\text{sam}} = 2f_0$. Will it be subject to aliasing? If we take $\phi_0 = 0$, we will get

$$x[n] = \cos(\pi n) = (-1)^n,$$

which seems to be alias free. However, if we take $\phi_0 = \pi/2$, we will get

$$x[n] = \sin(\pi n) = 0,$$

so the sampled signal vanishes completely! The answer is therefore that the choice $f_{\text{sam}} = 2f_0$ is *not* permitted in this case. Another argument for this conclusion is: Since the Fourier transform of $x(t)$ contains delta functions at $\omega = \pm 2\pi f_0$, sampling at $f_{\text{sam}} = 2f_0$ violates Nyquist's no-aliasing condition. □

Example 3.9 Consider the signal $x(t)$ whose Fourier transform is

$$X^{\text{F}}(\omega) = \begin{cases} \dfrac{1}{f_0}, & |\omega| \leq \omega_1, \\[2mm] \dfrac{0.5}{f_0}\left[1 + \cos\left(\dfrac{|\omega| - \omega_1}{2\alpha f_0}\right)\right], & \omega_1 < |\omega| \leq \omega_2, \\[2mm] 0, & |\omega| > \omega_2, \end{cases} \qquad (3.20)$$

where $0 < \alpha < 1$, and

$$\omega_1 = (1 - \alpha)\pi f_0, \quad \omega_2 = (1 + \alpha)\pi f_0. \tag{3.21}$$

This signal is band limited, its bandwidth being $(1 + \alpha)\pi f_0$, which is more than πf_0, but less than $2\pi f_0$. The shape of $X^F(\omega)$ is shown in Figure 3.6(a). This is called an *excess bandwidth, raised-cosine spectrum*, and α is called the *bandwidth excess*. For example, when $\alpha = 0.4$, we call it a 40 percent raised-cosine spectrum.

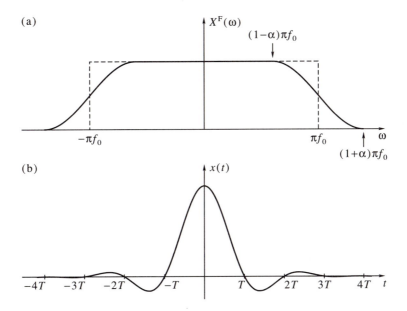

Figure 3.6 A raised-cosine signal: (a) the spectrum; (b) the waveform.

The inverse Fourier transform of $X^F(\omega)$ is given by

$$
\begin{aligned}
x(t) &= \frac{1}{\pi f_0} \int_0^{(1-\alpha)\pi f_0} \cos(\omega t)\,d\omega \\
&\quad + \frac{0.5}{\pi f_0} \int_{(1-\alpha)\pi f_0}^{(1+\alpha)\pi f_0} \left[1 + \cos\left(\frac{1}{2\alpha f_0}[\omega - (1 - \alpha)\pi f_0]\right)\right]\cos(\omega t)\,d\omega \\
&= \frac{\sin(\pi f_0 t)\cos(\pi \alpha f_0 t)}{\pi f_0 t[1 - (2\alpha f_0 t)^2]}.
\end{aligned}
\tag{3.22}
$$

The signal is shown in Figure 3.6(b). We observe that sampling at a rate $f_{sam} = f_0$ yields the discrete-time signal

$$x[n] = \delta[n] \quad \Rightarrow \quad X^f(\theta) = 1.$$

Thus the sampled signal is obviously aliased, since $f_{sam} = f_0$ is below the Nyquist rate. Moreover, the signal $x(t)$ is Nyquist-T for $T = 1/f_0$. Another way to arrive at this conclusion is to observe that in forming the sum

$$f_0 \sum_{k=-\infty}^{\infty} X^F(\omega - 2\pi k f_0),$$

the cosine shapes exactly add up to 1 at all intervals of overlap, so the result of the infinite sum is identically 1 at all frequencies.

Signals with raised-cosine spectra are useful in digital communication applications, thanks to their Nyquist-T property. They provide an example in which aliasing is not only harmless, but necessary for proper operation! □

The third and final case to be considered is that of a signal $x(t)$ whose bandwidth is not limited, as illustrated in Figure 3.7. In this case, the sum (3.10) includes an infinite number of nonzero terms, so the shape of the Fourier transform of the sampled signal must be distorted. Part b of the figure shows, in a thick solid line, the result of the infinite summation. The other lines (thin solid, two dashed, and two dotted) show five of the terms in the sum, corresponding to $k = -2, -1, 0, 1, 2$. The conclusion is that sampling of an infinite bandwidth signal always gives rise to aliasing, no matter how high the sampling rate.

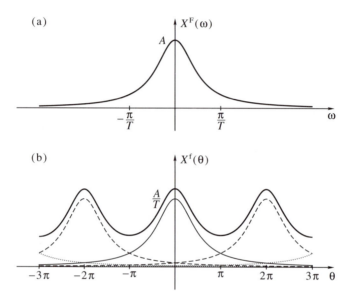

Figure 3.7 Sampling of a signal of an infinite bandwidth: (a) Fourier transform of the continuous-time signal; (b) Fourier transform of the sampled signal.

A famous theorem in the theory of Fourier transforms asserts that a signal of finite duration (i.e., which is identically zero outside a certain time interval) must have an infinite bandwidth.[3] Real-life signals always have finite duration, so we conclude that their bandwidth is always infinite. Sampling theory therefore implies that real-life signals must become aliased when sampled. Nevertheless, the bandwidth of a real-life signal is always *practically finite*, meaning that the percentage of energy outside a certain frequency range is negligibly small. It is therefore permitted, in most practical situations, to sample at a rate equal to twice the practical bandwidth, since then the effect of aliasing will be negligible.

Example 3.10 The practical bandwidth of a speech signal is about 4 kHz. Therefore, it is common to sample speech at 8–10 kHz. The practical bandwidth of a musical audio signal is about 20 kHz. Therefore, compact-disc digital audio uses a sampling rate of 44.1 kHz.   □

A common practice in sampling of a continuous-time signal is to filter the signal *before* it is passed to the sampler. The filter used for this purpose is an analog low-pass filter whose cutoff frequency is not larger than *half the sampling rate*. Such a filter is called an *antialiasing* filter.

In summary, the rules of safe sampling are:

> - Never sample below the Nyquist rate of the signal. To be on the safe side, use a safety factor (e.g., sample at 10 percent higher than the Nyquist rate).
> - In case of doubt, use an antialiasing filter before the sampler.

Example 3.11 This story happened in 1976 (a year after Oppenheim and Schafer's classic *Digital Signal Processing* was published, but sampling and its consequences were not yet common knowledge among engineers); its lesson is as important today as it was then. A complex and expensive electrohydraulic system had to be built as part of a certain large-scale project. The designer of the system constructed a detailed mathematical model of it, and this was given to a programmer whose task was to write a computer simulation program of the system. When the simulation was complete, the system was still under construction. The programmer then reported that, under certain conditions, the system exhibited nonsinusoidal oscillations at a frequency of about 8 Hz, as shown in Figure 3.8. This gave rise to a general concern, since such a behavior was judged intolerable. The designer declared that such oscillations were not possible, although high-frequency oscillations, at about 100 Hz, were possible. Further examination revealed the following: The simulation had been carried out at a rate of 1000 Hz, which was adequate. However, to save disk storage (which was expensive those days) and plotting time (which was slow), the simulation output had been stored and plotted at a rate of 100 Hz. In reality, the oscillations were at 108 Hz, but as a result of the choice of plotting rate, they were aliased and appeared at 8 Hz. When the simulation output was plotted again, this time at 1000 Hz, it showed the oscillations at their true frequency, see Figure 3.9.   □

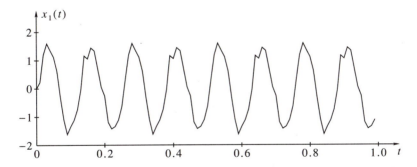

Figure 3.8 The apparent 8 Hz oscillations of the electrohydraulic system in Example 3.11.

3.4 Reconstruction

Suppose we are given a sampled signal $x[n]$ that is known to have been obtained from a band-limited signal $x(t)$ by sampling at the Nyquist rate (or higher). Since the Fourier transform of the sampled signal preserves the shape of the Fourier transform of the continuous-time signal, we should be able to reconstruct $x(t)$ exactly from its samples. How then do we accomplish such reconstruction? The answer is given by Shannon's reconstruction theorem.

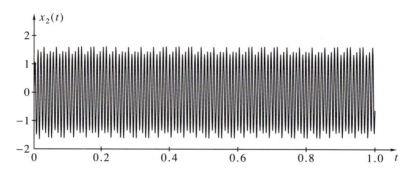

Figure 3.9 The true 108 Hz oscillations of the electrohydraulic system in Example 3.11.

Theorem 3.2 A band-limited signal $x(t)$ whose bandwidth is smaller than $\pm\pi/T$ can be exactly reconstructed from its samples $\{x(nT),\ -\infty < n < \infty\}$, using the formula

$$x(t) = \sum_{n=-\infty}^{\infty} x(nT)\text{sinc}\left(\frac{t-nT}{T}\right). \tag{3.23}$$

Proof As we did for the sampling theorem, we shall give two proofs of Shannon's theorem, one that is based on the impulse train and one that is not. Consider first $X_p^F(\omega)$, the Fourier transform of $x_p(t)$, and recall from (3.9) that in the nonaliased case it is related to $X^F(\omega)$ through

$$X^F(\omega) = TX_p^F(\omega)H^F(\omega), \tag{3.24}$$

where $H^F(\omega)$ is an ideal low-pass filter, that is,

$$H^F(\omega) = \text{rect}\left(\frac{\omega T}{2\pi}\right) \tag{3.25}$$

(see Figure 3.10). Therefore, $x(t)$ is given by the convolution

$$x(t) = T\int_{-\infty}^{\infty} h(\tau)x_p(t-\tau)d\tau. \tag{3.26}$$

The impulse response of the ideal low-pass filter is obtained from (2.35) as

$$h(t) = \frac{1}{T}\text{sinc}\left(\frac{t}{T}\right). \tag{3.27}$$

Therefore,

$$x(t) = \int_{-\infty}^{\infty} \text{sinc}\left(\frac{\tau}{T}\right)\left[\sum_{n=-\infty}^{\infty} x(nT)\delta(t-\tau-nT)\right]d\tau = \sum_{n=-\infty}^{\infty} x(nT)\text{sinc}\left(\frac{t-nT}{T}\right). \tag{3.28}$$

We now give a direct proof of (3.23). We have from (3.19)

$$x(t) = \frac{1}{2\pi}\int_{-\infty}^{\infty} X^F(\omega)e^{j\omega t}d\omega = \frac{T}{2\pi}\int_{-\pi/T}^{\pi/T} X^f(\omega T)e^{j\omega t}d\omega$$

$$= \frac{T}{2\pi}\int_{-\pi/T}^{\pi/T}\left[\sum_{n=-\infty}^{\infty} x(nT)e^{-j\omega nT}\right]e^{j\omega t}d\omega$$

$$= \frac{T}{2\pi}\sum_{n=-\infty}^{\infty} x(nT)\int_{-\pi/T}^{\pi/T} e^{j\omega(t-nT)}d\omega = \sum_{n=-\infty}^{\infty} x(nT)\text{sinc}\left(\frac{t-nT}{T}\right). \tag{3.29}$$

$\square$

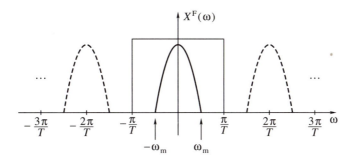

Figure 3.10 Reconstruction of a band-limited signal by an ideal low-pass filter (replicas shown by dashed lines will be eliminated by the filter).

Let us interpret the meaning of the reconstruction formula (3.23). Consider first a time point that coincides with one of the sampling points, say $t = n_0 T$. We have, by the properties of the sinc function,

$$\text{sinc} \left(\frac{n_0 T - nT}{T} \right) = \delta[n - n_0]. \tag{3.30}$$

The right side of (3.23) is indeed equal to $x(n_0 T)$, as it should be. This is true for all n_0, so we conclude that the reconstruction formula is correct at the sampling points. Now choose a point t located between $n_0 T$ and $(n_0 + 1)T$. Then none of the values of the sinc function will be zero in general. However, points closer to $n_0 T$ will contribute to the sum more than points farther from $n_0 T$, since the sinc is a decaying function (although not monotonically decaying). All these contributions, infinite in number, will add exactly to $x(t)$, as guaranteed by Shannon's formula. What if $x(t)$ has an irregular behavior between the two sampling points $n_0 T$ and $(n_0 + 1)T$, which the sinc function doesn't see? The answer is that the bandwidth of such a signal is necessarily higher than the limit π/T imposed by Shannon's theorem. Stated in different words: A signal whose bandwidth is limited to π/T must be fairly smooth—it cannot have more than "half an oscillation" between any two adjacent sampling points.

The Shannon reconstructor (also called *Shannon interpolator* or *sinc interpolator*) is a noncausal system. Reconstruction of $x(t)$ at any time t requires knowledge of the entire sequence $\{x(nT), -\infty < n < \infty\}$. Therefore, Shannon's reconstructor is important mainly from a theoretical viewpoint. In real-time signal processing, we must use a causal reconstructor, preferably one that is convenient to implement. Let $h(t)$ be the impulse response of the reconstructor. The reconstruction formula is then

$$\hat{x}(t) = \sum_{n=-\infty}^{\lfloor t/T \rfloor} x(nT)h(t - nT). \tag{3.31}$$

The impulse response $h(t)$ should be chosen such that $\hat{x}(t)$ is a good approximation of $x(t)$. The limits of the sum reflect the causality requirement (the upper limit is the largest integer not larger than t/T).

A simple device that realizes (3.31) is the *zero-order hold* (ZOH) circuit. A zero-order hold is a device that, when given an input sample $x[n]$, maintains a constant output equal to $x[n]$ during the interval $[nT, nT + T)$ and then resets its output to zero. When fed with the sequence $\{x[n], -\infty < n < \infty\}$, it yields the output

$$\hat{x}(t) = x[n], \quad nT \le t < nT + T, \quad \text{for all } n \in \mathbb{Z}. \tag{3.32}$$

The impulse response of the ZOH device is

$$h_{\text{zoh}}(t) = \begin{cases} 1, & 0 \le t < T, \\ 0, & \text{otherwise.} \end{cases} \tag{3.33}$$

The impulse response (3.33) is shown in Figure 3.11; the response $\hat{x}(t)$ to the input sequence $x(nT)$ is shown in Figure 3.12. The latter figure also shows, in a dashed line, the ideal waveform $x(t)$, which would be obtained by a Shannon reconstructor.

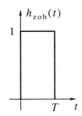

Figure 3.11 Impulse response of a zero-order hold.

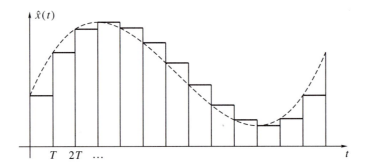

Figure 3.12 Time response of a zero-order hold. Staircase line: actual response $\hat{x}(t)$; dashed line: ideal response $x(t)$.

To understand to what extent $\hat{x}(t)$ approximates $x(t)$, let us compute the frequency response of the ZOH. As we recall, the frequency response of the ideal (Shannon) reconstructor is a perfect rectangle on $-\pi/T \le \omega \le \pi/T$, with zero phase; see (3.25). By analyzing how the frequency response of the ZOH deviates from the perfect rectangle, we will understand the nature of distortions introduced by the ZOH. We have from (3.33)

$$H_{\text{zoh}}^{\text{F}}(\omega) = \int_0^T e^{-j\omega t} dt = \frac{1 - e^{-j\omega T}}{j\omega} = T \operatorname{sinc}\left(\frac{\omega T}{2\pi}\right) e^{-j0.5\omega T}. \tag{3.34}$$

The magnitude and phase responses of $H_{\text{zoh}}^{\text{F}}(\omega)$ are shown in Figure 3.13. We observe the following differences with respect to the ideal low-pass filter:

1. The magnitude response at low frequencies is not flat, but decays gradually. Furthermore, it decreases to zero at $\omega = \pm 2\pi/T$, rather than at $\pm\pi/T$.

2. The magnitude response has nonvanishing ripple at high frequencies, so the reconstructed signal $\hat{x}(t)$ has undesired high-frequency energy. In the time domain, the high-frequency energy is apparent in the staircaselike form of the output created by the hold operation.

3. The phase of the response is not zero, but piecewise linear, with slope $-0.5T$.

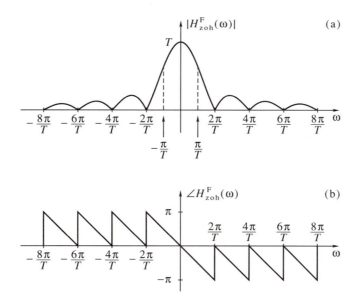

Figure 3.13 Frequency response of a zero-order hold: (a) magnitude; (b) phase.

The undesirable effects of the zero-order hold can be partly overcome by passing its output through a low-pass analog filter. The ideal magnitude response $|G^F(\omega)|$ of such a filter is shown in Figure 3.14. The magnitude response is the inverse of $|H^F_{zoh}(\omega)|$ in the frequency range $[-\pi/T, \pi/T]$ and zero elsewhere. We thus get from (3.34) that

$$|G^F(\omega)| = \left[\operatorname{sinc}\left(\frac{\omega T}{2\pi}\right) \right]^{-1} \operatorname{rect}\left(\frac{\omega T}{2\pi}\right). \tag{3.35}$$

Such a filter is impossible to implement, due its infinitely steep cutoff, so in practice an approximation must be used instead.

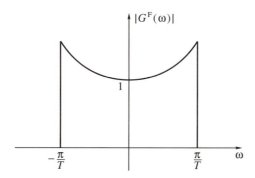

Figure 3.14 Magnitude response of an ideal reconstruction filter at the output of a ZOH.

Other reconstruction devices, more sophisticated than the ZOH, are sometimes used, but are not discussed here (see Problems 3.29 and 3.41 for two alternatives). We remark that non-real-time reconstruction offers much more freedom in choosing the reconstruction filter and allows for a greatly improved frequency response (compared with the one shown in Figure 3.13). Non-real-time signal processing involves sampling and storage of a finite (but potentially long) segment of a physical signal, followed by off-line processing of the stored samples. If such a signal needs to be reconstructed, it is perfectly legitimate, and even advisable, to use a noncausal filter for reconstruction.

We summarize our discussion of reconstruction by showing a complete typical DSP system, as depicted in Figure 3.15. The continuous-time signal $x(t)$ is passed through an antialiasing filter, then fed to the sampler. The resulting discrete-time signal $x[n]$ is then processed digitally as needed in the specific application. The discrete-time signal at the output $y[n]$ is passed to the ZOH and then low-pass filtered to give the final continuous-time signal $\hat{y}(t)$.

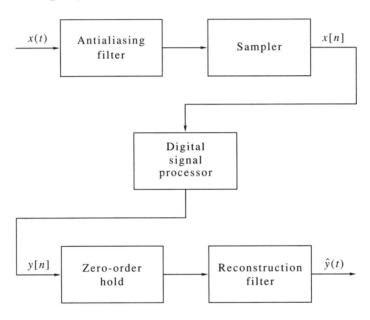

Figure 3.15 A typical digital signal processing system.

3.5 Physical Aspects of Sampling and Reconstruction*

So far we have described sampling and reconstruction as mathematical operations. We now describe electronic circuitry for implementing those two operations. We shall see that hardware limitations introduce imperfections to both sampling and reconstruction. These imperfections are chiefly *nonlinearities*, of which there are three major types:

1. Saturation, since voltages must be confined to a certain range, depending on the specific electronic circuit.

2. Quantization, since only a finite number of bits can be handled by the circuit.

3. Nonlinearities introduced by electronic components (tolerances of resistors, non-linearity of operational amplifiers, and so on).

Nonlinearities of these types appear in both sampling and reconstruction in a similar manner. Each of the two operations, sampling and reconstruction, also has its own characteristic imperfections. Sampling is prone to signal smearing, since it must be performed over finite time intervals. Switching delays make reconstruction prone to instantaneous deviations of the output signal (known to engineers as *glitches*).

3.5.1 Physical Reconstruction

Physical reconstruction is implemented using a device called *digital-to-analog converter*, or D/A. A D/A converter approximates zero-order hold operation. It accepts a discrete-time signal $x[n]$ in a form of a sequence of binary numbers. Each binary number is held fixed by a data register for a period of T seconds (the sampling interval). The binary number at the register's output is converted to a voltage waveform $\hat{x}(t)$. The voltage is approximately proportional to the present value of $x[n]$, and remains fixed for T seconds. When the next binary number $x[n + 1]$ appears at the register's output, $\hat{x}(t)$ changes accordingly. The result is approximately the staircase waveform shown in Figure 3.12. Figure 3.16 depicts this sequence of operations schematically.

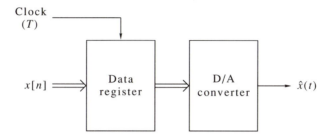

Figure 3.16 Schematic diagram of physical reconstruction by a D/A converter.

A simple electronic circuit for implementing a D/A converter is shown in Figure 3.17. The figure illustrates a 3-bit D/A, but it can be easily extended to any number of bits. This D/A is *unipolar*, that is, the binary number at its input is considered nonnegative and the voltage at its output has only one polarity (always negative in Figure 3.17, assuming that V_{ref} is positive). The signal $x[n]$ is represented by the binary number $b_0 b_1 \ldots b_{B-1}$, where B is the number of bits, 3 in this example. The most significant bit (MSB) is b_0, and the least significant bit (LSB) is b_{B-1}. It is convenient to think of this number as a fraction in the range $[0, 1 - 2^{-B}]$, that is, as representing the numerical value $\sum_{k=0}^{B-1} b_k 2^{-(k+1)}$. The circuit operates as follows:

1. Assuming that the gain of the operational amplifier is high (approximately infinity), its inverting input has nearly zero potential (*virtual ground* is the common term). Therefore, regardless of the positions of the switches, the bottom terminals of the $2R$ resistances of the resistor ladder behave as if they were grounded. Application of rules for parallel-series connection of resistors shows that the total resistance seen by the voltage source V_{ref} is $2R$.

2. Assuming that the voltage source is ideal (i.e., its internal resistance is zero), the current flowing from it is $I = V_{\text{ref}}/2R$. The current splits into two equal parts at the node, so $I_0 = I/2$. Continuing in this manner, we observe that $I_1 = I_0/2 = I/4$ and, in general,

$$I_k = 2^{-(k+1)}I, \quad 0 \le k \le B - 1. \tag{3.36}$$

3. Assume that the switches are ideal and that the kth switch is in the right position when $b_k = 1$ and in the left position is when $b_k = 0$. Then the current fed by the kth section to the operational amplifier's input is 0 or I_k, depending on whether b_k is 0 or 1. In other words, this current is $b_k I_k$. Therefore we get, by current balancing at the input of the operational amplifier (assuming it has an infinite

input resistance),

$$\frac{V_{\text{out}}}{2R} = -\sum_{k=0}^{B-1} b_k I_k = -I \sum_{k=0}^{B-1} b_k 2^{-(k+1)}, \tag{3.37}$$

$$V_{\text{out}} = -V_{\text{ref}} \sum_{k=0}^{B-1} b_k 2^{-(k+1)}. \tag{3.38}$$

As we see, the output voltage is proportional to the binary number at the input.[4]

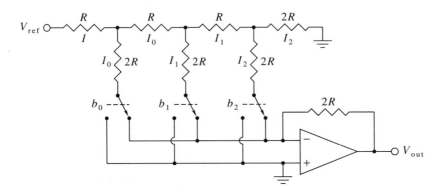

Figure 3.17 Schematic diagram of a digital-to-analog converter.

Practical D/A converters are often *bipolar*; their output voltage can be either positive or negative, depending on the sign of the binary number at the input. Table 3.1 illustrates, in its first two columns, the correspondence between binary numbers and voltages (relative to V_{ref}) for a 3-bit bipolar D/A. The binary word $00\dots0$ corresponds to the smallest (most negative) voltage and the word $11\dots1$ corresponds to the largest (most positive) voltage. This correspondence is called *offset binary*. The possible voltage values are symmetric with respect to 0; it is impossible to get zero voltage; finally, the largest possible absolute voltage is $(1 - 2^{-B})V_{\text{ref}}$. The absolute possible voltage represents the *saturation level* of the D/A. The voltage increment, also called the *quantization level*, is $2^{-(B-1)}V_{\text{ref}}$.

$8\frac{V_{\text{out}}}{V_{\text{ref}}}$	offset binary	two's-complement
-7	000	100
-5	001	101
-3	010	110
-1	011	111
1	100	000
3	101	001
5	110	010
7	111	011

Table 3.1 Correspondence between output voltage and binary representations (offset binary and two's-complement) in a bipolar D/A converter.

Numbers in a computer are usually represented in a *two's-complement* form.[5] The two's-complement representations of the eight possible voltages in the 3-bit case are shown in the third column of Table 3.1. As we see, two's-complement representation is obtained from offset binary representation by a simple rule: Invert the most significant bit and leave the other bits unchanged. Figure 3.18 illustrates the correspondence between voltages and binary numbers in the two's-complement case.

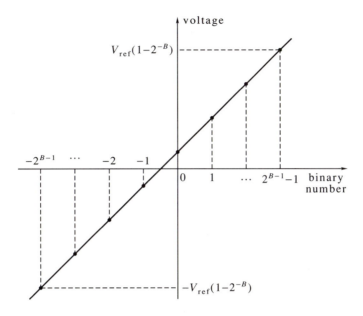

Figure 3.18 Correspondence between two's-complement binary numbers and voltages in a digital-to-analog converter.

The ratio between the maximum possible voltage and half the quantization level is called the *dynamic range* of the D/A. As we have seen, this number is $2^B - 1$, or nearly 2^B. The corresponding parameter in decibels is about $6B$ dB. Thus, a 10-bit D/A has a dynamic range of 60 dB. For example, high-fidelity music usually requires a dynamic range of 90 dB or more, so at least a 16-bit D/A is necessary for this purpose. D/A converters with high dynamic ranges are expensive to manufacture, since they impose tight tolerances on the analog components.

3.5.2 Physical Sampling

Physical sampling is implemented using a device called *analog-to-digital converter*, or A/D. The A/D device approximates point sampling. It accepts a continuous-time signal $x(t)$ in a form of an electrical voltage and produces a sequence of binary numbers $x[n]$, which approximate the corresponding samples $x(nT)$. Often the electrical voltage is not fed to the A/D directly, but through a device called *sample-and-hold*, or S/H; see Figure 3.19. Sample-and-hold is an analog circuit whose function is to measure the input signal value at the clock instant (i.e., at an integer multiple of T) and hold it fixed for a time interval long enough for the A/D operation to complete. Analog-to-digital conversion is potentially a slow operation, and variation of the input voltage during the conversion may disrupt the operation of the converter. The S/H prevents such disruption by keeping the input voltage constant during conversion. When the input voltage variation is slow relative to the speed of the A/D, the S/H is not needed and the input voltage may be fed directly to the A/D.

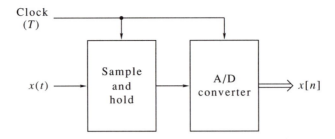

Figure 3.19 Schematic diagram of physical sampling by an A/D converter.

Similarly to D/A converters, A/D converters exhibit saturation and quantization. Figure 3.20 depicts two types of A/D response. Part a shows a *rounding* A/D; its binary output corresponding to an input voltage V_{in} is given by

$$x = \left\lfloor \frac{V_{in}}{V_{ref}} 2^{B-1} + 0.5 \right\rfloor, \tag{3.39}$$

where V_{ref} is the reference voltage to the A/D, B is the number of bits, and $\lfloor a \rfloor$ denotes the integer nearest to a from below. As we see from (3.39) and from the figure, the binary number can be up to half the quantization level above or below the ideal output (represented in the figure by a dashed line). The rounding A/D thus has a symmetric response, except near the positive and negative saturation levels. The reason for the asymmetry at the ends is that the number of possible values is necessarily even, being a power of 2. In two's-complement arithmetic, the maximum positive value is $2^{B-1} - 1$ and the maximum negative value is -2^{B-1}. Correspondingly, the positive saturation level is slightly smaller than the negative one.

Figure 3.20(b) shows a *truncating* A/D; its binary output corresponding to an input voltage V_{in} is given by

$$x = \left\lfloor \frac{V_{in}}{V_{ref}} 2^{B-1} \right\rfloor. \tag{3.40}$$

As we see from (3.40) and from the figure, the binary number is always less than the ideal output (represented in the figure by a dashed line) by up to one quantization level. The truncating A/D thus has an asymmetric response.

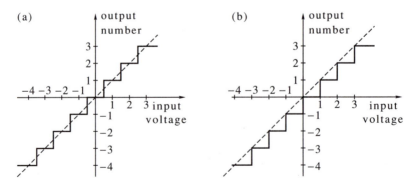

Figure 3.20 Quantization in analog-to-digital converter: (a) rounding; (b) truncation. Staircase lines show the actual responses; dashed lines show the ideal responses.

Figure 3.21 illustrates the result of sampling an analog signal with a rounding A/D. In this example the A/D is 4 bits, or 16 quantization levels (8 for positive values of the input signal). As we see, rounding results in an error that lies in the range plus or minus half the quantization level. In addition, we see how the extreme positive values of the input signal are chopped because of saturation. The error in the sampled values due to quantization (but not due to saturation) is called *quantization noise*.

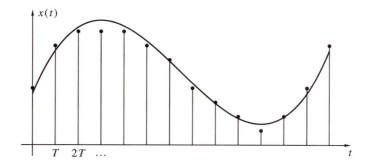

Figure 3.21 Quantization error in a rounding A/D.

Analog-to-digital converters can be implemented in several ways, depending on speed and cost considerations. The speed of an A/D converter is measured by the maximum number of conversions per second. The time for a single conversion is approximately the reciprocal of the speed. Usually, the faster the A/D, the more complex is the hardware and the costlier it is to build. Common A/D implementations include:

1. Successive approximation A/D; see Figure 3.22. This A/D builds the output bits in a feedback loop, one at a time, starting with the most significant bit (MSB). Feedback is provided by a D/A, which converts the output bits of the A/D to an electrical voltage. Initially the shift register MSB is set to 1, and all the other bits to 0. This sets the data register MSB to 1, so the D/A output becomes $0.5V_{\text{ref}}$. The comparator decides whether this is lower or higher than the input voltage V_{in}. If it is lower, the MSB retains the value 1. If it is higher, the MSB is reset to 0. At the next clock cycle, the 1 in the shift register shifts to the bit below the MSB, and sets the corresponding bit of the data register to 1. Again a comparison is made between the D/A output and V_{in}. If $V_{\text{in}} > V_{\text{fb}}$, the bit retains the value 1, otherwise it is reset to 0. This process is repeated B times, until all the data register bits are set to their proper values. Such an A/D is relatively inexpensive, requiring only a D/A, a comparator, a few registers, and simple logic. Its conversion speed is proportional to the number of bits, because of its serial operation. Successive approximation A/D converters are suitable for many applications, but not for ones in which speed is of prime importance.

   When a bipolar A/D converter is required, it is convenient to use offset-binary representation, since then the binary number is a monotone function of the voltage. The representation can be converted to two's-complement by inverting the MSB.

2. Flash A/D; see Figure 3.23 for a 3-bit example. This converter builds all output bits in parallel by directly comparing the input voltage with all possible output values. It requires $2^B - 1$ comparators. As is seen from Figure 3.23, the bottom comparators up to the one corresponding to the input voltage will be set to 1, whereas the ones above it will be set to 0. Therefore, we can determine that the quantized voltage is n quantization levels up from $-V_{\text{ref}}$ if the nth comparator is set to 1 and the $(n + 1)$st comparator is set to 0. This is accomplished by

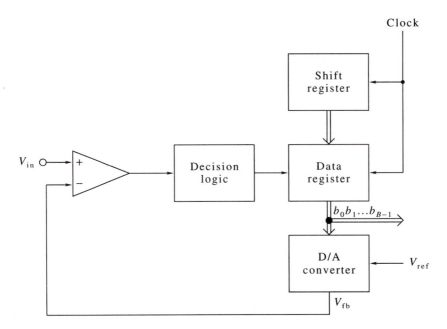

Figure 3.22 Schematic diagram of a successive approximation A/D converter.

the AND gates shown in the figure. The quantized voltage is $-V_{\text{ref}}$ if the bottom comparator is set to 0, and is $(1 - 2^{-(B-1)})V_{\text{ref}}$ if the top comparator is set to 1. The encoding logic then converts the gate outputs to the appropriate binary word. This scheme implements a truncating A/D; if a rounding A/D is required, it can be achieved by changing the reference voltages V_{ref} and $-V_{\text{ref}}$ to $(1 - 2^{-B})V_{\text{ref}}$ and $-(1 + 2^{-B})V_{\text{ref}}$, respectively. Another possibility is to change the bottom and top resistors to $0.5R$ and $1.5R$, respectively.

Flash A/D converters are the fastest possible, since all bits are obtained simultaneously. On the other hand, their hardware complexity grows exponentially with the number of bits, so they become prohibitively expensive for a large number of bits. Their main application is for conversion of video signals, since these require high speeds (on the order of 10^7 conversions per second), whereas the number of bits is typically moderate.

3. Half-flash A/D; see Figure 3.24. This A/D offers a compromise between speed and complexity. It uses two flash A/D converters, each for half the number of bits. The number of comparators is $2(2^{B/2} - 1)$, which is significantly less than the number of comparators in a flash A/D converter having the same number of bits. The $B/2$ most significant bits are found first, and then converted to analog using a $B/2$-bit D/A. The D/A output is subtracted from the input voltage and used, after being passed through a S/H, to find the $B/2$ least significant bits. The conversion time is about twice that of a full-flash A/D.

4. Sigma–delta A/D converter. This type of converter provides high resolutions (i.e., a large number of bits) with relatively simple analog circuitry. It is limited to applications in which the signal bandwidth is relatively low and speed is not a major factor. The theory of sigma–delta converters relies on concepts and techniques we have not studied yet. We therefore postpone the explanation of such converters to Section 14.7; see page 586.

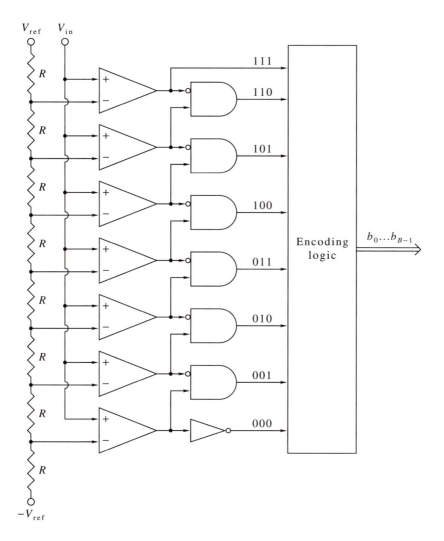

Figure 3.23 Schematic diagram of a flash A/D converter.

Now we discuss the circumstances under which an S/H is needed at the input of the A/D. For safe operation without S/H, the input voltage should not change by more than half the quantization level during the conversion time. Assume that the input voltage is a sinusoid of amplitude V_{ref} and frequency f. The greatest rate of change occurs when the sinusoid crosses zero, its value there being $2\pi f V_{ref}$. Let us denote by f_{ad} the number of conversions per second, so the duration of a single conversion is $1/f_{ad}$. Then the maximum amount of change during the conversion time is $2\pi f V_{ref}/f_{ad}$. This must be less than $2^{-B}V_{ref}$, so we arrive at the condition

$$f \le \frac{f_{ad}}{\pi 2^{B+1}}. \tag{3.41}$$

Condition (3.41) imposes an upper limit on the frequency of the input signal for which an S/H is not needed. If the frequency of the input signal exceeds the right side of (3.41), an S/H is necessary for safe operation.

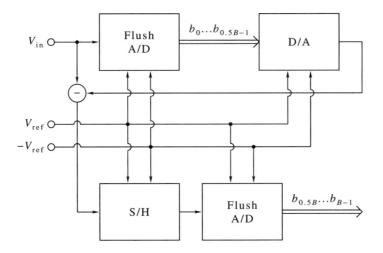

Figure 3.24 Schematic diagram of a half-flash A/D converter.

3.5.3 Averaging in A/D Converters

So far we have assumed ideal operation of the S/H circuit. In reality, S/H performs averaging of the continuous-time signal over a short, but finite period. Thus, a physical S/H can be mathematically described (at least approximately) by the relationship

$$x[n] = \frac{1}{\Delta} \int_{nT-\Delta}^{nT} x(t)dt, \tag{3.42}$$

where Δ is the averaging interval, assumed to be smaller than the sampling interval T (otherwise the A/D will not be able to operate in real time). The operation (3.42) is depicted in Figure 3.25.

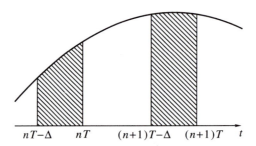

Figure 3.25 Averaging in A/D converter.

Intuitively, if $\Delta \ll T$, then physical sampling must be well approximated by point sampling. However, when Δ is a sizable fraction of T, we expect low-pass filtering effect that is not negligible. We now analyze this effect and express it in precise terms. Define the auxiliary signal $y(t)$ as the indefinite integral of $x(t)$, that is,

$$y(t) = \int^{t} x(\tau)d\tau + C, \tag{3.43}$$

where C is an arbitrary constant. Then we have from (3.42)

$$x[n] = \frac{1}{\Delta}\{y(nT) - y(nT - \Delta)\}. \tag{3.44}$$

The sequence $y(nT)$ is the point sampling of $y(t)$ and $y(nT - \Delta)$ is the point sampling of $y(t - \Delta)$. The Fourier transform of the former is $(j\omega)^{-1}X^F(\omega)$ and that of the latter is $(j\omega)^{-1}X^F(\omega)e^{-j\omega\Delta}$. Therefore we get, by linearity and by the sampling theorem (3.10),

$$X^f(\theta) = \frac{1}{T} \sum_{k=-\infty}^{\infty} \tilde{X}^F\left(\frac{\theta - 2\pi k}{T}\right), \tag{3.45}$$

where we define

$$\tilde{X}^F(\omega) = \frac{1 - e^{-j\omega\Delta}}{j\omega\Delta} X^F(\omega) = e^{-j0.5\omega\Delta}\mathrm{sinc}\left(\frac{\omega\Delta}{2\pi}\right) X^F(\omega). \tag{3.46}$$

As we see, sampling with averaging is equivalent to low-pass filtering followed by point sampling. The transfer function of the low-pass filter is identical to that of a ZOH, except that the hold interval is Δ. When Δ is small with respect to T, the attenuation and phase lag of the low-pass filter are negligible. At the other extreme, when Δ approaches T, we get the same magnitude attenuation and phase delay as described for the ZOH; see Figure 3.13. Recall, however, that if the signal $x(t)$ was passed through an antialiasing filter, its bandwidth does not exceed $\pm\pi/T$. Therefore, only the region $[-\pi/T, \pi/T]$ of Figure 3.13 is of interest in this case.

In summary, ideal point sampling is a good approximation of physical sampling if the averaging interval of the S/H is small with respect to the sampling interval. If not, there is an additional low-pass filtering effect that needs to be taken into account, as expressed by (3.46).

3.6 Sampling of Band-Pass Signals*

A continuous-time signal $x(t)$ is called *band pass* if its Fourier transform vanishes outside a certain frequency range that does not include $\omega = 0$: in other words, if there exist $0 < \omega_1 < \omega_2$ such that

$$X^F(\omega) = 0 \text{ for } |\omega| \leq \omega_1 \text{ and for } |\omega| \geq \omega_2. \tag{3.47}$$

Signals transmitted electromagnetically—such as radio, radar, or signals transmitted via optical fibers—are band pass. It is common to define the *bandwidth* of a band-pass signal as $\omega_2 - \omega_1$ [or $(\omega_2 - \omega_1)/2\pi$ in hertz]. For example, the bandwidth of commercial AM radio transmission is 10 kHz in the United States and 9 kHz in Europe and many other countries. The bandwidth of commercial FM radio transmission is about 180 kHz.

Band-pass signals are band limited to $\pm\omega_2$, so they can be sampled at a rate ω_2/π without being aliased. Since, however, the frequency support of the signal is only $2(\omega_2 - \omega_1)$, it appears to be wasteful to sample at such a rate if $\omega_2 \gg (\omega_2 - \omega_1)$. We now show that indeed, a band-pass signal can be sampled at a rate not much higher than $(\omega_2 - \omega_1)/\pi$ without being subject to aliasing.

Consider first the special case of ω_2 an integer multiple of the bandwidth, say

$$\omega_2 = L(\omega_2 - \omega_1). \tag{3.48}$$

Let us sample the signal at an interval

$$T = \frac{\pi}{\omega_2 - \omega_1} = \frac{\pi L}{\omega_2}. \tag{3.49}$$

As usual, denote the impulse-sampled signal by $x_p(t)$ and its Fourier transform by $X_p^F(\omega)$. Then, by the sampling theorem,

$$X_p^F(\omega) = \frac{1}{T} \sum_{k=-\infty}^{\infty} X^F\left(\omega - \frac{2\pi k}{T}\right) = \frac{1}{T} \sum_{k=-\infty}^{\infty} X^F(\omega - 2k(\omega_2 - \omega_1)). \tag{3.50}$$

Now, $X^F(\omega - 2k(\omega_2 - \omega_1))$ is nonzero only in the range

$$\omega_1 \leq |\omega - 2k(\omega_2 - \omega_1)| \leq \omega_2,$$

or

$$(2k - L)(\omega_2 - \omega_1) \leq \omega \leq (2k + 1 - L)(\omega_2 - \omega_1),$$
$$(2k + L - 1)(\omega_2 - \omega_1) \leq \omega \leq (2k + L)(\omega_2 - \omega_1).$$

This is illustrated in Figure 3.26 for even L (4 in this case) and odd L (3 in this case). As we see, the replicas corresponding to different values of k *do not overlap*. In particular, the replicas corresponding to $k = \pm L/2$ if L is even, or $k = \pm(L-1)/2$ if L is odd, appear in the interval

$$-(\omega_2 - \omega_1) \leq \omega \leq \omega_2 - \omega_1,$$

which corresponds to the interval $[-\pi, \pi]$ in the θ domain. The conclusion is that the sampled signal *is not aliased*, despite being sampled at a rate smaller than Nyquist's critical rate.

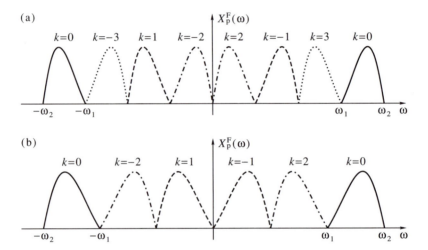

Figure 3.26 Sampling of a band-pass signal below the Nyquist rate when $(\omega_2 - \omega_1)$ is an integer multiple of the bandwidth: (a) $L = 4$; (b) $L = 3$.

Next consider the general case, where ω_2 is not an integer multiple of the bandwidth. The idea in this case is to extend the bandwidth of $x(t)$ artificially to the range

$$\omega_0 < |\omega| < \omega_2,$$

such that (1) $\omega_0 \leq \omega_1$ and (2) ω_2 is an integer multiple of $\omega_2 - \omega_0$, say

$$\omega_2 = L(\omega_2 - \omega_0). \tag{3.51}$$

We can then sample the signal at an interval

$$T = \frac{\pi}{(\omega_2 - \omega_0)} = \frac{\pi L}{\omega_2} \tag{3.52}$$

and the sampled signal will be alias free, by the same argument as before. Figure 3.27 illustrates the procedure of extending the bandwidth and sampling at the interval given by (3.52).

The integer factor L is calculated as follows. We have

$$\omega_0 = \frac{L - 1}{L}\omega_2 \leq \omega_1 \implies L \leq \frac{\omega_2}{\omega_2 - \omega_1}. \tag{3.53}$$

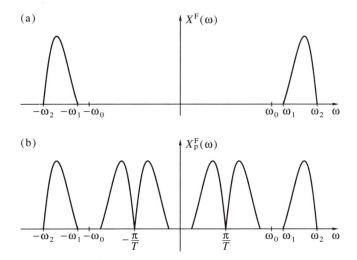

Figure 3.27 Sampling of a band-pass signal below the Nyquist rate in a general case: (a) Fourier transform of the continuous-time signal; (b) Fourier transform of the sampled signal.

But L must be an integer, so

$$L = \left\lfloor \frac{\omega_2}{\omega_2 - \omega_1} \right\rfloor. \tag{3.54}$$

In summary, the signal can be sampled at interval

$$T = \frac{\pi \lfloor \omega_2 / (\omega_2 - \omega_1) \rfloor}{\omega_2}. \tag{3.55}$$

This is only slightly different from $\pi / (\omega_2 - \omega_1)$ if $\omega_2 \gg (\omega_2 - \omega_1)$.

When implementing band-pass sampling, the following point is of extreme importance and should not be overlooked: Regardless of the sampling rate, the A/D converter must be speed compatible with the highest signal frequency ω_2, rather than with the sampling rate. Recall from our discussion of physical sampling that A/D conversion involves averaging. Unless the averaging interval Δ is much smaller than one period of ω_2, averaging will disrupt the information and the samples will be grossly distorted. Therefore, A/D converters for band-pass sampling are faster, hence costlier, than A/D converters for low-pass signals.

Example 3.12 Radio receivers, since the invention of the *superheterodyne*, convert their high-frequency input signal to an intermediate frequency, or IF. The great advantage of the superheterodyne is that the IF frequency is constant for the entire range of frequencies of the input signal. This makes it easy to amplify the signal and control its bandwidth; hence most of the amplification is done at the IF stage. The IF signal is, like the high-frequency input signal, modulated by the information signal, which is usually low frequency. Traditionally, the IF output is demodulated and low-pass filtered to provide the information signal.

Modern digital communication receivers often rely on digital processing of the information. At the time of writing, there is a growing interest in *direct IF sampling*, as an alternative to the procedure of demodulation followed by low-pass filtering and sampling. Direct IF sampling has the potential of eliminating costly and space-consuming analog hardware. For example, suppose that the IF frequency is 1 MHz and the information bandwidth is 38.4 kHz. In this case we have

$$L = \lfloor 1019.2/38.4 \rfloor = 26,$$

and the sampling frequency is

$$f_{\text{sam}} = \frac{2 \times 1019.2}{26} = 78.4 \,\text{kHz}.$$

The sample-and-hold time (or the conversion time for a flash A/D) should be about 0.1 microsecond or less in this case.   □

When sampling a band-pass signal, there is seldom a need to reconstruct it in the pass band (usually either there is no need for reconstruction at all, or reconstruction is needed in the base band). If band-pass reconstruction is needed, it can be performed by passing the sampled signal through an ideal band-pass filter with frequency response

$$H^{\text{F}}(\omega) = \begin{cases} T, & \omega_0 \leq |\omega| \leq \omega_2, \\ 0, & \text{otherwise,} \end{cases} \tag{3.56}$$

see Figure 3.28. The computation of the corresponding impulse response $h(t)$ is discussed in Problem 3.34.

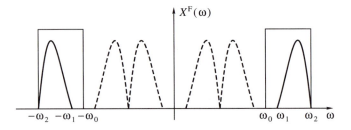

Figure 3.28 Reconstruction of a band-pass signal by an ideal band-pass filter (replicas shown by dashed lines will be eliminated by the filter).

Instead of extending the bandwidth on the left, we can also extend it on the right. You are asked to carry out the procedure and compare the resulting sampling rate with the one given in (3.55) (see Problem 3.32).

3.7   Sampling of Random Signals*

When a continuous-time random signal is sampled, the result is a discrete-time random signal. Our interest here is to learn how the parameters of the signal—mean, variance, covariance function, and PSD—are affected by sampling.

Let $x(t)$ be a continuous-time WSS random signal and $x[n] = x(nT)$ its point sampling. The mean and variance of the sampled signal are not affected by sampling, since

$$E(x[n]) = E(x(nT)) = \mu_x, \tag{3.57a}$$

$$E\{(x[n] - \mu_x)^2\} = E\{(x(nT) - \mu_x)^2\} = \gamma_x. \tag{3.57b}$$

The covariance sequence of $x[n]$ is derived as follows:

$$\kappa_x[m] = E\{(x[n+m] - \mu_x)(x[n] - \mu_x)\}$$
$$= E\{[x(nT + mT) - \mu_x][x(nT) - \mu_x]\} = \kappa_x(mT). \tag{3.58}$$

We conclude that:

1. Sampling of a WSS random signal yields a discrete-time WSS random signal.
2. The covariance sequence of the sampled signal is obtained by sampling the covariance function of the continuous-time signal at the same sampling interval.

The sampling theorem relations (3.10) and (3.58) immediately imply:

Theorem 3.3 The PSD of a sampled WSS signal $x[n]$ is related to that of the continuous-time WSS signal $x(t)$ by

$$K_x^f(\theta) = \frac{1}{T} \sum_{k=-\infty}^{\infty} K_x^F\left(\frac{\theta - 2\pi k}{T}\right). \tag{3.59}$$

□

Formula (3.59) can be regarded as a sampling theorem for random WSS signals. It implies that Nyquist's no-aliasing condition, as well as the aliasing phenomenon in case Nyquist's condition is violated, apply to the power spectra of random signals in the same manner as they apply to the Fourier transforms of nonrandom signals. One difference should be borne in mind, however. Whereas the right side of (3.10) is, in general, an infinite sum of complex-valued functions, the right side of (3.59) is a sum of real-valued nonnegative functions. The sampling theorem for random signals applies to the PSDs of the signals, not to their Fourier transforms.

White noise represents an exception to the sampling formulas (3.58) and (3.59). White noise cannot be point sampled. This is hardly surprising since, as we saw in Section 2.6, continuous-time white noise has infinite variance and does not exist as a physical entity. On the other hand, band-limited white noise has finite variance and its point sampling is well defined. The covariance and spectral characteristics of sampled band-limited white noise depend on the sampling rate relative to the noise bandwidth. An important special case occurs when the signal is sampled exactly at the Nyquist rate. In this case, the sampled signal becomes a discrete-time white noise. This can be seen either from (2.78) in the covariance domain, or from (2.77) in the frequency domain. The variance of the discrete-time white noise is then N_0/T, or $N_0 \omega_m/\pi$. This result is of importance when sampling signals accompanied by noise. Often the noise has large bandwidth, sometimes much larger than the signal bandwidth. If the sampling rate is chosen according to the signal bandwidth, the noise will be aliased and its variance may increase considerably after sampling, due to the summation in (3.59). In such cases it expedient to insert an antialiasing filter before the sampler. The antialiasing filter limits the bandwidth of the noise before the sampler. As a result, the sampled signal will be accompanied by discrete-time white noise whose variance is the smallest possible, since aliasing is prevented by the antialiasing filter. The following example illustrates such a case.

Example 3.13 Binary phase-shift keying (BPSK) is one of the simplest methods for transmitting digital information.[6] Suppose we are given a sequence of bits $b[n]$, appearing every T seconds. A non–return to zero (NRZ) signal $x(t)$ for this sequence is defined as

$$x(t) = \begin{cases} 1, & b[n] = 0, \\ -1, & b[n] = 1, \end{cases} \quad \text{for} \quad nT \le t < (n+1)T. \tag{3.60}$$

A typical waveform of an NRZ signal is shown in Figure 3.29(a). When such a waveform is used for modulating a sinusoidal carrier wave, the resulting high-frequency wave has the form

$$m(t) = x(t)\cos(\omega_0 t). \tag{3.61}$$

The signal $m(t)$ is a BPSK signal; it has phase 0° whenever the bit is 0 and phase 180° whenever the bit is 1; hence the name *binary phase-shift keying*. However, here we are interested in the signal $x(t)$, rather than in the modulated signal $m(t)$.

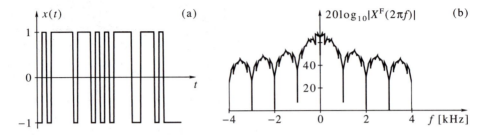

Figure 3.29 An NRZ signal: (a) waveform; (b) magnitude spectrum.

Figure 3.29(b) shows a typical magnitude spectrum of an NRZ signal. Here the bits appear at a rate of 1000 per second and the spectrum is shown in the range ±4 kHz. As we see, the magnitude decays rather slowly as the frequency increases. This behavior of the spectrum is problematic because communication systems usually require narrowing the spectrum as much as possible for a given bit rate.

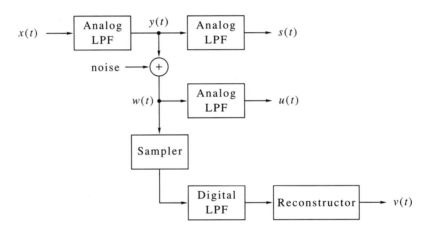

Figure 3.30 Block diagram of the system in Example 3.13.

A common remedy is to pass the NRZ signal through a low-pass filter (LPF) before it is sent to the modulator, as shown in the top part of Figure 3.30. The bandwidth of the filter is on the order of the bit rate, in our case 1 kHz. Figure 3.31(a) shows the result of passing the signal $x(t)$ through such a filter [denoted by $y(t)$], and Figure 3.31(b) shows the corresponding spectrum. As we see, the magnitude now decays much more rapidly as the frequency increases.

The received high-frequency signal is demodulated and then passed through another low-pass filter, as shown in the top part of Figure 3.30. This new filter is often identical or similar to the filter used in the transmitter. The signal at the output of this filter, denoted by $s(t)$, is used for detecting the transmitted bits.

In practice, the communication channel adds noise to the signal, resulting in a new signal $w(t)$, as shown in the middle part of Figure 3.30. When $w(t)$ is passed through the analog low-pass filter, it yields a signal $u(t)$ different from $s(t)$.

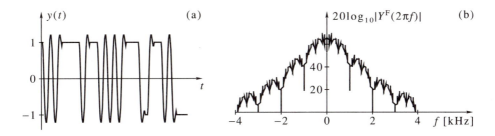

Figure 3.31 A filtered NRZ signal: (a) waveform; (b) magnitude spectrum.

To get to the point we wish to illustrate, let us assume that an engineer who was given the task of implementing the filter at the receiver decided to replace the traditional analog filter by a digital filter, as shown in the bottom part of Figure 3.30. The engineer decided to sample the demodulated signal at a rate of 8 kHz, judged to be more than enough for subsequent digital low-pass filtering to 1 kHz. The engineer also built the traditional analog filter for comparison and tested the two filters on real data provided by the receiver. The waveform $u(t)$ of the analog filter output is shown in Figure 3.32(a); the theoretical signal $s(t)$ is shown for comparison (dotted line). As we see, the real-life waveform at the output of the analog filter is quite similar to the theoretical one. However, the reconstructed output of the digital filter $v(t)$ was found to be distorted, as shown in Figure 3.32(b).

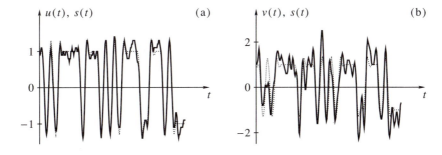

Figure 3.32 A received BPSK signal: (a) analog filtering; (b) digital filtering.

To find the source of the problem, the engineer recorded the waveform and the spectrum of the input signal to the two filters $w(t)$. Figure 3.33 shows the result. Contrary to the simulated signal, which is synthetic and smooth, the real-life signal is noisy. The noise has large bandwidth (about 100 kHz in this example), of which only a small part is shown in the figure. The analog filter attenuates most of this noise, retaining only the noise energy within ±1 kHz. This is why the signals $u(t)$ and $s(t)$ are similar. On the other hand, the noise energy is aliased in the sampling process, and appears to the digital filter as energy in the range ±4 kHz. The digital filter removes about 75 percent of this energy (the part outside 1 kHz), but the remaining 25 percent is enough to create the distortion seen in Figure 3.32(b).

The lesson of this example is that an analog antialiasing filter should have been inserted prior to the sampler, to remove the noise at frequencies higher than 4 kHz. With such a filter, the two systems would have performed approximately the same.

□

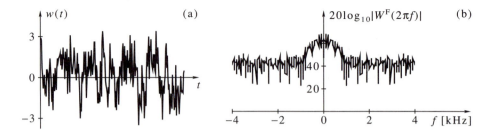

Figure 3.33 A noisy BPSK signal: (a) waveform; (b) magnitude spectrum.

3.8   Sampling in the Frequency Domain*

Suppose we are given a continuous-time signal $x(t)$ with a Fourier transform $X^F(\omega)$, and we wish to sample its Fourier transform, that is, to let

$$X^F[k] = X^F(k\Omega), \quad k \in \mathbb{Z},$$

for some $\Omega > 0$. How is the time signal corresponding to $X^F[k]$ related to $x(t)$, and under what conditions can we reconstruct $X^F(\omega)$ from its samples? We can answer these questions by invoking duality of the Fourier transform. Define $Y^F(\omega)$ as the impulse sampling of $X^F(\omega)$, that is,

$$Y^F(\omega) = \sum_{k=-\infty}^{\infty} X^F(k\Omega)\delta(\omega - k\Omega). \tag{3.62}$$

Then we have, by the dual of the sampling formula (3.9),

$$y(t) = \frac{1}{\Omega} \sum_{n=-\infty}^{\infty} x\left(t - \frac{2\pi n}{\Omega}\right). \tag{3.63}$$

As a result of the infinite summation, the signal corresponding to the sampled Fourier transform is periodic in t with period $2\pi/\Omega$. We therefore say that *sampling in the frequency domain gives rise to periodicity in the time domain.*

Formula (3.63) implies that $y(t)$ is *time aliased*; it is the signal obtained when shifting $x(t)$ by all integer multiples of $2\pi/\Omega$ and summing. In particular, if $x(t)$ is *time limited* to a duration not greater than $2\pi/\Omega$, then it can be recovered unambiguously from $y(t)$ because then the shifted replicas $x(t - 2\pi n/\Omega)$ do not overlap. The support of $x(t)$ does not have to be symmetric around $t = 0$; only its total duration matters.

When $x(t)$ is time limited to a duration $2\pi/\Omega$, we can recover $X^F(\omega)$ from its samples using the dual of Shannon's reconstruction formula

$$X^F(\omega) = \sum_{k=-\infty}^{\infty} X^F(k\Omega)\,\text{sinc}\left(\frac{\omega - k\Omega}{\Omega}\right). \tag{3.64}$$

Example 3.14 Suppose we wish to measure the impulse response of an electronic amplifier. Impulse response is difficult to measure directly because it is impossible to input an ideal impulse (i.e., a delta function) to a physical device. A convenient alternative is to measure the frequency response of the amplifier, and obtain the impulse response by inverse Fourier transform. Frequency response can be measured by feeding the amplifier with a sinusoidal input, and measuring the gain and phase shift of the output sinusoid. By repeating this process for a finite number of discrete frequencies, we get a sampling of the Fourier transform of the amplifier.

Suppose it is known a priori that the effective duration of the impulse response is no longer than 10 microseconds, and we are interested in a time resolution of 0.1

microsecond. How should we design the test procedure? By what we have said, the frequency increment should be 100 kHz or less (the reciprocal of the maximum duration) to avoid time aliasing. To meet the time resolution requirement, we should use frequencies up to 10 MHz (the reciprocal of the time resolution). In summary, the frequency response should be measured for the frequencies

$$\omega[k] = k\Omega = 2\pi k \cdot 10^5, \quad 0 \le k \le 100. \qquad \square$$

3.9 Summary and Complements

3.9.1 Summary

In this chapter we introduced the mathematical theory of sampling and reconstruction, as well as certain physical aspects of these two operations. The basic concept is point sampling of a continuous-time signal, which amounts to forming a sequence of regularly spaced time points (T seconds apart) and picking the signal values at these time points. An equivalent mathematical description of this operation is impulse sampling, which amounts to multiplying the continuous-time signal by an impulse train.

A fundamental result in sampling theory is the sampling theorem (3.10), which expresses the Fourier transform of a sampled signal as a function of the continuous-time signal. As the theorem shows, sampling leads to periodic replication of the Fourier transform. A major consequence of this replication is *aliasing*: The spectral shape of the sampled signal is distorted by high-frequency components, disguised as low-frequency components. The physical implication of aliasing is that high-frequency information is lost and low-frequency information is distorted.

An exception to the aliasing phenomenon occurs when the continuous-time signal is band limited and the sampling rate is at least twice the bandwidth. In such a case there is no aliasing: The Fourier transforms of the continuous-time signal and the sampled signal are equal up to a proportionality constant. This implies that all the information in the continuous-time signal is preserved in the sampled signal.

To prevent aliasing or to minimize its adverse effects, it is recommended to low-pass filter the continuous-time signal prior to sampling, such that the relative energy at frequencies above half the sampling rate will be zero or negligible. Such a low-pass filter is called an antialiasing filter.

A second fundamental result in sampling theory is the reconstruction theorem (3.23), which expresses the continuous-time signal as a function of the sampled signal values, provided that the signal has not been aliased during sampling. Ideal reconstruction is performed by an ideal low-pass filter whose cutoff frequency is half the sampling rate. Such an operation is physically impossible, so approximate reconstruction schemes are required. The simplest and most common approximate reconstructor is the zero-order hold (3.33). The zero-order hold has certain undesirable effects—nonuniform gain in the low-frequency range and nonzero gain in the high range. Both effects can be mitigated by an appropriate low-pass analog filter at the output of the zero-order hold.

We have devoted part of the chapter to physical circuits for sampling and reconstruction: analog-to-digital (A/D) and digital-to-analog (D/A) converters. Physical devices have certain undesirable effects on the signal, which cannot be completely avoided. The most prominent effect is quantization of the signal value to a finite number of bits. Other effects are smearing (or averaging), delays, and various nonlinearities.

Many applications involve band-pass or modulated signals. Such signals can often be sampled at a rate much lower than twice the highest frequency without being subject to aliasing. The lowest sampling rate at which a band-pass signal can be sampled without aliasing is typically slightly larger than twice its net bandwidth. Reconstruction of a sampled band-pass signal can be performed by an analog band-pass filter.

The last topic presented in this chapter is sampling in the frequency domain. The mathematical properties of sampling in the frequency domain are dual to those obtained for time-domain sampling. Sampling in the frequency domain leads to time aliasing in general. Time aliasing can be avoided if the signal has finite duration and the frequency sampling interval is smaller than the inverse of the duration.

3.9.2 Complements

1. [p. 48] Nyquist [1928] is credited with calling the attention of the engineering community to the lower limit $2f_m$ on the sampling frequency. Shannon [1949] provided a rigorous proof of the sampling theorem and the reconstruction formula (3.23). However, Whittaker [1915] preceded both, although his work was apparently unknown to electrical engineers of his time.

2. [p. 52] A linear time-invariant communication channel whose impulse response is a Nyquist-T signal can be used to transmit digital communication signals at a symbol rate $1/T$ without intersymbol interference. The channel whose impulse response is $h(t) = \text{sinc}(t/T)$ has the minimum bandwidth of all such channels. This property was discovered by Nyquist. The raised-cosine channel, whose impulse response is given by (3.22), also has the Nyquist-T property, but its bandwidth is larger than that of the sinc channel (for the same T). See Haykin [1994, Sec. 7.5] for a detailed discussion of this subject.

3. [p. 56] This property of the Fourier transform follows from the following argument. A finite duration signal $x(t)$ that has a Fourier transform $X^F(\omega)$ also possesses an analytic Laplace transform $X^L(s)$ on the entire complex plane. A theorem in complex function theory states that a function analytic in a domain has a countable number of zeros at most in the domain [Markushevich, 1977, Sec. 82, Theorem 17.3]. In particular, $X^F(\omega)$ has no more than a countable number of zeros on $-\infty < \omega < \infty$. Therefore, $X^F(\omega)$ cannot be band limited.

4. [p. 64] Sometimes a sample-and-hold circuit is inserted at the output of the D/A converter (see Section 3.5.2 for an explanation of the operation of this device). The sample-and-hold device helps overcome transients caused by unequal switching times of the D/A switches; it is particularly useful when the D/A operates at a high speed.

5. [p. 64] Two's-complement *integer* arithmetic is defined as follows: The range of numbers representable by B bits is from -2^{B-1} to $2^{B-1} - 1$. If x is a number in this range, it is represented by the positive number $x \bmod 2^B$, expressed as a B-bit binary number. Note that the MSB of positive numbers is 0 and that of negative numbers is 1. Two's-complement *fractional* arithmetic is defined similarly, except that the range of numbers is from -1 to $1 - 2^{-(B-1)}$, and a number x in this range is represented by the positive integer $(2^B x) \bmod 2^B$.

6. [p. 75] For further reading on binary phase-shift keying see, for example, Haykin [1994, Sec. 8.11].

3.10 Problems

3.1 We are given the signal

$$x[n] = \begin{cases} (-1)^m, & n = 2m, \\ 0, & n = 2m + 1. \end{cases}$$

Specify two possible continuous-time signals $x(t)$ from which $x[n]$ could have been obtained by sampling, one in which $x[n]$ is not aliased, and one in which it is aliased. Specify the sampling interval T in each case.

3.2 The signal

$$x(t) = e^{-0.02t^2} \mathrm{sinc}(t)$$

was sampled at interval T. It was then found that the Fourier transform of the sampled signal is

$$X^f(\theta) = 1.$$

What is the minimum T for which such a result is possible? If this is impossible for any T, explain why.

3.3 The signal $x(t)$ has the Fourier transform

$$X^F(\omega) = \frac{\pi}{\omega_m} \left[1 + \cos\left(\frac{\pi \omega}{\omega_m}\right) \right], \quad |\omega| \le \omega_m.$$

The signal is sampled at interval $T = 2\pi/\omega_m$. Find the sampled signal $x[n]$.

3.4 Let $x(t)$ be a band-limited signal, whose Fourier transform $X^F(\omega)$ is identically zero outside the interval $[-1.5\omega_0, 1.5\omega_0]$. Define $T = 2\pi/\omega_0$ and let

$$y(t) = \sum_{n=-\infty}^{\infty} x(t - nT).$$

(a) Prove that $y(t)$ has the form

$$y(t) = C_0 + C_1 \cos(\omega_0 t + \phi_0),$$

where C_0, C_1, ϕ_0 are constants.

(b) Express C_0, C_1, ϕ_0 it terms of $X^F(\omega)$.

3.5 Prove the following properties of a band-limited signal sampled at or above the Nyquist rate:

$$\text{Area conservation:} \quad \int_{-\infty}^{\infty} x(t)dt = T \sum_{n=-\infty}^{\infty} x(nT), \tag{3.65}$$

$$\text{Energy conservation:} \quad \int_{-\infty}^{\infty} |x(t)|^2 dt = T \sum_{n=-\infty}^{\infty} |x(nT)|^2. \tag{3.66}$$

3.6 Repeat Example 3.2 for the signal

$$x(t) = e^{-\alpha|t|}, \quad \alpha > 0, \quad -\infty < t < \infty,$$

and establish an infinite sum formula similar to (3.17) for this case.

3.7 Let $x(t)$ be a periodic signal with period T_0 and let $X^S[k]$ be its Fourier series coefficients. Find a general expression for $X^f(\theta)$, the Fourier transform of the sampled signal $x(nT)$, in terms of the $X^S[k]$. Hint: Remember that $X^f(\theta)$ must be defined first on $\theta \in [-\pi, \pi)$ and then extended periodically.

3.8 Let $x(t)$ be a continuous-time complex periodic signal with period T_0. The signal is band limited, such that its Fourier series coefficients $X^S[k]$ vanish for $|k| > 3$.

  (a) The signal is sampled at interval $T = T_0/N$, where N is integer. What is the minimum N that meets Nyquist's condition?

  (b) With this value of N, what is the minimum number of samples from which $X^S[k]$ can be computed? Explain how to perform the computation.

  (c) Instead of sampling as in part a, we sample at $T = T_0/5.5$. Plot the Fourier transform of the point-sampled signal as a function of θ. Is it possible to compute the $X^S[k]$ in this case? If so, what is the minimum number of samples and how can the computation be performed?

3.9 A continuous-time signal $x(t)$ is passed through a filter with impulse response $h(t)$, and then sampled at interval T; see Figure 3.34(a). The signal is band limited to $\pm w_1$, and the frequency response of the filter is band limited to $\pm w_2$. We wish to change the order of the operations: Sample the signal first and then pass the sampled signal through a digital filter; see Figure 3.34(b). We require that:

  • the impulse response of the digital filter be $Th(nT)$;

  • the outputs of the two systems be equal for any input signal $x(t)$ that meets the bandwidth restriction.

What is the condition on the sampling interval T to meet this requirement?

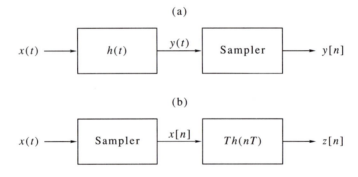

Figure 3.34 Pertaining to Problem 3.9.

3.10 Consider a continuous-time signal whose Fourier transform is

$$Y^F(w) = \begin{cases} \dfrac{1}{f_0}, & |w| \le w_1, \\ \dfrac{1}{f_0}\cos\left(\dfrac{|w| - w_1}{4\alpha f_0}\right), & w_1 < |w| \le w_2, \\ 0, & |w| > w_2, \end{cases} \qquad (3.67)$$

where w_1 and w_2 are defined in (3.21).

  (a) Compute the signal $y(t)$. Take care to compute separately for $t = 0$, for $t = \pm 1/(4\alpha f_0)$, and for all other t.

  (b) Assume that the signal is sampled at interval $T = 1/f_0$. Is the sampled signal alias free?

  (c) With the same sampling interval, is the signal Nyquist-T (as defined in Example 3.9)?

(d) Repeat parts b and c for the signal

$$x(t) = \{y * y\}(t).$$

Hint: This problem is related to Example 3.9.

3.11 The signal $x(t)$ has the Fourier transform

$$X^F(\omega) = \begin{cases} 1 - \frac{|\omega|T}{\pi}, & |\omega| \leq \frac{\pi}{T}, \\ 0, & \text{otherwise.} \end{cases}$$

Let $y(t)$ be the impulse sampling of $x(t)$ using the alternating impulse train $q_T(t)$ defined in Problem 2.24, that is,

$$y(t) = x(t)q_T(t).$$

Compute $Y^F(\omega)$ as a function of $X^F(\omega)$. Draw the shape of $Y^F(\omega)$.

3.12 Let $x(t)$ be a periodic square wave with period T_0, that is,

$$x(t) = \begin{cases} 1, & nT_0 \leq t < (n + 0.5)T_0, \\ -1, & (n + 0.5)T_0 \leq t < (n + 1)T_0, \end{cases} \quad \text{for} \quad -\infty < n < \infty.$$

(a) Show that the sampled signal $x(nT)$ is always aliased, regardless of the choice of T.

(b) We wish to sample $x(t)$ after passing it through an antialiasing filter. Let $y(t)$ be the output of the antialiasing filter. We require that (1) $y(nT)$ be nonaliased with respect to $y(t)$, (2) $y(t)$ contain at least 99 percent of the energy of $x(t)$, and (3) the frequency responses of $y(t)$ and $x(t)$ be identical in the range $\pm\pi/T$. What sampling interval T will meet these requirements, and what is the cutoff frequency of the antialiasing filter? Hint: Use Parseval's relationship for the Fourier series of $x(t)$.

3.13 We are given the continuous-time signal

$$x(t) = a\sin(\omega_0 t) + b\cos(2\omega_0 t).$$

The signal is sampled by impulse sampling at interval $T = 0.5\pi/\omega_0$.

(a) Show that the sampled signal $x_p(t)$ is periodic, find its period and compute its Fourier series coefficients $X_p^S[k]$.

(b) Suppose that a malfunction causes all samples

$$\{x(3mT), -\infty < m < \infty\}$$

to be lost and their values to be replaced by zero. Let $y(t)$ denote the resulting impulse-sampled signal. The signal $y(t)$ is passed through an ideal low-pass filter with frequency response

$$H^F(\omega) = T \operatorname{rect}\left(\frac{\omega}{3\omega_0}\right).$$

Find the signal $z(t)$ at the output of the filter.

3.14 Let

$$x(t) = \frac{\beta}{\beta^2 + t^2}, \quad \beta > 0, \quad -\infty < t < \infty.$$

The signal is sampled at interval T. Find $X^f(0)$ for the sampled signal $x[n]$. Hint: Find the Fourier transform of the signal $y(t) = 1/(\beta - jt)$ first, by duality to the signal $x_1(t)$ in Problem 2.3.

3.15 Let $x(t)$ be the continuous-time signal

$$x(t) = \begin{cases} 1, & |t| \le (N + 0.5)T, \\ 0, & \text{otherwise,} \end{cases}$$

where N is a positive integer.

(a) Find $X^f(\theta)$, the Fourier transform of the sampled signal $x(nT)$.

(b) Use the result of part a to derive a closed-form expression for the infinite sum

$$\sum_{k=-\infty}^{\infty} \text{sinc}\left[\frac{(N + 0.5)(\omega T - 2\pi k)}{\pi}\right].$$

3.16 We are given the signal $x(t)$ whose Fourier transform is

$$X^F(\omega) = \begin{cases} \cos\left(\frac{\pi\omega}{\omega_0}\right), & |\omega| \le \omega_0, \\ 0, & |\omega| > \omega_0. \end{cases}$$

Testing of the signal has revealed that an unwanted noise has been added to the signal in the frequency range $0.5\omega_0 < |\omega| \le \omega_0$. It was therefore decided to disregard the contents of the signal in this frequency range and to use only the information in the range $|\omega| \le 0.5\omega_0$.

(a) What is the minimum sampling frequency ω_{sam} for which there will be no aliasing in the range $|\omega| \le 0.5\omega_0$? Assume that antialiasing filtering before sampling is not permitted.

(b) After sampling at the rate found in part a, the signal is reconstructed using an ideal low-pass filter with cutoff frequency $0.5\omega_0$. Give an expression for the signal $\hat{x}(t)$ at the output of the reconstructor.

3.17 Figure 3.35 shows one period of a continuous-time periodic signal $x(t)$ with period $T_0 = 6$. The signal is sampled at interval $T = 1$, resulting in the discrete-time signal $x[n]$. The discrete-time signal is then reconstructed using the ideal reconstructor (3.23), resulting in the continuous-time signal $\hat{x}(t)$.

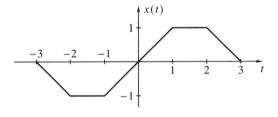

Figure 3.35 Pertaining to Problem 3.17.

(a) Is $\hat{x}(t)$ periodic?

(b) Is $\hat{x}(t)$ equal to $x(t)$? Answer without performing any calculations.

(c) Compute $\hat{x}(t)$.

3.18 The signal $x(t)$ is periodic, with period T_0. We are given that its Fourier series coefficients $X^S[k]$ are zero for $|k| > K$ for some given K. Let $N = 2K + 1$. The signal is sampled at interval $T = T_0/N$. Prove that $x(t)$ can be expressed in terms of its samples over one period, $\{x(nT),\ 0 \le n \le N - 1\}$, as follows:

$$x(t) = \frac{1}{N} \sum_{n=0}^{N-1} x(nT) D\left(\frac{t}{T_0} - \frac{n}{N}, N\right),$$ (3.68)

where

$$D(a, N) = \frac{\sin(0.5Na)}{\sin(0.5a)}.$$ (3.69)

Hint: Use Shannon's interpolation formula (3.23) first and then (2.157).

3.19 The signal $x(t)$ is sampled at the instants

$$t = nT,\ nT + a,\quad n \in \mathbb{Z},$$

where a is a constant, $0 < a < T$. Note that this is a nonuniform sampling. Let $q(t)$ be an impulse train consisting of Dirac delta functions at the sampling instants. Let

$$x_q(t) = x(t)q(t),$$

that is, $x_q(t)$ is the impulse sampling of $x(t)$.

(a) Write an explicit expression for $q(t)$.

(b) Express $X_q^F(\omega)$ as a function of $X^F(\omega)$. Bring the result to the form

$$X_q^F(\omega) = \frac{1}{T} \sum_{k=-\infty}^{\infty} A_k X^F(\omega - k\omega_{\text{sam}}).$$

Specify the values of A_k and ω_{sam}.

(c) Let $X^F(\omega)$ be given by

$$X^F(\omega) = \begin{cases} 1 - \dfrac{2|\omega|}{\omega_{\text{sam}}}, & -0.5\omega_{\text{sam}} \le \omega \le 0.5\omega_{\text{sam}}, \\ 0, & \text{otherwise,} \end{cases}$$

where ω_{sam} is as found in part b. Let $a = 0.25T$. Draw, on separate plots, the real and imaginary parts of $X_q^F(\omega)$ in the range $-2.5\omega_{\text{sam}} \le \omega \le 2.5\omega_{\text{sam}}$.

(d) Under the conditions of part c, what reconstruction filter is needed to reconstruct $x(t)$ from $x_q(t)$?

3.20 We are given a band-limited signal $x(t)$, whose Fourier transform is nonzero only for $|f| \le 8\,\text{kHz}$. The signal is sampled at interval T, resulting in the discrete-time signal $x[n]$. The signal $x[n]$ is passed through a nonlinear system, resulting in the discrete-time signal

$$y[n] = (x[n])^3 + 0.5x[n].$$

The signal $y[n]$ is reconstructed by an ideal low-pass filter. What is the minimal sampling rate $1/T$ that will ensure that the reconstructed signal will be

$$\hat{y}(t) = [x(t)]^3 + 0.5x(t)$$

up to a scale factor?

3.21 Recall Problem 3.7 and suppose we reconstruct $\hat{x}(t)$ from $x(nT)$ using Shannon's formula (3.23).

(a) Show that if T/T_0 is not a rational number, the reconstructed signal $\hat{x}(t)$ is not periodic.

(b) Show that if T/T_0 is rational, that is,

$$\frac{T}{T_0} = \frac{p}{q},$$

where p, q are coprime integers, then $\hat{x}(t)$ is periodic. Show that, in this case, $\hat{x}(t)$ is a finite sum of sinusoidal signals. Find the period of $\hat{x}(t)$ as a function of T, p, and q.

3.22 Let $x(t)$ be the signal

$$x(t) = 3\cos(100\pi t) + 2\sin(250\pi t).$$

(a) The signal is sampled at an interval $T_1 = 0.0025$ second, yielding the discrete-time signal $x[n]$. Now we reconstruct $\hat{x}(t)$ from $x[n]$ using an ideal reconstructor corresponding to $T_2 = 0.005$ second, that is,

$$\hat{x}(t) = \sum_{n=-\infty}^{\infty} x[n]\operatorname{sinc}\left(\frac{t - nT_2}{T_2}\right).$$

Give an explicit expression for $\hat{x}(t)$.

(b) Now let $T_1 = 0.005$ second and $T_2 = 0.0025$ second, and repeat part a.

3.23 Let $x(t)$ be band limited to $\pm\omega_m$ and let $T = \pi/\omega_m$. Define

$$y(t) = \sum_{n=-\infty}^{\infty} x(nT)\delta(t - 2nT).$$

(a) Find the Fourier transform of $y(t)$ as a function of $X^F(\omega)$. Hint: Write $y(t)$ as an impulse sampling of a signal related to $x(t)$, then use property (2.7) of the Fourier transform.

(b) Is the Fourier transform of $y(t)$ aliased? Explain both mathematically and based on physical reasoning.

3.24 Let $x(t)$ be band limited to $\pm\omega_m$ and let $T = \pi/\omega_m$. The signal $x(t)$ is sampled at interval T and then reconstructed according to the formula

$$z(t) = \sum_{n=-\infty}^{\infty} x(nT)\operatorname{sinc}\left(\frac{t - 2nT}{T}\right)$$

(note the factor 2 in the numerator of the argument of the sinc).

(a) Express $Z^F(\omega)$ in terms of $X^F(\omega)$ and plot the result. Hint: Express $z(t)$ as a convolution with a certain impulse-sampled signal and use the solution to Problem 3.23; then pass to the frequency domain.

(b) Give a closed-form expression for $z(t)$ if

$$x(t) = \cos(0.6\omega_m t).$$

(c) Show that the operation in part a can be also performed as follows:

- Form a new sequence $u[n]$ by inserting zeros between adjacent points of $x(nT)$, that is, define

$$u[n] = \begin{cases} x(0.5nT), & n \text{ even}, \\ 0, & n \text{ odd}. \end{cases}$$

- Reconstruct $z(t)$ from $u[n]$ by Shannon's formula (3.23).

3.25 The discrete-time signal

$$x[n] = \cos\left(\frac{2\pi n}{16}\right)$$

is reconstructed by a zero-order hold with $T = 1$, to give the continuous-time signal $y(t)$. This signal is sampled in the following manner:

$$z[n] = y((2n + 0.5)T).$$

Find an expression for $z[n]$.

3.26 The signal $x(t)$ has the Fourier transform

$$X^F(\omega) = \begin{cases} 1, & 0.25\pi \leq |\omega| \leq 1.25\pi, \\ 0, & \text{otherwise.} \end{cases}$$

The signal is sampled at interval $T = 1$ to give the discrete-time signal $x[n]$. The signal $x[n]$ is given to a Shannon reconstructor (with $T = 1$), yielding the continuous-time signal $y(t)$. Compute $y(t)$.

3.27 The signal

$$x(t) = \cos^3(0.4\pi t)$$

is sampled at interval $T = 1$ (point sampling) and then reconstructed by a Shannon reconstructor that uses $T = 0.5$. Compute the reconstructed signal.

3.28 The signal

$$x(t) = \cos(0.9\pi t)\cos(0.4\pi t)$$

is sampled at interval $T = 1$ second, yielding the discrete-time signal $x[n]$. The discrete-time signal is passed through an ideal high-pass filter, which eliminates all frequencies below 0.6π. The signal at the output of the filter, $y[n]$, is reconstructed by a Shannon reconstructor having $T = 1$ second. Find an expression for the signal $y(t)$ at the output of the reconstructor.

3.29 A *first-order hold* is a reconstructor that works as follows: At time $t = nT$ it computes the straight line connecting $(nT - T, x[nT - T])$ and $(nT, x[nT])$. During the interval $[nT, nT + T)$ it takes $\hat{x}(t)$ as the ordinate of the point on the straight line whose abscissa is t, see Figure 3.36.

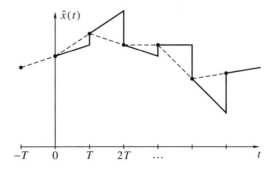

Figure 3.36 Pertaining to Problem 3.29.

(a) Compute and plot the impulse response of the first-order hold.

(b) Compute and plot the frequency response of the first-order hold (magnitude and phase).

(c) Compare the first-order hold with the zero-order hold and draw conclusions.

3.30 Let $x[n]$ be a discrete-time signal. Suppose that, because of a malfunction, all odd-numbered values are replaced by zero, and we get the signal

$$y[n] = \begin{cases} x[n], & n \text{ even}, \\ 0, & n \text{ odd}. \end{cases}$$

It is suggested to reconstruct $x[n]$ by a digital ZOH, which is defined by

$$\hat{x}[2m] = \hat{x}[2m + 1] = y[2m], \quad m \in \mathbb{Z}.$$

Show that the digital ZOH operation is equivalent to passing $y[n]$ through a digital linear time-invariant filter. Compute the magnitude and phase responses of this filter.

3.31 Consider the following reconstructor (note that it is not causal):

- Between every two points of the sampled signal, $x(nT)$ and $x(nT+T)$, we compute a midsample by linear interpolation, that is,

$$\hat{x}(nT + 0.5T) = 0.5[x(nT) + x(nT + T)].$$

- We interleave x and $\hat{x}$ in their order of their appearance; that is, we form the discrete-time signal

$$y[n] = \begin{cases} x(mT), & n = 2m, \\ \hat{x}(mT + 0.5T), & n = 2m + 1. \end{cases}$$

- We pass $y[n]$ through a ZOH (at interval $0.5T$) to get the reconstructed signal $\hat{y}(t)$.

(a) Find and plot the impulse response of the reconstructor.

(b) Compute the frequency response of the reconstructor.

3.32 Suppose that, instead of choosing w_0 to the left of the interval $[w_1, w_2]$, as we did in Section 3.6, we choose it to the right of the interval. Repeat the procedure described there for this case. Show that the resulting sampling interval is always smaller than the one given in (3.55).

3.33 Explain why in practice it is usually advisable, in sampling of band-pass signals, to extend the bandwidth to both left and right.

3.34 Write the impulse response $h(t)$ of the reconstruction filter (3.56).

3.35 Let $x(t)$ be a signal whose Fourier transform $X^F(w)$ is nonzero only on $3 \le |w| \le 9$. However, only the frequencies $5 \le |w| \le 7$ contain useful information whereas the other frequencies contain only noise.

(a) What is the smallest sampling rate that will enable exact reconstruction of the *useful* signal, if we do not perform any filtering on $x(t)$ before sampling?

(b) How will the answer to part a change if it is permitted to pass $x(t)$ through a filter before sampling?

3.36 $x(t)$ is a band-pass signal whose Fourier transform is nonzero only for $3.5 \le \omega \le 4.5$. Let

$$y(t) = [x(t)]^2.$$

What is the smallest rate at which $y(t)$ can be sampled such that it will be possible to reconstruct $y(t)$ exactly from its samples?

3.37 Let $x(t)$ be a complex band-limited signal on $[-\omega_m, \omega_m]$ and assume that $X^F(\omega_m) = X^F(-\omega_m) = 0$. Define

$$y(t) = x(t) \sum_{n=1}^{\infty} \lambda^n \sin(n\omega_0 t),$$

where λ is a given real number in the range $0 < \lambda < 1$.

(a) Express $Y^F(\omega)$ in terms of $X^F(\omega)$.

(b) What is the minimum value of ω_0 that enables exact reconstruction of $x(t)$ from $y(t)$?

(c) Suggest a procedure for reconstructing $x(t)$ from $y(t)$ when the condition on ω_0 is met.

3.38 A digital communication channel is given, capable of transmitting 19,200 bits per second. We wish to use the channel to transmit a band-limited analog signal $x(t)$, by sampling and digitizing. The magnitude of the analog signal is limited to $|x(t)| \le x_{max}$. The error between the digitized signal and $x(t)$ must not exceed $\pm 10^{-4} x_{max}$.

(a) What is the required number of bits of the A/D?

(b) What is the maximum bandwidth of the analog signal for which the channel can be used?

3.39 The signal $x(t) = \cos(\omega_0 t)$ is sampled by a nonideal sampler, as described in Section 3.5.3, with averaging interval Δ. The discrete-time signal $x[n]$ is then reconstructed, using an ideal (Shannon) reconstructor. Assume that the sampling interval T meets the Nyquist condition and that $\Delta < T$. Derive an expression for the reconstructed signal.

3.40 The purpose of this problem is to demonstrate that plots of signals sampled only slightly above the Nyquist rate can be misleading.

(a) Let

$$x(t) = \sin(0.98\pi t).$$

Sample the signal at interval $T = 1$ and plot it for $0 \le n \le 100$. Does the plot look like a sinusoid?

(b) Sample the signal at $T = 1/8$ and plot it for $0 \le n \le 800$. Then examine the signal details by plotting it for $0 \le n \le 100$. Does the plot look like a sinusoid now?

(c) Interpret what you have seen and draw conclusions.

3.41* Ben and Erik are given the continuous-time signal

$$x(t) = \cos(2\pi f_0 t + \phi_0).$$

They are told that $\phi_0 = 0.25\pi$, but they know nothing about f_0. Each is asked to sample the signal at a rate of his choice and report the value of f_0 based on the sampled signal. Ben uses a sampling frequency $f_{sam} = 150\,\text{Hz}$, and reports that $f_0 = 50\,\text{Hz}$. Erik uses a sampling frequency $f_{sam} = 240\,\text{Hz}$, and reports that $f_0 = 20\,\text{Hz}$.

(a) Is it possible, based on this information, to determine the true value of f_0? If not, what are all the possible values of f_0?

(b) If it is known that $f_0 < 1000\,\text{Hz}$, is it then possible to determine the true value of f_0?

3.42* We are given two band-pass signals: $x_1(t)$ is limited to the frequency range 1000–1350 Hz; $x_2(t)$ is limited to the frequency range 2000–2400 Hz. We are required to transmit *both* signals, in a digital form, over a single channel. The following scheme is proposed for this purpose; see Figure 3.37. Each signal is sampled separately, then the two discrete-time signals $x_1[n]$, $x_2[n]$ are combined to a single sequence $z[n]$ by a packetizer. The packetizer takes N_1 samples from the first signal and N_2 samples from the second signal as they become available, builds a sequence of $N_1 + N_2$ regularly spaced samples, and passes them to the channel. At the receiving end of the channel there is a distributor, which separates $z[n]$ back to the individual sequences $x_1[n]$, $x_2[n]$ and outputs each at regularly spaced intervals to the reconstructors. Finally, the reconstructors rebuild the continuous-time signals.

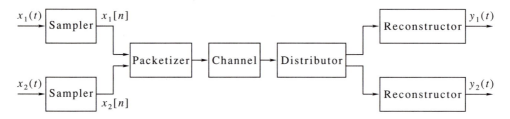

Figure 3.37 Pertaining to Problem 3.42.

(a) What are the minimum sampling rates for the two continuous-time signals?

(b) How many samples per second must the channel be capable of transmitting?

(c) What is the minimum packet size $N_1 + N_2$? How many packets are transmitted every second?

3.43* Let $x(t)$ be a real signal, band limited to $\pm\omega_m$. Consider the double-side-band modulated signal (see Problem 2.15)

$$y(t) = x(t)\cos(\omega_0 t),$$

where $\omega_0 \gg \omega_m$. The signal $y(t)$ is band-pass. DSB is often used in wireless communication. In receiving DSB signals, the most common scheme is to demodulate the signal first by

$$z(t) = 2y(t)\cos(\omega_0 t) = 2x(t)\cos^2(\omega_0 t) = x(t)[1 + \cos(2\omega_0 t)].$$

The signal $z(t)$ is then passed through a low-pass filter, which removes the high-frequency component $x(t)\cos(2\omega_0 t)$. The output of the filter is $x(t)$. This can be sampled at an interval $T \le \pi/\omega_m$ for digital processing.

(a) Show that $x(nT)$ can also be obtained by sampling $y(t)$ directly, without demodulating it first. This requires a careful choice of T. Find the largest T (i.e., the smallest sampling rate) for which this can be done, and explain why it works. Interpret your solution in the frequency domain.

(b) Suppose now that the receiver does not know ω_0 accurately and uses $\omega_0 + \Delta\omega$ instead, where $\Delta\omega$ is a small frequency error. Show that both schemes are prone to failure in this case, and explain exactly why and how they fail.

3.44* You are given the task of designing a digital receiver for radio transmission from a satellite. The transmitted signal is double side-band, as defined in Problem 3.43. The nominal carrier frequency is $f_0 = 160\,\text{MHz}$ and the information bandwidth is $f_m = 19.2\,\text{kHz}$. The satellite's orbit is at altitude $h = 300\,\text{km}$. The frequency generator at the satellite has relative accuracy $\pm 0.5 \times 10^{-6}$, and the one at the receiver has relative accuracy $\pm 3 \times 10^{-6}$. The receiver is required to work from horizon to horizon, that is, at all possible positions of the satellite relative to the receiver.

This problem deals with the front end of the receiver, which consists of a DSB demodulator, a low-pass filter, and a sampler; see Figure 3.38. The three parameters you have to determine are (1) the nominal frequency of the demodulator f_d (see the figure); (2) the sampling rate T; (3) the cutoff frequency of the low-pass filter f_c. Because of the inaccuracies of the various frequencies, you decide to demodulate the signal only partially, such that the Fourier transform remains band pass and its positive and negative parts do not overlap under any circumstances. You leave to the digital processor the task of handling the residual modulation.

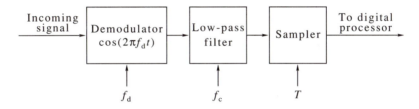

Figure 3.38 Pertaining to Problem 3.44.

Design the three parameters according to the supplied information, using the following hints and guidelines:

- If you have already solved Problem 3.43 you are aware of the inadequacy of the solution proposed there for the situation described here, so do not attempt it.
- You will need a few facts from your college physics education. In particular:
    (a) The Earth radius is $r_0 = 6370\,\text{km}$ and the Earth gravity at sea level is $g(0) = 9.81\,\text{m/s}^2$. The gravity at altitude h, denoted by $g(h)$, is inversely proportional to the square of the distance of a point at that altitude from the center of the Earth.
    (b) The orbital velocity of the satellite is the square root of the product of the gravity at its orbit by the radius of the orbit.
    (c) The Earth rotates once every 24 hours.
    (d) Doppler effect acts to change the frequency of an electromagnetic wave; look at a physics book to remind yourself of the formula.
- It is good engineering practice to insert reasonable safety margins at various places.

3.45* This problem introduces the concept of a *T-parent* of a discrete-time signal.

(a) Prove the following claim:

> Let $x[n]$ be a discrete-time signal whose Fourier transform $X^f(\theta)$ exists. Then, for every positive T, there exists a unique continuous-time signal $x(t)$ such that $X^F(\omega)$ is band limited to $|\omega| \leq \pi/T$ and $x[n] = x(nT)$.

The signal $x(t)$ can be called the *T-parent* of $x[n]$.

(b) Find the *T-parent* of the signal

$$x[n] = \begin{cases} 1, & n = -1, 0, 1, \\ 0, & \text{otherwise.} \end{cases}$$

(c) Generalize part a to any finite duration signal, that is, a signal satisfying

$$x[n] = 0, \quad n < n_1, \; n > n_2.$$

(d) Is the signal

$$x(t) = e^{-t/T}, \; t \geq 0$$

the *T-parent* of

$$x[n] = e^{-n}, \; n \geq 0?$$

Give reasons.

Chapter 4

The Discrete Fourier Transform

A discrete-time signal $x[n]$ can be recovered unambiguously from its Fourier transform $X^{\mathrm{f}}(\theta)$ through the inverse transform formula (2.95). This, however, requires knowledge of $X^{\mathrm{f}}(\theta)$ for all $\theta \in [-\pi, \pi]$. Knowledge of $X^{\mathrm{f}}(\theta)$ at a finite subset of frequencies is not sufficient, since the sequence $x[n]$ has an infinite number of terms in general. If, however, the signal has finite duration, say $\{x[n],\ 0 \le n \le N-1\}$, then knowledge of $X^{\mathrm{f}}(\theta)$ at N frequency points *may* be sufficient for recovering the signal, provided these frequencies are chosen properly. In other words, we may be able to sample the Fourier transform at N points and compute the signal from these samples. An intuitive justification of this claim can be given as follows: The Fourier transform is a linear operation. Therefore, the values of $X^{\mathrm{f}}(\theta)$ at N values of θ, say $\{\theta[k],\ 0 \le k \le N-1\}$, provide N linear equations at N unknowns: the signal values $\{x[n],\ 0 \le n \le N-1\}$. We know from linear algebra that such a system of equations has a unique solution if the coefficient matrix is nonsingular. Therefore, if the frequencies are chosen so as to satisfy this condition, the signal values can be computed unambiguously.

The sampled Fourier transform of a finite-duration, discrete-time signal is known as the *discrete Fourier transform* (DFT). The DFT contains a finite number of samples, equal to the number of samples N in the given signal. The DFT is perhaps the most important tool of digital signal processing. We devote most of this chapter to a detailed study of this transform, and to the closely related concept of circular convolution. In the next chapter we shall study fast computational algorithms for the DFT, known as fast Fourier transform algorithms. Then, in Chapter 6, we shall examine the use of the DFT for practical problems.

The DFT is but one of a large family of transforms for discrete-time, finite-duration signals. Common to most of these transforms is their interpretation as frequency domain descriptions of the given signal. The magnitude of the transform of a sinusoidal signal should be relatively large at the frequency of the sinusoid and relatively small at other frequencies. If the transform is linear, then the transform of a sum of sinusoids is the sum of the transforms of the individual sinusoids, so it should have relatively large magnitudes at the corresponding frequencies. Therefore, a standard way of understanding and interpreting a transform is to examine its action on a sinusoidal signal.

Among the many relatives of the DFT, the discrete cosine transform (DCT) has gained importance in recent years, because of its use in coding and compression of images. In recognition of its importance, we devote a section in this chapter to the

DCT. The discrete sine transform (DST), which is closely related to the DCT, is also discussed briefly.

4.1 Definition of the DFT and Its Inverse

Let the discrete-time signal $x[n]$ have finite duration, say in the range $0 \le n \le N - 1$. The Fourier transform of this signal is

$$X^f(\theta) = \sum_{n=0}^{N-1} x[n]e^{-j\theta n}. \tag{4.1}$$

Let us sample the frequency axis using a total of N equally spaced samples in the range $[0, 2\pi)$, so the sampling interval is $2\pi/N$; in other words, we use the frequencies

$$\theta[k] = \frac{2\pi k}{N}, \quad 0 \le k \le N - 1. \tag{4.2}$$

The result is, by definition, the discrete Fourier transform. Mathematically,

$$X^d[k] = \{\mathcal{D}x\}[k] = \sum_{n=0}^{N-1} x[n] \exp\left(-\frac{j2\pi kn}{N}\right), \quad 0 \le k \le N - 1. \tag{4.3}$$

Figure 4.1 illustrates the Fourier transform of a discrete-time signal and its DFT samples.

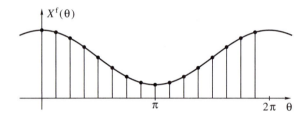

Figure 4.1 The relationship of the DFT to the Fourier transform. Solid line: Fourier transform; Circles: DFT samples (shown for $N = 16$).

In working with the DFT, it is common to use the notation

$$W_N = \exp\left(\frac{j2\pi}{N}\right). \tag{4.4}$$

Also, it is common to denote the DFT operation for a length-N signal by $\text{DFT}_N\{x[n]\}$. With these notations we can rewrite (4.3) as

$$X^d[k] = \text{DFT}_N\{x[n]\} = \sum_{n=0}^{N-1} x[n]W_N^{-kn}, \quad 0 \le k \le N - 1. \tag{4.5}$$

The sequence of integer powers of W_N, $\{W_N^n, -\infty < n < \infty\}$ is ubiquitous in the DFT world. This sequence is periodic with period N, since

$$W_N^N = \exp\left(\frac{j2\pi N}{N}\right) = e^{j2\pi} = 1 = W_N^0.$$

Figure 4.2 illustrates this sequence in the range $0 \le n \le N - 1$, for even and odd values of N. The magnitude of all the numbers in the sequence is 1, and the phases are equally spaced, starting at zero. The phase π appears if and only if N is even, and then it corresponds to $n = N/2$.

The formula given in the following lemma is a useful tool in deriving various DFT-related results.

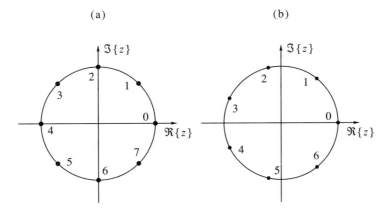

Figure 4.2 The sequence W_N^n in the complex plane: (a) even N; (b) odd N. The numbers indicate the values of n.

Theorem 4.1 (lemma)

$$\sum_{n=0}^{N-1} W_N^{kn} = N\delta[k \bmod N] = \begin{cases} 0, & (k \bmod N) \neq 0, \\ N, & (k \bmod N) = 0. \end{cases} \tag{4.6}$$

Proof The left side of (4.6) is a geometric series with parameter

$$q = W_N^k = \exp\left(\frac{j2\pi k}{N}\right).$$

When k is a multiple of N, say $k = mN$, we have $q = 1$ because

$$e^{j2\pi m} = 1, \quad \text{for all } m \in \mathbb{Z}.$$

Therefore, in this case, the sum evaluates to N. When k is not a multiple of N we have, by the summation formula for a geometric series,

$$\sum_{n=0}^{N-1} W_N^{kn} = \frac{W_N^{Nk} - 1}{W_N^k - 1} = \frac{1 - 1}{W_N^k - 1} = 0. \tag{4.7}$$

In summary,

$$\sum_{n=0}^{N-1} W_N^{kn} = N\delta[k \bmod N]. \tag{4.8}$$

$\square$

Example 4.1 The DFT of

$$x[n] = \begin{cases} 1, & n = 0, \\ 0, & 1 \leq n \leq N - 1, \end{cases} \tag{4.9}$$

is

$$X^{\mathrm{d}}[k] = 1, \quad 0 \leq k \leq N - 1. \tag{4.10}$$

$\square$

Example 4.2 The DFT of

$$x[n] = 1, \quad 0 \leq n \leq N - 1, \tag{4.11}$$

is

$$X^{\mathrm{d}}[k] = \sum_{n=0}^{N-1} W_N^{-kn} = \begin{cases} N, & k = 0, \\ 0, & 1 \le k \le N - 1. \end{cases} \tag{4.12}$$

□

Example 4.3 As we said at the beginning of this chapter, a good way of getting a feeling for a time-to-frequency transform is to examine its action on a sinusoidal signal. Since we are dealing with finite-duration signals, let us compute the DFT of a finite segment of a sinusoid:

$$x[n] = \cos(\theta_0 n + \phi_0), \quad 0 \le n \le N - 1, \tag{4.13}$$

where $0 < \theta_0 < \pi$. Expressing the cosine in terms of two complex exponentials, we get

$$
\begin{aligned}
X^{\mathrm{d}}[k] &= \sum_{n=0}^{N-1} \cos(\theta_0 n + \phi_0) \exp\left(-\frac{j 2\pi k n}{N}\right) \\
&= 0.5 e^{j\phi_0} \sum_{n=0}^{N-1} \exp\left[jn\left(\theta_0 - \frac{2\pi k}{N}\right)\right] + 0.5 e^{-j\phi_0} \sum_{n=0}^{N-1} \exp\left[-jn\left(\theta_0 + \frac{2\pi k}{N}\right)\right] \\
&= 0.5 e^{j\phi_0} \frac{1 - e^{j\theta_0 N}}{1 - \exp\left[j\left(\theta_0 - \frac{2\pi k}{N}\right)\right]} + 0.5 e^{-j\phi_0} \frac{1 - e^{-j\theta_0 N}}{1 - \exp\left[-j\left(\theta_0 + \frac{2\pi k}{N}\right)\right]}.
\end{aligned}
\tag{4.14}
$$

In particular, consider the case of θ_0 an integer multiple of $2\pi/N$, say

$$\theta_0 = \frac{2\pi m}{N}.$$

Then we get

$$X^{\mathrm{d}}[k] = \begin{cases} 0.5 N e^{j\phi_0}, & k = m, \\ 0.5 N e^{-j\phi_0}, & k = N - m, \\ 0, & \text{otherwise.} \end{cases} \tag{4.15}$$

The magnitude of the DFT indeed peaks at $k = m$, as well as at $k = N - m$. The phase of the DFT at $k = m$ is ϕ_0, the phase of the given sinusoidal signal. The DFT at all other frequencies is zero in this case. The conclusion is that, at least for sinusoids at particular frequencies—the integer multiples of $2\pi/N$—the DFT yields the expected result. Figure 4.3(a) illustrates the magnitude DFT for $N = 64$, $\theta_0 = 2\pi \cdot 15/64$, and $\phi_0 = 0$.

When θ_0 is not an integer multiple of $2\pi/N$, the DFT expression is less transparent. However, we still see from (4.14) that the magnitude of the denominator of the first term is minimized for k nearest to $N\theta_0/2\pi$, and the magnitude of the denominator of the second term is minimized for k nearest to $N(1 - \theta_0/2\pi)$. The numerators vary in magnitude between 0 and 2, but they are not identically zero if θ_0 is not an integer multiple of $2\pi/N$. Therefore, the magnitude DFT peaks for $k \approx N\theta_0/2\pi$, as well as for $k \approx N(1 - \theta_0/2\pi)$. Figure 4.3(b) illustrates the magnitude DFT for $N = 64$, $\theta_0 = 2\pi \cdot 15.25/64$, and $\phi_0 = 0$. As we see, the peak is at $k = 15$, which is the integer nearest to 15.25. The DFT values at all other frequencies are not zero, since the frequency is not an integer multiple of $2\pi/N$. Figure 4.3(c) illustrates the magnitude DFT for $N = 64$, $\theta_0 = 2\pi \cdot 15.5/64$, and $\phi_0 = 0$. In this case, since the frequency is exactly in the middle between $k = 15$ and $k = 16$, the peaks at these two values of k are equal.

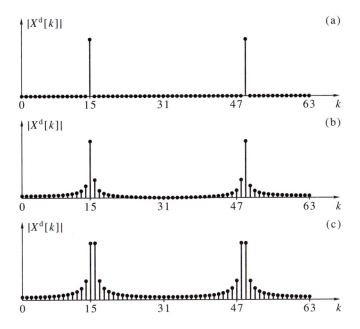

Figure 4.3 The magnitude DFT of a sinusoidal signal ($N = 64$): (a) $\theta_0 = 2\pi \cdot 15/64$; (b) $\theta_0 = 2\pi \cdot 15.25/64$; (c) $\theta_0 = 2\pi \cdot 15.5/64$.

In conclusion, the magnitude DFT of a sinusoidal signal indeed exhibits a peak at k approximately proportional to the frequency of the signal. If the frequency is an integer multiple of $2\pi/N$, we also get the phase of the sinusoid from the phase of the DFT at the peak point. □

The DFT values $\{X^d[k], 0 \le k \le N-1\}$ uniquely define the sequence $x[n]$ through the *inverse DFT formula* (IDFT):

Theorem 4.2 (inverse DFT)

$$x[n] = \{\mathcal{D}^{-1}X^d\}[n] = \text{IDFT}_N\{X^d[k]\} = \frac{1}{N}\sum_{k=0}^{N-1}X^d[k]W_N^{kn}, \quad 0 \le n \le N-1. \quad (4.16)$$

Proof

$$\frac{1}{N}\sum_{k=0}^{N-1}X^d[k]W_N^{kn} = \frac{1}{N}\sum_{k=0}^{N-1}\left[\sum_{m=0}^{N-1}x[m]W_N^{-km}\right]W_N^{kn} = \frac{1}{N}\sum_{m=0}^{N-1}x[m]\left[\sum_{k=0}^{N-1}W_N^{(n-m)k}\right]$$

$$= \frac{1}{N}\sum_{m=0}^{N-1}x[m]N\delta[(n-m)\bmod N] = x[n]. \quad (4.17)$$

□

The procedure `bfdft` in Program 4.1 illustrates possible MATLAB implementation of the DFT and its inverse. This is a brute-force implementation, obtained directly from the definition. The program computes the direct DFT if the variable `swtch` is 0, and the inverse DFT if it is nonzero. You are advised not to use this program for serious applications, since MATLAB offers a much more efficient implementation. Direct DFT in MATLAB is the function `fft` and inverse DFT is the function `ifft`. We shall learn about these implementations in Chapter 5.

Example 4.4 Let N be even, and

$$X^d[k] = \begin{cases} 1, & k = 0, \\ -1, & k = 0.5N, \\ 0, & \text{otherwise.} \end{cases}$$

The inverse DFT is given by

$$x[n] = N^{-1}(W_N^0 - W_N^{nN/2}) = N^{-1}[1 - (-1)^n],$$

since $W_N^{N/2} = e^{j\pi} = -1$. $\square$

A peculiarity of the DFT definition is the range of frequencies represented by the variable k. According to (4.2), the range $0 \le k \le \lfloor N/2 \rfloor$ corresponds to frequencies $0 \le \theta \le \pi$, whereas the range $\lceil N/2 \rceil \le k \le N-1$ corresponds to $\pi \le \theta < 2\pi$. The latter range is equivalent to $-\pi \le \theta < 0$. Therefore, the first $\lfloor N/2 \rfloor$ values of k correspond to positive frequencies and the remaining correspond to negative frequencies. The point $k = 0$ always corresponds to zero frequency and $k = N/2$ (for even N) corresponds to $\theta = \pm\pi$. When plotting $X^d[k]$ as a function of k from 0 to $N - 1$, we will see the positive frequencies on the left and the negative frequencies on the right, which is opposite to common custom. For example, low-pass frequency responses will appear as high pass, and vice versa. To avoid this confusion, one may wish to interchange the two halves of the DFT points before plotting them, that is, to rearrange the sequence (for even N) as

$$\{X^d[N/2], X^d[N/2 + 1], \ldots, X^d[N - 1], X^d[0], \ldots, X^d[N/2 - 1]\}.$$

Figure 4.4 illustrates the visual appearance of the DFT of a low-pass signal in the original order and after rearrangement.

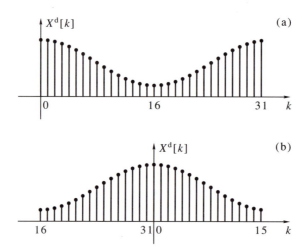

Figure 4.4 Rearrangement of the DFT: (a) index k in original order; (b) index k in a shifted order (shown for $N = 32$).

The following MATLAB expression interchanges the halves of a vector:

```
y = [x(ceil(length(x)/2)+1:length(x)), x(1:ceil(length(x)/2))];
```

the MATLAB function `fftshift` does the same operation.

The *frequency resolution* of the DFT is the spacing between two adjacent frequencies, that is, $\Delta\theta = 2\pi/N$. This number is also called the *frequency bin* of the DFT. The corresponding physical frequency resolution (or frequency bin), in hertz, is

$$\Delta f = 1/NT,$$

where T is the sampling interval. Note that the product NT is the total duration of the continuous-time signal from which the discrete-time signal was obtained by sampling. Therefore we get the following important relationship:

> The frequency resolution of the DFT is the inverse of the signal duration.[1]

Note, in particular, that the frequency resolution does not depend on the number of samples in the interval! This is perhaps counterintuitive and occasionally confusing, so it is worth bearing in mind.

So far we have defined the DFT for finite-duration signals. However, the definition remains unchanged if we consider discrete-time periodic signals instead, by using the DFT definition on *one period* of the signal. Thus, the DFT of a finite-duration signal $\{x[n], \ 0 \le n \le N - 1\}$ is identical to the DFT of its periodic extension $\tilde{x}[n]$. The same holds for the transformed sequence $X^{\mathrm{d}}[k]$; instead of limiting k to the range $0 \le k \le N - 1$, we can use its periodic extension for all $k \in \mathbb{Z}$. There is no need to change either of (4.5) or (4.16), since W_N^{kn} is periodic in both k and n (with period N in both cases). From now on we shall use the DFT and inverse DFT for the finite-duration sequences and their periodic extensions interchangeably.

The relationship

$$\tilde{x}[n] = \frac{1}{N} \sum_{k=0}^{N-1} X^{\mathrm{d}}[k] W_N^{kn}, \quad n \in \mathbb{Z}, \tag{4.18}$$

can be viewed as a Fourier series representation of the periodic signal $\tilde{x}[n]$. Clearly, a discrete-time periodic signal has no more than a finite number of harmonics, equal to the period N. Thus, the IDFT formula (4.16) fulfills the role of Fourier series of discrete-time periodic signals, and (4.3) is the formula for the coefficients. Compare the two formulas with formulas (2.39) and (2.40), given in Section 2.3 for continuous-time periodic signals.

4.2 Matrix Interpretation of the DFT*

The DFT is a linear operation, acting on N-dimensional vectors (the sequences $\{x[n], \ 0 \le n \le N - 1\}$), and producing N-dimensional vectors (the sequences $\{X^{\mathrm{d}}[k], \ 0 \le k \le N - 1\}$). Therefore, it can be conveniently represented by an $N \times N$ matrix. Define

$$\boldsymbol{F}_N = \begin{bmatrix} W_N^0 & W_N^0 & W_N^0 & \cdots & W_N^0 \\ W_N^0 & W_N^{-1} & W_N^{-2} & \cdots & W_N^{-(N-1)} \\ W_N^0 & W_N^{-2} & W_N^{-4} & \cdots & W_N^{-2(N-1)} \\ \vdots & \vdots & \vdots & \cdots & \vdots \\ W_N^0 & W_N^{-(N-1)} & W_N^{-2(N-1)} & \cdots & W_N^{-(N-1)^2} \end{bmatrix},$$

$$\boldsymbol{x}_N = \begin{bmatrix} x[0] \\ x[1] \\ \vdots \\ x[N-1] \end{bmatrix}, \quad \boldsymbol{X}_N^{\mathrm{d}} = \begin{bmatrix} X^{\mathrm{d}}[0] \\ X^{\mathrm{d}}[1] \\ \vdots \\ X^{\mathrm{d}}[N-1] \end{bmatrix}. \tag{4.19}$$

Then we have from (4.3)

$$X_N^d = F_N x_N. \tag{4.20}$$

The matrix F_N is called the *DFT matrix* of dimension N.

Example 4.5 The DFT matrices of dimensions 2, 3, 4 are as follows:

$$F_2 = \begin{bmatrix} 1 & 1 \\ 1 & -1 \end{bmatrix}, \quad F_3 = \begin{bmatrix} 1 & 1 & 1 \\ 1 & -0.5(1 + j\sqrt{3}) & -0.5(1 - j\sqrt{3}) \\ 1 & -0.5(1 - j\sqrt{3}) & -0.5(1 + j\sqrt{3}) \end{bmatrix}, \tag{4.21}$$

$$F_4 = \begin{bmatrix} 1 & 1 & 1 & 1 \\ 1 & -j & -1 & j \\ 1 & -1 & 1 & -1 \\ 1 & j & -1 & -j \end{bmatrix}. \tag{4.22}$$

DFT matrices are convenient for hand computation of DFTs of short lengths. For example, let $x[n]$ be the signal $\{1, 3, 0, -2\}$. Then

$$X_4^d = \begin{bmatrix} 1 & 1 & 1 & 1 \\ 1 & -j & -1 & j \\ 1 & -1 & 1 & -1 \\ 1 & j & -1 & -j \end{bmatrix} \begin{bmatrix} 1 \\ 3 \\ 0 \\ -2 \end{bmatrix} = \begin{bmatrix} 2 \\ 1 - 5j \\ 0 \\ 1 + 5j \end{bmatrix}.$$

Therefore $X^d[k] = \{2, 1 - 5j, 0, 1 + 5j\}$.   □

The DFT matrix has the following properties:

1. The elements on the first row and the first column are 1, since $W_N^0 = 1$.
2. It is symmetric, since its (k, n)th element, W_N^{-kn}, is symmetric in k and n.
3. If $\bar{F}_N$ is the complex conjugate of the DFT matrix, then

$$F_N \bar{F}_N' = N I_N, \tag{4.23}$$

where I_N is the $N \times N$ identity matrix. This follows from (4.6), since

$$(F_N \bar{F}_N')_{k,l} = \sum_{n=0}^{N-1} W_N^{-kn} W_N^{ln} = \sum_{n=0}^{N-1} W_N^{(l-k)n} = N\delta[(k - l) \bmod N]. \tag{4.24}$$

A complex $N \times N$ matrix Q satisfying $Q\bar{Q}' = I_N$ is called a *unitary matrix*. If Q is real, it is called an *orthonormal matrix*. It follows from (4.23) that the matrix $N^{-1/2} F_N$ is unitary. It is also symmetric (property 2), so it is a *symmetric unitary matrix*. The matrix $N^{-1/2} F_N$ is called the *normalized DFT matrix*.

We have from (4.16)

$$x_N = N^{-1} \bar{F}_N X_N^d, \tag{4.25}$$

so the inverse DFT operation can be described by the conjugate of the DFT matrix, with additional multiplication by N^{-1}.

The members of any set of N orthonormal vectors in an N-dimensional vector space form an orthonormal basis for the space. For example, the columns of the identity matrix I_N are orthonormal, so they form an orthonormal basis. This is called the *natural basis*. Let us denote these columns by $\{e_{N,n}, 0 \le n \le N - 1\}$. Then an arbitrary sequence $x[n]$, regarded as a vector x_N, can be expressed as

$$x_N = \sum_{n=0}^{N-1} x[n] e_{N,n}. \tag{4.26}$$

The numbers $\{x[n], \; 0 \le n \le N - 1\}$ are the coordinates of x_N in this basis. We emphasize that the $e_{N,n}$ can be viewed as a basis for either a real N-dimensional space (if the $x[n]$ are restricted to real sequences) or a complex N-dimensional space (if the $x[n]$ are allowed to be complex sequences). Since the vectors x_N are discrete-time signals, we identify the natural coordinates of a vector as temporal sequences. In other words, the nth natural coordinate of a vector is its value at time n.

Since the columns of the matrix $N^{-1/2}\bar{F}_N$ are orthonormal, we can regard them as basis vectors in a complex N-dimensional vector space. Let us denote these vectors by $\{N^{-1/2}f_{N,k}, \; 0 \le k \le N - 1\}$. With this point of view in mind, we can express the inverse DFT relationship as

$$x_N = \sum_{k=0}^{N-1} (N^{-1/2}X^{\mathrm{d}}[k])(N^{-1/2}f_{N,k}). \tag{4.27}$$

Therefore, the numbers $\{N^{-1/2}X^{\mathrm{d}}[k], \; 0 \le k \le N - 1\}$ are the coordinates of x_N in this basis. Since the basis is orthonormal, they are also the *projections* of the signal vector on the members of the basis.

We can now interpret Example 4.3 in geometrical terms, as follows. Since

$$\cos(\theta_0 n) = 0.5e^{j\theta_0 n} + 0.5e^{-j\theta_0 n},$$

it is a linear combination of two basis vectors when $\theta_0 = 2\pi m/N$. This explains why only two points in Figure 4.3(a) have nonzero ordinates. When $\theta_0 \ne 2\pi m/N$, both $e^{j\theta_0 n}$ and $e^{-j\theta_0 n}$ are close to basis vectors, but are not aligned with them. Therefore, $\cos(\theta_0 n)$ has nonzero projections on all basis vectors, as shown in Figure 4.3(b, c). Of those, the largest projections are on the basis vectors nearest to $\exp(j\theta_0 n)$ and $\exp(-j\theta_0 n)$, that is, on $\exp(j2\pi mn/N)$ and $\exp[j2\pi(N - m)n/N]$, where m is the integer nearest to $N\theta_0/2\pi$.

We summarize the preceding discussion as follows. The values of the DFT of a sequence $x[n]$ can be viewed as the coordinates of the sequence in a particular orthonormal basis (up to the constant factor $N^{-1/2}$). The vectors constituting this basis are the columns of the DFT matrix (again, up to a constant factor), therefore they are called the *DFT basis*. Figure 4.5 illustrates the DFT basis for $N = 8$. It is common to draw the basis vectors as staircase waveforms. The horizontal axis of the waveform is time, so the nth stair of the kth waveform indicates the nth coordinate of the kth basis vector in the natural basis. In this case we have 8 basis vectors, each having a real part and a complex part. For large N, the real-part waveforms will look more and more like continuous-time cosine functions at linearly increasing frequencies, whereas the imaginary-part waveforms will look like continuous-time sine functions.

4.3 Properties of the DFT

The discrete Fourier transform has properties similar to those of the usual Fourier transform. We now list the main properties of the DFT and their proofs.

1. **Linearity**

$$z[n] = ax[n] + by[n] \iff Z^{\mathrm{d}}[k] = aX^{\mathrm{d}}[k] + bY^{\mathrm{d}}[k], \quad a, b \in \mathbb{C}. \tag{4.28}$$

The proof follows immediately from (4.3).

2. **Periodicity**

$$X^{\mathrm{d}}[k] = X^{\mathrm{d}}[k + N]. \tag{4.29}$$

This is a direct result of the periodicity of W_N, as explained in Section 4.1.

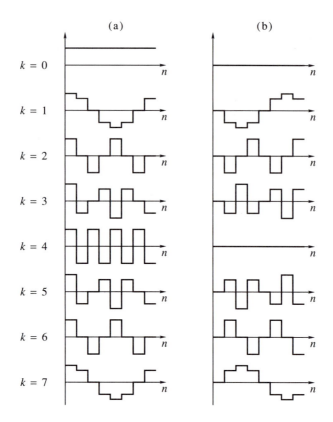

Figure 4.5 The DFT basis vectors for $N = 8$: (a) real part; (b) imaginary part.

3. **Circular shift**

$$y[n] = x[(n - m) \bmod N] \iff Y^{\mathrm{d}}[k] = W_N^{-km} X^{\mathrm{d}}[k], \quad m \in \mathbb{Z}. \tag{4.30}$$

Proof

$$Y^{\mathrm{d}}[k] = \sum_{n=0}^{N-1} x[(n - m) \bmod N] W_N^{-kn} = \sum_{n=0}^{N-1} x[(n - m) \bmod N] W_N^{-k(n-m+m)}$$

$$= W_N^{-km} \sum_{n=0}^{N-1} x[(n - m) \bmod N] W_N^{-k[(n-m) \bmod N]} = W_N^{-km} X^{\mathrm{d}}[k]. \tag{4.31}$$

In passing from the first to the second line we used the property

$$W_N^{-k(l \bmod N)} = W_N^{-kl}, \quad \text{for all } k, l \in \mathbb{Z}.$$

4. **Frequency shift (modulation)**

$$y[n] = W_N^{mn} x[n] \iff Y^{\mathrm{d}}[k] = X^{\mathrm{d}}[(k - m) \bmod N], \quad m \in \mathbb{Z}. \tag{4.32}$$

Proof

$$Y^{\mathrm{d}}[k] = \sum_{n=0}^{N-1} x[n] W_N^{mn} W_N^{-kn} = \sum_{n=0}^{N-1} x[n] W_N^{-n(k-m)}$$

$$= \sum_{n=0}^{N-1} x[n] W_N^{-n[(k-m) \bmod N]} = X^{\mathrm{d}}[(k - m) \bmod N]. \tag{4.33}$$

5. **Parseval's theorem**

$$\sum_{n=0}^{N-1} x[n]\bar{y}[n] = \frac{1}{N}\sum_{k=0}^{N-1} X^d[k]\bar{Y}^d[k], \tag{4.34}$$

and its special case

$$\sum_{n=0}^{N-1} |x[n]|^2 = \frac{1}{N}\sum_{k=0}^{N-1} |X^d[k]|^2. \tag{4.35}$$

Proof Using the vector notations (4.19) for the sequence and its DFT, we can express (4.34) as

$$\bar{y}'x = N^{-1}(\bar{Y}^d)'X^d. \tag{4.36}$$

Now, using the matrix representation of the DFT (4.20) and property (4.23) of the DFT matrix, we have

$$N^{-1}(\bar{Y}^d)'X^d = N^{-1}\bar{y}'\bar{F}'_N F_N x = \bar{y}'I_N x = \bar{y}'x, \tag{4.37}$$

which is the same as (4.36). The special case (4.35) is obtained by taking the sequence $y[n]$ equal to $x[n]$.

6. **Conjugation**

$$y[n] = \bar{x}[n] \iff Y^d[k] = \bar{X}^d[(N-k) \bmod N]. \tag{4.38}$$

Proof

$$Y^d[k] = \sum_{n=0}^{N-1} \bar{x}[n]W_N^{-kn} = \overline{\left[\sum_{n=0}^{N-1} x[n]W_N^{kn}\right]}$$

$$= \overline{\left[\sum_{n=0}^{N-1} x[n]W_N^{-[(N-k) \bmod N]n}\right]} = \bar{X}^d[(N-k) \bmod N]. \tag{4.39}$$

7. **Symmetry** If $x[n]$ is real valued then

$$X^d[(N-k) \bmod N] = \bar{X}^d[k], \tag{4.40a}$$

$$\Re\{X^d[(N-k) \bmod N]\} = \Re\{X^d[k]\}, \tag{4.40b}$$

$$\Im\{X^d[(N-k) \bmod N]\} = -\Im\{X^d[k]\}, \tag{4.40c}$$

$$|X^d[(N-k) \bmod N]| = |X^d[k]|, \tag{4.40d}$$

$$\sphericalangle X^d[(N-k) \bmod N] = -\sphericalangle X^d[k]. \tag{4.40e}$$

Proof Equality (4.40a) follows from

$$X^d[(N-k) \bmod N] = \sum_{n=0}^{N-1} x[n]W_N^{-[(N-k) \bmod N]n} = \sum_{n=0}^{N-1} x[n]W_N^{kn} = \bar{X}^d[k]. \tag{4.41}$$

The other four equalities follow from the first.

The conjugate symmetry property of real sequences requires clarification. Substitution of $k = 0$ gives $X^d[0] = \bar{X}^d[0]$. This is obvious, since

$$X^d[0] = \sum_{n=0}^{N-1} x[n], \tag{4.42}$$

which is real. So, $k = 0$ is its own conjugate symmetric. Substitution of $k = 0.5N$ (if N is even) gives $X^d[0.5N] = \bar{X}^d[0.5N]$. This is again obvious, since

$$X^d[0.5N] = \sum_{n=0}^{N-1} x[n]W_N^{-0.5Nn} = \sum_{n=0}^{N-1} (-1)^n x[n], \tag{4.43}$$

which is real. So, for even N, $k = 0.5N$ is its own conjugate symmetric. The other terms are not conjugate symmetric to themselves; for example, $k = 1$ is conjugate symmetric to $k = N - 1$, $k = 0.5N - 1$ is conjugate symmetric to $k = 0.5N + 1$, and so forth. If N is odd, $0.5N$ is not an integer, so the self-conjugate center point is missing. Figure 4.6 shows the symmetries in the odd and even cases.

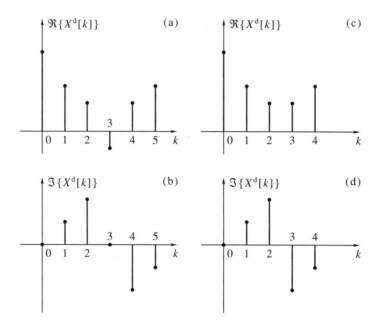

Figure 4.6 Symmetries in DFT of a real signal: (a) even N, real part; (b) even N, imaginary part; (c) odd N, real part; (d) odd N, imaginary part.

4.4   Zero Padding

The DFT of a length-N sequence is itself a length-N sequence, so it gives the frequency response of the signal at N points. Suppose we are interested in computing the frequency response at M equally spaced frequency points, where $M > N$. A simple device accomplishes this goal: We add $M - N$ zeros at the tail of the given sequence, thus forming a length-M sequence. The DFT of the new sequence has M frequency points. We prove that the values of the new DFT are indeed samples of the frequency response of the given signal at M equally spaced frequencies. Denote

$$x_a[n] = \begin{cases} x[n], & 0 \leq n \leq N - 1, \\ 0, & N \leq n \leq M - 1. \end{cases} \tag{4.44}$$

The operation of adding zeros to the tail of a sequence is called *zero padding*. The DFT of the zero-padded sequence $x_a[n]$ is given by

$$X_a^d[k] = \sum_{n=0}^{M-1} x_a[n] \exp\left(-\frac{j2\pi kn}{M}\right) = \sum_{n=0}^{N-1} x[n] \exp\left(-\frac{j2\pi kn}{M}\right) = X^f(\theta[k]), \tag{4.45}$$

where

$$\theta[k] = \frac{2\pi k}{M}, \quad 0 \leq k \leq M - 1. \tag{4.46}$$

As we see, $X_a^d[k]$ is indeed a sampling of $X^f(\theta)$ at M equally spaced frequency points in the range $[0, 2\pi)$.

The notation $\mathrm{DFT}_M\{x[n]\}$ is used for describing the zero-padded DFT operation. For example, if the length of $x[n]$ is 16, then $\mathrm{DFT}_{16}\{x[n]\}$ denotes the usual DFT, and $\mathrm{DFT}_{64}\{x[n]\}$ denotes the DFT of the sequence obtained by zero padding to length 64. The following MATLAB expression performs zero padding on a row vector:

```
y = [x, zeros(1,M-length(x))];
```

Figure 4.7 illustrates the zero-padding operation. Part a shows a signal of length $N = 8$, and part b shows the magnitude of its length-N DFT (note that we interchange the positive- and negative-index halves). Part c shows the signal obtained by zero padding to length $M = 32$, and part d shows the magnitude of the length-M DFT of the zero-padded signal.

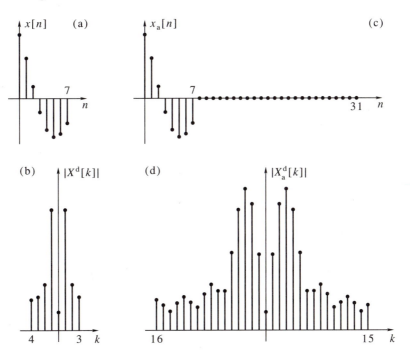

Figure 4.7 Increasing the DFT length by zero padding: (a) a signal of length 8; (b) the 8-point DFT of the signal (magnitude); (c) zero padding the signal to length 32; (d) the 32-point DFT of the zero-padded signal.

We can interpret the zero-padded DFT $X_a^d[l]$ as interpolation operation on $X^d[k]$. In particular, if M is an integer multiple of N, say $M = LN$, we have interpolation by a factor L. In this case, the points $X_a^d[kL]$ of the zero-padded DFT are identical to the corresponding points $X^d[k]$ of the conventional DFT. If M is not an integer multiple of N, most of the points $X^d[k]$ do not appear as points of $X_a^d[l]$ (Problem 4.21).

Zero padding is typically used for improving the visual continuity of plots of frequency responses. When plotting $X_a^d[k]$, we typically see more details than when plotting $X^d[k]$. However, the additional details do not represent additional information about the signal, since all the information is in the N given samples of $x[n]$. Indeed, computation of the inverse DFT of $X_a^d[k]$ gives the zero-padded sequence $x_a[n]$, which consists only of the $x[n]$ and zeros.

4.5   Zero Padding in the Frequency Domain*

The interpretation of zero padding given at the end of the preceding section raises an intriguing thought: If zero padding in the time domain provides interpolation in the frequency domain, then zero padding in the frequency domain must provide interpolation in the time domain. We should therefore be able to use zero padding of the DFT as a means of interpolating finite-duration, discrete-time signals. Implementation of this idea requires care, since we must preserve the symmetry properties of the zero-padded DFT. Let us consider, for simplicity, the case of odd N. The case of even N is left as an exercise (Problem 4.42). Also, we assume that M is an integer multiple of N, say $M = NL$, where L is the *interpolation factor*.

Assume we are given the DFT of a length-N sequence (where N is odd), and define the zero-padded DFT as

$$X_i^d[k] = \begin{cases} LX^d[k], & 0 \le k \le \frac{N-1}{2}, \\ LX^d[k - M + N], & M - \frac{N-1}{2} \le k \le M - 1, \\ 0, & \text{otherwise}, \end{cases} \tag{4.47}$$

for $M = LN$, $L > 0$. The new DFT has zero values at high frequencies; conjugate symmetry is preserved if possessed by $X^d[k]$. Let us now define $x_i[n]$ as the length-M IDFT of $X_i^d[k]$, that is,

$$x_i[n] = \frac{1}{M} \sum_{k=0}^{M-1} X_i^d[k] W_M^{nk}, \quad 0 \le n \le M - 1. \tag{4.48}$$

Then,

$$\begin{aligned} x_i[n] &= \frac{1}{N} \sum_{k=0}^{(N-1)/2} X^d[k] W_M^{nk} + \frac{1}{N} \sum_{k=M-(N-1)/2}^{M-1} X^d[k - M + N] W_M^{nk} \\ &= \frac{1}{N} \sum_{m=0}^{N-1} x[m] \left[\sum_{k=0}^{(N-1)/2} W_M^{nk} W_N^{-mk} + W_M^{-nN} \sum_{k=(N+1)/2}^{N-1} W_M^{nk} W_N^{-mk} \right] \\ &= \frac{1}{N} \sum_{m=0}^{N-1} x[m] \left[\frac{(W_M^n W_N^{-m})^{(N+1)/2} - 1}{W_M^n W_N^{-m} - 1} + \frac{1 - (W_M^n W_N^{-m})^{-(N-1)/2}}{W_M^n W_N^{-m} - 1} \right] \\ &= \frac{1}{N} \sum_{m=0}^{N-1} x[m] \frac{\sin[\pi(n - mL)/L]}{\sin[\pi(n - mL)/M]}. \end{aligned} \tag{4.49}$$

The right side of (4.49) bears similarity to Shannon's reconstruction formula

$$x(t) = \sum_{n=-\infty}^{\infty} x(nT) \text{sinc}\left(\frac{t - nT}{T}\right). \tag{4.50}$$

This similarity is not coincidental. Shannon's formula describes the reconstruction of a continuous-time signal from its samples. Formula (4.49) describes the reconstruction of a discrete-time signal from a signal obtained after resampling a discrete-time signal at an L-times slower rate. In Shannon's formula, the impulse response of the interpolating filter is a sinc function. Here it is the function $\sin(\pi n/L)/\sin(\pi n/M)$, which can be thought of as a discrete-time (periodic) sinc. Note, in particular, that for values of n that are integer multiples of L, we get from (4.49)

$$x_i[kL] = x[k],$$

so the interpolation becomes an identity at the time points of the original length-N signal.

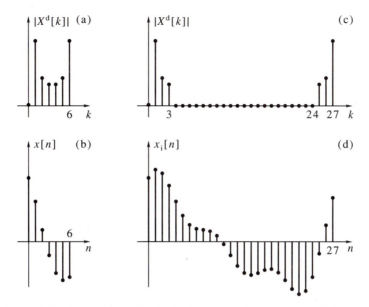

Figure 4.8 Interpolation by zero padding in the frequency domain: (a) DFT of a signal of length 7; (b) the time-domain signal; (c) zero padding the DFT to length 28; (d) the time-domain signal of the zero-padded DFT.

Figure 4.8 illustrates zero padding in the frequency domain. Part a shows the magnitude of the DFT of a signal of length $N = 7$, and part b shows the signal itself. Here we did not interchange the halves of the DFT, for better illustration of the operation (4.47). Part c shows the DFT after being zero padded to a length $M = 28$, and part d shows the inverse DFT of the zero-padded DFT. The result in part d exhibits two peculiarities:

1. The interpolated signal is rather wiggly. The wiggles, which are typical of this kind of interpolation, are introduced by the interpolating function $\sin(\pi n/L)/\sin(\pi n/M)$.

2. The last point of the original signal is $x[6]$, and it corresponds to the point $x_i[24]$ of the interpolated signal. The values $x_i[25]$ through $x_i[27]$ of the interpolated signal are actually interpolations involving the periodic extension of the signal. These values are not reliable, since they are heavily influenced by the initial values of $x[n]$ (note how $x[27]$ is pulled upward, closer to the value of $x[0]$).

Because of these phenomena, interpolation by zero padding in the frequency domain is not considered desirable and is not in common use (see, however, a continuation of this discussion in Problem 6.14).

4.6 Circular Convolution

You may have noticed that the convolution and multiplication properties were conspicuously missing from the list of properties of the DFT. This is because the DFT satisfies these properties only for a certain kind of convolution, which we now define.

Let $x[n]$ and $y[n]$ be two finite length sequences, of *equal* length N. We define their *circular convolution* as

$$z[n] = \{x \circledast y\}[n] = \sum_{m=0}^{N-1} x[m]y[(n-m) \bmod N], \quad 0 \le n \le N-1. \tag{4.51}$$

Other names for this operation are *cyclic convolution* and *periodic convolution*.

The circular convolution of two length-N sequences is itself a length-N sequence. It is convenient to think of a circular convolution as though the sequences are defined on points on a circle, rather than on a line. Take, for example, $N = 12$, and imagine the n coordinate as the hour on a watch, with 0 instead of 12. To perform the circular convolution, proceed as follows:

1. Spread the $x[n]$ clockwise, starting at the zero hour.
2. Spread the $y[n]$ counterclockwise, starting at the zero hour (i.e, the point $y[1]$ goes on the 11th hour, and so on).
3. To compute $z[n]$, rotate the sequence of $y[n]$s clockwise by n steps, then perform element-by-element multiplication of the two sequences, and sum.
4. Repeat for all n from 0 through 11.

Figure 4.9 illustrates this procedure using a 6-digit watch. In this figure we use $x[n] = y[n] = n$, so the numbers represent both indices and values of the two sequences. The value of n and the corresponding value of $z[n]$ is shown beneath each position of the watch.

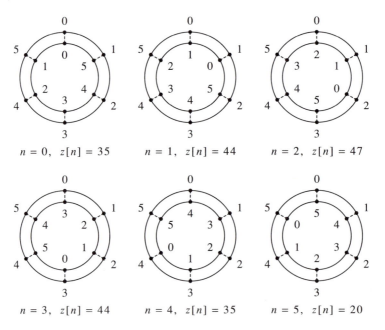

Figure 4.9 Illustration of circular convolution for $N = 6$. Outer circles: $x[m]$; inner circles: $y[n-m]$. The result of the convolution is the sum of products of sequence values near the ends of each dashed line.

Another useful way to look at circular convolution is to regard it as partial convolution of the periodic extensions of the two sequences. By "partial" we mean that summation is performed only over one period, not from $-\infty$ to ∞. Thus,

$$z[n] = \sum_{m=0}^{N-1} \tilde{x}[m]\tilde{y}[n-m]. \tag{4.52}$$

With this interpretation, the range of n is not necessarily limited to $[0, N - 1]$, but it can be any integer. Then the right side defines $\tilde{z}[n]$, the periodic extension of $z[n]$.

Circular convolution can be described as a matrix–vector multiplication:

$$
\begin{bmatrix} z[0] \\ z[1] \\ \vdots \\ z[N-2] \\ z[N-1] \end{bmatrix} = \begin{bmatrix} y[0] & y[N-1] & \cdots & y[2] & y[1] \\ y[1] & y[0] & \cdots & y[3] & y[2] \\ \vdots & \vdots & \ddots & \vdots & \vdots \\ y[N-2] & y[N-3] & \cdots & y[0] & y[N-1] \\ y[N-1] & y[N-2] & \cdots & y[1] & y[0] \end{bmatrix} \begin{bmatrix} x[0] \\ x[1] \\ \vdots \\ x[N-2] \\ x[N-1] \end{bmatrix}. \tag{4.53}
$$

Observe the arrangement of the elements of the sequence $y[n]$ in the matrix. A matrix having such a structure is called *circulant*, because its rows are consecutive circular shifts of the first row.

Example 4.6 Let us compute the circular convolution of the sequences

$$
x[n] = \{2, 3, -1, 5\}, \quad y[n] = \{1, 4, -2, -3\}.
$$

We can do the computation by the clock method, using (4.52), or through (4.53). Here we choose the latter:

$$
\begin{bmatrix} z[0] \\ z[1] \\ z[2] \\ z[3] \end{bmatrix} = \begin{bmatrix} 1 & -3 & -2 & 4 \\ 4 & 1 & -3 & -2 \\ -2 & 4 & 1 & -3 \\ -3 & -2 & 4 & 1 \end{bmatrix} \begin{bmatrix} 2 \\ 3 \\ -1 \\ 5 \end{bmatrix} = \begin{bmatrix} 15 \\ 4 \\ -8 \\ -11 \end{bmatrix}.
$$

□

The procedure `circonv` in Program 4.2 illustrates one of several ways to implement circular convolution in MATLAB. The program essentially relies on performing a conventional convolution of the two sequences and then folding the $(N - 1)$-long tail of the result on the head. This procedure is a good illustration of the relation between the two kinds of convolution and should be studied carefully (Problem 4.25).

Circular convolution is, similarly to conventional convolution, a commutative and associative operation, that is,

$$
x \circledast y = y \circledast x, \quad (x \circledast y) \circledast z = x \circledast (y \circledast z). \tag{4.54}
$$

The proof of these properties is left to Problem 4.28. Commutativity allows us to exchange the roles of x and y in the expression (4.53).

We can now state and prove the convolution and multiplication properties of the DFT.

Theorem 4.3 (convolution)

$$
z[n] = \{x \circledast y\}[n] \iff Z^{\mathrm{d}}[k] = X^{\mathrm{d}}[k] Y^{\mathrm{d}}[k]. \tag{4.55}
$$

Proof We have

$$
z[n] = \sum_{m=0}^{N-1} x[m] y[(n - m) \bmod N]. \tag{4.56}
$$

Using the linearity property (4.28) and the shift property (4.30), we can write

$$
Z^{\mathrm{d}}[k] = \sum_{m=0}^{N-1} x[m] W_N^{-mk} Y^{\mathrm{d}}[k] = Y^{\mathrm{d}}[k] \sum_{m=0}^{N-1} x[m] W_N^{-mk} = X^{\mathrm{d}}[k] Y^{\mathrm{d}}[k]. \tag{4.57}
$$

□

Theorem 4.4 (multiplication)

$$z[n] = x[n]y[n] \iff Z^{\mathrm{d}}[k] = \frac{1}{N}\{X^{\mathrm{d}} \circledast Y^{\mathrm{d}}\}[k]. \tag{4.58}$$

The proof follows from the previous one by duality: Start at right side and compute its IDFT using linearity and the modulation property (4.32). □

We restate the last two theorems, to emphasize their importance:

> The DFT of a circular convolution of two sequences is the product of the individual DFTs. The DFT of a product of two sequences is, up to a proportionality constant, the circular convolution of the individual DFTs.

Example 4.7 Consider again the sequences in Example 4.6. We have

$$X^{\mathrm{d}}[k] = \{9, 3 + j2, -7, 3 - j2\}, \quad Y^{\mathrm{d}}[k] = \{0, 3 - j7, -2, 3 + j7\}.$$

Therefore,

$$Z^{\mathrm{d}}[k] = \{0, 23 - j15, 14, 23 + j15\}.$$

Taking the inverse DFT of $Z^{\mathrm{d}}[k]$ gives

$$z[n] = \{15, 4, -8, -11\},$$

which is equal to the result in Example 4.6. □

The fact that it is circular, rather than conventional convolution that translates to multiplication in the frequency domain is a common source of confusion and mistakes. The following claim is often made by beginners: "Linear time-invariant filtering is equivalent to multiplication in the frequency domain. Therefore, we can perform linear filtering by computing the DFT of the input signal $X^{\mathrm{d}}[k]$, multiply by the DFT of the impulse response of the filter $H^{\mathrm{d}}[k]$ (which is presumably the frequency response of the filter), and compute the inverse DFT of the product." This claim is wrong, because the DFT is not the frequency response of either the signal or the filter. Rather, it is the sampling of the frequency response in the θ domain. Consequently, the inverse DFT of $H^{\mathrm{d}}[k]X^{\mathrm{d}}[k]$ is $\{x \circledast y\}[n]$, whereas the proper linear filtering operation is $\{x * y\}[n]$. Nevertheless, the idea behind the aforementioned claim is essentially correct, and frequency-domain operations with the DFT *can be used* for convolution and filtering. The proper way of doing this is elaborated in the next section.

4.7 Linear Convolution via Circular Convolution

Suppose we are given two finite-duration sequences having different lengths, say $\{x[n], \; 0 \le n \le N_1 - 1\}$, and $\{y[n], \; 0 \le n \le N_2 - 1\}$, and we wish to perform their (conventional) discrete-time convolution

$$z[n] = \sum_{m=m_1}^{m_2} x[m]y[n - m], \tag{4.59}$$

where

$$m_1 = \max\{0, n + 1 - N_2\}, \quad m_2 = \min\{N_1 - 1, n\}.$$

To avoid confusion, we will henceforth refer to conventional convolution as *linear convolution*.

Let us zero-pad $x[n]$ and $y[n]$ to a length $N = N_1 + N_2 - 1$ and denote the zero-padded sequences by $x_a[n]$, $y_a[n]$, respectively. Now we can express (4.59) as

$$z[n] = \sum_{m=0}^{n} x_a[m]y_a[n-m], \quad 0 \le n \le N - 1. \tag{4.60}$$

The zero-padded sequences have the same length; let us compute their circular convolution:

$$\{x_a \circledast y_a\}[n] = \sum_{m=0}^{N-1} x_a[m]y_a[(n-m) \bmod N]$$

$$= \sum_{m=0}^{n} x_a[m]y_a[n-m] + \sum_{m=n+1}^{N-1} x_a[m]y_a[n-m+N]. \tag{4.61}$$

In the second sum, the lengths of the two sequences $x[n]$ and $y[n]$ impose the following limits on the summation index m:

$$n + 1 \le m \le N_1 - 1,$$
$$N_1 + n \le m \le N + n.$$

The intersection of these two limits is empty, so the second term on the right side of (4.61) is zero. The first term is identical to the right side of (4.60), so the conclusion is that

$$\{x * y\}[n] = \{x_a \circledast y_a\}[n], \quad 0 \le n \le N - 1. \tag{4.62}$$

Therefore, we can perform linear convolution of two finite-length sequences by computing the circular convolution of the corresponding zero-padded sequences, provided zero padding is made to the sum of the lengths minus one. Figure 4.10 illustrates this procedure for $N_1 = 4$, $N_2 = 3$. In this figure we use $x[n] = y[n] = 1$ for convenience. The small solid circles (4 in the sequence $x[n]$, 3 in the sequence $y[n]$) represent the value 1. The small open circles represent the value 0 (i.e., the zero-padded points). As we see, the zero-valued points are properly aligned for each n so as to make the circular convolution equal to the desired linear convolution.

The circular convolution $\{x_a \circledast y_a\}$ can be performed by computing the DFTs of both sequences, multiplying the resulting vectors point by point, and then computing the inverse DFT of the result. The advantage of performing the convolution this way will become clear in the next chapter, when we study the fastOurier transform.

Example 4.8 We wish to compute the linear convolution of the two sequences

$$x[n] = \{2, 3\}, \quad y[n] = \{1, -4, 5\}.$$

Direct computation of the linear convolution gives

$$z[n] = \{2, -5, -2, 15\}.$$

Computation by circular convolution and DFT proceeds as follows. The zero-padded sequences are

$$x_a[n] = \{2, 3, 0, 0\}, \quad y_a[n] = \{1, -4, 5, 0\}.$$

The DFTs of the zero-padded sequences are

$$X_a^d[k] = \{5, 2 - j3, -1, 2 + j3\}, \quad Y_a^d[k] = \{2, -4 + j4, 10, -4 + j4\}.$$

Therefore,

$$Z_a^d[k] = \{10, 4 + j20, -10, 4 - j20\}.$$

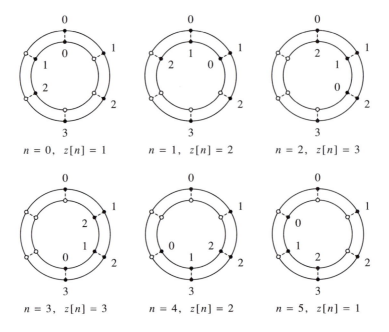

Figure 4.10 Illustration of linear convolution via circular convolution for $N = 6$. Outer circles: $x[m]$; inner circles: $y[n - m]$. Solid circles represent original sequence point; open circles represent zero-padded points.

Finally, taking the inverse DFT yields

$$z_a[k] = \{2, -5, -2, 15\},$$

which is the same as the result obtained by direct computation.   □

4.8   The DFT of Sampled Periodic Signals

We already mentioned, in Chapter 3, that no continuous-time signal can be simultaneously limited in time and in frequency; see page 56. To use the DFT, we need to sample a finite-duration signal (or to use a finite number of samples from an infinite-duration signal, which amounts to the same). But then the signal bandwidth must be infinite, so sampling will involve aliasing. The DFT cannot repair this aliasing, since it represents only sampling in the frequency domain of the Fourier transform of the sampled signal. The conclusion is that, in general, the DFT *does not* provide an exact description of a continuous-time signal in the frequency domain.

There is an exception to the preceding conclusion. When we sample a continuous-time *periodic* signal we can, under certain restrictions, get an exact frequency-domain description from the DFT of a finite number of samples. To see this, let $x(t)$ be a periodic signal with period T_0. Then the signal admits a Fourier series

$$x(t) = \sum_{m=-\infty}^{\infty} X^S[m] \exp\left(\frac{j2\pi mt}{T_0}\right). \tag{4.63}$$

Suppose we sample the signal an integer number of samples over a single period, say N samples. The corresponding sampling interval is $T = T_0/N$. The samples then make up the finite sequence

$$x[n] = x(nT) = \sum_{m=-\infty}^{\infty} X^S[m] \exp\left(\frac{j2\pi mn}{N}\right), \quad 0 \le n \le N - 1. \tag{4.64}$$

Write m as

$$m = k + lN, \quad 0 \le k \le N - 1, \; -\infty < l < \infty,$$

and substitute in (4.64) to get

$$x[n] = \sum_{k=0}^{N-1} \sum_{l=-\infty}^{\infty} X^S[k + lN] \exp\left(\frac{j2\pi kn}{N}\right) = \frac{1}{N} \sum_{k=0}^{N-1} \left(N \sum_{l=-\infty}^{\infty} X^S[k + lN]\right) W_N^{nk}. \tag{4.65}$$

This has the form of an inverse DFT, so we conclude that

$$X^d[k] = N \sum_{l=-\infty}^{\infty} X^S[k + lN]. \tag{4.66}$$

As we see, each DFT coefficient of the one-period sampled sequence is an infinite sum of Fourier coefficients of the continuous-time signal. This is again a manifestation of aliasing: In general, we cannot recover the individual Fourier coefficients $X^S[m]$ from the DFT coefficients $X^d[k]$. There exists, however, a special case in which there is no aliasing. Suppose the signal $x(t)$ is band limited and its nonzero Fourier coefficients are only $\{X^S[m], -M \le m \le M\}$. Note that there is no contradiction to the theorem quoted in the beginning of the section, since any periodic signal has an infinite duration. If we choose $N \ge 2M + 1$, we will get

$$X^d[k] = \begin{cases} NX^S[k], & 0 \le k \le M, \\ NX^S[k - N], & N - M \le k \le N - 1. \end{cases} \tag{4.67}$$

In this special case we can recover the $2M + 1$ nonzero Fourier coefficients from the corresponding values of the DFT coefficients. In summary, if a periodic band-limited signal is sampled an integer number of times N over one period, such that N is at least $2M + 1$ (where M is the index of the highest nonzero harmonic), the DFT coefficients of the sampled sequence are equal, up to a constant factor N, to the corresponding Fourier coefficients of the signal harmonics.

Example 4.9 Eliza and Beth are two engineers in a DSP company. Eliza, the senior, generates a continuous-time triangular signal whose fundamental frequency is 1 Hz; one period of this signal is shown in Figure 4.11. She samples the signal at interval T, an integer number of samples per period, and gives the values of the sampled signal to Beth, the junior. Eliza tells Beth that the signal is periodic, gives its period and the sampling interval T, and asks Beth to reconstruct the signal. Beth, after an hour's work, reports the following answer:

$$\hat{x}(t) = 0.25(\sqrt{2} + 2) \sin(2\pi t) + 0.25(\sqrt{2} - 2) \sin(6\pi t).$$

What was the value of T and how did Beth arrive at the result?

Since Beth knows both the period and the sampling interval, she knows the number of samples per period. It is obvious from her result that she was able to get only the first and third harmonics (recall that a symmetric triangular wave, like the one shown in Figure 4.11, has only odd harmonics). Thus, the number of samples per period is at least 7 (since with 6 samples or less she would not have found a third harmonic) at most 10 (since with 11 samples she would have found a fifth harmonic).

Let us try $T = 1/7$ first. With this sampling interval, the 7 samples of the triangular wave are

$$x[n] = \{0, 4/7, 6/7, 2/7, -2/7, -6/7, -4/7\}.$$

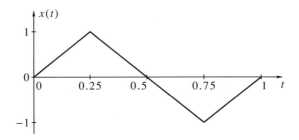

Figure 4.11 The signal in Example 4.9.

The DFT of $x[n]$ is

$$X^d[k] = \{0, -j2.8128, j0.0764, j0.2873, -j0.2873, -j0.0764, +j2.8128\}.$$

As we see, there is a second harmonic in this case, which was evidently introduced by aliasing. Since we know that Beth did not find a second harmonic, $T = 1/7$ is not the right answer.

Next let us try $T = 1/8$. With this sampling interval, the 8 samples of the triangular wave are

$$x[n] = \{0, 0.5, 1, 0.5, 0, -0.5, -1, -0.5\}.$$

The DFT of $x[n]$ is now

$$X^d[k] = \{0, -j(\sqrt{2} + 2), 0, -j(\sqrt{2} - 2), 0, j(\sqrt{2} - 2), 0, j(\sqrt{2} + 2)\}.$$

Now we have only the first and third harmonics, as found by Beth. The signal reconstructed from these harmonics is

$$\begin{aligned}
\hat{x}(t) &= -j0.125(\sqrt{2} + 2)e^{j2\pi t} - j0.125(\sqrt{2} - 2)e^{j6\pi t} \\
&\quad + j0.125(\sqrt{2} - 2)e^{-j6\pi t} + j0.125(\sqrt{2} + 2)e^{-j2\pi t} \\
&= 0.25(\sqrt{2} + 2)\sin(2\pi t) + 0.25(\sqrt{2} - 2)\sin(6\pi t),
\end{aligned}$$

which is precisely what Beth found; therefore $T = 1/8$.   □

4.9   The Discrete Cosine Transform*

The discrete cosine transform (DCT) is a kin of the DFT. It can be regarded as a discrete-time version of the Fourier cosine series, described in Section 2.5. Like the DFT, the DCT provides information about the signal in the frequency domain. Unlike the DFT, the DCT of a real signal is real valued. The transform is linear, so it can be expressed in a matrix–vector form

$$X_N^c = C_N x_N, \tag{4.68}$$

where x_N is the N-vector describing the signal, X_N^c is the N-vector describing the result of the transform, and C_N is a square nonsingular $N \times N$ matrix describing the transform itself. The matrix C_N is real valued.

Recall that, to obtain the Fourier cosine series, we extended the signal symmetrically around the origin. We wish to do the same in the discrete-time case. It turns out that symmetric extension of a discrete-time signal is not as obvious as in continuous time, and there are several ways of proceeding. Each symmetric extension gives rise to a different transform. In total there are four types of discrete cosine transform, as described next.

4.9.1 Type-I Discrete Cosine Transform

Let $x[n]$ be a discrete-time real signal of length N. A minimal symmetric extension of this signal involves adding $N - 2$ more points: $x[0]$ and $x[N - 1]$ remain unique, whereas $x[1]$ through $x[N - 2]$ are duplicated symmetrically around $x[N - 1]$. The points $x[0]$ and $x[N - 1]$ are not duplicated, but are multiplied by $2^{1/2}$. We thus define

$$x_1[n] = \begin{cases} 2^{1/2}x[n], & n = 0, N - 1 \\ x[n], & 1 \leq n \leq N - 2, \\ x[2N - n - 2], & N \leq n \leq 2N - 3. \end{cases} \quad (4.69)$$

Figure 4.12 illustrates this definition for $N = 10$.

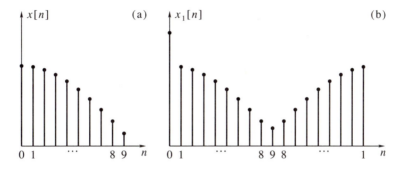

Figure 4.12 The symmetric extension used in the definition of DCT-I: (a) the given sequence; (b) the symmetrically extended sequence (for clarity, the time indices are marked not by the actual n but in a way that shows the symmetry).

The DFT of $x_1[n]$ is given by

$$X_1^d[k] = \sum_{n=0}^{2N-3} x_1[n]W_{2N-2}^{-nk} = 2^{1/2}x[0] + 2^{1/2}x[N - 1](-1)^k$$

$$+ \sum_{n=1}^{N-2} x[n]W_{2N-2}^{-nk} + \sum_{n=N}^{2N-3} x[2N - n - 2]W_{2N-2}^{-nk}$$

$$= 2^{1/2}x[0] + 2^{1/2}x[N - 1](-1)^k + \sum_{n=1}^{N-2} x[n](W_{2N-2}^{-nk} + W_{2N-2}^{nk})$$

$$= 2^{1/2}x[0] + 2^{1/2}x[N - 1](-1)^k + 2\sum_{n=1}^{N-2} x[n]\cos\left(\frac{\pi nk}{N - 1}\right)$$

$$= 2\sum_{n=0}^{N-1} a[n]x[n]\cos\left(\frac{\pi nk}{N - 1}\right), \quad 0 \leq k \leq 2N - 3, \quad (4.70)$$

where

$$a[n] = \begin{cases} 2^{-1/2}, & n = 0, N - 1, \\ 1, & \text{otherwise.} \end{cases} \quad (4.71)$$

The sequence $X_1^d[k]$ is real and satisfies $X_1^d[k] = X_1^d[2N - 2 - k]$. Therefore,

$$x_1[n] = \frac{1}{2N - 2}\sum_{k=0}^{2N-3} X_1^d[k]W_{2N-2}^{nk} = \frac{1}{2N - 2}X_1^d[0]$$

$$+ \frac{1}{2N - 2}X_1^d[N - 1](-1)^n + \frac{1}{N - 1}\sum_{k=1}^{N-2} X_1^d[k]\cos\left(\frac{\pi nk}{N - 1}\right). \quad (4.72)$$

The discrete cosine transform of type I (DCT-I) is defined in terms of the DFT of $x_1[n]$ as follows:

$$X^{\mathrm{cl}}[k] = \frac{1}{\sqrt{2(N-1)}}a[k]X_1^{\mathrm{d}}[k] = \sqrt{\frac{2}{N-1}} \sum_{n=0}^{N-1} a[k]a[n]x[n]\cos\left(\frac{\pi nk}{N-1}\right). \qquad (4.73)$$

The inverse transform is obtained from (4.72) as

$$x[n] = \sqrt{\frac{2}{N-1}} \sum_{k=0}^{N-1} a[k]a[n]X^{\mathrm{cl}}[k]\cos\left(\frac{\pi nk}{N-1}\right). \qquad (4.74)$$

We can express both DCT-I and its inverse as matrix–vector operations:

$$X_N^{\mathrm{cl}} = C_N^{\mathrm{I}}x_N, \quad x_N = C_N^{\mathrm{I}}X_N^{\mathrm{cl}}, \qquad (4.75)$$

where

$$[C_N^{\mathrm{I}}]_{k,n} = \sqrt{\frac{2}{N-1}}a[k]a[n]\cos\left(\frac{\pi nk}{N-1}\right), \quad 0 \le k,n \le N-1. \qquad (4.76)$$

C_N^{I} is called the DCT-I matrix. It is clear from the definition that this matrix is symmetric. It also follows from (4.75) that C_N^{I} is its own inverse, that is,

$$(C_N^{\mathrm{I}})^{-1} = C_N^{\mathrm{I}}. \qquad (4.77)$$

The DCT-I matrix is a symmetric orthonormal matrix. Recall, for comparison, that the normalized Fourier matrix $N^{-1/2}F_N$ is a symmetric unitary matrix, that is,

$$(N^{-1/2}F_N)^{-1} = N^{-1/2}\bar{F}_N. \qquad (4.78)$$

Example 4.10 Let us test the DCT-I transform on a sinusoidal waveform, as we did in Example 4.3 for the DFT. A sinusoidal waveform with an arbitrary initial phase can always be written as a linear combination of a cosine and a sine, because

$$\cos(\theta_0 n + \phi_0) = \cos\phi_0\cos(\theta_0 n) - \sin\phi_0\sin(\theta_0 n).$$

Since the DCT-I is a real transform, we need to consider the transforms of a cosine and a sine separately. Let the frequency be an integer multiple of $\pi/(N-1)$, which is matched to DCT-I as seen from (4.73). The signals to be tested are thus

$$x_{\cos}[n] = \cos\left(\frac{\pi mn}{N-1}\right), \quad x_{\sin}[n] = \sin\left(\frac{\pi mn}{N-1}\right).$$

Figure 4.13(a) shows $X_{\cos}^{\mathrm{cl}}[k]$ for $N = 32$ and $m = 10$; Figure 4.13(b) shows $X_{\sin}^{\mathrm{cl}}[k]$. As we see, the transform of the cosine signal exhibits a distinct peak at the right frequency, but is not free from ripple at other frequencies. The transform of the sine signal is not as informative. It is zero at the frequency of the signal, exhibits positive and negative peaks at the neighboring frequencies, and has a sizable ripple at other frequencies. For a sinusoidal signal of an arbitrary initial phase, the transform will be a weighted sum of the graphs shown in the two parts of Figure 4.13.   □

4.9.2   Type-II Discrete Cosine Transform

Type-II DCT is based on nonminimal symmetrization of the signal $x[n]$ which duplicates all N elements. Let us define

$$x_2[n] = \begin{cases} x[n], & 0 \le n \le N-1, \\ x[2N-1-n], & N \le n \le 2N-1. \end{cases} \qquad (4.79)$$

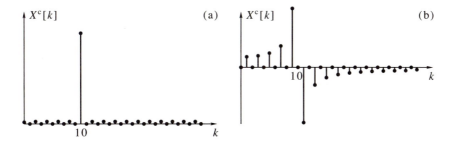

Figure 4.13 DCT-I of a sinusoidal waveform, $N = 32$: (a) pure cosine; (b) pure sine.

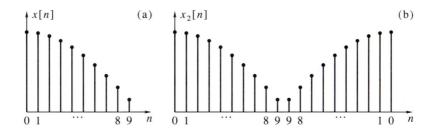

Figure 4.14 The symmetric extension used in the definition of DCT-II: (a) the given sequence; (b) the symmetrically extended sequence (for clarity, the time indices are marked not by the actual n but in a way that shows the symmetry).

Figure 4.14 illustrates this definition for $N = 10$. Part a shows the sequence $x[n]$ and part b the symmetrically extended sequence $x_2[n]$.

The DFT of $x_2[n]$ is given by

$$X_2^d[k] = \sum_{n=0}^{N-1} x[n]W_{2N}^{-nk} + \sum_{n=N}^{2N-1} x[2N-1-n]W_{2N}^{-nk}$$

$$= \sum_{n=0}^{N-1} x[n](W_{2N}^{-nk} + W_{2N}^{(n+1)k}), \quad 0 \le k \le 2N-1. \tag{4.80}$$

This sequence is neither real nor symmetric in k. However, if we multiply it by $W_{2N}^{-0.5k}$ we will get a real sequence, since

$$U[k] = W_{2N}^{-0.5k}X_2^d[k] = \sum_{n=0}^{N-1} x[n](W_{2N}^{-(n+0.5)k} + W_{2N}^{(n+0.5)k})$$

$$= 2 \sum_{n=0}^{N-1} x[n] \cos\left[\frac{\pi(n+0.5)k}{N}\right], \quad 0 \le k \le 2N-1. \tag{4.81}$$

We make two observations:

1. The sequence $U[k]$ satisfies the antisymmetry relation

    $$U[2N-k] = -U[k], \tag{4.82}$$

    since

    $$\cos\left[\frac{\pi(n+0.5)(2N-k)}{N}\right] = -\cos\left[\frac{\pi(n+0.5)k}{N}\right]. \tag{4.83}$$

2. $U[N] = 0$, since

    $$\cos\left[\frac{\pi(n+0.5)N}{N}\right] = \cos[\pi(n+0.5)] = 0. \tag{4.84}$$

The original sequence can be expressed in terms of $U[k]$ as follows:

$$x[n] = \frac{1}{2N} \sum_{k=0}^{2N-1} X_2^d[k] W_{2N}^{nk} = \frac{1}{2N} \sum_{k=0}^{2N-1} U[k] W_{2N}^{0.5k} W_{2N}^{nk}$$

$$= \frac{1}{2N} \sum_{k=0}^{N-1} U[k] W_{2N}^{(n+0.5)k} - \frac{1}{2N} \sum_{k=N+1}^{2N-1} U[2N-k] W_{2N}^{(n+0.5)k}$$

$$= \frac{1}{2N} U[0] + \frac{1}{2N} \sum_{k=1}^{N-1} U[k] (W_{2N}^{(n+0.5)k} + W_{2N}^{-(n+0.5)k})$$

$$= \frac{1}{2N} U[0] + \frac{1}{N} \sum_{k=1}^{N-1} U[k] \cos \left[\frac{\pi(n+0.5)k}{N} \right]. \tag{4.85}$$

The transform DCT-II is defined as

$$X^{c2}[k] = \frac{1}{\sqrt{2N}} b[k] U[k], \tag{4.86}$$

where

$$b[k] = \begin{cases} 2^{-1/2}, & k = 0, \\ 1, & k \neq 0. \end{cases} \tag{4.87}$$

Substitution of (4.81) and (4.87) in (4.86) gives

$$X^{c2}[k] = \sqrt{\frac{2}{N}} \sum_{n=0}^{N-1} b[k] x[n] \cos \left[\frac{\pi(n+0.5)k}{N} \right]. \tag{4.88}$$

Substitution of (4.86) and (4.87) in (4.85) gives

$$x[n] = \sqrt{\frac{2}{N}} \sum_{k=0}^{N-1} b[k] X^{c2}[k] \cos \left[\frac{\pi(n+0.5)k}{N} \right]. \tag{4.89}$$

Equations (4.88) and (4.89) are the DCT-II transform pair. The matrix–vector representation of the DCT-II transform pair is

$$X_N^{c2} = C_N^{II} x_N, \quad x_N = (C_N^{II})' X_N^{c2}, \tag{4.90}$$

where

$$[C_N^{II}]_{k,n} = \sqrt{\frac{2}{N}} b[k] \cos \left[\frac{\pi(n+0.5)k}{N} \right], \quad 0 \leq k, n \leq N-1. \tag{4.91}$$

It follows from (4.90) that C_N^{II} is a real orthonormal matrix, that is, it satisfies

$$(C_N^{II})^{-1} = (C_N^{II})'. \tag{4.92}$$

4.9.3 Type-III Discrete Cosine Transform

The type-III DCT is the inverse, or transpose, of the type-II transform. It is defined by

$$X^{c3}[k] = \sqrt{\frac{2}{N}} \sum_{n=0}^{N-1} b[n] x[n] \cos \left[\frac{\pi(k+0.5)n}{N} \right], \tag{4.93}$$

$$x[n] = \sqrt{\frac{2}{N}} \sum_{k=0}^{N-1} b[n] X^{c3}[k] \cos \left[\frac{\pi(k+0.5)n}{N} \right]. \tag{4.94}$$

The DCT-III matrix is

$$C_N^{III} = (C_N^{II})'. \tag{4.95}$$

Problem 4.44 discusses a direct construction of the DCT-III.

4.9.4 Type-IV Discrete Cosine Transform

Both DCT-II and DCT-III are not symmetrical in k and n. Type-IV DCT is, like DCT-I, symmetric. Its construction is similar to that of DCT-II. However, in addition to extending $x[n]$ symmetrically we modulate it as follows:

$$x_4[n] = \begin{cases} W_{2N}^{-0.5n}x[n], & 0 \le n \le N-1, \\ -W_{2N}^{-0.5n}x[2N-1-n], & N \le n \le 2N-1. \end{cases} \tag{4.96}$$

The DFT of $x_4[n]$ is given by

$$\begin{aligned} X_4^d[k] &= \sum_{n=0}^{N-1} x[n]W_{2N}^{-0.5n}W_{2N}^{-nk} - \sum_{n=N}^{2N-1} x[2N-1-n]W_{2N}^{-0.5n}W_{2N}^{-nk} \\ &= \sum_{n=0}^{N-1} x[n]W_{2N}^{-n(k+0.5)} - \sum_{n=0}^{N-1} x[n]W_{2N}^{-(2N-1-n)(k+0.5)} \\ &= \sum_{n=0}^{N-1} x[n](W_{2N}^{-n(k+0.5)} + W_{2N}^{(n+1)(k+0.5)}), \quad 0 \le k \le 2N-1. \end{aligned} \tag{4.97}$$

This sequence is neither real nor symmetric in k. However, if we multiply it by $W_{2N}^{-0.5(k+0.5)}$ we will get a real sequence, since

$$\begin{aligned} V[k] &= W_{2N}^{-0.5(k+0.5)}X_4^d[k] = \sum_{n=0}^{N-1} x[n](W_{2N}^{-(n+0.5)(k+0.5)} + W_{2N}^{(n+0.5)(k+0.5)}) \\ &= 2\sum_{n=0}^{N-1} x[n]\cos\left[\frac{\pi(n+0.5)(k+0.5)}{N}\right], \quad 0 \le k \le 2N-1. \end{aligned} \tag{4.98}$$

The sequence $V[k]$ satisfies the antisymmetry relation

$$V[2N-1-k] = -V[k], \tag{4.99}$$

since

$$\cos\left[\frac{\pi(n+0.5)(2N-1-k+0.5)}{N}\right] = -\cos\left[\frac{\pi(n+0.5)(k+0.5)}{N}\right]. \tag{4.100}$$

The original sequence can be expressed in terms of $V[k]$ as follows:

$$\begin{aligned} x[n] &= W_{2N}^{0.5n}\frac{1}{2N}\sum_{k=0}^{2N-1} X_4^d[k]W_{2N}^{nk} = W_{2N}^{0.5n}\frac{1}{2N}\sum_{k=0}^{2N-1} V[k]W_{2N}^{0.5(k+0.5)}W_{2N}^{nk} \\ &= \frac{1}{2N}\sum_{k=0}^{N-1} V[k]W_{2N}^{(n+0.5)(k+0.5)} - \frac{1}{2N}\sum_{k=N}^{2N-1} V[2N-1-k]W_{2N}^{(n+0.5)(k+0.5)} \\ &= \frac{1}{2N}\sum_{k=0}^{N-1} V[k](W_{2N}^{(n+0.5)(k+0.5)} + W_{2N}^{-(n+0.5)(k+0.5)}) \\ &= \frac{1}{N}\sum_{k=0}^{N-1} V[k]\cos\left[\frac{\pi(n+0.5)(k+0.5)}{N}\right]. \end{aligned} \tag{4.101}$$

The transform DCT-IV is defined as

$$X^{c4}[k] = \frac{1}{\sqrt{2N}}V[k] = \sqrt{\frac{2}{N}}\sum_{n=0}^{N-1} x[n]\cos\left[\frac{\pi(n+0.5)(k+0.5)}{N}\right]. \tag{4.102}$$

The inverse transform is obtained from (4.101) as

$$x[n] = \sqrt{\frac{2}{N}}\sum_{k=0}^{N-1} X^{c4}[k]\cos\left[\frac{\pi(n+0.5)(k+0.5)}{N}\right]. \tag{4.103}$$

Equations (4.102) and (4.103) are the DCT-IV transform pair. The matrix–vector representation of the DCT-IV transform pair is

$$X_N^{c4} = C_N^{IV} x_N, \quad x_N = C_N^{IV} X_N^{c4}, \tag{4.104}$$

where

$$[C_N^{IV}]_{k,n} = \sqrt{\frac{2}{N}} \cos\left[\frac{\pi(n+0.5)(k+0.5)}{N}\right], \quad 0 \le k, n \le N-1. \tag{4.105}$$

It follows from (4.104) that C_N^{IV} is a symmetric orthonormal matrix, that is, it satisfies

$$(C_N^{IV})^{-1} = C_N^{IV}. \tag{4.106}$$

4.9.5   Discussion

All four DCTs are real orthonormal transforms. As we saw in Section 4.2, the columns of the inverse of an orthonormal transform can be regarded as basis vectors for the N-dimensional vector space, and the components of the transform of a given signal are the coordinates of the signal in this basis. Thus, for any of the four transforms we have

$$x_N = \sum_{k=0}^{N-1} X^c[k] c_{N,k}, \tag{4.107}$$

where $c_{N,k}$ is the kth column of C_N'. Since the DCTs are real, their corresponding vectors form bases to the real N-dimensional vector space (unlike the DFT vectors, which form a basis for the complex N-dimensional space). Figures 4.15 and 4.16 show the basis vectors of the four transforms for $N = 8$. As we explained in Section 4.2, each vector is interpreted as a discrete-time signal, that is, its components are interpreted as points in time.

The DCT is not as useful as the DFT for frequency-domain signal analysis, due to its deficiencies in representing pure sinusoidal waveforms—recall Example 4.10. The main application of the DCT is in signal compression. We shall discuss this application in Section 14.1; see page 551.

4.10   The Discrete Sine Transform*

The discrete sine transform (DST) is similar to the DCT. It can be regarded as a discrete-time version of the Fourier sine series, which we described in Section 2.5. There are four types of DST, which differ in the way the given signal is extended antisymmetrically. Since their derivation is similar to the derivation of the corresponding DCTs, and since in practice they are less important than the DCTs, we skip the details and give only the DST definitions. We describe the transforms in terms of the (k, n)th element of the corresponding matrix, where n is the column index (corresponding to $x[n]$) and k is the row index (corresponding to $X^{si}[k]$, where $i = 1, 2, 3, 4$). In all cases $0 \le n, k \le N - 1$.

1. DST-I

$$[S_N^I]_{k,n} = \sqrt{\frac{2}{N+1}} \sin\left[\frac{\pi(k+1)(n+1)}{N+1}\right], \quad (S_N^I)^{-1} = S_N^I. \tag{4.108}$$

2. DST-II

$$[S_N^{II}]_{k,n} = \sqrt{\frac{2}{N}} c[k] \sin\left[\frac{\pi(k+1)(n+0.5)}{N}\right], \quad (S_N^{II})^{-1} = (S_N^{II})'. \tag{4.109}$$

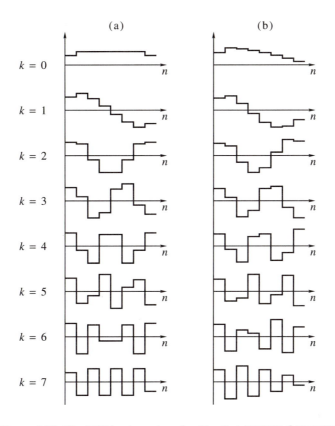

Figure 4.15 The DCT basis vectors for $N = 8$: (a) DCT-I; (b) DCT-II.

3. DST-III

$$[S_N^{III}]_{k,n} = \sqrt{\frac{2}{N}} c[n] \sin\left[\frac{\pi(k+0.5)(n+1)}{N}\right], \quad (S_N^{III})^{-1} = (S_N^{III})'. \tag{4.110}$$

4. DST-IV

$$[S_N^{IV}]_{k,n} = \sqrt{\frac{2}{N}} \sin\left[\frac{\pi(k+0.5)(n+0.5)}{N}\right], \quad (S_N^{IV})^{-1} = S_N^{IV}. \tag{4.111}$$

In these formulas we have defined

$$c[n] = \begin{cases} 2^{-1/2}, & n = N-1, \\ 1, & n \neq N-1. \end{cases} \tag{4.112}$$

4.11 Summary and Complement

4.11.1 Summary

In this chapter we introduced the discrete Fourier transform (4.3). The DFT is defined for discrete-time, finite-duration signals and is a uniform sampling of the Fourier transform of a signal, with a number of samples equal to the length of the signal. The signal can be uniquely recovered from its DFT through the inverse DFT formula (4.16). The DFT can be represented as a matrix–vector multiplication. The DFT matrix is unitary, except for a constant scale factor. The columns of the IDFT matrix can be interpreted

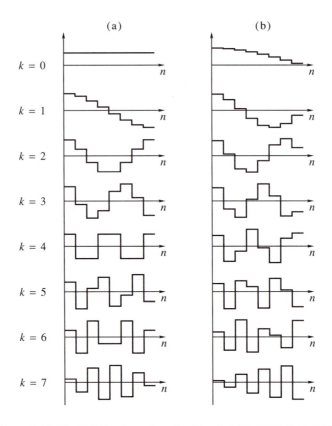

Figure 4.16 The DCT basis vectors for $N = 8$: (a) DCT-III; (b) DCT-IV.

as a basis for the N-dimensional vector space, and the DFT values can be interpreted as the coordinates of the signal in this basis. The DFT shares many of the properties of the usual Fourier transform.

The DFT of a signal of length N can also be defined at M frequency points, with $M > N$. This is done by padding the signal with $M - N$ zeros and computing the M-point DFT of the zero-padded signal. Zero padding is useful for refining the display of the frequency response of a finite-duration signal, thus improving its visual appearance. Frequency-domain zero padding allows interpolation of a finite-duration signal.

Closely related to the DFT is the operation of circular convolution (4.51). Circular convolution is defined between two signals of the same (finite) length, and the result has the same length. It can be thought of as a conventional (linear) convolution of the periodic extensions of the two signals over one period. The DFT of the circular convolution of two signals is the product of the individual DFTs. The DFT of the product of two signals is the circular convolution of the individual DFTs (up to a factor N^{-1}).

Circular convolution can be used for computing the linear convolution of two finite-duration signals, not necessarily of the same length. This is done by zero padding the two sequences to a common length, equal to the sum of the lengths minus 1, followed by a circular convolution of the zero-padded signals. The latter can be performed by multiplication of the DFTs and taking the inverse DFT of the product.

In general, the DFT does not give an exact picture of the frequency-domain characteristics of a signal, only an approximate one. An exception occurs when the signal is periodic and band limited, and sampling is at an integer number of samples per period, at a rate higher than the Nyquist rate. In this case the DFT values are equal, up to a constant factor, to the Fourier series coefficients of the signal.

In this chapter we also introduced the discrete cosine and sine transforms. Contrary to the DFT, these transforms are real valued (when applied to real-valued signals) and orthonormal. There are four types of each. The DCT has become a powerful tool for image compression applications. We shall describe the use of the DCT for compression in Section 14.1; see page 551. Additional material on the DCT can be found in Rao and Yip [1990].

4.11.2   Complement

1. [p. 99] *Resolution* here refers to the spacing of frequency points; it does not necessarily determine the *accuracy* to which the frequency of a sinusoidal signal can be determined from the DFT. We shall study the accuracy issue in Chapter 6.

4.12   MATLAB Programs

Program 4.1 Brute-force DFT and IDFT.

```
function y = bfdft(x,swtch);
% Synopsis: y = bfdft(x,swtch).
% Brute-force DFT or IDFT.
% Input parameters:
% x: the input vector
% swtch: 0 for DFT, 1 for IDFT.
% Output parameters:
% y: the output vector.

N = length(x);
x = reshape(x,1,N);
n = 0:N-1;
if (swtch), W = exp((j*2*pi/N)*n);
else, W = exp((-j*2*pi/N)*n), end
y = zeros(1,N);
for k = 0:N-1,
   y(k+1) = sum(x.*(W.^k));
end
if (swtch), y = (1/N)*y; end
```

Program 4.2 Circular convolution via linear convolution.

```
function z = circonv(x,y);
% Synopsis: z = circonv(x,y).
% Performs circular convolution by means of linear convolution.
% Input parameters:
% x, y; the two vectors to be convolved.
% Output parameter:
% z: the result of the circular convolution.

N = length(x);
if (length(y) ~= N),
   error('Vectors of unequal lengths in circonv');
end
z = conv(reshape(x,1,N),reshape(y,1,N));
z = z(1:N) + [z(N+1:2*N-1),0];
```

4.13 Problems

4.1 The sequence $x[n]$ has length N, and $X^d[k]$ is known to be real. What does this imply about $x[n]$?

4.2 If both $x[n]$ and $X^d[k]$ are real, what else can we say about these two sequences? If $x[n]$ is real and $X^d[k]$ is purely imaginary, what else can we say about $x[n]$?

4.3 Let $N = 4M$, where M is an integer. Let

$$X^d[k] = \begin{cases} 0.5, & k = M, \\ 0.5, & k = 3M, \\ 0, & \text{otherwise.} \end{cases}$$

(a) Is the signal $x[n]$ (the inverse DFT of $X^d[k]$) real?

(b) Compute $x[n]$ explicitly.

(c) Let

$$y[n] = (-1)^n x[2n], \quad 0 \le n \le 2M - 1.$$

Compute $Y^d[k]$.

4.4 We are given a real sequence $x[n]$ of length $4N$. Define $y[n]$ as

$$y[n] = x[(n - N) \bmod 4N] + x[n].$$

Also define

$$Z^d[k] = Y^d[k] + Y^d[4N - k].$$

Express $Z^d[k]$ as a function of $X^d[k]$ for each of the following four cases separately: $k \bmod 4 = 0, 1, 2, 3$.

4.5 The signal $x[n]$ is obtained by sampling the continuous-time signal $\sin(2\pi t)$ at interval $T = 0.125$ and taking 12 consecutive samples. At which values of k will $|X^d[k]|$ be maximum?

4.6 Let $x[n]$ be a real sequence of even-length N satisfying

$$x[n] = x[N - 1 - n].$$

Let

$$Y^d[k] = W_N^{-0.5k} X^d[k].$$

(a) Show that $Y^d[k]$ is real.

(b) Show that $Y^d[k]$ is antisymmetric modulo N, that is,

$$Y^d[k] = -Y^d[N - k].$$

4.7 Let $x[n]$ be the sequence

$$x[n] = \begin{cases} \alpha^n, & n \ge 0, \\ 0, & n < 0, \end{cases}$$

where α is real, $|\alpha| < 1$. We are given another sequence $y[n]$ of length N such that

$$Y^d[k] = X^f\left(\frac{2\pi k}{N}\right).$$

Give an explicit expression for $y[n]$.

4.8 Let $x[n]$ be a signal of length N and let $y[n]$ be the periodic extension of $x[n]$ to M periods, so the length of $y[n]$ is MN. Express Y^{d} in terms of X^{d}.

4.9 Examine the following claim: "Every real periodic discrete-time signal is necessarily a finite sum of discrete-time sinusoidal signals." Is the claim right or wrong? If it is right, explain why. If it is wrong, give a counterexample.

4.10 The sequences $x_1[n]$, $x_2[n]$ have length N each. Define the sequence $y[n]$ of length $3N$ as

$$\{y[n]\}_{n=0}^{3N-1} = \{x_1[0], 0, x_2[0], x_1[1], 0, x_2[1], \ldots, x_1[N-1], 0, x_2[N-1]\}.$$

Express Y^{d} in terms of X_1^{d} and X_2^{d}.

4.11 The sequences $x_1[n]$, $x_2[n]$ have length N each. Define the sequences $y_1[n]$, $y_2[n]$ as

$$y_1[n] = \begin{cases} x_1[n], & 0 \le n \le N-1, \\ 0, & N \le n \le 2N-1. \end{cases}$$

$$y_2[n] = \begin{cases} 0, & 0 \le n \le N-1, \\ x_2[n-N], & N \le n \le 2N-1. \end{cases}$$

Let

$$z[n] = x_1[n] + x_2[n].$$

Express Z^{d} in terms of Y_1^{d} and Y_2^{d}.

4.12 We are given a sequence $x[n]$ of length 6 and we know that

$$\{|X^{\mathrm{d}}[k]|, \ 0 \le k \le 5\} = \{12, 7, 3, 0, 3, 7\}.$$

Which of the following is correct? (1) The sequence is necessarily real; (2) the sequence is necessarily imaginary; (3) the sequence is necessarily complex; (4) the information is insufficient for conclusion. Explain your answer.

4.13 Let N be an even number.

(a) Let $f[n]$ be a sequence of length N. We are given that

$$F^{\mathrm{d}}[k] = (-1)^k, \quad 0 \le k \le N-1.$$

  Find $f[n]$.

(b) Let $x[n]$ be a sequence of length N and let

$$Y^{\mathrm{d}}[k] = \begin{cases} X^{\mathrm{d}}[k], & k \text{ even}, \\ 0, & k \text{ odd}. \end{cases}$$

  Express $y[n]$, the inverse DFT of $Y^{\mathrm{d}}[k]$, as a function of the sequence $x[n]$. Hint: Use the result of part a.

(c) Let $\tilde{x}[n]$ and $\tilde{y}[n]$ be the periodic extensions of $x[n]$ and $y[n]$, respectively, that is,

$$\tilde{x}[n] = x[n \bmod N], \quad \tilde{y}[n] = y[n \bmod N].$$

  Find a filter $h[n]$ such that

$$\tilde{y} = h * \tilde{x}.$$

(d) Return to part b and assume that N is odd, and that $x[n]$ is real valued. Is $y[n]$ real valued in general?

4.14 Suppose you wish to implement an inverse DFT, but your computer program (or hardware device) is capable of computing only a direct DFT. Show how to use the direct DFT to perform inverse DFT.

4.15 This problem explores the result of successive applications of the DFT operation.

(a) Let $x[n]$ have length N and let

$$y = \text{DFT}_N\{\text{DFT}_N\{x\}\}.$$

Express the sequence $y[n]$ in terms of the elements of the sequence $x[n]$ in the simplest possible form.

(b) Suppose we perform the operation DFT_N P times in cascade on the input sequence $x[n]$ and obtain the output sequence $w[n]$. What is the minimum value of P for which $w[n] = Ax[n]$, where A is a constant? What is the value of A?

4.16 The sequence $\{x[n], 0 \le n \le 4\}$ was obtained from a periodic signal $x(t)$ whose period is $T_0 = 1$ second by sampling at an interval $T = 0.2$ second. The DFT of the sequence is

$$X^d[k] = \{2, j, 1, 1, -j\}.$$

(a) Explain why this information is not sufficient to reconstruct $x(t)$ unambiguously.

(b) What additional information will make it possible to reconstruct $x(t)$ unambiguously? With this additional information, write the expression for $x(t)$.

4.17 This problem shows that the DFT values of a finite-duration signal uniquely determine the Fourier transform of the signal at all frequency points.

(a) We are given the DFT of a length-N sequence, $X^d[k]$. Express the Fourier transform of the sequence, $X^f(\theta)$, as a function of $\{X^d[k], 0 \le k \le N - 1\}$.

(b) Write a MATLAB program that implements the result in part a. The inputs to the program are the vector of $X^d[k]$ and a vector of frequencies θ at which the frequency response is to be calculated.

4.18 Define the transform $X^e[k]$ of the length-N sequence $x[n]$ as the samples of the Fourier transform of $x[n]$ at the frequency points

$$\theta[k] = \frac{\pi(2k + 1)}{N}, \quad 0 \le k \le N - 1.$$

Let $x[n], y[n], z[n]$ be three sequences of length N each, and suppose we know that

$$Z^e[k] = X^e[k]Y^e[k].$$

Express the sequence $z[n]$ in terms of the sequences $x[n], y[n]$. Hint: Write $X^e[k]$ as the DFT of a sequence related to $x[n]$.

4.19 Define the transform $X^g[k]$ of a real sequence $\{x[n], 0 \le n \le N - 1\}$ as

$$X^g[k] = \sum_{n=0}^{N-1} x[n]W_N^{-k(n+0.5)}. \tag{4.113}$$

(a) Compute $X^g[k]$ as a function of $X^d[k]$.

(b) Let $y[n]$ be the sequence

$$y[n] = \begin{cases} x[n], & 0 \le n \le N-1, \\ x[2N-n-1], & N \le n \le 2N-1. \end{cases}$$

Compute $Y^g[k]$ and express it as a function of $X^d[k]$ for even values of k. Is the result always real or complex in general?

4.20 A length-N signal $x[n]$ is zero padded to length $2N$ and the corresponding DFT $X_a^d[k]$ is computed. The length-N inverse DFT of the sequence of odd-index components $X_a^d[2l+1]$ is then computed. What is the result of this operation?

4.21 Let $X^d[k]$ be the DFT of a length-N signal $x[n]$, and $X_a^d[l]$ the DFT of the signal zero padded to length M. What points $0 \le k \le N-1$ have corresponding points $0 \le l \le M-1$ such that $X^d[k] = X_a^d[l]$:

(a) When M is an integer multiple of N?

(b) When M is not an integer multiple of N?

4.22 In Section 4.4 we saw how to use the DFT for computing the Fourier transform of a length-N sequence at M equally spaced frequency points when $M > N$. Suppose we wish to do this when $M < N$, that is, to compute

$$X^f(\theta[k]), \quad \theta[k] = \frac{2\pi k}{M}, \quad 0 \le k \le M-1.$$

(a) Show how to do this using DFT. Remark: At the time of writing, MATLAB did not perform this computation correctly. When called with an argument N shorter than the sequence length, `fft` gave the DFT of a truncated sequence, rather than the Fourier transform of the given sequence. Check your current version of MAT-LAB to see if this problem has been corrected!

(b) Illustrate the procedure you developed in part a for $N = 17$ and $M = 5$.

4.23 A continuous-time signal $x(t)$ is sampled at interval $T = 1/1.28$ millisecond, a total of 100 samples. We wish to compute the Fourier transform of the sampled signal at $f = 305$ Hz using DFT.

(a) Using DFT of the 100 samples, what is the frequency nearest to f at which the DFT is computed, and what is the corresponding k?

(b) What is the minimum number of zeros with which we must pad the 100 samples to obtain the DFT at $f = 305$ Hz exactly?

4.24 Pierre, a programmer in a DSP company, was instructed to write a computer program that zero-pads a sequence of length N to length LN (L a positive integer), then computes the DFT. Misunderstanding the concept of zero padding, he padded each point $x[n]$ of the given sequence by $L-1$ zeros. What did his program compute?

4.25 Show, by carrying out the necessary derivation, that Program 4.2 indeed performs circular convolution.

4.26 We are given the two sequences

$$x = \{1, -3, 1, 5\}, \quad w = \{7, -7, -9, -3\}.$$

Does there exist a sequence y of length 4 such that

$$x \circledast y = w?$$

If so, find y; if not, prove that such y does not exist. Hint: Solve using DFT.

4.27 Given that $x = \{2, -1\}$, $w = x * y$, and $w = \{6, -1, 7, -4\}$, compute the sequence y using DFT.

4.28 Prove the commutativity and associativity of circular convolution (4.54).

4.29 We are given two sequences $x_1[n]$, $x_2[n]$ of length 75 each. It is known that $x[n] = 0$ for all $0 \le n \le 7$ and for all $60 \le n \le 74$, but is nonzero otherwise; $x_2[n]$ is nonzero in general. Let

$$y[n] = \{x_1 \circledast x_2\}[n], \quad z[n] = \{x_1 * x_2\}[n].$$

For what values of n will $y[n \bmod 75] = z[n]$ hold?

4.30 The sequences $x[n]$, $y[n]$, $z[n]$ have length 3 each, and it is known that

$$z[n] = \{x \circledast y\}[n].$$

We are given that
$$x[n] = \{1, 1, 0\}, \quad Z^{\mathrm{d}}[k] = \{2, -W_3^2, -W_3\}.$$

Find the sequence $y[n]$.

4.31 In Section 4.7 we showed how to perform linear convolution of sequences of lengths N_1, N_2 by zero padding to length $N_1 + N_2 - 1$. Suppose, instead, that we zero-pad the two sequences to length $M > N_1 + N_2 - 1$. How then is the circular convolution of the zero-padded sequences related to the linear convolution?

4.32 Suggest a procedure for performing linear convolution of *three* sequences of lengths N_1, N_2, N_3 by DFT.

4.33 The *circular correlation* (or *circular cross-correlation*) of two discrete-time signals, both of length N, is defined as (cf. Problem 2.36)

$$z[n] = \sum_{m=0}^{N-1} x[(m + n) \bmod N]\bar{y}[m]. \qquad (4.114)$$

Express the circular correlation operation in the frequency domain.

4.34* Let
$$x(t) = t^3 + 1, \quad -\infty < t < \infty.$$

The signal is sampled at interval T, for $-N \le n \le N$, and shifted to the right by N, yielding a sequence of length $(2N + 1)$. Find $X^{\mathrm{d}}[0]$ of this sequence.

4.35* Let $x(t)$ be the signal

$$x(t) = a_1 \cos(2\pi f t + \phi_1) + a_2 \cos(2\pi \cdot 1.25 f t + \phi_2).$$

The frequency f is unknown, but it is known to lie in the range

$$1000\,\mathrm{Hz} < f < 1600\,\mathrm{Hz}$$

(with strict inequalities). The parameters a_1, a_2, ϕ_1, ϕ_2 are real and unknown.

(a) It is required to sample the signal such that for all f in the specified range it will be possible to reconstruct $x(t)$ from its samples unambiguously. What is the smallest sampling rate that meets this requirement?

(b) Is the signal $x(t)$ periodic? If so, what is its period? If not, explain why.

(c) Now suppose we know that $f = 1200\,\text{Hz}$, and we wish to find a_1 and a_2. Suggest a convenient way of doing it using DFT (not necessarily as efficiently as possible). Choose a sampling interval T and a number of sample N, and show the details of the calculation.

(d) Consider the following alternative to the operation in part c. Generate the signal

$$y(t) = x(t)\cos(2\pi 1000t).$$

Pass $y(t)$ through an ideal low-pass filter having a cutoff frequency of $1000\,\text{Hz}$. Let $z(t)$ denote the output of the filter. Now repeat part c for $z(t)$. What then do you choose for N and T?

4.36* Let $x[n]$ be a signal on $0 \le n \le N-1$ and $\tilde{x}[n]$ its periodic extension with period N. It is known that

$$X^{\mathrm{d}}[k] = 4\cos\left(\frac{2\pi kL}{N}\right) + 2$$

for some integer L in the range $0 < L < N$. Find the values of L for which $\tilde{x}[n]$ is periodic with period $N' < 0.5N$. Specify the constraint on N for this problem to have a solution.

4.37* Write F_5 in full, similarly to the matrices given in (4.21), (4.22).

4.38* Let N be a prime integer. Prove that each row of F_N, with the exception of the first, is a permutation of the sequence $\{W_N^{-n},\ 0 \le n \le N-1\}$ (i.e., each element of this sequence appears in the row exactly once).

4.39* Plot the basis vectors of the natural basis for $N = 8$, similarly to Figure 4.5.

4.40* This problem examines the representation of circular convolution by matrix notation.

(a) Denote by Y the matrix of elements $y[n]$ appearing in (4.53). Write a general expression for the (k, l)th element of this matrix.

(b) Let F_N be the DFT matrix, as given in (4.19). Derive a general expression for the (k, l)th element of the matrix product $F_N Y \bar{F}_N'$ and bring it to the simplest form.

(c) Return to (4.53) and use the property (4.23) of the DFT matrix to write it in the form

$$F_N z_N = N^{-1}(F_N Y \bar{F}_N')F_N x_N. \tag{4.115}$$

Justify the equivalence of this form to (4.53).

(d) Substitute the result of part b in (4.115), then write (4.115) in terms of the DFTs of the three sequences. What property of the DFT have you thus proved?

4.41* Write a MATLAB program that implements interpolation of a finite sequence $x[n]$ by zero padding in the frequency domain. The program should treat both even and odd lengths of $x[n]$ and not be limited to M which is an integer multiple of N.

4.42* Modify the definition of zero padding in the frequency domain (4.47) to the case of even N. You are not requested to derive (4.49) for this case. Hint: Take care to preserve the conjugate symmetry property of $X_i^d[k]$.

4.43* Repeat Example 4.10 for DCT-II. Match the signals to this transform by taking

$$x_{\cos}[n] = \cos\left[\frac{\pi m(n + 0.5)}{N - 1}\right], \quad x_{\sin}[n] = \sin\left[\frac{\pi m(n + 0.5)}{N - 1}\right].$$

Use $N = 32$ and $m = 10$. Plot the results, similarly to Figure 4.13. What is the main difference with respect to the results in Example 4.10?

4.44* Construct the DCT-III directly, similarly to the other three DCTs. Hint: Define $x_3[n]$ as

$$x_3[n] = \begin{cases} \sqrt{2}x[n], & n = 0, \\ x[n]W_{2N}^{-0.5n}, & 1 \le n \le N - 1, \\ 0, & n = N, \\ -x[2N - n]W_{2N}^{-0.5n}, & N + 1 \le n \le 2N - 1. \end{cases}$$

4.45* Write a MATLAB program that implements the four types of DCT.

4.46* Write a MATLAB program that implements the four types of DST.

4.47* Define a new discrete cosine transform, based on the following symmetric extension of $x[n]$:

$$x_5[n] = \begin{cases} 2^{1/2}x[0], & n = 0, \\ x[n], & 1 \le n \le N - 1, \\ x[2N - n - 1], & N \le n \le 2N - 2. \end{cases}$$

The transform is to be real and orthonormal. Derive the formulas for the transform and its inverse.

4.48* Define a new trigonometric function

$$\mathrm{cas}(\theta) = \cos\theta + \sin\theta = \sqrt{2}\cos(\theta - 0.25\pi) \tag{4.116}$$

(the notation *cas* is short for "cosine and sine"). For a real length-N sequence $x[n]$, define

$$X^h[k] = \{\mathcal{H}x\}[k] = \sum_{n=0}^{N-1} x[n]\mathrm{cas}\left(\frac{2\pi nk}{N}\right). \tag{4.117}$$

$X^h[k]$ is called the *discrete Hartley transform*, or DHT for short. The DHT is a *real transform*, that is, the transform of a real sequence is real.

(a) Prove that the function cas satisfies the orthogonality property

$$\sum_{n=0}^{N-1} \mathrm{cas}\left(\frac{2\pi kn}{N}\right)\mathrm{cas}\left(\frac{2\pi nl}{N}\right) = N\delta[k - l], \quad 0 \le k,l \le N - 1. \tag{4.118}$$

(b) Explain why (4.118) implies that the inverse DHT is

$$x[n] = \frac{1}{N}\sum_{k=0}^{N-1} X^h[k]\mathrm{cas}\left(\frac{2\pi nk}{N}\right). \tag{4.119}$$

From this we conclude that the same computer program that computes the DHT can be used for computing the inverse DHT.

(c) Express the DHT in terms of the DFT of $x[n]$ (assuming that this sequence is real).

(d) Express the DFT of $x[n]$ in terms of the DHT.

4.49* This problem explores certain properties of the DHT.

(a) Establish the following shift property of the DHT:

$$y[n] = x[(n - m) \bmod N] \implies$$

$$Y^h[k] = X^h[k] \cos\left(\frac{2\pi mk}{N}\right) + X^h[(N - k) \bmod N] \sin\left(\frac{2\pi mk}{N}\right). \qquad (4.120)$$

(b) Use (4.120) for expressing the DHT of the circular convolution of two sequences $x[n], y[n]$ as a function of the DHTs of the two sequences. The formula looks more complicated than the corresponding formula for the DFTs, but it does not require more computations. Explain why.

Chapter 5

The Fast Fourier Transform

The invention of the fast Fourier transform, by Cooley and Tukey in 1965, was a major breakthrough in digital signal processing and, in retrospect, in applied science in general. Until then, practical use of the discrete Fourier transform was limited to problems in which the number of data to be processed was relatively small. The difficulty in applying the Fourier transform to real problems is that, for an input sequence of length N, the number of arithmetic operations in direct computation of the DFT is proportional to N^2. If, for example, $N = 1000$, about a million operations are needed. In the 1960s, such a number was considered prohibitive in most applications.

Cooley and Tukey's discovery was that when N, the DFT length, is a composite number (i.e., not a prime), the DFT operation can be decomposed to a number of DFTs of shorter lengths. They showed that the total number of operations needed for the shorter DFTs is smaller than the number needed for direct computation of the length-N DFT. Each of the shorter DFTs, in turn, can be decomposed and performed by yet shorter DFTs. This process can be repeated until all DFTs are of prime lengths—the prime factors of N. Finally, the DFTs of prime lengths are computed directly. The total number of operations in this scheme depends on the factorization of N into prime factors, but is usually much smaller than N^2. In particular, if N is an integer power of 2, the number of operations is on the order of $N \log_2 N$. For large N this can be smaller than N^2 by many orders of magnitude. Immediately, the discrete Fourier transform became an immensely practical tool. The algorithms discovered by Cooley and Tukey soon became known as the *fast Fourier transform*, or FFT. This name should not mislead you: FFT algorithms are just computational schemes for computing the DFT; they are not new transforms!

Since Cooley and Tukey's pioneering work, there have been enormous developments in FFT algorithms, and fast algorithms in general.[1] Today this is unquestionably one of the most highly developed areas of digital signal processing.

This chapter serves as an introduction to FFT. We first explain the general principle of DFT decomposition, show how it reduces the number of operations, and present a recursive implementation of this decomposition. We then discuss in detail the most common special case of FFT: the radix-2 algorithms. Next we present the radix-4 FFT, which is more efficient than radix-2 FFT. Finally we discuss a few FFT-related topics: FFTs of real sequences, linear convolutions using FFT, and the chirp Fourier transform algorithm for computing the DFT at a selected frequency range.

5.1   Operation Count

Before entering the main topic of this chapter, let us discuss the subject of operation count. It is common, in evaluating the computational complexity of a numerical algorithm, to count the number of real multiplications and the number of real additions. By "real" we mean either fixed-point or floating-point operations, depending on the specific computer and the way arithmetic is implemented on it. Subtraction is considered equivalent to addition. Divisions, if present, are counted separately. Other operations, such as loading from memory, storing in memory, indexing, loop counting, and input–output, are usually not counted, since they depend on the specific architecture and the implementation of the algorithm. Such operations represent overhead: They should not be ignored, and their contribution to the total load should be estimated (at least roughly) and taken into account.

Modern-day DSP microprocessors typically perform a real multiplication *and* a real addition in a single machine cycle, so the traditional adage that "multiplication is much more time-consuming than addition" has largely become obsolete. The operation is $y = ax + b$, and is usually called MAC (for multiply/accumulate). When an algorithm such as FFT is to be implemented on a machine equipped with MAC instruction, it makes more sense to count the *maximum* of the number of additions and the number of multiplications (rather than their sum).

The DFT operation is, in general, a multiplication of a complex $N \times N$ matrix by a complex N-dimensional vector. Therefore, if we do not make any attempt to save operations, it will require N^2 complex multiplications and $N(N-1)$ complex additions. However, the elements of the DFT matrix on the first row and on the first column are 1. It is therefore possible to eliminate $2N - 1$ multiplications, ending up with $(N - 1)^2$ complex multiplications and $N(N - 1)$ complex additions. Each complex multiplication requires four real multiplications and two real additions; each complex addition requires two real additions. Therefore, straightforward computation of the DFT requires $4(N - 1)^2$ real multiplications and $4(N - 0.5)(N - 1)$ real additions. If the input vector is real, the number of operations can be reduced by half.

The preceding operation count ignores the need to evaluate the complex numbers W_N^{-nk}. Usually, it is assumed that these numbers are computed off line and stored. If this is not true, the computation of these numbers must be taken into account. More on this will be said in Section 5.2.3.

5.2   The Cooley–Tukey Decomposition

5.2.1   Derivation of the CT Decomposition

The Cooley–Tukey (CT) decomposition of the discrete Fourier transform is based on the factorization of N, the DFT length, as a product of numbers smaller than N. Let us thus assume that N is not a prime, so it can be written as $N = PQ$, where both factors are greater than 1. Such a factorization is usually not unique; however, all we need right now is the existence of one such factorization.

We are given the DFT formula

$$X^{\mathrm{d}}[k] = \sum_{n=0}^{N-1} x[n]W_N^{-nk}. \tag{5.1}$$

We begin by dividing the range of integers 0 to $N - 1$ in two different ways. For the time index n, division is into Q intervals of length P each. For the frequency index k, division is into P intervals of length Q each. We then express the variables n and k in the form

$$n = Pq + p, \quad 0 \le q \le Q - 1, \ 0 \le p \le P - 1, \tag{5.2a}$$

$$k = Qs + r, \quad 0 \le s \le P - 1, \ 0 \le r \le Q - 1. \tag{5.2b}$$

Figure 5.1 illustrates this division for $P = 3$ and $Q = 2$.

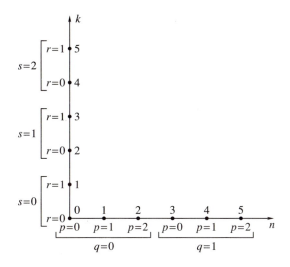

Figure 5.1 Cooley–Tukey decomposition of n and k in the DFT: $P = 3$, $Q = 2$.

The product nk appearing in the exponent of W_N in the DFT formula can be written in terms of p, q, r, s as

$$nk = (Qs + r)(Pq + p) = Nsq + Qsp + Prq + rp. \tag{5.3}$$

Therefore,

$$W_N^{-nk} = W_N^{-Nsq} W_N^{-Qsp} W_N^{-Prq} W_N^{-rp}. \tag{5.4}$$

We have the relationships (Problem 5.1)

$$W_N^N = 1, \quad W_N^Q = W_P, \quad W_N^P = W_Q. \tag{5.5}$$

Therefore,

$$W_N^{-nk} = W_P^{-sp} W_Q^{-rq} W_N^{-rp}. \tag{5.6}$$

Substitution of (5.6) in (5.1) gives

$$X^{\mathrm{d}}[Qs + r] = \sum_{p=0}^{P-1} \sum_{q=0}^{Q-1} x[Pq + p] W_P^{-sp} W_Q^{-rq} W_N^{-rp}$$

$$= \sum_{p=0}^{P-1} W_N^{-rp} \left[\sum_{q=0}^{Q-1} x[Pq + p] W_Q^{-rq} \right] W_P^{-sp}. \tag{5.7}$$

Define the following P sequences, each of length Q:

$$x_p[q] = x[Pq + p], \quad 0 \le q \le Q - 1, \ 0 \le p \le P - 1. \tag{5.8}$$

Each of the $x_p[q]$ is called a *decimated* sequence, because it is obtained by choosing one out of every P elements of $x[n]$, such that the selected elements are uniformly spaced. The DFT of $x_p[q]$ is

$$X_p^d[r] = \sum_{q=0}^{Q-1} x_p[q] W_Q^{-rq}. \tag{5.9}$$

Substitution of (5.9) in (5.7) gives

$$X^d[Qs + r] = \sum_{p=0}^{P-1} W_N^{-rp} X_p^d[r] W_P^{-sp}. \tag{5.10}$$

Let us define, for each $0 \le r \le Q - 1$, the length-P sequence

$$y_r[p] = W_N^{-rp} X_p^d[r], \quad 0 \le p \le P - 1. \tag{5.11}$$

Substitution of (5.11) in (5.10) gives

$$X^d[Qs + r] = \sum_{p=0}^{P-1} y_r[p] W_P^{-sp}. \tag{5.12}$$

Equation (5.12), together with the auxiliary definitions (5.8), (5.9), (5.11), is the Cooley–Tukey decomposition of the DFT. The decomposition can be implemented by the following procedure:

1. Form the P decimated sequences $x_p[q]$ and compute the Q-point DFT of each.

2. Multiply each output of each DFT by the corresponding complex number W_N^{-rp}. These numbers are called *twiddle factors*.

3. For each r, compute the P-point DFT of the sequence $y_r[p]$ to get $X^d[Qs + r]$.

Figure 5.2 illustrates the CT decomposition for $N = 6$, $P = 3$, $Q = 2$. The three decimated sequences are $\{x[0], x[3]\}$, $\{x[1], x[4]\}$, and $\{x[2], x[5]\}$. Each of these sequences is transformed using a DFT$_2$ operation. The outputs of the three DFTs are multiplied by the six twiddle factors. Then the order of the numbers is changed, to yield two sequences of three numbers each, $\{y_0[0], y_0[1], y_0[2]\}$ and $\{y_1[0], y_1[1], y_1[2]\}$. Finally, each of the two sequences is transformed using a DFT$_3$ operation to obtain the DFT of the given sequence $x[n]$.

Example 5.1 Let us continue to explore the details of the Cooley–Tukey decomposition for $N = 6$, $P = 3$, $Q = 2$. For convenience, we express the DFT operations as matrix-vector multiplications. Using the definition (5.9) of the sequences $X_p^d[r]$ and the block diagram shown in Figure 5.2, we see that

$$\begin{bmatrix} X_0^d[0] \\ X_0^d[1] \\ X_1^d[0] \\ X_1^d[1] \\ X_2^d[0] \\ X_2^d[1] \end{bmatrix} = \begin{bmatrix} W_2^0 & 0 & 0 & W_2^0 & 0 & 0 \\ W_2^0 & 0 & 0 & W_2^{-1} & 0 & 0 \\ 0 & W_2^0 & 0 & 0 & W_2^0 & 0 \\ 0 & W_2^0 & 0 & 0 & W_2^{-1} & 0 \\ 0 & 0 & W_2^0 & 0 & 0 & W_2^0 \\ 0 & 0 & W_2^0 & 0 & 0 & W_2^{-1} \end{bmatrix} \begin{bmatrix} x[0] \\ x[1] \\ x[2] \\ x[3] \\ x[4] \\ x[5] \end{bmatrix}. \tag{5.13}$$

The operation of multiplication by the twiddle factors can be described by

$$\begin{bmatrix} y_0[0] \\ y_0[1] \\ y_0[2] \\ y_1[0] \\ y_1[1] \\ y_1[2] \end{bmatrix} = \begin{bmatrix} W_6^0 & 0 & 0 & 0 & 0 & 0 \\ 0 & 0 & W_6^0 & 0 & 0 & 0 \\ 0 & 0 & 0 & 0 & W_6^0 & 0 \\ 0 & W_6^0 & 0 & 0 & 0 & 0 \\ 0 & 0 & 0 & W_6^{-1} & 0 & 0 \\ 0 & 0 & 0 & 0 & 0 & W_6^{-2} \end{bmatrix} \begin{bmatrix} X_0^d[0] \\ X_0^d[1] \\ X_1^d[0] \\ X_1^d[1] \\ X_2^d[0] \\ X_2^d[1] \end{bmatrix}. \tag{5.14}$$

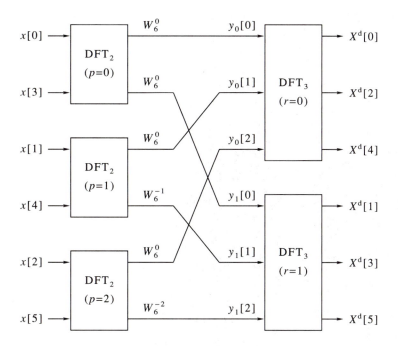

Figure 5.2 Cooley–Tukey DFT algorithm, illustrated for $P = 3$, $Q = 2$.

Finally,

$$
\begin{bmatrix} X^{\mathrm{d}}[0] \\ X^{\mathrm{d}}[1] \\ X^{\mathrm{d}}[2] \\ X^{\mathrm{d}}[3] \\ X^{\mathrm{d}}[4] \\ X^{\mathrm{d}}[5] \end{bmatrix} = \begin{bmatrix} W_3^0 & W_3^0 & W_3^0 & 0 & 0 & 0 \\ 0 & 0 & 0 & W_3^0 & W_3^0 & W_3^0 \\ W_3^0 & W_3^{-1} & W_3^{-2} & 0 & 0 & 0 \\ 0 & 0 & 0 & W_3^0 & W_3^{-1} & W_3^{-2} \\ W_3^0 & W_3^{-2} & W_3^{-4} & 0 & 0 & 0 \\ 0 & 0 & 0 & W_3^0 & W_3^{-2} & W_3^{-4} \end{bmatrix} \begin{bmatrix} y_0[0] \\ y_0[1] \\ y_0[2] \\ y_1[0] \\ y_1[1] \\ y_1[2] \end{bmatrix}. \tag{5.15}
$$

Denoting the matrices in (5.13), (5.14), (5.15) by $G_{6,1}$, $G_{6,2}$, $G_{6,3}$, respectively, and combining these three equations, we can write

$$
X_6^{\mathrm{d}} = G_{6,3} G_{6,2} G_{6,1} x_6. \tag{5.16}
$$

Comparing this result with (4.20), we arrive at the following decomposition of the DFT$_6$ matrix F_6:

$$
F_6 = G_{6,3} G_{6,2} G_{6,1}. \tag{5.17}
$$

We can now count the number of arithmetic operations needed for computing a 6-point DFT by CT decomposition and compare it with the number of operations in a direct computation. In (5.13) there are no multiplications, since $W_2^0 = 1$ and $W_2^{-1} = -1$. There is one addition per row, or a total of 6 additions (subtractions and additions are counted together). In (5.14) there are 2 multiplications, by W_6^{-1} and W_6^{-2}, and no additions. Finally, in (5.15) there are 8 multiplications and 12 additions. In summary, the DFT computation by CT decomposition requires 10 complex multiplications and 18 complex additions. For comparison, direct multiplication by F_6 requires 25 multiplications and 30 additions. Therefore, the CT decomposition reduces the number of both multiplications and additions. This reduction is achieved by replacing a large DFT by a number of small DFTs, and is a special case of a general property of the CT decomposition. □

The matrix decomposition derived in Example 5.1 for F_6 can be generalized to any composite N. The CT decomposition implies a decomposition of F_N to a product of three matrices: The rightmost describes the P Q-point DFTs, the center describes the multiplication by the twiddle factors, and the leftmost describes the Q P-point DFTs. Saving in numerical operations results from the CT decomposition because each of the factors in the matrix decomposition is sparse: The number of nonzero entries is a small percentage of the total number of entries in the matrix. Therefore, multiplication by each such matrix requires a number of operations much smaller than the square of its dimension.

We finally observe that the CT decomposition applies to the inverse DFT equally well. The necessary changes are:

1. Replacing the small DFTs by inverse DFTs.
2. Replacing the twiddle factors W_N^{-rp} by W_N^{rp}.

5.2.2   Recursive CT Decomposition and Its Operation Count

The number of operations needed for computing a length-N DFT via the CT decomposition is computed as follows. The procedure requires P DFTs of length Q, then the multiplications by the twiddle factors, and finally Q DFTs of length P. Suppose we compute each of the DFTs directly. A P-point DFT then requires $(P-1)^2$ complex multiplications and $P(P-1)$ complex additions. Similarly, a Q-point DFT requires $(Q-1)^2$ complex multiplications and $Q(Q-1)$ complex additions. The total number of twiddle factors is PQ. However, $P+Q-1$ of these factors are 1, so $(P-1)(Q-1)$ complex multiplications are required for the twiddle factors. Therefore, the total number of complex multiplications is $N(P+Q-3)+1$ and the total number of complex additions is $N(P+Q-2)$. By comparison, the corresponding numbers in a direct computation of an N-point DFT are $(N-1)^2$ and $N(N-1)$. Therefore, if $P+Q \ll N$, we have a significant reduction in the number of operations. This is the principle of operation of Cooley–Tukey FFT algorithms.

Obviously, we do not have to stop at one stage of the decomposition. As long as either P or Q (or both) is composite, we can factor it and use the CT decomposition for each of the shorter DFTs. In general, we can continue this procedure recursively until we are left only with DFTs of prime lengths, the prime factors of N. If we carry out the CT decomposition until we are left with DFTs whose lengths are all prime numbers, the resulting algorithm is called a *mixed radix* FFT. In the special case of N an integer power of a single prime number, it is called a *prime radix* FFT. If, say, $N = P^r$, the algorithm is called a *radix-P* FFT.

The total number of operations in recursive implementation of the CT decomposition can be computed as follows. Let the number of complex additions in an N-point DFT be $\mathcal{A}_c(N)$ and the number of complex multiplications be $\mathcal{M}_c(N)$. Then we get by counting as before,

$$\mathcal{A}_c(N) = P\mathcal{A}_c(Q) + Q\mathcal{A}_c(P), \tag{5.18a}$$

$$\mathcal{M}_c(N) = P\mathcal{M}_c(Q) + Q\mathcal{M}_c(P) + (P-1)(Q-1). \tag{5.18b}$$

We can substitute in these equalities in a recursive manner, until both arguments P and Q are prime. Then we substitute the number of multiplications or additions for the DFTs of corresponding primes.

Example 5.2 Let $N = 30 = 2 \times 3 \times 5$. We have

$$\mathcal{A}_c(2) = 2, \quad \mathcal{A}_c(3) = 6, \quad \mathcal{A}_c(5) = 20,$$

$$\mathcal{M}_c(2) = 0, \quad \mathcal{M}_c(3) = 4, \quad \mathcal{M}_c(5) = 16.$$

Therefore,

$$\mathcal{A}_c(6) = 6 + 12 = 18, \quad \mathcal{M}_c(6) = 8 + 2 = 10,$$

and finally,

$$\mathcal{A}_c(30) = 120 + 90 = 210, \quad \mathcal{M}_c(6) = 96 + 50 + 20 = 166.$$

For comparison, direct computation of a 30-point DFT requires 841 complex multiplications and 870 complex additions. □

5.2.3 Computation of the Twiddle Factors

In counting the operations for direct DFT and for Cooley–Tukey FFT, we ignored the need to compute the twiddle factors W_N^{-pr}. There are only N distinct twiddle factors, since the sequence W_N^{-n} is periodic with period N. There are two common approaches to twiddle factor computation:

1. In special-purpose applications (which are often in real time), the length N is usually fixed, but the FFT must be performed repeatedly on different input sequences. In this case, the most efficient solution is to precompute and store the N twiddle factors in a lookup table. Then, each time one of them is needed, we compute the corresponding position in the lookup table and retrieve its value. Specifically, denote

$$W[n] = W_N^{-n}, \quad 0 \le n \le N - 1. \tag{5.19}$$

   Then

$$W_N^{-rp} = W_N^{-(rp \bmod N)} = W[rp \bmod N]. \tag{5.20}$$

   Note that the factors appearing in the small FFTs are also available from the table, for example,

$$W_Q^{-m} = W_N^{-Pm} = W[Pm \bmod N].$$

2. In general purpose implementations (such as computer subroutines in mathematical software), the length N usually varies. In this case precomputing and storage is not practical, and the solution is to compute the twiddle factors either prior to the algorithm itself or on the fly. The simplest way is to compute W_N^{-1} directly by cosine and sine, and then use the recursion $W_N^{-(m+1)} = W_N^{-m} W_N^{-1}$. Note, however, that this may be subject to roundoff error accumulation if N is large or the computer word length is small. The opposite (and most conservative) approach is to compute each twiddle factor directly by the appropriate trigonometric functions. Various other schemes have been devised, but are not discussed here.

5.2.4 Computation of the Inverse DFT

So far we discussed only the direct DFT. As we saw in Section 4.1, the inverse DFT differs from the direct DFT in two respects: (1) instead of negative powers of W_N, it uses positive powers; (2) there is an additional division of each output value by N. Every FFT algorithm can be thus modified to perform inverse DFT, by using positive powers of W_N as twiddle factors and multiplying each component of the output (or of the input) by N^{-1}. This entails N extra multiplications, so the computational load increases only slightly. Most FFT computer programs are written so that they can be switched from direct to inverse FFT by an input flag. MATLAB offers an exception: It uses two different calls, `fft` and `ifft`, for the two operations.

5.2.5   Time Decimation and Frequency Decimation

As a special case of the CT procedure, consider choosing P as the *smallest* prime factor of the length of the DFT at each step of the recursion. Then the next step of the recursion needs to be performed for the P DFTs of length Q, whereas in the Q DFTs of length P no further computational savings are possible. The algorithms thus obtained are called *time-decimated* FFTs. Of special importance is the radix-2, time-decimated FFT, to be studied in Section 5.3. In a dual manner, consider choosing Q as the smallest prime factor of the length of the DFT at each step of the recursion. Then the next step of the recursion needs to be performed for the Q DFTs of length P, whereas in the P DFTs of length Q no further computational savings are possible. The algorithms thus obtained are called *frequency-decimated* FFTs. Of special importance is the radix-2 frequency-decimated FFT, also studied in Section 5.3.

5.2.6   MATLAB Implementation of Cooley–Tukey FFT

Programs 5.1, 5.2, and 5.3 implement a frequency-decimated FFT algorithm in MATLAB. The implementation is recursive and is based on the Cooley–Tukey decomposition. The main program, `edufft`, prepares the sequence W_N^{-n} or W_N^{n}, depending on whether direct or inverse FFT is to be computed. It then calls `ctrecur` to do the actual FFT computation, and finally normalizes by N^{-1} in the case of inverse FFT. The program `ctrecur` first tries to find the smallest possible prime factor of N. If such a factor is not found, that is, if N is a prime, it calls `primedft` to compute the DFT. Otherwise it sets this prime factor to Q, sets P to N/Q, and then starts the Cooley–Tukey recursion. Figure 5.3 illustrates how the program performs the recursion (for $Q = 3$, $P = 4$). The first step is to arrange the input vector (shown in part a) in a matrix whose rows are the decimated sequences of length Q (shown in part b). Then, since Q is a prime, `primedft` is called to compute the DFT of each row. The next step is to multiply the results by the twiddle factors (shown in part c). The recursion is now invoked, that is, `ctrecur` calls itself for the columns (shown in part d). Finally, the result is rearranged as a vector (shown in part e) and the program exits.

The powers of W_N are computed only once, in `edufft`, and are passed as arguments to the other routines, thus eliminating redundant operations. Even so, the program runs very slow if N is highly composite. This is due to the high overhead of MATLAB in performing recursions and passing parameters. Therefore, this program should be regarded as educational, given here only for illustration, rather than for use in serious applications. The MATLAB routines `fft` and `ifft` should be used in practice, because they are implemented efficiently in an internal code.

5.3   Radix-2 FFT

Suppose the length of the input sequence is an integer power of 2, say $N = 2^r$. We can then choose $P = 2$, $Q = N/2$ and continue recursively until the entire DFT is built out of 2-point DFTs. This is a special case of time-decimated FFT. In a dual manner we can start with $Q = 2$, $P = N/2$, and continue recursively. This is a special case of frequency-decimated FFT. Both are called *radix-2* FFTs.

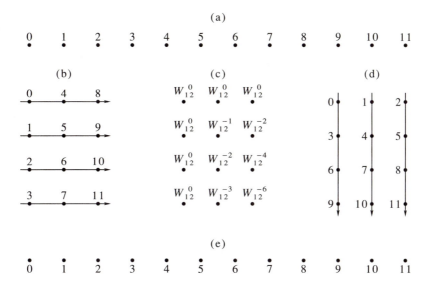

Figure 5.3 Illustration of the steps in the recursive computation of Cooley–Tukey FFT, shown for $Q = 3, P = 4$: (a) the input vector in its natural order; (b) rearrangement of the input vector as a matrix and performing DFT on each row; (c) element-by-element multiplication by the twiddle factors; (d) performing DFT on each column and reindexing of the elements of the matrix; (e) rearrangement of the output as a vector in natural order.

Because of their simplicity of implementation and high computational efficiency, radix-2 FFTs are the most commonly used FFT algorithms. To count their number of operations, observe first that a 2-point DFT is given by

$$\begin{bmatrix} X^d[0] \\ X^d[1] \end{bmatrix} = \begin{bmatrix} 1 & 1 \\ 1 & -1 \end{bmatrix} \begin{bmatrix} x[0] \\ x[1] \end{bmatrix}. \tag{5.21}$$

Thus, a 2-point DFT requires two additions and *no* multiplications. Therefore, the recursive formulas for the number of operations are

$$\mathcal{A}_c(N) = 0.5N\mathcal{A}_c(2) + 2\mathcal{A}_c(N/2) = N + 2\mathcal{A}_c(N/2), \tag{5.22a}$$

$$\mathcal{M}_c(N) = 0.5N\mathcal{M}_c(2) + 2\mathcal{M}_c(N/2) + 0.5N - 1 = 2\mathcal{M}_c(N/2) + 0.5N - 1. \tag{5.22b}$$

The solutions of these difference equations can be verified to be (Problem 5.21)

$$\mathcal{A}_c(N) = N\log_2 N, \tag{5.23a}$$

$$\mathcal{M}_c(N) = 0.5N(\log_2 N - 2) + 1. \tag{5.23b}$$

The corresponding count of real operations is

$$\mathcal{A}_r(N) = 3N\log_2 N - 2N + 2, \tag{5.24a}$$

$$\mathcal{M}_r(N) = 2N(\log_2 N - 2) + 4. \tag{5.24b}$$

The operation count is the same for both types of radix-2 FFT (time decimated and frequency decimated).

As an example of the computational load, consider $N = 1024$. Direct computation of the DFT would require about 1024^2, or a million complex operations of each kind. By comparison, a radix-2 FFT algorithm requires about 10×1024 complex additions and 4×1024 complex multiplications. Therefore, in this case there is a 100-fold reduction in the number of additions, and 250-fold reduction in the number of multiplications.

5.3.1 The 2-Point DFT Butterfly

The basic building block of radix-2 FFT algorithms is the 2-point DFT. Figure 5.4 illustrates a 2-point DFT. In this figure and in the subsequent ones, we use the following conventions:

1. A line with an arrow indicates signal flow.
2. A circle around a + sign, with two or more lines leading to it, indicates addition.
3. A constant number above a line indicates multiplication of the signal flowing in that line by the constant number.

Because of its visual shape, the 2-point DFT operation is called a *butterfly*.

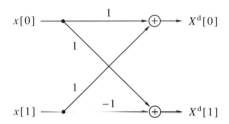

Figure 5.4 The butterfly of a 2-point DFT.

5.3.2 Time-Decimated Radix-2 FFT

The time-decimated radix-2 FFT is obtained by applying the CT decomposition recursively, each time taking $P = 2$ and $Q = N/2$. Figure 5.5 illustrates the flow of computation in the case of $N = 2^3 = 8$. As we see, the main feature of this algorithm is that all internal DFT operations are of length 2; that is, they are replicas of the butterfly shown in Figure 5.4. There are $\log_2 N$ main sections (three in this case), cascaded from left to right. They correspond to the $\log_2 N$ stages in the recursion. In each section there are $0.5N$ butterflies, so the total number of butterflies is $0.5N \log_2 N$. The figure is best read from right to left, although the flow of computation is from left to right. The output variables are ordered in their natural sequential order for convenience, from $X^d[0]$ to $X^d[N-1]$. They are obtained by combining the outputs of two DFTs of length $0.5N$ each (at the next-to-last section), after multiplying them by the N twiddle factors. Note that, in the case $N = 8$, five of the twiddle factors have value 1. In general there will be $0.5N + 1$ such factors. Next, the output values of each of the two DFTs of length $0.5N$ are obtained by combining the outputs of two DFTs of length $0.25N$ each (at the second-to-last section), after multiplying them by the $0.5N$ twiddle factors. Now three of four twiddle factors have value 1. Finally, the first section shows the $0.5N$ DFTs of length 2 that operate on the input sequence. Since this is the first section, all twiddle factors are identically 1.

   As we see from Figure 5.5, the input variables are supplied in a permuted order, rather than in their natural order. The permutation is formed by recursively moving the even indices to the beginning and the odd indices to the end. This is precisely the time-decimation operation. A little reflection reveals that the permutation can be mathematically described as follows:

1. Express each index n in the range $[0, N-1]$ as an r-digit binary number, say

$$(n)_2 = n_{r-1} n_{r-2} \cdots n_1 n_0.$$

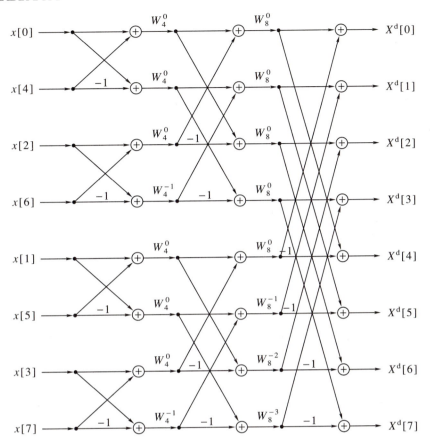

Figure 5.5 Time-decimated radix-2 FFT, shown for $N = 8$. All branches not marked otherwise have gains 1.

2. Reverse the order of bits to form

$$(\tilde{n})_2 = n_0 n_1 \cdots n_{r-2} n_{r-1}.$$

3. Replace $x[n]$ in the input sequence by $x[\tilde{n}]$.

This operation is called *bit reversal*. In the time-decimated radix-2 FFT, bit reversal is performed on the input sequence before starting the butterfly and twiddle factor operations. The output sequence will then be obtained in its natural order.

Example 5.3 For $N = 8$, the binary representations of the numbers 0 through 7 are 000, 001, 010, 011, 100, 101, 110, 111. After bit reversal, they change to 000, 100, 010, 110, 001, 101, 011, 111. Converting back to decimal, we get the sequence 0, 4, 2, 6, 1, 5, 3, 7. This agrees with the ordering of the input sequence in Figure 5.5. □

Another interesting observation can be made from Figure 5.5: Once the output variables of each section have been computed, there is no more need for the input variables. Therefore, we can perform the entire algorithm with N storage words, plus a few temporary variables, plus storage for the twiddle factors. This use of storage is called *in-place computation*, and it is effective if N is large. When in-place computation is used, the output sequence overwrites the input sequence. Therefore, the input sequence must be saved if needed for further use.

5.3.3 Frequency-Decimated Radix-2 FFT

The frequency-decimated radix-2 FFT is obtained by applying the CT decomposition recursively, each time taking $P = N/2$ and $Q = 2$. The frequency-decimated algorithm is the dual of the time-decimated algorithm, and the two have similar properties. Figure 5.6 illustrates the flow of computation in the case of $N = 2^3 = 8$. As we see, the input sequence is now given in its natural order, whereas the output appears in a bit-reversed order. The appearance of the sections is reversed, and the twiddle factors change positions. The total numbers of butterflies, twiddle factors, and operations are identical to those of the time-decimated algorithm. Practically, there is no reason to prefer one over the other, and the choice between the two is usually made arbitrarily.

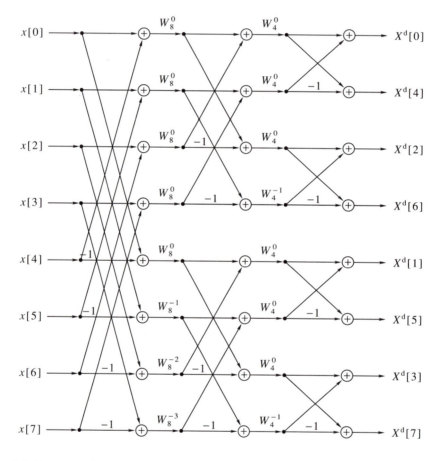

Figure 5.6 Frequency-decimated radix-2 FFT, shown for $N = 8$. All branches not marked otherwise have gains 1.

5.3.4 Signal Scaling in Radix-2 FFT*

When implementing FFT algorithms in fixed point, it is necessary to scale the input signal, the output signal, and the intermediate signals at the various sections so as to make the best use of the computer's accuracy. If the algorithm is implemented in floating point (such as in MATLAB), scaling is almost never a problem. However, many real-time signal processing systems operate in fixed point, because fixed-point hardware is usually faster and cheaper than floating-point hardware. It is therefore expedient to scale the signals in such a way that they will be represented by as many significant bits as possible, while simultaneously avoiding overflows.

Overflows are potentially harmful to the operation of a numerical algorithm if not detected and handled properly. For example, consider an implementation that can handle only numbers in the range $(-1, 1)$. If the numbers 0.85 and 0.2 are to be added, the theoretical result is 1.05. This result, however, causes overflow. In most fixed-point implementations, such an overflow will cause the result to appear as -0.95 (or nearly so). The difference between the true result and the apparent one is equal to the full dynamic range of the signal! Scaling, possibly accompanied by overflow detection and handling, is aimed at solving this kind of problems.

Consider the scaling problem in radix-2 FFT. Assume that the computer represents numbers as fractions in the range $[-1, 1]$, and that the input signal is scaled such that $|x[n]| \leq 1$ for all n. If the input signal is real, it means that $-1 \leq x[n] \leq 1$. If the input signal is complex, it means that $(\Re\{x[n]\})^2 + (\Im\{x[n]\})^2 \leq 1$. Now, since $|W_N| = 1$, we have

$$|X^d[k]| = \left| \sum_{n=0}^{N-1} x[n] W_N^{-kn} \right| \leq \sum_{n=0}^{N-1} |x[n]| \cdot |W_N^{-kn}| = \sum_{n=0}^{N-1} |x[n]| \leq N. \qquad (5.25)$$

This shows that fixed-point DFT is not safe from overflow: The output signal can be as large as N times the largest magnitude of the input signal. Indeed, this happens if $x[n] = \exp(j2\pi n m_0/N)$ for some integer m_0 (Problem 5.16).

Looking more closely into the inner structure of the radix-2 FFT, we observe the following:

1. Multiplication of a signal by a twiddle factor preserves the absolute value of the signal, since the absolute value of any twiddle factor is 1. Therefore, this operation does not cause overflow.

2. A butterfly operation can lead to overflow, since it involves one addition and one subtraction. However, if the absolute values of the inputs to the butterfly are less than 1, the absolute values of the outputs are less than 2. Therefore, we can eliminate overflow at a butterfly operation if we force both inputs to have absolute values less than 0.5.

There are two common approaches to elimination of overflow in butterfly operations:

1. The entire array is multiplied by 0.5 before each stage of the FFT is executed. Recall that the algorithm uses in-place computation. Therefore, the input vector is multiplied by 0.5 before performing the first stage, then the resulting array is multiplied by 0.5 before performing the second stage, and so on. There are $\log_2 N$ stages, so the output array is scaled by $0.5^{\log_2 N}$, that is, by N^{-1}, with respect to the input. Clearly, $|X^d[k]| \leq 1$ at the output. Multiplication by 0.5 can be conveniently performed by shifting the fixed-point number one bit to the right (with sign extension). We remark that this scaling is superior to one-time normalization of the input by N^{-1}, because it makes better use of the computer word length at the intermediate stages (i.e., bits are lost successively, one per stage). Even so, the loss of significant bits in this scheme can be excessive. For example, if $N = 1024$, we lose 10 bits of accuracy along the way.

2. The array is not multiplied by 0.5 before the stage. However, overflow detection is inserted at each butterfly (most computers can test for overflow in the hardware). If overflow is detected, the entire array at the present stage is multiplied by 0.5

and the offending overflow is fixed as well. A counter keeps track of the number of times this happens. If overflow is detected at m stages (necessarily $m \leq \log_2 N$), it means that the output is scaled by 0.5^m with respect to the input. This scheme is called *block floating point*. It provides variable scaling, depending on the input signal, and is nearly optimal for any given signal. For example, the FFT of the signal $\{x[n] = 1, \ 0 \leq n \leq N - 1\}$ will be scaled by N^{-1}, as in the first scheme; however, the FFT of $x[n] = \delta[n]$ will not be scaled at all.

Example 5.4 We illustrate block floating-point scaling for time-decimated radix-2 FFT with $N = 8$. The input signal in this example is

$$x[n] = 0.65^{n+1}, \quad 0 \leq n \leq 7.$$

We assume that the computer is accurate to ± 0.0001 and uses truncation arithmetic (for comparison, a 16-bit fixed-point computer has accuracy about three times better). Table 5.1 shows the signal values during the computation. The first column gives the input signal, in bit-reversed order. The second, third, and fourth columns give the outputs of the three stages (see Figure 5.5). The fourth column is the output signal $X^d[k]$. At the first output of the second stage there is overflow, so the entire vector is multiplied by 0.5. Therefore, the output signal is scaled by 0.5 in this case.   □

Input	1st stage output	2nd stage output[†]	3rd stage output
0.6500	0.7660	0.5448	0.8989
0.1160	0.5340	$0.2670 - j0.1128$	$0.3378 - j0.2873$
0.2746	0.3236	0.2212	$0.2212 - j0.1438$
0.0490	0.2256	$0.2670 + j0.1128$	$0.1962 - j0.0617$
0.4225	0.4979	0.3541	0.1907
0.0754	0.3471	$0.1735 - j0.0733$	$0.1962 + j0.0617$
0.1785	0.2103	0.1438	$0.2212 + j0.1438$
0.0318	0.1467	$0.1735 + j0.0733$	$0.3378 + j0.2873$

Table 5.1 The signals in Example 5.4; [†]in this column there is scaling by 0.5.

5.4 Radix-4 Algorithms*

As we saw in the preceding section, the computational efficiency of radix-2 algorithms is largely a result of the absence of multiplications in the 2×2 butterflies. In this section we investigate another radix whose corresponding butterflies can be implemented without multiplications: radix-4. The resulting algorithms will turn out to be more efficient than radix-2 algorithms.

Consider the 4-point DFT operation. By our discussion of radix-2 FFT, it can be computed with eight complex additions and one complex multiplication. However, the twiddle factor appearing in the multiplication is W_4^{-1}, which is equal to $-j$. Now, multiplication of the complex number $a + jb$ by $-j$ gives $b - ja$, which amounts to replacing the real part by the imaginary part, and replacing the imaginary part by the negative of the real part. Therefore, no numerical multiplication is needed. To reiterate: A 4-point DFT can be performed with eight complex additions and no multiplications. At the point where multiplication by $-j$ is called for, we replace it by the exchange operation $a + jb \longrightarrow b - ja$. When writing computer code in a low-level language (e.g., in C or Assembler), the exchange operation can be hard-coded, so it does not require extra machine time. Figure 5.7 shows the basic 4-point DFT building block.

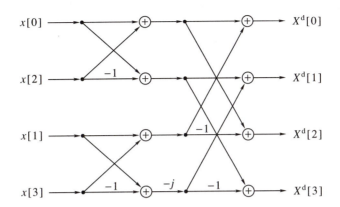

Figure 5.7 Multiplication-free 4-point DFT building block.

We can use the multiplication-free property of the 4-point DFT to construct radix-4 FFTs that are more efficient than radix-2 FFTs if N is an integer power of 4. The recursive formulas for the number of operations are

$$\mathcal{A}_c(N) = 0.25N\mathcal{A}_c(4) + 4\mathcal{A}_c(N/4) = 2N + 4\mathcal{A}_c(N/4), \tag{5.26a}$$

$$\mathcal{M}_c(N) = 0.25N\mathcal{M}_c(4) + 4\mathcal{M}_c(N/4) + 3(0.25N - 1)$$
$$= 4\mathcal{M}_c(N/4) + 0.75N - 3. \tag{5.26b}$$

The solutions of these difference equations can be verified to be (Problem 5.21)

$$\mathcal{A}_c(N) = 2N \log_4 N = N \log_2 N, \tag{5.27a}$$

$$\mathcal{M}_c(N) = 0.75N \log_4 N - N + 1 = 0.375N \log_2 N - N + 1. \tag{5.27b}$$

The corresponding counts of real operations are

$$\mathcal{A}_r(N) = 2.75N \log_2 N - 2N + 2, \tag{5.28a}$$

$$\mathcal{M}_r(N) = 1.5N \log_2 N - 4N + 4. \tag{5.28b}$$

The operation count is the same for both types of radix-4 FFT (time-decimated and frequency-decimated). Instead of bit reversal, we must perform reversal of the base-4 digits of the indices.

As we see, radix-4 FFT achieves about 25 percent reduction in the number of multiplications, as well as a small reduction in the number of additions. This reduction is achieved provided N is an integer power of 4 (i.e., an even power of 2), and assuming that the computer program handles the multiplications by $-j$ as we have described. If N is an odd power of 2, we have the option of doing radix-2 at one section and radix-4 at the remaining sections.

5.5 DFTs of Real Sequences

The FFT algorithms that we have described operate on complex sequences. If the input sequence is real, no computational advantage is gained, since after the first section all variables become complex. On the other hand, we know that direct computation of the DFT permits 50 percent reduction in multiplications if the input sequence is real, since the product of a real number and a complex one requires only two real multiplications. A question arises as to whether there is a way to achieve a similar reduction with FFT. The answer is a conditional yes: We can compute the FFTs of *two* real sequences in a number of operations only slightly higher than that needed for a single complex FFT. Frequently we must compute the FFTs of many real sequences successively. We can then combine them in pairs, perform a single complex FFT for each pair, thus reducing the total count of operations by almost 50 percent.

The basic idea is to build a single complex sequence using the two real sequences as its real and imaginary parts. If the input sequences are $x[n]$ and $y[n]$, we form

$$z[n] = x[n] + jy[n] \tag{5.29}$$

and compute $Z^d[k]$, the FFT of $z[n]$. The next step is to extract $X^d[k]$ and $Y^d[k]$ from $Z^d[k]$. We have

$$x[n] = 0.5(z[n] + \bar{z}[n]), \tag{5.30a}$$

$$y[n] = j0.5(\bar{z}[n] - z[n]). \tag{5.30b}$$

Therefore, by the conjugation property of the DFT (4.38),

$$X^d[k] = 0.5(Z^d[k] + \bar{Z}^d[N - k]), \tag{5.31a}$$

$$Y^d[k] = j0.5(\bar{Z}^d[N - k] - Z^d[k]). \tag{5.31b}$$

As we see, the operation overhead is $2N$ additions and $2N$ multiplications. However, the multiplications are by 0.5, so they can be performed with right shifts.

Problem 5.18 explores the computation of FFTs of real sequences further.

5.6   Linear Convolution by FFT

In Section 4.7 we showed how to perform linear convolution of two finite-duration sequences using circular convolution. The method involved zero padding the two sequences to a length equal to the sum of the individual lengths minus 1. Now, equipped with the fast Fourier transform, we further develop this idea.

Let the two sequences be $\{x[n],\ 0 \le n \le N_1 - 1\}$ and $\{y[n],\ 0 \le n \le N_2 - 1\}$. Define N to be the smallest power of 2 not smaller than $N_1 + N_2 - 1$ and let $x_a[n], y_a[n]$ be the corresponding zero-padded sequences. Then, by Problem 4.31, the convolution $\{x * y\}[n]$ can be obtained by performing $\{x_a \circledast y_a\}[n]$ and retaining the first $N_1 + N_2 - 1$ elements of the result. Furthermore, using the convolution property of the DFT, we can obtain the latter by the operation

$$z_a[n] = \text{IDFT}\{X_a^d[k]Y_a^d[k]\}. \tag{5.32}$$

The total number of operations, assuming use of a radix-2 FFT, is three times the number of operations for a single N-point FFT: Two for the direct DFTs and one for the inverse DFT. This amounts to about $6N \log_2 N$ real multiplications and $9N \log_2 N$ real additions (less if the input sequences are real and we transform them together as described in Section 5.5). We also need N complex multiplications, or $4N$ real multiplications, to compute the product of the FFTs. By comparison, direct computation of the convolution requires about $N_1 N_2$ multiplications and a similar number of additions (Problem 5.14). Therefore, judging by the number of multiplications, it is preferable to perform the convolution by FFT whenever

$$N_1 N_2 > (N_1 + N_2 - 1)[6 \log_2 (N_1 + N_2 - 1) + 4]. \tag{5.33}$$

In digital signal processing it commonly happens that one of the two sequences, say $y[n]$, is known *a priori* and has a fixed length N_2, whereas the other sequence, $x[n]$, is known only in real time, and its length is not fixed and is potentially much larger than N_2. In Chapter 9 we shall learn more about such situations. For now we concern ourselves with the problem of computing the linear convolution $x * y$ under these conditions. Using zero padding is possible, but may not be advisable, for two reasons: (1) The sequence $y[n]$ will have to be padded by many zeros, resulting in many unnecessary computations, and (2) the DFT will have to be performed on very long sequences, which may be inconvenient or impossible. A better approach is to split the long sequence $x[n]$ to segments, each of length N_1 which is of the same order as N_2, convolve $y[n]$ with each of the segments separately, and properly add up the partial results to form the desired convolution $x * y$. This method is known as *overlap-add* (OLA) convolution; we now describe its details.

Let us write the sequence $x[n]$ as

$$x[n] = \sum_i x_i[n], \tag{5.34}$$

where

$$x_i[n] = \begin{cases} x[n], & N_1 i \leq n \leq N_1(i+1) - 1, \\ 0, & \text{otherwise.} \end{cases} \tag{5.35}$$

We have left the range of the index i unspecified on purpose, to emphasize that the length of $x[n]$ need not be known in advance. If this length is not an integer multiple of N_1, we pad $x[n]$ with zeros to make it so. By linearity of the convolution operator we have

$$z[n] = \{x * y\}[n] = \sum_i \{x_i * y\}[n]. \tag{5.36}$$

The convolution $\{x_i * y\}$ has length $N_1 + N_2 - 1$, and it is nonzero for the range of time points

$$N_1 i \leq n \leq N_1 i + N_1 + N_2 - 2.$$

We can write $\{x_i * y\}$ as a sum of two sequences, say

$$z_i[n] = \{x_i * y\}[n] = u_i[n] + v_i[n], \tag{5.37}$$

where $u_i[n]$ is nonzero in the range

$$N_1 i \leq n \leq N_1(i+1) - 1,$$

and $v_i[n]$ is nonzero in the range

$$N_1(i+1) \leq n \leq N_1(i+1) + N_2 - 2.$$

The range of definition of $u_i[n]$ coincides with that of $x_i[n]$. On the other hand, the range of definition of $v_i[n]$ coincides with the initial part of the range of $x_{i+1}[n]$. Therefore, the proper sequence of operations, implied by (5.36) and (5.37), is

1. Zero-pad $x_i[n]$ and $y[n]$ to a length $N_1 + N_2 - 1$.
2. Perform the circular convolution $\{x_i \otimes y\}$, which is equal to the linear convolution $\{x_i * y\}$. The circular convolution is performed by FFT, as explained earlier.
3. Use $u_i[n]$ as a partial result for the ith stage. Add to its initial part the sequence $v_{i-1}[n]$ from the previous stage. Save $v_i[n]$ for the $(i+1)$st stage.

Figure 5.8 illustrates the various sequences, their proper timings, and the way they are combined. The signal $y[n]$ in this example has length $N_2 = 4$ and its values are

$\{0.1, 0.5, 0.25, 0.15\}$. The length chosen for the sequences $x_i[n]$ is $N_1 = 5$. Accordingly, the lengths of $z_i[n]$, $u_i[n]$ and $v_i[n]$ are 8, 5, and 3, respectively. The reason for the name *overlap-add* should be now clear: The tail $v_i[n]$ of the convolution $\{x_i * y\}[n]$ overlaps with the head of $\{x_{i+1} * y\}[n]$, and must be added to it.

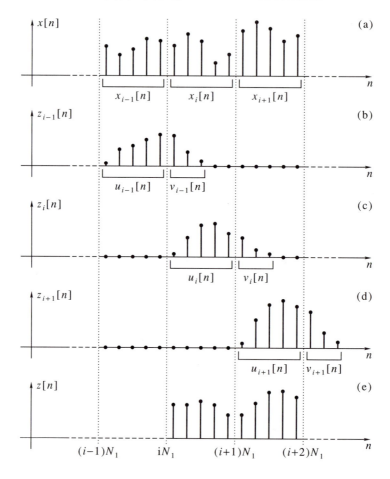

Figure 5.8 Illustration of overlap-add convolution: (a) the input signal $x[n]$; (b), (c), (d) the three output segments $z_{i-1}[n]$ through $z_{i+1}[n]$; (e) a segment of the output signal $z[n]$ corresponding to $x_i[n]$ and $x_{i+1}[n]$.

The procedure ola in Program 5.4 is a MATLAB implementation of the overlap-add method. The length of the sequence x is assumed to be much greater than the length of y. The FFT of y is computed first and used throughout. Then the program loops over the segments, and finally handles the tail of the sequence x, which may be shorter than the other segments.

Linear convolution of a long sequence $x[n]$ by a fixed-length sequence $y[n]$ can also be performed by a method called *overlap-save*, which is dual to overlap-add. The details of the overlap-save method are left as an exercise (Problem 5.23).

We now discuss the optimal choice of the FFT length N in the overlap-add method. We assume that the sequences in question are real. As we saw in Section 5.5, this enables the simultaneous computation of the FFTs of two consecutive segments (Program 5.4 does not work this way, however). Since we have the freedom to choose N_1, let us choose it such that $N_1 + N_2 - 1$ is a power of 2. The FFT of the zero-padded $y[n]$ needs to be computed only once, so we ignore the operations it requires. For each pair of segments of N_1 output points we need $4(N_1 + N_2 - 1)\log_2(N_1 + N_2 - 1)$ multiplications for the two FFTs (a direct one and an inverse one), as well as $4(N_1 + N_2 - 1)$ multiplications for the product of the FFTs. It is convenient, in this application, to count the number of operations per sample of the signal $x[n]$. Using the overlap-add method for convolution is more efficient than direct convolution when the number of operations per sample is smaller. We thus get the criterion

$$2\left(1 + \frac{N_2 - 1}{N_1}\right)[1 + \log_2(N_1 + N_2 - 1)] < N_2. \tag{5.38}$$

Table 5.2 shows the optimal choice of the parameter $N_1 + N_2 - 1$ for various values of N_2. The optimum is defined as the value for which the left side of (5.38) is minimal, under the constraint that this parameter is an integer power of 2. For $N_2 < 19$ the inequality (5.38) does not hold, meaning that direct convolution is more efficient than overlap-add.

Range of N_2	Optimal $N_1 + N_2 - 1$
19–26	128
27–47	256
48–86	512
87–158	1024

Table 5.2 Optimal choice of the segment length as a function of the length of $y[n]$ for OLA convolution of real sequences by radix-2 FFT.

5.7 DFT at a Selected Frequency Range

5.7.1 The Chirp Fourier Transform

Since the FFT is an implementation of the DFT, it provides a frequency resolution of $2\pi/N$, where N is the length of the input sequence. If this resolution is not sufficient in a given application, we have the option of zero padding the input sequence, as described in Section 4.4. However, this may be unduly expensive in operation count if the required resolution is much higher than $2\pi/N$. In practice, one is seldom interested in high resolution over the entire frequency band. More often one is interested only in a relatively small part of the band. For example, suppose we wish to determine the frequency θ_0 of a sinusoid to a good accuracy from N data points. Performing FFT on the data will enable determination of θ_0 to an accuracy $\pm\pi/N$. If we could compute $X^f(\theta)$ at high resolution in an interval of width $2\pi/N$ around the maximum point of the FFT, we might be able to improve the accuracy of θ_0. A special technique, called *chirp Fourier transform*, accomplishes this.[2]

Suppose we are given a sequence $\{x[n],\ 0 \le n \le N - 1\}$ and we wish to compute

$$X^f(\theta[k]) = \sum_{n=0}^{N-1} x[n]e^{-j\theta[k]n}, \tag{5.39}$$

where $\Delta\theta$ is the desired frequency resolution and

$$\theta[k] = \theta_0 + k\Delta\theta, \quad 0 \le k \le K - 1. \tag{5.40}$$

Figure 5.9 illustrates the desired operation, showing $X^f(\theta)$ and the desired range of computation, as well as a magnification of the desired range and the desired transform values $\{X^f(\theta[k]), 0 \le k \le K - 1\}$.

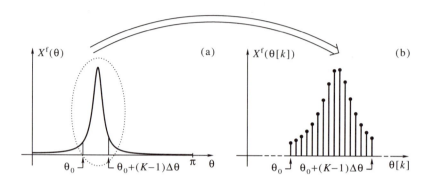

Figure 5.9 Chirp Fourier transform: (a) Full range of $X^f(\theta)$; (b) selected range of $X^f(\theta)$ and the sampled values $X^f(\theta[k])$.

Direct computation of (5.39) would require NK complex multiply–add operations. This can be high if K is large: for example, if it is of the same order as N. The chirp Fourier transform, which we now describe, reduces the number of operations to a small multiple of $(N + K) \log_2(N + K)$.

Define $W = e^{j\Delta\theta}$. Then we can rewrite (5.39) as

$$X^f(\theta[k]) = \sum_{n=0}^{N-1} x[n] e^{-j\theta_0 n} W^{-nk}. \tag{5.41}$$

Let us now use the identity

$$nk = 0.5[n^2 + k^2 - (k - n)^2]$$

in (5.41) to get

$$X^f(\theta[k]) = W^{-0.5k^2} \sum_{n=0}^{N-1} x[n] e^{-j\theta_0 n} W^{-0.5n^2} W^{0.5(k-n)^2}. \tag{5.42}$$

Define the two sequences

$$g[n] = x[n] e^{-j\theta_0 n} W^{-0.5n^2}, \quad h[n] = W^{0.5n^2}. \tag{5.43}$$

Then we have from (5.42)

$$X^f(\theta[k]) = W^{-0.5k^2} \{g * h\}[k]. \tag{5.44}$$

The sequence $g[n]$ is finite, having the same length as $x[n]$, whereas $h[n]$ is infinite. However, since we are interested in $X^f(\theta[k])$ only on a finite set of ks, we need only the finite segment $\{h[n], -(N - 1) \le n \le (K - 1)\}$. We can therefore summarize the chirp Fourier transform algorithm as follows:

1. Find a number L that is an integer power of 2 and is not smaller than $N + K - 1$.

2. Form the sequence $g[n]$ as defined in (5.43) and zero-pad it to a length L.

3. Form the sequence $\{h[n], 0 \le n \le K - 1\}$, as defined in (5.43); concatenate to it the sequence $\{h[n], -(L - K) \le n \le -1\}$; note that the sequence thus formed is a circular shift of $h[n]$.

4. Perform the operation

$$\text{IFFT}\{\text{FFT}\{g[n]\} \cdot \text{FFT}\{h[n]\}\}.$$

Multiply the first K terms of the resulting sequence by $W^{-0.5k^2}$ to get the sequence $\{X^f(\theta[k]), 0 \le k \le K - 1\}$.

Figure 5.10 shows a block diagram of the chirp Fourier transform algorithm, and the procedure `chirpf` in Program 5.5 gives its MATLAB implementation.

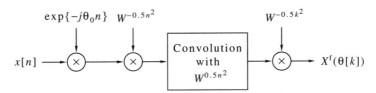

Figure 5.10 Block diagram of the chirp Fourier transform algorithm.

The computational complexity of the chirp Fourier transform algorithm is proportional to $L \log_2 L$, which is close to $(N + K) \log_2 (N + K)$, plus a small multiple of N and another small multiple of K for the auxiliary sequences.

5.7.2 Zoom FFT*

The so-called *zoom FFT* is an alternative to chirp Fourier transform for computing the DFT of a sequence at a selected frequency range. Suppose we are interested in computing the DFT at the points

$$k_0 \le k \le k_0 + K - 1$$

for some k_0 and K. This problem is similar to the one posed in (5.39), (5.40) and depicted in Figure 5.9, with $\theta_0 = 2\pi k_0/N$ and $\Delta\theta = 2\pi/N$. Suppose further that the sequence $x[n]$ has length $N = KL$ for some integer L. The sequence may be zero padded if we want the frequency resolution $\Delta\theta = 2\pi/N$ to be better than the resolution corresponding to the given signal length.

Let the index n be expressed as

$$n = l + mL, \quad 0 \le l \le L - 1, 0 \le m \le K - 1.$$

Then

$$X^d[k] = \sum_{n=0}^{N-1} x[n] W_N^{-kn} = \sum_{m=0}^{K-1} \sum_{l=0}^{L-1} x[l + mL] W_N^{-k(l+mL)}$$

$$= \sum_{l=0}^{L-1} \left[\sum_{m=0}^{K-1} x[l + mL] W_K^{-km} \right] W_N^{-kl}. \tag{5.45}$$

Let

$$X_l^d[r] = \sum_{m=0}^{K-1} x[l + mL] W_K^{-rm} \tag{5.46}$$

be the DFT of the decimated sequence $\{x[l + mL], 0 \le m \le K - 1\}$. Then, since $X_l^d[r]$ is periodic in r with period K, we have

$$X_l^d[k] = X_l^d[k \bmod K].$$

We therefore conclude from (5.45), (5.46) that $X^d[k]$ can be computed in the desired frequency range using the following procedure:

1. Compute the K-point DFTs of the L decimated sequences $x[l + mL]$, as given in (5.46).

2. For each k in the desired range let

$$X^d[k] = \sum_{l=0}^{L-1} X_l^d[k \bmod K] W_N^{-kl}. \tag{5.47}$$

The first step of this procedure requires about $0.5LK(\log_2 K - 2)$ complex multiplications (assuming that K is an integer power of 2), and the second requires LK complex multiplications (we cannot do this step by FFT). Therefore, the total number of complex multiplications is about $0.5N \log_2 K$. In addition, we need all the numbers $\{W_N^n,\ 0 \le n \le N - 1\}$. This is more efficient than N-point FFT if $L > 4$, and much more efficient if $L \gg K$. Also, it is usually convenient, if N is large, to perform several short FFTs, rather than one large FFT.

5.8   Summary and Complements

5.8.1   Summary

In this chapter we introduced fast Fourier transform algorithms, in particular the Cooley–Tukey algorithms. The CT decomposition of the DFT enables reducing the number of arithmetic operations whenever the length of the DFT is a composite number. The CT decomposition can be performed recursively, until only prime-length DFTs are left to be performed, and these are computed directly.

Of particular importance are radix-2 FFT algorithms, for the case in which the DFT length is an integer power of 2. The two main algorithms are the time decimated and the frequency decimated. The computational complexity of these algorithms is proportional to $N \log_2 N$. The radix-4 FFT algorithm is more efficient than radix-2 algorithms in multiplication count, but not in addition count.

An important use of FFT is for performing linear convolutions. We presented the overlap-add method for performing a linear convolution between a fixed-length sequence and a sequence of an indefinite length. Finally, we presented variations of the FFT that enable computation of the DFT at a selected frequency range: the chirp Fourier transform and the zoom FFT.

We have not treated FFT algorithms other than the ones based on the CT decomposition. Two alternative approaches, important in certain applications, are as follows:

1. Prime factor FFT [Good, 1958; Thomas, 1963; Burrus, 1988]. These algorithms are useful when the DFT length is a product of different primes, each appearing once in the factorization. It is possible, in this case, to perform the DFT without twiddle factor multiplications.

2. Winograd Fourier transform algorithms, based on performing DFTs of small prime numbers by convolutions [Winograd, 1978]. Winograd algorithms require a number of multiplications proportional to N (as opposed to $N \log_2 N$). However, they require considerably more additions than Cooley–Tukey FFT.

5.8.2   Complements

1. [p. 133] The term *fast algorithms* usually refers to computational schemes that implement certain calculations in a relatively small number of operations. An example from computer science: Direct sorting of N numbers requires about N^2

comparisons; sorting by the *quicksort* algorithm requires about $N \log_2 N$ comparisons on the average, so it is a fast algorithm.

2. [p. 151] A signal of the form

$$x(t) = e^{j0.5\alpha t^2}, \quad \alpha > 0,$$

is called a *chirp*. It is a complex signal whose magnitude is constant and whose frequency increases linearly in time. A chirp signal has the following property: If its duration is the interval $[0, T_0]$, its bandwidth is about αT_0. Therefore, bandwidth increases simultaneously with the duration. In radar and sonar applications, the signal duration determines the resolution to which the velocity of a target can be determined and the signal bandwidth determines the resolution to which the range can be determined. A chirp signal allows simultaneous control of the velocity resolution (through the parameter T_0) and the range resolution (through the parameter α). This property is the reason for its importance in radar and sonar applications. For further reading on this subject, see Rihaczek [1985].

We finally remark that the chirp Fourier transform has nothing to do with this particular application; it got its name because it uses the discrete-time chirp signal $W^{0.5n^2}$.

5.9   MATLAB Programs

Program 5.1 FFT and IFFT, by Cooley–Tukey frequency decimation recursion.

```
function X = edufft(x,swtch);
% Synopsis: X = edufft(x,swtch).
% An "educational" FFT algorithm, based on Cooley-Tukey frequency
% decimation procedure. Runs slow due to MATLAB recursion overhead.
% Input parameters:
% x: the input vector
% swtch: 0 for FFT, 1 for IFFT.
% Output:
% X: the output vector.

N = length(x); x = reshape(x,1,N);
if (swtch), W = exp((j*2*pi/N)*(0:N-1));
else, W = exp((-j*2*pi/N)*(0:N-1)); end
X = ctrecur(x,W); if (swtch), X = (1/N)*X; end
```

Program 5.2 Cooley–Tukey frequency decimation recursion.

```
function X = ctrecur(x,W);
% Synopsis: X = ctrecur(x,W).
% Cooley-Tukey frequency-decimation recursion.
% Input parameters:
% x: the input vector, assumed to be a row
% W: vector of powers of W, assumed to be a row.
% Output:
% y: the output row vector.

N = length(x); Q = N;
for i = 2:floor(sqrt(N)),
   if (rem(N,i) == 0), Q = i; break, end, end

if (Q == N), X = primedft(x,W);
else,
   P = N/Q; tmp = reshape(x,P,Q);
   for p = 0:P-1,
   tmp(p+1,:) = primedft(tmp(p+1,:),W(1:P:N));
   if (p > 0), tmp(p+1,2:Q) = tmp(p+1,2:Q).*W(rem(p*(1:Q-1),N)+1); end
   end
   for q = 1:Q, tmp(:,q) = (ctrecur(tmp(:,q).',W(1:Q:N))).'; end
   X = reshape(tmp.',1,N);
end
```

Program 5.3 Brute-force DFT or conjugate DFT for prime length.

```
function y = primedft(x,W);
% Synopsis: y =primedft(x,W).
% Brute-force DFT or conjugate DFT in case of prime length.
% Input parameters:
% x: the input vector, assumed to be a row
% W: vector of powers of W, assumed to be a row.
% Output:
% y: the output row vector.

N = length(x); n = 1:N-1; y = zeros(1,N); y(1) = sum(x);
for k = 1:N-1,
   y(k+1) = x(1) + sum(x(2:N).*W(rem(k*n,N)+1)); end
```

Program 5.4 Overlap-add convolution.

```
function z = ola(x,y,N);
% Synopsis: z = ola(x,y,N).
% Computes the convolution z = x*y by the overlap-add method.
% Input parameters:
% x: the long input sequence
% y: the short  input sequence
% N: length of the FFT.
% Output parameters:
% z: the output sequence.

N2 = length(y); lx = length(x); y = reshape(y,1,N2);
x = reshape(x,1,lx); z = zeros(1,N2+lx-1);
lz = length(z); rflag = 0; N1 = N-N2+1;
if (any(imag(x)) | any(imag(y))), rflag = 1; end
nframe = floor(lx/N1); ltail = lx - N1*nframe;
Y = fft([y, zeros(1,N-N2)]);
for k = 1:nframe,
   nst = (k-1)*N1;
   temp = ifft(fft([x(1,nst+1:nst+N1), zeros(1,N-N1)]).*Y);
   if (rflag), temp = real(temp); end
   z(1,nst+1:nst+N) = z(1,nst+1:nst+N) + temp;
end
if (ltail > 0),
   nst = nframe*N1+1; temp = [x(1,nst:lx), zeros(1,N-ltail)];
   temp = ifft(fft(temp).*Y);
   if (rflag), temp = real(temp); end
   z(1,nst:lz) = z(1,nst:lz) + temp(1,1:N2+ltail-1);
end
```

Program 5.5 The chirp Fourier transform algorithm.

```
function X = chirpf(x,theta0,dtheta,K);
% Synopsis: X = chirpf(x,theta0,dtheta,K).
% Computes the chirp Fourier transform on a frequency
% interval.
% Input parameters:
% x   : the input vector
% theta0 : initial frequency (in radians)
% dtheta : frequency increment (in radians)
% K   : number of points on the frequency axis.
% Output:
% X: the chirp Fourier transform of x.

N = length(x); x = reshape(x,1,N); n = 0:N-1;
g = x.*exp(-j*(0.5*dtheta*n+theta0).*n);
L = 1; while (L < N+K-1), L = 2*L; end
g = [g, zeros(1,L-N)];
h = [exp(j*0.5*dtheta*(0:K-1).^2), ...
    exp(j*0.5*dtheta*(-L+K:-1).^2)];
X = ifft(fft(g).*fft(h));
X = X(1:K).*exp(-j*0.5*dtheta*(0:K-1).^2);
```

5.10 Problems

5.1 Explain why the relationships in (5.5) are true.

5.2 Write MATLAB statements that compute the integers p, q for given n, P, and Q, as defined in (5.2a).

5.3 In our count of operations for FFT algorithms, we assumed that a complex multiplication requires four real multiplications and two real additions. However, it is also possible to perform a complex multiplication with three real multiplications and three real additions. To see this, suppose $a = a_{\rm r} + ja_{\rm i}$ is an intermediate complex number appearing in the algorithm and $W = W_{\rm r} + jW_{\rm i}$ is a twiddle factor used to multiply a. Then

$$aW = (a_{\rm r}W_{\rm r} - a_{\rm i}W_{\rm i}) + j(a_{\rm r}W_{\rm i} + a_{\rm i}W_{\rm r})$$
$$= [(a_{\rm r} - a_{\rm i})W_{\rm i} + a_{\rm r}(W_{\rm r} - W_{\rm i})] + j[(a_{\rm r} - a_{\rm i})W_{\rm i} + a_{\rm i}(W_{\rm r} + W_{\rm i})].$$

The numbers $(W_{\rm r} - W_{\rm i})$ and $(W_{\rm r} + W_{\rm i})$ can be prepared off line, since the W are themselves prepared off line. Then we are left with three multiplications and three additions that must be performed on line, since they depend on a.
Compute the total number of operations in the radix-2 algorithm when this scheme is used.

5.4 Verify (5.17) by direct multiplication of the three matrices.

5.5 Suppose we wish to implement a 27-point DFT by radix-3 time-decimated FFT. Write down explicitly the proper ordering of the 27 input data.

5.6 Write a MATLAB program `bitrev` that performs bit reversal. The calling sequence of the program should be

$$n = \texttt{bitrev(r)}$$

where r is a positive integer. The output n is a vector of length 2^r whose components are the integers from 0 to $2^r - 1$ in a bit-reversed order. Hint: It is convenient to implement the program recursively. Verify that the program does not call itself more times than necessary.

5.7 Let $N = p^r$, where p is a prime. Derive recursive expressions for the numbers of complex multiplications and complex additions in a radix-p time-decimated FFT, similarly to the derivation of the case $p = 2$. Then solve the recursions and find explicit expressions.

5.8 Let the sequence $x[n]$ have length $N = 2^{2r+1}$. Consider the following three methods for computing $X^{\rm d}[k]$:

(a) By radix-2 FFT.

(b) Decomposition of N to $N = PQ$, $Q = 2$, $P = 4^r$, then using the CT decomposition such that the length-P DFTs are performed by radix-4 FFT.

(c) Zero padding the sequence to length 4^{r+1}, performing radix-4 FFT on the zero-padded sequence, then discarding the odd-index points of the transform.

Compute the number of complex multiplications and the number of complex additions for each of the methods and conclude which one is the most efficient.

5.9 Let the sequence $x[n]$ have length $N = 3 \times 4^r$. Suppose we are interested in the DFT of $x[n]$ at N or more equally spaced frequency points. Consider the following two methods of computation:

(a) Decomposition of N to $N = PQ$, $Q = 3$, $P = 4^r$, then using the CT decomposition such that the length-P DFTs are performed by radix-4 FFT.

(b) Zero padding the sequence to length 4^{r+1} and performing radix-4 FFT on the zero-padded sequence.

Compute the number of complex multiplications for each of the methods and conclude which one is the most efficient (when only multiplications are of concern).

5.10 We are given a sequence of length $N = 240$.

(a) Compute the number of complex multiplications and the number of complex additions needed in a full (i.e., recursive) CT decomposition.

(b) Compute the corresponding numbers of operations if the sequence is zero-padded to length 256 before the FFT.

5.11 It is required to compute the DFT of a sequence of length $N = 24$. Zero padding is permitted. Find the number of complex multiplications needed for each of the following solutions and state which is the most efficient:

(a) Cooley–Tukey recursive decomposition of 24 into primes.

(b) Zero padding to $M = 25$ and using radix-5 FFT.

(c) Zero padding to $M = 27$ and using radix-3 FFT.

(d) Zero padding to $M = 32$ and using radix-2 FFT.

(e) Zero padding to $M = 64$ and using radix-4 FFT.

5.12 Write down the twiddle factors $\{W_8^n, \ 0 \le n \le 7\}$ explicitly. Then show that multiplication of any complex number by W_8^n requires either no multiplications, or two multiplications at most.

5.13 Consider the sequences given in Problem 4.10. Let $\mathcal{M}_{c1}$ be the number of complex multiplications in zero padding $y[n]$ to the nearest power of 2, then performing radix-2 FFT on the zero-padded sequence. Let $\mathcal{M}_{c2}$ be the number of complex multiplications in computing the radix-2 FFTs of $x_1[n]$ and $x_2[n]$ first (zero padding as necessary), then computing the DFT of $y[n]$ using the result of Problem 4.10.

(a) If $N = 2^r$, $r \ge 1$, show that $\mathcal{M}_{c2} \le \mathcal{M}_{c1}$.

(b) If N is not an integer power of 2, does it remain true that $\mathcal{M}_{c2} \le \mathcal{M}_{c1}$? If so, prove it; if not, give a counterexample.

5.14 Count the number of multiplications and additions needed for linear convolution of sequences of lengths N_1, N_2, if computed directly. Avoid multiplications and additions of zeros.

5.15 We are given two real sequences $x[n]$, $y[n]$ of length 2^r each. Compute the number of real multiplications needed for the linear convolution of the two sequences, first by direct computation, and then by using radix-2 FFT in the most efficient manner.

5.16 Assume that the input signal to a DFT satisfies $|x[n]| \leq 1$ for all $0 \leq n \leq N - 1$. Show that the largest possible value of any of the $|X^d[k]|$ is N. Find all sequences $x[n]$ for which $|X^d[k]| = N$ for some k.

5.17 In Section 5.5 we saw how to compute the DFTs of two real sequences by one complex DFT, thus saving about 50 percent of the operations. Show how to do the reverse operation: Suppose we are given the DFTs $X^d[k]$, $Y^d[k]$ of two real sequences, and show how to compute the signals $x[n]$, $y[n]$ by one complex IDFT.

5.18 In Section 5.5 we showed how to compute the DFTs of two real sequences simultaneously with almost 50 percent saving in computations. Suppose, however, that we need the DFT of only *one* real sequence $x[n]$ of even length. Show that this can also be done at only slightly more than 50 percent of the number of operations in a complex FFT. Hint: Form the complex input sequence

$$y[n] = x[2n] + jx[2n + 1], \quad 0 \leq n \leq 0.5N - 1,$$

then use ideas from Section 5.5 and from the time-decimated radix-2 FFT. Count the number of operations needed to obtain the final result.

5.19* A discrete-time periodic signal $x[n]$ with period N is given at the input of a linear time invariant filter

$$H^z(z) = \sum_{l=0}^{L-1} h[l]z^{-l}.$$

Let $y[n]$ denote the output signal.

(a) Show that $y[n]$ is periodic and find its period.

(b) Suppose that $N = 2^r$, and that $2r < L < N$. Show how to compute in an efficient manner the samples of $y[n]$ over one period.

(c) Repeat part b when L is an integer multiple of N, say $L = MN$. Show that the number of operations is only slightly larger than that in part b.

(d) Repeat part b when L is larger than N, but not necessarily an integer multiple of N.

5.20* Obtain a factorization of F_8 to a product of five matrices from the time-decimated radix-2 FFT; use Figure 5.5.

5.21* Equations such as (5.22) can be solved by converting them to linear difference equations in the variable $r = \log_2 N$. With this change of variable, they become

$$\mathcal{A}_c[r] = 2\mathcal{A}_c[r - 1] + 2^r, \tag{5.48a}$$

$$\mathcal{M}_c[r] = 2\mathcal{M}_c[r - 1] + 2^{r-1} - 1. \tag{5.48b}$$

These difference equations can be solved using techniques studied in Chapter 7. If you know how to solve linear difference equations, continue to solve this problem now. If not, defer it until you have studied Chapter 7.

(a) Solve the difference equations and verify (5.23).

(b) Use the same technique to obtain and solve difference equations for radix-4 FFT, and verify the radix-4 operation count (5.27).

5.22* The purpose of this problem is to present the principle of operation of *split-radix FFT*. Here we derive the time-decimated version.

(a) Assume that N, the length of the input sequence, is an integer multiple of 4. Decimate the input sequence $x[n]$ as follows. First take the even-index elements $\{x[2m], 0 \le m \le 0.5N - 1\}$ and assume we have computed their $0.5N$-DFT. Denote the result by $\{F^d[k], 0 \le k \le 0.5N - 1\}$. Next take the elements $\{x[4m + 1], 0 \le m \le 0.25N - 1\}$ and assume we have computed their $0.25N$-DFT. Denote the result by $\{G^d[k], 0 \le k \le 0.25N - 1\}$. Finally take the elements $\{x[4m+3], 0 \le m \le 0.25N - 1\}$ and assume we have computed their $0.25N$-DFT. Denote the result by $\{H^d[k], 0 \le k \le 0.25N-1\}$. Show that $X^d[k], 0 \le k \le N-1$ can be computed in terms of $F^d[k], G^d[k], H^d[k]$.

(b) Draw the butterfly that describes the result of part a. Count the number of complex operations in the butterfly, and the number of operations for twiddle factor multiplications. Remember that multiplication by j does not require actual multiplication, only exchange of the real and imaginary parts.

(c) Take $N = 16$ and count the total number of butterflies of the type you obtained in part b that are needed for the computation. Note that eventually N ceases to be an integer multiple of 4 and then we must use the usual 2×2 butterflies. Count the number of those as well. Also, count the number of twiddle factor multiplications.

(d) Repeat part c for $N = 32$. Attempt to draw a general conclusion about the multiple of $N \log_2 N$ that appears in the count of complex multiplications.

Split-radix FFT is more efficient than radix-2 or radix-4 FFT. However, it is more complicated to program, hence it is less widely used.

5.23* The purpose of this problem is to develop the overlap-save method of linear convolution, in a manner similar to the overlap-add method. We denote the long sequence by $x[n]$, the fixed-length sequence by $y[n]$, and the length of $y[n]$ by N_2. We take N as a power of 2 greater than N_2, and denote $N_1 = N - N_2 + 1$. So far everything is the same as in the case of overlap-add.

(a) Show that, if $y[n]$ is zero-padded to length N and circular convolution is performed between the zero-padded sequence and a length-N segment of $x[n]$, then the last N_1 points of the result are identical to corresponding N_1 points of the linear convolution, whereas the first $N_2 - 1$ points are different. Specify the range of indices of the linear convolution thus obtained.

(b) Break the input sequence $x[n]$ into overlapping segments of length N each, where the overlap is $N_2 - 1$. Denote the ith segment by $\{x_i[n], 0 \le n \le N - 1\}$. Express $x_i[n]$ in terms of a corresponding point of $x[n]$.

(c) Show how to discard parts of the circular convolutions $\{x_i \circledast y\}[n]$ and patch the remaining parts together so as to obtain the desired linear convolution $\{x * y\}[n]$.

5.24* Program the zoom FFT (see Section 5.7.2) in MATLAB. The inputs to the program are the sequence $x[n]$, and initial index k_0, and the number of frequency points K. Hints: (1) Zero-pad the input sequence if its length is not a product of K; (2) use the MATLAB feature of performing FFTs of all columns of a matrix simultaneously.

5.25* Using the material in Section 4.9, write a MATLAB program fdct that computes the four DCTs using FFT. The calling sequence of the program should be

$$X = \text{fdct}(x, \text{typ})$$

where x is the input vector, typ is the DCT type (from 1 to 4), and X is the output vector.

Chapter 6

Practical Spectral Analysis

We illustrate the topic of this chapter by an example from musical signal processing. Suppose we are given a recording of, say, the fourth Symphony of Brahms, as a digitized waveform. We want to perform spectral analysis of the audio signal. Why would we want to do that? For the sake of the story, assume we want to use the signal for reconstructing the full score of the music, note by note, instrument by instrument. We hasten to say, lest you develop misconceptions about the power of digital signal processing, that such a task is still beyond our ability (in 1996 at least). However, the future may well prove such tasks possible.

Let us do a few preliminary calculations. The music is over 40 minutes long, or about 2500 seconds. Compact-disc recordings are sampled at 44.1 kHz and are in stereo. However, to simplify matters, assume we combine the two channels into a single monophonic signal by summing them at each time point. Our discrete-time signal then has a number of samples N on the order of 10^8. So, are we to compute the DFT of a sequence one hundred million samples long? This appears to be both unrealistic and useless. It is unrealistic because speed and storage requirements for a hundred million point FFT are too high by today's standards. It is useless because all we will get as a result is a wide-band spectrum, covering the range from about 20 Hz to about 20 kHz, and including all notes of all instruments, with their harmonics, throughout the symphony. The percussion instruments, which are naturally wide band and noiselike, will contribute to the unintelligible shape of the spectrum. True, the frequency resolution will be excellent—on the order of 0.4×10^{-3} Hz—but there is no apparent use for such resolution. The human ear is far from being able to perceive frequency differences that are fractions of millihertz, and distances between adjacent notes in music scales are three to six orders of magnitude higher.

If not a full-length DFT, then what? A bit of reflection will tell us that what we really want is *a sequence of short DFTs*. Each DFT will exhibit the spectrum of the signal during a relatively short interval. Thus, for example, if the violins play the note E during a certain time interval, the spectrum should exhibit energies at the frequency of the note E (329.6 Hz) and at the characteristic harmonics of the violin. In general, if the intervals are short enough, we may be able to track the note-to-note changes in the music. If the frequency resolution is good enough, we may be able to identify the musical instrument(s) playing at each time interval. This is the essence of spectral analysis. The human ear-brain is certainly an excellent spectrum analyzer. A trained musician can identify individual notes and instruments in very complex musical compositions.

Continuing our example, each DFT would have a length of, say, 4096, corresponding to a time interval of about 92.9 milliseconds. The frequency resolution will be about 11 Hz, which is quite adequate. There are about 26,000 distinct intervals in the symphony $[= (40 \times 60)/(92.9 \times 10^{-3})]$, so we will need to compute 26,000 FFTs of length 4096. We may choose to be conservative and overlap the intervals to a certain extent (say 50 percent), to ensure that nothing important that may occur at the border between adjacent intervals is missed. Then we will need about 52,000 FFTs of length 4096. With today's technology, the entire computation can be done in considerably less time than the music itself.

We complete our Brahms example by showing what some of the music looks like in the time and frequency domains. The fourth movement of the symphony starts with eight chords, each about 1.7 seconds. Table 6.1 gives the frequencies of the individual notes in each chord.[1] Figures 6.1 and 6.2 show eight segments of the signal, one for each chord, of length 92.9 milliseconds. Figures 6.3 and 6.4 show the magnitudes of the Fourier transforms of the eight segments, in the range 0 to 2000 Hz. We also mark, on each spectrum, the frequencies of the notes of the chord, taken from Table 6.1, as well as all their harmonics up to 2000 Hz. As we see, the note frequencies and their harmonics fall on peaks of the corresponding spectra in most cases.

Bar No.	1	2	3	4	5	6	7	8
note 1	130.8	110.0	82.4	130.8	92.5	98.0	87.3	82.4
note 2	261.6	220.0	164.8	261.6	164.8	164.8	123.5	164.8
note 3	329.6	261.6	246.9	329.6	185.0	196.0	174.6	207.6
note 4	440.0	370.0	329.6	440.0	277.2	329.6	220.0	246.9
note 5	523.2	440.0	392.0	523.2	329.6	493.9	246.9	329.6
note 6	659.2	523.2	493.9	659.2	466.2	987.8	311.1	415.2
note 7	—	740.0	659.2	880.0	554.4	—	440.0	493.9
note 8	—	—	784.0	—	932.4	—	493.9	659.2

Table 6.1 Frequencies (in Hz) of the notes in the first 8 bars of the fourth movement of Brahms's Symphony No. 4.

Dividing a signal having long duration into short segments and performing spectral analysis on each segment is known as *short-time spectral analysis*. Practical short-time spectral analysis requires more than just DFT (or FFT) computations. In this chapter we study the principal technique used for short-time spectral analysis, that of *windowing*. We first describe the basic windowing operation and interpret it in the time and frequency domains. We then list the most common windows and examine their properties. Finally, we apply the windowing technique to the problem of measuring the frequencies of sinusoidal signals, first without and then with additive noise. Short-time spectral analysis is by no means limited to sinusoidal signals, however. In Section 13.1 we shall study the use of short-time spectral analysis to random signals; see page 513.

6.1   The Effect of Rectangular Windowing

Assume we are given a long, possibly infinite-duration signal $y[n]$. We pick a relatively short segment of $y[n]$, say

$$x[n] = \begin{cases} y[n], & 0 \le n \le N - 1, \\ 0, & \text{otherwise.} \end{cases} \tag{6.1}$$

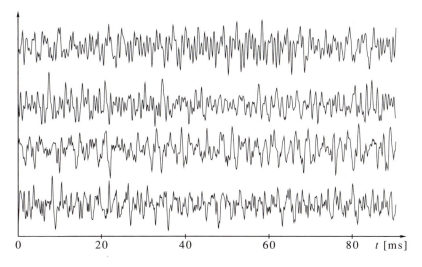

Figure 6.1 Brahms's Symphony No. 4, fourth movement, bars 1 through 4 (top to bottom), a 92.9-millisecond segment of each bar.

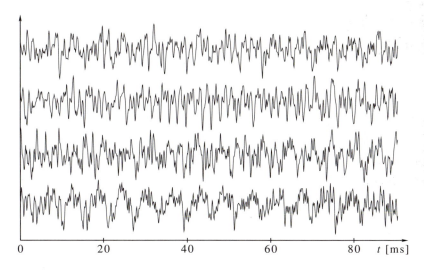

Figure 6.2 Brahms's Symphony No. 4, fourth movement, bars 5 through 8 (top to bottom), a 92.9-millisecond segment of each bar.

We can describe the operation of getting $x[n]$ from $y[n]$ as a multiplication of $y[n]$ by a *rectangular window*, that is,

$$x[n] = y[n]w_{\mathrm{r}}[n], \quad \text{where } w_{\mathrm{r}}[n] = \begin{cases} 1, & 0 \leq n \leq N-1, \\ 0, & \text{otherwise.} \end{cases} \tag{6.2}$$

How is the Fourier transform of a rectangular-windowed signal related to that of the original signal? Before we give a general answer, let us look at an example. Consider an exponential signal and its Fourier transform,

$$y[n] = \begin{cases} a^n, & n \geq 0, \\ 0, & n < 0, \end{cases} \quad Y^{\mathrm{f}}(\theta) = \frac{1}{1 - ae^{-j\theta}}, \tag{6.3}$$

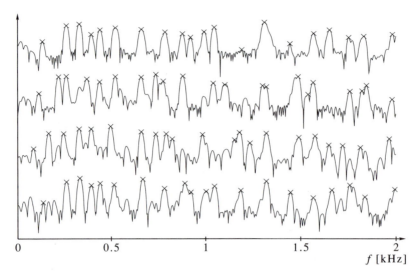

Figure 6.3 Spectra of the signals shown in Figure 6.1. x's denote frequencies of chord notes and their harmonics.

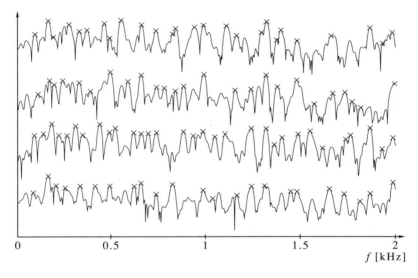

Figure 6.4 Spectra of the signals shown in Figure 6.2. x's denote frequencies of chord notes and their harmonics.

assuming $|a| < 1$. Now consider the Fourier transform of the rectangular-windowed signal,

$$X^f(\theta) = \sum_{n=0}^{N-1} a^n e^{-j\theta n} = \frac{1 - a^N e^{-j\theta N}}{1 - a e^{-j\theta}}. \tag{6.4}$$

Figure 6.5 shows the magnitudes of the Fourier transforms of the original and the rectangular-windowed signals, with $N = 16$ and $a = 0.9$. The transform of $y[n]$ is smooth, that of $x[n]$ is wiggly, but the latter approximately follows the former.

Now return to (6.2); recall that multiplication in the time domain translates to

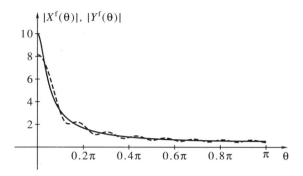

Figure 6.5 The Fourier transform (magnitude) of an exponential sequence. Solid line: unwindowed; dashed line: windowed.

convolution in the frequency domain, so

$$X^{\mathrm{f}}(\theta) = \frac{1}{2\pi} \{Y^{\mathrm{f}} * W_{\mathrm{r}}^{\mathrm{f}}\}(\theta), \tag{6.5}$$

where $W_{\mathrm{r}}^{\mathrm{f}}(\theta)$ is the Fourier transform of the rectangular window, given by

$$W_{\mathrm{r}}^{\mathrm{f}}(\theta) = \sum_{n=0}^{N-1} e^{-j\theta n} = \frac{1 - e^{-j\theta N}}{1 - e^{-j\theta}} = \frac{\sin(0.5\theta N)}{\sin(0.5\theta)} e^{-j0.5\theta(N-1)}. \tag{6.6}$$

The function

$$D(\theta, N) = \frac{\sin(0.5\theta N)}{\sin(0.5\theta)} \tag{6.7}$$

is called the *Dirichlet kernel*. It is shown in Figure 6.6 for $N = 40$. The main properties of the Dirichlet kernel are as follows:

1. Its maximum value is N, occurring at $\theta = 0$.
2. Its zeros nearest to the origin are at $\theta = \pm 2\pi/N$. The region between the two zeros is called the *main lobe* of the Dirichlet kernel.
3. There are additional zeros at $\{\theta = 2m\pi/N, \ m = \pm 2, \pm 3, \ldots\}$. Between every pair of adjacent zeros there is a maximum or a minimum, approximately at $\theta = (2m + 1)\pi/N$. The regions between these zeros are called *side lobes*.
4. The highest side lobe (in absolute value) occurs at $\theta = \pm 3\pi/N$ and its value (for large N) is approximately $2N/3\pi$. The ratio of the highest side lobe to the main lobe is $2/(3\pi)$, or about -13.5 dB.

Had $W_{\mathrm{r}}^{\mathrm{f}}(\theta)$ been an impulse function $\delta(\theta)$, we would have obtained $X^{\mathrm{f}}(\theta) = Y^{\mathrm{f}}(\theta)$. Since $W_{\mathrm{r}}^{\mathrm{f}}(\theta)$ is not an impulse, we get distortions. These distortions are of two kinds:

1. Smearing and, as a consequence, loss of resolution due to the finite width of the main lobe. Any impulselike feature in $Y^{\mathrm{f}}(\theta)$ (such as one resulting from a periodic component in $y[n]$) will have a width approximately $\pm 2\pi/N$ in the spectrum of the rectangular-windowed signal $x[n]$. Two periodic components in $y[n]$ whose frequency separation is less than $2\pi/N$ will blend with each other and may appear as one component in $X^{\mathrm{f}}(\theta)$.
2. Side-lobe interferences. Suppose $y[n]$ contains a strong sinusoidal component at frequency θ_0 and weak components at other frequencies. The side lobes of the strong component may mask the main lobe of the weak component. This masking is worst when the frequency of the weak component differs from θ_0 by an odd multiple of π/N.

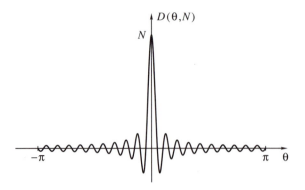

Figure 6.6 The Dirichlet kernel for $N = 40$.

In summary, rectangular windowing has undesirable side effects, so it is generally not a good way of performing short-time spectral analysis. The solution to this problem is presented in the next section.

6.2   Windowing

As we have seen, the undesirable side effects of rectangular windowing are due to the convolution of the Fourier transform of the original signal with the Dirichlet kernel. Suppose, instead, that we wish to obtain $X^{\mathrm{f}}(\theta)$ as the convolution of $Y^{\mathrm{f}}(\theta)$ with another function $W^{\mathrm{f}}(\theta)$. The time-domain multiplication property of the Fourier transform tells us how to do it: We must multiply $y[n]$ by the sequence $w[n]$, the inverse Fourier transform of $W^{\mathrm{f}}(\theta)$. Thus, we must perform the operation

$$x[n] = y[n]w[n], \tag{6.8}$$

and then we will get

$$X^{\mathrm{f}}(\theta) = \frac{1}{2\pi}\{Y^{\mathrm{f}} * W^{\mathrm{f}}\}(\theta). \tag{6.9}$$

However, the sequence $w[n]$ cannot be arbitrary. First, it must have finite duration, else it will not select a finite segment of $y[n]$. Second, its length N must agree with the desired length of the finite segment we wish to analyze. Third, the elements of $w[n]$ should be nonnegative, because it is not reasonable to invert the polarities of the signal values. Within these limitations, we would like $W^{\mathrm{f}}(\theta)$ to have:

1. A main lobe that is as narrow as possible.

2. Side lobes that are as low as possible.

The sequence $w[n]$ is called a *window*, and the operation (6.8) is called *windowing*. Windowing amounts to truncating the signal $y[n]$ to a finite length N while reshaping it through multiplication. The special case of rectangular windowing corresponds to truncation without reshaping. The transform $W^{\mathrm{f}}(\theta)$ of the window is called the *kernel function* of the window [hence the name Dirichlet kernel for (6.7)]. Figure 6.7 depicts a hypothetical desired shape of a kernel. The kernel of a good window should be a good approximation to a delta function, since then the convolution operation (6.9) will not distort $Y^{\mathrm{f}}(\theta)$ too much. However, the inverse Fourier transform of $W^{\mathrm{f}}(\theta) = 2\pi\delta(\theta)$ is $w[n] \equiv 1$, and this does not have a finite duration. Choosing a window always involves trade-off between the width of the main lobe and the level of the side lobes. In general, the narrower the main lobe, the higher the side lobes, and vice versa. The rectangular window has the narrowest possible main lobe of all windows of the same length, but its side lobes are the highest. The side-lobe level of the rectangular window, which is $-13.5\,\text{dB}$, is undesired in most applications. We are almost always ready to increase the main-lobe width (and degrade the frequency resolution) in order to reduce the side-lobe level.

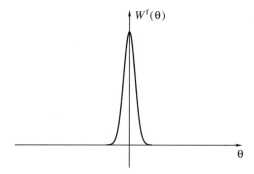

Figure 6.7 A hypothetical window kernel function.

To reiterate:

> The main parameters of a window by which we judge its suitability to a given task are (1) its main-lobe width relative to the width of the main lobe of a rectangular window of the same length N and (2) its side-lobe level relative to the amplitude of the main lobe.

6.3 Common Windows

6.3.1 Rectangular Window

The rectangular window was discussed previously, so we just repeat its main properties for completeness. The width of its main lobe is $4\pi/N$ and its side-lobe level is $-13.5\,\text{dB}$. Figure 6.8 depicts a rectangular window with $N = 41$, in the time and frequency domains. The magnitude response of the kernel function is shown in decibels, for better visualization of the side-lobe level.

6.3.2 Bartlett Window

Bartlett's method of side-lobe level reduction is based on squaring of the kernel function of the rectangular window, thereby reducing the side-lobe level by a factor of 2 (in dB). Squaring in the frequency domain is equivalent to convolving a rectangular window with itself in the time domain. Suppose that the desired window length N is odd and let $w_r[n]$ be a rectangular window of length $(N + 1)/2$. Define

$$w_t[n] = \frac{2}{N+1}\{w_r * w_r\}[n] = 1 - \frac{|2n - N + 1|}{N+1}, \quad 0 \le n \le N - 1. \tag{6.10}$$

The corresponding kernel function is then

$$W_t^f(\theta) = \frac{2}{N+1}D^2(\theta, 0.5(N+1))e^{-j0.5\theta(N-1)} = \frac{2\sin^2[0.25\theta(N+1)]}{(N+1)\sin^2(0.5\theta)}e^{-j0.5\theta(N-1)}. \tag{6.11}$$

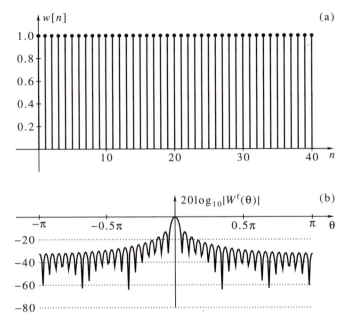

Figure 6.8 A rectangular window, $N = 41$: (a) time-domain plot; (b) frequency-domain magnitude plot.

This window is called the *Bartlett window* (after its discoverer) or a *triangular window* (owing to its shape). Figure 6.9 depicts the Bartlett window in the time and frequency domains, for $N = 41$. As is clear from its construction, the side-lobe level of the Bartlett window is -27 dB. The width of the main lobe is $8\pi/(N+1)$, which is nearly twice that of a rectangular window of the same length.

When N is even, we can obtain the Bartlett window by convolving a rectangular window of length $N/2$ with one of length $(N+2)/2$. The time-domain window function is again (6.10) and the kernel function is (see Problem 6.2)

$$
\begin{aligned}
W_t^f(\theta) &= \frac{2}{(N+1)} D(\theta, 0.5N) D(\theta, 0.5(N+2)) e^{-j0.5\theta(N-1)} \\
&= \frac{2\sin(0.25\theta N)\sin[0.25\theta(N+2)]}{(N+1)\sin^2(0.5\theta)} e^{-j0.5\theta(N-1)}.
\end{aligned}
\tag{6.12}
$$

6.3.3 Hann Window

Whereas the Bartlett window achieves side-lobe level reduction by squaring, the Hann window achieves a similar effect by superposition. The kernel function of the Hann window is obtained by adding three Dirichlet kernels, shifted in frequency so as to yield partial cancellation of their side lobes. Figure 6.10 illustrates the construction of the Hann window. The two side kernels are shifted by $\pm 2\pi/(N-1)$ with respect to the one at the center. The center kernel has magnitude 0.5 and the ones at the sides have magnitudes 0.25 each. Figure 6.10 also shows the sum of the three kernels, in which attenuation of the side lobes is apparent.

The kernel function of the Hann window is given by

$$
\begin{aligned}
W_{hn}^f(\theta) &= \left[0.5 D(\theta, N) + 0.25 D\left(\theta - \frac{2\pi}{N-1}, N\right) + 0.25 D\left(\theta + \frac{2\pi}{N-1}, N\right) \right] e^{-j0.5\theta(N-1)} \\
&= 0.5 W_r^f(\theta) - 0.25 W_r^f\left(\theta - \frac{2\pi}{N-1}\right) - 0.25 W_r^f\left(\theta + \frac{2\pi}{N-1}\right).
\end{aligned}
\tag{6.13}
$$

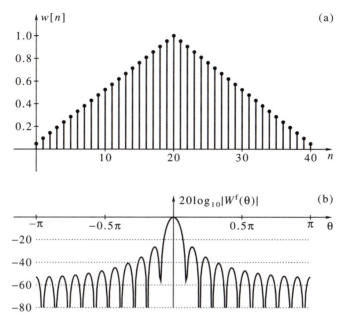

Figure 6.9 Bartlett window, $N = 41$: (a) time-domain plot; (b) frequency-domain magnitude plot.

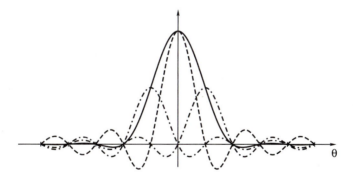

Figure 6.10 Construction of the Hann window from three Dirichlet kernels. Dashed line: center kernel; dot-dashed lines: shifted kernels; solid line: the sum.

By the modulation property of the Fourier transform we get that the window sequence corresponding to (6.13) is

$$w_{\mathrm{hn}}[n] = 0.5 - 0.25 \exp\left(\frac{j2\pi n}{N-1}\right) - 0.25 \exp\left(-\frac{j2\pi n}{N-1}\right)$$

$$= 0.5\left[1 - \cos\left(\frac{2\pi n}{N-1}\right)\right], \quad 0 \le n \le N-1. \tag{6.14}$$

The Hann window is also known in the literature as the "Hanning" window, for reasons explained in the next section. It is also called a cosine window. Figure 6.11 depicts the Hann window in the time and frequency domains, for $N = 41$. The side-lobe level of this window is $-32\,\mathrm{dB}$ and the width of the main lobe is $8\pi/N$. The main lobe width is the same as that of the Bartlett window, but the side-lobe level is lower.

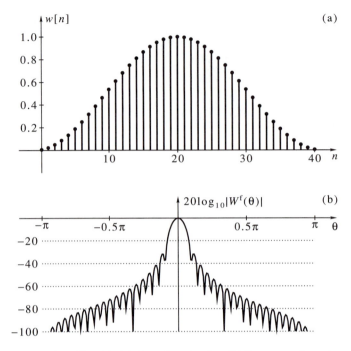

Figure 6.11 Hann window, $N = 41$: (a) time-domain plot; (b) frequency-domain magnitude plot.

The Hann window has a peculiar property: Its two end points are zero. When applied to a signal $y[n]$, it effectively deletes the points $y[0]$ and $y[N-1]$. This suggests increasing the window length by 2 with respect to the desired N and deleting the two end points. The modified Hann window thus obtained is

$$w_{\text{hn}}[n] = 0.5\left\{1 - \cos\left[\frac{2\pi(n+1)}{N+1}\right]\right\}, \quad 0 \le n \le N-1; \tag{6.15}$$

the kernel function changes accordingly.

6.3.4 Hamming Window

The Hamming window is obtained by a slight modification of the Hann window, which amounts to choosing different magnitudes for the three Dirichlet kernels. The window and kernel function are

$$w_{\text{hm}}[n] = 0.54 - 0.46\cos\left(\frac{2\pi n}{N-1}\right), \quad 0 \le n \le N-1, \tag{6.16}$$

$$W_{\text{hm}}^{\text{f}}(\theta) = 0.54W_{\text{r}}^{\text{f}}(\theta) - 0.23W_{\text{r}}^{\text{f}}\left(\theta - \frac{2\pi}{N-1}\right) - 0.23W_{\text{r}}^{\text{f}}\left(\theta + \frac{2\pi}{N-1}\right). \tag{6.17}$$

Figure 6.12 shows the three replicas and their sum for the Hamming window. As we see, the side lobes of the sum are lower than those of the Hann window. Hamming got the numbers 0.54 and 0.46 by trial and error, seeking to minimize the amplitude of the highest side lobe.

Figure 6.13 depicts the Hamming window in the time and frequency domains, for $N = 41$. The side-lobe level of this window is -43 dB; the main-lobe width is $8\pi/N$, the same as that of the Bartlett and Hann windows. A peculiar property of this window is that the highest side lobe is not the one nearest to the main lobe. Its end points are not zero, so its length need not be increased by 2, as in the case of the Hann window.

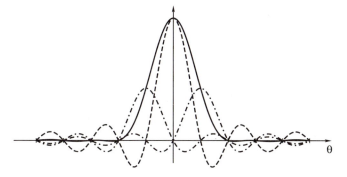

Figure 6.12 Construction of the Hamming window from three Dirichlet kernels. Dashed line: center kernel; dot-dashed lines: shifted kernels; solid line: the sum.

The Hamming window is also called a raised-cosine window. When Blackman and Tukey described the Hann and Hamming windows in their classical book of 1958, they nicknamed the former "Hanning," which though probably intended as a pun, was its official name for many years. Recently, however, the window has been properly renamed *Hann* after its inventor.

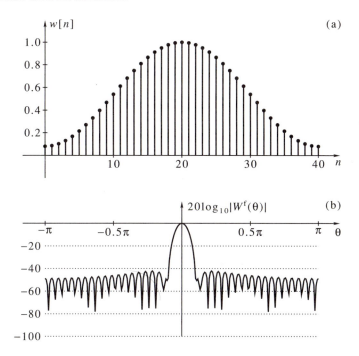

Figure 6.13 Hamming window, $N = 41$: (a) time-domain plot; (b) frequency-domain magnitude plot.

6.3.5 Blackman Window

The Hamming window has the lowest possible side-lobe level among all windows based on three Dirichlet kernels. The Blackman window uses five Dirichlet kernels, thus reducing the side-lobe level still further. The Blackman window and its kernel function are

$$w_b[n] = 0.42 - 0.5\cos\left(\frac{2\pi n}{N-1}\right) + 0.08\cos\left(\frac{4\pi n}{N-1}\right), \quad 0 \le n \le N-1, \qquad (6.18)$$

$$W_b^f(\theta) = 0.42 W_r^f(\theta) - 0.25 W_r^f\left(\theta + \frac{2\pi}{N-1}\right) - 0.25 W_r^f\left(\theta - \frac{2\pi}{N-1}\right)$$
$$+ 0.04 W_r^f\left(\theta + \frac{4\pi}{N-1}\right) + 0.04 W_r^f\left(\theta - \frac{4\pi}{N-1}\right). \qquad (6.19)$$

Figure 6.14 depicts the Blackman window in the time and frequency domains, for $N = 41$. The side-lobe level of the Blackman window is -57 dB and the width of the main lobe is $12\pi/N$. As in the case of the Hann window, the two end points of the Blackman window are zero, so in practice we can increase N by 2 and remove the two end points. The weights of the Blackman window are not optimized to minimize the side-lobe level. Such optimization was performed by Harris [1978] and is discussed in Problem 6.8.

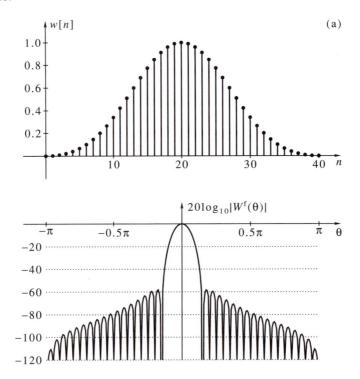

Figure 6.14 Blackman window, $N = 41$: (a) time-domain plot; (b) frequency-domain magnitude plot.

6.3.6   Kaiser Window

The windows described so far are considered as classical. They have been derived based on intuition and educated guesses. Modern windows are based on optimality criteria; they aim to be best in a certain respect, while meeting certain constraints. Different optimality criteria give rise to different windows. Of special importance in this category are the following optimality criteria:

1. *Dolph's criterion:* Minimize the width of the main lobe of the kernel, under the constraint that the window length be fixed and the side-lobe level not exceed a given maximum value. The window values depend on the length and the permitted side-lobe level.

2. *Kaiser's criterion:* Minimize the width of the main lobe of the kernel, under the constraint that the window length be fixed and the energy in the side lobes not exceed a given percentage of the total energy. The energy in the side lobes is defined as the integral of the square magnitude of the kernel function over the range $[-\pi, \pi]$, excluding the interval of the main lobe. The window values depend on the length and the permitted energy in the side lobes.

Of the windows based on these two criteria, the Kaiser window is much more popular than the Dolph window. Kaiser's criterion gives rise to a family of windows that has become, perhaps, the most widely used for modern digital signal processing [Kuo and Kaiser, 1966]. We now describe the Kaiser window.

The solution to Kaiser's optimization problem is described in terms of the *modified Bessel function* of order zero. This function is given by the infinite power series

$$I_0(x) = \sum_{k=0}^{\infty} \left(\frac{x^k}{2^k k!} \right)^2. \tag{6.20}$$

Using this function, the Kaiser window is given by

$$w_k[n] = \frac{I_0\left[\alpha \sqrt{1 - \left(\frac{|2n-N+1|}{N-1} \right)^2} \right]}{I_0[\alpha]}, \quad 0 \le n \le N - 1. \tag{6.21}$$

The parameter α is used for controlling the main-lobe width and the side-lobe level. In general, higher α leads to a wider main lobe and lower side lobes. Figure 6.15 shows the dependence of the main-lobe width (as a multiple of $2\pi/N$) and the side-lobe level on α. The jagged appearance of the upper graph is due to the finite window length used for the computation. In theory (for window length approaching infinity), the graph should be smooth.

Figure 6.16 depicts the Kaiser window in the time and frequency domains, for $N = 41$ and $\alpha = 12$. In this case, the main-lobe width is $16\pi/N$ and the side-lobe level is $-90\,\text{dB}$.

6.3.7   Dolph Window*

As we have said, Dolph's criterion (also called the Dolph–Chebyshev criterion) for window design is the minimization of the main-lobe width under the constraint that the window length be fixed and the side-lobe level not exceed a given maximum value [Dolph, 1946]. The mathematical theory underlying the Dolph window is beyond the scope of this book, so we describe the window, but not its derivation.

The kernel function of the Dolph window is

$$W_d^f(\theta) = \begin{cases} C \cos\{(N-1)\arccos[\beta\cos(0.5\theta)]\}, & |\beta\cos(0.5\theta)| \le 1, \\ C \cosh\{(N-1)\text{arccosh}[\beta\cos(0.5\theta)]\}, & |\beta\cos(0.5\theta)| > 1, \end{cases} \tag{6.22}$$

where

$$\beta = \cosh[(N-1)^{-1}\text{arccosh}(10^{-0.05\alpha})], \tag{6.23}$$

and α is the side-lobe level in decibels (a negative number). The parameter C is a normalization constant, chosen to make the middle coefficient of the window equal to

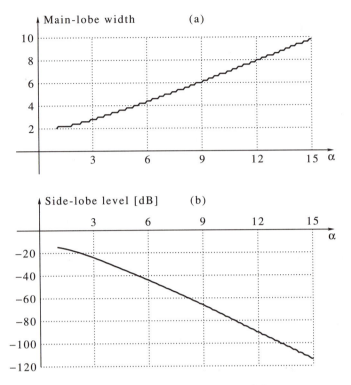

Figure 6.15 Properties of the Kaiser window as a function of the parameter α: (a) main-lobe width, as a multiple of $2\pi/N$; (b) side-lobe level.

1. The functions *cosh* and *arccosh* are

$$\cosh(x) = 0.5(e^x + e^{-x}), \quad \mathrm{arccosh}(x) = \log_e[x + (x^2 - 1)^{1/2}].$$

The window sequence $w_d[n]$ is obtained by sampling $W_d^f(\theta)$ at the N points (where N is the desired window length)

$$\theta[k] = \frac{2\pi k}{N}, \quad 0 \le k \le N - 1,$$

and computing the inverse DFT of the result. The procedure is straightforward when N is odd, but needs care when N is even. For even N, we want to get a window that is symmetric around the fractional point $0.5(N - 1)$. This requires shifting to the left by half a sample, which is equivalent to multiplying $W_d^f(\theta[k])$ by the sequence $\exp(j\pi k/N)$ before the inverse DFT. Problem 4.6 provides the theoretical justification for this procedure. We finally must swap the two halves of the inverse DFT to get the sequence $w_d[n]$ (for both even and odd N).

Figure 6.17 illustrates the Dolph window in the time and frequency domains for $N = 41$ and $\alpha = -60$ dB. As we see, the side-lobe level is constant for all side lobes. Because of this property, the Dolph window is called an *equiripple window*.

The Dolph window has a certain idiosyncrasy: The two end points, $w_d[0]$ and $w_d[N - 1]$, are sometimes larger in magnitude than points nearer to the center. This is not noticeable in Figure 6.17, but becomes more pronounced as N increases. Many engineers are reluctant to use a window that emphasizes the end points of the data segment (although there is no theoretical objection to this), so the Dolph window is not popular. Another reason for the rare use of this window is its high sensitivity to coefficient accuracy, which prevents its use in computers having short word lengths.

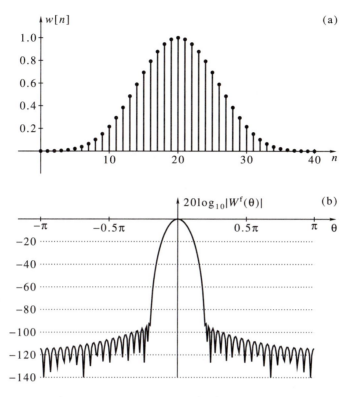

Figure 6.16 Kaiser window, $N = 41$, $\alpha = 12$: (a) time-domain plot; (b) frequency-domain magnitude plot.

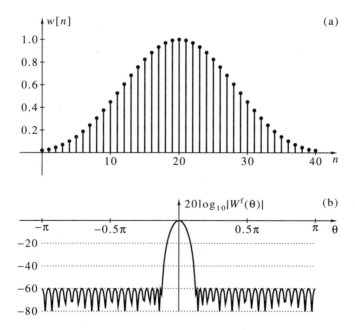

Figure 6.17 Dolph window, $N = 41$, $\alpha = -60$ dB: (a) time-domain plot; (b) frequency-domain magnitude plot.

6.3.8   MATLAB Implementation of Common Windows

The MATLAB Signal Processing Toolbox contains the functions boxcar, bartlett, hanning, hamming, blackman, kaiser, and chebwin, which generate the seven window sequences described in this section. Here we also include, in Program 6.1, MATLAB implementations of these windows. The procedure window accepts the length N and the window's name as input arguments. In the case of Kaiser or Dolph windows, it accepts the parameter α as a third argument. The output is the desired window sequence. The Dolph window is implemented in dolph, shown in Program 6.2.

The following differences exist between Program 6.1 and the aforementioned MATLAB routines:

1. Whereas the MATLAB output is a column vector, ours is a row vector.

2. The output of the MATLAB command bartlett(N) is a vector of N elements, the first and last of which are zero. When these two zeros are deleted, the remaining vector is identical to the one obtained by entering the command window(N-2,'bart').

3. The output of the command window(N,'hann') is a vector of N elements, the first and last of which are zero. When these two zeros are deleted, the remaining vector is identical to the output of the MATLAB command hanning(N).

6.4   Frequency Measurement

One of the most important applications of the DFT is the measurement of frequencies of periodic signals, in particular sinusoidal signals. Sinusoidal signals are prevalent in science and engineering and the need to measure the frequency of a sinusoidal signal arises in numerous applications. The Fourier transform is a natural tool for this purpose. Practical signals are measured over a finite time interval and the Fourier transform can be computed only on a discrete set of frequencies. The implications of these two restrictions on the theory and practice of frequency measurement are explored in this section.

It is convenient, from a pedagogical point of view, to deal with complex exponential signals first, and proceed to real-valued sinusoids later. In practice, real-valued sinusoids are much more common. However, in certain applications (such as radar and communication), complex signals appear naturally. Therefore, the treatment of complex exponential signals is useful not only as an introduction to the case of real signals, but also for its own merit.

6.4.1   Frequency Measurement for a Single Complex Exponential

The simplest case of a signal for which frequency measurement is a meaningful problem is a single complex exponential signal. Suppose we are given the signal

$$y(t) = Ae^{j(\omega_0 t + \phi_0)}, \tag{6.24}$$

and we wish to measure the frequency ω_0. We sample the signal at interval T and collect N consecutive data points, thus getting the discrete-time signal

$$y[n] = Ae^{j(\theta_0 n + \phi_0)}, \quad 0 \le n \le N-1, \tag{6.25}$$

where $\theta_0 = \omega_0 T$. We assume that $-\pi < \omega_0 T < \pi$, so measurement of θ_0 implies unambiguous measurement of ω_0.

The Fourier transform of the sampled signal is given by

$$Y^{\mathrm{f}}(\theta) = Ae^{j\phi_0} \sum_{n=0}^{N-1} e^{-j(\theta-\theta_0)n} = Ae^{-j[0.5(\theta-\theta_0)(N-1)-\phi_0]}D(\theta - \theta_0, N). \qquad (6.26)$$

In particular, since $D(0, N) = N$, evaluation of the Fourier transform at the frequency $\theta = \theta_0$ gives

$$Y^{\mathrm{f}}(\theta_0) = NAe^{j\phi_0}, \quad \text{therefore} \quad |Y^{\mathrm{f}}(\theta_0)| = NA. \qquad (6.27)$$

Furthermore, since $|D(\theta, N)| < N$ for all $\theta \neq 0$, the point $\theta = \theta_0$ is the unique global maximum of $|Y^{\mathrm{f}}(\theta)|$ on $-\pi < \theta < \pi$. We therefore conclude that θ_0 can be obtained by finding the point of global maximum of $|Y^{\mathrm{f}}(\theta)|$ in the frequency range $(-\pi, \pi)$. To reiterate:

> The magnitude of the Fourier transform of a finite segment of a complex exponential signal $y[n] = Ae^{j(\theta_0 n + \phi_0)}$ exhibits a unique global maximum at the frequency $\theta = \theta_0$, thus enabling the determination of θ_0 from $|Y^{\mathrm{f}}(\theta)|$.

In practice, it is not possible to find the global maximum of $|Y^{\mathrm{f}}(\theta)|$ exactly, since we cannot evaluate this function at an infinite number of frequency points. A first approximation can be obtained by computing the DFT of $y[n]$ and searching for the point of maximum of $Y^{\mathrm{d}}[k]$. The index k_0 for which $Y^{\mathrm{d}}[k]$ is maximized yields a corresponding frequency $\theta[k_0] = 2\pi k_0/N$ and this serves as the measured value of θ_0 (subtracting 2π if $k_0 \geq 0.5N$). Better approximations can be obtained, if necessary, by either zero padding the sequence $y[n]$ or using a chirp Fourier transform, as explained in Section 5.7. In summary, measurement of the frequency of a single complex exponential can be accomplished by simple DFT or by a DFT-based algorithm, depending on the desired accuracy.

6.4.2 Frequency Measurement for Two Complex Exponentials

Proceeding to a more difficult problem, we now consider a signal consisting of two complex exponentials, that is,

$$y(t) = A_1 e^{j(\omega_1 t + \phi_1)} + A_2 e^{j(\omega_2 t + \phi_2)}. \qquad (6.28)$$

Our aim is to measure the frequencies ω_1 and ω_2. As before, we sample the signal at interval T and collect N measurements of the sampled signal to obtain

$$y[n] = A_1 e^{j(\theta_1 n + \phi_1)} + A_2 e^{j(\theta_2 n + \phi_2)}, \quad 0 \leq n \leq N - 1, \qquad (6.29)$$

where $\theta_i = \omega_i T$, $i = 1, 2$; we assume that $-\pi < \omega_i T < \pi$.

The Fourier transform of the sampled signal is given by

$$\begin{aligned}
Y^{\mathrm{f}}(\theta) &= A_1 e^{j\phi_1} \sum_{n=0}^{N-1} e^{-j(\theta-\theta_1)n} + A_2 e^{j\phi_2} \sum_{n=0}^{N-1} e^{-j(\theta-\theta_2)n} \\
&= A_1 e^{-j[0.5(\theta-\theta_1)(N-1)-\phi_1]}D(\theta - \theta_1, N) \\
&\quad + A_2 e^{-j[0.5(\theta-\theta_2)(N-1)-\phi_2]}D(\theta - \theta_2, N).
\end{aligned} \qquad (6.30)$$

In particular, evaluation of the Fourier transform at the frequency $\theta = \theta_1$ gives

$$Y^{\mathrm{f}}(\theta_1) = NA_1 e^{j\phi_1} + A_2 e^{-j[0.5(\theta_1-\theta_2)(N-1)-\phi_2]}D(\theta_1 - \theta_2, N). \qquad (6.31)$$

We observe the following:

1. If $A_2 = 0$, that is, if only one complex sinusoid is present, then $|Y^f(\theta_1)| = NA_1$, as before. Therefore we can find θ_1 from the maximum point of $|Y^f(\theta)|$.

2. If $A_2 \neq 0$, but

$$|A_2 D(\theta_1 - \theta_2, N)| \ll NA_1, \tag{6.32}$$

we still have a good chance of finding a local maximum of $|Y^f(\theta)|$ near θ_1, since the behavior of $|Y^f(\theta)|$ near θ_1 will not be much perturbed by the second term in (6.31). However, this local maximum is not necessarily a global maximum any more, because the second term in (6.31) affects the global behavior of $|Y^f(\theta)|$.

3. Condition (6.32) will hold if $|\theta_2 - \theta_1| \geq 2\pi/N$, and if A_2 is not much larger than A_1. Otherwise, (6.32) may fail, and we may be unable to find a local maximum of $|Y^f(\theta)|$ near θ_1.

4. The discussion is symmetric with respect to the two sinusoidal components. We can find another local maximum of $|Y^f(\theta)|$ near θ_2, provided the interference from the Dirichlet kernel centered at θ_1 is sufficiently small in the vicinity of θ_2.

Example 6.1 Figure 6.18 shows the behavior of $|Y^f(\theta)|$ in the range of θ of interest (in dB below the maximum) for several choices of θ_1, θ_2, A_1, and A_2 (in all cases $N = 64$). In parts a, b, and c the amplitudes of the two complex exponentials are equal. In part a, the difference $\theta_2 - \theta_1$ is $2\pi/N$. As we see, the two peaks are well separated, indicating the presence of two complex exponentials. However, the peaks appear to repel each other, causing their locations to deviate slightly from the true θ_1 and θ_2. This phenomenon is called *bias*; it typically occurs when the frequencies are close to each other. In part b, the frequency difference is $1.5\pi/N$. Now the two peaks start to merge with each other, but are still distinct; bias is again apparent. In part c, the frequency difference is π/N. In this case there is only one peak, so the two complex exponentials cannot be distinguished from each other.

In parts d, e, and f the frequencies are the same as in parts a, b, and c, respectively, but $A_2 = 0.25A_1$. There are still two distinguishable peaks when $\theta_2 - \theta_1 = 2\pi/N$ (part d); however, the weaker component is considerably biased. When the frequency difference decreases, the weaker component becomes invisible.   □

Condition (6.32) may be hard to satisfy since the side lobes of the Dirichlet kernel are relatively high; the highest side lobe is only 13.5 dB lower than the main lobe. For example, if $|\theta_2 - \theta_1| = 3\pi/N$ and A_2 is larger than A_1 by more than 13.5 dB, the Dirichlet kernel centered at θ_2 will interfere with the one centered at θ_1 and may prohibit measurement of θ_1. Because of the slow rate of decay of the side lobes of the Dirichlet kernel, however, the problem does not completely disappear, even if the distance between the two frequencies is large with respect to $2\pi/N$.

Windowing alleviates the problems we have described. Suppose we multiply $y[n]$ by a window $w[n]$ prior to computing the Fourier transform and denote

$$x[n] = y[n]w[n].$$

Then (6.30) changes to

$$X^f(\theta) = A_1 e^{j\phi_1} \sum_{n=0}^{N-1} w[n]e^{-j(\theta-\theta_1)n} + A_2 e^{j\phi_2} \sum_{n=0}^{N-1} w[n]e^{-j(\theta-\theta_2)n}$$

$$= A_1 e^{j\phi_1} W^f(\theta - \theta_1, N) + A_2 e^{j\phi_2} W^f(\theta - \theta_2, N), \tag{6.33}$$

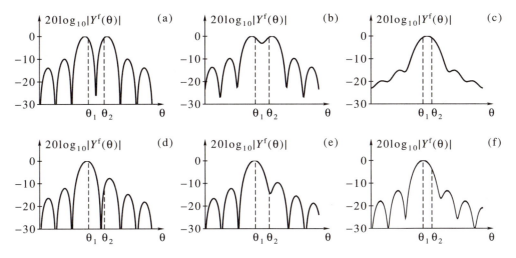

Figure 6.18 The Fourier transform of two complex exponentials in Example 6.1: (a) $A_2 = A_1$, $\theta_2 - \theta_1 = 2\pi/N$; (b) $A_2 = A_1$, $\theta_2 - \theta_1 = 1.5\pi/N$; (c) $A_2 = A_1$, $\theta_2 - \theta_1 = \pi/N$; (d) $A_2 = 0.25A_1$, $\theta_2 - \theta_1 = 2\pi/N$; (e) $A_2 = 0.25A_1$, $\theta_2 - \theta_1 = 1.5\pi/N$; (f) $A_2 = 0.25A_1$, $\theta_2 - \theta_1 = \pi/N$.

where $W^f(\theta, N)$ is the kernel function of the window. In particular,

$$X^f(\theta_1) = A_1 e^{j\phi_1} W^f(0, N) + A_2 e^{j\phi_2} W^f(\theta_1 - \theta_2, N). \tag{6.34}$$

We have $W^f(0, N) = \sum_{n=0}^{N-1} w[n]$, and this sum is approximately proportional to N. Suppose that

$$|A_2 W^f(\theta_1 - \theta_2, N)| \ll A_1 \sum_{n=0}^{N-1} w[n]; \tag{6.35}$$

then we have a good chance of finding a local maximum of $|X^f(\theta)|$ near θ_1. Condition (6.35) holds if $|\theta_2 - \theta_1|$ is greater than the width of the main lobe of the kernel function, and if $20\log_{10}(A_1/A_2)$ is larger than the side-lobe level. We can increase the chances of meeting the latter condition by choosing a window with an extremely low side-lobe level. For example, if we use a Kaiser window with side-lobe level of $-80\,\text{dB}$, we will be able to handle sinusoidal components whose amplitudes differ by up to four orders of magnitude. However, this comes at the price of making the frequency separation condition more difficult to meet, due to the widening of the main lobe. We may be able to compensate for this by increasing the number of samples N, depending on the application.

Example 6.2 Figure 6.19 shows the behavior of the windowed DFT $|X^f(\theta)|$ for two complex exponentials, using the Hann window. The amplitudes A_1, A_2 are the same as in Example 6.1. The frequency differences are $8\pi/N$, $6\pi/N$, and $4\pi/N$. As we see, the two peaks are well separated when the frequency difference is $8\pi/N$, and are still distinguishable when it is $6\pi/N$. The bias is considerably smaller than in the case of unwindowed DFT. However, the two complex exponentials become indistinguishable when the frequency difference is $4\pi/N$, whereas with unwindowed DFT they would be well separated (as we recall from Example 6.1). $\qquad\square$

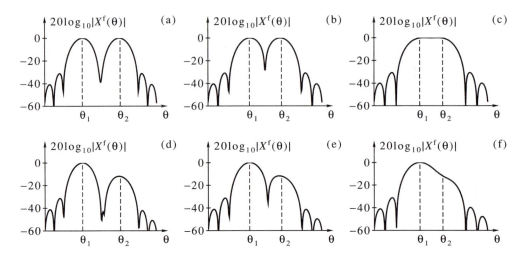

Figure 6.19 The windowed Fourier transform of two complex exponentials in Example 6.2: (a) $A_2 = A_1$, $\theta_2 - \theta_1 = 8\pi/N$; (b) $A_2 = A_1$, $\theta_2 - \theta_1 = 6\pi/N$; (c) $A_2 = A_1$, $\theta_2 - \theta_1 = 4\pi/N$; (d) $A_2 = 0.25A_1$, $\theta_2 - \theta_1 = 8\pi/N$; (e) $A_2 = 0.25A_1$, $\theta_2 - \theta_1 = 6\pi/N$; (f) $A_2 = 0.25A_1$, $\theta_2 - \theta_1 = 4\pi/N$.

6.4.3   Frequency Measurement for Real Sinusoids

We now generalize the frequency measurement problem from 2 to M signal components. We also change from complex exponentials to real-valued sinusoidal signals. The given signal is

$$y(t) = \sum_{k=1}^{M} A_k \cos(\omega_k t + \phi_k). \tag{6.36}$$

Our aim is to measure the frequencies $\{\omega_k, \ 1 \leq k \leq M\}$. Sometimes the amplitudes A_k and the phases ϕ_k are also of interest, but we do not consider them here. Since we are now dealing with real sinusoids, we need to consider only positive frequencies.

We sample the signal (6.36) at interval T and collect N measurements to get

$$y[n] = \sum_{k=1}^{M} A_k \cos(\theta_k n + \phi_k), \quad 0 \leq n \leq N - 1, \tag{6.37}$$

where $\theta_k = \omega_k T$, $\ 1 \leq k \leq M$; we assume that $-\pi < \omega_k T < \pi$.

To compute the Fourier transform of the signal $y[n]$, we express it in the form

$$y[n] = \sum_{k=1}^{M} [0.5 A_k e^{j\phi_k} e^{j\theta_k n} + 0.5 A_k e^{-j\phi_k} e^{-j\theta_k n}]. \tag{6.38}$$

Then,

$$\begin{aligned}
Y^{\mathrm{f}}(\theta) &= \sum_{k=1}^{M} 0.5 A_k e^{j\phi_k} \left[\sum_{n=0}^{N-1} e^{-j(\theta - \theta_k)n} \right] + \sum_{k=1}^{M} 0.5 A_k e^{-j\phi_k} \left[\sum_{n=0}^{N-1} e^{-j(\theta + \theta_k)n} \right] \\
&= \sum_{k=1}^{M} 0.5 A_k e^{-j[0.5(\theta - \theta_k)(N-1) - \phi_k]} D(\theta - \theta_k, N) \\
&\quad + \sum_{k=1}^{M} 0.5 A_k e^{-j[0.5(\theta + \theta_k)(N-1) + \phi_k]} D(\theta + \theta_k, N). \tag{6.39}
\end{aligned}$$

Let us examine the behavior of $Y^{\mathrm{f}}(\theta)$ at one of the sinusoid frequencies themselves, say at $\theta = \theta_m$:

$$Y^f(\theta_m) = 0.5NA_m e^{j\phi_m} + \sum_{\substack{k=1 \\ k \neq m}}^{M} 0.5A_k e^{-j[0.5(\theta_m - \theta_k)(N-1) - \phi_k]} D(\theta_m - \theta_k, N)$$

$$+ \sum_{k=1}^{M} 0.5A_k e^{-j[0.5(\theta_m + \theta_k)(N-1) + \phi_k]} D(\theta_m + \theta_k, N). \tag{6.40}$$

Suppose that all the following conditions hold:

1. All frequencies $\{\theta_k,\ k \neq m\}$ are far from θ_m by $2\pi/N$ at least.

2. The frequency θ_m is not smaller than π/N and not higher than $\pi(1 - 1/N)$.

3. All $\{A_k, k \neq m\}$ are either smaller or not much larger than A_m.

Then the first term in (6.40) will dominate the sum, so $|Y^f(\theta)|$ can be expected to have a local maximum near $\theta = \theta_m$. If we assume that the conditions hold for *all* θ_m, then $|Y^f(\theta)|$ can be expected to exhibit M local maxima in the range $\theta \in (0, \pi)$ near the true values of the θ_m. Of course, the total number of local maxima will usually be larger than M. However, the M maxima of interest correspond to the *main lobes* of the M Dirichlet kernels, so they can be expected to be the largest. We can state the principle of measuring frequencies of real sinusoids by the Fourier transform as follows: Find all local maxima of $|Y^f(\theta)|$ in the range $\theta \in (0, \pi)$ and take the frequencies of the M largest maxima as the measured values of $\{\theta_k\}$.

As in the case of two complex exponentials, the side lobes of the Dirichlet kernel are likely to interfere and prevent this procedure from working. We therefore modify it by applying a window prior to the Fourier transform, that is, we form

$$x[n] = y[n]w[n].$$

Then (6.39) changes to

$$\begin{aligned} X^f(\theta) &= \sum_{n=0}^{N-1} y[n]w[n]e^{-j\theta n} \\ &= \sum_{k=1}^{M} 0.5A_k e^{j\phi_k} \left[\sum_{n=0}^{N-1} w[n]e^{-j(\theta - \theta_k)n} \right] + \sum_{k=1}^{M} 0.5A_k e^{-j\phi_k} \left[\sum_{n=0}^{N-1} w[n]e^{-j(\theta + \theta_k)n} \right] \\ &= \sum_{k=1}^{M} 0.5A_k e^{j\phi_k} W^f(\theta - \theta_k, N) + \sum_{k=1}^{M} 0.5A_k e^{-j\phi_k} W^f(\theta + \theta_k, N). \end{aligned} \tag{6.41}$$

In particular,

$$X^f(\theta_m) = 0.5W^f(0, N)A_m e^{j\phi_m} + \sum_{k \neq m} 0.5A_k e^{j\phi_k} W^f(\theta_m - \theta_k, N)$$

$$+ \sum_{k=1}^{M} 0.5A_k e^{-j\phi_k} W^f(\theta_m + \theta_k, N). \tag{6.42}$$

As we see from (6.42), $|X^f(\theta)|$ is likely to have its M highest local maxima near the $\{\theta_k\}$ if the following conditions are satisfied for each θ_m:

1. All frequencies $\{\theta_k,\ k \neq m\}$ are far from θ_m at least by half the width of the main lobe of $W^f(\theta, N)$.

2. The frequency θ_m is not smaller than half the width of the main lobe of $W^f(\theta, N)$ and not larger than π minus half the width of the main lobe of $W^f(\theta, N)$.

3. The quantities $\{20 \log_{10} A_k\}$ differ from each other by no more than the side-lobe level of the kernel function.

In summary, we state the principle of frequency measurement by Fourier transform as follows:

> Choose a window whose side-lobe level reflects the largest expected ratio between the strongest and weakest sinusoidal components. Multiply the sampled signal by the window and compute $|X^f(\theta)|$. Find all local maxima of $|X^f(\theta)|$ in the range $\theta \in (0, \pi)$. Obtain $\{\theta_k\}$ as the frequencies of the M largest maxima.

6.4.4   Practice of Frequency Measurement

Practical frequency measurement using DFT consists, as a minimum, of the following three steps:

1. Multiplication of the sampled sequence by a window.

2. Computation of the DFT, usually through FFT.

3. Search for the local maxima of the absolute value of the DFT and selection of the maxima of interest.

Additional steps are necessary if measurement of the amplitudes and phases is required as well, but we shall not discuss them here.

As we have explained, the choice of a window requires knowledge of the nature of the signal. Specifically, we need to know the maximum ratio between different A_k, or at least an estimate of the maximum ratio. It is also useful to know the minimum possible separation between frequencies of different components, the distance of the lowest frequency from zero, and the distance of the highest frequency from π. These enable us to tell whether the required resolution can be achieved for the given signal. Without such prior knowledge, the best we can do is to try the procedure on the given signal, but there is no guarantee that it will succeed.

When using basic (windowed) DFT, the accuracy of the measured frequencies is limited by the number of samples. In certain applications, the number of samples may be large, but we may not be able to afford to use them all in a single DFT operation because the number of operations is prohibitively large. In other applications, the number of samples is limited and we must make do with the available amount of data.

As was said before, we have at least two means of interpolating between the DFT points: zero padding and the chirp Fourier transform. The former is recommended when the total number of operations, after zero padding, is not too large. The latter is an attractive alternative when zero padding is too costly in computations. It is often advantageous to repeat the measurement procedure twice in the following manner. In the first pass we compute the DFT at the basic resolution provided by the number of samples N and find the local maxima of interest. In the second pass we compute the DFT at a higher resolution around the local maxima found in the first pass, using the chirp Fourier transform algorithm, to improve the accuracy of the local maxima.

A convenient procedure for finding the M largest local maxima of the DFT is as follows. Suppose we are given the DFT sequence $X^d[k]$. Define

$$\Delta[k] = |X^d[k]| - |X^d[k-1]|, \quad 1 \le k \le N - 1. \tag{6.43}$$

Find all occurrences of k_l such that $\Delta[k_l] \ge 0$ and $\Delta[k_l + 1] < 0$. Each k_l is a point of local maximum (at the resolution of the given sequence). Now take the subsequence $|X^d[k_l]|$ and sort it in a decreasing order of magnitude. The indices of the first M points of the sorted sequence, multiplied by $2\pi/N$, are the frequencies of local maxima.

The procedure `maxdft` in Program 6.3 implements frequency measurement in MAT-LAB. The program contains two main parts. In the first part, the M largest local maxima of the DFT (where M is chosen by the user) are found to an accuracy $\pm\pi/N$. In the second part, the routine `chirpf` (described in Section 5.7) is invoked for each of the local maxima, to find the maximum point to an accuracy $\pm 2\pi/N^2$. The procedure `locmax` in Program 6.4 implements the local maxima search. We note that the input vector must be real (the program needs modification in order to work properly for complex vectors) and that the DFT is optionally windowed.

Example 6.3 Consider the discrete-time signal

$$y[n] = \sin(2\pi \cdot 0.1992n) + 0.005 \sin(2\pi \cdot 0.25n), \quad 0 \leq n \leq 127. \tag{6.44}$$

The frequencies were selected such that the stronger component is in the middle between the points $k = 25$ and $k = 26$ of the DFT, and the weaker component is at $k = 32$ exactly. The amplitude difference is 46 dB in this case.

Figure 6.20 shows the magnitude of the DFT with three windows: rectangular, Bartlett, and Hann. Figure 6.21 continues to show three more windows: Hamming, Blackman, and Kaiser with $\alpha = 10$. As we see, the rectangular and Bartlett windows mask the weaker component and, except for a certain irregularity in the side lobes, we cannot discern anything. The Hann window does show the weaker component, because at a distance of 6.5 units of $2\pi/N$ its side lobes are already low enough. The Hamming window, perhaps unexpectedly, has poorer performance than the Hann window in this case, because of the relative flatness of its side lobes (look again at Figure 6.13!). The Blackman and Kaiser windows both perform well; the weaker component is clearly visible and easily distinguishable from the stronger one. In the Kaiser window, the two components are better separated than in the Blackman window.   □

6.5   Frequency Measurement of Signals in Noise*

We now generalize the discussion in Section 6.4 to sinusoidal signals measured with additive white noise. Noise is present to one degree or another in almost all real-life applications. Noise is often broad band and becomes white (or nearly so) after prefiltering and sampling. Proper understanding of the effect of noise on frequency measurement is therefore crucial to practical spectral analysis. The basic method of frequency measurement in the presence of noise is the same as when there is no noise: multiplication of the data vector by a window, computation of the magnitude of the DFT, search for maximum (or several local maxima), followed by an optional fine search, by either zero padding or the chirp Fourier transform. Noise affects this procedure in two respects:

1. It masks the peaks in the magnitude of the DFT (in addition to masking caused by side lobes, which we have already encountered), thus making them more difficult to identify. The problem is then to find the peaks belonging to the sinusoidal signals in the DFT when the DFT contains many noise peaks. This is an example of a general problem known as *signal detection.*

2. It causes the point of maximum to deviate from the true frequency, thus introducing errors to the measured frequencies. Since noise always introduces errors, it is common to refer to frequency measurement in noise as *frequency estimation.* The terms *estimation* and *estimates* imply randomness of the measured parameter(s) due to randomness in the signal.

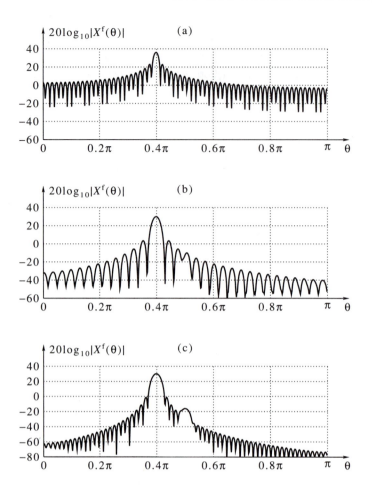

Figure 6.20 Frequency measurement of two sinusoids by windowed DFT: (a) rectangular window; (b) Bartlett window; (c) Hann window.

Our goal in this section is to analyze these two effects quantitatively. Some of the mathematical derivations are based only on the material in Chapter 2; these will be carried out in detail. Other results rely on tools beyond the scope of this book; for these we shall give only the final formulas.

6.5.1   Signal Detection

To simplify the derivations, we examine first the case of a single complex exponential signal. A real sinusoid has two complex exponential components, which interfere with each other and complicate the analysis, so we defer its discussion until later. Since we assume that the signal is complex valued, we also assume that the noise is complex valued for compatibility. Our signal model is thus

$$y[n] = s[n] + v[n] = Ae^{j\theta_0 n} + v[n], \quad 0 \le n \le N - 1, \tag{6.45}$$

where $v[n]$ is discrete-time complex white noise,[2] with zero mean and variance y_v.

The windowed signal is

$$x[n] = y[n]w[n] = Ae^{j\theta_0 n}w[n] + v[n]w[n].$$

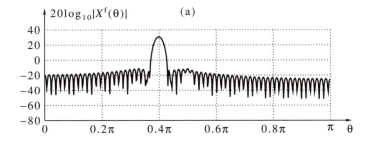

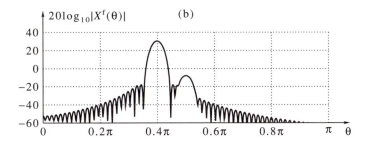

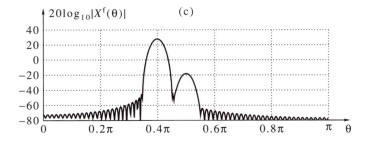

Figure 6.21 Frequency measurement of two sinusoids by windowed DFT: (a) Hamming window; (b) Blackman window; (c) Kaiser window, $\alpha = 10$.

The Fourier transform of the windowed signal is given by

$$X^{\mathrm{f}}(\theta) = A \sum_{n=0}^{N-1} w[n]e^{j(\theta_0-\theta)n} + \sum_{n=0}^{N-1} w[n]v[n]e^{-j\theta n}. \tag{6.46}$$

Since the elements of the window are always real and nonnegative, the value of θ at which the transform of $s[n]w[n]$ is maximized is $\theta = \theta_0$. The corresponding value of $X^{\mathrm{f}}(\theta_0)$ is

$$X^{\mathrm{f}}(\theta_0) = A \sum_{n=0}^{N-1} w[n] + \sum_{n=0}^{N-1} w[n]v[n]e^{-j\theta_0 n}. \tag{6.47}$$

Let us define

$$CG = \frac{1}{N} \sum_{n=0}^{N-1} w[n]. \tag{6.48}$$

The parameter CG is called the *coherent gain* of the window. Using this parameter, we can express (6.47) as

$$X^{\mathrm{f}}(\theta_0) = NA \cdot CG + \sum_{n=0}^{N-1} w[n]v[n]e^{-j\theta_0 n}. \tag{6.49}$$

The coherent gain depends on the window type, but is nearly independent of the window length N. Therefore, the gain of the complex exponential at the maximum point of the Fourier transform is approximately linear in N.

As we saw in the preceding section, frequency measurement is based on the magnitude of the Fourier transform. Here it will be more convenient to work with the square magnitude, so let us derive an expression for $|X^{\mathrm{f}}(\theta_0)|^2$. We have

$$|X^{\mathrm{f}}(\theta_0)|^2 = (NA \cdot CG)^2 + 2NA \cdot CG \cdot \Re\left\{\sum_{n=0}^{N-1} w[n]v[n]e^{-j\theta_0 n}\right\}$$
$$+ \sum_{n=0}^{N-1}\sum_{m=0}^{N-1} w[n]v[n]w[m]\bar{v}[m]e^{-j\theta_0(n-m)}. \tag{6.50}$$

The square magnitude of the Fourier transform is a random variable; its mean is given by

$$E(|X^{\mathrm{f}}(\theta_0)|^2) = (NA \cdot CG)^2 + 2NA \cdot CG \cdot \Re\left\{\sum_{n=0}^{N-1} w[n]E(v[n])e^{-j\theta_0 n}\right\}$$
$$+ \sum_{n=0}^{N-1}\sum_{m=0}^{N-1} w[n]w[m]E(v[n]\bar{v}[m])e^{-j\theta_0(n-m)}$$
$$= (NA \cdot CG)^2 + \gamma_v \sum_{n=0}^{N-1}\sum_{m=0}^{N-1} w[n]w[m]\delta[n-m]e^{-j\theta_0(n-m)}$$
$$= (NA \cdot CG)^2 + \gamma_v \sum_{n=0}^{N-1} w^2[n]. \tag{6.51}$$

The first term on the right side of (6.51) comes from the complex exponential signal, whereas the second comes from the noise. We define the *output signal-to-noise ratio* (OSNR) of the windowed Fourier transform as the ratio between the two, that is,

$$\mathrm{SNR_o} = \frac{(NA \cdot CG)^2}{\gamma_v \sum_{n=0}^{N-1} w^2[n]}. \tag{6.52}$$

We also define the *input signal-to-noise ratio* (ISNR) as

$$\mathrm{SNR_i} = \frac{A^2}{\gamma_v}. \tag{6.53}$$

The *processing gain* of the windowed Fourier transform is, by definition, the ratio between the OSNR and N times the ISNR. We get from (6.52), (6.53)

$$\mathrm{PG} = \frac{\mathrm{SNR_o}}{N \cdot \mathrm{SNR_i}} = \frac{N^2(CG)^2}{N \sum_{n=0}^{N-1} w^2[n]} = \frac{\left(\sum_{n=0}^{N-1} w[n]\right)^2}{N \sum_{n=0}^{N-1} w^2[n]}. \tag{6.54}$$

Similarly to the coherent gain, the processing gain depends on the window type, but it is nearly independent of the window length N. Therefore, $\mathrm{SNR_o}$ is approximately linear in N.

Practice shows that when the OSNR is about 14 dB or greater (or 25 in absolute value), the presence of the complex exponential can be detected reliably,[3] and its frequency estimated to a good accuracy. Using the definition of processing gain, we get the following rule of thumb:

If the amplitude of the complex exponential A, the noise variance γ_v, and the number of samples N satisfy

$$\frac{NA^2 \cdot \text{PG}}{\gamma_v} \geq 25, \tag{6.55}$$

there is high probability that the maximum magnitude of the Fourier transform will be near the true frequency; it is then safe to use the windowed Fourier transform for frequency estimation.

The discrete-time signal $y[n]$ is often obtained by antialias filtering and sampling of a continuous-time signal accompanied by broad-band noise. Recall from Section 3.7 that, in this case, the variance of the discrete-time noise is given by $\gamma_v = N_0/T$, where N_0 is the power density of the continuous-time noise, in units of power per hertz, and T is the sampling interval. Then the input SNR is given by

$$\text{SNR}_i = \frac{A^2 T}{N_0}. \tag{6.56}$$

Correspondingly, the rule of thumb (6.55) can be expressed as

$$\frac{DA^2 \cdot \text{PG}}{N_0} \geq 25, \tag{6.57}$$

where $D = NT$ is the measurement interval of the continuous-time signal.

In our analysis so far, we have assumed that the windowed Fourier transform is computed at all frequencies θ in order to find the maximum point. In practice this does not happen, of course, since only a finite number of frequency points can be computed. In the simplest case we compute the windowed DFT of the signal at a resolution of $2\pi/N$. If zero padding or a chirp Fourier transform is used, the resolution is better. Let us denote by $\Delta\theta$ the distance between two adjacent frequencies at which $|X^f(\theta)|$ is computed (this is the frequency resolution of the algorithm being used). Then, if θ_0 is not an integer multiple of $\Delta\theta$, the use of the coherent gain [as defined in (6.48)] and the processing gain [as defined in (6.54)] are not justified any more. Instead, both parameters will depend on the deviation of θ_0 from the nearest integer multiple of $\Delta\theta$. In the worst case, θ_0 is exactly in the middle between two integer multiples of $\Delta\theta$. Accordingly, we define the *worst-case coherent gain* as

$$\text{CG}(\Delta\theta) = \frac{1}{N} \left| \sum_{n=0}^{N-1} w[n] e^{j0.5\Delta\theta n} \right|, \tag{6.58}$$

and the *worst-case processing gain* as

$$\text{PG}(\Delta\theta) = \frac{\left| \sum_{n=0}^{N-1} w[n] e^{j0.5\Delta\theta n} \right|^2}{N \sum_{n=0}^{N-1} w^2[n]}. \tag{6.59}$$

The rule of thumb (6.55) [or (6.57)] should be modified in this case by substituting $\text{PG}(\Delta\theta)$ for PG. We emphasize again that $\Delta\theta$ depends on the algorithm: For simple DFT it is $2\pi/N$, for zero-padded DFT it is $2\pi/L$ (where L is the length after zero padding), and for the chirp Fourier transform it is on the order of $2\pi/N^2$. In the case of the chirp Fourier transform, the loss due to $\Delta\theta$ is usually negligible and $\text{PG}(0)$ can be used for most practical purposes.

Table 6.2 summarizes the various parameters of the windows described in Section 6.3. SLL is the side-lobe level, as defined in Section 6.3. $\text{CG}(0)$ and $\text{CG}(2\pi/N)$ are the best- and worst-case coherent gains, where the worst case corresponds to simple DFT. $\text{PG}(0)$ and $\text{PG}(2\pi/N)$ are the corresponding processing gains. The parameter J_w

will be discussed later; for now it should be ignored. The values shown in the table are for large N. For short windows, the values differ slightly.

As we see from Table 6.2, all gains are actually losses, since they are all negative in dB. We must remember, however, that these values are normalized by N and that there is always an additional gain of N, as seen in (6.49) and (6.55). Windows with lower side-lobe levels have smaller best-case gains in general, but the difference between the best case and the worst case is also smaller. The rectangular window is the only one whose best-case gains are 0 dB, but its worst-case values are inferior. The worst-case processing gain of all windows is between -3 and -4 dB. The procedure cpgains in Program 6.5 implements the computation of the coherent gain and processing gain, as well as the parameter J_w defined in Section 6.5.2.

Window	SLL	CG(0)	$CG\left(\frac{2\pi}{N}\right)$	PG(0)	$PG\left(\frac{2\pi}{N}\right)$	J_w
rectangular	-13.5	0	-3.92	0	-3.92	1.00
Bartlett	-27	-6.01	-7.84	-1.25	-3.07	1.59
Hann	-32	-6.03	-7.45	-1.77	-3.19	2.35
Hamming	-43	-5.36	-7.11	-1.35	-3.10	1.65
Blackman	-57	-7.54	-8.64	-2.38	-3.47	3.15
Kaiser, $\alpha = 4$	-30	-4.39	-6.45	-0.96	-3.02	1.41
Kaiser, $\alpha = 8$	-58	-7.22	-8.40	-2.22	-3.40	2.82
Kaiser, $\alpha = 12$	-90	-8.93	-9.75	-3.03	-3.85	4.70
Dolph, $\alpha = -40$	-40	-4.60	-6.67	-1.15	-3.21	1.79
Dolph, $\alpha = -60$	-60	-6.40	-7.82	-1.81	-3.24	2.16
Dolph, $\alpha = -80$	-80	-7.66	-8.74	-2.41	-3.50	3.13

Table 6.2 Windows and their parameters (all in dB except for J_w).

6.5.2 Frequency Estimation

Let us now discuss the accuracy of the frequency estimated from the maximum point of the Fourier transform when noise is present. To isolate the effect of noise from that of sampling at a finite number of frequencies, we assume that $|X^f(\theta)|$ is computed at arbitrarily high resolution, for example, by the chirp Fourier transform. Let us denote by $\hat{\theta}_0$ the frequency where $|X^f(\theta)|$ is maximum. Then $\hat{\theta}_0$ is close, but not identical, to the true frequency θ_0. The difference $\hat{\theta}_0 - \theta_0$ is a random variable, depending on the noise, the amplitude of the complex exponential, and the number of samples. The derivation of the mean and variance of this random variable relies on mathematical tools beyond the scope of this book, so we give only the final results. These results are valid when the inequality (6.55) [or (6.57)] is satisfied.

For a given window $w[n]$, define

$$J_w = \frac{N^3 \sum_{n=0}^{N-1} [n - 0.5(N-1)]^2 w^2[n]}{12 \left(\sum_{n=0}^{N-1} [n - 0.5(N-1)]^2 w[n] \right)^2}. \tag{6.60}$$

This parameter is nearly independent of N for large N, depending only on the window type. Using this parameter, we have

$$E(\hat{\theta}_0 - \theta_0)^2 \approx \frac{6\gamma_v J_w}{A^2 N^3} = \frac{6 J_w}{\text{SNR}_i \cdot N^3}. \tag{6.61}$$

In addition,

$$E(\hat{\theta}_0 - \theta_0) \approx 0. \tag{6.62}$$

The parameter J_w is between 1 and 4 for most practical windows. Its smallest value, 1, is obtained for a rectangular window, that is, for an unwindowed Fourier transform. This case represents the best accuracy (i.e., the smallest mean-square error) that can be obtained for given SNR and number of samples.

When the discrete-time signal $y[n]$ is obtained by antialias filtering and sampling of a continuous-time signal accompanied by broad-band noise, the input SNR is given by (6.56). In this case we are interested in the error in ω, rather than the error in θ. Substituting $\omega = \theta/T$ and $D = NT$, we get from (6.61)

$$E(\hat{\omega}_0 - \omega_0)^2 \approx \frac{6N_0 J_w}{A^2 D^3}. \tag{6.63}$$

The mean-square error in the frequency f is given by

$$E(\hat{f}_0 - f_0)^2 \approx \frac{6N_0 J_w}{(2\pi)^2 A^2 D^3}. \tag{6.64}$$

Substitution of (6.57) in (6.63) leads to the inequality

$$[E(\hat{f}_0 - f_0)^2]^{1/2} \leq \sqrt{\frac{6J_w \cdot PG}{100\pi^2}} \cdot \frac{1}{D}. \tag{6.65}$$

The left side of (6.65) represents the root-mean-square (RMS) frequency error in hertz. As we recall from Section 4.1, the quantity $1/D$ (the reciprocal of the measurement interval) represents the basic resolution of the DFT in hertz; see page 99. The numerical factor multiplying $1/D$ in (6.65) is smaller than 1 for all common windows. For example, with PG = 0.5 and $J_w = 4$, this factor is about 0.11. We are thus led to the conclusion that *the RMS frequency error due to noise is smaller than the basic DFT resolution.* Furthermore, if the left side of (6.57) is much larger than 25, this factor will be much smaller than 1. It therefore pays to compute the maximum point of $|X^f(\theta)|$ at a resolution much better than the basic DFT resolution. Zero padding is not practical for this purpose, since it would require a high padding ratio. On the other hand, the chirp Fourier transform is useful. An alternative to the chirp Fourier transform for this purpose is discussed in Problem 6.20.

6.5.3 Detection and Frequency Estimation for Real Sinusoids

Having treated the case of a single complex exponential, let us now proceed to see the effect of noise on a single real sinusoid. The measured signal is now

$$\begin{aligned} y[n] = s[n] + v[n] &= A\cos(\theta_0 n + \phi_0) + v[n] \\ &= 0.5Ae^{j(\theta_0 n + \phi_0)} + 0.5Ae^{-j(\theta_0 n + \phi_0)} + v[n]. \end{aligned} \tag{6.66}$$

Note that the noise $v[n]$ is now real valued. The Fourier transform of the windowed signal $x[n]$ at $\theta = \theta_0$ is

$$X^f(\theta_0) = 0.5NAe^{j\phi_0} \cdot CG + 0.5A \sum_{n=0}^{N-1} w[n]e^{-j(2\theta_0 n + \phi_0)} + \sum_{n=0}^{N-1} w[n]v[n]e^{-j\theta_0 n}. \tag{6.67}$$

Let us assume that the number of samples N is such that $2\theta_0$ and $2(\pi - \theta_0)$ are both larger than the main-lobe width of $W^f(\theta)$, the kernel function of the window. As we saw in Section 6.4, this guarantees that only a side lobe of the negative spectral component interfere with the positive one. In other words, the second term on the right side of (6.67) is negligible, resulting in

$$X^f(\theta_0) \approx 0.5NAe^{j\phi_0} \cdot CG + \sum_{n=0}^{N-1} w[n]v[n]e^{-j\theta_0 n}. \tag{6.68}$$

Comparing this expression with (6.49), we see that the main change is the factor 0.5, resulting from splitting the real sinusoid into two spectral lines, one positive and one negative. The constant phase term $e^{j\phi_0}$ is immaterial, since it disappears when taking the absolute value. Therefore, all results derived for a complex exponential, and the conclusions drawn from them hold for the real case, the only change being the replacement of A by $0.5A$. In particular, formula (6.52) changes to

$$\text{SNR}_o = \frac{(NA \cdot CG)^2}{4\gamma_v \sum_{n=0}^{N-1} w^2[n]}. \tag{6.69}$$

The input SNR of a real sinusoid is defined as

$$\text{SNR}_i = \frac{A^2}{2\gamma_v}. \tag{6.70}$$

Note that, although $(0.5A)^2 = 0.25A^2$, there is a division only by 2 in the definition, since the average power of a sinusoid is half its square amplitude. The processing gain is now defined as

$$\text{PG} = \frac{\text{SNR}_o}{0.5N \cdot \text{SNR}_i}, \tag{6.71}$$

so its expression in terms of the window coefficients (6.54) remains unchanged. The rules of thumb (6.55), (6.57) change to

$$\frac{NA^2 \cdot \text{PG}}{\gamma_v} \geq 100, \qquad \frac{DA^2 \cdot \text{PG}}{N_0} \geq 100. \tag{6.72}$$

The frequency error expressions change to

$$E(\hat{\theta}_0 - \theta_0)^2 \approx \frac{24\gamma_v J_w}{A^2 N^3}, \qquad E(\hat{\theta}_0 - \theta_0) \approx 0, \tag{6.73a}$$

$$E(\hat{f}_0 - f_0)^2 \approx \frac{24N_0 J_w}{(2\pi)^2 A^2 D^3}, \qquad E(\hat{f}_0 - f_0) \approx 0. \tag{6.73b}$$

We emphasize again that these approximations are not as good as the ones in the case of a complex exponential, due to the interference term from the negative spectral line, which we have neglected.

Finally, we consider the frequency estimation problem for the M-sinusoids model (6.37), modified to include additive white noise $v[n]$, that is,

$$y[n] = \sum_{k=1}^{M} A_k \cos(\theta_k n + \phi_k) + v[n], \quad 0 \leq n \leq N - 1. \tag{6.74}$$

Assume that the following conditions are satisfied:

1. The frequencies and the amplitudes are such that, for each k, the kth complex exponential $0.5A_k e^{j(\theta_k n + \phi_k)}$ is only weakly disturbed by the other complex exponentials.

2. The inequalities (6.72) hold for all A_k.

In this case, the approximate expressions (6.73) apply to each k individually, that is,

$$E(\hat{\theta}_k - \theta_k)^2 \approx \frac{24\gamma_v J_w}{A_k^2 N^3}, \qquad E(\hat{\theta}_k - \theta_k) \approx 0, \tag{6.75a}$$

$$E(\hat{f}_k - f_k)^2 \approx \frac{24N_0 J_w}{(2\pi)^2 A_k^2 D^3}, \qquad E(\hat{f}_k - f_k) \approx 0. \tag{6.75b}$$

Great care should be taken, however, in using these approximations, since they tend to be rather crude, especially when the number of sinusoids is large. Unfortunately, accurate analysis of the M-sinusoid problem is difficult and few general results are known. Therefore, it is good practice to perform computer simulations and field tests, and to avoid relying on the approximate formulas exclusively.

Example 6.4 Consider again the discrete-time signal in Example 6.3,

$$y[n] = \sin(2\pi \cdot 0.1992n) + 0.005 \sin(2\pi \cdot 0.25n), \quad 0 \le n \le 127. \tag{6.76}$$

Here we wish to examine the windowed Fourier transform of this signal when measured with white noise. Since the Kaiser window with $\alpha = 10$ was proved to be the best among all windows tested in Example 6.3, we use only this window here. First, we choose the noise standard deviation such that (6.55) will be satisfied with equality for the weaker of the two sinusoids. We have

$$N = 128, \quad A_2 = 0.005, \quad \text{PG} = 0.538,$$

which gives $\gamma_v^{1/2} = 0.00415$. Figure 6.22(a) shows a typical DFT magnitude for these parameters. Comparing this figure with Figure 6.21(c), we see that the DFT floor has been raised and looks noisy. The peak corresponding to the first (strong) sinusoid has not suffered any discernible degradation. The peak corresponding to the second (weak) sinusoid is still the second largest, but is only a few dB above some noise peaks.

Figure 6.22(b) shows what happens when the noise standard deviation is doubled. Now inequality (6.55) is not satisfied any more. As we see, the peak corresponding to the weak sinusoid is no longer the second in magnitude. Thus, the noise masks this sinusoid and renders it undetectable with high probability.

Figure 6.22(c) shows the effect of raising the noise level such that (6.55) holds with equality for the strong sinusoid. The corresponding value of $\gamma_v^{1/2}$ is 0.83. The weak sinusoid is now completely lost in the noise. The strong sinusoid still gives the largest peak, but some noise peaks appear only a few dB below it. Increasing the noise level still further would mask the strong sinusoid as well. □

Example 6.5 We continue Example 6.4 and examine the accuracy of the frequency estimates. The parameter J_w of a Kaiser window with $\alpha = 10$ is 3.79. The RMS of the frequency error $\hat{\theta}_2 - \theta_2$ is, according to (6.73), approximately 5.47×10^{-3}. To confirm this value, we perform a sequence of random simulations. (Simulations of this kind are called *Monte-Carlo simulations*, in honor of Europe's unofficial gambling capital.) Each simulation adds a different noise realization to the signal $y[n]$ (using the MATLAB function randn), and then the frequencies $\hat{\theta}_1, \hat{\theta}_2$ are found from the largest local maxima of the windowed Fourier transform. After L such simulations have been performed, we compute the *empirical mean error*

$$\langle \hat{\theta}_2 - \theta_2 \rangle = \frac{1}{L} \sum_{l=1}^{L} (\hat{\theta}_{2,l} - \theta_2), \tag{6.77}$$

and the *empirical RMS error*

$$\langle (\hat{\theta}_2 - \theta_2)^2 \rangle^{1/2} = \left[\frac{1}{L} \sum_{l=1}^{L} (\hat{\theta}_{2,l} - \theta_2)^2 \right]^{1/2}, \tag{6.78}$$

where $\{\hat{\theta}_{2,l}, \ 1 \le l \le L\}$ are the results of the individual random runs. We do not do the same for $\hat{\theta}_1$, since its accuracy, according to (6.73), is much higher than the accuracy

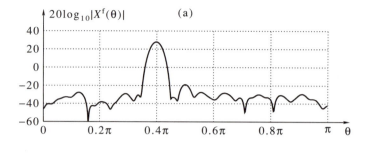

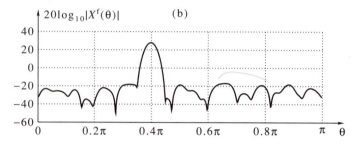

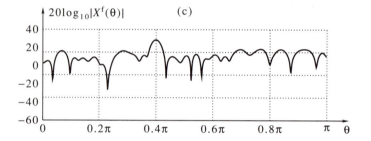

Figure 6.22 Frequency measurement of two sinusoids in noise by Kaiser-windowed DFT: (a) $y_v^{1/2} = 0.00415$; (b) $y_v^{1/2} = 0.0083$; (c) $y_v^{1/2} = 0.83$.

of the chirp Fourier transform algorithm in this case. In this test we use $L = 1000$. The results are:

$$\langle \hat{\theta}_2 - \theta_2 \rangle = -8.64 \times 10^{-4}, \quad \langle (\hat{\theta}_2 - \theta_2)^2 \rangle^{1/2} = 5.27 \times 10^{-3}.$$

As we see, the empirical RMS is close to the theoretical RMS, which is 5.47×10^{-3}, and the empirical mean is smaller than the empirical RMS by almost an order of magnitude. Therefore, the computer simulations are in good agreement with the theoretical formulas in this case.

Next we repeat the same experiment with $y_v^{1/2} = 0.83$. As we saw in Example 6.4, only the frequency θ_1 is of interest in this case. Formula (6.73) predicts an RMS error of 5.47×10^{-3} for $\hat{\theta}_1$ in this case. The empirical mean and RMS errors obtained from the simulations are

$$\langle \hat{\theta}_1 - \theta_1 \rangle = -8.78 \times 10^{-5}, \quad \langle (\hat{\theta}_1 - \theta_1)^2 \rangle^{1/2} = 5.61 \times 10^{-3}.$$

Again, there is good agreement between the empirical and the theoretical RMS error. Moreover, the empirical mean is smaller than the empirical RMS by almost two orders of magnitude.   □

6.6 Summary and Complements

6.6.1 Summary

This chapter was devoted to practical aspects of spectral analysis, in particular short-time spectral analysis. Short-time spectral analysis is needed (1) when the data sequence is naturally short or (2) when it is long, but splitting it into short segments and analyzing each segment separately makes more sense physically.

The main tool in short-time spectral analysis is windowing. Windowing is the combined operation of selecting a fixed-length segment of the data and shaping the signal in the segment by multiplication, as expressed by (6.8). The equivalent operation in the frequency domain is convolution with the window's kernel function (its Fourier transform), as expressed by (6.9).

The two main parameters by which we judge the suitability of a window for a given task are the width of the main lobe of its kernel function and the side-lobe level relative to the main lobe. The main lobe acts to smear the Fourier transform of the signal (through frequency-domain convolution); therefore it should be as narrow as possible. The side lobes lead to interference between signal components at different frequencies, so they should be as low as possible. These two requirements contradict each other; therefore, the selection of a window for a particular application involves a trade-off between the two parameters.

The rectangular window has the narrowest main lobe (for a given length of the window), but the highest side lobes. Because of its high side lobes, it is rarely used. Common windows, arranged in a decreasing order of side-lobe level, are Bartlett (6.10), Hann (6.14), Hamming (6.16), and Blackman (6.18). Two window types that enable tuning of the side-lobe level via an additional parameter are the Kaiser window (6.21) and the Dolph window (6.22). The former approximately minimizes the side-lobe energy for a given width of the main lobe. The latter is an equiripple window; its side lobes have a flat, tunable level. Of the two, the Kaiser window is more commonly used.

We have demonstrated the use of spectral analysis for sinusoidal frequency measurement, first without noise and then in the presence of noise. Sinusoidal frequency measurement accuracy depends on the length of the data interval, the separation between the frequencies of the various sinusoids, the relative amplitudes of the sinusoids, and the window used. If noise is present, accuracy also depends on the signal to noise ratio. Special forms of the DFT, such as zero-padded DFT or chirp Fourier transform, can be used for increasing the accuracy.

6.6.2 Complements

1. [p. 164] The notes of the eight chords are shown in the following table; see Section 14.3 for an explanation of the relationships between notes and their frequencies.

Bar No.	1	2	3	4	5	6	7	8
note 1	C	A	E	C	F#	E	B	E
note 2	E	C	G	E	A#	G	D#	G#
note 3	A	F#	B	A	C#	B	F	B
note 4	—	—	—	—	E	—	A	—

2. [p. 186] A discrete-time complex white noise is $v[n] = v_r[n] + jv_i[n]$, where

(a) Each of $v_r[n]$, $v_i[n]$ is white noise and they have zero means and equal variances.

(b) $v_r[n]$, $v_i[n]$ are uncorrelated, that is, $E(v_r[n]v_i[m]) = 0$ for all n, m.

The covariance sequence of $v[n]$ is defined by

$$\kappa_v[m] = E(v[n+m]\bar{v}[n]) = \gamma_v \delta[m] = (\gamma_{v_r} + \gamma_{v_i})\delta[m].$$

3. [p. 188] It is common to measure the reliability of signal detection by two criteria:

   (a) The probability that a nonexistent signal will be falsely detected. This is called the *false alarm probability* and is denoted by P_{fa}.

   (b) The probability that an existing signal will not be detected. This is called the *miss probability* and is denoted by P_{miss}. Its complement, $P_{det} = 1 - P_{miss}$, is called the *detection probability*.

These two criteria always conflict with each other; decreasing the probability of false alarm increases the probability of miss and vice versa. Quantitative analysis of these probabilities for the problem of detecting sinusoids in noise is beyond the level of this book; see, for example, Van Trees [1968].

6.7 MATLAB Programs

Program 6.1 Generation of the seven common window types.

```
function  w = window(N,name,alpha);
% Synopsis: w = window(N,name,alpha).
% Generates a window.
% input parameters:
% N: the window length
% name: the window name inside quotes, as follows:
%       'rect' for rectangular
%       'bart' for Bartlett
%       'hann' for Hann
%       'hamm' for Hamming
%       'blac' for Blackman
%       'kais' for Kaiser
%       'dolp' for Dolph
% alpha: the Kaiser window parameter for name = 'kais',
%        the side-lobe level (in dB) for name = 'dolp',
%        not used otherwise.
% Output:
% w: the window (a row vector)

n = 0:N-1;
if (name(1:4) == 'rect'),
   w = ones(1,N);
elseif (name(1:4) == 'bart'),
   w = 1-(1/(N+1))*abs(2*n-N+1);
elseif (name(1:4) == 'hann'),
   w = 0.5*(1-cos((2*pi/(N-1))*n));
elseif (name(1:4) == 'hamm'),
   w = 0.54-0.46*cos((2*pi/(N-1))*n);
elseif (name(1:4) == 'blac'),
   w = 0.42-0.5*cos((2*pi/(N-1))*n) + 0.08*cos((4*pi/(N-1))*n);
elseif (name(1:4) == 'kais'),
   w = 1/(besseli(0,alpha))* ...
       besseli(0,alpha*sqrt(1-((1/(N-1))*abs(2*n-N+1)).^2));
elseif (name(1:4) == 'dolp'),
   w = dolph(N,alpha);
else,
   error('Unrecognized window name in WINDOW')
end
```

Program 6.2 Implementation of Dolph window.

```
function w = dolph(N,sll);
% Synopsis: w = dolph(N,sll).
% Computes a Dolph-Chebyshev window.
% Input parameters:
% N: the window length
% sll: the side-lobe level, in dB (negative number)
% Output:
% w: the window.

b = 10^(-sll/20); b = cosh(log(b+sqrt(b^2-1))/(N-1));
c = b*cos((pi/N)*(0:N-1)); ind1 = find(abs(c) <= 1);
ind2 = find(abs(c) > 1); w = zeros(1,N);
w(ind1) = cos((N-1)*acos(c(ind1)));
w(ind2) = cosh((N-1)*log(c(ind2)+sqrt(c(ind2).^2-1))); w = real(w);
if (rem(N,2) == 0), w = w.*exp((j*pi/N)*(0:N-1)); end
w = fftshift(real(ifft(w))); w = (1/w(floor(N+1)/2))*w;
```

Program 6.3 Computation of the *M* largest maxima of the DFT of a real vector.

```
function [theta,val] = maxdft(x,M,name,alpha);
% Synopsis: [theta,val] = maxdft(x,M,name,alpha).
% Finds the M largest maxima of the DFT of the real vector x.
% Input parameters:
% x: the input vector
% M: number of local maxima to be found
% name: an optional window for x; one of the names in window.m
% alpha: needed if name = 'kaiser'.
% Output parameters:
% theta: vector of thetas at the local maxima
% val: vector of corresponding values of DFT(x).

N = length(x); x = reshape(x,1,N);
if (nargin == 3), x = x.*window(N,name);
elseif (nargin == 4), x = x.*window(N,name,alpha); end
X = abs(fft(x)); X = X(1:floor((N+1)/2));
[y,ind] = locmax(X); ind = ind(1:M);
theta = (2*pi/N)*(ind-1); val = zeros(1,M);
for m = 1:M,
   X = chirpf(x,theta(m)-(2*pi/N),(2*pi/N^2),2*N+1);
   [y,ind] = max(abs(X));
   theta(m) = theta(m)-(2*pi/N) + (ind-1)*2*pi/N^2;
   val(m) = y;
end
```

Program 6.4 Search for the local maxima of a vector and their indices.

```
function [y,ind] = locmax(x);
% Synopsis: [y,ind] = locmax(x).
% Finds all local maxima in a vector and their locations,
% sorted in decreasing maxima values.
% Input:
% x: the input vector.
% Output parameters:
% y: the vector of local maxima values
% ind: the corresponding vector of indices of the input vector x.

n = length(x); x = reshape(x,1,n);
xd = x(2:n)-x(1:n-1);
i = find(xd(1:n-2) > 0.0 & xd(2:n-1) < 0.0) + 1;
if (x(1) > x(2)), i = [1,i]; end
if (x(n) > x(n-1)), i = [i,n]; end
[y,ind] = sort(x(i)); ind = fliplr(ind);
ind = i(ind); y = x(ind);
```

Program 6.5 The coherent gain and processing gain of a window.

```
function [cg,pg,jw] = cpgains(w,dtheta);
% Synopsis: [cg,pg,jw] = cpgains(w,dtheta).
% Computes the coherent gain and the processing gain of
% a given window as a function of the frequency deviation.
% Also computes the parameter Jw (see text).
% Input parameters:
% w: the window sequence (of length N)
% dtheta: the frequency deviation:
%    0 gives the best-case gains
%    2*pi/N gives the worst case for N-point DFT
%    2*pi/M gives the worst case for M-point zero-padded DFT
% Output parameters:
% cg: the coherent gain, in dB
% pg: the processing gain, in dB
% jw: the parameter Jw.

N = length(w); w = reshape(w,1,N);
cg = (1/N)*abs(sum(w.*exp(j*0.5*dtheta*(0:N-1))));
pg = N*cg^2/sum(w.*w);
cg = 20*log10(cg); pg = 10*log10(pg);
n = (0:N-1)-0.5*(N-1);
jw = (N^3/12)*(sum((w.*n).^2)/(sum(w.*(n.^2)))^2);
```

6.8 Problems

6.1 We have seen that the largest side lobe of the rectangular window has attenuation $-13.5\,\text{dB}$ with respect to the main lobe. What is approximately the attenuation of the smallest side lobe? Note that the result depends on N.

6.2 Let N be even. Show that the convolution of a rectangular window of length $N/2$ and a rectangular window of length $(N + 2)/2$ gives the Bartlett window (6.10) for even-length N, up to a constant scale factor. Then show that the kernel function of this window is (6.12).

6.3 A window $w[n]$ is generated by convolving a rectangular window of length N with a Bartlett window of length $2N - 1$. What are the length, main lobe width, and side-lobe level of $w[n]$?

6.4 Recall that the Bartlett window was obtained by convolving a rectangular window $w_r[n]$ with itself. Suppose that we generate a window $w[n]$ by K-fold convolution of $w_r[n]$ with itself. Describe the properties of $w[n]$.

6.5 Let $x[n]$ be a discrete-time signal on $0 \le n \le N - 1$ and $X^d[k]$ its DFT. Define

$$Y_1^d[k] = 0.5X^d[k] - 0.25X^d[(k - 1) \bmod N] - 0.25X^d[(k + 1) \bmod N]. \quad (6.79)$$

(a) Show that $Y_1^d[k]$ represents a windowing operation in the time domain. How is this window related to the Hann window? Hint: Use (4.32).

(b) Define

$$Y_2^d[k] = 0.42X^d[k] - 0.25X^d[(k - 1) \bmod N] - 0.25X^d[(k + 1) \bmod N]$$
$$+ 0.04X^d[(k - 2) \bmod N] + 0.04X^d[(k + 2) \bmod N]. \quad (6.80)$$

Show that $Y_2^d[k]$ represents a windowing operation in the time domain. To which of the windows you have learned is this window related?

(c) Discuss possible implementation advantages of the windows in parts a and b.

(d) Can the idea proposed in this problem be applied to other windows you know?

6.6 Consider the window

$$w[n] = \sin^4\left(\frac{\pi n}{N - 1}\right).$$

Explore the properties of this window: main-lobe width and side-lobe level. How is this window related to the Hann window?

6.7 We are given 128 samples of the signal

$$x[n] = \sin\left(2\pi\frac{6.3}{128}n\right) + 0.001 \sin\left(2\pi\frac{56}{128}n\right).$$

(a) Explain why a rectangular window is not adequate for detecting the second component. Hint: Problem 6.1 is relevant here.

(b) Of the Hann and Hamming windows, guess which one is better in this case for detecting the second component, then test your guess on a computer.

6.8 Change the coefficients $0.42, 0.5, 0.08$ in the Blackman window to 0.42323, 0.49755, 0.07922, respectively. Using MATLAB, find the side-lobe level of this window. This is called the *Blackman–Harris window*; see Harris [1978].

6.9 When using windowing *and* zero padding for a length-N sequence $x[n]$, which of the following two is the proper procedure: (1) zero-pad first to length M, then use a window of length M or (2) use a length-N window on $x[n]$, then zero-pad to length M? Give reasons. If you are in doubt, experiment with MATLAB.

6.10 A communication system receives 2000 samples of the signal $\cos(\theta_0 n + \phi_0)$. For certain reasons, however, the signal is chopped in the following manner: After every 80 signal points there are 20 erased points, that is, points containing zeros. In MATLAB, such a signal is generated by the command

```
x = cos(theta0*(0:1999)+phi0).* ...
    kron(ones(1,20),[ones(1,80), zeros(1,20)]);
```

It is required to measure the frequency θ_0, using windowed DFT. Consider the following windowing options:

(a) A single window of length 2000 applied to the entire data sequence.

(b) 20 windows of length 80 each, applied to the individual chops.

(c) Both a and b.

Explore these three options with MATLAB. Decide which is the proper one to use and give reasons.

6.11 The signal $x(t) = \cos(2\pi t)$ is sampled at interval $T = 0.01$ second, and $N = 100$ samples are collected. The frequency of the signal is then measured from a windowed DFT (without zero padding). Of the windows you have studied, which ones will show that the signal has nonzero frequency and which will mislead us to think that the frequency is zero? For the Kaiser window, the answer depends on the parameter α. Answer first without trying it out; then verify your answers with MATLAB. Finally, repeat the experiment with $N = 200$.

6.12 This problem introduces another measure of performance of windows, called *roll-off rate*. The roll-off rate measures how fast the side lobes of the window decay as $|\theta|$ increases. It can be shown that, for large values of N, the peaks of the magnitude response $|W^f(\theta)|$ of a window are approximately tangent to a curve of the form

$$A(\theta) = \frac{K}{|\theta|^r},$$

where r is an integer. The larger r, the faster the decay of the side lobes. The quantity $6r$ is called the roll-off rate, and is measured in dB/octave. The reason for this definition is that

$$20\log_{10} A(\theta) = 20\log_{10} K - 20r\log_{10}|\theta|;$$

therefore, $20\log_{10} A(\theta)$ decreases by $6r$ dB each time $|\theta|$ is doubled. A frequency is said to be an *octave* higher than another frequency if it is two times higher. This term comes from music, where the interval between two notes bearing the same name is an integer number of octaves; that is, the frequency ratio is an integer power of 2.

A convenient way of visualizing the roll-off rate is to plot $|W^f(\theta)|$ and $A(\theta)$ on a log-log scale. Then $A(\theta)$ appears as a straight line tangent to the peaks of $|W^f(\theta)|$. In the following exercises, use windows of length $N = 128$, zero-pad to 1024 for plotting, and use the MATLAB command `loglog` for plotting.

(a) For a rectangular window, $K = 2$ and $r = 1$. Therefore, the roll-off rate is 6 dB/octave. Convince yourself that this is so by plotting $|W^f(\theta)|$ and $A(\theta)$.

(b) Guess the roll-off rate of the Bartlett window, then confirm your guess by examining the plots.

(c) The roll-off rate of the Hann window is 18 dB/octave. Convince yourself that this is so by plotting $|W^f(\theta)|$ and $A(\theta)$. Find K by experimenting.

(d) Guess the roll-off rate of the Hamming window, then confirm your guess by examining its plot.

(e) Find the roll-off rates of the Blackman and Kaiser windows by experimenting with their plots.

(f) What is the roll-off rate of the Dolph window? Answer without relying on a computer.

6.13 Discuss whether windowing is needed for frequency measurement of a *single* real sinusoid.

6.14* Recall the procedure of signal interpolation by zero padding in the frequency domain, studied in Section 4.5 and implemented in Problem 4.41. Suggest how to use windowing for improving the appearance of the interpolated signal, modify the program you have written for Problem 4.41, experiment, and report the results.

6.15* A student who has studied Section 6.5 proposed the following idea: "We know that, since $J_w > 1$ for all windows except the rectangular, the RMS errors of the frequency estimates $\hat{\theta}_m$ are larger for any window than for a rectangular window. We cannot work with a rectangular window exclusively, because of the problem of cross-interference. However, we can have the best of both worlds if we use different windows for the two estimation steps. During the coarse phase we will use a window with low side-lobe level, so that we can reliably identify the spectral peaks. Then, when performing the fine step of frequency estimation, we will revert to a rectangular window to get the best possible accuracy."

What is wrong with this idea?

6.16* A student who has studied Sections 6.4 and 6.5 proposed the following idea: "We know that any window other than the rectangular widens the main lobe by a factor of 2 at least. Therefore, when using a windowed DFT for sinusoid detection, we do not need all N frequency points, but we can do equally well with the $N/2$ even-index points. Since the main lobe of the window is at least $\pm 4\pi/N$, the sinusoid will show in at least one of the even-index DFT points. The even-index points can be implemented with *one* DFT of length $N/2$, by slightly modifying the radix-2 frequency-decimated FFT algorithm. This will save about half the number of operations, without degrading our ability to detect the sinusoidal signal."

(a) Explain the proposed modification of the radix-2 frequency-decimated FFT algorithm.

(b) What is wrong with this idea? Hint: Examine the worst-case processing gain of the windows you have studied under the new conditions.

6.17* An NRZ signal $x(t)$ (recall its description in Example 3.13) having bit interval $T = 5 \times 10^{-4}$ second [see (3.60) for the definition of T] is modulated by a complex carrier $e^{j2\pi f_0 t}$ to give the complex signal

$$y(t) = x(t)e^{j2\pi f_0 t}.$$

The carrier frequency f_0 is known to be in the range $|f_0| < 1000\,\text{Hz}$. It is required to estimate f_0 from a finite measured interval of the modulated BPSK signal.

(a) Explain why the measurement of f_0 from the magnitude of the DFT of $y(t)$ (with or without a window) is likely to be problematic. If you are in doubt, experiment with MATLAB.

(b) It is proposed to estimate f_0 by the following procedure:

    i. Sample the signal and square the result; that is, compute

$$z[n] = y^2(nT).$$

    ii. Estimate f_0 from the magnitude of the DFT of $z[n]$.

Explain why this method works and determine the sampling interval necessary for unambiguous measurement of f_0.

(c) Suppose that discrete-time white noise $v[n]$ is added to $y(nT)$. Let γ_v be the variance of the white noise. Assume that a rectangular window is used. Compute the quantity SNR_o which is, by definition, the ratio of the signal power to the noise power at the frequency of maximum of the magnitude of the DFT. Assume that the input SNR is relatively high, so the term $v^2[n]$ can be neglected in the expansion of $z[n]$.

(d) Draw conclusions from part c about the loss in SNR due to the squaring operation used in part b.

The method presented in this problem is useful for carrier frequency estimation of BPSK signals but is limited to cases of relatively high input SNR, for the reasons you have discovered in parts c and d.

6.18* We wish to detect the presence of a single complex exponential in white noise using windowed DFT, as discussed in Section 6.5.1. Suppose we know a priori that the frequency θ_0 of the complex exponential is less in magnitude than a certain θ_c, where $0 < \theta_c < \pi$. It is proposed to filter the signal $y[n]$ by an ideal low-pass filter having frequency response

$$H^f(\theta) = \text{rect}\left(\frac{\theta}{2\theta_c}\right)$$

and use the filtered signal for the windowed DFT. The variance of the noise will be reduced by this filter, whereas the complex exponential will not be affected. Therefore, it is argued that the processing gain of the windowed DFT will be increased.

(a) Show that the noise term in $E(|X^f(\theta_0)|^2)$ is given by

$$\frac{\gamma_v}{2\pi} \int_{-\theta_c-\theta_0}^{\theta_c-\theta_0} |W^f(\theta)|^2 d\theta,$$

where $W^f(\theta)$ is the kernel function of the window. Hint: The second term on the third line of (6.51) changes to

$$\sum_{n=0}^{N-1}\sum_{m=0}^{N-1} w[n]w[m]\kappa[n-m]e^{-j\theta_0(n-m)}, \tag{$*$}$$

where $\kappa[n-m]$ is the covariance sequence of the discrete-time noise at the output of the low-pass filter. Use (2.135) to express $\kappa[n-m]$ as

$$\kappa[n-m] = \frac{\gamma_v}{2\pi} \int_{-\theta_c}^{\theta_c} e^{j\theta(n-m)} d\theta.$$

Substitute this expression in $(*)$ and interchange the order of the integration and the two summations.

(b) Show that $\theta_c = \pi$ gives the result in (6.51).

(c) Explain why, if the window is good and $|\theta_0|$ is not close to θ_c, the proposed idea helps little in increasing the processing gain of the windowed DFT.

6.19* This problem discusses the effect of *sampling jitter* on the spectrum of a sampled signal. Sampling jitter is the name given to random perturbations of the sampling instants. Ideal uniform sampling is defined by the sequence of sampling instants $t[n] = nT$. Sampling jitter is characterized by a sequence of sampling instants $t[n] = nT + \varepsilon[n]$, where $\varepsilon[n]$ is a sequence of random variables resulting from the nonideal operation of the sampling hardware. Here we illustrate the effect of sampling jitter on a sinusoidal signal $x(t) = \cos(\omega_0 t)$.

(a) Choose $\omega_0 = 0.5\pi$; generate 512 samples of $x(t)$, sampled uniformly at $T = 1$. Compute the windowed DFT of the sampled signal, using the Blackman window, and plot the magnitude response in dB.

(b) Generate a sequence of 512 uniformly distributed random variables in the range $\pm\varepsilon_{max}$, where $\varepsilon_{max} = 0.1$, and use it for generating the sequence of sampling instants $t[n]$. Now generate the sampled sequence $x(t[n])$, compute its windowed DFT using the Blackman window, and superimpose the magnitude response on the one in part a. Repeat this experiment for $\varepsilon_{max} = 0.01, 0.001$. Report what you have observed.

(c) Analyze the jitter effect in mathematical terms. Hint: Assume that $|\omega_0\varepsilon[n]| \ll 1$ and expand $x(nT + \varepsilon[n])$ in first-order Taylor series around nT.

6.20* The following procedure for estimating the local peaks of a windowed Fourier transform is a useful alternative to the chirp Fourier transform algorithm:

(a) Find the index k_0 of the windowed DFT (without zero padding) that corresponds to the local peak of interest.

(b) Compute $20\log_{10}|X^f(\theta)|$ at the three points

$$\theta_a = \frac{2\pi(k_0 - 1)}{N}, \quad \theta_b = \frac{2\pi k_0}{N}, \quad \theta_c = \frac{2\pi(k_0 + 1)}{N}.$$

(c) Find a parabola that passes through the three points

$$\{(\theta_x, 20\log_{10}|X^f(\theta_x)|), \ x = a, b, c\}$$

(there is always a unique such parabola).

(d) Take $\hat{\theta}$ as the maximum point of the parabola.

The advantage of this procedure is in that it requires only a few operations. Its disadvantage is in that it is likely to give less accurate results than the chirp Fourier transform.

(a) Derive a closed-form formula for $\hat{\theta}$ in terms of the three points (abscissas and ordinates). Hint: Derive the coefficients of the parabola first, then the maximum point as a function of the coefficients.

(b) Write a MATLAB procedure to implement the computation.

(c) Change the MATLAB program `maxdft` such that it will use this procedure instead of `chirpf`.

(d) Repeat Example 6.4 with this procedure.

Chapter 7

Review of z-Transforms and Difference Equations

The z-transform fulfills, for discrete-time signals, the same need that the Laplace transform fulfills for continuous-time signals: It enables us to replace operations on signals by operations on complex functions. Like the Laplace transform, the z-transform
is mainly a tool of convenience, rather than necessity. Frequency-domain analysis,
as we have seen, can be dealt with both theoretically and practically without the z-
transform. However, for certain operations the convenience of using the z-transform
(or the Laplace transform in the continuous-time case) outweighs the burden of having
to learn yet another tool. This is especially true when dealing with linear filtering and
linear systems in general. Applications of the z-transform in linear system analysis
include:

1. Time-domain interpretation of LTI systems responses.
2. Stability testing.
3. Block-diagram manipulation of systems consisting of subsystems connected in
   cascade, parallel, and feedback.
4. Decomposition of systems into simple building blocks.
5. Analysis of systems and signals that do not possess Fourier transforms (e.g.,
   unstable LTI systems).

In this chapter we give the necessary background on the z-transform and its relation
to linear systems.[1] We pay special attention to *rational systems*, that is, systems that
can be described by difference equations. We shall use the bilateral z-transform almost
exclusively. This is in contrast with continuous-time systems, where the unilateral
Laplace transform is usually emphasized. The unilateral z-transform will be dealt with
briefly, in connection to the solution of difference equations with initial conditions.

Proper understanding of the material in this chapter requires certain knowledge of
complex function theory, in particular analytic functions and their basic properties. If
you are not familiar with this material, you can still use the main results given here,
especially the ones concerning rational systems, and take their derivations as a matter
of belief.

7.1   The z-Transform

The *bilateral* (or *two-sided*) *z-transform* of a sequence $x[n]$ is the complex-valued function[2]

$$\{Zx\}(z) = X^z(z) = \sum_{n=-\infty}^{\infty} x[n]z^{-n}, \tag{7.1}$$

where z belongs to a subset of the complex numbers $\mathbb{C}$, called the *region of convergence* (ROC) of the transform. By definition, the region of convergence is the set of all z such that the right side of (7.1) is absolutely convergent, that is,[3]

$$\sum_{n=-\infty}^{\infty} |x[n]z^{-n}| < \infty. \tag{7.2}$$

To determine the shape of the region of convergence, let us write the transform variable z in a polar form, that is, as $z = re^{j\theta}$, where r and θ are real, $r \geq 0$ and $-\pi \leq \theta < \pi$. We then get

$$\sum_{n=-\infty}^{\infty} |x[n]z^{-n}| = \sum_{n=-\infty}^{\infty} |x[n]| \cdot |re^{j\theta}|^{-n} = \sum_{n=-\infty}^{\infty} |x[n]|r^{-n}. \tag{7.3}$$

Define[4]

$$R_1 = \inf\left\{r \geq 0 : \sum_{n=-\infty}^{\infty} |x[n]|r^{-n} < \infty\right\}, \tag{7.4a}$$

$$R_2 = \sup\left\{r \geq 0 : \sum_{n=-\infty}^{\infty} |x[n]|r^{-n} < \infty\right\}. \tag{7.4b}$$

Then we get that the region of convergence includes the annulus

$$R_1 < |z| < R_2, \tag{7.5}$$

provided this annulus is nonempty.

The numbers R_1 and R_2 depend on the sequence $x[n]$, as seen from (7.4).[5] The region of convergence may include the circle $|z| = R_1$ or the circle $|z| = R_2$ or both, but there is no general criterion for testing these possibilities.[6] In any case, we call the set (7.5) the *interior* of the region and the union of the two circles the *boundary* of the region. Figure 7.1 illustrates the region of convergence of the z-transform.

Recall that a function $f(z)$ is *analytic* in a domain in the complex plane if it is differentiable at every point of the domain. A fundamental property of the z-transform, which we quote without proof, is

Theorem 7.1 The z-transform is an analytic function of z in the interior of its region of convergence.   □

The sequence $x[n]$ can be computed from its z-transform as given by the following theorem:

Theorem 7.2 (inverse z-transform)

$$x[n] = \frac{1}{2\pi j} \oint X^z(z)z^{n-1} dz, \tag{7.6}$$

where the complex integration is over any simple contour in the interior of the region of convergence of $X^z(z)$ that circles the point $z = 0$ exactly once in the counterclockwise direction.

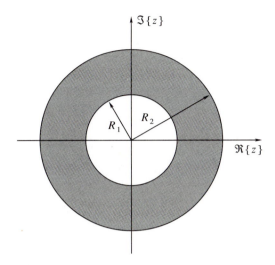

Figure 7.1 The region of convergence of the z-transform.

Proof The proof relies on the following result from complex function theory:

$$\frac{1}{2\pi j}\oint z^k dz = \delta[k+1] = \begin{cases} 1, & k = -1, \\ 0, & k \in \mathbb{Z},\ k \neq -1, \end{cases} \tag{7.7}$$

where the integral is over any simple contour that circles the point $z = 0$ exactly once in the counterclockwise direction. We get from (7.6) and (7.7)

$$\frac{1}{2\pi j}\oint X^z(z)z^{n-1}dz = \frac{1}{2\pi j}\oint \left[\sum_{k=-\infty}^{\infty} x[k]z^{-k}\right]z^{n-1}dz$$

$$= \sum_{k=-\infty}^{\infty} x[k]\left[\frac{1}{2\pi j}\oint z^{n-k-1}dz\right]$$

$$= \sum_{k=-\infty}^{\infty} x[k]\delta[n-k] = x[n]. \tag{7.8}$$

For the substitution made in the derivation to be valid, the contour must lie in the interior of the region of convergence of $X^z(z)$. $\qquad\square$

Table 7.1 gives the z-transforms of a few common sequences. In this table, $\upsilon[n]$ denotes the discrete-time, unit-step function. Methods of deriving the transforms in this table are discussed later in this chapter.

Example 7.1 The following special cases illustrate various points related to the definition of the z-transform and its region of convergence.

1. The z-transform of $x[n] = \delta[n]$ is obviously $X^z(z) = 1$. Here the z-transform is defined on the entire complex plane.

2. Let

$$x[n] = \begin{cases} a^n, & n \geq 0, \\ 0, & n < 0. \end{cases} \tag{7.9}$$

$x[n]$	$X^z(z)$	ROC
$\delta[n]$	1	all z
$\upsilon[n]$	$\dfrac{1}{1 - z^{-1}}$	$|z| > 1$
$a^n \upsilon[n]$	$\dfrac{1}{1 - az^{-1}}$	$|z| > |a|$
$-a^n \upsilon[-(n+1)]$	$\dfrac{1}{1 - az^{-1}}$	$|z| < |a|$
$na^n \upsilon[n]$	$\dfrac{az^{-1}}{(1 - az^{-1})^2}$	$|z| > |a|$
$n^2 a^n \upsilon[n]$	$\dfrac{az^{-1}(1 + az^{-1})}{(1 - az^{-1})^3}$	$|z| > |a|$
$a^n \cos(\theta_0 n)\,\upsilon[n]$	$\dfrac{1 - a\cos\theta_0\,z^{-1}}{1 - 2a\cos\theta_0\,z^{-1} + a^2 z^{-2}}$	$|z| > |a|$
$a^n \sin(\theta_0 n)\,\upsilon[n]$	$\dfrac{a\sin\theta_0\,z^{-1}}{1 - 2a\cos\theta_0\,z^{-1} + a^2 z^{-2}}$	$|z| > |a|$

Table 7.1 Common sequences and their z-transforms.

Then

$$X^z(z) = \sum_{n=0}^{\infty} a^n z^{-n} = \sum_{n=0}^{\infty} (az^{-1})^n = \frac{1}{1 - az^{-1}}, \quad |z| > |a|. \qquad (7.10)$$

In this example $R_1 = |a|$ and $R_2 = \infty$.

3. Let

$$x[n] = \begin{cases} -a^n, & n < 0, \\ 0, & n \geq 0. \end{cases} \qquad (7.11)$$

Then

$$X^z(z) = -\sum_{n=-\infty}^{-1} a^n z^{-n} = -\sum_{n=-\infty}^{-1} (az^{-1})^n = -\sum_{n=1}^{\infty} (za^{-1})^n$$

$$= -\frac{za^{-1}}{1 - za^{-1}} = \frac{1}{1 - az^{-1}}, \quad |z| < |a|. \qquad (7.12)$$

In this example $R_1 = 0$ and $R_2 = |a|$. The z-transforms of (7.9) and (7.11) have the same functional form, but disjoint regions of convergence. As we see, the region of convergence is an indispensable part of the transform: Without knowledge of this region, we cannot tell whether the function $1/(1 - az^{-1})$ comes from the sequence (7.9) or from (7.11).

4. Let

$$x[n] = a^{|n|}, \quad -\infty < n < \infty. \qquad (7.13)$$

Then

$$X^z(z) = \sum_{n=-\infty}^{-1} a^{-n} z^{-n} + \sum_{n=0}^{\infty} a^n z^{-n} = \frac{az}{1 - az} + \frac{1}{1 - az^{-1}}$$

$$= \frac{1 - a^2}{(1 - az)(1 - az^{-1})}, \quad |a| < |z| < |a|^{-1}. \qquad (7.14)$$

Here $R_1 = |a|$ and $R_2 = |a|^{-1}$ provided $|a| < 1$; otherwise the region of convergence is empty. This example shows that there exist sequences that have no z-transforms.

5. Let

$$x[n] = \frac{1}{|n| + 1}, \quad -\infty < n < \infty. \tag{7.15}$$

We have

$$\sum_{n=-\infty}^{\infty} |x[n]| = \sum_{n=-\infty}^{\infty} \frac{1}{|n| + 1} = \infty,$$

so (7.2) does not converge for $|z| = 1$. This implies lack of convergence for $|z| < 1$, since then

$$\sum_{n=0}^{\infty} |x[n]z^{-n}| > \sum_{n=0}^{\infty} |x[n]| = \infty,$$

as well as lack of convergence for $|z| > 1$, since then

$$\sum_{n=-\infty}^{0} |x[n]z^{-n}| > \sum_{n=-\infty}^{0} |x[n]| = \infty.$$

In summary, (7.2) does not converge for any z, so $x[n]$ does not have a z-transform. Note the difference between this example and the preceding one: There the sequence had exponential growth for both positive and negative n when $|a| > 1$. Here the sequence decays to zero for both positive and negative n, but it does not decay fast enough.

6. Let

$$x[n] = \frac{1}{n^2 + 1}, \quad -\infty < n < \infty. \tag{7.16}$$

Then (see Problem 2.49),

$$\sum_{n=-\infty}^{\infty} |x[n]| = \frac{\pi^2}{3} - 1 < \infty, \tag{7.17}$$

so the z-transform exists on the circle $|z| = 1$. However, the sum (7.2) diverges for all $|z| \neq 1$ in this case. The interior of the region of convergence of $X^z(z)$ is therefore empty, and $X^z(z)$ is not analytic anywhere.

7. Let $x[n]$ be a finite-duration sequence; that is, $x[n] = 0$ for $n < n_1$ and for $n > n_2$. Then

$$X^z(z) = \sum_{n=n_1}^{n_2} x[n]z^{-n}. \tag{7.18}$$

The sum is finite, so it converges for all $0 < |z| < \infty$; therefore $R_1 = 0$ and $R_2 = \infty$.

8. A sequence $x[n]$ satisfying $x[n] = 0$ for all $n < 0$ is called a *causal* sequence. For such a sequence,

$$X^z(z) = \sum_{n=0}^{\infty} x[n]z^{-n}. \tag{7.19}$$

Let R_1 be as defined in (7.4a). Then

$$|z| > R_1 \implies \sum_{n=0}^{\infty} |x[n]z^{-n}| = \sum_{n=0}^{\infty} |x[n]| \cdot |z|^{-n} < \infty, \tag{7.20}$$

so z is in the region of convergence for all $|z| > R_1$, meaning that $R_2 = \infty$.

9. A sequence $x[n]$ satisfying $x[n] = 0$ for all $n > 0$ is called an *anticausal* sequence. For such a sequence,

$$X^z(z) = \sum_{n=-\infty}^{0} x[n]z^{-n}. \tag{7.21}$$

Let R_2 be as defined in (7.4b). Then

$$|z| < R_2 \implies \sum_{n=-\infty}^{0} |x[n]z^{-n}| = \sum_{n=-\infty}^{0} |x[n]| \cdot |z|^{-n} < \infty, \qquad (7.22)$$

so z is in the region of convergence for all $|z| < R_2$, meaning that $R_1 = 0$.

10. Suppose we are given the z-transform

$$X^z(z) = \frac{z(z + 1.2)}{(z - 0.4)(z - 2)}. \qquad (7.23)$$

What is the corresponding sequence $x[n]$? Since the region of convergence is not given, there is no unique answer. Let us, however, explore the various possibilities. At the points $z = 0.4$ and $z = 2$ the function $X^z(z)$ is singular. Since the z-transform is an analytic function in the interior of its region of convergence, the circles $|z| = 0.4$ and $|z| = 2$ cannot be in the interior of the region of convergence. Nevertheless, they can belong to the boundary of the region. Now, since the region of convergence is always of the form (7.5), there are only three possibilities:

$$0 < |z| < 0.4 \quad \text{(case I)},$$
$$0.4 < |z| < 2 \quad \text{(case II)},$$
$$2 < |z| < \infty \quad \text{(case III)}.$$

This is illustrated in Figure 7.2.

To find $x[n]$ in the three cases, we express $X^z(z)$ in the form

$$X^z(z) = \frac{2z}{z - 2} - \frac{z}{z - 0.4} = \frac{2}{1 - 2z^{-1}} - \frac{1}{1 - 0.4z^{-1}}. \qquad (7.24)$$

Now parts 2, 3, 8, and 9 of Example 7.1 lead us to conclude that

$$x[n] = \begin{cases} \begin{cases} -2 \times 2^n + 0.4^n, & n < 0, \\ 0, & n \geq 0, \end{cases} & \text{(case I)}, \\[2em] \begin{cases} -2 \times 2^n, & n < 0, \\ -0.4^n, & n \geq 0, \end{cases} & \text{(case II)}, \\[2em] \begin{cases} 2 \times 2^n - 0.4^n, & n \geq 0, \\ 0, & n < 0, \end{cases} & \text{(case III)}. \end{cases} \qquad (7.25)$$

$\square$

7.2 Properties of the z-Transform

In this section we list the main properties of the z-transforms. Most of these properties are shared by Fourier transforms of sequences, as listed in Section 2.7. However, in the case of the z-transform we must also specify the region of convergence at which the property holds. This was not a problem in the case of the Fourier transform, since the range of argument θ of the Fourier transform is always the entire real line.

The sequences $x[n]$, $y[n]$, $w[n]$ mentioned next have z-transforms $X^z(z)$, $Y^z(z)$, $W^z(z)$ and radii of convergence (R_{x1}, R_{x2}), (R_{y1}, R_{y2}), (R_{w1}, R_{w2}), respectively. In each case, the region of convergence may include either boundary or both.

1. **Linearity**

$$w[n] = ax[n] + by[n] \implies W^z(z) = aX^z(z) + bY^z(z), \quad a, b \in \mathbb{C}, \qquad (7.26)$$

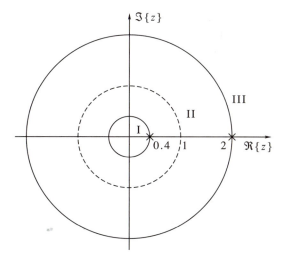

Figure 7.2 Possible regions of convergence in Example 7.1, part 10.

where

$$R_{w1} \le \max\{R_{x1}, R_{y1}\}, \quad R_{w2} \ge \min\{R_{x2}, R_{y2}\}.$$

The proof of (7.26) is easy and will be skipped.

2. **Time shift**

$$w[n] = x[n - m] \iff W^z(z) = z^{-m} X^z(z), \quad m \in \mathbb{Z}. \tag{7.27}$$

The region of convergence of $W^z(z)$ is identical to that of $X^z(z)$ with one exception: If $x[n]$ is anticausal then $X^z(z)$ exists on $z = 0$, but $W^z(z)$ may not exist on $z = 0$ if m is positive.

Proof

$$W^z(z) = \sum_{n=-\infty}^{\infty} x[n - m]z^{-n} = \sum_{n=-\infty}^{\infty} x[n]z^{-(n+m)} = z^{-m} X^z(z). \tag{7.28}$$

3. **Multiplication by a geometric series**

$$w[n] = a^n x[n] \iff W^z(z) = X^z\left(\frac{z}{a}\right), \quad a \in \mathbb{C}, \ a \neq 0, \tag{7.29}$$

where

$$R_{w1} = |a| R_{x1}, \quad R_{w2} = |a| R_{x2}.$$

Proof

$$W^z(z) = \sum_{n=-\infty}^{\infty} a^n x[n]z^{-n} = \sum_{n=-\infty}^{\infty} x[n]\left(\frac{z}{a}\right)^{-n} = X^z\left(\frac{z}{a}\right). \tag{7.30}$$

4. **Time reversal**

$$w[n] = x[-n] \iff W^z(z) = X^z(z^{-1}), \tag{7.31}$$

where

$$R_{w1} = (R_{x2})^{-1}, \quad R_{w2} = (R_{x1})^{-1}.$$

Proof

$$W^z(z) = \sum_{n=-\infty}^{\infty} x[-n]z^{-n} = \sum_{n=-\infty}^{\infty} x[n]z^n = X^z(z^{-1}). \tag{7.32}$$

5. Differentiation

$$w[n] = nx[n] \iff W^z(z) = -z\frac{dX^z(z)}{dz}, \tag{7.33}$$

$$R_{w1} = R_{x1}, \quad R_{w2} = R_{x2}.$$

Proof

$$-z\frac{dX^z(z)}{dz} = -z\frac{d}{dz}\sum_{n=-\infty}^{\infty} x[n]z^{-n} = -z\sum_{n=-\infty}^{\infty}(-n)x[n]z^{-(n+1)}$$

$$= \sum_{n=-\infty}^{\infty} nx[n]z^{-n} = W^z(z). \tag{7.34}$$

6. Time-domain convolution

$$w[n] = \{x * y\}[n] \iff W^z(z) = X^z(z)Y^z(z), \tag{7.35}$$

where

$$R_{w1} \le \max\{R_{x1}, R_{y1}\}, \quad R_{w2} \ge \min\{R_{x2}, R_{y2}\}.$$

Proof

$$W^z(z) = \sum_{n=-\infty}^{\infty}\left[\sum_{m=-\infty}^{\infty} x[m]y[n-m]\right]z^{-(n-m)-m}$$

$$= \sum_{m=-\infty}^{\infty} x[m]z^{-m}\sum_{k=-\infty}^{\infty} y[k]z^{-k} = X^z(z)Y^z(z). \tag{7.36}$$

7. Time-domain multiplication

$$w[n] = x[n]y[n] \iff W^z(z) = \frac{1}{2\pi j}\oint X^z(u)Y^z\left(\frac{z}{u}\right)u^{-1}du, \tag{7.37}$$

where

$$R_{w1} \le R_{x1}R_{y1}, \quad R_{w2} \ge R_{x2}R_{y2}.$$

Proof Using (7.7), we get

$$\frac{1}{2\pi j}\oint X^z(u)Y^z\left(\frac{z}{u}\right)u^{-1}du$$

$$-\frac{1}{2\pi j}\oint\left[\sum_{n=-\infty}^{\infty} x[n]u^{-n}\right]\left[\sum_{k=-\infty}^{\infty} y[k](z/u)^{-k}\right]u^{-1}du$$

$$= \sum_{n=-\infty}^{\infty} x[n]\sum_{k=-\infty}^{\infty} y[k]z^{-k}\left[\frac{1}{2\pi j}\oint u^{k-n-1}du\right]$$

$$= \sum_{n=-\infty}^{\infty} x[n]\sum_{k=-\infty}^{\infty} y[k]z^{-k}\delta[k-n] = \sum_{n=-\infty}^{\infty} x[n]y[n]z^{-n} = w[n]. \tag{7.38}$$

The contour of integration must be such that

$$R_{x1} < |u| < R_{x2} \text{ and } R_{y1} < |z/u| < R_{y2}.$$

Such a contour is guaranteed to exist if

$$R_{x1}R_{y1} < |z| < R_{x2}R_{y2},$$

hence the region of convergence of $W^z(z)$.

7.3   Transfer Functions

Recall that a discrete-time LTI system is completely characterized, as far as its input–output relationships are concerned, by its impulse response sequence $h[n]$. The z-transform $H^z(z)$ of the impulse response, if exists, is called the *transfer function* of the system. Not every LTI system possesses a transfer function, since not every sequence has a z-transform. However, systems of practical interest usually have transfer functions, which is to say that $H^z(z)$ exists on a subset of the complex plane.

Of particular interest are LTI systems that are stable in the sense of having the bounded-input, bounded-output (BIBO) property. An LTI system is BIBO-stable if its response to any bounded input signal is a bounded signal. Let $x[n]$ be any input to the system and $y[n]$ the corresponding output. Then the system is BIBO-stable if

$$|x[n]| \le B_1 \text{ for all } n \in \mathbb{Z} \implies$$

$$\text{there exists } B_2 \text{ such that } |y[n]| \le B_2 \text{ for all } n \in \mathbb{Z}. \tag{7.39}$$

The BIBO-stability of an LTI system can be determined from the impulse response of the system, as stated in the following theorem.

Theorem 7.3 A discrete-time LTI system is BIBO-stable if and only if its impulse response satisfies

$$\sum_{n=-\infty}^{\infty} |h[n]| < \infty. \tag{7.40}$$

Proof Suppose the impulse response of the system satisfies (7.40) and let $x[n]$ be bounded by B_1. Then,

$$|y[n]| = \left| \sum_{m=-\infty}^{\infty} h[m]x[n-m] \right| \le \sum_{m=-\infty}^{\infty} |h[m]|\,|x[n-m]|$$

$$\le B_1 \sum_{m=-\infty}^{\infty} |h[m]| < \infty. \tag{7.41}$$

Therefore, by taking $B_2 = B_1 \sum_{m=-\infty}^{\infty} |h[m]|$, we deduce that the system is BIBO-stable. This proves sufficiency of the condition (7.40). To prove necessity, assume that (7.40) does not hold; that is, the left side is infinite. Consider the input sequence

$$x[n] = \begin{cases} \dfrac{h[-n]}{|h[-n]|}, & h[-n] \ne 0, \\ 0, & h[-n] = 0. \end{cases} \tag{7.42}$$

This sequence is bounded in magnitude by 1. The output at time $n = 0$ is

$$y[0] = \sum_{m=-\infty}^{\infty} h[m]x[-m] = \sum_{m=-\infty}^{\infty} |h[m]| = \infty, \tag{7.43}$$

so the output is not bounded. This proves that (7.40) is necessary for BIBO-stability.□

Corollary An LTI system is BIBO-stable if and only if the region of convergence of its transfer function includes the circle $|z| = 1$.
Proof Substitution $|z| = 1$ makes the left sides of (7.40) and (7.2) equal (with the obvious change of $x[n]$ to $h[n]$).   □

The circle $|z| = 1$ is called the *unit circle*. It is therefore common to say that an LTI system is stable if and only if the region of convergence of its transfer function includes the unit circle.

Another important property that an LTI system may possess is causality. A system is said to be *causal* if, for any time n, the response $y[n]$ depends only on past and present inputs $\{x[m], \ -\infty < m \le n\}$, but not on future inputs $\{x[m], \ m > n\}$.

Theorem 7.4 An LTI system is causal if and only if its impulse response $h[n]$ is a causal sequence.

Proof Decompose the system's output as follows:

$$y[n] = \sum_{m=-\infty}^{-1} h[m]x[n-m] + \sum_{m=0}^{\infty} h[m]x[n-m]. \tag{7.44}$$

The system is causal if and only if the first term is identically zero for an arbitrary input signal; but this is possible if and only if $h[m] = 0$ for all $m < 0$.   □

As we saw in Example 7.1, part 8, the region of convergence of the transfer function of a causal system is of the form $R_1 < |z| < \infty$. It may also include the boundary $|z| = R_1$, but there is no general criterion for testing this.

Referring to Example 7.1, part 10, and assuming that $x[n]$ is the impulse response of an LTI system, we see that:

1. The system is not stable and not causal in case I.

2. The system is stable, but not causal, in case II.

3. The system is causal, but not stable, in case III.

The region of convergence of the transfer function $H^z(z)$ of an LTI system that is both causal and stable must include the unit circle and all points outside it. In particular, $H^z(z)$ may possess no singularities in that region. An alternative statement of this property is:

> All the singularities of the transfer function of a stable and causal LTI system must be inside the unit circle.

7.4   Systems Described by Difference Equations

7.4.1   Difference Equations

A differential equation relates a function $y(t)$ and a few of its derivatives to another function $x(t)$ and a few of its derivatives. Similarly, a *difference equation* relates a sequence $y[n]$ and a few of its time shifts to another sequence $x[n]$ and a few of its time shifts.

Linear differential equations with constant coefficients are of special importance in the theory of continuous-time LTI systems. Similarly, linear difference equations with constant coefficients are of special importance in the theory of discrete-time LTI systems. A linear difference equation with constant coefficients is, by definition, the following relationship between two sequences $x[n]$ and $y[n]$:

$$y[n] = -a_1 y[n-1] - \cdots - a_p y[n-p] + b_0 x[n] + b_1 x[n-1] + \cdots + b_q x[n-q]$$

$$= -\sum_{i=1}^{p} a_i y[n-i] + \sum_{i=0}^{q} b_i x[n-i]. \tag{7.45}$$

The numbers p and q are called, respectively, the *denominator* and *numerator orders* of the difference equation. The constants $\{a_1, a_2, \ldots, a_p\}$ are called the *denominator coefficients*, and the constants $\{b_0, b_1, \ldots, b_q\}$ are called the *numerator coefficients* (the reason for these names will soon become clear). We assume that $a_p \neq 0$ and $b_q \neq 0$, because otherwise we can delete zero-valued a coefficients or b coefficients and reduce p or q accordingly.

The difference equation (7.45) represents a causal relationship between the input sequence $x[n]$ and the output sequence $y[n]$. In other words, $y[n]$ depends on its own past values and on past and present (but not future) values of $x[n]$. A system whose input–output relationship obeys a difference equation such as (7.45) is linear and time invariant; see Problem 7.11.

The transfer function of a system described by a difference equation can be computed as follows. Take the z-transform of (7.45) and use the linearity and time-shift properties of the z-transform to obtain

$$Y^z(z) = -\sum_{i=1}^{p} a_i z^{-i} Y^z(z) + \sum_{i=0}^{q} b_i z^{-i} X^z(z) = -Y^z(z) \sum_{i=1}^{p} a_i z^{-i} + X^z(z) \sum_{i=0}^{q} b_i z^{-i},$$

$$(7.46)$$

hence

$$H^z(z) = \frac{Y^z(z)}{X^z(z)} = \frac{b_0 + b_1 z^{-1} + \cdots + b_q z^{-q}}{1 + a_1 z^{-1} + \cdots + a_p z^{-p}} = \frac{b(z)}{a(z)}. \qquad (7.47)$$

A transfer function that has the form (7.47) is called *rational*, since it is a ratio of two polynomials in the variable z^{-1}, namely $a(z)$ and $b(z)$. The polynomials are called the *numerator* and *denominator polynomials* of the transfer function, for obvious reasons.

It is occasionally useful to express (7.47) in positive powers of z. To this end, let b_{q-r} be the first nonzero numerator coefficient. Thus, for example, $r = q$ if $b_0 \neq 0$, etc. Then we can write from (7.47)

$$H^z(z) = \begin{cases} \dfrac{z^{p-q}(b_{q-r}z^r + \cdots + b_q)}{z^p + a_1 z^{p-1} + \cdots + a_p}, & p \geq q, \\[3mm] \dfrac{b_{q-r}z^r + \cdots + b_q}{z^{q-p}(z^p + a_1 z^{p-1} + \cdots + a_p)}, & p < q. \end{cases} \qquad (7.48)$$

Now the numerator and the denominator on the right side are polynomials in z. If $p \geq q$, the denominator and numerator degrees are p and $p - q + r$, respectively. If $p < q$, the denominator and numerator degrees are q and r, respectively. In any case, since $r \leq q$, the numerator degree is not larger than the denominator degree.

A rational transfer function in positive powers of the variable (s in continuous time, z in discrete time) whose numerator degree is not larger than its denominator degree is called *proper*. The transfer function in (7.48) is proper because the system it represents is causal.[7] If the denominator and numerator degrees are equal, the transfer function is said to be *exactly proper*. This happens when $r = q$, that is, when $b_0 \neq 0$. The transfer function is exactly proper if $y[n]$ depends on the present value of $x[n]$, not only on its past values. If $b_0 = 0$, then $y[n]$ depends only on past values of $x[n]$. In this case the numerator degree of (7.48) will be strictly less than the denominator degree. Such a transfer function is called *strictly proper*.

7.4.2 Poles and Zeros

According to the fundamental theorem of algebra, a polynomial $P(z)$ of degree n has exactly n roots, including multiplicities. Thus, the polynomial can be written as a product of first-order factors

$$P(z) = \prod_{i=1}^{n} (z - \lambda_i). \qquad (7.49)$$

The roots λ_i can be either real or complex. However, if the coefficients of the polynomial are real, complex roots come in conjugate pairs; in other words, if λ is a root of multiplicity m, then $\bar{\lambda}$ is also a root of multiplicity m.

Let n_d, n_n denote, respectively, the degrees of the denominator and numerator polynomials in (7.48), in either of the two cases. Let $\{\alpha_i, \ 1 \le i \le n_d\}$ be the roots of the denominator polynomial in (7.48) and $\{\beta_i, \ 1 \le i \le n_n\}$ the roots of the numerator polynomial. Then we can write (7.48) as

$$H^z(z) = b_{q-r} z^{p-q} \frac{\prod_{i=1}^{r}(z - \beta_i)}{\prod_{i=1}^{p}(z - \alpha_i)} = b_{q-r} \frac{\prod_{i=1}^{n_n}(z - \beta_i)}{\prod_{i=1}^{n_d}(z - \alpha_i)}. \tag{7.50}$$

The numbers α_i are called the *poles* of the system, and the numbers β_i are called the *zeros*. These numbers can be real or complex; however, complex poles or zeros come in conjugate pairs if the coefficients of the difference equation (7.45) are real. Of the poles, exactly p are nonzero, because of the condition $a_p \ne 0$. There are $q-p$ additional poles at $z = 0$ if $q > p$. Of the zeros, exactly r are nonzero, because of the condition $b_q \ne 0$. There are $p - q$ additional zeros at $z = 0$ if $p > q$.

Two polynomials are said to be *coprime* if no polynomial of degree 1 or more is a factor of both. Thus, two polynomials are coprime if and only if they have no root in common. If the numerator and denominator polynomials of a rational transfer function are coprime, the corresponding system is said to be *minimal*. The reason for this name is that the system cannot be described by a difference equation having smaller p and q. If the two polynomials do have a common factor, the system is said to be *nonminimal*. A nonminimal system can be brought to a minimal form by canceling the greatest common factor of the numerator and denominator polynomials. In dealing with rational transfer functions, we assume that the function is given in a minimal form unless stated otherwise.

7.4.3 Partial Fraction Decomposition

Another useful form of a rational transfer function is as a *partial fraction decomposition*. Partial fraction decomposition can be derived from either (7.47) (with negative powers of z) or (7.48) (with positive powers of z). It will turn out to be more useful to work with negative powers of z. Consider first the case where $q \ge p$. Then we can write (7.47) in the form

$$H^z(z) = c_0 + \cdots + c_{q-p} z^{-(q-p)} + \frac{d_0 + \cdots + d_{p-1} z^{-(p-1)}}{1 + a_1 z^{-1} + \cdots + a_p z^{-p}} = c(z) + \frac{d(z)}{a(z)}. \tag{7.51}$$

The coefficients of the new polynomials can be obtained by equating coefficients of powers of z^{-1} in the equality

$$c(z)a(z) + d(z) = b(z). \tag{7.52}$$

We have $q + 1$ equations in $q + 1$ unknowns: the numbers $\{c_0, \ldots, c_{q-p}, d_0, \ldots, d_{p-1}\}$. These equations can be written explicitly as follows:

$$\sum_{i=0}^{\min\{q-p,k\}} c_i a_{k-i} + d_k = b_k, \quad 0 \le k \le p - 1, \tag{7.53a}$$

$$\sum_{i=k-p}^{\min\{q-p,k\}} c_i a_{k-i} = b_k, \quad p \le k \le q. \tag{7.53b}$$

It can be shown that this set of equations is always nonsingular, so there is always a unique solution.

The case $q < p$ is a special case of (7.51), with $c(z) = 0$ and $d(z) = b(z)$. Thus, by computing the partial fraction decomposition of $d(z)/a(z)$ we shall cover all cases. Assume, further, that all the poles of the system are *simple*; that is, no two of them are equal. Then we have

$$\frac{d(z)}{a(z)} = \frac{d_0 + d_1 z^{-1} + \cdots + d_{p-1} z^{-(p-1)}}{\prod_{i=1}^{p}(1 - \alpha_i z^{-1})} = \sum_{i=1}^{p} \frac{A_i}{1 - \alpha_i z^{-1}}. \tag{7.54}$$

The coefficient A_k can be found by multiplying (7.54) by $(1 - \alpha_k z^{-1})$ and then substituting $z = \alpha_k$. This gives

$$A_k = \frac{d_0 + d_1 \alpha_k^{-1} + \cdots + d_{p-1}\alpha_k^{-(p-1)}}{\prod_{i \neq k}(1 - \alpha_i \alpha_k^{-1})} = \frac{d_0 \alpha_k^{p-1} + d_1 \alpha_k^{p-2} + \cdots + d_{p-1}}{\prod_{i \neq k}(\alpha_k - \alpha_i)}. \tag{7.55}$$

Since the α_i are all distinct, the denominator is nonzero. Therefore, A_k is well defined.

In summary, the partial fraction decomposition of $H^z(z)$ is given by

$$H^z(z) = c_0 + \cdots + c_{q-p} z^{-(q-p)} + \sum_{k=1}^{p} \frac{A_k}{1 - \alpha_k z^{-1}}, \tag{7.56}$$

with A_k as in (7.55). We remark that (7.56) can be generalized to transfer functions of systems with multiple poles. However, such systems are rare in digital signal processing applications, so this generalization is of little use to us and we shall not discuss it here.

The procedure `tf2pf` in Program 7.1 implements the partial fraction decomposition (7.56) in MATLAB. The first part of the program computes $c(z)$ by solving the linear equations (7.53). The matrix `temp` is the coefficient matrix of the linear equations. It is built using the MATLAB function `toeplitz`. A *Toeplitz* matrix is a matrix whose elements are equal along the diagonals. Such a matrix is uniquely defined by its first column and first row. You are encouraged to learn more about this function by typing `help toeplitz`. The second part of the program computes the α_k and A_k. The MATLAB function `residue` can also be used for partial fraction decomposition. However, `residue` was programmed to deal with polynomials expressed in positive powers of the argument, so it is more suitable for transfer functions in the s domain. It can be adapted for use with negative powers, but this requires care.

The procedure `pf2tf` in Program 7.2 implements the inverse operation, that is, the conversion of partial fraction decomposition to a transfer function. It iteratively brings the partial fractions under a common denominator, using convolution to perform multiplication of polynomials. It takes the real part of the results at the end, since it implicitly assumes that the transfer function is real.

7.4.4 Stability of Rational Transfer Functions

A causal LTI system is stable if and only if it possesses no singularities in the domain $|z| \geq 1$. For a rational system, this is equivalent to saying that all poles are inside the unit circle. Testing the stability of a rational LTI system by explicitly computing its poles may be inconvenient, however, if the order p of the denominator polynomial is high. There exist several tests for deciding whether a polynomial $a(z)$ is stable, that is, has all its roots inside the unit circle, without explicitly finding the roots. The best known are the *Jury test* and the *Schur-Cohn test*. Here we describe the latter.

Let there be given a monic pth-order polynomial in powers of z^{-1}; *monic* means that the coefficient of z^0 is 1. Denote the polynomial as

$$a_p(z) = 1 + a_{p,1} z^{-1} + \cdots + a_{p,p} z^{-p} \tag{7.57}$$

(the reason for labeling the coefficients with two indices will become clear presently). Since $a_{p,p}$ is, up to a sign, the product of all roots of $a_p(z)$, the condition $|a_{p,p}| < 1$ is necessary for stability of $a_p(z)$. This condition is not sufficient, however. For example, the polynomial $1 + 4z^{-1} + 0.5z^{-2}$ is not stable. The Schur–Cohn test begins by testing whether $|a_{p,p}| < 1$. If this condition is violated, the polynomial is not stable. If it holds, the polynomial is reduced to a degree $p - 1$ by computing

$$a_{p-1}(z) = (1 - a_{p,p}^2)^{-1}[a_p(z) - a_{p,p}z^{-p}a_p(z^{-1})]. \tag{7.58}$$

The coefficients of the reduced-degree polynomial are

$$a_{p-1,k} = (1 - a_{p,p}^2)^{-1}[a_{p,k} - a_{p,p}a_{p,p-k}], \quad 0 \le k \le p - 1. \tag{7.59}$$

The polynomial $a_{p-1}(z)$ is monic since $a_{p-1,0}$, as given by (7.59), is equal to 1. The Schur–Cohn theorem asserts that, given that $|a_{p,p}| < 1$, the polynomial $a_p(z)$ is stable if and only if $a_{p-1}(z)$ is stable. Since $a_{p-1}(z)$ has degree lower than that of $a_p(z)$, we can repeat the test recursively. At the mth step we are given a polynomial

$$a_{p-m}(z) = 1 + a_{p-m,1}z^{-1} + \cdots + a_{p-m,p-m}z^{-(p-m)}. \tag{7.60}$$

We first test if $|a_{p-m,p-m}| < 1$. If not, we conclude that the initial polynomial $a_p(z)$ is not stable and terminate the recursion. If yes, we perform the degree-reduction step

$$a_{p-m-1}(z) = (1 - a_{p-m,p-m}^2)^{-1}[a_{p-m}(z) - a_{p-m,p-m}z^{-(p-m)}a_{p-m}(z^{-1})], \tag{7.61}$$

or, equivalently,

$$a_{p-m-1,k} = (1 - a_{p-m,p-m}^2)^{-1}[a_{p-m,k} - a_{p-m,p-m}a_{p-m,p-m-k}], \ 0 \le k \le p - m - 1. \tag{7.62}$$

We then decrease m by 1 and repeat. If the procedure does not terminate prior to $m = p$, we conclude that the initial polynomial $a_p(z)$ is stable.

A proof of the Schur–Cohn test is beyond the scope of this book. The procedure sctest in Program 7.3 implements the test. In this procedure, the conversion of the polynomial to monic (by dividing by the leading coefficient) is performed first, thus enabling the procedure to handle nonmonic polynomials. Next the magnitude of the last coefficient is tested. If it is not smaller than 1, the procedure terminates with a negative result. Otherwise the reduced-degree polynomial is computed and the program calls itself recursively. There is no explicit multiplication by $(1 - a_{p-m,p-m}^2)^{-1}$, since the conversion to monic at the beginning takes care of this. If the degree of the input polynomial is zero, the program exits with positive result.

Example 7.2 Consider the second-order polynomial

$$a_2(z) = 1 + a_1z^{-1} + a_2z^{-2}.$$

The first step in the Schur–Cohn test gives the necessary condition for stability

$$|a_2| < 1.$$

The reduced-degree polynomial is

$$a_1(z) = 1 + \frac{a_1(1 - a_2)}{1 - a_2^2}z^{-1} = 1 + \frac{a_1}{1 + a_2}z^{-1}.$$

The second step in the test gives the additional condition

$$-1 < \frac{a_1}{1 + a_2} < 1.$$

In summary, $a_2(z)$ is stable if and only if

$$-1 < a_2 < 1, \quad 1 + a_1 + a_2 > 0, \quad 1 - a_1 + a_2 > 0. \tag{7.63}$$

Figure 7.3 shows the so-called triangle of stability of the coefficients of a second-order polynomial. □

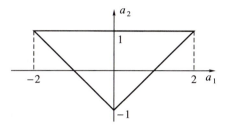

Figure 7.3 The triangle of stability of a second-order polynomial; points inside the triangle correspond to a stable polynomial.

7.4.5 The Noise Gain of Rational Transfer Functions*

In Section 2.9 we introduced the noise gain of an LTI system. The noise gain measures the ratio between the output and input variances of the system when the input is white noise. Here we show how to compute the noise gain of a system whose transfer function is rational and stable.

By Parseval's theorem,

$$\text{NG} = \frac{1}{2\pi} \int_{-\pi}^{\pi} |H^{\text{f}}(\theta)|^2 d\theta = \sum_{n=-\infty}^{\infty} |h[n]|^2. \tag{7.64}$$

Therefore, in the special case of a transfer function whose denominator polynomial is $a(z) = 1$, we get

$$\text{NG} = \frac{1}{2\pi} \int_{-\pi}^{\pi} |b(e^{j\theta})|^2 d\theta = \sum_{k=0}^{q} b_k^2. \tag{7.65}$$

In this case, the noise gain can be easily computed from the numerator coefficients.

When $a(z) \neq 1$, the sum in (7.64) is infinite, so it cannot be computed exactly in a direct manner. An indirect procedure can be used for computing the noise gain exactly, as shown in the following derivation. Let

$$G^{\text{f}}(\theta) = |H^{\text{f}}(\theta)|^2 = \sum_{n=-\infty}^{\infty} g[n]e^{-j\theta n}. \tag{7.66}$$

We have from (2.137) that the noise gain is

$$\text{NG} = \frac{1}{2\pi} \int_{-\pi}^{\pi} |H^{\text{f}}(\theta)|^2 d\theta = \frac{1}{2\pi} \int_{-\pi}^{\pi} G^{\text{f}}(\theta) d\theta = g[0]. \tag{7.67}$$

However, (7.66) also implies that

$$G^z(z) = \sum_{n=-\infty}^{\infty} g[n]z^{-n} = H^z(z)H^z(z^{-1}) = \frac{b(z)b(z^{-1})}{a(z)a(z^{-1})}. \tag{7.68}$$

We assume for now that the numerator and denominator polynomials have equal degrees, that is, $p = q$. To find the constant coefficient in the series decomposition of (7.68), it is convenient to express the right side in the form

$$\frac{b(z)b(z^{-1})}{a(z)a(z^{-1})} = \frac{c(z)}{a(z)} + \frac{c(z^{-1})}{a(z^{-1})}, \tag{7.69}$$

where $c(z)$ is a polynomial of degree p in powers of z^{-1}. Such a polynomial exists if and only if

$$a(z)c(z^{-1}) + a(z^{-1})c(z) = b(z)b(z^{-1}). \tag{7.70}$$

This equation can be written explicitly in terms of the coefficients of the three polynomials as

$$(1 + \cdots + a_p z^{-p})(c_0 + \cdots + c_p z^p) + (1 + \cdots + a_p z^p)(c_0 + \cdots + c_p z^{-p})$$
$$= (b_0 + b_1 z^{-1} + \cdots + b_p z^{-p})(b_0 + b_1 z + \cdots + b_p z^p). \tag{7.71}$$

Equation (7.71) holds if and only if the coefficients of z^i are equal on both sides for all $-p \le i \le p$. Because of the symmetry in this equation, however, we need look only at $0 \le i \le p$. We then get that (7.71) is equivalent to the set of linear equations

$$\left\{ \begin{bmatrix} 1 & a_1 & \cdots & a_p \\ 0 & 1 & \cdots & a_{p-1} \\ \vdots & \vdots & \cdots & \vdots \\ 0 & 0 & \cdots & 1 \end{bmatrix} + \begin{bmatrix} 1 & a_1 & \cdots & a_p \\ a_1 & a_2 & \cdots & 0 \\ \vdots & \vdots & \cdots & \vdots \\ a_p & 0 & \cdots & 0 \end{bmatrix} \right\} \begin{bmatrix} c_0 \\ c_1 \\ \vdots \\ c_p \end{bmatrix} = \begin{bmatrix} b_0 & b_1 & \cdots & b_p \\ 0 & b_0 & \cdots & b_{p-1} \\ \vdots & \vdots & \cdots & \vdots \\ 0 & 0 & \cdots & b_0 \end{bmatrix} \begin{bmatrix} b_0 \\ b_1 \\ \vdots \\ b_p \end{bmatrix}. \tag{7.72}$$

The coefficient matrix in (7.72) is nonsingular if $a(z)$ is a stable polynomial (we shall not prove this here). Therefore, (7.72) has a unique solution for the coefficients c_k. This implies that the decomposition (7.69) exists and is unique.

Now that we have established the existence and uniqueness of (7.69), let us denote by $f[n]$ the impulse response of $c(z)/a(z)$, so

$$\frac{c(z)}{a(z)} = \sum_{n=0}^{\infty} f[n]z^{-n}. \tag{7.73}$$

Then we have by symmetry,

$$\frac{c(z^{-1})}{a(z^{-1})} = \sum_{n=0}^{\infty} f[n]z^{n}. \tag{7.74}$$

Therefore we can write, from (7.68), (7.69),

$$\sum_{n=-\infty}^{\infty} g[n]z^{-n} = \sum_{n=0}^{\infty} f[n](z^{-n} + z^n). \tag{7.75}$$

In particular,

$$g[0] = \text{NG} = 2f[0]. \tag{7.76}$$

From (7.74), however, when substituting $z = 0$, we see that $f[0] = c_0$. Therefore,

$$\text{NG} = 2c_0. \tag{7.77}$$

In summary, the noise gain can be computed by solving the linear equations (7.72); then $2c_0$ is the desired result.

The procedure `nsgain` in Program 7.4 implements the noise-gain computation in MATLAB. The case $p = 0$ is handled first, using (7.65). The case $p \ne q$ (but $p \ne 0$) is handled by extending one of the polynomials to degree $\max\{p, q\}$ by adding zero-valued coefficients. The coefficient matrix of the linear equations is built by calling the functions `toeplitz` and `hankel`. We have already commented on Toeplitz matrices when discussing partial fraction decomposition. A *Hankel* matrix is a matrix whose elements are equal along the antidiagonals. Such a matrix is uniquely defined by its first column and last row. You are encouraged to learn more about this function by typing `help hankel`.

7.5 Inversion of the z-Transform

The general formula for the inverse z-transform is (7.6). However, this formula is seldom used for actual computation of the inverse z-transform. In this section we describe common methods for inverse z-transform calculation.

If the region of convergence includes the unit circle, we can use the unit circle as the contour of integration. Substitution of $z = e^{j\theta}$ and $dz = je^{j\theta}d\theta$ in (7.6) gives

$$x[n] = \frac{1}{2\pi} \int_{-\pi}^{\pi} X^z(e^{j\theta})e^{j\theta n}d\theta = \frac{1}{2\pi} \int_{-\pi}^{\pi} X^f(\theta)e^{j\theta n}d\theta, \tag{7.78}$$

so the inverse z-transform coincides with an inverse Fourier transform. If the region of convergence does not include the unit circle, we can take a circle with radius r (greater or smaller than 1, depending on the region of convergence) and substitute $z = re^{j\theta}$ and $dz = jre^{j\theta}d\theta$ to get

$$x[n] = \frac{1}{2\pi} \int_{-\pi}^{\pi} X^z(re^{j\theta})r^n e^{j\theta n}d\theta = r^n \frac{1}{2\pi} \int_{-\pi}^{\pi} X^z(re^{j\theta})e^{j\theta n}d\theta. \tag{7.79}$$

Therefore, the inverse z-transform can be computed by evaluating the inverse Fourier transform of $X^z(re^{j\theta})$ and multiplying the result by the sequence r^n.

Sometimes $X^z(z)$ can be expanded as a known power series (e.g., a series available in standard tables). The following example illustrates such a case.

Example 7.3 Let

$$X^z(z) = -\log_e(1 - az^{-1}) = az^{-1} + \frac{1}{2}a^2 z^{-2} + \frac{1}{3}a^3 z^{-3} + \cdots, \quad |z| > |a|. \tag{7.80}$$

This series is given in most standard series tables. Therefore,

$$x[n] = \begin{cases} \frac{1}{n}a^n, & n \geq 1, \\ 0, & n < 1. \end{cases} \tag{7.81}$$

$\square$

A general and powerful tool for inverse z-transform computation is the Cauchy residue theorem. If a function $f(z)$ is analytic in a domain in the complex plane except at a single point z_0 in the domain, the *residue* of $f(z)$ at z_0 is

$$\text{res}\{f(z) \text{ at } z_0\} = \frac{1}{2\pi j} \oint f(z)dz, \tag{7.82}$$

where the contour of integration lies in the domain of analyticity of $f(z)$ and circles z_0 once in the counterclockwise direction. If $f(z)$ has a finite number of singularities inside the contour, say $\{z_k, 1 \leq k \leq K\}$, and is analytic otherwise on the contour and inside it, then

$$\frac{1}{2\pi j} \oint f(z)dz = \sum_{k=1}^{K} \text{res}\{f(z) \text{ at } z_k\}. \tag{7.83}$$

Equation (7.83) is the Cauchy residue theorem.

If $f(z)$ is singular at z_0 and there exists an integer m such that

$$g(z) \triangleq f(z)(z - z_0)^m$$

is analytic at z_0, then z_0 is said to be a *pole* of order m. A singularity point that is not an mth-order pole for any m is called an *essential singularity*. The residue of $f(z)$ at an mth-order pole z_0 is given by

$$\text{res}\{f(z) \text{ at } z_0\} = \frac{1}{(m-1)!} \frac{d^{m-1}g(z)}{dz^{m-1}} \bigg|_{z=z_0}. \tag{7.84}$$

The Cauchy residue theorem can be used for inverse z-transform computation as follows. Assume that $X^z(z)$ has only a finite number of poles and no essential singularities in the region $0 \le |z| \le R_1$ (where R_1 is the inner radius of the region of convergence). Then, according to (7.83),

$$x[n] = \frac{1}{2\pi j} \oint X^z(z) z^{n-1} dz = \sum_{k=1}^{K_n} \mathrm{res}\{X^z(z) z^{n-1} \text{ at } z_{k,n}\}, \qquad (7.85)$$

where $\{z_{k,n}, 1 \le k \le K_n\}$ are the residues of $X^z(z) z^{n-1}$ in the region $0 \le |z| \le R_1$. The poles of $X^z(z) z^{n-1}$ in this region are the poles of $X^z(z)$, and possibly a pole at $z = 0$. The residues can be computed from (7.84).

Example 7.4 Let

$$X^z(z) = \frac{z}{(z-a)^2}, \qquad |z| > |a|. \qquad (7.86)$$

Then

$$X^z(z) z^{n-1} = \frac{z^n}{(z-a)^2}, \qquad |z| > |a|. \qquad (7.87)$$

For $n \ge 0$, the only singular point is $z = a$, which is a second-order pole. At this point we have $g(z) = z^n$, so

$$\left. \frac{dg(z)}{dz} \right|_{z=a} = \left. nz^{n-1} \right|_{z=a} = na^{n-1}. \qquad (7.88)$$

For $n < 0$, the two singular points are $z = a$ (a second-order pole, as before) and $z = 0$, which is a pole of order $m = -n$. At the point $z = 0$ we have $g(z) = (z-a)^{-2}$, so

$$\frac{1}{(m-1)!} \left. \frac{d^{m-1}g(z)}{dz^{m-1}} \right|_{z=0} = \frac{m!}{(m-1)!} (a-z)^{-(m+1)} \Big|_{z=0}$$
$$= ma^{-(m+1)} = -na^{n-1}. \qquad (7.89)$$

The residue at $z = a$ is na^{n-1}, as before. Therefore, the sum of the two residues is zero. In summary,

$$x[n] = \begin{cases} na^{n-1}, & n \ge 0, \\ 0, & n < 0. \end{cases} \qquad (7.90)$$

$\square$

Our main interest here is the case of $X^z(z)$ rational and, in particular, $X^z(z)$ is the transform of a *causal* sequence. If all the poles of $X^z(z)$ are distinct, it can be written as a partial fraction decomposition [cf. (7.56)]

$$X^z(z) = \sum_{i=0}^{q-p} c_i z^{-i} + \sum_{k=1}^{p} \frac{A_k}{1 - \alpha_k z^{-1}}. \qquad (7.91)$$

Therefore we have, by the linearity and time-shift properties, and by Example 7.1, part 2,

$$x[n] = \sum_{i=0}^{q-p} c_i \delta[n-i] + \sum_{k=1}^{p} A_k \alpha_k^n, \qquad n \ge 0. \qquad (7.92)$$

The poles α_k can be real or complex. If the coefficients of the rational function $X^z(z)$ are real, complex poles must appear in conjugate pairs. Assume, for example, that α_1 and α_2 are conjugates of each other, say

$$\alpha_1 = \rho e^{j\phi}, \qquad \alpha_2 = \rho e^{-j\phi}.$$

Then A_1 and A_2 must also be conjugate of each other, say

$$A_1 = a + jb, \quad A_2 = a - jb.$$

Then

$$A_1\alpha_1^n + A_2\alpha_2^n = (a + jb)\rho^n e^{jn\phi} + (a - jb)\rho^n e^{-jn\phi}$$
$$= \rho^n(a + jb)[\cos(n\phi) + j\sin(n\phi)] + \rho^n(a - jb)[\cos(n\phi) - j\sin(n\phi)]$$
$$= 2\rho^n[a\cos(n\phi) - b\sin(n\phi)]. \qquad (7.93)$$

This conversion can be performed for all complex pairs in (7.92), thus bringing it to a purely real form.

The procedure invz in Program 7.5 implements the computation of the inverse z-transform of a rational causal transfer function. Inverse z-transform computation by partial fraction decomposition can be generalized to the case of repeated poles, as well as to the case of rational noncausal transfer functions. These generalizations are not discussed here.

When learning the z-transform, you should get acquainted with the characteristic shapes of sequences associated with different pole locations. The shape is mainly determined by (1) whether the pole is real or complex, and (2) whether it is inside, on, or outside the unit circle. Figure 7.4 illustrates the sequences associated with real poles for six different cases: $\alpha > 1$, $\alpha = 1$, $0 < \alpha < 1$, $-1 < \alpha < 0$, $\alpha = -1$, and $\alpha < -1$. As we see, the sequences are divergent when $|\alpha| > 1$, convergent when $|\alpha| < 1$, and have constant amplitude when $|\alpha| = 1$. When $\alpha < 0$, the sequence is oscillatory. This is in contrast with continuous-time signals, which are never oscillatory when the corresponding pole is real.

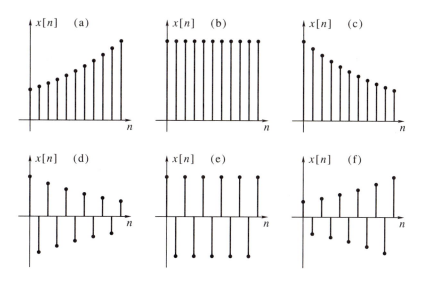

Figure 7.4 The sequences associated with a single real pole at $z = \alpha$ in the z-transform: (a) $\alpha > 1$, (b) $\alpha = 1$, (c) $0 < \alpha < 1$, (d) $-1 < \alpha < 0$, (e) $\alpha = -1$, (f) $\alpha < -1$.

Figure 7.5 illustrates the sequences associated with complex poles for six different cases: $|\alpha| > 1$ and $\sphericalangle\alpha < 0.5\pi$, $|\alpha| = 1$ and $\sphericalangle\alpha < 0.5\pi$, $|\alpha| < 1$ and $\sphericalangle\alpha < 0.5\pi$, $|\alpha| < 1$ and $\sphericalangle\alpha > 0.5\pi$, $|\alpha| = 1$ and $\sphericalangle\alpha > 0.5\pi$, and finally $|\alpha| > 1$ and $\sphericalangle\alpha > 0.5\pi$. The convergence-divergence behavior is the same as for real poles. However, in the case $\alpha = 1$, the constant amplitude property is not evident, because a discrete-time sinusoidal signal is not necessarily periodic. The sequences are always oscillatory and in general appear quite irregular.

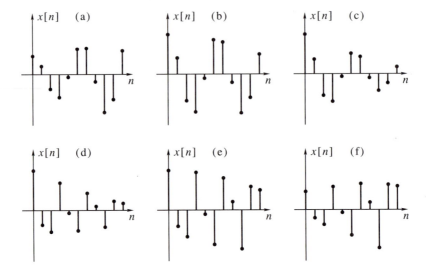

Figure 7.5 The sequences associated with a pair of complex poles at $z = \alpha, z = \bar{\alpha}$ in the z-transform: (a) $|\alpha| > 1$, $\measuredangle\alpha < 0.5\pi$, (b) $|\alpha| = 1$, $\measuredangle\alpha < 0.5\pi$, (c) $|\alpha| < 1$, $\measuredangle\alpha < 0.5\pi$, (d) $|\alpha| < 1$, $\measuredangle\alpha > 0.5\pi$, (e) $|\alpha| = 1$, $\measuredangle\alpha > 0.5\pi$, (f) $|\alpha| > 1$, $\measuredangle\alpha > 0.5\pi$.

7.6 Frequency Responses of Rational Transfer Functions

The frequency response of a BIBO-stable system is obtained from the transfer function by the substitution $z = e^{j\theta}$. For rational transfer functions, it is especially convenient to use the pole–zero factorization (7.50) of the transfer function. We have in this case

$$H^f(\theta) = b_{q-r}e^{j\theta(p-q)}\frac{\prod_{i=1}^{r}(e^{j\theta} - \beta_i)}{\prod_{i=1}^{p}(e^{j\theta} - \alpha_i)}. \tag{7.94}$$

Since the magnitude of a product of complex numbers is the product of the magnitudes of the factors, we get from (7.94)

$$|H^f(\theta)| = b_{q-r}\frac{\prod_{i=1}^{r}|e^{j\theta} - \beta_i|}{\prod_{i=1}^{p}|e^{j\theta} - \alpha_i|}. \tag{7.95}$$

Similarly, since the phase of a product of complex numbers is the sum of the phases of the factors, we get

$$\measuredangle H^f(\theta) = \theta(p - q) + \sum_{i=1}^{r}\measuredangle(e^{j\theta} - \beta_i) - \sum_{i=1}^{p}\measuredangle(e^{j\theta} - \alpha_i). \tag{7.96}$$

Figure 7.6 illustrates how to determine the magnitude and phase of the frequency response from the pole-zero plot of the transfer function. Each frequency θ is represented by a point on the unit circle. The transfer function represented by this figure is

$$H^z(z) = \frac{1 + 0.2z^{-1}}{[1 - 0.867z^{-1}][1 - (0.067 + j0.867)z^{-1}][1 - (0.067 - j0.867)z^{-1}]}.$$

The complex numbers $(e^{j\theta} - \beta_i)$ are represented by vectors pointing from the zero locations β_i to the point on the unit circle. Similarly, the complex numbers $(e^{j\theta} - \alpha_i)$

are represented by vectors pointing from the pole locations α_i to the point on the unit circle. To compute the magnitude response at a specific θ, we must form the product of magnitudes of all vectors pointing from the zeros, then form the product of magnitudes of all vectors pointing from the poles, divide the former by the latter, and finally multiply by the constant factor b_{q-r}. To compute the phase response at a specific θ, we must form the sum of angles of all vectors pointing from the zeros, then form the sum of angles of all vectors pointing from the poles, subtract the latter from the former, and finally add the linear-phase term $\theta(p - q)$. For the transfer function represented by Figure 7.6 we get the frequency response shown in Figure 7.7.

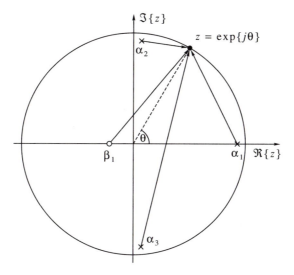

Figure 7.6 Using the pole–zero plot to obtain the frequency response.

The graphical procedure leads to the following observations:

1. A real pole near $z = 1$ results in a high DC gain, whereas a real pole near $z = -1$ results in a high gain at $\theta = \pi$.

2. Complex poles near the unit circle result in a high gain near the frequencies corresponding to the phase angles of the poles.

3. A real zero near $z = 1$ results in a low DC gain, whereas a real zero near $z = -1$ results in a low gain at $\theta = \pi$.

4. Complex zeros near the unit circle result in a low gain near the frequencies corresponding to the phase angles of the poles.

Examination of the pole–zero pattern of the transfer function thus permits rapid, although coarse, estimation of the general nature of the frequency response.

A MATLAB implementation of frequency response computation of a rational system does not require a pole–zero factorization, but can be performed directly on the polynomials of the transfer function. The procedure `frqresp` in Program 7.6 illustrates this computation. The program has three modes of operation. In the first mode, only the polynomial coefficients and the desired number of frequency points K are given. The program then selects K equally spaced points on the interval $[0, \pi]$ and performs the computation by dividing the zero-padded FFTs of the numerator and the denominator. In the second mode, the program is given a number of points and a frequency interval. The program then selects K equally spaced points on the given interval and

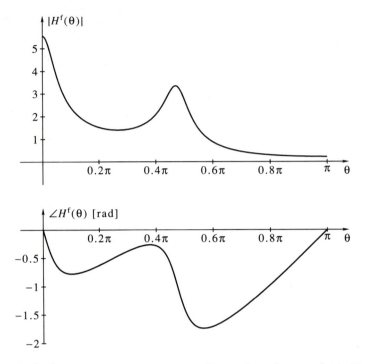

Figure 7.7 The frequency response corresponding to the pole–zero plot in Figure 7.6.

performs the computation by dividing the chirp Fourier transforms of the numerator and the denominator. In the third mode, the program is given a vector of individual frequencies, not necessarily equally spaced, and performs the computation point by point. We remark that the MATLAB function `freqz`, located in the Signal Processing Toolbox, performs a similar computation.

7.7 The Unilateral z-Transform

The *unilateral*, or *one-sided*, z-transform of a sequence $x[n]$ is defined as

$$X_+^z(z) = \sum_{n=0}^{\infty} x[n]z^{-n}. \qquad (7.97)$$

The definition applies to general sequences, whether causal or not. Equality of $X^z(z)$ and $X_+^z(z)$ holds if and only if the sequence $x[n]$ is causal.

The region of convergence of the unilateral z-transform includes the open annulus

$$R_1 < |z| < \infty, \qquad (7.98)$$

where

$$R_1 = \inf\left\{r \geq 0 : \sum_{n=0}^{\infty} |x[n]|r^{-n} < \infty\right\}. \qquad (7.99)$$

In addition, it may or may not include the boundary $|z| = R_1$.

The inverse unilateral z-transform formula is identical to that of the bilateral transform, (7.6). However, it yields only the values of $x[n]$ for $n \geq 0$. The unilateral z-transform provides no information on values of $x[n]$ for $n < 0$.

The unilateral z-transform shares certain, but not all properties of the bilateral z-transform. Of special interest to us here are its two shift properties, given in the following theorem.

Theorem 7.5 Let

$$y[n] = x[n + m], \quad m > 0, \tag{7.100a}$$

$$u[n] = x[n - m], \quad m > 0. \tag{7.100b}$$

Then,

$$Y_+^z(z) = z^m X_+^z(z) - \sum_{i=0}^{m-1} x[i] z^{m-i}, \tag{7.101a}$$

$$U_+^z(z) = z^{-m} X_+^z(z) + \sum_{i=-m}^{-1} x[i] z^{-(m+i)}. \tag{7.101b}$$

Proof of (7.101a)

$$Y_+^z(z) = \sum_{n=0}^{\infty} x[n+m] z^{-n} = z^m \sum_{n=0}^{\infty} x[n+m] z^{-(n+m)} = z^m \sum_{i=m}^{\infty} x[i] z^{-i}$$

$$= z^m \left[\sum_{i=0}^{\infty} x[i] z^{-i} - \sum_{i=0}^{m-1} x[i] z^{-i} \right] = z^m X_+^z(z) - \sum_{i=0}^{m-1} x[i] z^{m-i}. \tag{7.102}$$

Proof of (7.101b)

$$U_+^z(z) = \sum_{n=0}^{\infty} x[n-m] z^{-n} = z^{-m} \sum_{n=0}^{\infty} x[n-m] z^{-(n-m)} = z^{-m} \sum_{i=-m}^{\infty} x[i] z^{-i}$$

$$= z^{-m} \left[\sum_{i=-m}^{-1} x[i] z^{-i} + \sum_{i=0}^{\infty} x[i] z^{-i} \right] = z^{-m} X_+^z(z) + \sum_{i=-m}^{-1} x[i] z^{-(m+i)}. \tag{7.103}$$

□

Two other properties of the unilateral z-transform are given by the following theorems.

Theorem 7.6 (initial value theorem) The initial value $x[0]$ of the sequence $x[n]$ can be obtained from its unilateral z-transform by

$$x[0] = \lim_{z \to \infty} X_+^z(z). \tag{7.104}$$

Proof Take the limits of both sides of (7.97) as $z \to \infty$ and observe that the only term on the right side that does not go to zero is $x[0]$. □

Theorem 7.7 (final value theorem) Assume that $X_+^z(z)$ is such that $(z-1)X_+^z(z)$ exists on the unit circle. Then

$$\lim_{n \to \infty} x[n] = \lim_{z \to 1} (z-1) X_+^z(z). \tag{7.105}$$

Proof Observe that if $(z-1)X_+^z(z)$ exists on the unit circle, then either $X_+^z(z)$ exists on the unit circle, or $X_+^z(z)$ has a simple pole at $z = 1$, which is canceled by the factor $(z-1)$. In the former case the sequence $\{x[n], n \geq 0\}$ is stable, so necessarily $\lim_{n \to \infty} x[n] = 0$. Indeed, in this case $\lim_{z \to 1}(z-1)X_+^z(z) = 0$. In the latter case $X_+^z(z)$ admits the additive decomposition

$$X_+^z(z) = \frac{A}{z-1} + B(z), \tag{7.106}$$

where $B(z)$ exists on the unit circle, so it is the transform of a stable sequence. We get the constant A from (7.106), multiplying by $(z - 1)$ and taking the limit $z \to 1$:

$$A = \lim_{z \to 1}(z - 1)X_+^z(z). \qquad (7.107)$$

On the other hand, the inverse z-transform of (7.106) gives

$$x[n] = A + b[n], \quad n \geq 1. \qquad (7.108)$$

Taking the limit as $n \to \infty$ and using the stability of $b[n]$ gives

$$A = \lim_{n \to \infty} x[n]. \qquad (7.109)$$

$\square$

An important application of the final value theorem is for computing the final (or steady-state) value of the response of a stable and causal LTI system to a unit-step input. The z-transform of a unit step is $1/(1 - z^{-1})$, so the steady-state response to a unit step of a system whose transfer function is $H^z(z)$ is

$$\lim_{n \to \infty} y[n] = \lim_{z \to 1} \frac{(z - 1)H^z(z)}{(1 - z^{-1})} = \lim_{z \to 1} zH^z(z) = H^z(1). \qquad (7.110)$$

The quantity $H^z(1)$ is called the *DC gain* of the system. Although $H^z(1)$ may exist for any $H^z(z)$, it is meaningful only for a stable and causal transfer function.

The main use of the unilateral z-transform is for solving difference equations with initial conditions. Consider the general difference equation

$$y[n] = -\sum_{i=1}^{p} a_i y[n - i] + \sum_{i=0}^{q} b_i x[n - i]. \qquad (7.111)$$

and assume that we are given the initial conditions

$$\{y[-p], y[-p + 1], \ldots, y[-1]\}.$$

Assume also that the input sequence $x[n]$ is causal, so $x[n] = 0$ for $n < 0$. Because the difference equation is linear, we can express its solution as

$$y[n] = y_{zir}[n] + y_{zsr}[n], \qquad (7.112)$$

where

1. $y_{zir}[n]$ is the *zero-input response*, or the *natural response*, defined as the solution of (7.111) when the initial conditions are as given above, while the input $x[n]$ is identically zero.

2. $y_{zsr}[n]$ is the *zero-state response*, or the *forced response*, defined as the solution of (7.111) when the initial conditions are zero and the input $x[n]$ is present.

Computation of $y_{zsr}[n]$ requires only the techniques we studied in Section 7.5, since

$$Y_{zsr}^z(z) = \frac{b(z)X^z(z)}{a(z)}. \qquad (7.113)$$

If $X^z(z)$ is rational, we can invert the right side of (7.113) using partial fraction decomposition. If $X^z(z)$ is not rational but its numerical values are known, we can compute $y_{zsr}[n]$ numerically using, for example, the MATLAB function `filter`.

Computation of $y_{zir}[n]$ involves solving the homogeneous equation

$$y[n] = -\sum_{i=1}^{p} a_i y[n - i], \quad n \geq 0, \qquad (7.114)$$

subject to the initial conditions

$$\{y[-p], y[-p + 1], \ldots, y[-1]\}.$$

The solution can be obtained using the unilateral z-transform, as follows. We transform (7.114) and use the shift property (7.101) to get

$$Y_+^z(z) = -\sum_{i=1}^{p} a_i \left[z^{-i} Y_+^z(z) + \sum_{l=-i}^{-1} y[l] z^{-(i+l)} \right]. \tag{7.115}$$

Therefore,

$$Y_+^z(z) = -\frac{\sum_{i=1}^{p} a_i \sum_{l=-i}^{-1} y[l] z^{-(i+l)}}{1 + \sum_{i=1}^{p} a_i z^{-i}}. \tag{7.116}$$

This can also be written, after rearrangement of the numerator, as

$$Y_+^z(z) = -\frac{\sum_{i=0}^{p-1} \left[\sum_{l=i-p}^{-1} y[l] a_{i-l} \right] z^{-i}}{a(z)}. \tag{7.117}$$

Finally, $y_{\text{zir}}[n]$ is obtained by computing the inverse z-transform of (7.117).

The procedure numzir in Program 7.7 implements the computation of the numerator of (7.117). The zero-input response can then be computed by calling invz, or the partial fraction decomposition of (7.117) be computed by calling tf2pf, as needed.

7.8   Summary and Complements

7.8.1   Summary

In this chapter we introduced the z-transform of a discrete-time signal (7.1) and discussed its use to discrete-time linear system theory. The z-transform is a complex function of a complex variable. It is defined on a domain in the complex plane called the region of convergence. The ROC usually has the form of an annulus. The inverse z-transform is given by the complex integration formula (7.6).

The z-transform of the impulse response of a discrete-time LTI system is called the transfer function of the system. Of particular interest are systems that are stable in the BIBO sense. For such systems, the ROC of the transfer function includes the unit circle. If the system is also causal, the ROC includes the unit circle and all that is outside it. Therefore, all the singularities of the transfer function of a stable and causal system must be inside the unit circle.

Of special importance are causal LTI systems whose transfer functions are rational functions of z. Such a system can be described by a difference equation (7.45). Its transfer function is characterized by the numerator and denominator polynomials. The roots of these two polynomials are called zeros and poles, respectively. For a stable system, all the poles are inside the unit circle. The stability of a rational transfer function can be tested without explicitly finding the poles, but by means of the Schur-Cohn test, which requires only simple rational operations. A rational transfer function whose poles are simple (i.e., of multiplicity one) can be expressed by partial fraction decomposition (7.56).

Several methods exist for computation of inverse z-transforms. Contour integration is the most general, but usually the least convenient method. The Cauchy residue theorem and power series expansions are convenient in certain cases. Partial fraction decomposition is the preferred method for computing inverse z-transforms of rational functions.

The frequency response of an LTI system can be computed from its transfer function by substituting $z = e^{j\theta}$. This is especially convenient when the transfer function is given in a factored form. Regions of low magnitude response (in the vicinity of zeros) and high magnitude response (in the vicinity of poles) can be determined by visual examination of the pole–zero map of the system.

The unilateral z-transform (7.97) is also useful, mainly for solving difference equations with initial conditions. The solution of such an equation is conveniently expressed as a sum of two terms: the zero-input response and the zero-state response. The former is best obtained by the unilateral z-transform, whereas the latter can be done using the bilateral z-transform.

7.8.2 Complements

1. [p. 205] The earliest references on sampled-data systems, which paved the way to the z-transforms, are by MacColl [1945] and Hurewicz [1947]. The z-transform was developed independently by a number of people in the late 1940s and early 1950s. The definitive reference in the western literature is by Ragazzini and Zadeh [1952] and that in the eastern literature is Tsypkin [1949, 1950]. Barker [1952] and Linvill [1951] have proposed definitions similar to the z-transform. Jury [1954] has invented the modified z-transform, which we shall mention later in this book (Problem 10.35).

2. [p. 206] In complex function theory, the z-transform is a special case of a Laurent series: It is the Laurent series of $X^z(z)$ around the point $z_0 = 0$. The inverse z-transform formula is the inversion formula of Laurent series; see, for example, Churchill and Brown [1984].

3. [p. 206] The region of convergence of the z-transform is defined as the set of all complex numbers z such that the sum of the absolute values of the series converges, as seen in (7.2). Why did we require absolute convergence when the z-transform is defined as a sum of complex numbers, as seen in (7.1)? The reason is that the value of an infinite sum such as (7.1) can potentially depend on the order in which the elements of the series are added. In general, a series $\sum_{n=1}^{\infty} a_n$ may converge (that is, yield a finite result) even when $\sum_{n=1}^{\infty} |a_n| = \infty$. However, in such a case the value of $\sum_{n=1}^{\infty} a_n$ will vary if the order of terms is changed. Such a series is said to be *conditionally convergent*. On the other hand, if the sum of absolute values is finite, the sum will be independent of the order of summation. Such a series is said to be *absolutely convergent*. In the z-transform we are summing a two-sided sequence, so we do not want the sum to depend on the order of terms. The requirement of absolute convergence (7.2) eliminates the problem and guarantees that (7.1) be unambiguous.

4. [p. 206] Remember that (1) the *infimum* of a nonempty set of real numbers $\mathbb{S}$, denoted by $\inf\{\mathbb{S}\}$, is the largest number having the property of being smaller than all members of $\mathbb{S}$; (2) the *supremum* of a nonempty set of real numbers $\mathbb{S}$, denoted by $\sup\{\mathbb{S}\}$, is the smallest number having the property of being larger than all members of $\mathbb{S}$. The infimum is also called *greatest lower bound*; the supremum is also called *least upper bound*. Every nonempty set of real numbers has a unique infimum and a unique supremum. If the set is bounded from above, its supremum is finite; otherwise it is defined as ∞. If the set is bounded from below, its infimum is finite; otherwise it is defined as $-\infty$.

5. [p. 206] The *Cauchy-Hadamard theorem* expresses the radii of convergence R_1 and R_2 explicitly in terms of the sequence values:

$$R_1 = \limsup_{n \to \infty} |x[n]|^{1/n}, \quad R_2 = \liminf_{n \to \infty} |x[-n]|^{-1/n}. \tag{7.118}$$

6. [p. 206] The *extended complex plane* is obtained by adding a single point $z = \infty$ to the conventional complex plane $\mathbb{C}$. The point $z = \infty$ has modulus (magnitude) larger than that of any other complex number; its argument (phase) is undefined. By comparison, the point $z = 0$ has modulus smaller than that of any other complex number and undefined argument. The region of convergence may be extended to include the point $z = \infty$ if and only if the sequence is causal; see the discussion in Example 7.1, part 8.

7. [p. 215] In continuous-time systems, properness is related to realizability, not causality. For example, the continuous-time transfer function $H^L(s) = s$ is not proper. It represents pure differentiation, which is a causal, but not a physically realizable operation.

7.9   MATLAB Programs

Program 7.1 Partial fraction decomposition of a rational transfer function.

```
function [c,A,alpha] = tf2pf(b,a);
% Synopsis: [c,A,alpha] = tf2pf(b,a).
% Partial fraction decomposition of b(z)/a(z). The polynomials are in
% negative powers of z. The poles are assumed distinct.
% Input parameters:
% a, b: the input polynomials
% Output parameters:
% c: the free polynomial; empty if deg(b) < deg(a)
% A: the vector of gains of the partial fractions
% alpha: the vector of poles.

% Compute c(z) and d(z)
p = length(a)-1; q = length(b)-1;
a = (1/a(1))*reshape(a,1,p+1);
b = (1/a(1))*reshape(b,1,q+1);
if (q >= p), % case of nonempty c(z)
   temp = toeplitz([a,zeros(1,q-p)]',[a(1),zeros(1,q-p)]);
   temp = [temp,[eye(p); zeros(q-p+1,p)]];
   temp = temp\b';
   c = temp(1:q-p+1)'; d = temp(q-p+2:q+1)';
else
   c = []; d = [b,zeros(1,p-q-1)];
end

% Compute A and alpha
alpha = cplxpair(roots(a)).'; A = zeros(1,p);
for k = 1:p,
   temp = prod(alpha(k)-alpha(find(1:p ~= k)));
   if (temp == 0), error('Repeated roots in TF2PF');
   else, A(k) = polyval(d,alpha(k))/temp; end
end
```

Program 7.2 Conversion of partial fraction decomposition to a rational transfer function.

```
function [b,a] = pf2tf(c,A,alpha);
% Synopsis: [b,a] = pf2tf(c,A,alpha).
% Conversion of partial fraction decomposition to the form b(z)/a(z).
% The polynomials are in negative powers of z.
% Input parameters:
% c: the free polynomial; empty if deg(b) < deg(a)
% A: the vector of gains of the partial fractions
% alpha: the vector of poles.
% Output parameters:
% a, b: the output polynomials

p = length(alpha); d = A(1); a = [1,-alpha(1)];
for k = 2:p,
   d = conv(d,[1,-alpha(k)]) + A(k)*a;
   a = conv(a,[1,-alpha(k)]);
end
if (length(c) > 0),
   b = conv(c,a) + [d,zeros(1,length(c))];
else, b = d; end
a = real(a); b = real(b);
```

Program 7.3 The Schur–Cohn stability test.

```
function s = sctest(a);
% Synopsis: s = sctest(a).
% Schur-Cohn stability test.
% Input:
% a: coefficients of polynomial to be tested.
% Output:
% s: 1 if stable, 0 if unstable.

n = length(a);
if (n == 1), s = 1; % a zero-order polynomial is stable
else,
   a = reshape((1/a(1))*a,1,n); % make the polynomial monic
   if (abs(a(n)) >= 1), s = 0; % unstable
   else,
      s = sctest(a(1:n-1)-a(n)*fliplr(a(2:n))); % recursion
   end
end
```

Program 7.4 Computation of the noise gain of a rational transfer function.

```
function ng = nsgain(b,a);
% Synopsis: ng = nsgain(b,a).
% Computes the noise gain of a rational system b(z)/a(z).
% Input parameters:
% b, a: the numerator and denominator coefficients.
% Output:
% ng: the noise gain

p = length(a)-1; q = length(b)-1; n = max(p,q);
if (p == 0), ng = sum(b.^2); return, end
a = [reshape(a,1,p+1),zeros(1,n-p)];
b = [reshape(b,1,q+1),zeros(1,n-q)];
mat = toeplitz([1; zeros(n,1)],a) + ...
      hankel(a',[a(n+1),zeros(1,n)]);
vec = toeplitz([b(1); zeros(n,1)],b)*b';
vec = mat\vec; ng = 2*vec(1);
```

Program 7.5 The inverse z-transform of a rational transfer function.

```
function x = invz(b,a,N);
% Synopsis: x = invz(b,a,N).
% Computes first N terms of the inverse z-transform
% of the rational transfer function b(z)/a(z).
% The poles are assumed distinct.
% Input parameters:
% b, a: numerator and denominator input polynomials
% N: number of points to be computed
% Output:
% x: the inverse sequence.

[c,A,alpha] = tf2pf(b,a);
x = zeros(1,N);
x(1:length(c)) = c;
for k = 1:length(A),
    x = x+A(k)*(alpha(k)).^(0:N-1);
end
x = real(x);
```

Program 7.6 Frequency response of a rational transfer function.

```
function H = frqresp(b,a,K,theta);
% Synopsis: H = frqresp(b,a,K,theta).
% Frequency response of b(z)/a(z) on a given frequency interval.
% Input parameters:
% b, a: numerator and denominator polynomials
% K: the number of frequency response points to compute
% theta: if absent, the K points are uniformly spaced on [0, pi];
%        if present and theta is a 1-by-2 vector, its entries are
%        taken as the end points of the interval on which K evenly
%        spaced points are placed; if the size of theta is different
%        from 2, it is assumed to be a vector of frequencies for which
%        the frequency response is to be computed, and K is ignored.
% Output:
% H: the frequency response vector.

if (nargin == 3),
   H = fft(b,2*K-2)./fft(a,2*K-2); H = H(1:K);
elseif (length(theta) == 2),
   t0 = theta(1); dt = (theta(2)-theta(1))/(K-1);
   H = chirpf(b,t0,dt,K)./chirpf(a,t0,dt,K);
else
   H = zeros(1,length(theta));
   for i = 1:length(theta),
      H(i) = sum(b.*exp(-j*theta(i)*(0:length(b)-1)))/ ...
             sum(a.*exp(-j*theta(i)*(0:length(a)-1)));
   end
end
```

Program 7.7 Computation of the numerator of (7.117).

```
function b = numzir(a,yinit);
% Synopsis: b = numzir(a,yinit).
% Compute the numerator polynomial for finding the zero-input
% response of the homogeneous equation a(z)y(z) = 0.
% Input parameters:
% a: the coefficient polynomial of the homogeneous equation
% yinit: the vector of y[-1], y[-2], ..., y[-p].
% Output:
% b: the numerator.

p = length(a) - 1; a = fliplr(reshape(-a(2:p+1),1,p));
b = conv(reshape(yinit,1,p),a); b = fliplr(b(1:p));
```

7.10   Problems

7.1 Derive the z-transforms of

$$x_1[n] = \begin{cases} \cos(\theta_0 n), & n \geq 0, \\ 0, & n < 0, \end{cases} \qquad x_2[n] = \begin{cases} \sin(\theta_0 n), & n \geq 0, \\ 0, & n < 0, \end{cases}$$

$$x_3[n] = \begin{cases} n\cos(\theta_0 n), & n \geq 0, \\ 0, & n < 0, \end{cases} \qquad x_4[n] = \begin{cases} n\sin(\theta_0 n), & n \geq 0, \\ 0, & n < 0. \end{cases}$$

Specify the region of convergence of the transforms.

7.2 Find the z-transform of

$$x[n] = \begin{cases} n, & 1 \leq n \leq N, \\ 2N - n, & N + 1 \leq n \leq 2N - 1, \\ 0, & \text{otherwise.} \end{cases}$$

7.3 Use the differentiation property (7.33) to find the z-transform of the sequences

$$x_1[n] = \begin{cases} na^n, & n \geq 0, \\ 0, & n < 0, \end{cases} \qquad x_2[n] = \begin{cases} n^2 a^n, & n \geq 0, \\ 0, & n < 0. \end{cases}$$

Specify the region of convergence of the transforms.

7.4 Show that the z-transform of the sequence

$$x[n] = \begin{cases} \dfrac{(n+1)(n+2)\cdots(n+m)a^n}{m!}, & n \geq 0, \\ 0, & n < 0, \end{cases}$$

(where $m = 1, 2, \ldots$) is

$$X^z(z) = \frac{1}{(1 - az^{-1})^{m+1}}.$$

Hint: Use induction.

7.5 Can the function $X^z(z) = \bar{z}$ be the z-transform of any sequence $x[n]$? If so, what is $x[n]$? If not, why not?

7.6 Let $x[n]$ be a finite-duration signal on $0 \leq n \leq N - 1$.

(a) Express $X^d[k]$, the DFT of the signal, as a function of $X^z(z)$.

(b) Express $X^z(z)$ as a function of $X^d[k]$.

7.7 Let $x[n]$ be a symmetric sequence, that is, $x[n] = x[-n]$ for all n.

(a) Prove that $x[n]$ possesses a z-transform in a region in the complex plane if and only if $\sum_{n=-\infty}^{\infty} |x[n]| < \infty$.

(b) Let

$$X^z(z) = \frac{z^{-1}}{(1 - 0.5z^{-1})(1 - 2z^{-1})}.$$

What is the region of convergence of $X^z(z)$ if $x[n]$ is symmetric and what is $x[n]$?

7.8 Give an example of a causal sequence that does not have a z-transform.

7.9 Let $X^z(z)$ be the z-transform of $x[n]$. What is the z-transform of $n^2 x[n]$?

7.10 Let $x[n]$, $y[n]$ be causal sequences. Prove that

$$\frac{1}{2\pi j} \oint X^z(z) Y^z(z) z^{-1} dz = \left[\frac{1}{2\pi j} \oint X^z(z) z^{-1} dz\right] \cdot \left[\frac{1}{2\pi j} \oint Y^z(z) z^{-1} dz\right].$$

7.11 Prove that a system obeying a difference equation such as (7.45) is linear and time invariant. Assume that $y[n]$ is identically zero when $x[n]$ is identically zero.

7.12 We are given a system whose transfer function is

$$H^z(z) = (1 - Az^{-3}) \sum_{k=0}^{2} \frac{b_k}{1 - W_3^k z^{-1}}, \quad \text{where } W_3 = e^{j2\pi/3}.$$

(a) For what value(s) of A will the system be BIBO-stable?

(b) For A found in part a:

    i. Compute the impulse response of the system $h[n]$.

    ii. Compute $H^f(\theta)$ for $\theta = 0, 2\pi/3, 4\pi/3$.

    iii. Express the relation between $\{b_0, b_1, b_2\}$ and $\{h[0], h[1], h[2]\}$ in matrix-vector form. Interpret this equation.

7.13 The signal $y(t)$ obeys a differential equation

$$\frac{dy(t)}{dt} = -\alpha y(t) + x(t),$$

where $\alpha > 0$. The signal is sampled at interval T. Prove that the sampled signal $y[n]$ obeys a difference equation

$$y[n] = ay[n-1] + u[n].$$

Express a and $u[n]$ in terms of α, T, and $x[n]$.

7.14 An LTI system has the transfer function

$$H^z(z) = \sum_{k=2}^{10} \left(\frac{1}{1 - kz^{-1}} + \frac{1}{1 - k^{-1}z^{-1}}\right).$$

Find the region of convergence and the impulse response of the system if it is known that the system is (a) causal (but not necessarily stable) and (b) stable (but not necessarily causal).

7.15 A causal LTI system has four zeros at $z = \pm 0.5 \pm j0.5$ and two poles at $z = \pm j0.5$. The system has no other poles or zeros, except possibly at $z = 0$.

(a) What is the minimum number of poles or zeros that the system must have at $z = 0$?

(b) Draw an approximate magnitude response plot of the system.

(c) Define

$$g[n] = (-1)^n h[n],$$

where $h[n]$ is the impulse response of the given system. Find the poles and zeros of $G^z(z)$.

7.16 A stable LTI system has the transfer function

$$H^z(z) = \frac{3(1 - z^{-1})}{(1 - 0.5z^{-1})(1 - 2z^{-1})}.$$

Find the impulse response of the system.

7.17 The *step response* of an LTI system is its response to the unit-step signal $u[n]$.

(a) Explain how to compute the step response of an LTI system obeying a difference equation. Assume that all the poles of the system are simple, and none of them is at $z = 1$.

(b) Compute the step response of the system

$$H^z(z) = \frac{2 + 2.7z^{-1} - 0.36z^{-2}}{1 + 0.5z^{-1} - 0.36z^{-2}}.$$

7.18 We are given a causal LTI system whose transfer function is

$$H^z(z) = \frac{1 + z^{-1} + z^{-2}}{(1 - 0.8z^{-1})(1 + 0.5z^{-1})}.$$

The impulse response $h[n]$ of the system is used for generating a new impulse response $h_1[n]$ according to

$$h_1[n] = y^n h[n],$$

where y is constant.

(a) Write the transfer function $H_1^z(z)$ and the corresponding difference equation.

(b) Is $H_1^z(z)$ stable if $|y| < 1$? If $|y| > 1$?

(c) Generalize to an arbitrary causal, rational, and stable LTI system $H^z(z)$ of orders p, q.

7.19 Find the difference equation of the LTI system whose impulse response is

$$h[n] = \begin{cases} a^{n-1} \cos(\theta_0 n), & n \geq 1, \\ 0, & \text{otherwise.} \end{cases}$$

7.20 We are given a causal LTI system obeying the difference equation

$$y[n] = 2y[n-1] - y[n-2] + x[n-1].$$

The input $x[n]$ is generated from the output of the system and from an external input $v[n]$ according to

$$x[n] = K(v[n] - 0.5v[n-1] - y[n] + 0.5y[n-1]),$$

where K is constant. Find all values of K for which the transfer function from $v[n]$ to $y[n]$ is stable.

7.21 Bob, Nick, and Dave are given a causal and stable LTI system and are told that its transfer function is

$$H^z(z) = \frac{b_0 + b_1 z^{-1}}{1 + a_1 z^{-1}}.$$

They are asked to find the parameters b_0, b_1, a_1. Bob feeds a unit impulse to the system and reports that $h[0] = 1$. Nick feeds a unit step to the system and reports that the DC gain is 4. Dave feeds the signal $\cos(\pi n/3)$ to the system and reports that the amplitude of the sinusoidal signal at the output is 2. Based on this information, what are the values of b_0, b_1, a_1?

7.22 The response of a causal LTI system to the input signal.

$$x[n] = 0.5^n, \quad n \geq 0$$

is known to be

$$y[n] = 0.25^n, \quad n \geq 0.$$

Find the impulse response of the system.

7.23 A causal LTI system is described by the difference equation

$$y[n] = -0.5y[n-1] + x[n] + x[n-1].$$

The input to the system is

$$x[n] = \begin{cases} 1, & n \text{ even}, \quad n \geq 0, \\ 0, & \text{otherwise}. \end{cases}$$

What is the output of the system?

7.24 This problem discusses the computation of the coefficients of a difference equation from a finite set of impulse response values.

(a) Show that the impulse response of the system described by the difference equation (7.45) satisfies

$$h[n] = -a_1 h[n-1] - a_2 h[n-2] - \ldots - a_p h[n-p], \quad \text{for all } n > q. \quad (7.119)$$

(b) Deduce from (7.119) that the denominator coefficients can be obtained from the impulse response coefficients by solving the linear equations

$$\begin{bmatrix} h[q] & h[q-1] & \cdots & h[q-p+1] \\ h[q+1] & h[q] & \cdots & h[q-p+2] \\ \vdots & \vdots & \cdots & \vdots \\ h[q+p-1] & h[q+p-2] & \cdots & h[q] \end{bmatrix} \begin{bmatrix} a_1 \\ a_2 \\ \vdots \\ a_p \end{bmatrix} = - \begin{bmatrix} h[q+1] \\ h[q+2] \\ \vdots \\ h[q+p] \end{bmatrix}.$$

$$(7.120)$$

It can be shown that the coefficient matrix of this set of equations is nonsingular (hence has a unique solution) if and only if the polynomials $a(z), b(z)$ are coprime.

(c) Show that the numerator coefficients can be obtained from the formula

$$\begin{bmatrix} b_0 \\ b_1 \\ \vdots \\ b_q \end{bmatrix} = \begin{bmatrix} h[0] & 0 & \cdots & 0 \\ h[1] & h[0] & \cdots & 0 \\ \vdots & \vdots & \cdots & \vdots \\ h[q] & h[q-1] & \cdots & h[q-p] \end{bmatrix} \begin{bmatrix} 1 \\ a_1 \\ \vdots \\ a_p \end{bmatrix}. \quad (7.121)$$

7.25 Let $G^z(z)$ and $H^z(z)$ be the causal LTI systems

$$G^z(z) = \frac{1}{1 + a_1 z^{-1} + a_2 z^{-2}}, \quad H^z(z) = \frac{1 - 0.25z^{-1}}{1 - 0.5z^{-1}}.$$

Find a_1, a_2 such that $\{g[0], g[1], g[2]\}$ will be equal to $\{h[0], h[1], h[2]\}$, respectively. For a_1, a_2 you have found, is $g[3]$ equal to $h[3]$ as well? Hint: Use Problem 7.24.

7.26 Write a MATLAB program that implements the Schur–Cohn test without recursive calls.

7.27 Find the inverse z-transforms of the functions

$$X_1^z(z) = \exp(z^{-1}), \quad X_2^z(z) = \cos(z^{-1}), \quad X_3^z(z) = \sin(z^{-1}),$$

where the ROC is $z \neq 0$ in all cases.

7.28 Find the causal sequence $x[n]$ whose Fourier transform is

$$X^f(\theta) = \frac{1}{e^{j\theta} + \frac{1}{6} - \frac{1}{6}e^{-j\theta}}.$$

7.29 An LTI system has the transfer function

$$H^z(z) = \frac{a_p + a_{p-1}z^{-1} + \cdots + a_1 z^{-(p-1)} + z^{-p}}{1 + a_1 z^{-1} + \cdots + a_{p-1}z^{-(p-1)} + a_p z^{-p}}.$$

Find $|H^f(\theta)|$, the magnitude response of the system, as a function of θ and the coefficients $a_1, \ldots, a_p$. Interpret the result.

7.30 A causal LTI has the z-transform

$$H^z(z) = \frac{z^p - (1 - \varepsilon)}{z^p + (1 - \varepsilon)},$$

where p is odd and $0 < \varepsilon \ll 1$.

(a) Compute the poles and the zeros of the system. Is the system stable?

(b) Draw the pole–zero map of the system for $p = 5$.

(c) Draw an approximate magnitude response plot of the system for $p = 5$ and $\varepsilon = 0.1$.

7.31 The program `frqresp` may be sensitive to roundoff errors if the order of the $H^z(z)$ is large. Write a MATLAB program `frqrspzp` that computes the frequency response $H^f(\theta)$ from the pole–zero factorization (7.50). As we shall see in Chapter 10, digital filters are often designed to give the poles and zeros directly, so the program you are asked to write here will be useful later.

7.32 Let $x[n]$ be a periodic signal with period N, and $y[n]$ be one period of $x[n]$, that is

$$y[n] = \begin{cases} x[n], & 0 \le n \le N - 1, \\ 0, & \text{otherwise.} \end{cases}$$

(a) Express $X^z_+(z)$ in terms of $Y^z_+(z)$ and state its region of convergence (note that these are unilateral z-transforms).

(b) Find $X^z_+(z)$ in the special case where $N = 6$ and

$$y[n] = \begin{cases} 1, & 0 \le n \le 2, \\ -1, & 3 \le n \le 5. \end{cases}$$

7.33 A signal $x[n]$ is fed to two LTI systems, yielding outputs $y_1[n]$, $y_2[n]$, respectively. The LTI systems obey the difference equations

$$y_1[n] = 0.5y_1[n - 1] + x[n], \quad y_2[n] = 0.5y_2[n - 1] - 2x[n - 1].$$

The signal $w[n]$ is generated according to

$$w[n] = y_1[n] + y_2[n]$$

and fed to a third LTI system obeying the difference equation

$$y_3[n] = 2.5y_3[n - 1] - y_3[n - 2] + w[n].$$

(a) Find the response of $y_3[n]$ to a unit impulse at the input $x[n]$.

(b) Is the transfer function from $x[n]$ to $y_3[n]$ BIBO-stable?

(c) Find the DC gain of the system.

(d) Assume that $y_1[n]$ and $y_2[n]$ are zero for $n < 0$, but $y_3[-1] = y_3[-2] = 1$. Find the zero-input response of $y_3[n]$.

(e) Reconcile the result in part d with your answer to part b.

7.34* This problem explores the existence of a z-transform of the sinc function.

(a) Does the sequence $\text{sinc}(0.5n)$ possess a z-transform? If so, find it and specify the region of convergence. If not, explain why.

(b) Generalize part a to $\text{sinc}(rn)$, where r is a rational number in the range $(0, 1)$.

The result of part b holds for any $r \in (0, 1)$, whether rational or not. However, the proof of the irrational case is much more difficult.

7.35* Repeat Problem 7.34 for the sequence $\text{sinc}^2(0.5n)$, then for $\text{sinc}^2(rn)$.

7.36* This problem discusses the z-transform of a subsequence of a given sequence.

(a) The causal signal $x[n]$ has z-transform

$$X^z(z) = \frac{3}{1 - 0.4z^{-1} - 0.32z^{-2}}.$$

Define

$$y[n] = x[2n], \quad n \geq 0.$$

Find $Y^z(z)$, the z-transform of $y[n]$.

(b) Repeat part a for

$$X^z(z) = \frac{2(1 - 0.5z^{-1})}{1 - z^{-1} + 0.5z^{-2}}.$$

(c) Let $X^z(z)$ be a causal rational function and let $y[n]$ be related to $x[n]$ as in part a. Assume that the poles of $X^z(z)$ are simple. Is $Y^z(z)$ a rational function?

(d) Repeat part c if

$$y[n] = x[Mn], \quad n \geq 0,$$

where M is a fixed positive integer.

(e) If $y[n] = x[2n]$ and

$$Y^z(z) = \frac{1}{1 - 0.25z^{-1}},$$

what is $x[n]$?

7.37* Compute the noise gain of the filter

$$H^z(z) = \frac{b_0}{1 + a_1z^{-1}}$$

as a function of b_0 and a_1.

Chapter 8

Introduction to Digital Filters

It is hard to give a formal definition of the term *filtering*. The electrical engineer often thinks of filtering as changing the frequency domain characteristics of the given (input) signal. Of course, from a purely mathematical point of view, a frequency-domain operation often has a corresponding time-domain interpretation and vice versa. However, electrical engineers are trained, by tradition, to think in the frequency domain. This way of thinking has proved its effectiveness. We have already seen this when discussing spectral analysis and its applications (in Chapters 4 through 6), and we shall see it again in this chapter and the ones to follow.

Examples of filtering operations include:

1. Noise suppression. This operation is necessary whenever the signal of interest has been contaminated by noise. Examples of signals that are typically noisy include:

   (a) Received radio signals.

   (b) Signals received by imaging sensors, such as television cameras or infrared imaging devices.

   (c) Electrical signals measured from the human body (such as brain, heart, or neurological signals).

   (d) Signals recorded on analog media, such as analog magnetic tapes.

2. Enhancement of selected frequency ranges. Examples of signal enhancement include:

   (a) Treble and bass control or graphic equalizers in audio systems. These typically serve to increase the sound level at high and low frequencies, to compensate for the lower sensitivity of the ear at those frequencies, or for special sound effects.

   (b) Enhancement of edges in images. Edge enhancement improves recognition of objects in an image, whether recognition by a human eye or by a computer. It is essentially an amplification of the high frequencies in the Fourier transform of the image: Edges are sharp transitions in the image brightness, and we know from Fourier theory that sharp transitions in a signal appear as high-frequency components in the frequency domain.

3. Bandwidth limiting. In Section 3.3 we learned about bandwidth limiting as a means of aliasing prevention in sampling. Bandwidth limiting is also useful in communication applications. A radio or a television signal transmitted over a

specific channel is required to have a limited bandwidth, to prevent interference with neighboring channels. Thus, amplitude modulation (AM) radio is limited to ± 5 kHz (in the United States) or to ± 4.5 kHz (in Europe and other countries) around the carrier frequency. Frequency modulation (FM) radio is limited to about ± 160 kHz around to the carrier frequency. Bandwidth limiting is accomplished by attenuating frequency components outside the permitted band below a specified power level (measured in dB with respect to the power level of the transmitted signal).

4. Removal or attenuation of specific frequencies. For example:

   (a) Blocking of the DC component of a signal.

   (b) Attenuation of interferences from the power line. Such interferences appear as sinusoidal signals at 50 or 60 Hz, and are common in measurement instruments designed to measure (and amplify) weak signals.

5. Special operations. Examples include:

   (a) Differentiation. Differentiation of a continuous-time signal is described in the time and frequency domains as

   $$y(t) = \frac{dx(t)}{dt}, \quad Y^{\mathrm{F}}(\omega) = j\omega X^{\mathrm{F}}(\omega). \tag{8.1}$$

   (b) Integration. Integration of a continuous-time signal is described in the time and frequency domains as

   $$y(t) = \int_{-\infty}^{t} x(\tau)d\tau, \quad Y^{\mathrm{F}}(\omega) = \frac{X^{\mathrm{F}}(\omega)}{j\omega} + \pi X^{\mathrm{F}}(0)\delta(\omega). \tag{8.2}$$

   (c) Hilbert transform. The continuous-time Hilbert filter has an impulse response

   $$h(t) = \frac{1}{\pi t} \tag{8.3}$$

   and a frequency response

   $$H^{\mathrm{F}}(\omega) = -j\,\mathrm{sign}(\omega). \tag{8.4}$$

   The Hilbert transform of a signal $x(t)$ is defined as its convolution with $h(t)$, that is, as $y(t) = \{x * h\}(t)$.[1]

These operations can be approximated by digital filters operating on the sampled input signal, using methods described in Chapters 9 and 10.

8.1 Digital and Analog Filtering

Analog filtering is performed on continuous-time signals and yields continuous-time signals. It is implemented using operational amplifiers, resistors, and capacitors. Theoretically, the frequency range of an analog filter is infinite. In practice, it is always limited, depending on the application and the technology. For example, common operational amplifiers operate up to a few hundred kilohertz. Special amplifiers operate up to a few hundred megahertz. Very high frequencies can be handled by special devices, such as microwave and surface acoustic wave (SAW) devices. Analog filters suffer from sensitivity to noise, nonlinearities, dynamic range limitations, inaccuracies due to variations in component values, lack of flexibility, and imperfect repeatability.

Digital filtering is performed on discrete-time signals and yields discrete-time signals. It is usually implemented on a computer, using operations such as addition, multiplication, and data movement. Sometimes, special-purpose hardware is used for carrying out similar operations. The frequency range is always finite, and is limited to one-half of the sampling rate. Digital filters are accurate to any desired degree (determined by the computer word length), highly linear (except for quantization effects, which will be studied in Chapter 11), flexible (especially if implemented in software), and perfectly repeatable; moreover, they do not suffer from internal noise and have a practically unlimited dynamic range if implemented in floating point. There exist, also, digital filters that are completely unlike analog filters, and enable operations that cannot be realized (or are difficult to realize) by analog means. These include pure time delays, time-varying filters, and adaptive filters. On the other hand, digital filtering requires interface to the physical world (A/D and D/A conversion); care must be taken to avoid (or minimize) aliasing, as well, and operating frequency range is sometimes limited by the available computational speed.

In this book we study only linear time-invariant (LTI) filters. Analog LTI filters are usually specified by their s-domain transfer function. The transfer function of an analog filter is rational, causal, and proper, that is,

$$H^L(s) = \frac{b_0 s^q + b_1 s^{q-1} + \cdots + b_q}{s^p + a_1 s^{p-1} + \cdots + a_p}, \qquad (8.5)$$

where $q \le p$. Except in rare cases, the filter is required to be stable. Digital LTI filters are usually specified by their z-domain transfer function. The transfer function is usually rational and causal, that is,

$$H^z(z) = \frac{b_0 + b_1 z^{-1} + \cdots + b_q z^{-q}}{1 + a_1 z^{-1} + \cdots + a_p z^{-p}}. \qquad (8.6)$$

Here we do not require that $q \le p$. As in the case of analog filters, usually the filter is required to be stable.

Digital filters for which $p \ge 1$ (and $a_p \ne 0$), are called *infinite impulse response* (IIR) filters, because their corresponding impulse response sequence $h[n]$ has an infinite duration—it never dies out completely. To see this, recall from (7.56) that $h[n]$ can be expressed as

$$h[n] = c_0 \delta[n] + \cdots + c_{q-p} \delta[n - q + p] + \sum_{k=1}^{p} A_k \alpha_k^n, \quad n \ge 0, \qquad (8.7)$$

where α_k are the poles of the transfer function $H^z(z)$. Therefore, for $n > q - p$, $h[n]$ is a linear combination of geometric series α_k^n. Note that the impulse response $h(t)$ of a rational analog filter also has an infinite duration. The transfer function $H^L(s)$ admits a partial fraction decomposition

$$H^L(s) = c_0 + \sum_{k=1}^{p} \frac{A_k}{s - \lambda_k}, \qquad (8.8)$$

where λ_k are the poles of $H^L(s)$ (assuming that they are distinct); therefore,

$$h(t) = c_0 \delta(t) + \sum_{k=1}^{p} A_k e^{\lambda_k t}, \quad t \ge 0. \qquad (8.9)$$

Digital filters for which $p = 0$ are called *finite impulse response* (FIR) filters. Their transfer function is

$$H^z(z) = b_0 + b_1 z^{-1} + \cdots + b_q z^{-q}. \qquad (8.10)$$

The impulse response $h[n]$ of an FIR filter is

$$h[n] = \begin{cases} b_n, & 0 \le n \le q, \\ 0, & \text{otherwise.} \end{cases} \qquad (8.11)$$

As we see, the impulse response is nonzero only for a finite number of samples, hence the name *finite impulse response*. FIR filters are characteristic of the discrete time domain. Analog FIR filters are possible, but they are difficult to implement and are rarely used.[2]

8.2   Filter Specifications

Before a digital filter can be designed and implemented, we need to specify its performance requirements. A typical filter should pass certain frequencies and attenuate other frequencies; therefore, we must define exactly the frequencies in question, as well as the required gains and attenuations. There are four basic filter types, as illustrated in Figure 8.1:

1. *Low-pass filters* are designed to pass low frequencies, from zero to a certain cutoff frequency θ_0, and to block high frequencies. We encountered analog low-pass filters when we discussed antialiasing filters and reconstruction filters in Sections 3.3 and 3.4.

2. *High-pass filters* are designed to pass high frequencies, from a certain cutoff frequency θ_0 to π, and to block low frequencies.[3]

3. *Band-pass filters* are designed to pass a certain frequency range $[\theta_1, \theta_2]$, which does not include zero, and to block other frequencies. We encountered analog band-pass filters when we discussed reconstruction of band-pass signals in Section 3.6.

4. *Band-stop filters* are designed to block a certain frequency range $[\theta_1, \theta_2]$, which does not include zero, and to pass other frequencies.

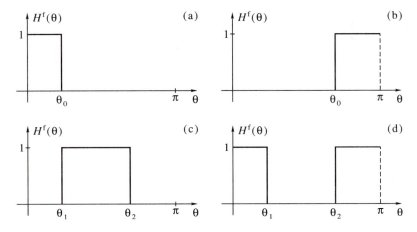

Figure 8.1 Ideal frequency responses of the four basic filter types: (a) low-pass; (b) high-pass; (c) band-pass; (d) band-stop.

The frequency responses shown in Figure 8.1 are ideal. Frequency responses of practical filters are not shaped in straight lines. The response of practical filters varies continuously as a function of the frequency: It is neither exactly 1 in the pass bands, nor exactly 0 in the stop bands. In this section we define the terms and parameters used for digital filter specifications. We assume that the filters are real, so their magnitude response is symmetric and their phase response is antisymmetric; recall properties

(2.107d, e) of the Fourier transform of a real sequence. Therefore, we shall usually show only the positive frequency range, from 0 to π.

8.2.1 Low-Pass Filter Specifications

A low-pass (LP) filter is designed to pass low frequencies, from zero to a certain cutoff frequency θ_p, with approximately unity gain. The frequency range $[0, \theta_p]$ is called the *pass band* of the filter. High frequencies, from a certain frequency θ_s up to π, are to be attenuated. The frequency range $[\theta_s, \pi]$ is called the *stop band* of the filter. The frequency range (θ_p, θ_s), between the pass band and the stop band, is called the *transition band*. The exact behavior of the frequency response in the transition band is usually of little importance.

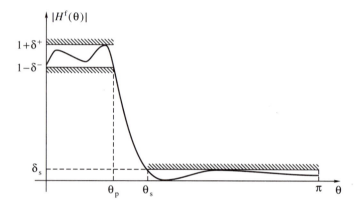

Figure 8.2 Specification of a low-pass filter.

Figure 8.2 provides a graphical description of the specifications of a low-pass filter. The hatched areas in the pass band and in the stop band indicate forbidden magnitude values in these bands. In the transition band there are no forbidden values, but it is usually required that the magnitude decrease monotonically in this band. The mathematical specification of a low-pass filter is

$$1 - \delta^- \leq |H^f(\theta)| \leq 1 + \delta^+, \quad 0 \leq \theta \leq \theta_p, \tag{8.12a}$$

$$0 \leq |H^f(\theta)| \leq \delta_s, \qquad \theta_s \leq \theta \leq \pi. \tag{8.12b}$$

The parameters δ^+ and δ^- are the positive and negative tolerances of the magnitude response in the pass band. The desired (or nominal) magnitude response in the pass band is 1. The parameter δ_s is the tolerance of the magnitude response in the stop band. The desired (or nominal) magnitude response in the stop band is 0.

The quantity $\max\{\delta^+, \delta^-\}$ is called the *pass-band ripple*. Another useful quantity is

$$A_p = \max\{20 \log_{10}(1 + \delta^+), -20 \log_{10}(1 - \delta^-)\}. \tag{8.13}$$

This parameter is the pass-band ripple in dB. A convenient approximation to A_p is obtained by noting that

$$\log_e(1 \pm \delta) \approx \pm \delta,$$

hence

$$20 \log_{10}(1 \pm \delta) = 20 \log_{10} e \cdot \log_e(1 \pm \delta) \approx \pm 8.6859 \delta. \tag{8.14}$$

Substitution of (8.14) in (8.13) for δ^+ and δ^- thus gives

$$A_p \approx 8.6859 \max\{\delta^+, \delta^-\}. \tag{8.15}$$

It is common, in digital filter design, to use different pass-band tolerances for IIR and FIR filters. For IIR filters, it is common to use $\delta^+ = 0$, and denote δ^- as δ_p. For FIR filters, it is common to use $\delta^+ = \delta^-$, and denote their common value as δ_p. Thus, the value 1 is the maximum pass-band gain for IIR filters but the midrange pass-band gain for FIR filters.

The quantity δ_s is called the *stop-band attenuation*. Another useful quantity is

$$A_s = -20 \log_{10} \delta_s. \tag{8.16}$$

This parameter is the stop-band attenuation in dB.

The frequency response specification we have described concerns the magnitude only and ignores the phase. We shall discuss the phase response in Section 8.4; for now we continue to ignore it.

Example 8.1 A typical application of low-pass filters is noise attenuation. Consider, for example, a signal sampled at $f_{sam} = 20$ kHz. Suppose that the signal is band limited to 1 kHz, but discrete-time white noise is also present at the sampled signal, and the SNR is 10 dB. We wish to attenuate the noise in order to improve the SNR as much as is reasonably possible, without distorting the magnitude of the signal by more than 0.1 dB at any frequency.

Since the noise is white, its energy is distributed uniformly at all frequencies up to 10 kHz (remember that we are already in discrete time). The best we can theoretically do is pass the signal through an ideal low-pass filter having cutoff frequency 1 kHz. This will leave 10 percent of the noise energy (from zero up to 1 kHz) and remove the remainder. Since the noise energy is decreased by 10 dB, the SNR at the output of an ideal filter will be 20 dB.

Suppose now that, due to implementation limitations, the digital filter cannot have a transition bandwidth less than 200 Hz. In this case, the noise energy in the range 1 kHz to 1.2 kHz will also be partly passed by the filter. The response of the filter in the transition band decreases monotonically, so we assume that the average noise gain in this band is about 0.5. Therefore, the filter now leaves about 11 percent, or -9.6 dB, of the noise energy. Thus, the SNR at the output of the filter cannot be better than 19.6 dB.

Finally, consider the contribution to the output noise energy made by the stop-band characteristics of the filter. The stop band is from 1.2 kHz to 10 kHz. Since the stop-band attenuation of a practical filter cannot be infinite, we must accept an output SNR less than 19.6 dB, say 19.5 dB. Thus, the total noise energy in the filter's output must not be greater than 11.22 percent of its input energy. Out of this, 11 percent is already lost to the pass band and the transition band, so the stop band must leave no more than 0.22 percent of the noise energy. The total noise gain in the stop band is not higher than $(8.8/10)\delta_s^2$, and this should be equal to 0.0022. Therefore, we get that $\delta_s = 0.05$, or $A_s = 26$ dB. In summary, the digital filter specifications should be

$$\theta_p = 0.1\pi, \quad \theta_s = 0.12\pi, \quad A_p = 0.1 \text{ dB}, \quad A_s = 26 \text{ dB}. \qquad \Box$$

8.2.2 High-Pass Filter Specifications

Next we consider high-pass (HP) filters. A high-pass filter is designed to pass high frequencies, from a certain cutoff frequency θ_p to π, with approximately unity gain.

The frequency range $[\theta_p, \pi]$ is the pass band of the filter. Low frequencies, from 0 to a certain frequency θ_s, are to be attenuated. The frequency range $[0, \theta_s]$ is the stop band of the filter. The frequency range (θ_s, θ_p) is the transition band.

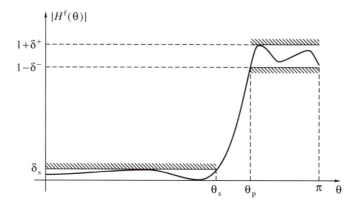

Figure 8.3 Specification of a high-pass filter.

Figure 8.3 provides a graphical description of the specifications of a high-pass filter. As before, the hatched areas in the pass band and in the stop band indicate forbidden magnitude values in these bands. The mathematical specification of a high-pass filter is

$$0 \le |H^f(\theta)| \le \delta_s, \qquad 0 \le \theta \le \theta_s, \qquad (8.17\text{a})$$

$$1 - \delta^- \le |H^f(\theta)| \le 1 + \delta^+, \quad \theta_p \le \theta \le \pi. \qquad (8.17\text{b})$$

The meaning of the parameters δ^+, δ^-, and δ_s is the same as for low-pass filters. The ripple and attenuation parameters A_p and A_s are defined as in (8.13) and (8.16), respectively. As in the case of low-pass filters, it is common to define $\delta^+ = 0$, $\delta^- = \delta_p$ for IIR filters and $\delta^+ = \delta^- = \delta_p$ for FIR filters.

Example 8.2 Electrical signals measured from the human brain are mostly in the frequency range from 0 to 30 Hz under normal conditions. The level of EEG signals (when measured by electrodes attached to the scalp) is up to about 200 μV. When the brain is stimulated by the senses, signals at other frequencies may appear, depending on the stimulus. Such signals are called *evoked potentials*. In this example we consider *brain stem auditory evoked potentials*: signals generated in the brain stem in response to sound. The level of such signals is relatively low, on the order of 1 μV. Their spectrum depends on the frequency of the sound signal but can contain energy from about 30 Hz to a few kilohertz. To analyze such signals, it is necessary to attenuate the normal brain signals (background activity). This can be accomplished by means of a high-pass filter.

Suppose that we are interested in analyzing brain stem auditory evoked potentials in the frequency range 50 to 1000 Hz. We pass the signal through an antialiasing filter and sample at 2500 Hz. The sampled signal is then passed through a digital high-pass filter. A filter suitable for this purpose should have a stop band from 0 to 30 Hz and a pass band from 50 to 1250 Hz. A reasonable value of A_s for this application is 50 dB, and a reasonable value of A_p is 0.2 dB. In summary, the filter specifications are

$$\theta_s = 0.024\pi, \quad \theta_p = 0.04\pi, \quad A_p = 0.2\,\text{dB}, \quad A_s = 50\,\text{dB}. \qquad \square$$

8.2.3 Band-Pass Filter Specifications

A band-pass (BP) filter is designed to pass signals in a certain frequency range, from $\theta_{p,1}$ to $\theta_{p,2}$. Outside this range, it must attenuate the signal below a specified level. Figure 8.4 provides a graphical description of the specifications of a band-pass filter. The mathematical specification of a band-pass filter is

$$0 \leq |H^f(\theta)| \leq \delta_{s,1}, \qquad 0 \leq \theta \leq \theta_{s,1}, \tag{8.18a}$$

$$1 - \delta^- \leq |H^f(\theta)| \leq 1 + \delta^+, \quad \theta_{p,1} \leq \theta \leq \theta_{p,2}, \tag{8.18b}$$

$$0 \leq |H^f(\theta)| \leq \delta_{s,2}, \qquad \theta_{s,2} \leq \theta \leq \pi. \tag{8.18c}$$

Note that the attenuations in the two stop bands, $\delta_{s,1}$ and $\delta_{s,2}$, are not necessarily equal. As in the case of low-pass and high-pass filters, it is common to define $\delta^+ = 0$, $\delta^- = \delta_p$ in IIR filters, and $\delta^+ = \delta^- = \delta_p$ in FIR filters. The pass-band ripple is $\max\{\delta^+, \delta^-\}$, and the corresponding dB value is

$$A_p \approx 8.6859 \max\{\delta^+, \delta^-\}.$$

There are two stop-band attenuation parameters, and their corresponding dB values are

$$A_{s,1} = -20 \log_{10} \delta_{s,1}, \quad A_{s,2} = -20 \log_{10} \delta_{s,2}.$$

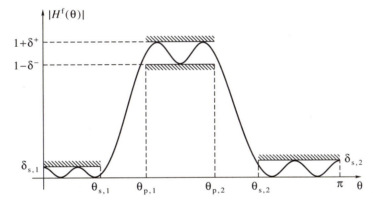

Figure 8.4 Specification of a band-pass filter.

Example 8.3 A certain BPSK communication system (see Example 3.13) is designed to share five channels, each having bandwidth 4 kHz, in a single receiver. This is accomplished by modulating each channel with a different carrier frequency, using double-side-band (DSB) modulation. The carrier frequencies are 10 kHz apart. The received signal, which contains all five channels, is demodulated such that the carrier frequencies after demodulation are 5, 15, 25, 35, 45 kHz. The spectrum of the demodulated signal is shown in Figure 8.5. Between every two adjacent channels there is a 2-kHz-wide band, called a *guard band,* that is free of any signal.

The signal shown in Figure 8.5 is sampled at $f_{sam} = 100$ kHz, using a 12-bit A/D converter. The digital signal is to be split into the five individual channels using five band-pass filters. We wish to prescribe the specifications of the five filters.

We first note that the guard bands can be used as transition bands for the filters, since no signal is expected in these bands. Therefore, the band-edge frequencies of the

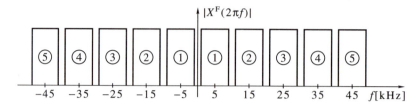

Figure 8.5 The spectrum of the signal in Example 8.3; the numbers in circles correspond to the five channels.

filters are as shown in Table 8.1. The frequencies are given in kilohertz for convenience (multiplication by $2000\pi/f_{\text{sam}}$ will convert them to the θ domain). Note, in particular, that the first filter is low pass and the fifth is high pass.

Next we observe that the dynamic range of the digital signal is 72.25 dB, since $20\log_{10} 2^{12} = 72.25$. Moreover, the practical dynamic range is typically 6 dB lower, since a 12-bit A/D usually does not provide more than 11 true bits.[4] It is therefore reasonable to require stop-band attenuation of 66 dB. This guarantees that even the strongest possible signal in one channel will be below the least significant bit (hence undetectable) in the other channels.

The pass-band ripple cannot be directly inferred from the given information. However, BPSK systems are not highly sensitive to amplitude distortions. For the purpose of this example, we require that the pass-band ripple be 0.5 dB for all five filters. Table 8.1 summarizes the specifications of the filters.   □

Channel	1	2	3	4	5
$f_{\text{s},1}$ [kHz]	—	9	19	29	39
$f_{\text{p},1}$ [kHz]	—	11	21	31	41
$f_{\text{p},2}$ [kHz]	9	19	29	39	—
$f_{\text{s},2}$ [kHz]	11	21	31	41	—
A_{s} [dB]	66	66	66	66	66
A_{p} [dB]	0.5	0.5	0.5	0.5	0.5

Table 8.1 The filter specifications in Example 8.3.

8.2.4 Band-Stop Filter Specifications

A band-stop (BS) filter is designed to attenuate signals in a certain frequency range, from $\theta_{\text{s},1}$ to $\theta_{\text{s},2}$. Outside this range, it must pass the signal within a specified tolerance. Figure 8.6 provides a graphical description of the specifications of a band-stop filter. The mathematical specification of a band-stop filter is

$$1 - \delta_1^- \leq |H^{\text{f}}(\theta)| \leq 1 + \delta_1^+, \qquad 0 \leq \theta \leq \theta_{\text{p},1}, \tag{8.19a}$$

$$0 \leq |H^{\text{f}}(\theta)| \leq \delta_{\text{s}}, \qquad \theta_{\text{s},1} \leq \theta \leq \theta_{\text{s},2}, \tag{8.19b}$$

$$1 - \delta_2^- \leq |H^{\text{f}}(\theta)| \leq 1 + \delta_2^+, \qquad \theta_{\text{p},2} \leq \theta \leq \pi. \tag{8.19c}$$

The tolerance parameters δ_1^-, δ_2^-, δ_1^+, δ_2^+ are permitted to be different in general. There are two pass-band ripple parameters, $\max\{\delta_1^+,\delta_1^-\}$ and $\max\{\delta_2^+,\delta_2^-\}$, and their corresponding dB values are

$$A_{p,1} \approx 8.6859 \max\{\delta_1^+, \delta_1^-\}, \quad A_{p,2} \approx 8.6859 \max\{\delta_2^+, \delta_2^-\}.$$

The stop-band attenuation is δ_s and its corresponding dB value is $A_s = -20 \log_{10} \delta_s$.

For IIR filters it is common to define $\delta_1^+ = \delta_2^+ = 0$, $\delta_1^- = \delta_{p,1}$, $\delta_2^- = \delta_{p,2}$; for FIR filters it is common to define $\delta_1^+ = \delta_1^- = \delta_{p,1}$, $\delta_2^+ = \delta_2^- = \delta_{p,2}$.

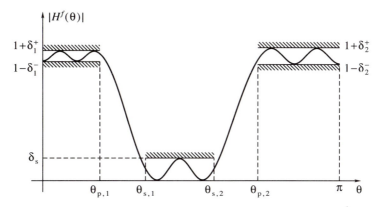

Figure 8.6 Specification of a band-stop filter.

Example 8.4 A certain physiological signal is measured by electrodes, amplified, sampled at 200 Hz, and processed digitally. Signals measured from the human body are low level, so considerable amplification is necessary to bring them to the operating level of the A/D. In such circumstances, it is difficult to completely isolate the analog circuitry from power-line interferences. Suppose that examination of the analog voltage at the A/D input reveals a power-line signal component (60 Hz in the United States) at amplitude approximately equal to that of the signal of interest. The proposed solution to this problem is to pass the digital signal through a band-stop filter, designed to attenuate the power-line interference with minimal distortion of the signal of interest. Since the power-line frequency has short-term fluctuations from its nominal value (although its long-term average value is highly stable), it is decided to use a stop band from 59 to 61 Hz. The permitted interference amplitude after filtering is required to be 1 percent or less of the amplitude of the signal of interest. This implies stop-band attenuation of 40 dB. Suppose further that the spectrum of the signal is approximately flat in the range ±100 Hz, and we do not want the filter to cut more than 5 percent of the signal energy. Thus, the stop band and the transition band together should not span more than 5 Hz. We therefore define the two pass-band edge frequencies as 57.5 and 62.5 Hz. We define the pass-band ripple as 0.1 dB, so that the amplitude distortion of the signal in the pass band will be less than 1 percent. In summary, the filter specifications in this example are

$$\theta_{p,1} = 0.575\pi, \ \theta_{s,1} = 0.59\pi, \ \theta_{s,2} = 0.61\pi, \ \theta_{p,2} = 0.625\pi,$$

$$A_{p,1} = A_{p,2} = 0.1 \, \text{dB}, \ A_s = 40 \, \text{dB}. \qquad \square$$

8.2.5 Multiband Filters

The four filter types we have described are the most commonly used, but they are far from encompassing the full generality of LTI filtering. Multiband filters generalize these four types in that they allow for different gains or attenuations in different frequency bands. A *piecewise-constant multiband filter* is characterized by the following parameters:

1. A division of the frequency range $[0, \pi]$ to a finite union of intervals. Some of these intervals are pass bands, some are stop bands, and the remaining are transition bands.

2. A desired gain and a permitted tolerance for each pass band.

3. An attenuation threshold for each stop band.

Suppose that we have K_p pass bands, and let $\{[\theta_{p,l,k}, \theta_{p,h,k}], 1 \leq k \leq K_p\}$ denote the corresponding frequency intervals. Similarly, suppose that we have K_s stop bands, and let $\{[\theta_{s,l,k}, \theta_{s,h,k}], 1 \leq k \leq K_s\}$ denote the corresponding frequency intervals (where l and h stand for "low" and "high," respectively). Let $\{C_k, 1 \leq k \leq K_p\}$ denote the desired gains in the pass band, and $\{\delta_k^-, \delta_k^+, 1 \leq k \leq K_p\}$ the pass-band tolerances. Finally, let $\{\delta_{s,k}, 1 \leq k \leq K_s\}$ denote the stop-band attenuations. Then the multiband filter specification is

$$C_k - \delta_k^- \leq |H^f(\theta)| \leq C_k + \delta_k^+, \quad \theta_{p,l,k} \leq \theta \leq \theta_{p,h,k}, \quad 1 \leq k \leq K_p, \quad (8.20a)$$

$$0 \leq |H^f(\theta)| \leq \delta_{s,k}, \quad \theta_{s,l,k} \leq \theta \leq \theta_{s,h,k}, \quad 1 \leq k \leq K_s. \quad (8.20b)$$

Figure 8.7 illustrates the specifications of a six-band filter having three pass bands and three stop bands.

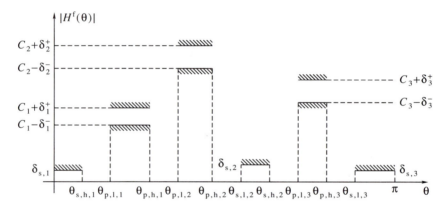

Figure 8.7 Specification of a multiband filter.

Multiband filters are not necessarily suitable for any filtering problem that may come up in a particular application. Sometimes the required gain behavior of the filter is too complex to be faithfully described by a piecewise-constant approximation. Then the engineer must rely on experience and understanding of the problem at hand. The field of closed-loop feedback control is a typical example. Filters used for closed-loop control are called *controllers*, or *compensators*. They are carefully tailored to the controlled system, their gain is continuously varying (rather than piecewise constant), and their phase behavior is of extreme importance. Consequently, compensator design for control systems is a discipline by itself, and the techniques studied in this book are of little use for it.

8.3 The Magnitude Response of Digital Filters

A digital filter designed to meet given specifications must have its magnitude response lying in the range $[1 - \delta^-, 1 + \delta^+]$ in the pass band, and in the range $[0, \delta_s]$ in the stop band. The exact magnitude response of the filter in each of these ranges is of secondary importance. For practical filters, the magnitude response in each band typically has one of two forms:

1. A monotone response, either increasing or decreasing.

2. An oscillating, or rippling, response.

A rippling response in the pass band is typically such that it reaches the tolerance limits $1 - \delta^-, 1 + \delta^+$ several times in the band. Similarly, a rippling stop-band response typically reaches the limits $0, \delta_s$ several times in the band. Such a response is called *equiripple*. In Chapter 10 we shall encounter filters that are monotone in both bands, filters that are equiripple in both bands, and filters that are monotone in one band and equiripple in the other.

If you are experienced with analog filters, you are probably familiar with asymptotic Bode diagrams of the magnitude responses of such filters. Digital filters do not have asymptotic Bode diagrams, for the following reasons:

1. In an analog filter, the frequency response of a first-order factor $(s - \lambda_k)$ is $(j\omega - \lambda_k)$, which is a rational function of ω. In a digital filter, on the other hand, the frequency response of a first-order factor $(1 - \alpha_k z^{-1})$ is $(1 - \alpha_k e^{-j\theta})$, which is not a rational function of θ.

2. It is meaningless to seek approximations as θ tends to infinity, since θ is limited to π.

Therefore, the digital filter designer usually relies on exact magnitude response plots, computed by programs such as `frqresp`, described in Section 7.6.

8.4 The Phase Response of Digital Filters

In most filtering applications, the magnitude response of the filter is of primary concern. However, the phase response may also be important in certain applications. It turns out that the phase response of practical filters cannot be made arbitrary, but is subject to certain restrictions. In this section we study the properties of the phase of digital filters. We restrict our discussion to filters that are *real, causal, stable, and rational*. We use the abbreviation RCSR for such filters. Certain results in this section hold for more general classes of filters, but we shall not need these generalizations here.

8.4.1 Phase Discontinuities

Suppose we are given an RCSR filter whose transfer function is $H^z(z)$. The frequency response of the filter can be expressed as

$$H^f(\theta) = H_r^f(\theta) + jH_i^f(\theta) = |H^f(\theta)|e^{j\psi(\theta)}, \tag{8.21}$$

where $H_r^f(\theta)$, $H_i^f(\theta)$, $|H^f(\theta)|$, $\psi(\theta)$ are, respectively, the real part, imaginary part, magnitude, and phase of the frequency response. Then

$$|H^f(\theta)| = \sqrt{[H_r^f(\theta)]^2 + [H_i^f(\theta)]^2}, \tag{8.22}$$

$$\psi(\theta) = \begin{cases} \arctan2\{H_i^f(\theta), H_r^f(\theta)\}, & H^f(\theta) \neq 0, \\ \text{undefined}, & H^f(\theta) = 0, \end{cases} \quad (8.23)$$

where the value of $\arctan2(y, x)$ is, by definition, the unique angle $\alpha \in (-\pi, \pi]$ for which[5]

$$\cos \alpha = \frac{x}{(x^2 + y^2)^{1/2}}, \quad \sin \alpha = \frac{y}{(x^2 + y^2)^{1/2}}.$$

Since $H^z(z)$ is RCSR, the frequency response $H^f(\theta)$ is a continuous function of θ. As we see from (8.23), this implies continuity of $\psi(\theta)$, except in two cases:

1. At points θ_0 where $H_i^f(\theta_0) = 0$ and $H_r^f(\theta) < 0$ we have $\psi(\theta_0) = \pi$ because of our definition of the range of $\psi(\theta)$. Therefore, there will be a jump of 2π if $H_i^f(\theta_0^-)$ or $H_i^f(\theta_0^+)$ (or both) are negative, because then $\psi(\theta_0^+)$ or $\psi(\theta_0^-)$ (or both) will be $-\pi$.

2. At points θ_0 where $H^f(\theta_0) = 0$ the phase is not defined, so it is not continuous either.

For rational transfer functions, the number of discontinuity points of the phase $\psi(\theta)$ in the range $-\pi < \theta \leq \pi$ is necessarily finite. This follows immediately from the following lemma.

Theorem 8.1 (lemma) If $H^z(z)$ is RCSR, the imaginary part of $H^f(\theta)$ is zero only on a finite number of points in the range $-\pi < \theta \leq \pi$.

Proof We have, with $z = e^{j\theta}$,

$$H_i^f(\theta) = \frac{1}{2j}[H^f(\theta) - H^f(-\theta)] = \frac{1}{2j}\left[\frac{b(z)}{a(z)} - \frac{b(z^{-1})}{a(z^{-1})}\right]$$

$$= \frac{b(z)a(z^{-1}) - b(z^{-1})a(z)}{2ja(z)a(z^{-1})}. \quad (8.24)$$

Therefore, $H_i^f(\theta) = 0$ if and only if the numerator on the right side is zero. The numerator is a polynomial in mixed (positive and negative) powers of z. Such a polynomial has only a finite number of zeros, and only a subset of those are on the unit circle. Therefore, the number of points in the range $-\pi < \theta \leq \pi$ for which $H_i^f(\theta) = 0$ is finite. In particular, the number of points for which $H_i^f(\theta) = 0$ and $H_r^f(\theta) \leq 0$, that is, the discontinuity points of the phase, is finite. $\square$

8.4.2 Continuous-Phase Representation

As we have seen, the phase of $H^f(\theta)$ is not defined when $H^f(\theta) = 0$. This leads to the following question: Can we *define* the phase $\psi(\theta_0)$ at points where $H^f(\theta_0) = 0$ so as to make it continuous at such points? As the following example shows, the answer is "not always."

Example 8.5 Let $H^z(z) = 1 - z^{-1}$. The corresponding frequency response is

$$H^f(\theta) = 1 - e^{-j\theta} = (e^{j0.5\theta} - e^{-j0.5\theta})e^{-j0.5\theta} = 2j\sin(0.5\theta)e^{-j0.5\theta}$$

$$= 2\sin(0.5\theta)e^{j(0.5\pi - 0.5\theta)}. \quad (8.25)$$

Therefore,

$$|H^f(\theta)| = 2|\sin(0.5\theta)|, \quad \psi(\theta) = \begin{cases} 0.5\pi - 0.5\theta, & 0 < \theta < \pi, \\ -0.5\pi - 0.5\theta, & -\pi < \theta < 0. \end{cases} \quad (8.26)$$

In particular, at the point $\theta_0 = 0$,

$$\psi(\theta_0^-) = -0.5\pi, \quad \psi(\theta_0^+) = 0.5\pi. \tag{8.27}$$

Figure 8.8 shows the magnitude and phase responses of $H^f(\theta)$. As we see, the phase response jumps by π radians when passing through $\theta = 0$, so in this case it is impossible to define $\psi(0)$ so as to make the phase continuous.   □

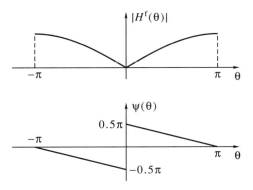

Figure 8.8 The magnitude and phase responses of $H^f(\theta)$ in Example 8.5.

Example 8.5 is an illustration of a more general phenomenon, as the next theorem shows:

Theorem 8.2 Let $H^z(z)$ be an RCSR transfer function. Then at any discontinuity point of the phase response, the phase jumps by either π or 2π radians.

Proof As we have seen, discontinuity of the phase can occur only at points θ_0 where either $H_i^f(\theta_0) = 0$ and $H_r^f(\theta_0) < 0$, or $H^f(\theta_0) = 0$. We have already seen that the jump is 2π in the former case, so it only remains to examine the latter. Since $H^z(z)$ is rational, this can happen only if $H^z(z)$ has a zero at the point $z = e^{j\theta_0}$. Let m be the multiplicity of this zero. Then $H^z(z)$ must have the form

$$H^z(z) = H_1^z(z)(1 - e^{j\theta_0}z^{-1})^m, \tag{8.28}$$

where $H_1^z(z)$ is nonzero at $z = e^{j\theta_0}$. Therefore, as in Example 8.5,

$$\begin{aligned}
H^f(\theta) &= H_1^f(\theta)[1 - e^{-j(\theta-\theta_0)}]^m \\
&= H_1^f(\theta)\{2\sin[0.5(\theta - \theta_0)]\}^m e^{j[0.5m\pi - 0.5(\theta-\theta_0)m]}. \tag{8.29}
\end{aligned}$$

Since $H_1^f(\theta_0) \neq 0$, the local behavior of $\psi(\theta)$ around θ_0 is not affected by the first factor. If m is even, the sine term does not change sign, so the phase remains continuous. If m is odd, there is a sign reversal in the sine term, so there is a jump of π radians in the phase.   □

A close examination of the proof of Theorem 8.2 reveals that discontinuities at which the phase jumps by π are the result of the magnitude being positive *by definition*. Suppose that we relax the definition of magnitude and permit it to be positive or negative, requiring only that it be real. As the next theorem shows, this allows us to express $H^f(\theta)$ in terms of a modified, continuous-phase function.

Theorem 8.3 The frequency response of an RCSR transfer function $H^z(z)$ can be expressed as

$$H^f(\theta) = A(\theta)e^{j\phi(\theta)}, \tag{8.30}$$

where $A(\theta)$ is real, but not necessarily positive, and $\phi(\theta)$ is continuous.

Proof The proof is by construction. Let θ_1 be the first discontinuity point to the right of $-\pi$. Define

$$A(\theta) = |H^f(\theta)|, \quad \phi(\theta) = \psi(\theta), \quad -\pi < \theta < \theta_1,$$

$$A(\theta_1) = \lim_{\theta \uparrow \theta_1} A(\theta), \quad \phi(\theta_1) = \lim_{\theta \uparrow \theta_1} \phi(\theta).$$

The notation $\lim_{\theta \uparrow \theta_1}$ means limit as θ approaches θ_1 from below. Next let θ_2 be the first discontinuity point to the right of θ_1 and define

$$A(\theta) = (-1)^k |H^f(\theta)|, \quad \phi(\theta) = \psi(\theta) + k\pi, \quad \theta_1 < \theta < \theta_2,$$

where k is an integer (positive or negative), chosen so as to preserve continuity of $\phi(\theta)$ at θ_1; k is even in the case of 2π discontinuity, and odd in the case of π discontinuity. We continue this way until we exhaust all discontinuity points [whose number is necessarily finite, according to the lemma (Theorem 8.1)]. Each time we pass a discontinuity, we set the values of $A(\theta)$ and $\phi(\theta)$ as the limits from the left of the functions constructed up to that discontinuity. We then preserve continuity on the right either by adding an even multiple of π to $\phi(\theta)$ while leaving the sign of $A(\theta)$ unchanged or by adding an odd multiple of π to $\phi(\theta)$ while reversing the sign of $A(\theta)$. After all discontinuity points have been exhausted, $A(\theta)$ and $\phi(\theta)$ are defined on $\theta \in (-\pi, \pi)$ such that $\phi(\theta)$ is continuous and $A(\theta)$ is real, but not necessarily positive. Finally, the values of $A(-\pi)$ and $\phi(-\pi)$ are defined by continuity from the right, and those of $A(\pi)$ and $\phi(\pi)$ by continuity from the left.   □

The form (8.30) is called a *continuous-phase representation* of the frequency response, $A(\theta)$ is called the *amplitude function*, and $\phi(\theta)$ the *continuous phase*. Bear in mind the difference between the magnitude $|H^f(\theta)|$, which is always real and nonnegative, and the amplitude $A(\theta)$, which can assume any real value. Continuous-phase representation is not unique, since we can replace $A(\theta)$ by $-A(\theta)$ and $\phi(\theta)$ by $\phi(\theta) + \pi$. Also, we can obviously add to $\phi(\theta)$ an arbitrary integer multiple of 2π. We can make the representation unique by imposing the additional condition

$$0 \le \phi(0) < \pi.$$

Example 8.6 The continuous-phase representation of $H^f(\theta)$ given in Example 8.5 is obviously

$$A(\theta) = 2\sin(0.5\theta), \quad \phi(\theta) = 0.5\pi - 0.5\theta. \tag{8.31}$$

The amplitude and phase functions are shown in Figure 8.9. As we see, $\phi(\theta)$ is continuous.   □

8.4.3   Linear Phase

Consider the filter

$$H^z(z) = z^{-L}, \quad H^f(\theta) = e^{-j\theta L}, \tag{8.32}$$

where L is an integer. Both magnitude and amplitude functions of this filter are 1 at all frequencies, and the phase response is $\phi(\theta) = -L\theta$, which is a linear function of

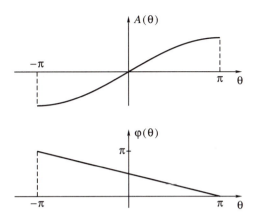

Figure 8.9 The amplitude and continuous-phase responses of $H^f(\theta)$ in Example 8.6.

the frequency. This filter represents a pure delay; that is, the response to an input signal $x[n]$ is the signal $x[n-L]$. In particular, the filter does not distort the form of the input signal.

Are there other filters, beside the pure delay filter, that do not distort signals applied to their input except for a delay? As the next example shows, the answer is affirmative.

Example 8.7 Consider a filter whose frequency response obeys

$$H^f(\theta) = \begin{cases} e^{-j\theta L}, & \theta_1 \le |\theta| \le \theta_2, \\ \text{arbitrary}, & \text{otherwise}, \end{cases} \tag{8.33}$$

where L is a positive integer and θ_1, θ_2 are fixed frequencies. Such a filter cannot be RCSR, but this is immaterial for our present discussion. Let $x[n]$ be a signal whose Fourier transform is nonzero only for $\theta_1 \le |\theta| \le \theta_2$. Then the signal at the output of the filter is

$$y[n] = \frac{1}{2\pi} \int_{-\pi}^{\pi} H^f(\theta) X^f(\theta) e^{j\theta n} d\theta = \frac{1}{2\pi} \int_{-\pi}^{\pi} X^f(\theta) e^{j\theta(n-L)} d\theta = x[n-L], \tag{8.34}$$

so the filter $H^f(\theta)$ acts as a pure delay filter for signals limited to the band $\theta_1 \le |\theta| \le \theta_2$. □

How will the result in Example 8.7 change if we let the proportionality factor of the phase be noninteger? The answer is given in the next example.

Example 8.8 Let

$$H^f(\theta) = \begin{cases} e^{-j\theta(L+\delta)}, & \theta_1 \le |\theta| \le \theta_2, \\ \text{arbitrary}, & \text{otherwise}, \end{cases} \tag{8.35}$$

where L, θ_1, θ_2 are as in Example 8.7 and δ is a proper fraction. For a signal $x[n]$ whose bandwidth is limited to $\theta_1 \le |\theta| \le \theta_2$ we get

$$y[n] = \frac{1}{2\pi} \int_{-\pi}^{\pi} X^f(\theta) e^{j\theta(n-L-\delta)} d\theta = \frac{1}{2\pi} \int_{-\pi}^{\pi} \left[\sum_{m=-\infty}^{\infty} x[m] e^{-j\theta m} \right] e^{j\theta(n-L-\delta)} d\theta$$

$$= \sum_{m=-\infty}^{\infty} x[m] \left[\frac{1}{2\pi} \int_{-\pi}^{\pi} e^{j\theta(n-m-L-\delta)} d\theta \right] = \sum_{m=-\infty}^{\infty} x[m] \operatorname{sinc}(n-m-L-\delta). \tag{8.36}$$

This time, the filter acts as a *sinc interpolator*, and $y[n]$ can be interpreted as the interpolated value of the continuous-time signal $x(t)$ at time $(n - L - \delta)T$. In summary, a filter that has a pure linear-phase characteristic over a certain frequency band, such that the phase proportionality factor is noninteger, acts as an *interpolated-delay* operator for signals in that frequency band.   □

A digital filter whose frequency response admits a continuous-phase representation

$$H^{\mathrm{f}}(\theta) = A(\theta)e^{-j\theta\tau_{\mathrm{p}}} \tag{8.37}$$

is said to have *linear phase*. The proportionality factor τ_{p} is called the *phase delay* of the filter and is measured in *samples* (a dimensionless number). There are no restrictions on the characteristics of the amplitude function $A(\theta)$. However, for practical filters, $A(\theta)$ will usually contain pass bands and stop bands, as we have explained in Section 8.2. In the pass bands, $A(\theta) \approx 1$ (up to the specified pass-band ripple); therefore, the filter acts as an approximate pure delay filter for signals in the pass band. In the stop bands, $A(\theta) \approx 0$ (up to the specified stop-band attenuation); therefore, the phase characteristic is immaterial in these bands. If the phase delay τ_{p} is an integer, there is a delay of τ_{p} samples; if it is fractional, there is an interpolated delay of τ_{p} samples. In any case, the output signal is almost a distortion-free copy of the input signal, as long as the input signal is in the pass band of the filter.

The concept of phase delay is defined also for filters that do not necessarily have linear phase. In general, the phase delay is defined as

$$\tau_{\mathrm{p}}(\theta) = -\frac{\phi(\theta)}{\theta}. \tag{8.38}$$

A linear-phase filter is therefore characterized by a constant phase delay.

8.4.4 Generalized Linear Phase

Suppose that the phase function is not strictly proportional to the frequency, but includes an additive constant term, as in the following example.

Example 8.9 Let

$$H^{\mathrm{f}}(\theta) = \begin{cases} e^{j\phi_0 - j\theta(L+\delta)}, & \theta_1 \le \theta \le \theta_2, \\ e^{-j\phi_0 - j\theta(L+\delta)}, & -\theta_2 \le \theta \le -\theta_1, \\ \text{arbitrary}, & \text{otherwise}, \end{cases} \tag{8.39}$$

where L is an integer, δ a proper fraction, ϕ_0 a fixed angle, and θ_1, θ_2 are fixed frequencies (again, such a filter cannot be RCSR). Consider an input signal of the form $x[n]\cos(\theta_{\mathrm{c}}n)$. Such a signal is called *amplitude modulated*, and $x[n]$ itself is the *modulating signal* or the *envelope*. The frequency θ_{c} is called the *carrier frequency*. Here we assume that

1. The modulating signal is band limited to $|\theta| \le \theta_0$ for some θ_0.

2. The following relationship holds among the frequencies of interest:

$$\theta_1 \le \theta_{\mathrm{c}} - \theta_0 < \theta_{\mathrm{c}} + \theta_0 \le \theta_2.$$

The Fourier transform of the modulated signal is given by

$$0.5X^{\mathrm{f}}(\theta - \theta_{\mathrm{c}}) + 0.5X^{\mathrm{f}}(\theta + \theta_{\mathrm{c}}).$$

Therefore, the Fourier transform of the signal $y[n]$ at the output of the filter is

$$Y^f(\theta) = 0.5X^f(\theta - \theta_c)e^{j\phi_0 - j\theta(L+\delta)} + 0.5X^f(\theta + \theta_c)e^{-j\phi_0 - j\theta(L+\delta)}$$
$$= 0.5X^f(\theta - \theta_c)e^{-j(\theta - \theta_c)(L+\delta)}e^{j\phi_0 - j\theta_c(L+\delta)}$$
$$+ 0.5X^f(\theta + \theta_c)e^{-j(\theta + \theta_c)(L+\delta)}e^{-j\phi_0 + j\theta_c(L+\delta)}. \qquad (8.40)$$

By the modulation property of the Fourier transform, $y[n]$ is given by

$$y[n] = 0.5e^{j[\theta_c n + \phi_0 - \theta_c(L+\delta)]}w[n] + 0.5e^{-j[\theta_c n + \phi_0 - \theta_c(L+\delta)]}w[n]$$
$$= w[n]\cos[\theta_c n + \phi_0 - \theta_c(L+\delta)], \qquad (8.41)$$

where $w[n]$ is the inverse Fourier transform of $X^f(\theta)e^{-j\theta(L+\delta)}$. As we saw in Example 8.8, $w[n]$ is an exact delay of $x[n]$ if $\delta = 0$, or an interpolated delay of $x[n]$ if $\delta \neq 0$. In conclusion, the filter's output is a modulated signal such that its envelope is a delayed version of the envelope of the input to the filter. The constant phase ϕ_0 affects the carrier, but not the envelope of the modulated signal at the output. □

A digital filter whose frequency response admits a continuous-phase representation

$$H^f(\theta) = A(\theta)e^{j(\phi_0 - \theta\tau_g)} \qquad (8.42)$$

is said to have *generalized linear phase* (GLP). The slope τ_g is called the *group delay* of the filter and is measured in samples. There are no restrictions of the characteristics of the amplitude function $A(\theta)$. However, for practical filters $A(\theta)$ will usually contain pass bands and stop bands. In the pass bands, $A(\theta) \approx 1$ (up to the specified pass-band ripple); therefore, the filter acts as an approximate pure delay filter for the envelopes of modulated signals in the pass band. In the stop bands, $A(\theta) \approx 0$ (up to the specified stop-band attenuation); therefore, the phase characteristic is immaterial in these bands. If the group delay τ_g is an integer, there is a delay of the envelope by τ_g samples; if it is fractional, there is an interpolated delay of τ_g samples. In any case, the envelope of the output signal is almost a distortion-free copy of the envelope of the input signal, as long as the modulated signal is in the pass band of the filter.

The concept of group delay is defined also for filters that do not necessarily have generalized linear phase. In general, if the phase function $\phi(\theta)$ is differentiable, the group delay is defined as

$$\tau_g(\theta) = -\frac{d\phi(\theta)}{d\theta}. \qquad (8.43)$$

A GLP filter is therefore characterized by a constant group delay.

The group delay of a filter with frequency response $H^f(\theta)$ can be computed as follows:

$$\tau_g(\theta) = -\frac{d}{d\theta}\arctan2\{H_i^f(\theta), H_r^f(\theta)\} = \frac{\dfrac{dH_r^f(\theta)}{d\theta}H_i^f(\theta) - \dfrac{dH_i^f(\theta)}{d\theta}H_r^f(\theta)}{[H_r^f(\theta)]^2 + [H_i^f(\theta)]^2}. \qquad (8.44)$$

This computation is particularly simple if the filter is FIR, since then

$$H^f(\theta) = \sum_{n=0}^{N} h[n]e^{-j\theta n}, \qquad \frac{dH^f(\theta)}{d\theta} = -j\sum_{n=0}^{N} nh[n]e^{-j\theta n}. \qquad (8.45)$$

If the filter is a rational IIR, its group delay is the difference between the group delay of the numerator and that of the denominator (Problem 8.13). Each of the two is FIR, so its group delay can be computed using (8.45). The procedure `grpdly` in Program 8.1 illustrates this computation. The input parameters of the program are in the same format as the parameters of Program 7.6 (which computes the frequency response of an RCSR filter). The program first determines if the filter is FIR. If so, it computes the group delay using (8.44) and (8.45). If not, it calls itself recursively twice, once for the numerator and once for the denominator, and subtracts the results.

8.4.5 Restrictions on GLP Filters

We now show that periodicity and realness impose restrictions on the parameters ϕ_0 and τ_g of a generalized linear-phase filter. Consider first the implications of periodicity. Since $H^f(\theta)$ is periodic with period 2π, we get from (8.42)

$$A(\theta)e^{j(\phi_0-\theta\tau_g)} = A(\theta + 2\pi)e^{j(\phi_0-\theta\tau_g-2\pi\tau_g)}. \tag{8.46}$$

Therefore,

$$A(\theta) = A(\theta + 2\pi)e^{-j2\pi\tau_g}. \tag{8.47}$$

But, since $A(\cdot)$ is real valued, $2\tau_g$ must be an integer. We therefore have two cases:

1. The group delay is an integer, say $\tau_g = M$. In this case we get

   $$A(\theta) = A(\theta + 2\pi), \tag{8.48}$$

   so the amplitude function is periodic with period 2π.

2. The group delay has a fractional part 0.5, say $\tau_g = M + 0.5$. In this case we get

   $$A(\theta) = -A(\theta + 2\pi), \tag{8.49}$$

   so the amplitude function does not have a period of 2π. However, in this case we still have

   $$A(\theta) = A(\theta + 4\pi), \tag{8.50}$$

   so the amplitude function is periodic with period 4π.

Consider next the implications of the assumption that the filter has real impulse response. In this case, the frequency response satisfies the conjugate symmetry condition

$$H^f(-\theta) = \bar{H}^f(\theta), \tag{8.51}$$

so

$$A(-\theta)e^{j(\phi_0+\theta\tau_g)} = A(\theta)e^{-j(\phi_0-\theta\tau_g)}. \tag{8.52}$$

This gives

$$e^{j2\phi_0} = \frac{A(\theta)}{A(-\theta)}, \tag{8.53}$$

from which it follows that $e^{j2\phi_0}$ is necessarily real. As before, we have two cases:

1. $\phi_0 = 0$ and $A(\theta) = A(-\theta)$, so $A(\theta)$ is a symmetric function. In this case, the filter has constant phase delay $\tau_p = \tau_g$.

2. $\phi_0 = \pi/2$ and $A(\theta) = -A(-\theta)$, so $A(\theta)$ is an antisymmetric function.

In summary, real digital filters with generalized linear phase come in four types:

Type-I filters have integer group delay $\tau_g = M$ and initial phase $\phi_0 = 0$, so they have constant phase delay.

Type-II filters have fractional group delay $\tau_g = M + 0.5$ and initial phase $\phi_0 = 0$, so they also have constant phase delay.

Type-III filters have integer group delay $\tau_g = M$ and initial phase $\phi_0 = \pi/2$, so their phase delay is not constant.

Type-IV filters have fractional group delay $\tau_g = M + 0.5$ and initial phase $\phi_0 = \pi/2$, so their phase delay is not constant.

8.4.6 Restrictions on Causal GLP Filters

Causality imposes further restrictions on the possible form of digital GLP filters. To find these restrictions, observe from (8.42) that a GLP filter (causal or not) satisfies

$$H^f(\theta)e^{-j(\phi_0 - \theta\tau_g)} = A(\theta). \tag{8.54}$$

Consider first the case $\phi_0 = 0$. We then get from (8.54)

$$h[2\tau_g - n] = \frac{1}{2\pi}\int_{-\pi}^{\pi} H^f(\theta)e^{j\theta(2\tau_g - n)}d\theta = \frac{1}{2\pi}\int_{-\pi}^{\pi} A(\theta)e^{j\theta(\tau_g - n)}d\theta. \tag{8.55}$$

Taking the conjugate of (8.55) and using the fact that $A(\theta)$ is real, we get

$$h[2\tau_g - n] = \frac{1}{2\pi}\int_{-\pi}^{\pi} A(\theta)e^{j\theta(n - \tau_g)}d\theta = \frac{1}{2\pi}\int_{-\pi}^{\pi} H^f(\theta)e^{j\theta n}d\theta = h[n]. \tag{8.56}$$

Now assume that the filter is causal, so $h[n] = 0$ for $n < 0$. Then, by (8.56), $h[n] = 0$ for $n > 2\tau_g$. Thus, the filter has a finite impulse response, its order N is equal to $2\tau_g$, and it satisfies the symmetry condition

$$h[n] = h[N - n]. \tag{8.57}$$

Next consider the case $\phi_0 = \pi/2$. We then get from (8.54) that

$$h[2\tau_g - n] = \frac{1}{2\pi}\int_{-\pi}^{\pi} H^f(\theta)e^{j\theta(2\tau_g - n)}d\theta = \frac{j}{2\pi}\int_{-\pi}^{\pi} A(\theta)e^{j\theta(\tau_g - n)}d\theta. \tag{8.58}$$

Taking the conjugate of (8.58) and using the fact that $A(\theta)$ is real, we get

$$h[2\tau_g - n] = -\frac{j}{2\pi}\int_{-\pi}^{\pi} A(\theta)e^{j\theta(n - \tau_g)}d\theta = -\frac{1}{2\pi}\int_{-\pi}^{\pi} H^f(\theta)e^{j\theta n}d\theta = -h[n]. \tag{8.59}$$

We again get, by (8.59), that $h[n] = 0$ for $n > 2\tau_g$. Thus, the filter has a finite impulse response, its order N is equal to $2\tau_g$, and it satisfies the antisymmetry condition

$$h[n] = -h[N - n]. \tag{8.60}$$

In summary, we have proved the following theorem:

Theorem 8.4 A linear phase RCSR filter is necessarily FIR. The phase or group delay of such a filter is half its order; it satisfies either the symmetry relationship (8.57) or the antisymmetry relationship (8.60). In the former case the phase delay is constant, while in the latter the group delay is constant. □

Theorem 8.4 implies that, if we look for an RCSR filter having linear phase (either exact or generalized), there is no sense in considering an IIR filter, but we must restrict ourselves to an FIR filter.[6] This property of digital FIR filters is one of the main reasons they are more commonly used than digital IIR filters.

8.4.7 Minimum-Phase Filters*

Let $H^z(z)$ be an RCSR filter. Stability implies that all poles are inside the unit circle. However, there is no constraint on the location of zeros, as far as stability is concerned.

We show that, in general, there is freedom in choosing the zero locations such that the magnitude response will remain fixed, but the phase response will vary. Then we shall discuss how this freedom can be exploited.

Let β be a zero of $H^z(z)$, so $(1 - \beta z^{-1})$ is a factor of the transfer function. Then we can write

$$H^z(z) = (1 - \beta z^{-1})H_0^z(z). \tag{8.61}$$

Let us replace this factor by

$$\bar{\beta} - z^{-1} = \bar{\beta}[1 - \bar{\beta}^{-1}z^{-1}].$$

We then get the new transfer function

$$H_1^z(z) = \bar{\beta}[1 - \bar{\beta}^{-1}z^{-1}]H_0^z(z). \tag{8.62}$$

We have

$$\bar{\beta} - e^{-j\theta} = -e^{-j\theta}(1 - \bar{\beta}e^{j\theta}) = -e^{-j\theta}\overline{(1 - \beta e^{-j\theta})}. \tag{8.63}$$

We get that $(\bar{\beta} - e^{-j\theta})$ and $(1 - \beta e^{-j\theta})$ have the same magnitude, implying that $H^z(z)$ and $H_1^z(z)$ have the same magnitude response. However, the phase responses of the two filters are different in general. Thus, the replacement of zeros of $H^z(z)$ by their conjugate inverses is a means of changing the phase response without affecting the magnitude response. Note that, if β is complex and we replace the zero at β by a zero at $\bar{\beta}^{-1}$, we must simultaneously replace the zero at $\bar{\beta}$ by a zero at β^{-1}; otherwise $H_1^z(z)$ will not be real.

The following theorem shows how the group delay of the filter is affected by zero replacement.

Theorem 8.5 Suppose $|\beta| < 1$, so the zero at β is inside the unit circle. Then the group delay of $H_1^z(z)$ is larger than the group delay of $H^z(z)$ at all frequencies.

Proof Let

$$\beta = \beta_r + j\beta_i, \quad \zeta = \arctan2(\beta_i, \beta_r).$$

Then the phase contributed by the factor $(1 - \beta z^{-1})$ is

$$\phi_\beta(\theta) = \angle\{1 - (\beta_r + j\beta_i)(\cos\theta - j\sin\theta)\}$$
$$= \arctan2(\beta_r \sin\theta - \beta_i \cos\theta, 1 - \beta_r \cos\theta - \beta_i \sin\theta)$$
$$= \arctan2\{\sin(\theta - \zeta), |\beta|^{-1} - \cos(\theta - \zeta)\}. \tag{8.64}$$

The contribution of $\phi_\beta(\theta)$ to the group delay is given by

$$\tau_{g,\beta}(\theta) = -\frac{d\phi_\beta(\theta)}{d\theta} = \frac{\sin^2(\theta - \zeta) + \cos^2(\theta - \zeta) - |\beta|^{-1}\cos(\theta - \zeta)}{\sin^2(\theta - \zeta) + \cos^2(\theta - \zeta) + |\beta|^{-2} - 2|\beta|^{-1}\cos(\theta - \zeta)}$$
$$= \frac{|\beta| - \cos(\theta - \zeta)}{|\beta| + |\beta|^{-1} - 2\cos(\theta - \zeta)}. \tag{8.65}$$

Replacing β by $\bar{\beta}^{-1}$ leaves ζ unchanged, and it also leaves the denominator of (8.65) unchanged. However, the numerator of (8.65) increases, since $|\beta|$ is replaced by $|\beta|^{-1}$, which is larger. Therefore, replacing β by $\bar{\beta}^{-1}$ increases the group delay. If β is complex, the same proof holds for the replacement of $\bar{\beta}$ by β^{-1}. $\qquad \square$

A filter $H^z(z)$ for which the group delay is the minimum possible at all frequencies, of all filters having the same magnitude response as $H^z(z)$, is called *minimum phase*. The term *minimum group delay* is perhaps more accurate, but the former is the one in common use. Theorem 8.5 implies that an RCSR filter $H^z(z)$ is minimum phase if and only if it has no zeros outside the unit circle. This is because any zero outside the unit circle can be replaced by its conjugate inverse, thus reducing the group delay without affecting the magnitude response. A practical procedure for designing a minimum-phase filter meeting a prescribed magnitude specification is given as follows:

1. Design a filter meeting the magnitude specification, ignoring the phase.

2. Find the zeros of the filter.

3. Replace all zeros outside the unit circle by their conjugate inverses, and adjust the constant gain accordingly.

There is no need to apply the procedure to zeros *on* the unit circle, because this will not change the group delay.

Example 8.10 Consider the four filters

$$H_1^z(z) = \frac{(1 - 0.5z^{-1})(1 - 0.2z^{-1})}{1 - z^{-1} + 0.5z^{-2}}, \quad H_2^z(z) = \frac{0.5(1 - 2z^{-1})(1 - 0.2z^{-1})}{1 - z^{-1} + 0.5z^{-2}},$$

$$H_3^z(z) = \frac{0.2(1 - 0.5z^{-1})(1 - 5z^{-1})}{1 - z^{-1} + 0.5z^{-2}}, \quad H_4^z(z) = \frac{0.1(1 - 2z^{-1})(1 - 5z^{-1})}{1 - z^{-1} + 0.5z^{-2}}.$$

These filters have identical magnitude responses. The first is minimum phase, since both its zeros are inside the unit circle. The fourth is maximum phase, since both its zeros are outside the unit circle. The other two filters are mixed phase. Figure 8.10 shows the group delays of the four filters. As we see, $H_1^z(z)$ indeed has the smallest group delay at all frequencies and $H_4^z(z)$ has the largest group delay. The two other filters have group delays between the two extremes. Note that their graphs intersect, so one is not always smaller than the other. □

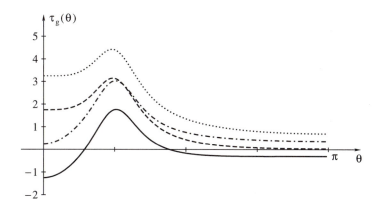

Figure 8.10 The group delays of the four filters in Example 8.10. Solid line: $H_1^z(z)$, dashed line: $H_2^z(z)$, dot-dashed line: $H_3^z(z)$, dotted line: $H_4^z(z)$.

8.4.8 All-Pass Filters*

A filter $H^z(z)$ is called *all pass* if its magnitude response is identically 1 (or, more generally, a positive constant) at all frequencies. The phase response of an all-pass filters is not restricted and is allowed to vary arbitrarily as a function of the frequency.

Example 8.11 Let

$$H^z(z) = \frac{\bar{a} - z^{-1}}{1 - az^{-1}}, \tag{8.66}$$

where $|a| < 1$; this is a stable IIR filter. We have

$$H^f(\theta) = \frac{\bar{a} - e^{-j\theta}}{1 - ae^{-j\theta}} = -e^{-j\theta}\frac{1 - \bar{a}e^{j\theta}}{1 - ae^{-j\theta}}. \tag{8.67}$$

The numerator and denominator on the right side are conjugates of each other; therefore

$$|H^f(\theta)| = 1, \tag{8.68}$$

so $H^z(z)$ is an all-pass filter. This result is not coincidental, but occurs because the zero is the inverse of the pole.   □

In general, a rational filter is all pass if and only if it has the same number of poles and zeros (including multiplicities), and each zero is the conjugate inverse of a corresponding pole. In other words, the transfer function of the filter must be a product of first-order all-pass filters, that is,

$$H^z(z) = \prod_{k=1}^{p} \frac{\bar{\alpha}_k - z^{-1}}{1 - \alpha_k z^{-1}}. \tag{8.69}$$

If the filter is stable then, since all the poles are inside the unit circle, all the zeros must be outside the unit circle. See Problem 7.29 for another (but mathematically equivalent) characterization of all-pass filters.

It follows from the foregoing discussion of minimum phase filters that an RCSR filter $H^z(z)$ that is not minimum phase is related to its minimum-phase companion $H^z_{mp}(z)$ by

$$H^z(z) = H^z_{mp}(z)H^z_{ap}(z), \tag{8.70}$$

where $H^z_{ap}(z)$ is an all-pass filter whose order is equal to the total number of zeros of $H^z(z)$ outside the unit circle. These zeros are precisely the zeros of $H^z_{ap}(z)$, whereas their conjugate inverses are the poles of $H^z_{ap}(z)$.

8.5   Digital Filter Design Considerations

A typical design process of a digital filter involves four steps:

1. Specification of the filter's response, as discussed in the preceding section. The importance of this step cannot be overstated. Often a proper specification is the key to the success of the system of which the filter is a part. Therefore, this task is usually entrusted to senior engineers, who rely on experience and engineering common sense. The remaining steps are then often put in the hands of relatively junior engineers.

2. Design of the transfer function of the filter. The main goal here is to meet (or surpass) the specification with a filter of *minimum complexity*. For LTI filters, minimum complexity is usually synonymous with minimum order. Design methods are discussed in detail in Chapters 9 and 10.

3. Verification of the filter's performance by analytic means, simulations, and testing with real data when possible.

4. Implementation by hardware, software, or both. Implementation is discussed in Chapter 11.

As we have mentioned, there are two classes of LTI digital filters: IIR and FIR. Design techniques for these two classes are radically different, so we briefly discuss each of them individually.

8.5.1 IIR Filters

Analog rational filters necessarily have an infinite impulse response. Good design techniques of analog IIR filters have been known for decades. The design of digital IIR filters is largely based on analog filter design techniques. A typical design procedure of a digital IIR filter thus involves the following steps:

1. Choosing a method of transformation of a given analog filter to a digital filter having approximately the same frequency response.

2. Transforming the specifications of the digital IIR filter to equivalent specifications of an analog IIR filter such that, after the transformation from analog to digital is carried out, the digital IIR filter will meet the specifications.

3. Designing the analog IIR filter according to the transformed specifications.

4. Transforming the analog design to an equivalent digital filter.

The main advantages of this design procedure are convenience and reliability. It is convenient, since analog filter design techniques are well established and the properties of the resulting filters are well understood. It is reliable, since a good analog design (i.e., one that meets the specifications at minimum complexity) is guaranteed to yield a good digital design. The main drawback of this method is its limited generality. Design techniques of analog filters are practically limited to the four basic types (LP, HP, BP, and BS) and a few others. More general filters, such as multiband, are hard to design in the analog domain and require considerable expertise.

Beside analog-based methods, there exist design methods for digital IIR filters that are performed in the digital domain directly. Typically they belong to one of two classes:

1. Methods that are relatively simple, requiring only operations such as solutions of linear equations. Since these give rise to filters whose quality is often mediocre, they are not popular.

2. Methods that are accurate and give rise to high-quality filters, but are complicated and hard to implement; consequently they are not popular either.

Direct digital design techniques for IIR filters are not studied in this book.

8.5.2 FIR Filters

Digital FIR filters cannot be derived from analog filters, since rational analog filters cannot have a finite impulse response. So, why bother with such filters? A full answer to this question will be given in Chapter 9, when we study FIR filters in detail. Digital FIR filters have certain unique properties that are not shared by IIR filters (whether analog or digital), such as:

1. They are inherently stable.

2. They can be designed to have a linear phase or a generalized linear phase.

3. There is great flexibility in shaping their magnitude response.

4. They are convenient to implement.

These properties are highly desirable in many applications and have made FIR filters far more popular than IIR filters in digital signal processing. On the other hand, FIR filters have a major disadvantage with respect to IIR filters: The relative computational complexity of the former is higher than that of the latter. By "relative" we mean that an FIR filter meeting the *same* specifications as a given IIR filter will require many more operations per unit of time.

Design methods for FIR filters can be divided into two classes:

1. Methods that require only relatively simple calculations. Chief among those is the *windowing* method, which is based on concepts similar to those studied in Section 6.2. Another method in this category is *least-squares design*. These methods usually give good results, but are not optimal in terms of complexity.

2. Methods that rely on numerical optimization and require sophisticated software tools. Chief among those is the *equiripple* design method, which is guaranteed to meet the given specifications at a minimum complexity.

8.6   Summary and Complements

8.6.1   Summary

In this chapter we introduced the subject of digital filtering. In general, filtering means shaping the frequency-domain characteristics of a signal. The most common filtering operation is attenuation of signals in certain frequency bands and passing signals in other frequency bands with only little distortion. In the case of digital filters, the input and output signals are in discrete time. The two basic types of digital filter are infinite impulse response and finite impulse response. The former resemble analog filters in many respects, whereas the latter are unique to the discrete-time domain.

Simple filters are divided into four kinds, according to their frequency response characteristics: low pass, high pass, band pass, and band stop. Each of these kinds is characterized by its own set of specification parameters. A typical set of specification parameters includes (1) band-edge frequencies and (2) ripple and attenuation tolerances. The task of designing a filter amounts to finding a transfer function $H^z(z)$ (IIR or FIR), such that the corresponding frequency response $H^f(\theta)$ will meet or surpass the specifications, preferably at minimum complexity.

The frequency response of a rational digital filter can be written in a continuous-phase representation (8.30). If the phase function $\phi(\theta)$ in this representation is exactly linear, signals in the pass band of the filter are passed to the output almost free of distortion. If the phase function is linear up to an additive constant, the envelope of a modulated signal in the pass band is passed to the output almost free of distortion.

In the class of real, causal, stable, and rational digital filters, only FIR filters can have linear phase. Such filters come in four types: The impulse response can be either symmetric or antisymmetric, and the order can be either even or odd. Symmetric impulse response yields exact linear phase, whereas antisymmetric one yields generalized linear phase.

The phase response of a digital rational filter is not completely arbitrary, but is related to the magnitude response. For a given magnitude response, there exists a unique filter whose zeros are all inside or on the unit circle (the poles must be inside the unit circle in any case, for stability). Such a filter is called minimum phase. Any other filter having the same magnitude response has the following property: Each of its zeros is collocated either with a zero of the minimum-phase filter, or with the conjugate inverse of a zero of the minimum-phase filter.

8.6.2 Complements

1. [p. 243] Since $1/\pi t$ is discontinuous at $t = 0$ and its limits at the two sides of the discontinuity do not exist, the convolution $\{x * h\}(t)$ needs to be defined carefully. The standard definition is

$$\{x * h\}(t) = \lim_{\varepsilon \to 0} \left[\int_{-\infty}^{-\varepsilon} \frac{x(t-\tau)}{\pi\tau} d\tau + \int_{\varepsilon}^{\infty} \frac{x(t-\tau)}{\pi\tau} d\tau \right]. \tag{8.71}$$

The right side is called the *Cauchy principal value* of the integral. The problem does not arise in the discrete-time Hilbert transform.

2. [p. 245] Surface acoustic wave (SAW) filters and switched-capacitor filters are an exception to this statement: They implement z-domain transfer functions by essentially analog means.

3. [p. 245] *Analog* high-pass filters are designed to pass high frequencies, from a certain cutoff frequency ω_0 to infinity. In this sense, digital filters are fundamentally different from analog filters. A similar remark holds for band-stop filters.

4. [p. 250] A bit is typically lost due to nonlinearities, noise, etc.

5. [p. 254] The function *arctan2*, as defined here, is identical to the MATLAB function atan2.

6. [p. 261] We have proved Theorem 8.4 for RCSR filters. The theorem holds for real causal stable filters if we replace the rationality assumption by the less restrictive assumption that the z-transform of the filter exists on an annulus whose interior contains the unit circle. The proof of the extended version of the theorem is beyond the scope of this book. Without this assumption, the theorem is not valid. It was shown by Clements and Pease [1989] that real causal linear-phase IIR filters do exist, but their z-transform does not exist anywhere, except possibly on the unit circle itself.

8.7   MATLAB Program

Program 8.1 Group delay of a rational transfer function.

```
function D = grpdly(b,a,K,theta);
% Synopsis: D = grpdly(b,a,K,theta).
% Group delay of b(z)/a(z) on a given frequency interval.
% Input parameters:
% b, a: numerator and denominator polynomials
% K: the number of frequency response points to compute
% theta: if absent, the K points are uniformly spaced on
%        [0, pi]; if present and theta is a 1-by-2 vector,
%        its entries are taken as the end points of the
%        interval on which K evenly spaced points are
%        placed; if the size of theta is different from 2,
%        it is assumed to be a vector of frequencies for
%        which the group delay is to be computed, and K is
%        ignored.
% Output:
% D: the group delay vector.

a = reshape(a,1,length(a)); b = reshape(b,1,length(b));
if (length(a) == 1), % case of FIR
   bd = -j*(0:length(b)-1).*b;
   if (nargin == 3),
      B = frqresp(b,1,K); Bd = frqresp(bd,1,K);
   else,
      B = frqresp(b,1,K,theta); Bd = frqresp(bd,1,K,theta);
   end
   D = (real(Bd).*imag(B)-real(B).*imag(Bd))./abs(B).^2;
else % case of IIR
   if (nargin == 3), D = grpdly(b,1,K)-grpdly(a,1,K);
   else, D = grpdly(b,1,K,theta)-grpdly(a,1,K,theta); end
end
```

8.8 Problems

8.1 Repeat Example 8.1 with the following changes:

- The signal bandwidth is $250\,\text{Hz}$.
- The transition bandwidth is $50\,\text{Hz}$.
- The average noise gain in the transition band is 0.25.
- The permitted pass-band ripple is 0.05 dB.

Specify the achievable output SNR and the required filter parameters.

8.2 The digital low-pass filter $H^z(z)$ is known to meet the specifications

$$\theta_p = 0.25\pi, \quad \theta_s = 0.35\pi, \quad \delta_p = 0.02, \quad \delta_s = 0.005.$$

Define a new filter $G^z(z)$ by

$$g[n] = \begin{cases} 2h[n], & n \text{ even,} \\ 0, & n \text{ odd.} \end{cases}$$

(a) Is $G^z(z)$ LP, HP, BP, or BS? Explain.

(b) What specifications is $G^z(z)$ guaranteed to meet?

8.3 A digital filter usually has some of its zeros on the unit circle. Discuss the relationship between the locations of the zeros on the unit circle and the nature of the frequency response of the filter (LP, HB, BP, BS). Pay special attention to zeros at $z = 1$ and at $z = -1$.

8.4 A digital LTI filter has the frequency response

$$H^f(\theta) = 4 \left(1 - \frac{|\theta|}{\pi} \right) e^{-j\theta}.$$

The continuous-time signal $x(t) = \cos(\pi t)$ is sampled at interval $T = 0.25$ and fed to the filter. Find the signal $y[n]$ at the output of the filter.

8.5 This problem explores the existence of zeros of a digital filter at $z = \pm 1$.

(a) Find a necessary and sufficient condition on the coefficients $h[n]$ of an FIR filter for the transfer function $H^z(z)$ to have a zero at $z = 1$. Repeat for a zero at $z = -1$.

(b) Fill the eight cells in Table 8.2 with yes or no, and explain.

Filter type	I	II	III	IV
Must have zero at $z = 1$				
Must have zero at $z = -1$				

Table 8.2 Pertaining to Problem 8.5.

8.6 If an FIR filter $H^z(z)$ has a zero at $z = 1$, its transfer function can be expressed as $H^z(z) = (1 - z^{-1})F^z(z)$; if it has a zero at $z = -1$, its transfer function can be expressed as $H^z(z) = (1 + z^{-1})F^z(z)$; if it has zeros at both locations, its transfer function can be expressed as $H^z(z) = (1 - z^{-2})F^z(z)$. In all cases, $F^z(z)$ is FIR. Use your answer to Problem 8.5 for expressing $H^z(z)$ of each of the four filter types in the applicable form. For each type, express the coefficients of $H^z(z)$ in terms of $f[n]$, the coefficients of the filter $F^z(z)$.

8.7 We are given a transfer function of a digital filter in a factored form

$$H^z(z) = b_0 \frac{\prod_{i=1}^{q}(1 - \beta_i z^{-1})}{\prod_{i=1}^{p}(1 - \alpha_i z^{-1})}.$$

Define the filter $G^z(z)$ by replacing all poles and zeros of $H^z(z)$ by their negatives, that is,

$$\alpha_i \longrightarrow -\alpha_i, \quad \beta_i \longrightarrow -\beta_i.$$

(a) Express the frequency response $G^f(\theta)$ in terms of $H^f(\theta)$.

(b) Suppose that all poles and zeros of $H^z(z)$ are on the imaginary axis. Can such a filter be low pass? high pass? band pass? band stop? Give reasons.

(c) Let

$$H^z(z) = \frac{(z - 1)^2}{z^2 - 1.212436z + 0.49}.$$

Write the corresponding transfer function $G^z(z)$. Is $G^z(z)$ low pass, high pass, band pass, or band stop?

8.8 Let

$$H^z(z) = 1 - 1.5z^{-1} + 0.5z^{-2}.$$

Find $A(\theta)$ and $\phi(\theta)$ in the continuous-phase representation of $H^f(\theta)$. Hint: Factor $H^z(z)$ to its zeros.

8.9 Is it possible for a minimum-phase FIR filter to have linear phase? If so, give an example. If not, explain why.

8.10 Suppose that the FIR filter $H^z(z)$ has all its zeros on the unit circle. Show that the filter has linear phase. Hint: Write $H^z(z)$ in a factored form

$$H^z(z) = h[0](1 - z^{-1})^{N_1}(1 + z^{-1})^{N_2} \prod_{k=1}^{N_3}(1 - 2\cos\zeta_k z^{-1} + z^{-2}),$$

where N_1 is the number of zeros at $z = 1$, N_2 is the number of zeros at $z = -1$, and N_3 is the number of complex conjugate pairs of zeros on the unit circle. Then show that each of the factors has linear phase. Finally show that the product has linear phase.

8.11 A noncausal filter is called a *half-band filter* if its frequency response satisfies

$$H^f(\theta) + H^f(\theta - \pi) = c, \tag{8.72}$$

where c is constant. The equivalent z-domain property is

$$H^z(z) + H^z(-z) = c. \tag{8.73}$$

Such a filter is called *zero phase, half-band* if, in addition to (8.72), it is symmetric, that is, $h[n] = h[-n]$. Figure 8.11 illustrates the frequency response of a zero-phase, half-band filter.

(a) Prove that a necessary and sufficient condition for a filter to be half-band is

$$h[2n] = \begin{cases} 0.5c, & n = 0, \\ 0, & n \neq 0. \end{cases} \tag{8.74}$$

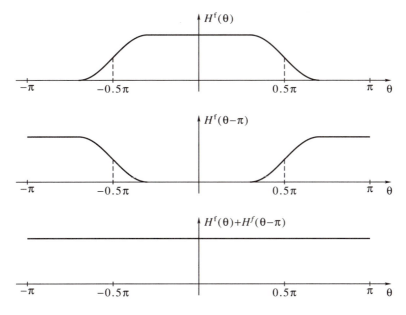

Figure 8.11 The frequency response of a half-band filter.

(b) What is $H^f(0.5\pi)$ for a zero-phase, half-band filter?

(c) A *causal half-band filter* is a causal filter satisfying

$$H^f(\theta) + H^f(\theta - \pi) = ce^{-j2\theta M} \tag{8.75}$$

for some positive integer M. If, in addition to (8.75), the filter is a symmetric FIR filter (that is, $h[n] = h[N - n]$), it is called *linear phase, half-band*. What is a necessary and sufficient condition for a causal filter to be half-band?

(d) Let $h[n]$ be the impulse response of a linear-phase, half-band FIR filter of order $4M$. Show that the order of the filter can be reduced to $4M - 2$ without changing the amplitude response. What will be the group delay of the reduced-order filter?

8.12 This problem introduces the concept of *deconvolution*.

(a) The signal $x[n]$ is passed through the RCSR IIR filter $H^z(z) = 1/a(z)$, and the signal $y[n]$ is obtained. We wish to recover the signal $x[n]$ by passing $y[n]$ through an RCSR filter $G^z(z)$. Under what conditions is this possible, and what is $G^z(z)$?

(b) Repeat part a when the filter is FIR, $H^z(z) = b(z)$.

(c) Repeat part a when the filter is $H^z(z) = b(z)/a(z)$.

The operation addressed in this problem is called deconvolution because it reverses the convolution operation of the filter.

8.13 Explain why the group delay of a rational IIR filter is the difference between the group delay of the numerator and that of the denominator.

8.14 Use MATLAB for computing the impulse responses of the four filters in Example 8.10. Plot them and observe the values of the early impulse response coefficients (the ones nearest to $n = 0$) relative to the coefficients at later times. Compare the behavior of the four filters in this respect, and reach a conclusion about the special nature of the impulse response of the minimum-phase filter $H_1^z(z)$.

8.15 We are given the four filters

$$H_1^z(z) = 1 + \frac{3}{16}z^{-2} - \frac{1}{64}z^{-4}, \quad H_2^z(z) = \frac{1}{4} + \frac{63}{64}z^{-2} - \frac{1}{16}z^{-4},$$

$$H_3^z(z) = -\frac{1}{16} + \frac{63}{64}z^{-2} + \frac{1}{4}z^{-4}, \quad H_4^z(z) = -\frac{1}{64} + \frac{3}{16}z^{-2} + z^{-4}.$$

(a) Compute the zeros of each of the four filters.

(b) Compute and plot the frequency response (magnitude and phase) of each of the four filters. Learn the MATLAB function unwrap and use it for the phase plots.

(c) Plot the group delay of each of the four filters.

(d) State your conclusions from this exercise.

8.16 Show that $H^z(z)$ in (8.69) and the filter in Problem 7.29 are mathematically equivalent in the following sense:

(a) If $H^z(z)$ in (8.69) is expanded, its numerator and denominator polynomials will be related as shown in Problem 7.29 (except for possible sign reversal).

(b) If $H^z(z)$ in Problem 7.29 is factored, its poles and zeros will satisfy the relationship shown in (8.69) (except for possible sign reversal).

8.17 Give an example of an analog FIR filter. Hint: Such a filter cannot have a rational transfer function. We encountered an analog FIR filter earlier in this book.

8.18 We are given the filter

$$H^z(z) = \frac{z^{-1} + 0.5}{1 + 0.5z^{-1}}$$

and the two input signals

$$x_1[n] = \sin(0.1\pi n), \quad x_2[n] = 3\cos(0.4\pi n).$$

Let $y_1[n]$, $y_2[n]$ be the output signals corresponding to $x_1[n]$, $x_2[n]$.

(a) What are the amplitudes of $y_1[n]$, $y_2[n]$? Answer with the minimum amount of computations.

(b) What are the phases of $y_1[n]$, $y_2[n]$ relative to those of $x_1[n]$, $x_2[n]$, respectively? Compute to four decimal digits.

(c) Is $H^z(z)$ a distortion-free filter? Explain.

8.19* The purpose of this problem is to derive a continuous-phase representation of an RCSR transfer function $H^z(z)$ from the pole–zero factorization of the transfer function. We saw in Section 7.4.2 that $H^z(z)$ can be written as

$$H^z(z) = b_{q-r}z^{p-q}\frac{\prod_{k=1}^{r}(z - \beta_k)}{\prod_{k=1}^{p}(z - \alpha_k)}. \tag{8.76}$$

(a) Divide the zeros $\{\beta_k, 1 \le k \le r\}$ into three sets:

- The first set, $\{\beta_{i,k}, 1 \le k \le r_i\}$, contains all the zeros inside the unit circle.
- The second set, $\{\beta_{u,k}, 1 \le k \le r_u\}$, contains all the zeros on the unit circle.
- The third set, $\{\beta_{o,k}, 1 \le k \le r_o\}$, contains all the zeros outside the unit circle.

Show that $H^z(z)$ can be written as

$$H^z(z) = b_{q-r} \prod_{k=1}^{r_o} (-\beta_{o,k}) z^{-q+r_i}$$

$$\cdot \frac{\left[\prod_{k=1}^{r_i}(1 - \beta_{i,k}z^{-1})\right]\left[\prod_{k=1}^{r_o}(1 - \beta_{o,k}^{-1}z)\right]\left[\prod_{k=1}^{r_u}(z - \beta_{u,k})\right]}{\prod_{k=1}^{p}(1 - \alpha_k z^{-1})}. \tag{8.77}$$

(b) Show that, since $|\alpha_k| < 1$, the factor $(1 - \alpha_k e^{-j\theta})$ of $H^f(\theta)$ has a positive real part, hence its phase is continuous and lies in the range $(-0.5\pi, 0.5\pi)$. Since $|\beta_{i,k}| < 1$, the factor $(1 - \beta_{i,k}e^{-j\theta})$ also has this property.

(c) Show that, since $|\beta_{o,k}| > 1$, the factor $(1 - \beta_{o,k}^{-1}e^{j\theta})$ of $H^f(\theta)$ has a positive real part, hence its phase is continuous and lies in the range $(-0.5\pi, 0.5\pi)$.

(d) Show that, since $|\beta_{u,k}| = 1$, the factor $(e^{j\theta} - \beta_{u,k})$ of $H^f(\theta)$ can be written as a product of a real (but not necessarily positive) function of θ and a linear-phase factor.

(e) Show that the constant factor in (8.77) is real and that z^{-q+r_i} provides a linear-phase factor.

(f) Show how to combine the results of parts b through e to obtain a continuous-phase representation of $H^f(\theta)$. This provides an alternative proof of Theorem 8.3.

8.20* Discrete-time white noise is given at the input of an FIR filter. What special property does the covariance sequence of the output have? Hint: Recall Problem 2.39.

8.21* Let $\{x[n], 0 \le n \le N - 1\}$ be a *fixed* real signal, and let $v[n]$ be a discrete-time white noise with zero mean and $\gamma_v = 1$. We feed $x[n] + v[n]$ to an LTI filter having an impulse response $h[n]$. Let $y_1[N - 1]$ denote the response of the filter to $x[n]$ and $y_2[N - 1]$ the response to $v[n]$, both at time $n = N - 1$. Define the output signal-to-noise ratio of the filter as

$$\text{SNR}_o = \frac{|y_1[N - 1]|^2}{E(|y_2[N - 1]|^2)}.$$

We are looking for a filter $h[n]$ that will maximize SNR_o.

(a) Show that $h[n]$ that maximizes SNR_o is

$$h[n] = cx[N - 1 - n], \quad 0 \le n \le N - 1,$$

where c is an arbitrary nonzero constant. Hint: Use the Cauchy–Schwarz inequality in form (2.145).

(b) What is the output SNR of the optimal filter?

(c) Solve this problem if the signal and the noise are $\{x[n], v[n], -\infty < n \le N - 1\}$. In what sense is the filter that you have found different from the one in part a? Hint: Use the Cauchy–Schwarz inequality in form (2.146).

The filter presented in this problem is called the *matched filter* for the signal $x[n]$. It is of great importance in signal detection, for example in radar and sonar applications. We shall encounter it again in Section 14.4.

8.22* This problem introduces single-side-band-modulated signals and filtering of such signals.

(a) The signal

$$x[n] = \cos(\theta_m n) \cdot \cos(\theta_c n), \quad \theta_m \ll \theta_c$$

is passed through an ideal band-pass filter whose band-edge frequencies are θ_c and $\theta_c + 2\theta_m$. What is the signal $y[n]$ at the output of the filter?

(b) How will the answer to part a change if the filter is not ideal, but has the specifications

$$\theta_{s,1} = \theta_c, \quad \theta_{p,1} = \theta_c + 0.5\theta_m, \quad \theta_{p,2} = \theta_c + 1.5\theta_m, \quad \theta_{s,2} = \theta_c + 2\theta_m,$$

$$A_s = 60\,\text{dB}, \quad A_p = 0.087\,\text{dB}.$$

(c) Repeat for the signal

$$x[n] = s[n] \cos \theta_c,$$

where $s[n]$ is a real band-pass signal in the frequency range $[0.5\theta_m, 1.5\theta_m]$.

Signals of this kind are an example of *single-side-band* (SSB) modulation. In communication, they are used for reducing the bandwidth of the modulated signal by half, while preserving the information in the modulating signal $s[n]$.

Chapter 9

Finite Impulse Response Filters

In this chapter we study the structure, properties, and design methods of digital FIR filters. Digital FIR filters have several favorable properties, thanks to which they are popular in digital signal processing. First and foremost of those is the linear-phase property, which, as we saw in Chapter 8, provides distortionless response (or nearly so) for signals in the pass band. We therefore begin this chapter with an expanded discussion of the linear-phase property, and study its manifestations in the time and frequency domains.

The simplest design method for FIR filters is impulse response truncation, so this is the first to be presented. This method is not quite useful by itself, since it has undesirable frequency-domain characteristics. However, it serves as a necessary introduction to the second design method to be presented—windowing. We have already encountered windows in Section 6.2, in the context of spectral analysis. Here we shall learn how windows can be used for mitigating the adverse effects of impulse response truncation in the same way as they mitigate the effects of signal truncation in spectral analysis.

The windowing design method, although simple and convenient, is not optimal. By this we mean that, for given pass-band and stop-band specifications, their order is not the minimum possible. We present two design methods based on optimality criteria. The first is *least-squares design*, which minimizes an integral-of-square-error criterion in the frequency domain. The second is *equiripple design*, which minimizes the maximum ripple in each band. Equiripple design is intricate in its basic theory and details of implementation. Fortunately, there exist well-developed computer programs for this purpose, which take much of the burden of the design from the individual user. We shall therefore devote relatively little space to this topic, presenting only its principles.

9.1 Generalized Linear Phase Revisited

Practical FIR filters are usually designed to have a linear phase, either exact or generalized. We shall omit the modifiers "exact" and "generalized," calling filters having either constant phase delay or constant group delay *linear-phase filters*. The transfer function of an FIR filter is usually expressed in terms of its impulse response coefficients, that is,

$$H^z(z) = h[0] + h[1]z^{-1} + \ldots + h[N]z^{-N}. \tag{9.1}$$

Although FIR filters are a special case of rational filters, we do not use the polynomial notation $b(z)$ for these filters. We use N to denote the *order* of the filter; in some books N is used for the *length* of the filter, so watch out for differences resulting from this notation. An Nth-order filter has $N+1$ nonzero coefficients, so its length is $N+1$. Thus, if the order is even, the length is odd, and vice versa.

In this section we expand the discussion of linear-phase filters and derive expressions for the frequency responses of the four filter types. Then we discuss the restrictions on the zero locations resulting from the linear-phase property.

9.1.1   Type-I Filters

Type-I filters have even order, $N = 2M$, and initial phase $\phi_0 = 0$. Consequently, they have constant phase delay. Their impulse response satisfies the symmetry condition

$$h[n] = h[N - n]. \tag{9.2}$$

Their frequency response is derived as follows:

$$
\begin{aligned}
H^{\mathrm{f}}(\theta) &= \sum_{n=0}^{2M} h[n]e^{-j\theta n} = e^{-j\theta M} \sum_{n=0}^{2M} h[n]e^{j\theta(M-n)} \\
&= e^{-j\theta M}\left\{ h[M] + \sum_{n=0}^{M-1} h[n]e^{j\theta(M-n)} + \sum_{n=M+1}^{2M} h[n]e^{j\theta(M-n)} \right\} \\
&= e^{-j\theta M}\left\{ h[M] + \sum_{n=0}^{M-1} h[n]e^{j\theta(M-n)} + \sum_{n=0}^{M-1} h[N-n]e^{-j\theta(M-n)} \right\} \\
&= e^{-j\theta M}\left\{ h[M] + 2 \sum_{n=0}^{M-1} h[n]\cos[\theta(M-n)] \right\}. \tag{9.3}
\end{aligned}
$$

In passing from the second to the third line, we substituted $n' = N - n$ in the second term and then renamed n' as n. In the fourth line we used Euler's formula for $\cos[\theta(M - n)]$.

The amplitude function of a type-I filter can be written in the form

$$A_{\mathrm{I}}(\theta) = \sum_{n=0}^{M} g[n]\cos(\theta n), \tag{9.4}$$

where

$$g[n] = \begin{cases} h[M], & n = 0, \\ 2h[M - n], & 1 \le n \le M. \end{cases} \tag{9.5}$$

The amplitude function is symmetric in θ and has period 2π. Figure 9.1 shows typical impulse and amplitude responses of a type-I filter.

Example 9.1 The filter

$$H^z(z) = 1 + 0.5z^{-1} - 0.3z^{-2} + 1.2z^{-3} - 0.3z^{-4} + 0.5z^{-5} + z^{-6}$$

is type I. Its amplitude function is

$$A(\theta) = 1.2 - 0.6\cos\theta + \cos(2\theta) + 2\cos(3\theta). \qquad \square$$

9.1.2   Type-II Filters

Type-II filters have odd order, $N = 2M + 1$, and initial phase $\phi_0 = 0$. Like type-I filters, they have constant phase delay. Their impulse response satisfies the symmetry

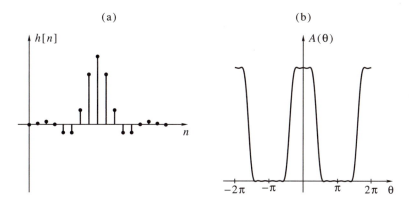

Figure 9.1 Type-I FIR filter: (a) impulse response; (b) amplitude response.

condition (9.2). Their frequency response is derived as follows:

$$H^f(\theta) = \sum_{n=0}^{2M+1} h[n]e^{-j\theta n} = e^{-j\theta(M+0.5)} \sum_{n=0}^{2M+1} h[n]e^{j\theta(M+0.5-n)}$$

$$= e^{-j\theta(M+0.5)} \left\{ \sum_{n=0}^{M} h[n]e^{j\theta(M+0.5-n)} + \sum_{n=M+1}^{2M+1} h[n]e^{j\theta(M+0.5-n)} \right\}$$

$$= e^{-j\theta(M+0.5)} \left\{ \sum_{n=0}^{M} h[n]e^{j\theta(M+0.5-n)} + \sum_{n=0}^{M} h[N-n]e^{-j\theta(M+0.5-n)} \right\}$$

$$= e^{-j\theta(M+0.5)} \left\{ 2 \sum_{n=0}^{M} h[n] \cos[\theta(M + 0.5 - n)] \right\}. \tag{9.6}$$

The amplitude function of a type-II filter can be written in the form

$$A_{\mathrm{II}}(\theta) = \cos(0.5\theta) \sum_{n=0}^{M} g[n]\cos(\theta n). \tag{9.7}$$

This is proved as follows:

$$\cos(0.5\theta) \sum_{n=0}^{M} g[n]\cos(\theta n) = \cos(0.5\theta) \sum_{n=0}^{M} g[M-n]\cos[\theta(M-n)]$$

$$= 0.5 \sum_{n=0}^{M} g[M-n]\{\cos[\theta(M + 0.5 - n)] + \cos[\theta(M - 0.5 - n)]\}$$

$$= 0.5 \sum_{n=1}^{M} \{g[M-n] + g[M+1-n]\}\cos[\theta(M + 0.5 - n)]$$

$$+ 0.5g[M]\cos[\theta(M + 0.5)] + 0.5g[0]\cos(0.5\theta). \tag{9.8}$$

Thus we can make (9.8) equal to $A_{\mathrm{II}}(\theta)$ by choosing

$$2h[n] = \begin{cases} 0.5g[M], & n = 0, \\ 0.5\{g[M+1-n] + g[M-n]\}, & 1 \le n \le M - 1, \\ 0.5g[1] + g[0], & n = M. \end{cases} \tag{9.9}$$

The coefficients $g[n]$ can be computed from the impulse response $h[n]$ by inverting the relationships (9.9). This yields the iterative formulas

$$g[M] = 4h[0], \tag{9.10a}$$

$$g[M - n] = 4h[n] - g[M + 1 - n], \quad 1 \le n \le M - 1, \tag{9.10b}$$

$$g[0] = 2h[M] - 0.5g[1]. \tag{9.10c}$$

The amplitude function of a type-II filter is symmetric in θ and has period 4π. As we see from (9.7), $A_{\mathrm{II}}(\pi) = 0$, because $\cos(0.5\pi) = 0$. Therefore, type-II filters are not suitable for high-pass or band-stop filters. This is in contrast with type-I filters, which are suitable for filters of all kinds. Figure 9.2 shows typical impulse and amplitude responses of a type-II filter.

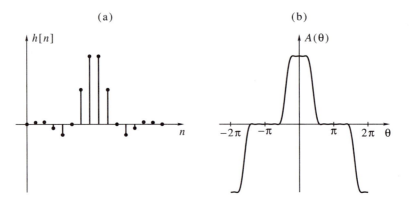

Figure 9.2 Type-II FIR filter: (a) impulse response; (b) amplitude response.

Example 9.2 The filter

$$H^z(z) = 0.4 + 0.6z^{-1} + 1.5z^{-2} + 1.5z^{-3} + 0.6z^{-4} + 0.4z^{-5}$$

is type II. Its amplitude function is

$$A(\theta) = \cos(0.5\theta)[2.6 + 0.8\cos\theta + 1.6\cos(2\theta)]. \qquad \square$$

9.1.3 Type-III Filters

Type-III filters have even order, $N = 2M$, and initial phase $\phi_0 = \pi/2$. Consequently, they have constant group delay, but not constant phase delay. Their impulse response satisfies the antisymmetry condition

$$h[n] = -h[N - n]. \tag{9.11}$$

Their frequency response is derived similarly to (9.3) (note that $h[M] = 0$, because of the antisymmetry condition):

$$
\begin{aligned}
H^f(\theta) &= e^{-j\theta M}\left\{\sum_{n=0}^{M-1} h[n]e^{j\theta(M-n)} + \sum_{n=M+1}^{2M} h[n]e^{j\theta(M-n)}\right\} \\
&= e^{-j\theta M}\left\{\sum_{n=0}^{M-1} h[n]e^{j\theta(M-n)} + \sum_{n=0}^{M-1} h[N-n]e^{-j\theta(M-n)}\right\} \\
&= e^{j(0.5\pi - \theta M)}\left\{2\sum_{n=0}^{M-1} h[n]\sin[\theta(M-n)]\right\}. \tag{9.12}
\end{aligned}
$$

The amplitude function of a type-III filter can be written in the form

$$A_{\text{III}}(\theta) = \sin\theta \sum_{n=0}^{M-1} g[n]\cos(\theta n). \qquad (9.13)$$

This is proved as follows:

$$\sin\theta \sum_{n=0}^{M-1} g[n]\cos(\theta n) = \sin\theta \sum_{n=1}^{M} g[M-n]\cos[\theta(M-n)]$$

$$= 0.5 \sum_{n=1}^{M} g[M-n]\{\sin[\theta(M+1-n)] - \sin[\theta(M-1-n)]\}$$

$$= 0.5 \sum_{n=0}^{M-1} g[M-1-n]\sin[\theta(M-n)] - 0.5 \sum_{n=2}^{M} g[M+1-n]\sin[\theta(M-n)]$$

$$+ 0.5g[0]\sin\theta. \qquad (9.14)$$

Thus we can make (9.14) equal to $A_{\text{III}}(\theta)$ by choosing

$$2h[n] = \begin{cases} 0.5g[M-1], & n = 0, \\ 0.5g[M-2], & n = 1, \\ 0.5\{g[M-1-n] - g[M+1-n]\}, & 2 \le n \le M-2, \\ g[0] - 0.5g[2], & n = M-1. \end{cases} \qquad (9.15)$$

The coefficients $g[n]$ can be computed from the impulse response $h[n]$ by inverting the relationships (9.15). This gives the iterative formulas

$$g[M-1] = 4h[0], \qquad (9.16a)$$

$$g[M-2] = 4h[1], \qquad (9.16b)$$

$$g[M-n] = 4h[n-1] + g[M+2-n], \quad 3 \le n \le M-1, \qquad (9.16c)$$

$$g[0] = 2h[M-1] + 0.5g[2]. \qquad (9.16d)$$

The amplitude function of a type-III filter is antisymmetric in θ and has period 2π. As we see from (9.13), $A_{\text{III}}(0) = A_{\text{III}}(\pi) = 0$, because $\sin 0 = \sin\pi = 0$. Therefore, type-III filters are not suitable for low-pass, high-pass, or band-stop filters. They can be used for band-pass filters, but since they do not have exact linear phase, only constant group delay, they are not common for this application. Type III is sometimes used for differentiators and Hilbert transformers, as we shall illustrate later. Figure 9.3 shows typical impulse and amplitude responses of a type-III filter.

Example 9.3 The filter

$$H^z(z) = 1 + 0.5z^{-1} - 0.3z^{-2} + 0.3z^{-4} - 0.5z^{-5} - z^{-6}$$

is type III. Its amplitude function is

$$A(\theta) = \sin\theta[1.4 + 2\cos\theta + 4\cos(2\theta)]. \qquad \square$$

9.1.4 Type-IV Filters

Type-IV filters have odd order, $N = 2M + 1$, and initial phase $\phi_0 = \pi/2$. Like type-III filters, they have constant group delay, but not constant phase delay. Their impulse response satisfies the antisymmetry condition (9.11). Their frequency response is derived similarly to (9.6):

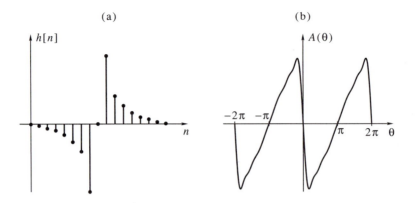

Figure 9.3 Type-III FIR filter: (a) impulse response; (b) amplitude response.

$$H^f(\theta) = e^{-j\theta(M+0.5)} \left\{ \sum_{n=0}^{M} h[n]e^{j\theta(M+0.5-n)} + \sum_{n=0}^{M} h[N-n]e^{-j\theta(M+0.5-n)} \right\}$$

$$= e^{j[0.5\pi - \theta(M+0.5)]} \left\{ 2 \sum_{n=0}^{M} h[n]\sin[\theta(M+0.5-n)] \right\}. \tag{9.17}$$

The amplitude function of a type-IV filter can be written in the form

$$A_{IV}(\theta) = \sin(0.5\theta) \sum_{n=0}^{M} g[n]\cos(\theta n). \tag{9.18}$$

This is proved as follows:

$$\sin(0.5\theta) \sum_{n=0}^{M} g[n]\cos(\theta n) = \sin(0.5\theta) \sum_{n=0}^{M} g[M-n]\cos[\theta(M-n)]$$

$$= 0.5 \sum_{n=0}^{M} g[M-n]\{\sin[\theta(M+0.5-n)] - \sin[\theta(M-0.5-n)]\}$$

$$= 0.5 \sum_{n=1}^{M} \{g[M-n] - g[M+1-n]\} \sin[\theta(M+0.5-n)]$$

$$+ 0.5g[M]\sin[\theta(M+0.5)] + 0.5g[0]\sin(0.5\theta). \tag{9.19}$$

Thus we can make (9.19) equal to $A_{IV}(\theta)$ by choosing

$$2h[n] = \begin{cases} 0.5g[M], & n = 0, \\ 0.5\{g[M-n] - g[M+1-n]\}, & 1 \le n \le M-1, \\ g[0] - 0.5g[1], & n = M. \end{cases} \tag{9.20}$$

The coefficients $g[n]$ can be computed from the impulse response $h[n]$ by inverting the relationships (9.20). This gives the iterative formulas

$$g[M] = 4h[0], \tag{9.21a}$$

$$g[M-n] = 4h[n] + g[M+1-n], \quad 1 \le n \le M-1, \tag{9.21b}$$

$$g[0] = 2h[M] + 0.5g[1]. \tag{9.21c}$$

The amplitude function of a type-IV filter is antisymmetric in θ and has period 4π. As we see from (9.18), $A_{IV}(0) = 0$, because $\sin 0 = 0$. Therefore, type-IV filters are not suitable for low-pass or band-stop filters. They can be used for high-pass and band-pass filters, but this is not common. Similarly to type-III filters, they are useful for differentiators and Hilbert transformers. Figure 9.4 shows typical impulse and amplitude responses of a type-IV filter.

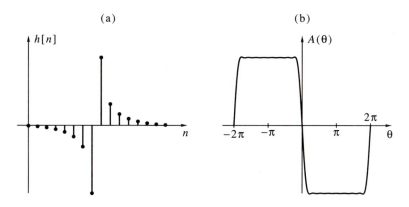

Figure 9.4 Type-IV FIR filter: (a) impulse response; (b) amplitude response.

Example 9.4 The filter

$$H^z(z) = 0.4 + 0.6z^{-1} + 1.5z^{-2} - 1.5z^{-3} - 0.6z^{-4} - 0.4z^{-5}$$

is type IV. Its amplitude function is

$$A(\theta) = \sin(0.5\theta)[5 + 4\cos\theta + 1.6\cos(2\theta)]. \qquad \square$$

9.1.5 Summary of Linear-Phase Filter Types

As we have seen, the frequency responses of linear-phase FIR filters of all four types are of the form

$$H^f(\theta) = A(\theta)e^{j(\phi_0 - 0.5\theta N)}, \tag{9.22}$$

where the amplitude response has the form

$$A(\theta) = F(\theta)G(\theta), \quad G(\theta) = \sum_{k=0}^{K} g[k]\cos(\theta k), \tag{9.23}$$

and $F(\theta)$ is one of the functions 1, $\cos(0.5\theta)$, $\sin\theta$, $\sin(0.5\theta)$, depending on the type. Table 9.1 summarizes the properties and parameters of the four types as a function of the order N and the impulse response coefficients $h[n]$.

9.1.6 Zero Locations of Linear-Phase Filters

The zeros of a linear-phase FIR filter cannot be completely arbitrary, but must satisfy certain constraints. Let $H^z(z)$ be the transfer function of an Nth-order, linear-phase filter with real coefficients. Then

$$H^z(z^{-1}) = \sum_{n=0}^{N} h[n]z^n = z^N \sum_{n=0}^{N} h[n]z^{-(N-n)} = z^N \sum_{n=0}^{N} h[N-n]z^{-n} = \pm z^N H^z(z), \tag{9.24}$$

where the sign of the right side is positive for a symmetric filter and negative for an antisymmetric filter. Let β be a zero of $H^z(z)$, so $H^z(\beta) = 0$. Then we get from (9.24)

Type	I	II	III	IV
Order	even	odd	even	odd
Symmetry of $h[n]$	symmetric	symmetric	anti-symmetric	anti-symmetric
Symmetry of $A(\theta)$	symmetric	symmetric	anti-symmetric	anti-symmetric
Period of $A(\theta)$	2π	4π	2π	4π
ϕ_0	0	0	0.5π	0.5π
$F(\theta)$ in (9.23)	1	$\cos(0.5\theta)$	$\sin\theta$	$\sin(0.5\theta)$
K in (9.23)	$N/2$	$(N-1)/2$	$(N-2)/2$	$(N-1)/2$
$g[n]$ in (9.23)	see (9.5)	see (9.10)	see (9.16)	see (9.21)
$H^f(0)$	arbitrary	arbitrary	0	0
$H^f(\pi)$	arbitrary	0	0	arbitrary
Uses	LP, HP, BP, BS, Multiband filters	LP, BP	Differentiators, Hilbert transformers	

Table 9.1 Properties and parameters of the four FIR filter types.

that $H^z(\beta^{-1}) = 0$ (note that necessarily $\beta \neq 0$, since a minimal FIR filter cannot have a zero at the origin). This shows that, whenever β is a zero of $H(z)$, so is β^{-1}. Therefore, the zeros of a linear-phase filter must obey the following restrictions:

1. A zero $\beta = 1$ can appear any number of times, because it is self reciprocal.
2. A zero $\beta = -1$ can appear any number of times, because it is self reciprocal.
3. If there is a zero $\beta = e^{j\zeta}$, where $\zeta \neq 0$, $\zeta \neq \pi$, then $\bar\beta = e^{-j\zeta}$ is also a zero, and the two are the reciprocals of each other. Such pairs can appear any number of times.
4. If there is a zero $\beta = r$, where r is real and $|r| \neq 1$, then $\beta^{-1} = r^{-1}$ must also be a zero. Such pairs can appear any number of times.
5. If there is a zero $\beta = re^{j\zeta}$, where $r \neq 1$, $\zeta \neq 0$, $\zeta \neq \pi$, then

$$\bar\beta = re^{-j\zeta}, \quad \beta^{-1} = r^{-1}e^{-j\zeta}, \quad \bar\beta^{-1} = r^{-1}e^{j\zeta}$$

must also be zeros. Such quadruples can appear any number of times.

Figure 9.5 illustrates the possible zero locations of a linear-phase FIR filter in the z-plane. The numbers in the figure match the numbers in the preceding list.

Recalling the discussion in Section 8.4.7, we conclude that a linear-phase filter is not minimum phase in general, since it has zeros outside the unit circle. The only exception occurs when *all* the zeros of the filter are on the unit circle, but such a case is rare in practice. We also recall that a minimum-phase filter can be obtained from any linear-phase filter by replacing the zeros outside the unit circle by their conjugate inverses. Such a procedure will leave the magnitude response unchanged. The minimum-phase filter obtained by this procedure has the following property: All its zeros inside the unit circle have multiplicity 2 at least.

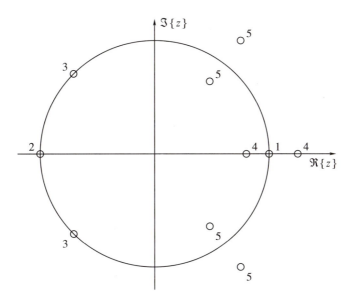

Figure 9.5 Possible zero locations of a linear-phase FIR filter.

Example 9.5 Let

$$H^z(z) = 1 - 0.5z^{-1} + 0.25z^{-2} + 0.875z^{-3} - 0.5z^{-4} + 0.25z^{-5} - 0.125z^{-6}.$$

We wish to find a linear-phase filter $H_1^z(z)$ of minimal order such that the zeros of $H^z(z)$ will be a subset of the zeros of $H_1^z(z)$. We find, using the function `roots` of MATLAB, that the zeros of $H^z(z)$ are

$$\beta_1 = -1, \quad \beta_2 = 0.5, \quad \beta_{3,4} = 0.5(1 \pm j\sqrt{3}), \quad \beta_{5,6} = \pm j0.5.$$

The filter $H_1^z(z)$ must include, in addition to these zeros, zeros at

$$\beta_7 = 2, \quad \beta_{8,9} = \pm j2.$$

Note that β_1 need not be duplicated because it is self reciprocal, and $\beta_{3,4}$ need not be duplicated because they are the reciprocals of each other. Thus, $H_1^z(z)$ is given by

$$H_1(z) = H^z(z)(1 - 2z^{-1})(1 + 4z^{-2}) = 1 - 2.5z^{-1} + 5.25z^{-2} - 9.625z^{-3}$$
$$+ 2.75z^{-4} + 2.75z^{-5} - 9.625z^{-6} + 5.25z^{-7} - 2.5z^{-8} + z^{-9}.$$

Next we wish to find a minimum-phase filter $H_2^z(z)$ having the same magnitude response as $H_1^z(z)$. The filter $H_2^z(z)$ includes $H^z(z)$ as a factor, because $H^z(z)$ is minimum phase, as well as the conjugate inverses of $\beta_{7,8,9}$, because these zeros are outside the unit circle. Therefore,

$$H_2(z) = K \cdot H^z(z)(1 - 0.5z^{-1})(1 + 0.25z^{-2})$$
$$= K(1 - z^{-1} + 0.75z^{-2} + 0.5z^{-3} - 0.8125z^{-4} + 0.6875z^{-5}$$
$$- 0.4844z^{-6} + 0.1875z^{-7} - 0.0625z^{-8} + 0.0156z^{-9}).$$

The factor K should be chosen so as to make the DC gains of the two filters equal. This is equivalent to saying that the sum of the coefficients must be equal. We get from this condition that $K = -8$.   □

9.2   FIR Filter Design by Impulse Response Truncation

9.2.1   Definition of the IRT Method

Ideal frequency responses, such as the ones shown in Figure 8.1, have infinite impulse responses. However, by Parseval's theorem, the impulse response $h[n]$ has finite energy. Truncating the impulse response of the ideal filter on both the right and the left thus yields a finite impulse response whose associated frequency response approximates that of the ideal filter. Furthermore, by shifting the truncated impulse response to the right (i.e., by delaying it), we can make it causal. This is the basic idea of the impulse response truncation (IRT) design method.

The phase response of an ideal filter is usually either identically zero, in which case the impulse response is symmetric, or identically $\pi/2$, in which case the impulse response is antisymmetric. In either case, the larger (in magnitude) impulse response coefficients are the ones whose indices n are close to zero. It is therefore reasonable to truncate the impulse response symmetrically around $n = 0$, before shifting it for causality. The filter thus obtained will be symmetric (in the case of zero phase) or antisymmetric (in the case of phase $\pi/2$), so it will have linear phase. This, however, limits the filter's length to an odd number, hence its order to an even number, hence its type to I or III. An alternative approach, which frees us from the constraint of even order, is to incorporate a linear-phase factor into the ideal response, that is, a factor $e^{-j0.5\theta N}$, in which N reflects the desired order after truncation. Then, after the impulse response has been computed, it is truncated to the range $0 \le n \le N$. The filter thus obtained is causal, has linear phase, and approximates the ideal frequency response. Its order can be even or odd, depending on the choice of N in the linear-phase factor. Thus, the FIR filter can be of any of the four types.[1] In summary, the impulse response truncation method consists of the following steps:

───────────┤ **Impulse-response truncation FIR filter design procedure** ├───────────

1. Write the desired amplitude response $A_d(\theta)$, according to the filter class: low pass, high pass, etc.

2. Choose the filter's phase characteristic: integer or fractional group delay, initial phase 0 or $\pi/2$.

3. Choose the filter's order N. The ideal desired frequency response, including the phase, can now be written as

$$H_d^f(\theta) = A_d(\theta)e^{j(\mu 0.5\pi - 0.5\theta N)}, \qquad (9.25)$$

   where $\mu = 0$ for a symmetric filter, and $\mu = 1$ for an antisymmetric one.

4. Compute the impulse response of the ideal filter, using the inverse Fourier transform formula

$$h_d[n] = \frac{1}{2\pi} \int_{-\pi}^{\pi} A_d(\theta)e^{j(\mu 0.5\pi - 0.5\theta N)} e^{j\theta n} d\theta. \qquad (9.26)$$

5. Truncate the impulse response by taking

$$h[n] = \begin{cases} h_d[n], & 0 \le n \le N, \\ 0, & \text{otherwise.} \end{cases} \qquad (9.27)$$

In most practical cases, the form of $A_d(\theta)$ is simple, so the integral can be computed in a closed form. This is illustrated in the following sections, which discuss the design of the most common kinds of FIR filters.

9.2.2 Low-Pass, High-Pass, and Band-Pass Filters

Let the desired amplitude response be

$$A_d(\theta) = \begin{cases} 1, & \theta_1 \le |\theta| \le \theta_2, \\ 0, & \text{otherwise.} \end{cases} \tag{9.28}$$

This is the ideal amplitude response of a band-pass filter. A low-pass filter is obtained as a special case, by taking $\theta_1 = 0$. Similarly, a high-pass filter is obtained as a special case, by taking $\theta_2 = \pi$. For these three kinds of filter it is reasonable to require constant phase delay, rather than constant group delay. This amounts to choosing $\mu = 0$, that is, filters of type I or II. Therefore, the ideal frequency response, including the linear-phase term, is

$$H_d^f(\theta) = \begin{cases} e^{-j0.5\theta N}, & \theta_1 \le |\theta| \le \theta_2, \\ 0, & \text{otherwise.} \end{cases} \tag{9.29}$$

The desired impulse response is computed as

$$
\begin{aligned}
h_d[n] &= \frac{1}{2\pi} \int_{-\theta_2}^{-\theta_1} e^{j\theta(n-0.5N)} d\theta + \frac{1}{2\pi} \int_{\theta_1}^{\theta_2} e^{j\theta(n-0.5N)} d\theta \\
&= \frac{\theta_2}{\pi} \operatorname{sinc}\left[\frac{\theta_2(n-0.5N)}{\pi}\right] - \frac{\theta_1}{\pi} \operatorname{sinc}\left[\frac{\theta_1(n-0.5N)}{\pi}\right].
\end{aligned}
\tag{9.30}$$

In the special case of a low-pass filter, we get

$$h_d[n] = \frac{\theta_2}{\pi} \operatorname{sinc}\left[\frac{\theta_2(n-0.5N)}{\pi}\right]. \tag{9.31}$$

In the special case of a high-pass filter, we get

$$h_d[n] = \delta[n-0.5N] - \frac{\theta_1}{\pi} \operatorname{sinc}\left[\frac{\theta_1(n-0.5N)}{\pi}\right], \tag{9.32}$$

provided N is even. Recall that odd N (i.e., type II) is not suitable for high-pass filters.

Example 9.6 Design a band-pass filter with band-edge frequencies 0.2π and 0.6π, and order $N = 40$. Figure 9.6 shows the impulse and amplitude responses of the filter. The oscillatory nature of the amplitude response, which results from the impulse response truncation, is discussed in detail later.   □

9.2.3 Multiband Filters

The ideal amplitude response of a multiband filter is a superposition of band-pass amplitude responses. Suppose that the desired frequency response has K bands, where the band-edge frequencies of the kth band are $\theta_{1,k}$, $\theta_{2,k}$ and its gain is C_k. Then,

$$A_d(\theta) = \sum_{k=1}^{K} A_{d,k}(\theta), \tag{9.33}$$

where

$$A_{d,k}(\theta) = \begin{cases} C_k, & \theta_{1,k} \le |\theta| \le \theta_{2,k}, \\ 0, & \text{otherwise.} \end{cases} \tag{9.34}$$

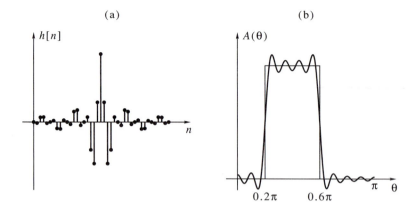

Figure 9.6 Band-pass filter designed by the IRT method, of order $N = 40$: (a) impulse response; (b) amplitude response.

Using superposition, we get from (9.30)

$$h_d[n] = \sum_{k=1}^{K} \frac{C_k}{\pi} \left\{ \theta_{2,k} \text{sinc} \left[\frac{\theta_{2,k}(n - 0.5N)}{\pi} \right] - \theta_{1,k} \text{sinc} \left[\frac{\theta_{1,k}(n - 0.5N)}{\pi} \right] \right\}. \quad (9.35)$$

A band-stop filter can be designed as a special case of a multiband filter having two bands and parameters $C_1 = C_2 = 1$, $\theta_{1,1} = 0$, $\theta_{2,2} = \pi$. Recall that type-II filters (odd N) are not suitable for band-stop filters.

The procedure `firdes` in Program 9.1 implements the design of a general multiband FIR filter by impulse response truncation. The program accepts the desired filter order N and the filter specifications. The specifications are entered as a $K \times 3$ matrix. The kth row of the matrix describes the kth pass band. Its first two entries are the lower and upper band-edge frequencies, and its third entry is the gain of the band. The program accepts a third optional window, which should be ignored for now (we shall discuss this parameter in Section 9.3). The program can handle the three single-band filter types as special cases.

Example 9.7 Design a two-band filter with $N = 80$, $\theta_{1,1} = 0.2\pi$, $\theta_{2,1} = 0.4\pi$, $\theta_{1,2} = 0.7\pi$, $\theta_{2,2} = 0.8\pi$, $C_1 = 1$, $C_2 = 0.5$. Figure 9.7 shows the impulse response and the amplitude response of the resulting filter. As we see, the shape of the amplitude response approximates the desired low-pass characteristic, but the undesirable oscillations are again apparent.   □

9.2.4 Differentiators

The frequency response of an ideal analog differentiator is

$$H^F(\omega) = j\omega.$$

A digital differentiator can approximate an analog differentiator only over a limited range of frequencies, determined by the sampling rate. The ideal frequency response of a digital differentiator is therefore

$$H^f(\theta) = j\frac{\theta}{T}, \quad -\pi \le \theta \le \pi.$$

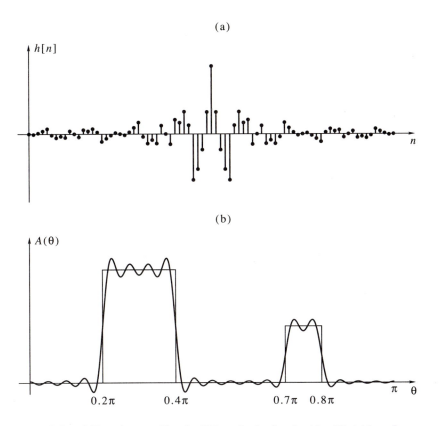

Figure 9.7 Multiband filter designed by the IRT method, of order $N = 80$: (a) impulse response; (b) frequency response.

The phase response of an ideal differentiator is 0.5π at all frequencies. Therefore, either a type-III or a type-IV filter are the natural choices in this case. We include a linear-phase term, so the desired response is

$$H_d^f(\theta) = \frac{\theta}{T} e^{j(0.5\pi - 0.5\theta N)}. \tag{9.36}$$

The impulse response is therefore

$$h_d[n] = \frac{j}{2\pi T} \int_{-\pi}^{\pi} \theta e^{j\theta(n-0.5N)} d\theta = \begin{cases} \dfrac{(-1)^{(n-0.5N)}}{(n-0.5N)T}, & N \text{ even}, \ n \neq 0.5N, \\ 0, & N \text{ even}, \ n = 0.5N, \\ \dfrac{(-1)^{(n-0.5N+0.5)}}{\pi(n-0.5N)^2 T} & N \text{ odd}. \end{cases} \tag{9.37}$$

We observe that the magnitude of $h_d[n]$ decays in proportion to $|n - 0.5N|^{-1}$ for even N, but in proportion to $|n - 0.5N|^{-2}$ for odd N. This has a significant influence on the oscillations of the frequency response, as is illustrated next.

Example 9.8 Design a differentiator by the IRT method, first of order $N = 16$, and then of order $N = 15$, with $T = 1$ in both cases. The coefficients are obtained by straightforward substitution in (9.37) and are shown in Figure 9.8. The amplitude responses are shown in Figure 9.9. As we see, the coefficients decay much more rapidly at the ends in the case of odd N than in the case of even N. Correspondingly, the amplitude response in the former case is much smoother, although the difference in

the orders is only 1. The conclusion is that an odd-order (i.e., type-IV) filter is preferred
to an even-order filter for FIR differentiators.   □

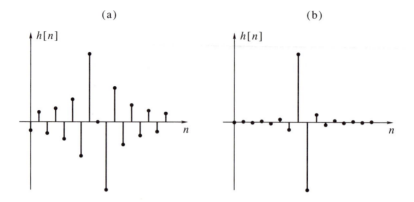

Figure 9.8 Impulse response of a differentiator designed by the IRT method: (a) $N = 16$; (b)
$N = 15$.

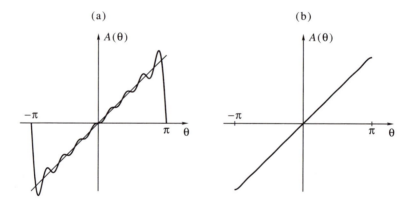

Figure 9.9 Amplitude response of a differentiator designed by the IRT method: (a) $N = 16$; (b)
$N = 15$.

9.2.5 Hilbert Transformers

The frequency response of an ideal analog Hilbert transformer is

$$H^{\mathrm{F}}(\omega) = \begin{cases} -j, & \omega > 0, \\ 0, & \omega = 0, \\ j, & \omega < 0. \end{cases}$$

We provide a brief description of Hilbert transformers here, to encourage interest in
them; however, we shall not explore their applications.[2]

Suppose that $x(t)$ is a real-valued signal; then its Fourier transform satisfies the
conjugate symmetry property $X^{\mathrm{F}}(\omega) = \bar{X}^{\mathrm{F}}(-\omega)$. Let $y(t)$ be the signal obtained by
passing $x(t)$ through an ideal Hilbert transformer. Then the Fourier transform of $y(t)$

is given by

$$Y^F(\omega) = H^F(\omega)X^F(\omega) = \begin{cases} -jX^F(\omega), & \omega > 0, \\ 0, & \omega = 0, \\ jX^F(\omega), & \omega < 0. \end{cases}$$

Let $z(t)$ be the complex signal

$$z(t) = x(t) + jy(t). \tag{9.38}$$

Then the Fourier transform of $z(t)$ is

$$Z^F(\omega) = [1 + jH^F(\omega)]X^F(\omega) = \begin{cases} 2X^F(\omega), & \omega > 0, \\ X^F(\omega), & \omega = 0, \\ 0, & \omega < 0. \end{cases}$$

The signal $z(t)$ is called the *analytic signal* of $x(t)$. Its frequency response is equal to that of $x(t)$ in the positive frequency range (except for a factor 2), and is zero in the negative frequency range. Thus, the spectrum of the analytic signal occupies only half the bandwidth and is free of the redundancy that exists in the spectrum of the original signal. It follows that the analytic signal can be sampled at half the sampling rate of the original signal. The analytic signal is complex: Its real part is equal to the original signal, and its imaginary part is obtained by passing the original signal through a Hilbert transformer.

As in the case of a differentiator, a digital Hilbert transformer can approximate an analog Hilbert transformer only over a limited range of frequencies, determined by the sampling rate. The ideal frequency response of a digital Hilbert transformer, including a linear-phase term, is

$$H_d^f(\theta) = \begin{cases} -e^{j(0.5\pi - 0.5\theta N)}, & \theta > 0, \\ 0, & \theta = 0, \\ e^{j(0.5\pi - 0.5\theta N)}, & \theta < 0. \end{cases} \tag{9.39}$$

As we see, $\phi_0 = 0.5\pi$ for a Hilbert transformer, so either type-III or type-IV filter is appropriate. The impulse response corresponding to (9.39) is given by

$$h_d[n] = \begin{cases} \dfrac{1 - \cos[(n - 0.5N)\pi]}{\pi(n - 0.5N)}, & n \neq 0.5N, \\ 0, & n = 0.5N. \end{cases} \tag{9.40}$$

The procedure `diffhilb` in Program 9.2 implements the design of an FIR differentiator or an FIR Hilbert transformer by impulse response truncation. The program accepts the filter type (differentiator or Hilbert) and the desired filter order N. The program accepts a third optional window, which should be ignored for now (we shall discuss this parameter in Section 9.3).

Example 9.9 Design a Hilbert transformer by the IRT method, first of order $N = 16$, and then of order $N = 15$. The coefficients are obtained by substitution in (9.40) and shown in Figure 9.10. The amplitude responses are shown in Figure 9.11. As we see, the amplitude responses are similar near $\theta = 0$, but the response of the odd-order (type-IV) filter is better behaved near $\theta = \pm\pi$. □

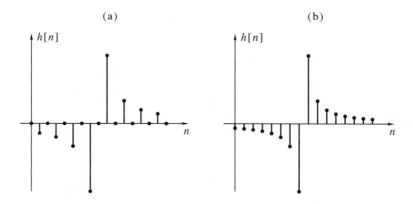

Figure 9.10 Impulse response of a Hilbert transformer designed by the IRT method: (a) $N = 16$; (b) $N = 15$.

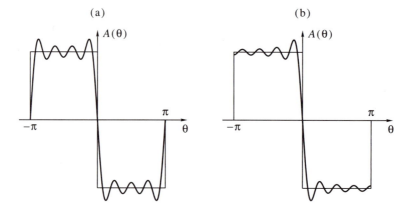

Figure 9.11 Amplitude response of a Hilbert transformer designed by the IRT method: (a) $N = 16$; (b) $N = 15$.

9.2.6 Optimality of the IRT Method

A measure of the error between the frequency response of the ideal filter $H_d^f(\theta)$ and that of the designed filter $H^f(\theta)$ is the integral of the square error, defined by

$$\varepsilon^2 = \frac{1}{2\pi} \int_{-\pi}^{\pi} |H_d^f(\theta) - H^f(\theta)|^2 d\theta. \tag{9.41}$$

Filters designed by the impulse response truncation method are optimal in the sense of minimizing the integral of the square error, as we now prove.

Theorem 9.1 For a given ideal frequency response $H_d^f(\theta)$ and a given order N, the filter obtained by impulse response truncation has a minimum integral of square error among all causal FIR filters of the given order.

Proof Let $\{h[n],\ 0 \le n \le N\}$ be the coefficients of any causal FIR filter of order N. By Parseval's theorem, ε^2 can be expressed as

$$\varepsilon^2 = \sum_{n=-\infty}^{\infty} (h_d[n] - h[n])^2 = \sum_{n=-\infty}^{-1} h_d^2[n] + \sum_{n=N+1}^{\infty} h_d^2[n] + \sum_{n=0}^{N} (h_d[n] - h[n])^2. \tag{9.42}$$

The first two terms on the right side of (9.42) are determined only by $H_d(\theta)$ and are not affected by the choice of filter's parameters. The third term is nonnegative, so ε^2 is minimized if and only if this term is zero. But this happens if and only if

$$h[n] = h_d[n], \quad 0 \le n \le N,$$

which is precisely the IRT design criterion. □

Despite its optimality in the sense of integral-of-square-error, IRT is not considered a good method, because of the oscillatory nature of the frequency response near discontinuity points of the desired response $H_d^f(\theta)$. As N increases, the oscillations become more localized, therefore ε^2 decreases. However, the magnitude of the oscillations relative to the size of the discontinuity does not decrease to zero, but tends to a finite limit. This property, known as the *Gibbs phenomenon*, is explained next.

9.2.7 The Gibbs Phenomenon

Consider the amplitude response of an IRT low-pass filter having band-edge frequency 0.5π, for different orders N. Figure 9.12 illustrates the responses for $N = 10$ and $N = 40$. The filter of the larger order has a narrower transition band, as expected. However, the pass-band and stop-band tolerance parameters, δ_p and δ_s, are approximately the same for both, and their value is about 0.09. It turns out that the value 0.09 is characteristic of the IRT design method, and is almost independent of the filter's order and shape. This phenomenon is named in honor of J. W. Gibbs,[3] and is mathematically explained as follows.

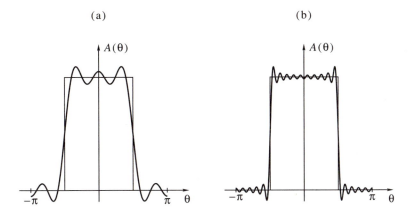

Figure 9.12 Amplitude response of a low-pass filter designed by the IRT method: (a) $N = 10$; (b) $N = 40$.

Let $A_d(\theta)$ be the amplitude response of an ideal low-pass filter with band-edge frequency θ_0. Denote for convenience $L = N + 1$ (the parameter L is the length of the sequence $h[n]$), and assume that

$$L \gg \frac{2\pi}{\theta_0} \quad \text{and} \quad L \gg \frac{2\pi}{\pi - \theta_0}. \tag{9.43}$$

The truncated impulse response is related to the ideal impulse response by

$$h[n] = h_d[n]w_r[n], \tag{9.44}$$

where $w_r[n]$ is a rectangular window of length L. Therefore, the amplitude response of the truncated filter is the convolution of $A_d(\theta)$ with the Dirichlet kernel $D(\theta, L)$, that is

$$A(\theta) = \frac{1}{2\pi} \int_{-\pi}^{\pi} A_d(\lambda) D(\theta - \lambda, L) d\lambda = \frac{1}{2\pi} \int_{-\theta_0}^{\theta_0} D(\theta - \lambda, L) d\lambda. \qquad (9.45)$$

We ignore the linear-phase terms in both $H_d^f(\theta)$ and $H^f(\theta)$, since they do not affect the behavior of the amplitude response.

Let us now concentrate on values of θ in the vicinity of θ_0. Specifically, we require that $(L|\theta - \theta_0|)/(2\pi)$ be a small number, say, not greater than 5. Assume first that $\theta < \theta_0$ and compute the integral in (9.45) as follows:

$$\frac{1}{2\pi} \int_{-\theta_0}^{\theta_0} D(\theta - \lambda, L) d\lambda = \frac{1}{2\pi} \int_{-\theta_0}^{\theta} D(\theta - \lambda, L) d\lambda + \frac{1}{2\pi} \int_{\theta}^{\theta_0} D(\theta - \lambda, L) d\lambda$$

$$= \frac{1}{2\pi} \int_{0}^{\theta + \theta_0} D(\mu, L) d\mu + \frac{1}{L\pi} \int_{0}^{0.5L(\theta_0 - \theta)} D\left(\frac{2\mu}{L}, L\right) d\mu, \qquad (9.46)$$

where we substituted $\mu = \theta - \lambda$ in the first integral, and $\mu = 0.5(\lambda - \theta)L$ in the second (we also used the symmetry of the Dirichlet kernel). Since θ is close to θ_0, and by assumption (9.43), the first integral is approximately equal to the integral of the Dirichlet kernel from 0 to π, the value of which is π. In the second integral, the argument of the Dirichlet kernel is a small number, so

$$D\left(\frac{2\mu}{L}, L\right) = \frac{\sin \mu}{\sin(\mu/L)} \approx \frac{L \sin \mu}{\mu}. \qquad (9.47)$$

Substitution of these approximations in (9.46) gives

$$A(\theta) \approx 0.5 + \frac{1}{\pi} \int_{0}^{0.5L(\theta_0 - \theta)} \frac{\sin \mu}{\mu} d\mu. \qquad (9.48)$$

The function

$$\text{Si}(x) = \int_{0}^{x} \frac{\sin \mu}{\mu} d\mu \qquad (9.49)$$

is called the *sine integral*. Using this function in (9.48), we get

$$A(\theta) \approx 0.5 + \frac{1}{\pi} \text{Si}[0.5(\theta_0 - \theta)L], \quad \theta < \theta_0. \qquad (9.50)$$

Using a similar derivation, we obtain

$$A(\theta) \approx 0.5 - \frac{1}{\pi} \text{Si}[0.5(\theta - \theta_0)L], \quad \theta > \theta_0. \qquad (9.51)$$

The sine integral is shown in Figure 9.13. The salient properties of this function are as follows:

1. For small values of x, $\text{Si}(x) \approx x$.

2. $\lim_{x \to \infty} \text{Si}(x) = 0.5\pi$.

3. The global maximum is at $x = \pi$, and its value is approximately $1.179 \times 0.5\pi$.

4. After the function has reached its global maximum, it oscillates around its limiting value in a decaying manner.

The properties of the sine integral explain the Gibbs phenomenon. As we see from (9.50) and (9.51), $A(\theta)$ takes the value 0.5 at $\theta = \theta_0$. At $\theta = \theta_0 - 2\pi/L$ it takes the value $0.5 + \pi^{-1}\text{Si}(\pi)$, which is approximately 1.0895. Similarly, at $\theta = \theta_0 + 2\pi/L$ it takes the value $0.5 - \pi^{-1}\text{Si}(\pi)$, which is approximately -0.0895. Figure 9.14 shows the details of the amplitude function $A(\theta)$ in the vicinity of θ_0. As we see, both the

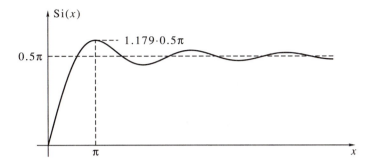

Figure 9.13 The sine integral function.

stop-band attenuation and the pass-band ripple are about 0.09, not depending on the filter's order. The width of the transition band is smaller than $4\pi/L$, so it decreases as the filter's order increases. The upper limit on the width of the transition band is the width of the main lobe of the Dirichlet kernel.

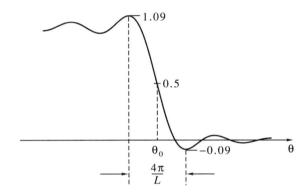

Figure 9.14 The Gibbs phenomenon.

The practical implication of the preceding discussion is that the IRT design method is suitable only for filters whose tolerances are not smaller than 0.09, or about 21 dB in the stop band and 0.75 dB in the pass band. Practical filters are almost always required to have smaller tolerances, so the IRT method is not suitable for their design. In the next section we shall see how the Gibbs phenomenon can be mitigated with the aid of windows.

9.3 FIR Filter Design Using Windows

In Chapter 6 we used windows for reducing the side-lobe interference resulting from truncating an infinite-duration signal to a finite length. We recall that windowing performs convolution in the frequency domain; hence it can be used to attenuate the Gibbs oscillations seen in the amplitude response of FIR filters. Windowing is applied to an impulse response of an FIR filter in the same way as to a finite-duration signal. Specifically, let $w[n]$ denote a window of length $L = N + 1$. The window design method of FIR filters is then as follows:

| **Windowed FIR filter design procedure** |

---

1. Define the ideal frequency response $H_d^f(\theta)$ as in the IRT method; see (9.25). For each pair $\{\theta_p, \theta_s\}$ (depending on the individual bands), take the corresponding band-edge frequency of the ideal response as the midpoint, that is, $0.5(\theta_p + \theta_s)$.

2. Obtain the ideal impulse response $h_d[n]$ as in the IRT method.

3. Compute the coefficients of the filter by

$$h[n] = \begin{cases} w[n]h_d[n], & 0 \le n \le N, \\ 0, & \text{otherwise.} \end{cases} \qquad (9.52)$$

Windows are always symmetric, that is, they satisfy

$$w[n] = w[N - n],$$

for either even or odd N. Therefore, the windowed impulse response $h[n]$ is either symmetric or antisymmetric, in agreement with $h_d[n]$. It follows that the FIR filter has linear phase, and the window does not affect its type.

The frequency response of the FIR filter is the convolution of that of the ideal filter with the kernel function of the window, that is,

$$H^f(\theta) = \frac{1}{2\pi}\{W^f * H_d^f\}(\theta) = \frac{1}{2\pi}\int_{-\pi}^{\pi} W^f(\lambda)H_d^f(\theta - \lambda)d\lambda. \qquad (9.53)$$

For a given ideal frequency response $H_d^f(\theta)$, the properties of the actual response— stop-band attenuation, pass-band ripple, and transition bandwidth—depend on the chosen window. As we learned in Section 6.2, a window is characterized by two primary parameters:

1. The width of the main lobe, equal to $4\pi\eta/L$, where η is a proportionality constant characteristic of the window. For example, recall from Section 6.2 that $\eta = 1$ for the rectangular window, $\eta = 2$ for the Bartlett, Hann, and Hamming windows, etc.

2. The magnitude of the highest side lobe relative to the main lobe.

The effect of the window on the amplitude response $A(\theta)$ can be analyzed similarly to the analysis of the Gibbs phenomenon. Let $A_w(\theta)$ be the amplitude function of the kernel $W^f(\theta)$ and define

$$\Gamma_w(x) = \frac{1}{L}\int_0^x A_w\left(\frac{2\mu}{L}\right)d\mu. \qquad (9.54)$$

Suppose that $A_d(\theta)$ is discontinuous at θ_0 such that $A_d(\theta_0^-) = 1$, $A_d(\theta_0^+) = 0$. Then we have, using the same derivation that has led to (9.50), (9.51),

$$A(\theta) \approx \begin{cases} 0.5 + \frac{1}{\pi}\Gamma_w[0.5(\theta_0 - \theta)L], & \theta < \theta_0, \\ 0.5 - \frac{1}{\pi}\Gamma_w[0.5(\theta - \theta_0)L], & \theta > \theta_0. \end{cases} \qquad (9.55)$$

Let us examine the shape of (9.55) for the windows studied in Section 6.2. Consider first the Bartlett window and recall that the amplitude response of its kernel is nonnegative [cf. (6.11)]. Therefore, the function $\Gamma_w(x)$ is monotone nondecreasing in this case. This implies that $A(\theta)$ is monotone in the vicinity of the discontinuity point, as illustrated in Figure 9.15. Since $A(\theta)$ is monotone, the transition bandwidth and the tolerance parameters are not well defined as separate entities, only jointly. Nevertheless, it is common to regard $8\pi/L$ as the transition bandwidth of the filter and, correspondingly, the values $\delta_s = \delta_p = 0.05$ as the tolerance parameters; see Figure 9.15. The respective dB values are

$$A_s \approx 26\,\text{dB}, \quad A_p \approx 0.45\,\text{dB}.$$

As we see, the Bartlett window offers only a small improvement over the rectangular window in terms of the tolerance parameters, whereas its transition bandwidth is twice as large. Therefore, the use of the Bartlett window for FIR filter design is rare. It can be recommended only if a monotone amplitude response is desired and if the tolerance specifications are not tight.

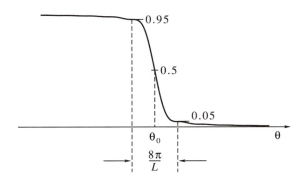

Figure 9.15 The amplitude response of an FIR filter based on the Bartlett window near a discontinuity point.

Let us now proceed to the Hann and Hamming windows, whose kernel functions are given by (6.13) and (6.17), respectively. The details of the amplitude functions of the corresponding FIR filters are shown in Figures 9.16 and 9.17. In these figures, as in the ones to follow, we omitted the transition band for better display of the pass-band and stop-band ripples. As we see, the transition bandwidth is slightly less than $8\pi/L$ for both windows. The tolerance parameters of the Hann window are $\delta_p = \delta_s = 0.0063$, and the respective dB values are

$$A_s \approx 44\,\text{dB}, \quad A_p \approx 0.055\,\text{dB}.$$

The tolerance parameters of the Hamming window are $\delta_p = \delta_s = 0.0022$, and the respective dB values are

$$A_s \approx 53\,\text{dB}, \quad A_p \approx 0.019\,\text{dB}.$$

The Hamming window has the advantage of smaller tolerances, whereas the Hann window has the advantage of faster decay of the ripple away from the discontinuity.

The details of the amplitude function of an FIR filter based on the Blackman window are shown in Figure 9.18. The kernel function of the Blackman window, given in (6.19), has a peculiar property: Its first side lobe is nonnegative, but subsequent side lobes have alternating signs. As a result, the transition bandwidth is $12\pi/L$, but the extremal points are at $\theta_0 \pm 8\pi/L$, as shown in Figure 9.18. The tolerance parameters of the Blackman window are $\delta_p = \delta_s = 0.0002$, and the respective dB values are

$$A_s \approx 74\,\text{dB}, \quad A_p \approx 0.0017\,\text{dB}.$$

Finally, let us examine filters based on the Kaiser window. The properties of a Kaiser window—main-lobe width and side-lobe level—can be controlled by the window

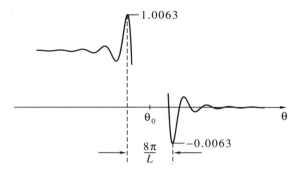

Figure 9.16 The amplitude response of an FIR filter based on the Hann window near a discontinuity point.

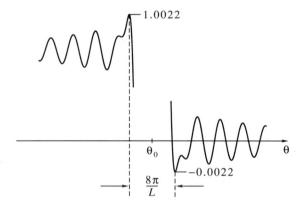

Figure 9.17 The amplitude response of an FIR filter based on the Hamming window near a discontinuity point.

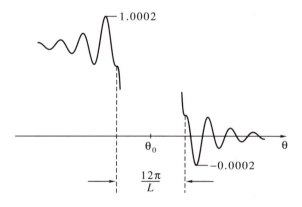

Figure 9.18 The amplitude response of an FIR filter based on the Blackman window near a discontinuity point.

parameter α. In the context of filter design, α is used for controlling the transition bandwidth and the tolerance parameters. Figure 9.19 shows the details of the amplitude response of a filter based on a Kaiser window with $\alpha = 10$. The transition bandwidth is about $13\pi/L$ in this case. The tolerance parameters are $\delta_p = \delta_s = 0.00001$, and the corresponding dB values are

$$A_s \approx 100 \, \text{dB}, \quad A_p \approx 0.000087 \, \text{dB}.$$

Table 9.2 summarizes the main properties of the windows we have described.

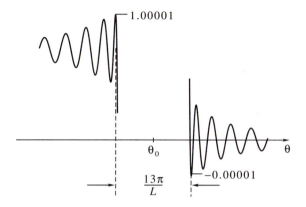

Figure 9.19 The amplitude response of an FIR filter based on the Kaiser window, with $\alpha = 10$, near a discontinuity point.

Window	eqn.	main-lobe width	side-lobe level, dB	δ_p, δ_s	pass-band ripple A_p [dB]	stop-band attn. A_s [dB]
rectangular	(6.2)	$4\pi/L$	-13.5	0.09	0.75	21
Bartlett	(6.10)	$8\pi/L$	-27	0.05	0.45	26
Hann	(6.14)	$8\pi/L$	-32	0.0063	0.055	44
Hamming	(6.16)	$8\pi/L$	-43	0.0022	0.019	53
Blackman	(6.18)	$12\pi/L$	-57	0.0002	0.0017	74
Kaiser	(6.21)	Depend on α				

Table 9.2 Windows and their properties.

For FIR filter design using the Kaiser window, there exist empirical formulas that give the parameter α and the order N as a function of the transition bandwidth $|\theta_p - \theta_s|$ and the tolerance parameters δ_p, δ_s, as follows:

$$A = -20 \log_{10}(\min\{\delta_p, \delta_s\}), \tag{9.56a}$$

$$\alpha = \begin{cases} 0.1102(A - 8.7), & A > 50, \\ 0.5842(A - 21)^{0.4} + 0.07886(A - 21), & 21 < A \le 50, \\ 0, & A \le 21, \end{cases} \tag{9.56b}$$

$$N = \frac{A - 7.95}{2.285|\theta_p - \theta_s|}. \tag{9.56c}$$

Note that if $A \le 21$ dB, a rectangular window is sufficient; therefore $\alpha = 0$ in this case. The order N should be rounded up to an even or odd integer, according to the desired filter type. Since the formulas are empirical, the filter is not guaranteed to meet the specifications. It is therefore necessary to check the response of the resulting filter and, if found unsatisfactory, to increase N or α (or both) and repeat the design.

The Kaiser window is better than the other windows we have mentioned, since for given tolerance parameters its transition band is always narrower. For this reason, and because of the convenience of controlling the filter tolerances via the parameter α, the Kaiser window has become the preferred choice in window-based filter design.

Program 9.1 can be used for window design of multiband FIR filters, by multiplying the IRT filter by a window. The window is entered as a third optional parameter, in which case it must have length equal to that of the filter. The same applies to Program 9.2 for differentiators and Hilbert transformers design.

The procedures firkais, kaispar, verspec in Programs 9.3, 9.4, and 9.5 implement Kaiser window FIR filter design according to given specifications. The program is limited to the six basic filter types: low pass, high pass, band pass, band stop, differentiator, and Hilbert transformer. The program accepts the filter type, the parity of the order (even or odd), the band-edge frequencies, and the tolerance parameters. It operates as follows:

1. Initial guesses for N and α are obtained by calling kaispar. This routine implements Kaiser's formulas (9.56). The requested parity is honored, except when the filter type is high pass or band stop, whereupon an even-order filter is forced regardless of the input parameter.

2. The filter is designed by calling firdes or diffhilb, according to the desired type.

3. The filter is tested against the specifications, by calling verspec. The test can result in three possible outcomes: 0 means that the filter meets the specifications; 1 means that N needs to be increased; 2 means that α needs to be increased. In the first case the program exits; in the second it increases N by 2 (to preserve the parity) and repeats the procedure; in the third it increases α by 0.05 and repeats the procedure.

4. The routine verspec computes the magnitude response on intervals near the edges of the bands, by calling frqresp. Each interval starts at a band-edge frequency, stretches over half the transition band in a direction away from the transition region, and contains 100 test points. If any band-edge frequency does not meet the specifications, the output is set to 1 to indicate a need for order increase. If there is a deviation from the specifications any interval, the output is set to 2 to indicate a need for increasing α. Otherwise the output is set to 0.

9.4   FIR Filter Design Examples

We now illustrate the window-based FIR design procedure by a few examples.

Example 9.10 Design a type-I low-pass filter according to the specifications:

$$\theta_p = 0.2\pi, \quad \theta_s = 0.3\pi, \quad \delta_p = \delta_s = 0.01.$$

The required stop-band attenuation is 40 dB, so the Hann window should be adequate. The band-edge frequency of the ideal response is the midpoint between θ_p and θ_s, that is, 0.25π. The transition bandwidth is $8\pi/(N+1) = 0.1\pi$, so the filter's order is chosen as $N = 80$. However, as we recall from Section 6.2, the Hann window has the property $w_{hn}[0] = w_{hn}[N] = 0$, so the actual order is only $N = 78$.

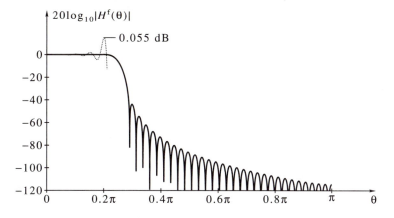

Figure 9.20 The magnitude response of the filter based on the Hann window in Example 9.10.

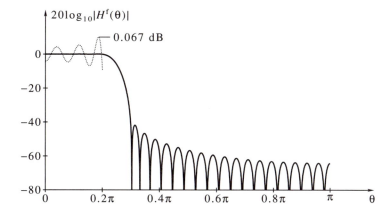

Figure 9.21 The magnitude response of the filter based on the Kaiser window in Example 9.10.

The magnitude response of the filter is shown in Figure 9.20. The figure also shows, as a dotted line, the details of the pass-band response at an expanded scale. For comparison, we design a filter based on a Kaiser window that meets the same specification. According to (9.56), the window parameter is $\alpha = 3.395$, and the order is $N = 46$. The magnitude response of the Kaiser window filter is shown in Figure 9.21. As we see, both filters meet the specifications, and the filter based on the Kaiser window does so at a considerably smaller order. However, the ripple decays more rapidly in the filter based on the Hann window.   □

Example 9.11 Design a type-II band-pass filter according to the specifications:

$$\theta_{s,1} = 0.1\pi, \quad \theta_{p,1} = 0.25\pi, \quad \theta_{p,2} = 0.6\pi, \quad \theta_{s,2} = 0.8\pi,$$

$$\delta_{s,1} = 0.005, \quad \delta_{s,2} = 0.0025, \quad \delta_p = 0.005.$$

The stop-band attenuation is determined by the smaller of the two tolerances, giving 52 dB. Therefore, a Hamming window is adequate here. The design transition bandwidth is the narrower of the two specified widths, so $8\pi/(N+1) = 0.15\pi$, which gives $N = 53$. Since the actual filter will have its two transition bands equal, we take the band-edge frequencies of the ideal response as 0.175π and 0.675π.

The magnitude response of the filter is shown in Figure 9.22. For comparison, we design a filter based on a Kaiser window that meets the same specification. According to (9.56), the window parameter is $\alpha = 4.772$ and the order is $N = 41$. The magnitude response of the Kaiser window filter is shown in Figure 9.23. As we see, both filters meet the specifications, and the filter based on the Kaiser window does so at a smaller order. In general, the magnitude responses of the two filters are quite similar.   □

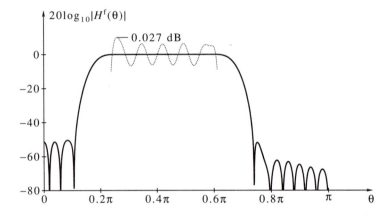

Figure 9.22 The magnitude response of the filter based on the Hamming window in Example 9.11.

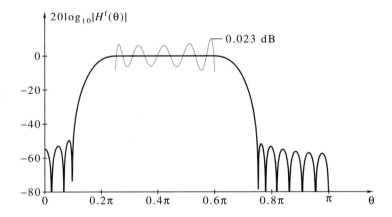

Figure 9.23 The magnitude response of the filter based on the Kaiser window in Example 9.11.

Example 9.12 Design a type-I high-pass filter according to the specifications:

$$\theta_s = 0.7\pi, \quad \theta_p = 0.8\pi, \quad \delta_s = 0.0002, \quad \delta_p = 0.001.$$

The required stop-band attenuation is 74 dB, so the Blackman window should be adequate, although marginal. The transition bandwidth is $12\pi/(N+1) = 0.1\pi$, so we choose the filter's order as $N = 120$. Like the Hann window, the Blackman window has the property $w_b[0] = w_b[N] = 0$, so the actual filter order is only $N = 118$. The band-edge frequency of the ideal response is 0.75π.

The magnitude response of the filter is shown in Figure 9.24. For comparison, we design a Kaiser window filter that meets the same specifications. According to (9.56), the window parameter is $\alpha = 7.196$ and the order is $N = 92$. The magnitude response of the Kaiser window filter is shown in Figure 9.25. As we see, both filters meet the specifications, and the filter based on the Kaiser window does so at a smaller order. However, the ripple decays more rapidly in the filter based on the Blackman window.□

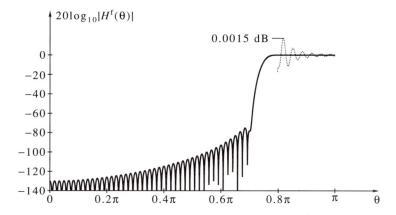

Figure 9.24 The magnitude response of the filter based on the Blackman window in Example 9.12.

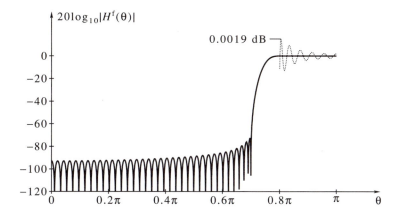

Figure 9.25 The magnitude response of the filter based on the Kaiser window in Example 9.12.

Example 9.13 Design a type-I band-stop filter according to the specifications:

$$\theta_{p,1} = 0.3\pi, \quad \theta_{s,1} = 0.4\pi \quad , \theta_{s,2} = 0.6\pi, \quad \theta_{p,2} = 0.7\pi,$$

$$\delta_s = 0.00001, \quad \delta_{p,1} = \delta_{p,2} = 0.0002.$$

The stop-band attenuation is 100 dB, so only a Kaiser window is adequate in this case. According to (9.56), the window parameter is $\alpha = 10.061$, and the order is $N = 130$. The magnitude response of the filter is shown in Figure 9.26.   □

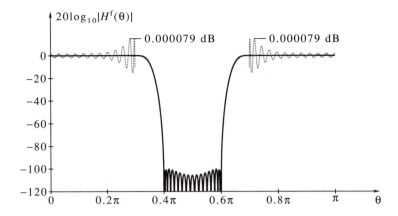

Figure 9.26 The magnitude response of the filter based on the Kaiser window in Example 9.13.

Example 9.14 A *digital raised-cosine* filter has an ideal frequency response like the one shown in Example 3.9. In particular, if $f_0 T = 0.5$ (where T is the sampling interval), then

$$H^f(\theta) = \begin{cases} 2, & |\theta| \le \theta_1, \\ 1 + \cos\left(\dfrac{|\theta| - \theta_1}{\alpha}\right), & \theta_1 < |\theta| \le \theta_2, \\ 0, & |\theta| > \theta_2, \end{cases} \qquad (9.57)$$

where $0 < \alpha < 1$, and

$$\theta_1 = 0.5(1 - \alpha)\pi, \quad \theta_2 = 0.5(1 + \alpha)\pi. \qquad (9.58)$$

In this special case, the filter is zero-phase, half-band, as defined in Problem 8.11, and its frequency response is as shown in Figure 8.11. Its noncausal impulse response is obtained from (3.22) as

$$h_d[n] = \begin{cases} \dfrac{\sin(0.5\pi n)\cos(0.5\pi\alpha n)}{0.5\pi n[1 - (\alpha n)^2]}, & n \ne 0, \\ 1, & n = 0. \end{cases} \qquad (9.59)$$

As we see, the even-index coefficients are zero except $h_d[0]$, as necessary for half-band filters [cf. (8.74)].

A causal FIR filter can be obtained from $h_d[n]$ by shifting it $2M$ points to the right, truncating to $0 \le n \le 4M$, and applying a window. The resulting filter will have a frequency response different from (9.57), but it will still be a half-band filter, because (8.74) continues to hold when multiplied by a window. The FIR filter will have zero coefficients in positions 0 and $4M$. Therefore, we can reduce the order to $4M - 2$ if we define

$$h[n] = \begin{cases} h_d[n - 2M + 1]w[n], & 0 \le n \le 4M - 2, \\ 0, & \text{otherwise,} \end{cases} \qquad (9.60)$$

where $w[n]$ is any window of length $4M - 1$.

Consider a specific example. Let $M = 10$, so the filter's order is $N = 38$. Let the bandwidth excess be $\alpha = 0.4$. Figure 9.27 shows the magnitude response of a rectangularly windowed FIR filter. This result is perhaps surprising: Instead of the familiar 21 dB stop-band attenuation, which we would expect from a rectangular window, we get attenuation of nearly 60 dB. However, the value 21 dB results from points

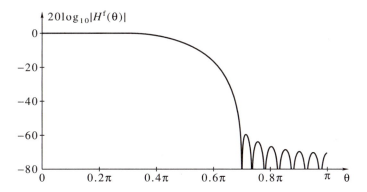

Figure 9.27 The magnitude response of a digital, raised-cosine, half-band filter; $N = 38$, rectangular window.

of discontinuity of the ideal frequency response. The raised-cosine response (9.57) is continuous and has a continuous first derivative at all points; only its second derivative is discontinuous at $\theta = 0.5(1 \pm \alpha)\pi$. The Gibbs phenomenon occurs whenever the frequency response or one of its derivatives is discontinuous, but such that the higher the order of the discontinuous derivative, the less pronounced the ripple around the discontinuity point. Another way to see this is from the rate of decay of the impulse response coefficients as a function of n. For a discontinuous function, the rate is $|n|^{-1}$, for discontinuous first derivative the rate is $|n|^{-2}$, and for discontinuous second derivative it is $|n|^{-3}$. High decay rate acts, in a sense, like a window to mitigate the Gibbs effect. As we see from (9.59), the rate of decay is $|n|^{-3}$ for the raised-cosine function.

In the range $0 \le \theta \le 0.7\pi$, the magnitude response of the FIR filter is very close to that of the ideal filter; it is plotted in Figure 9.27, but is not visible, because the two graphs coincide.

The conclusion from the preceding discussion is that a raised-cosine, half-band filter can be designed with a rectangular window if the required stop-band attenuation is 60 dB or less. If the attenuation is higher, another window should be employed. A window would not affect the half-band property, but would increase the transition band. It is recommended, in this case, to choose a Kaiser window and try different values of α, until the specifications are met. $\square$

9.5 Least-Squares Design of FIR Filters*

FIR filter design based on windows is simple and robust, yielding filters with good performance. However, in two respects it is not optimal:

1. The resulting pass-band and stop-band parameters δ_p, δ_s are equal (or almost so), even if we do not require them to be equal a priori. Often, the specification is more stringent in the stop band than in the pass band (i.e., $\delta_s < \delta_p$), so we obtain an unnecessarily high accuracy in the pass band.

2. The ripple of windows is not uniform in either the pass band or the stop band, but decays as we move away from the discontinuity points, according to the side-lobe pattern of the window (except for the Dolph window, which is rarely used for filter design). By allowing more freedom in the ripple behavior, we may be able to reduce the filter's order, thereby reducing its complexity.

In this section we present a design method that approximates the desired amplitude response $A_d(\theta)$ by a linear-phase FIR amplitude response function $A(\theta)$ according to an optimality criterion. The optimality criterion is a modified version of the frequency-domain integral of square error (9.41): To each frequency θ we assign a *weight* $V(\theta)$. The weight is a nonnegative number, reflecting the relative importance of the error between $A_d(\theta)$ and $A(\theta)$ at the particular frequency θ. For example, if we wish to design a low-pass filter, we can assign a constant weight δ_p^{-1} to all frequencies in the pass band, a different constant weight δ_s^{-1} to all frequencies in the stop band, and zero weight to all frequencies in the transition band. Then, the smaller the tolerance, the larger the weight given to the error in the corresponding band.

After choosing a weight function $V(\theta)$, we define the weighted frequency-domain error as

$$E(\theta) = V(\theta)[A_d(\theta) - A(\theta)], \tag{9.61}$$

and the integral of weighted square frequency-domain error as

$$\varepsilon^2 = \int_0^\pi E^2(\theta) d\theta \tag{9.62}$$

(because of symmetry, it is sufficient to integrate only over the positive frequencies). We assume that order N of the filter to be designed and its type are known. Under these assumptions, we can express $A(\theta)$ as in (9.23):

$$A(\theta) = F(\theta)G(\theta), \quad G(\theta) = \sum_{k=0}^K g[k]\cos(\theta k), \tag{9.63}$$

where K and $F(\theta)$ depend on the filter type, as given in Table 9.1. Substitution of (9.63) in (9.61) gives

$$E(\theta) = V(\theta)\left[A_d(\theta) - F(\theta)\sum_{k=0}^K g[k]\cos(\theta k)\right]. \tag{9.64}$$

Designing the FIR filter now reduces to determining the coefficient values $g[k]$ that would minimize ε^2 in (9.62).

To find the optimal coefficients, we differentiate ε^2 with respect to all $g[k]$ and equate the derivatives to zero. In general, this gives a local extremum point. However, only one extremum is possible in this case: the global minimum of the function.[4] Differentiation of (9.62) thus yields

$$\frac{\partial \varepsilon^2}{\partial g[m]} = \int_0^\pi 2E(\theta)\frac{\partial E(\theta)}{\partial g[m]}d\theta, \tag{9.65}$$

where

$$\frac{\partial E(\theta)}{\partial g[m]} = -V(\theta)F(\theta)\cos(\theta m). \tag{9.66}$$

Substitution of (9.64) and (9.66) in (9.65) gives

$$-2\int_0^\pi V^2(\theta)F(\theta)\cos(\theta m)\left[A_d(\theta) - F(\theta)\sum_{k=0}^K g[k]\cos(\theta k)\right]d\theta = 0, \quad 0 \le m \le K. \tag{9.67}$$

We therefore get a set of $K + 1$ equations in $K + 1$ unknowns

$$\sum_{k=0}^K c_{m,k} g[k] = d_m, \quad 0 \le m \le K, \tag{9.68}$$

where

$$c_{m,k} = \int_0^\pi V^2(\theta)F^2(\theta)\cos(\theta m)\cos(\theta k)d\theta, \tag{9.69a}$$

$$d_m = \int_0^\pi V^2(\theta)F(0)A_{\mathrm{d}}(\theta)\cos(\theta m)d\theta. \qquad (9.69b)$$

Solution of the linear equations gives the desired coefficients $g[k]$.

The least-squares design procedure for FIR filters is then as follows:

Least-squares FIR filter design procedure

1. Choose the filter order N and its type; these determine K and $F(\theta)$.
2. Choose the weight function $V(\theta)$. As we have said, a common choice for $V(\theta)$ is

$$V(\theta) = \begin{cases} (\delta_{\mathrm{p},l})^{-1}, & \theta \text{ in the } l\text{th pass band,} \\ (\delta_{\mathrm{s},l})^{-1}, & \theta \text{ in the } l\text{th stop band,} \\ 0, & \theta \text{ in any of the transition bands.} \end{cases} \qquad (9.70)$$

3. Compute the coefficients $c_{m,k}$ and d_m as given by (9.69). Since $A_{\mathrm{d}}(\theta)$ and $V(\theta)$ are typically piecewise constant, often the integrals can be computed in a closed form. However, it is more common to approximate these integrals by sums. The frequency range $[0, \pi]$ is sampled uniformly, at a number of points I proportional to the filter's order (say $I = 16N$ points). Then $c_{m,k}$ and d_m are approximated by

$$c_{m,k} \approx \frac{\pi}{I} \sum_{i=0}^{I-1} V^2(\theta_i)F^2(\theta_i)\cos(\theta_i m)\cos(\theta_i k), \qquad (9.71a)$$

$$d_m \approx \frac{\pi}{I} \sum_{i=0}^{I-1} V^2(\theta_i)F(\theta_i)A_{\mathrm{d}}(\theta_i)\cos(\theta_i m). \qquad (9.71b)$$

4. Solve the set of equations (9.68).
5. Compute the coefficients $h[n]$ of the filter, using one of (9.5), (9.9), (9.15), (9.20), according to the filter type.

The procedures `firls` and `firlsaux` in Programs 9.6 and 9.7 implement least-squares design of the four basic frequency responses in MATLAB. The program `firls` accepts the filter's order N, the desired frequency response characteristic (LP, HP, BP, BS), the band-edge frequencies, and the ripple tolerances. It uses piecewise-constant weighting and samples the frequency range $[0, \pi]$ at $16N$ points. It then implements the procedure in a straightforward manner. The program `firlsaux` is an auxiliary program used by `firls`.

Least-squares design often leads to filters of orders lower than those of filters based on windows, especially when there is a large difference between the tolerances δ_{p} and δ_{s}. This design method is highly flexible. The amplitude function $A(\theta)$ can have an almost arbitrary shape, and is not required to be expressed by a mathematical formula: We need only its numerical values on a sufficiently dense grid. There is much flexibility in choosing the weighting function, allowing much freedom in shaping the frequency response. The computational requirements are modest, and programming the method is straightforward. The main drawbacks of the least-squares method are as follows:

1. Meeting the specifications is not guaranteed a priori, and trial and error is often required. To help the design procedure meet the specifications at the desired band-edge frequencies, it is often advantageous to set the transition bands

slightly narrower than needed (e.g., increasing θ_p and decreasing θ_s for a low-pass filter). Also, it is often necessary to experiment with the weights until satisfactory results are achieved.

2. Occasionally, the resulting frequency response may be peculiar. For example, the transition-band response may be nonmonotonic, or the ripple may be irregular. In such cases, changing the weighting function usually solves the problem.

Example 9.15 Recall Example 9.10, but assume that the pass-band tolerance is $\delta_p = 0.1$ instead of 0.01. The Hann and Kaiser designs cannot benefit from this re-laxation, so they remain as designed in Example 9.10. A least-squares design gives a filter of order $N = 33$. Figure 9.28 shows the magnitude response of the filter; here it was obtained by artificially decreasing the transition band to $[0.21\pi, 0.29\pi]$.   □

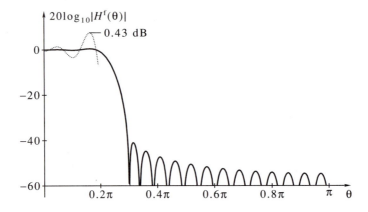

Figure 9.28 The magnitude response of the filter in Example 9.15.

9.6   Equiripple Design of FIR Filters*

The least-squares criterion, presented in the preceding section, often is not entirely satisfactory. A better criterion is the minimization of the maximum error at each band. This criterion leads to an *equiripple filter*, that is, a filter whose amplitude response oscillates uniformly between the tolerance bounds of each band. The design method we study in this section is optimal in the sense of minimizing the maximum magnitude of the ripple in all bands of interest, under the constraint that the filter order N be fixed. A computational procedure for solving this mathematical optimization problem was developed by Remez [1957] and is known as the *Remez exchange algorithm*. The algorithm in common use is by Parks and McClellan [1972a,b] and is known as the *Parks–McClellan algorithm*.

9.6.1   Mathematical Background

We begin as in Section 9.5, by defining the weighted frequency domain error (9.64). The desired amplitude response $A_d(\theta)$ and the weighting function $V(\theta)$ are assumed to be specified on a *compact* subset of $[0, \pi]$; that is, a set $\mathbb{S}$ that is a finite union of closed intervals. These intervals correspond to the pass bands and stop bands, and the complement of $\mathbb{S}$ in $[0, \pi]$ is the union of transition bands, at which the response is not specified. The weighting function fulfills the same role as in least-squares design and is chosen in a similar manner (i.e., proportional to the inverse of the tolerance at each band). We wish to find the coefficients $g[k]$ that minimize the maximum absolute weighted error $|E(\theta)|$ over the entire set $\mathbb{S}$.

Since $F(\theta)$ is a fixed nonnegative function, depending only on the filter type, we can rewrite (9.64) as

$$E(\theta) = \tilde{V}(\theta)[\tilde{A}_d(\theta) - G(\theta)], \quad \theta \in \mathbb{S}, \tag{9.72}$$

where

$$\tilde{V}(\theta) = V(\theta)F(\theta), \quad \tilde{A}_d(\theta) = \frac{A_d(\theta)}{F(\theta)}. \tag{9.73}$$

Then, the optimization problem to be solved is to minimize the nonnegative quantity ε defined by

$$\varepsilon = \max_{\theta \in \mathbb{S}} |E(\theta)| \tag{9.74}$$

over all choices of real coefficients $\{g[k], \ 0 \le k \le K\}$.

The solution of the optimization problem is based on the following theorem by Remez [1957], known as the *alternation theorem*.

Theorem 9.2 The function $G(\theta)$ is optimal in the minimax sense if and only if there exist at least $K + 2$ frequencies $\{\theta_i, \ 0 \le i \le K + 1\}$ in $\mathbb{S}$ (arranged in increasing order) such that

$$|E(\theta_i)| = \varepsilon, \quad 0 \le i \le K + 1, \tag{9.75a}$$

$$E(\theta_{i+1}) = -E(\theta_i), \quad 0 \le i \le K. \tag{9.75b}$$

$\square$

The theorem essentially says that the absolute error achieves its maximum $K + 2$ times or more, and does so at alternating signs (hence the name *alternation theorem*). In other words, the optimal approximation is equiripple. The reason for having at least $K + 2$ extrema can be intuitively explained as follows. The function $\cos(\theta k)$ is known to be kth-order polynomial in $\cos \theta$, called the kth-order Chebyshev polynomial,[5] and denoted by $T_k(\cos \theta)$. Therefore,

$$G(\theta) = \sum_{k=0}^{K} g[k]T_k(\cos \theta), \tag{9.76}$$

so $G(\theta)$ is a Kth-order polynomial in $\cos \theta$. Therefore, $G(\theta)$ has $K - 1$ extrema. If $\mathbb{S}$ consists of B disjoint bands (where B must be at least 2), there may be up to $2B$ additional extrema at the end points of the bands. The alternation theorem says that for the optimal solution, at least three end points must be extrema of the error function. The total number of extrema may be larger, up to a maximum of $K-1+2B$. For example, in the case of a low-pass filter $B = 2$, so there may be either $K + 2$ or $K + 3$ extrema.

9.6.2 The Remez Exchange Algorithm

The Remez exchange algorithm exploits the alternation theorem in the following manner. Assume, temporarily, that we *know* the frequencies $\{\theta_i, \ 0 \le i \le K + 1\}$ of the extremal points. Then the $K + 2$ parameters $\{g[k], \ 0 \le k \le K\}$ and ε can be obtained by solving the set of linear equations

$$E(\theta_i) = \tilde{V}(\theta_i)[\tilde{A}_d(\theta_i) - G(\theta_i)] = (-1)^i\varepsilon, \quad 0 \le i \le K + 1, \tag{9.77}$$

or, equivalently,

$$\sum_{k=0}^{K} g[k]\cos(\theta_i k) + \frac{(-1)^i \varepsilon}{\tilde{V}(\theta_i)} = \tilde{A}_d(\theta_i), \quad 0 \le i \le K+1. \tag{9.78}$$

The Remez exchange algorithm proceeds according to the following steps.

1. Choose the order N of the filter. The following empirical formula, proposed by Kaiser, can be used for estimating N:

$$N = \frac{-20\log_{10}\sqrt{\delta_p \delta_s} - 13}{2.32|\theta_p - \theta_s|}. \tag{9.79}$$

   The parameter K is derived from N according to filter type; see Table 9.1.

2. Choose initial values for $\{\theta_i, \ 0 \le i \le K+1\}$. This can be done, for example, by dividing the set $\mathbb{S}$ uniformly.

3. Solve the linear equations (9.78) and obtain $\{g[k], \ 0 \le k \le K\}$ and ε.

4. Find all the extremal points of $E(\theta)$, $\theta \in \mathbb{S}$. Both local extrema and end-point extrema must be examined. If there are more than $K+2$ such points, retain those corresponding to the $K+2$ largest values of $|E(\theta)|$. Denote the points thus found by $\{\tilde{\theta}_i\}$.

5. If $|\theta_i - \tilde{\theta}_i| < \mu$, where μ is a small user-chosen number (e.g., 10^{-4}), go to step 6. If not, replace $\{\theta_i\}$ by $\{\tilde{\theta}_i\}$, and go back to step 3 (this is the *exchange step*).

6. Get the coefficients $g[k]$ for the last values of the frequencies, and compute the frequency response. If it meets the specifications, go to step 7. If not, increase the order N (thereby increasing K) and go back to step 2.

7. Convert from the coefficients $g[k]$ to the coefficients $h[n]$, using one of the formulas (9.5), (9.9), (9.15), (9.20), according to the filter type.

Remarks

1. It turns out that ε in (9.78) can be computed explicitly, yielding

$$\varepsilon = \frac{\sum_{i=0}^{K+1} \eta_i \tilde{A}_d(\theta_i)}{\sum_{i=0}^{K+1} \frac{(-1)^i \eta_i}{\tilde{V}(\theta_i)}}, \tag{9.80}$$

   where

$$\eta_i = \prod_{\substack{k=0 \\ k \ne i}}^{K+1} [\cos\theta_i - \cos\theta_k]^{-1}. \tag{9.81}$$

2. Step 3 is performed by computing $E(\theta)$ on a dense grid in $\mathbb{S}$. The number of grid points is chosen by the user; a common choice is $16N$.

3. To compute $E(\theta)$ on the grid, it is necessary to evaluate $G(\theta)$ for each grid point. This can be done without solving the system of equations (9.78). Instead, we can use the Lagrange interpolation formula

$$G(\theta) = \frac{\sum_{i=0}^{K+1} G(\theta_i)\beta_i[\cos\theta - \cos\theta_i]^{-1}}{\sum_{i=0}^{K+1} \beta_i[\cos\theta - \cos\theta_i]^{-1}}, \tag{9.82}$$

   where

$$\beta_i = \prod_{\substack{k=0 \\ k \ne i}}^{K+1} [\cos\theta_i - \cos\theta_k]^{-1}. \tag{9.83}$$

The values $G(\theta_i)$ are known, since by (9.77)

$$G(\theta_i) = \tilde{A}_d(\theta_i) - \frac{(-1)^i \varepsilon}{\tilde{V}(\theta_i)}. \tag{9.84}$$

Only after the algorithm stops, we need to solve (9.78) once and obtain the final values of $g[k]$.

We do not include a program for equiripple FIR filter design in this book, due to the considerable complexity of such a program. If you have access to the MATLAB Signal Processing Toolbox, you can use the program remez for this purpose. The design examples given in Section 9.6.3 were prepared using this program.

9.6.3   Equiripple FIR Design Examples

Example 9.16 Consider the low-pass filter whose specifications were given in Example 8.1. An equiripple FIR filter that meets these specifications has order $N = 146$. Figure 9.29 shows the magnitude response of the filter. Figure 9.30(a) shows an input signal to the filter, consisting of a sinusoid at frequency $f_0 = 500\,\text{Hz}$ (i.e., in the middle of the pass band) and white noise at SNR 10 dB. Figure 9.30(b) shows the corresponding output signal. We choose a segment of the output signal delayed by 73 samples with respect to the input signal, to compensate for the phase delay of the filter. As we see, the output signal is considerably cleaner than the input signal, as a result of the noise attenuation performed by the filter. However, the amplitude of the output signal is not constant. This effect is due to the residual noise: The noise at the filter's output is not white any more, but approximately band limited, and its bandwidth is the same as the pass band of the filter.   □

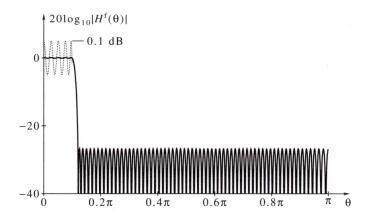

Figure 9.29 Magnitude response of the filter in Example 9.16.

Example 9.17 Consider the high-pass filter whose specifications were given in Example 8.2. An equiripple FIR filter that meets these specifications has order $N = 248$. Figure 9.31 shows the magnitude response of the filter.   □

Example 9.18 Consider the five filters whose specifications were given in Example 8.3. Here we design the band-pass filter for the second channel. An equiripple FIR filter

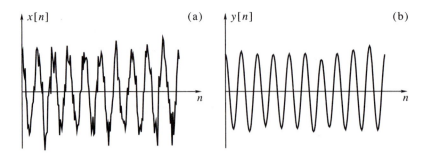

Figure 9.30 Input (a) and output (b) signals in Example 9.16.

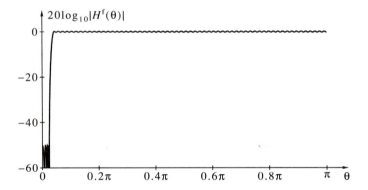

Figure 9.31 Magnitude response of the filter in Example 9.17.

that meets these specifications has order $N = 116$. Figure 9.32 shows the magnitude response of the filter.   □

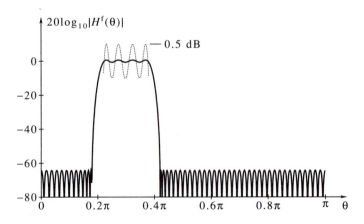

Figure 9.32 Magnitude response of the filter in Example 9.18.

Example 9.19 Consider the band-stop filter whose specifications were given in Example 8.4. An equiripple FIR filter that meets these specifications has order $N = 272$. Figure 9.33 shows the magnitude response of the filter. To illustrate the operation of this filter, we feed it with a simulated physiological signal $x[n]$, as shown in

Figure 9.34, part a. The response to this signal (delayed by 136 samples to account for the phase delay) is shown in part b. As we see, the response is almost identical to the input, since the filter eliminates only a negligible percentage of the energy. Part c shows the same signal as in part a, with an added 60 Hz sinusoidal interference, whose energy is equal to that of the signal. Part d shows the response to the signal in part c. As we see, the filter eliminates the sinusoidal interference almost completely, and the signal in part d is almost identical to those shown in parts a and b.   □

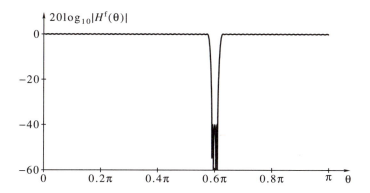

Figure 9.33 Magnitude response of the filter in Example 9.19.

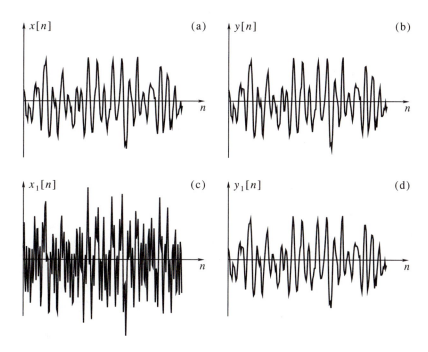

Figure 9.34 Signals in Example 9.19: (a) input signal; (b) output signal; (c) input signal with sinusoidal interference; (d) response to signal with sinusoidal interference.

9.7 Summary and Complements

9.7.1 Summary

This chapter was devoted to the properties and design of FIR filters. FIR filters are almost always designed so as to have linear phase (exact or generalized). There are four types of linear-phase filter: Types I and II have even symmetry of the impulse response coefficients, and even and odd orders, respectively; types III and IV have odd symmetry of the coefficients, and even and odd orders, respectively. A type-I filter is suitable for LP, HP, BP, and BS filters, as well as for multiband filters. Type II, on the other hand, is limited to LP and BP filters. Types III and IV are used mainly for differentiators and Hilbert transformers.

The zero locations of linear-phase FIR filters are not arbitrary, since for every zero at $z = \beta$, there must be zero at $z = \beta^{-1}$. This implies that linear-phase filters usually have zeros both inside and outside (as well as on) the unit circle.

The simplest design method for FIR filters is impulse response truncation. This method is based on computing the (infinite) impulse response of an ideal filter that has the desired frequency response by the inverse Fourier transform integral, and then truncating the impulse response to a finite length. It is applicable to LP, HP, BP, and BS filters, as well as to multiband filters, differentiators, and Hilbert transformers.

The IRT method is mathematically identical to truncating the Fourier series of a periodic function, hence it shares similar properties: On one hand, it is optimal in the sense of minimizing the integral of frequency-domain square error; on the other hand, it suffers from the Gibbs phenomenon. Windowing of the IRT filter attenuates the Gibbs oscillations, thereby reducing the pass-band ripple and increasing the stop-band attenuation. The tolerances of an FIR filter obtained by windowing depend on the window, but we always get $\delta_p \approx \delta_s$ in this design method. The width of the transition band(s) depends on the window and on the order of the filter. Of the various windows, the Kaiser window is most commonly used for filter design. The parameter α is determined by the specified ripple, whereas the order is determined by both the ripple and the specified transition bandwidth.

The least-squares design method is a convenient alternative to the windowing method when the tolerance parameters δ_p, δ_s differ widely, or when the desired amplitude response has a nonstandard shape (such as when it is only tabulated, not defined by a formula). Least-squares design requires trial and error in setting up the specification parameters and the weighting function.

Equiripple design is an optimal method, in the sense of providing the minimum-order filter that meets a given set of specifications. However, this design method requires sophisticated software tools, which are not easy to develop. Such software tools are available for standard filter types (e.g., in the Signal Processing Toolbox of MATLAB). Although the equiripple principle (in the form of the alternation theorem) applies to general amplitude responses and general weighting functions, this method is rarely used in its full generality, due to the absence of widely accessible software tools.

We summarize the subject by reiterating the main advantages and disadvantages of FIR filters:

1. Advantages:

   (a) Linear phase.

   (b) Inherent stability.

   (c) Flexibility in achieving almost any desired amplitude response.

     (d) Existence of convenient design techniques and sophisticated design tools.

     (e) Low sensitivity to finite word length effects (to be studied in Chapter 11).

2. Disadvantages:

     (a) High complexity of implementation, since large orders are needed to achieve tight tolerances and narrow transition bands.

     (b) Large delays, which may be undesirable in certain applications.

9.7.2 Complements

1. [p. 284] Many textbooks on digital signal processing teach the design of even-order filters first, assuming zero phase in the design process and then shifting the impulse response to make it causal. Then, special tricks are employed in designing odd-order filters. The method presented here avoids the need to learn such tricks.

2. [p. 288] See almost any book on communication systems for Hilbert transforms, analytic signals, and their applications. Also see the example in Section 14.4 of this book.

3. [p. 291] Josiah Willard Gibbs (1839–1903), a distinguished physicist, did not discover the phenomenon (it was the physicist Albert Abraham Michelson who did), but offered a mathematical explanation for it.

4. [p. 304] The solution of (9.67) is the unique global minimizer of ε^2 because the matrix of second partial derivatives of ε^2 with respect to the $g[k]$ is positive definite.

5. [p. 307] We shall discuss the Chebyshev polynomials in detail in Section 10.3, when we present Chebyshev filters.

9.8   MATLAB Programs

Program 9.1 Design of a multiband FIR filter.

```
function h = firdes(N,spec,win);
% Synopsis: h = firdes(N,spec,win).
% Design of a general multiband FIR filter by truncated
% impulse response, with optional windowing.
% Input parameters:
% N: the filter order (the number of coefficients is N+1)
% spec: a table of K rows and 3 columns, a row to a band:
%        spec(k,1) is the low cutoff frequency,
%        spec(k,2) is the high cutoff frequency,
%        spec(k,3) is the gain.
% win: an optional window of length N+1.
% Output:
% h: the impulse response coefficients.

flag = rem(N,2); [K,m] = size(spec);
n = (0:N) - N/2; if (~flag), n(N/2+1) = 1; end, h = zeros(1,N+1);
for k = 1:K,
   temp = (spec(k,3)/pi)*(sin(spec(k,2)*n) - sin(spec(k,1)*n))./n;
   if (~flag), temp(N/2+1) = spec(k,3)*(spec(k,2) - spec(k,1))/pi; end
   h = h + temp; end
if  nargin == 3, h = h.*reshape(win,1,N+1); end
```

Program 9.2 Design of FIR differentiators and Hilbert transformers.

```
function h = diffhilb(typ,N,win);
% Synopsis: h = diffhilb(typ,N,win).
% Design of an FIR differentiator or an FIR Hilbert transformer
% by truncated impulse response, with optional windowing.
% Input parameters:
% typ: 'd' for differentiator, 'b' for Hilbert
% N: the filter order (the number of coefficients is N+1)
% win: an optional window.
% Output:
% h: the filter coefficients.

flag = rem(N,2); n = (0:N)-(N/2); if (~flag), n(N/2+1) = 1; end
if (typ == 'd'),
   if (~flag), h = ((-1).^n)./n; h(N/2+1) = 0;
   else, h = ((-1).^(round(n+0.5)))./(pi*n.^2); end
elseif (typ == 'b'),
   h = (1-cos(pi*n))./(pi*n); if (~flag), h(N/2+1) = 0; end, end
if  nargin == 3, h = h.*win; end
```

Program 9.3 Kaiser window FIR filter design according to prescribed specifications.

```
function h = firkais(typ,par,theta,deltap,deltas);
% Synopsis: h = firkais(typ,par,theta,deltap,deltas).
% Designs an FIR filter of one of the six basic types by
% Kaiser window, to meet prescribed specifications.
% Input parameters:
% typ: the filter type:
%       'l','h','p','s' for LP, HP, BP, BS, respectively,
%       'b' for Hilbert transformer, 'd' for differentiator
% par: 'e' for even order (type I or III),
%       'o' for odd order (type II or IV)
% theta: vector of band-edge frequencies in increasing order.
% deltap: one or two pass-band tolerances
% deltas: one or two stop-band tolerances; not needed for
%         typ = 'b' or typ = 'd'
% Output:
% h: the filter coefficients.

if (nargin == 4), deltas = deltap; end
if (typ == 'p' | typ == 's'),
   if (length(deltap) == 1), deltap = deltap*[1,1]; end
   if (length(deltas) == 1), deltas = deltas*[1,1]; end
end
[N,alpha] = kaispar(typ,par,theta,deltap,deltas);
while (1),
   if (alpha == 0), win = window(N+1,'rect');
   else, win = window(N+1,'kais',alpha); end
   if (typ=='l'), h = firdes(N,[0,mean(theta),1],win);
   elseif (typ=='h'), h = firdes(N,[mean(theta),pi,1],win);
   elseif (typ=='p'),
   h = firdes(N,[mean(theta(1:2)),mean(theta(3:4)),1],win);
   elseif (typ=='s'),
   h = firdes(N, ...
       [0,mean(theta(1:2)),1; mean(theta(3:4)),pi,1],win);
   elseif (typ=='b' | typ=='d'),
   h = diffhilb(typ,N,win); end
   res = verspec(h,typ,theta,deltap,deltas);
   if (res==0), break;
   elseif (res==1), N = N+2;
   else, alpha = alpha+0.05; end
end
```

Program 9.4 Computation of N and α to meet the specifications of a Kaiser window FIR filter.

```
function [N,alpha] = kaispar(typ,par,theta,deltap,deltas);
% Synopsis: kaispar(typ,par,theta,deltap,deltas).
% Estimates parameters for FIR Kaiser window filter design.
% Input parameters: see description in firkais.m.
% Output parameters:
% N: the filter order
% alpha: the Kaiser window parameter.

A = -20*log10(min([deltap,deltas]));
if (A > 50), alpha = 0.1102*(A-8.7);
elseif (A > 21), alpha = 0.5842*(A-21)^(0.4)+0.07886*(A-21);
else, alpha = 0; end
if (typ == 'b'), dt = theta;
elseif (typ == 'd'), dt = (pi-theta);
else,
    if (length(theta) == 2), dt = theta(2)-theta(1);
    else, dt = min(theta(2)-theta(1),theta(4)-theta(3)); end
end
N = ceil((A-7.95)/(2.285*dt)); Npar = rem(N,2);
oddpermit = (par=='o') & (typ~='h') & (typ~='s');
if (Npar ~= oddpermit), N = N+1; end
```

Program 9.5 Verification that a given FIR filter meets the specifications.

```
function res = verspec(h,typ,t,dp,ds);
% Synopsis: res = verspec(h,typ,t,dp,ds).
% Verifies that an FIR filter meets the design specifications.
% Input parameters:
% h: the FIR filter coefficients
% other parameters: see description in firkais.m.
% Output:
% res: 0: OK, 1: increase order, 2: increase alpha.

if (typ=='l'), ntest = 1;
Hp = abs(frqresp(h,1,100,[max(0,1.5*t(1)-0.5*t(2)),t(1)]));
Hs = abs(frqresp(h,1,100,[t(2),min(pi,1.5*t(2)-0.5*t(1))]));
elseif (typ=='h'), ntest = 1;
Hp = abs(frqresp(h,1,100,[t(2),min(pi,1.5*t(2)-0.5*t(1))]));
Hs = abs(frqresp(h,1,100,[max(0,1.5*t(1)-0.5*t(2)),t(1)]));
elseif (typ=='p'), ntest = 2;
Hp1 = abs(frqresp(h,1,100,[t(2),min(t(3),1.5*t(2)-0.5*t(1))]));
Hs1 = abs(frqresp(h,1,100,[max(0,1.5*t(1)-0.5*t(2)),t(1)]));
Hp2 = abs(frqresp(h,1,100,[max(t(2),1.5*t(3)-0.5*t(4)),t(3)]));
Hs2 = abs(frqresp(h,1,100,[t(4),min(pi,1.5*t(4)-0.5*t(3))]));
Hp = [Hp1; Hp2]; Hs = [Hs1; Hs2];
elseif (typ=='s'), ntest = 2;
Hp1 = abs(frqresp(h,1,100,[max(0,1.5*t(1)-0.5*t(2)),t(1)]));
Hs1 = abs(frqresp(h,1,100,[t(2),min(t(3),1.5*t(2)-0.5*t(1))]));
Hp2 = abs(frqresp(h,1,100,[t(4),min(pi,1.5*t(4)-0.5*t(3))]));
Hs2 = abs(frqresp(h,1,100,[max(t(2),1.5*t(3)-0.5*t(4)),t(3)]));
Hp = [Hp1; Hp2]; Hs = [Hs1; Hs2];
elseif (typ=='b'), ntest = 1;
Hp = abs(frqresp(h,1,100,[t,2*t])); Hs = zeros(1,100);
end

res = 0;
for i = 1:ntest,
   if(max(abs(Hp(i,1)-1),abs(Hp(i,100)-1)) > dp(i)), res = 1;
   elseif(max(Hs(i,1),Hs(i,100)) > ds(i)), res = 1; end
end
if (res), return, end
for i = 1:ntest,
   if(max(abs(Hp(i,:)-1)) > dp(i)), res = 2;
   elseif(max(Hs(i,:)) > ds(i)), res = 2; end
end
```

Program 9.6 Least-squares design of linear-phase FIR filters.

```
function h = firls(N,typ,theta,deltap,deltas);
% Synopsis: h = firls(N,typ,theta,deltap,deltas).
% Designs an FIR filter of one of the four basic types by
% least-squares.
% Input parameters:
% N: the filter order
% typ: the filter type:
%      'l','h','p','s' for LP, HP, BP, BS, respectively,
% theta: vector of band-edge frequencies in increasing order.
% deltap: one or two pass-band tolerances
% deltas: one or two stop-band tolerances.
% Output:
% h: the filter coefficients.

thetai = (pi/(32*N)) + (pi/(16*N))*(0:(16*N-1));
if (rem(N,2)),
   F = cos(0.5*thetai); K = (N-1)/2;
else,
   F = ones(1,16*N); K = N/2;
end
if (typ == 'p' | typ == 's'),
   if (length(deltap) == 1), deltap = deltap*[1,1]; end
   if (length(deltas) == 1), deltas = deltas*[1,1]; end
end
[V,Ad] = firlsaux(typ,theta,deltap,deltas,thetai);
carray = cos(thetai'*(0:K)).*((F.*V)'*ones(1,K+1));
darray = (V.*Ad)';
g = (carray\darray)';
if (rem(N,2)),
   h = 0.25*[g(K+1),fliplr(g(3:K+1))+fliplr(g(2:K))];
   h = [h,0.25*g(2)+0.5*g(1)];
   h = [h,fliplr(h)];
else,
   h = [0.5*fliplr(g(2:K+1)),g(1),0.5*g(2:K+1)];
end
```

Program 9.7 An auxiliary subroutine for `firls`.

```
function [V,Ad] = firlsaux(typ,theta,deltap,deltas,thetai);
% Synopsis: [V,Ad] = firlsaux(typ,theta,deltap,deltas,thetai).
% An auxiliary function for FIRLS.
% Input parameters: see firls.m
% Output parameters:
% V, Ad: variables needed in firls.m

ind1 = find(thetai < theta(1));
ind3 = find(thetai > theta(length(theta)));
if (typ == 'p' | typ == 's'),
ind2 = find(thetai > theta(2) & thetai < theta(3)); end
V = zeros(1,length(thetai)); Ad = zeros(1,length(thetai));
if (typ == 'l'),
   Ad(ind1) = ones(1,length(ind1));
   Ad(ind3) = zeros(1,length(ind3));
   V(ind1) = (1/deltap)*ones(1,length(ind1));
   V(ind3) = (1/deltas)*ones(1,length(ind3));
elseif (typ == 'h'),
   Ad(ind1) = zeros(1,length(ind1));
   Ad(ind3) = ones(1,length(ind3));
   V(ind1) = (1/deltas)*ones(1,length(ind1));
   V(ind3) = (1/deltap)*ones(1,length(ind3));
elseif (typ == 'p'),
   Ad(ind1) = zeros(1,length(ind1));
   Ad(ind2) = ones(1,length(ind2));
   Ad(ind3) = zeros(1,length(ind3));
   V(ind1) = (1/deltas(1))*ones(1,length(ind1));
   V(ind2) = (1/deltap(1))*ones(1,length(ind2));
   V(ind3) = (1/deltas(2))*ones(1,length(ind3));
elseif (typ == 's'),
   Ad(ind1) = ones(1,length(ind1));
   Ad(ind2) = zeros(1,length(ind2));
   Ad(ind3) = ones(1,length(ind3));
   V(ind1) = (1/deltap(1))*ones(1,length(ind1));
   V(ind2) = (1/deltas(1))*ones(1,length(ind2));
   V(ind3) = (1/deltap(2))*ones(1,length(ind3));
end
```

9.9   Problems

9.1 Show that a type-II filter has a factor $(1 + z^{-1})$ in its transfer function, a type-III filter has a factor $(1 - z^{-2})$, and a type-IV filter has a factor $(1 - z^{-1})$.

9.2 Let $H^z(z)$ be a general FIR filter (not necessarily linear phase).

(a) Show that $H^z(z)$ can be expressed as a sum of two FIR filters

$$H^z(z) = G_1^z(z) + G_2^z(z),$$

where $G_1^z(z)$ and $G_2^z(z)$ have linear phase (exact or generalized) and are of the same order as $H^z(z)$ or less.

(b) Of what types are $G_1^z(z)$ and $G_2^z(z)$ when N is even? When N is odd?

(c) Express $|H^f(\theta)|$ in terms of the amplitude functions of $G_1^z(z)$ and $G_2^z(z)$.

9.3 The filters in Figure 9.35 are FIR, with linear phase, real and causal. Their orders are N_1, N_2, and $N_3 = N_1 + N_2$, respectively. Each of the three filters can be either symmetric or antisymmetric, so there are eight cases in total. For which of these eight cases will the equivalent filter have linear phase (either exact or generalized)? In each case state whether the equivalent filter is symmetric or antisymmetric.

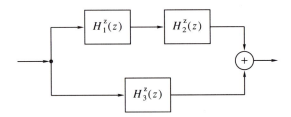

Figure 9.35 Pertaining to Problem 9.3.

9.4 The impulse response of a linear-phase FIR filter starts at the values

$$h[0] = 1, \quad h[1] = 3, \quad h[2] = -2.$$

For each of the four filter types, find the coefficients of the smallest order FIR filter that satisfies this condition.

9.5 Let $H^z(z)$ be the transfer function of a linear-phase FIR filter with real coefficients. The filter is known to have zeros in the following locations:

$$\beta_1 = 1, \quad \beta_2 = 0.5e^{j\pi/3}, \quad \beta_3 = -5, \quad \beta_4 = j.$$

What is the lowest possible order of the filter? What is the transfer function of the filter of the lowest order, and what is its type?

9.6 Suppose that $H^z(z)$ is a linear-phase FIR filter. Can a filter whose transfer function is $1/H^z(z)$ be causal and stable? Can $1/H^z(z)$ be stable if it is not required to be causal?

9.7 This problem explores the preservation of linear phase in cascade and parallel connections of FIR filters.

(a) Show that a cascade connection of two (generalized) linear-phase filters has (generalized) linear phase. Find the order of the equivalent filter, its amplitude function, and its initial phase as a function of the types of the two filters.

(b) Is the same true for a parallel connection? What if the two filters in parallel have the same order?

9.8 In Section 9.1.6 we saw that the zeros of a linear-phase filter must satisfy certain symmetry conditions, divided into five possible cases. Show that the converse is also true, that is: If every zero of an FIR filter satisfies one of the five symmetry conditions, then the filter has linear phase.

9.9 A real, causal, linear-phase FIR filter of order N is known to satisfy

$$\sum_{n=0}^{N} h[n] \cdot (0.8)^{-n} e^{-j\pi n/3} = 0.$$

Find the order and the coefficients of the filter of minimal order satisfying these conditions. Assume that $h[0] = 1$.

9.10 A real FIR filter is required to block all real sinusoidal signals of frequencies $\theta_k = 0.125 k\pi$, for all integer k. Subject to this requirement, the filter should have a minimal order. There are no other requirements from the filter. Specify the order and the transfer function of the filter. Does the filter have exact linear phase, generalized linear phase, or no linear phase at all?

9.11 You are given a fourth-order FIR filter $H_1^z(z)$ that has a double zero at $\beta_{1,3} = 0.5 + j0.75$, and a double zero at $\beta_{2,4} = 0.5 - j0.75$. Suggest a linear-phase FIR filter $H_2^z(z)$ whose magnitude response $|H_2^f(\theta)|$ is as close as possible to $|H_1^f(\theta)|$.

9.12 Let
$$x[n] = 0.3 \cos(0.9273n + 0.25\pi) - 0.4 \sin(\pi n - 0.3).$$
Find an FIR filter $h[n]$ of minimal order that satisfies
$$\{h * x\}[n] = 0, \quad n \in \mathbb{Z}.$$

9.13 A type-II FIR filter of order 5 is known to satisfy
$$H^z(1) = 6, \quad H^z(0.5 - 0.5j) = 0.$$
Find $H^z(-j)$.

9.14 The DFT of the coefficients of a causal FIR filter is $\{6, -1 - j, 0, -1 + j\}$. State as many properties of the filter as you can think of.

9.15 You are given two causal filters whose frequency responses are
$$H_1^f(\theta) = \frac{1}{1 - 0.5e^{-j\theta}}, \quad H_2^f(\theta) = \left(\frac{\sin 2\theta}{\sin 0.5\theta} e^{-j1.5\theta}\right)^2.$$
Let
$$H^f(\theta) = \frac{1}{2\pi} \{H_1^f * H_2^f\}(\theta).$$
Compute the coefficients of $H^f(\theta)$.

9.16 Let

$$x_1[n] = 0.2\text{sinc}(0.2n), \quad y_1[n] = \delta[n] - 0.6\text{sinc}(0.6n),$$
$$x_2[n] = 0.6\text{sinc}(0.6n), \quad y_2[n] = \delta[n] - 0.2\text{sinc}(0.2n).$$

Compute $\{x_1 * y_1\}[n]$ and $\{x_2 * y_2\}[n]$.

9.17 Design a symmetric FIR filter with group delay $N/2$ according to the ideal magnitude response

$$|H_d^f(\theta)| = \begin{cases} 1, & 0 \leq |\theta| \leq \frac{\pi}{3}, \\ 0, & \frac{\pi}{3} < |\theta| < \frac{2\pi}{3}, \\ 0.5, & \frac{2\pi}{3} \leq |\theta| \leq \pi. \end{cases}$$

(a) Compute $h_d[n]$.

(b) If the filter is designed with the Hamming window and its order is $N = 40$, what are the values of

$$\theta_{p,1}, \; \theta_{s,1}, \; \theta_{s,2}, \; \theta_{p,2}, \; \delta_{p,1}, \; \delta_s, \; \delta_{p,2}?$$

(c) Suppose we want to have $\delta_{p,1} = \delta_{p,2} = 0.01$, and $\delta_s = 0.005$. Is it possible to achieve this with a Hamming window filter of order $N = 39$?

9.18 Consider the following filter specifications:

$$\theta_p = 0.3\pi, \; \theta_s = 0.5\pi, \; \delta_p = 0.0025, \; \delta_s = 0.0021.$$

(a) Design a low-pass FIR type-I filter $H^z(z)$, using the Kaiser window, according to these specifications. Give the order N and the parameter α (there is no need to write down the coefficient values).

(b) Let $\tilde{H}^z(z)$ be the noncausal filter

$$\tilde{H}^z(z) = z^{N/2} H^z(z).$$

Define

$$\tilde{G}_1^z(z) = [\tilde{H}^z(z)]^2,$$
$$\tilde{G}_2^z(z) = 2\tilde{H}^z(z) - \tilde{G}_1^z(z).$$

Is $\tilde{G}_1^z(z)$ low pass, high pass, band pass, or band stop? What about $\tilde{G}_2^z(z)$?

(c) Compute the pass-band ripple and the stop-band attenuation of $\tilde{G}_1^z(z)$ and $\tilde{G}_2^z(z)$.

(d) Now assume that $\tilde{H}^z(z)$ is not available, only $H^z(z)$. We wish to realize two causal filters $G_1^z(z)$ and $G_2^z(z)$ having the same amplitude response as $\tilde{G}_1^z(z)$ and $\tilde{G}_2^z(z)$, respectively, using copies of $H^z(z)$ as building blocks (and other elements as needed). Show how to do this.

9.19 We are given the system shown in Figure 9.36. The signal $x(t)$ is limited to $|f| \leq 0.5\,\text{Hz}$, and the sampling interval is $T = 1$ second. The filter is FIR of even-order N. We wish to design the filter such that the output will satisfy

$$y[n] \approx x(n - L - 0.25),$$

where L is a given positive integer.

(a) What is the ideal frequency response $H_d^f(\theta)$ of the filter?

(b) Choose an appropriate N (which depends on L), and design the filter by the IRT method.

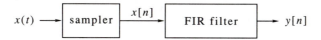

Figure 9.36 Pertaining to Problem 9.19.

(c) Does $H^z(z)$ have exact linear phase, generalized linear phase, or no linear phase at all?

9.20 Figure 9.37 shows the ideal magnitude and phase responses of a filter that is a differentiator at low frequencies and high pass at high frequencies.

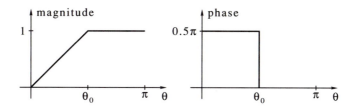

Figure 9.37 Pertaining to Problem 9.20.

(a) Compute the desired impulse response $h_d[n]$. Does the truncated impulse response have linear phase? Explain why or why not.

(b) Design an FIR filter of order $N = 128$ and having $\theta_0 = 0.5\pi$, using: (1) a rectangular window; (2) a Hamming window; (3) a Kaiser window with $\alpha = 6$; (4) a Kaiser window with $\alpha = 12$. Compute and plot the magnitude and phase responses of the four filters.

9.21 We are given a low-pass type-I FIR filter $h[n]$, with given parameters θ_p, θ_s, δ_p, δ_s. Define
$$g[n] = (-1)^{N/2}\delta[n - 0.5N] - (-1)^n h[n].$$

(a) What type is the filter $g[n]$ and what is the nature of its frequency response?

(b) Express the tolerance and band-edge parameters of the filter $g[n]$ in terms of the parameters of $h[n]$.

9.22 Consider a digital filter whose frequency response is like the right side of (3.67), with $f_0 T = 0.5$ and $\theta = \omega T$.

(a) Compute the ideal impulse response $h_d[n]$. Is it a half-band filter?

(b) Is
$$g_d[n] \triangleq \{h_d * h_d\}[n]$$
a half-band filter?

(c) Suggest a causal FIR filter $h[n]$ that will approximate this frequency response. Compute and plot the magnitude response of the FIR filter for $N = 38$ and $\alpha = 0.4$.

(d) Is
$$g[n] \triangleq \{h * h\}[n]$$
a half-band filter?

9.23 Suggest a definition of a *quarter-band filter.* Write a necessary and sufficient condition for a noncausal filter to be quarter band, and then for a causal filter. Suggest a procedure for designing a quarter-band FIR filter. Show the magnitude response of a quarter-band, raised-cosine filter with α, N, and window of your choice.

9.24 It is required to design a linear-phase FIR filter according to the desired amplitude response

$$A_d(\theta) = \cos(0.5\theta)[1 + \cos\theta].$$

The order of the filter N is required to be no larger than 8. Subject to this requirement, the filter must have the smallest possible integral of square error ε^2; see its definition (9.41). Find the filter's coefficients $h[n]$ and the associated value of ε^2. Hint: Solve without computing any integrals.

9.25 Find an FIR filter $H^z(z)$ whose amplitude response is

$$A(\theta) = \sin(2\theta).$$

You are free to choose the phase response $\phi(\theta)$.

9.26 Let $w[n]$ be an arbitrary window, and define a new window $v[n] = \{w * w\}[n]$. Show that the function $\Gamma_v(x)$ corresponding to $v[n]$ is monotone, similar to that of the Bartlett window.

9.27 Suggest a procedure for designing a low-pass, half-band filter using windows. Hint: What is the ideal frequency response?

9.28 Suppose we wish to design a low-pass filter with band-edge frequencies θ_p, θ_s. Instead of using the boxcarlike amplitude response (9.28) for $A_d(\theta)$, we can use the function shown in Figure 9.38.

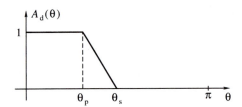

Figure 9.38 Pertaining to Problem 9.28.

(a) Compute the ideal impulse response $h_d[n]$ corresponding to $A_d(\theta)$ shown in the figure.

(b) Use MATLAB for designing a low-pass, linear-phase filter using the IRT method with $h_d[n]$ found in part a. Take $N = 20$, $\theta_p = 0.2\pi$, $\theta_s = 0.4\pi$. Give the filter coefficients $h[0]$ through $h[10]$ to four decimal digits as an answer.

(c) Plot the frequency response of the filter and use it for finding the pass-band ripple and the stop-band attenuation.

(d) Suggest an extension of this idea to band-pass filters (there is no need to work out the details for this case).

9.29 Suppose we wish to form the analytic signal corresponding to a given real signal $x[n]$, as in (9.38), where $y[n]$ is the Hilbert transform of $x[n]$. Since $y[n]$ is not given, we generate it by passing $x[n]$ through a linear-phase FIR Hilbert transformer, as explained in Section 9.2.5. Explain why we must delay $x[n]$ by $N/2$ samples before combining it with $y[n]$ (where N is the order of the Hilbert transformer). Hence state whether it is more convenient to use a type-III or type-IV transformer in this case.

9.30 A linear-phase FIR filter of an even-order N, whose initial conditions are set to zero, is fed with an input signal of length L. As we know, this results in a signal of length $L + N$. It is often desired to retain only L output values, the same as the number of input values. Consider the following three options:

- Deleting the first N output points.
- Deleting the last N output points.
- Deleting $N/2$ points from the beginning and $N/2$ points from the end of the output signal.

(a) Which of the three is used by the MATLAB function `filter`? Try before you give an answer.

(b) Which of the three makes more sense to you? Give reasons. Hint: Design a low-pass filter using `firdes`, and feed it with the signal $x[n] = 1$, $0 \leq n \leq L - 1$. Try the three options with MATLAB, then form an opinion.

(c) Will your answer to part b be different if $h[n]$ is a minimum-phase (rather than linear-phase) filter?

9.31 Design an even-order band-pass Hilbert transformer whose ideal frequency response is

$$H_d^f(\theta) = \begin{cases} -je^{-j0.5\theta N}, & 0.25\pi \leq \theta \leq 0.75\pi, \\ je^{-j0.5\theta N}, & -0.75\pi \leq \theta \leq -0.25\pi, \\ 0, & \text{otherwise.} \end{cases}$$

The transition bands of the filter should not be wider than 0.1π and its ripple should not be higher than 0.01 in any of the stop bands. As an answer, give a table of the coefficients, accurate to five decimal digits.

9.32* The signal $x_1(t)$ is low pass in the frequency range $|\omega| < 2\pi \cdot 20$, and the magnitude of its Fourier transform is known to be not larger than 1 in this range. The signal $x_2(t)$ is band pass in the frequency range $2\pi \cdot 40 < |\omega| < 2\pi \cdot 60$, and the magnitude of its Fourier transform is known to be not larger than 1 in this range. We measure the signal

$$y(t) = x_1(t) + x_2(t)$$

and sample it at frequency f_{sam}. The discrete-time signal $y[n]$ is filtered by a digital low-pass filter, whose aim is:

- to attenuate the component resulting from $x_2(t)$, such that the magnitude of its Fourier transform in the θ domain will not exceed 0.001;
- to pass the component resulting from $x_1(t)$, with tolerance ± 0.001.

The filter is to be designed by means of a Kaiser window. Two options are considered for the sampling frequency: (1) $f_{sam} = 120\,\text{Hz}$; (2) $f_{sam} = 100\,\text{Hz}$.

Which sampling rate will lead to a filter of lower order? Hints: Do not neglect the aliasing in case 1; use (9.56c) for the order of a Kaiser window filter.

9.33* Let $H^z(z)$ be a type-I, Nth-order, low-pass FIR filter with tolerances δ_p, δ_s. Define

$$G^z(z) = [3z^{-N/2} - 2H^z(z)][H^z(z)]^2.$$

What are the tolerances of $G^z(z)$? What is its order? Suggest a possible use of this idea.

9.34* Suppose we wish to design a differentiator using least-squares design.

(a) Explain why it makes sense to use the weighting function

$$W(\theta) = \delta \cdot \theta,$$

where δ is a small number.

(b) Write a MATLAB program for least-squares type-IV differentiator design. Use parts of `firls` as needed.

(c) Use the program to design a differentiator with $\delta = 0.01$. Find and report the order N that meets this specification.

9.35* Suppose we design an equiripple low-pass filter $G^z(z)$ according to the following specifications:

- Odd-order N, that is, type II.
- $\theta_p < \pi$ and a corresponding δ_p.
- $\theta_s = \pi$, that is, no stop band (and therefore no δ_s).

Such a filter is called *one-band filter.* Define

$$H^z(z) = 0.5[G^z(z^2) + z^{-N}].$$

(a) Show that $H^z(z)$ is a half-band filter.

(b) What is the order of $H^z(z)$?

(c) What are the pass-band and stop-band tolerances of $H^z(z)$, and what are the band-edge frequencies?

(d) Carry out the design for $\theta_p = 0.9\pi$, $\delta_p = 0.1$, and $N = 11$. Draw the amplitude response plots of $G^z(z)$ and $H^z(z)$.

This method of equiripple half-band filter design was developed by Vaidyanathan and Nguyen [1987].

9.36* The following method, called *frequency sampling,* is proposed for FIR filter design:

- Sample the desired frequency response $H_d^f(\theta)$ at the $N + 1$ points

$$\theta_k = \frac{2\pi k}{N + 1}, \quad 0 \le k \le N.$$

- Let the filter coefficients $h[n]$ be the inverse DFT of $\{H_d^f(\theta_k), 0 \le k \le N\}$.

It is clear from the construction of $h[n]$ that its frequency response coincides with the desired one. This, however, does not imply proximity of $H^f(\theta)$ to $H_d^f(\theta)$ at other frequencies.

(a) Prove that for frequencies different from θ_k,

$$H^f(\theta) = \frac{1}{N + 1} \sum_{k=0}^{N} H_d^f(\theta_k) \frac{1 - e^{-j\theta(N+1)}}{1 - e^{-j[\theta - 2\pi k/(N+1)]}}.$$

(b) Compute $h[n]$ in the special case of a low-pass filter with desired cutoff frequency θ_c. Be careful to maintain the conjugate symmetry property of $H_d^f(\theta_k)$; otherwise the impulse response $h[n]$ will not be real valued.

(c) Choose $N = 63$, $\theta_c = 0.3\pi$. Compute the impulse response and the frequency response of the filter. Plot the magnitude response.

(d) Conclude from part c on the properties of the frequency sampling design method.

Chapter 10

Infinite Impulse Response Filters

As we said in Chapter 8, the most common design method for digital IIR filters is based on designing an analog IIR filter and then transforming it to an equivalent digital filter. Accordingly, this chapter includes two main topics: analog IIR filter design and analog-to-digital transformations of IIR filters.

Well-developed design methods exist for analog low-pass filters. We therefore discuss such filters first. The main classes of analog low-pass filters are (1) Butterworth filters; (2) Chebyshev filters, of which there are two kinds; and (3) elliptic filters. These filters differ in the nature of their magnitude responses, as well as in their respective complexity of design and implementation. Familiarity with all classes helps one to choose the most suitable filter class for a specific application.

The design of analog filters other than low pass is based on frequency transformations. Frequency transformations enable obtaining a desired high-pass, band-pass, or band-stop filter from a prototype low-pass filter of the same class. They are discussed after the sections on low-pass filter classes.

The next topic in this chapter is the transformation of a given analog IIR filter to a similar digital filter, which could be implemented by digital techniques. Similarity is required in both magnitude and phase responses of the filters. Since the frequency response of an analog filter is defined for $-\infty < \omega < \infty$, whereas that of a digital filter is restricted to $-\pi \leq \theta < \pi$ (beyond which it is periodic), the two cannot be made identical. We shall therefore be concerned with similarity over a limited frequency range, usually the low frequencies.

Of the many transformation methods discussed in the literature, we shall restrict ourselves to three: the impulse invariant method, the backward difference method, and the bilinear transform. The first two are of limited applicability, but their study is pedagogically useful. The third is the best and most commonly used method of analog-to-digital filter transformation, so this is the one we emphasize.

IIR filter design usually concentrates on the magnitude response and regards the phase response as secondary. The next topic in this chapter explores the effect of phase distortions of digital IIR filters. We show that phase distortions due to variable group delay may be significant, even when the pass-band ripple of the filter is low.

The final topic discussed in this chapter is that of analog systems interfaced to a digital environment, also called sampled-data systems. This topic is marginal to digital signal processing but has great importance in related fields (such as digital control), and its underlying mathematics is well suited to the material in this chapter.

10.1 Analog Filter Basics

Analog filters are specified in a manner similar to digital filters; the main difference is that frequencies are specified in the ω domain (in rad/s), rather than in the θ domain. Thus, a low-pass analog filter is specified in terms of its pass-band edge frequency ω_p, stop-band edge frequency ω_s, pass-band ripple δ_p, and stop-band attenuation δ_s. For analog filters, the pass-band magnitude response is usually required to be in the range $[1 - \delta_p, 1]$. This is mainly a matter of convenience, since the filter's gain can be easily adjusted to make the pass-band ripple symmetrical with respect to 1 (see Problem 10.4). Also recall the definitions

$$A_p = -20\log_{10}(1 - \delta_p) \approx 8.6859\delta_p, \tag{10.1a}$$

$$A_s = -20\log_{10}\delta_s. \tag{10.1b}$$

It will be convenient to introduce the following auxiliary parameters, which are functions of the basic specification parameters:

$$d = \left[\frac{(1 - \delta_p)^{-2} - 1}{\delta_s^{-2} - 1}\right]^{1/2} = \left(\frac{10^{0.1A_p} - 1}{10^{0.1A_s} - 1}\right)^{1/2}, \tag{10.2}$$

$$k = \frac{\omega_p}{\omega_s}. \tag{10.3}$$

The parameter d is called the *discrimination factor*; the parameter k is called the *selectivity factor*. As we shall see later, these are the primary design parameters for analog filters.

Two other parameters of analog filters are the following:

1. The -3 dB frequency (also called the *cutoff frequency*) ω_{3db}, defined as the frequency at which the magnitude response of the filter is $1/\sqrt{2}$ of its nominal value at the pass band.

2. The asymptotic attenuation at high frequencies. This parameter is determined by the difference between the denominator and numerator degrees. It follows from (8.5) that

   $$H^F(\omega) \approx b_0(j\omega)^{q-p} \tag{10.4}$$

   for large enough ω (e.g., larger than about 5 times the largest absolute value of all poles and zeros). Then from (10.4) we can write

   $$20\log_{10}|H^F(\omega)| \approx 20\log_{10}|b_0| - 20(p - q)\log_{10}\omega. \tag{10.5}$$

   It is common to express this by saying that the asymptotic attenuation is $20(p - q)$ dB/decade.

The -3 dB frequency is occasionally useful for digital filters as well. The asymptotic attenuation is not defined for a digital filter, because the digital frequency θ is of interest only in the range $[-\pi, \pi]$.

Classical analog filters are constructed from a desired square-magnitude response function. This function is usually of the form (except for Chebyshev filter of the second kind, to be studied in Section 10.3.2)

$$|H^F(\omega)|^2 = \frac{1}{1 + \Lambda\left(\frac{\omega}{\omega_0}\right)}, \tag{10.6}$$

where $\Lambda(\cdot)$ is either a polynomial or a rational function, called the *attenuation function*, and ω_0 is a reference frequency. The attenuation function is nonnegative and is designed to be small in the pass band and large in the stop band. If $\Lambda(\cdot)$ is monotone, so is $|H^F(\omega)|^2$. If $\Lambda(\cdot)$ is oscillatory in one of the bands (or both), $|H^F(\omega)|^2$ exhibits ripple in that band. By constraining $\Lambda(\cdot)$ to be polynomial or rational, we impose on $H^L(s)$ to be rational, thus guaranteeing that it be realizable. The order of the polynomial, or the rational function, determines the order of $H^L(s)$. In addition, the attenuation function may depend on other parameters, used for tuning the behavior of $|H^F(\omega)|^2$.

The square-magnitude function $|H^F(\omega)|^2$ determines the product $H^L(s)H^L(-s)$, but it does not uniquely determine the transfer function $H^L(s)$. The set of poles of $H^L(s)H^L(-s)$ consists of the poles of $H^L(s)$ and their mirror images with respect to the imaginary axis in the s plane. The requirement that $H^L(s)$ be stable determines this transfer function uniquely: $H^L(s)$ must include all poles of $H^L(s)H^L(-s)$ on the left half of the s plane and only those.

The analog filters that we shall study in the following sections are all constructed in a similar manner, differing only in the choice of attenuation functions. Because of the different attenuation functions, the filters differ in their response characteristics.

10.2   Butterworth Filters

A low-pass Butterworth filter is defined in terms of its square-magnitude frequency response

$$|H^F(\omega)|^2 = \frac{1}{1 + \left(\frac{\omega}{\omega_0}\right)^{2N}}, \qquad (10.7)$$

where N is an integer (which will soon be seen to be the filter's order), and ω_0 is a parameter. Figure 10.1 illustrates the square magnitude of the frequency response for $N = 3$.

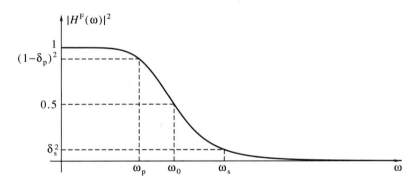

Figure 10.1 The square magnitude of the frequency response of a low-pass Butterworth filter, $N = 3$.

The salient properties of a low-pass Butterworth filter are as follows:

1. The magnitude response is a monotonically decreasing function of ω.

2. The maximum gain occurs at $\omega = 0$ and is $|H^F(0)| = 1$.

3. We have $|H^F(\omega_0)| = \sqrt{0.5}$, so the -3 dB point is ω_0 rad/s.

4. The asymptotic attenuation at high frequencies is $20N$ dB/decade.

5. The square magnitude of the frequency response satisfies

$$\left.\frac{\partial^k |H^F(\omega)|^2}{\partial \omega^k}\right|_{\omega=0} = 0, \quad 1 \le k \le 2N - 1. \qquad (10.8)$$

Thus, the magnitude response is nearly constant at low frequencies (the constant being 1). Because of this property, low-pass Butterworth filters are said to be *maximally flat at DC*.

The transfer function of Butterworth filter can be obtained by the following calculation. By substituting $s = j\omega$, or $\omega = s/j$, we get from (10.7)

$$H^L(s)H^L(-s) = \frac{1}{1 + \left(\frac{s}{j\omega_0}\right)^{2N}} = \frac{1}{1 + (-1)^N \left(\frac{s}{\omega_0}\right)^{2N}}. \tag{10.9}$$

The right side of (10.9) has $2N$ poles, which are the $2N$ complex roots of $(-1)^{(N+1)}$, multiplied by ω_0. The poles are given by the explicit expression

$$s_k = \omega_0 \exp\left[j\frac{(N+1+2k)\pi}{2N}\right], \quad 0 \le k \le 2N - 1. \tag{10.10}$$

Of those, the poles corresponding to $0 \le k \le N - 1$ are on the left side of the complex plane. These must correspond to $H^L(s)$, since we require $H^L(s)$ to be stable. The remaining roots are the poles of $H^L(-s)$. Figure 10.2 illustrates the pole locations of Butterworth filters of odd and even orders.

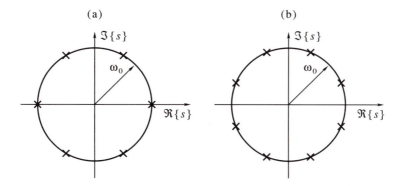

Figure 10.2 The poles of a low-pass Butterworth filter: (a) $N = 3$; (b) $N = 4$. The N poles of $H^L(s)$ are on the left side of the s plane and those of $H^L(-s)$ are on the right side.

A Butterworth filter has no zeros, but its constant coefficient in the numerator has to be adjusted so that $H^F(0) = 1$. In summary, Butterworth filter of order N is given by

$$H^L(s) = \prod_{k=0}^{N-1} \frac{-s_k}{s - s_k}, \tag{10.11}$$

where

$$s_k = \omega_0 \cos\left[\frac{\pi}{2} + \frac{(2k+1)\pi}{2N}\right] + j\omega_0 \sin\left[\frac{\pi}{2} + \frac{(2k+1)\pi}{2N}\right]$$
$$= -\omega_0 \sin\left[\frac{(2k+1)\pi}{2N}\right] + j\omega_0 \cos\left[\frac{(2k+1)\pi}{2N}\right], \quad 0 \le k \le N - 1. \tag{10.12}$$

Table 10.1 gives the coefficients of the polynomials $a(s)$ in the transfer function

$$H^L(s) = \frac{1}{a(s)} = \frac{1}{s^N + a_1 s^{N-1} + \cdots + a_{N-1}s + 1} \tag{10.13}$$

of an Nth-order LP Butterworth filter having $\omega_0 = 1$, for $2 \le N \le 6$. Such a filter is said to be *normalized*.

N	a_1	a_2	a_3	a_4	a_5
2	1.4142				
3	2.0000	2.0000			
4	2.6131	3.4142	2.6131		
5	3.2361	5.2361	5.2361	3.2361	
6	3.8637	7.4641	9.1416	7.4641	3.8637

Table 10.1 Coefficients of the denominator polynomials of normalized low-pass Butterworth filters of orders 2 through 6.

Designing a Butterworth filter to meet given specifications proceeds as follows. First we compute the discrimination factor d and the selectivity factor k from the given values of δ_p, δ_s, ω_p, and ω_s, using the formulas (10.2) and (10.3). Then we observe from (10.7) and Figure 10.1 that

$$\frac{1}{1 + \left(\frac{\omega_p}{\omega_0}\right)^{2N}} \geq (1 - \delta_p)^2, \qquad \frac{1}{1 + \left(\frac{\omega_s}{\omega_0}\right)^{2N}} \leq \delta_s^2. \tag{10.14}$$

This gives

$$\left(\frac{\omega_p}{\omega_0}\right)^{2N} \leq (1 - \delta_p)^{-2} - 1, \qquad \left(\frac{\omega_s}{\omega_0}\right)^{2N} \geq \delta_s^{-2} - 1. \tag{10.15}$$

Dividing these two expressions gives

$$\left(\frac{\omega_s}{\omega_p}\right)^{2N} \geq \frac{\delta_s^{-2} - 1}{(1 - \delta_p)^{-2} - 1}. \tag{10.16}$$

Using the definitions of d (10.2) and k (10.3) gives

$$\left(\frac{1}{k}\right)^{2N} \geq \left(\frac{1}{d}\right)^2. \tag{10.17}$$

Finally, taking the logarithms and dividing again gives

$$N \geq \frac{\log_e(1/d)}{\log_e(1/k)}. \tag{10.18}$$

The right side of (10.18) will not be an integer in general, so in practice we take N as the smallest integer larger than $\log_e(1/d) / \log_e(1/k)$. The frequency ω_0 can be chosen anywhere in the interval (Problem 10.3)

$$\omega_p \left[(1 - \delta_p)^{-2} - 1\right]^{-1/2N} \leq \omega_0 \leq \omega_s \left[\delta_s^{-2} - 1\right]^{-1/2N}. \tag{10.19}$$

In summary, the design procedure for a low-pass Butterworth filter is as follows:

Low-pass Butterworth filter design procedure

1. Compute d and k as a function of δ_p, δ_s, ω_p, and ω_s, using (10.2) and (10.3).
2. Compute N, using (10.18), and round upward to the nearest integer.
3. Choose ω_0, using (10.19).
4. Compute the poles s_k, using (10.12).
5. Compute the transfer function $H^L(s)$, using (10.11).

Example 10.1 Design a low-pass Butterworth filter to meet the specifications:

$$\delta_p = 0.001, \quad \delta_s = 0.001, \quad \omega_p = 1\,\text{rad/s}, \quad \omega_s = 2\,\text{rad/s}.$$

The selectivity and discrimination factors are

$$d = 4.4755 \times 10^{-5}, \quad k = 0.5.$$

We get

$$N \geq 14.45 \implies N = 15.$$

Therefore,

$$1.2301\,\text{rad/s} \leq \omega_0 \leq 1.2619\,\text{rad/s}.$$

The poles of the resulting filter (with $\omega_0 = 1.2301$) are

$$s_{0,14} = -0.1286 \pm j1.2234, \quad s_{1,13} = -0.3801 \pm j1.1699,$$
$$s_{2,12} = -0.6150 \pm j1.0653, \quad s_{3,11} = -0.8231 \pm j0.9141,$$
$$s_{4,10} = -0.9952 \pm j0.7230, \quad s_{5,9} = -1.1238 \pm j0.5003,$$
$$s_{6,8} = -1.2032 \pm j0.2558, \quad s_7 = -1.2301.$$

The magnitude response of the filter is shown in Figure 10.3. Part a shows the response in the frequency range 0 to 2 rad/s; part b shows the pass band at an expanded scale. The filter meets the pass-band specification exactly, but its transition band is slightly narrower than the specification.   □

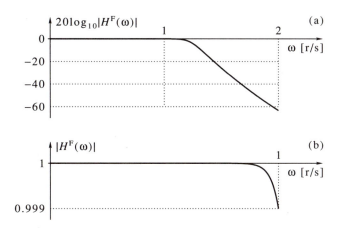

Figure 10.3 Magnitude response of the Butterworth filter in Example 10.1: (a) full range; (b) pass-band details.

10.3 Chebyshev Filters

Chebyshev has noted that $\cos(N\alpha)$ is a polynomial of degree N in $\cos\alpha$. Furthermore, $\cosh(N\alpha)$ is *the same* polynomial (of degree N) in $\cosh\alpha$. Correspondingly, Chebyshev polynomial of degree N is defined as[1]

$$T_N(x) = \begin{cases} \cos(N\arccos x), & |x| \leq 1, \\ \cosh(N\,\text{arccosh}\,x), & |x| > 1. \end{cases} \tag{10.20}$$

Chebyshev polynomials can be constructed by the following recursive formula:

Theorem 10.1

$$T_N(x) = 2xT_{N-1}(x) - T_{N-2}(x), \quad T_0(x) = 1, \ T_1(x) = x. \tag{10.21}$$

Proof We have the following trigonometric identities:

$$\cos(N\alpha) = \cos[(N-1)\alpha]\cos\alpha - \sin[(N-1)\alpha]\sin\alpha, \tag{10.22a}$$

$$\cos[(N-2)\alpha] = \cos[(N-1)\alpha]\cos\alpha + \sin[(N-1)\alpha]\sin\alpha. \tag{10.22b}$$

Adding these two identities gives

$$\cos(N\alpha) + \cos[(N-2)\alpha] = 2\cos[(N-1)\alpha]\cos\alpha, \tag{10.23}$$

which is identical to (10.21). □

The main properties of Chebyshev polynomials are as follows:

1. For $|x| \le 1$, $|T_N(x)| \le 1$, and it oscillates between -1 and $+1$ a number of times proportional to N. This is obvious from the definition (10.20).

2. For $|x| > 1$, $|T_N(x)| > 1$, and it is monotonically increasing in $|x|$. This is also obvious from the definition (10.20).

3. Chebyshev polynomials of odd orders are odd functions of x (i.e., they contain only odd powers of x) and Chebyshev polynomials of even orders are even functions of x (i.e., they contain only even powers of x). The proof is left as an exercise (Problem 10.7).

4. $T_N(0) = \pm 1$ for even N, and $T_N(0) = 0$ for odd N. The proof is again left as an exercise (Problem 10.8).

5. $|T_N(\pm 1)| = 1$ for all N. This follows from the definition (10.20).

Figure 10.4 shows the graphs of Chebyshev polynomials of orders 1 through 4.

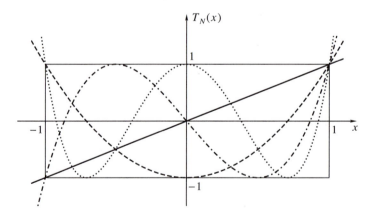

Figure 10.4 Chebyshev polynomials. Solid line: $N = 1$, dashed line: $N = 2$, dot-dashed line: $N = 3$, dotted line: $N = 4$.

The main property of Chebyshev polynomials—being oscillatory in the range $|x| \le 1$ and monotone outside it—is used for constructing filters that are equiripple in either the pass band or the stop band. Thus, the magnitude response oscillates between the permitted minimum and maximum values in the band a number of times depending on the filter's order. The equiripple property provides sharper transition between the pass band and the stop band, compared with that obtained when the magnitude response is monotone. As a result, the order of a Chebyshev filter needed to achieve given specifications is usually smaller than that of a Butterworth filter.

There are two kinds of Chebyshev filter: The first is equiripple in the pass band and monotonically decreasing in the stop band, whereas the second is monotonically decreasing in the pass band and equiripple in the stop band. The second kind is also called *inverse Chebyshev*.

10.3.1 Chebyshev Filter of the First Kind

Chebyshev filter of the first kind, or Chebyshev-I for short, is defined by the square-magnitude frequency response

$$|H^{\mathrm{F}}(\omega)|^2 = \frac{1}{1 + \varepsilon^2 T_N^2\left(\frac{\omega}{\omega_0}\right)}, \tag{10.24}$$

where N is an integer (which will soon be seen to be the filter's order), and ω_0 and ε are parameters. Figure 10.5 illustrates the square magnitudes of the frequency responses for $N = 2$ and $N = 3$.

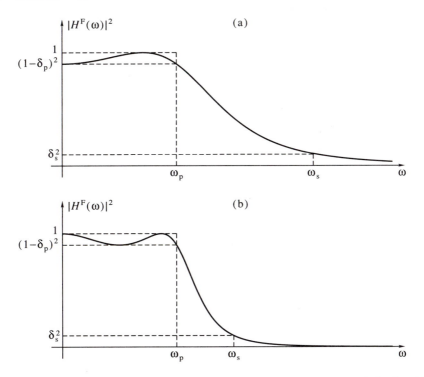

Figure 10.5 The square magnitude of the frequency response of a low-pass Chebyshev filter of the first kind: (a) $N = 2$; (b) $N = 3$.

The salient properties of a low-pass Chebyshev-I filter are as follows:

1. For $0 \le \omega \le \omega_0$ we have, by the properties of Chebyshev polynomials,

$$\frac{1}{1 + \varepsilon^2} \le |H^{\mathrm{F}}(\omega)|^2 \le 1. \tag{10.25}$$

2. From property 4 of Chebyshev polynomials we get that

$$|H^F(0)|^2 = \begin{cases} 1/(1+\varepsilon^2), & N \text{ even}, \\ 1, & N \text{ odd}. \end{cases} \tag{10.26}$$

3. For $\omega > \omega_0$, the response is monotonically decreasing, because of the monotone behavior of $T_N(x)$ for $|x| > 1$. Furthermore, since $T_N(x)$ is an Nth-order polynomial, the asymptotic attenuation of the filter is $20N$ dB/decade, same as that of Butterworth filter of the same order.

The poles of an Nth-order low-pass Chebyshev-I filter are given by the formula

$$s_k = -\omega_0 \sinh\left(\frac{1}{N}\operatorname{arcsinh}\frac{1}{\varepsilon}\right)\sin\left[\frac{(2k+1)\pi}{2N}\right]$$

$$+ j\omega_0\cosh\left(\frac{1}{N}\operatorname{arcsinh}\frac{1}{\varepsilon}\right)\cos\left[\frac{(2k+1)\pi}{2N}\right], \quad 0 \le k \le N-1. \tag{10.27}$$

The poles are located on an ellipse whose principal radii are $\sinh(N^{-1}\operatorname{arcsinh}\varepsilon^{-1})$ (horizontal) and $\cosh(N^{-1}\operatorname{arcsinh}\varepsilon^{-1})$ (vertical). The proof of (10.27) is not given here. Figure 10.6 illustrates the pole locations of a Chebyshev-I filter for odd and even N. The following identities can be used for computing the inverse hyperbolic functions:

$$\operatorname{arcsinh}(x) = \log_e(x + \sqrt{x^2+1}), \quad \operatorname{arccosh}(x) = \log_e(x + \sqrt{x^2-1}). \tag{10.28}$$

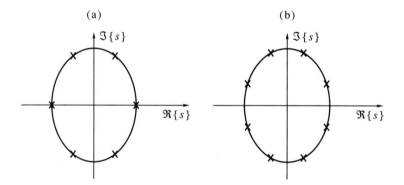

Figure 10.6 The poles of a low-pass Chebyshev filter of the first kind: (a) $N = 3$; (b) $N = 4$. The N poles of $H^L(s)$ are on the left side of the s plane and those of $H^L(-s)$ are on the right side.

A low-pass Chebyshev filter of the first kind has no zeros. Thus, the formula for the transfer function $H^L(s)$ is

$$H^L(s) = H_0 \prod_{k=0}^{N-1} \frac{-s_k}{s - s_k}, \tag{10.29}$$

where

$$H_0 = \begin{cases} (1+\varepsilon^2)^{-1/2}, & N \text{ even}, \\ 1, & N \text{ odd}. \end{cases} \tag{10.30}$$

Designing a Chebyshev-I filter to meet given specifications proceeds as follows. First, we compute the discrimination factor d and the selectivity factor k from the given values of δ_p, δ_s, ω_p, and ω_s, using (10.2) and (10.3). Then we observe from (10.24) that

$$\frac{1}{1 + \varepsilon^2 T_N^2\left(\frac{\omega_p}{\omega_0}\right)} \ge (1 - \delta_p)^2, \tag{10.31}$$

$$\frac{1}{1 + \varepsilon^2 T_N^2 \left(\frac{\omega_s}{\omega_0}\right)} \leq \delta_s^2. \tag{10.32}$$

Equation (10.31) can be satisfied with equality by choosing

$$\omega_0 = \omega_p, \quad \varepsilon = [(1 - \delta_p)^{-2} - 1]^{1/2}. \tag{10.33}$$

Then (10.32) can be satisfied by choosing

$$T_N(1/k) = \cosh[N \operatorname{arccosh}(1/k)] \geq \frac{1}{d}. \tag{10.34}$$

The inequality (10.34) leads to

$$N \geq \frac{\operatorname{arccosh}(1/d)}{\operatorname{arccosh}(1/k)}. \tag{10.35}$$

In practice, we take N as the smallest integer larger than the right side of (10.35). We note that the right side of (10.35) is always smaller than the right side of (10.18); see Problem 10.10. Thus, the order of a Chebyshev filter meeting given specifications will be smaller than that of Butterworth filter meeting the same specifications, or equal to it at most. In summary, the design procedure for a low-pass Chebyshev-I filter is as follows:

Low-pass Chebyshev-I filter design procedure
---

1. Compute d and k as a function of δ_p, δ_s, ω_p, and ω_s, using (10.2) and (10.3).
2. Compute N, using (10.35), and round upward to the nearest integer.
3. Compute ω_0 and ε, using (10.33).
4. Compute the poles s_k, using (10.27).
5. Compute H_0, using (10.30).
6. Compute the transfer function $H^L(s)$, using (10.29).

Example 10.2 Design a low-pass Chebyshev-I filter to meet the same specifications as in Example 10.1, therefore having the same selectivity and discrimination factors. We get

$$\varepsilon = 0.04475, \quad \omega_0 = 1 \operatorname{rad/s}, \quad N \geq 8.13 \implies N = 9.$$

The poles of the resulting filter are

$$s_{0,8} = -0.0755 \pm j1.0739, \quad s_{1,7} = -0.2175 \pm j0.9444,$$
$$s_{2,6} = -0.3332 \pm j0.7009, \quad s_{3,5} = -0.4087 \pm j0.3730,$$
$$s_4 = -0.4349.$$

The magnitude response of the filter is shown in Figure 10.7. Part a shows the response in the frequency range 0 to 2 rad/s; part b shows the pass band at an expanded scale. The filter meets the pass-band specification exactly, but its transition band is slightly narrower than the specification. □

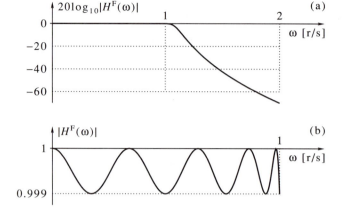

Figure 10.7 Magnitude response of the Chebyshev filter in Example 10.2: (a) full range; (b) pass-band details.

10.3.2 Chebyshev Filter of the Second Kind

Chebyshev filter of the second kind, or Chebyshev-II for short, is defined by the square-magnitude frequency response

$$|H^F(\omega)|^2 = 1 - \frac{1}{1 + \varepsilon^2 T_N^2\left(\frac{\omega_0}{\omega}\right)} = \frac{\varepsilon^2 T_N^2\left(\frac{\omega_0}{\omega}\right)}{1 + \varepsilon^2 T_N^2\left(\frac{\omega_0}{\omega}\right)}, \tag{10.36}$$

where N is the filter's order, and ω_0 and ε are parameters. Figure 10.8 illustrates the square magnitudes of the frequency responses for $N = 2$ and $N = 3$.

The salient properties of a low-pass Chebyshev-II filter are as follows:

1. For $\omega \geq \omega_0$ we have, by the properties of Chebyshev polynomials,

$$0 \leq |H^F(\omega)|^2 \leq \frac{\varepsilon^2}{1 + \varepsilon^2}. \tag{10.37}$$

2. $|H^F(0)|^2 = 1$ for all N, ω_0, and $\varepsilon > 0$.

3. For $0 \leq \omega < \omega_0$, the response is monotonically decreasing, because of the monotone behavior of $T_N(x)$ for $|x| > 1$.

4. In the limit as ω tends to infinity we have

$$\lim_{\omega \to \infty} |H^F(\omega)|^2 = \frac{\varepsilon^2 T_N^2(0)}{1 + \varepsilon^2 T_N^2(0)} = \begin{cases} \frac{\varepsilon^2}{1 + \varepsilon^2}, & N \text{ even,} \\ 0, & N \text{ odd.} \end{cases} \tag{10.38}$$

5. Since, for even N, $|H^F(\omega)|^2$ approaches a nonzero constant as ω tends to infinity, the asymptotic attenuation is 0 dB/decade for even N. For odd N we can find the asymptotic attenuation as follows. Near $x = 0$, Chebyshev polynomial of an odd order is approximately given by $T_N(x) \approx K_N x$, where K_N is constant. Therefore we have for large ω,

$$|H^F(\omega)|^2 \approx \frac{\varepsilon^2 K_N^2 \omega_0^2}{\omega^2}. \tag{10.39}$$

Thus, for odd N, the asymptotic attenuation is 20 dB/decade.

The poles of the transfer function $H^L(s)$ of a low-pass Chebyshev-II filter are inversely proportional to those of a Chebyshev-I filter of the same order. Denoting the

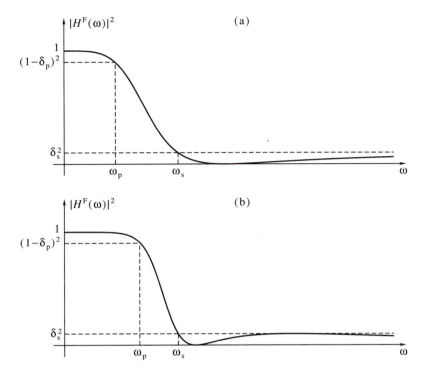

Figure 10.8 The square magnitude of the frequency response of a low-pass Chebyshev filter of the second kind: (a) $N = 2$; (b) $N = 3$.

poles by $\{v_k,\ 0 \le k \le N - 1\}$, we have

$$v_k = \frac{\omega_0^2}{s_k}, \quad 0 \le k \le N - 1, \tag{10.40}$$

where the s_k are given by (10.27). The poles of a Chebyshev filter of the second kind are *not* located on an ellipse.

Contrary to filters of the first kind, Chebyshev filters of the second kind have zeros. As is clear from the definition of $|H^{\mathrm{F}}(\omega)|^2$, the zeros are on the imaginary axis, at the frequencies for which $T_N(\omega_0/\omega) = 0$. The zeros of $T_N(x)$ are easily seen from the definition to be at $\{\cos[(2k + 1)\pi/2N],\ 0 \le k \le N - 1\}$. Therefore, the zeros of $H^{\mathrm{L}}(s)$ are at

$$u_k = \frac{j\omega_0}{\cos\left[\dfrac{(2k + 1)\pi}{2N}\right]}, \quad 0 \le k \le N - 1. \tag{10.41}$$

When N is even, there are N finite zeros. When N is odd, there are only $N - 1$ zeros, since $k = (N - 1)/2$ gives a zero at infinity.

In summary, the transfer function $H^{\mathrm{L}}(s)$ is given by

$$H^{\mathrm{L}}(s) = \prod_{k=0}^{N-1} \frac{v_k(s - u_k)}{u_k(s - v_k)}, \tag{10.42}$$

where $(s - u_k)/u_k$ is replaced by -1 if $u_k = \infty$.

Designing a Chebyshev filter of the second kind to meet given specifications proceeds as follows. First we compute the discrimination factor d and the selectivity factor k from the given values of δ_{p}, δ_{s}, ω_{p}, and ω_{s}, using (10.2) and (10.3). Then we observe from (10.36) that

$$\frac{\varepsilon^2 T_N^2\left(\frac{\omega_0}{\omega_p}\right)}{1 + \varepsilon^2 T_N^2\left(\frac{\omega_0}{\omega_p}\right)} \geq (1 - \delta_p)^2, \tag{10.43}$$

$$\frac{\varepsilon^2 T_N^2\left(\frac{\omega_0}{\omega_s}\right)}{1 + \varepsilon^2 T_N^2\left(\frac{\omega_0}{\omega_s}\right)} \leq \delta_s^2. \tag{10.44}$$

Equation (10.44) can be satisfied with equality by choosing

$$\omega_0 = \omega_s, \quad \varepsilon = (\delta_s^{-2} - 1)^{-1/2}. \tag{10.45}$$

Then we can satisfy (10.43) by choosing

$$T_N(1/k) = \cosh[N \operatorname{arccosh}(1/k)] \geq \frac{1}{d}, \tag{10.46}$$

hence

$$N \geq \frac{\operatorname{arccosh}(1/d)}{\operatorname{arccosh}(1/k)}. \tag{10.47}$$

In practice, we take N as the smallest integer larger than the right side of (10.47). For a given set of specifications, both kinds of Chebyshev filter will have the same order. Therefore, the choice between the two kinds depends only on the desired ripple characteristics (i.e., whether ripple is permitted in the pass band or in the stop band). In summary, the design procedure for a low-pass Chebyshev-II filter is as follows:

─────────────┤ **Low-pass Chebyshev-II filter design procedure** ├─────────────

1. Compute d and k as a function of δ_p, δ_s, ω_p, and ω_s, using (10.2) and (10.3).
2. Compute N, using (10.47), and round upward to the nearest integer.
3. Compute ω_0 and ε, using (10.45).
4. Compute the s_k, using (10.27).
5. Compute the poles v_k, using (10.40).
6. Compute the zeros u_k, using (10.41).
7. Compute the transfer function $H^L(s)$, using (10.42), where $(s - u_k)/u_k$ is replaced by -1 if $u_k = \infty$.

Example 10.3 Design a low-pass Chebyshev-II filter to meet the same specifications as in Example 10.1, therefore having the same selectivity and discrimination factors. We get

$$\varepsilon = 0.001, \quad \omega_0 = 2\,\text{rad/s}, \quad N \geq 8.13 \implies N = 9.$$

The poles of the resulting filter are

$$v_{0,8} = -0.1762 \pm j1.4520, \quad v_{1,7} = -0.5750 \pm j1.4770,$$
$$v_{2,6} = -1.1069 \pm j1.3496, \quad v_{3,5} = -1.7533 \pm j0.9273,$$
$$v_4 = -2.1084.$$

The zeros of the filter are

$$u_{0,8} = \pm j2.0308, \quad u_{1,7} = \pm j2.3094, \quad u_{2,6} = \pm j3.1114, \quad u_{3,5} = \pm j5.8476.$$

The magnitude response of the filter is shown in Figure 10.9. Part a shows the response in the frequency range 0 to 4 rad/s; part b shows the pass band at an expanded scale. The filter meets the stop-band specification exactly, but its transition band is narrower than the specification, so its tolerance in the pass band is much narrower than the specification. □

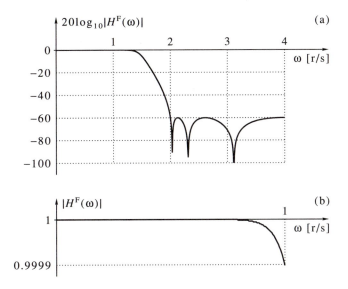

Figure 10.9 Magnitude response of the Chebyshev filter in Example 10.3: (a) full range; (b) pass-band details.

10.4 Elliptic Filters*

Elliptic filters are equiripple in both the pass band and the stop band. They achieve the minimal possible order for given specifications. A low-pass elliptic filter is defined by the square-magnitude response

$$|H^{\mathrm{F}}(\omega)|^2 = \frac{1}{1 + \varepsilon^2 R_N^2 \left(\frac{\omega}{\omega_0}\right)}, \tag{10.48}$$

where $R_N(x)$ is a *Chebyshev rational function* of degree N. The main properties of this function are as follows:

1. It is an even function of x for even N, and an odd function of x for odd N (similar to Chebyshev polynomials).

2. In the range $-1 \leq x \leq 1$, the function oscillates between -1 and 1 and all its zeros are in this range. Therefore, $|H^{\mathrm{F}}(\omega)|^2$ oscillates between 1 and $1/(1 + \varepsilon^2)$ for $0 \leq |\omega| \leq \omega_0$.

3. In the range $1 < |x| < \infty$, $|R_N(x)|$ oscillates between $1/d$ and ∞, where d is a design parameter (later identified with the discrimination factor d). As a result, $|H^{\mathrm{F}}(\omega)|^2$ oscillates between 0 and $1/(1 + \varepsilon^2/d^2)$ in the range $\omega_0 < |\omega| < \infty$.

Figure 10.10 shows the graphs of typical Chebyshev rational functions of orders 2 through 5, in the range $-1 \leq x \leq 1$.

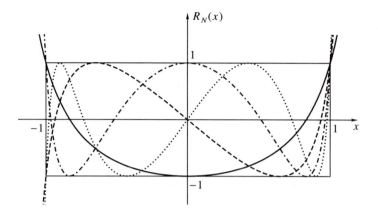

Figure 10.10 Chebyshev rational functions. Solid line: $N = 2$, dashed line: $N = 3$, dot-dashed line: $N = 4$, dotted line: $N = 5$.

Figure 10.11 illustrates the square magnitudes of the frequency responses of a low-pass elliptic filter for $N = 2$ and $N = 3$. Comparing this with Figure 10.5, we see that an elliptic filter indeed has a narrower transition band than a Chebyshev filter having the same order and tolerances.

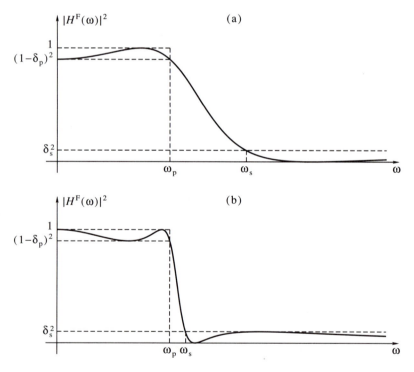

Figure 10.11 The square magnitude of the frequency response of a low-pass elliptic filter: (a) $N = 2$; (b) $N = 3$.

Before we characterize the Chebyshev rational functions, we state a few results on elliptic integrals and Jacobi elliptic functions. We give here the necessary notations and definitions of elliptic integrals and functions, but omit the proofs. A detailed discussion of this topic is given in Antoniou [1993].

The *elliptic integral of the first kind* is defined as

$$u(\phi, m) = \int_0^\phi (1 - m \sin^2 x)^{-1/2} dx, \tag{10.49}$$

where $0 < m < 1$. When we substitute $\phi = 0.5\pi$, we get the *complete elliptic integral of the first kind*

$$K(m) = u(0.5\pi, m) = \int_0^{0.5\pi} (1 - m \sin^2 x)^{-1/2} dx. \tag{10.50}$$

The complete elliptic integral of the first kind is computed in MATLAB by the function ellipke. It is a monotone increasing function of m, tending to 0.5π as m approaches 0, and tending to ∞ as m approaches 1.

The Jacobi elliptic sine function is obtained from the elliptic integral of the first kind by the following two operations:

1. Invert the function $u(\phi, m)$; that is, express ϕ as a function of u, with m as a parameter, say $\phi(u, m)$.

2. The Jacobi elliptic sine function is then

$$\text{sn}(u, m) = \sin[\phi(u, m)]. \tag{10.51}$$

The Jacobi elliptic sine function is a periodic function, with period $4K(m)$. It is computed in MATLAB by the function ellipj (which also computes two other Jacobi elliptic functions, of no interest to us here).

A Chebyshev rational function has the form

$$R_N(x) = Cx^{(N \bmod 2)} \prod_{l=1}^{\lfloor N/2 \rfloor} \frac{x^2 - z_l^2}{x^2 - p_l^2}, \quad \text{where} \quad C = \prod_{l=1}^{\lfloor N/2 \rfloor} \frac{1 - z_l^2}{1 - p_l^2}. \tag{10.52}$$

In the present case, the parameters N, $\{z_l, p_l, 0 \le l \le N/2\}$ are derived from the selectivity factor k and the discrimination factor d of the filter to be designed according to the following formulas:

$$N = \frac{K(k^2)K(1 - d^2)}{K(1 - k^2)K(d^2)}, \tag{10.53}$$

$$z_l = \begin{cases} \text{sn}[2lK(k^2)/N], & N \text{ odd}, \\ \text{sn}[(2l - 1)K(k^2)/N], & N \text{ even}, \end{cases} \qquad p_l = (kz_l)^{-1}. \tag{10.54}$$

After we have computed the poles and the zeros of $R_N(x)$, we can determine $H^L(s)$ from (10.48). To this end, we substitute $\omega = -js$ and $\omega_0 = 1$. We get

$$R_N(-js) = C(-js)^{(N \bmod 2)} \prod_{l=1}^{\lfloor N/2 \rfloor} \frac{s^2 + z_l^2}{s^2 + p_l^2}. \tag{10.55}$$

Substitution of (10.55) in (10.48) gives

$$H^L(s)H^L(-s) = \frac{\prod_{l=1}^{\lfloor N/2 \rfloor}(s^2 + p_l^2)^2}{\prod_{l=1}^{\lfloor N/2 \rfloor}(s^2 + p_l^2)^2 + (C\varepsilon)^2(-s^2)^{(N \bmod 2)} \prod_{l=1}^{\lfloor N/2 \rfloor}(s^2 + z_l^2)^2}. \tag{10.56}$$

Therefore, the zeros of $H^L(s)$ are $\{jp_l, -jp_l, 1 \le l \le \lfloor N/2 \rfloor\}$ and the poles are the subset of N roots of the denominator on the right side of (10.56) whose real part is negative.

The following steps summarize the design procedure of a low-pass elliptic filter having $\omega_p = 1$ and meeting given specifications:

| **Low-pass elliptic filter design procedure** |

---

1. Compute the discrimination factor d and the selectivity factor k, using (10.2) and (10.3).

2. Compute

$$\varepsilon = [(1 - \delta_p)^{-2} - 1]^{1/2}. \tag{10.57}$$

3. Compute the order N from (10.53). Usually N will be noninteger. Therefore, we have to round N upward to the nearest integer, then search for a new k (which will be larger than the given k) such that (10.53) is satisfied exactly with the integer N. This guarantees that the pass-band specifications will be met exactly, whereas the stop-band specifications will be exceeded. The search for new k is performed numerically, as explained in Section 10.5.

4. Compute z_l, p_l, and C as given by (10.54) and (10.52).

5. Compute the zeros of $H^L(s)$ as

$$u_{2l-1} = jp_l, \ u_{2l} = -jp_l, \quad 1 \le l \le \lfloor N/2 \rfloor. \tag{10.58}$$

6. Compute the poles of $H^L(s)$ as follows:

   (a) Solve the polynomial equation

$$\prod_{l=1}^{\lfloor N/2 \rfloor} (s^2 + p_l^2)^2 + (\varepsilon C)^2 (-s^2)^{(N \bmod 2)} \prod_{l=1}^{\lfloor N/2 \rfloor} (s^2 + z_l^2)^2 = 0. \tag{10.59}$$

   (b) Take the poles v_l of $H^L(s)$ as the subset of roots with negative real parts (there are always exactly N such roots).

7. Let

$$H_0 = \begin{cases} (1 + \varepsilon^2)^{-1/2} \dfrac{\prod_{l=1}^{N}(-v_l)}{\prod_{l=1}^{N}(-u_l)}, & N \text{ even}, \\[3mm] \dfrac{\prod_{l=1}^{N}(-v_l)}{\prod_{l=1}^{N-1}(-u_l)}, & N \text{ odd}. \end{cases} \tag{10.60}$$

8. Finally, the transfer function of the elliptic filter is given by

$$H^L(s) = H_0 \frac{\prod_{l=1}^{2\lfloor N/2 \rfloor}(s - u_l)}{\prod_{l=1}^{N}(s - v_l)}. \tag{10.61}$$

Example 10.4 Design a low-pass elliptic filter to meet the same specifications as in Example 10.1, therefore having the same selectivity and discrimination factors. We get

$$\varepsilon = 0.04475, \quad \omega_0 = 1 \, \text{rad/s}, \quad k = 0.5486, \quad N = 6.$$

The poles of the resulting filter are

$$v_{1,2} = -0.1259 \pm j0.8386, \quad v_{3,4} = -0.4995 \pm j0.8914,$$
$$v_{5,6} = -1.2454 \pm j0.5955.$$

The zeros of the filter are

$$u_{1,2} = \pm j1.0353, \quad u_{3,4} = \pm j1.4142, \quad u_{5,6} = \pm j3.8637.$$

The magnitude response of the filter is shown in Figure 10.12. Part a shows the response in the frequency range 0 to 4 rad/s; part b shows the pass band at an expanded scale. The filter meets the pass-band specification exactly, but its transition band is slightly narrower than the specification. $\qquad\square$

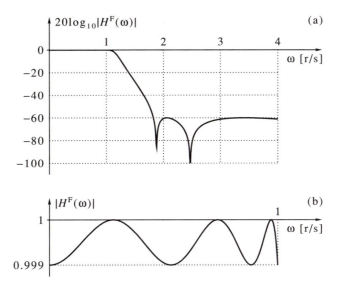

Figure 10.12 Magnitude response of the elliptic filter in Example 10.4: (a) full range; (b) pass-band details.

10.5 MATLAB Programs for Analog Low-Pass Filters

Table 10.2 summarizes the main properties and design equations of the four analog filters described in the preceding sections.

Filter	Butterworth	Chebyshev-I	Chebyshev-II	Elliptic
Pass band	monotone	equiripple	monotone	equiripple
Stop band	monotone	monotone	equiripple	equiripple
Order	(10.18)	(10.35)	(10.47)	(10.53)
ω_0	(10.19)	(10.33)	(10.45)	1
ε	N/A	(10.33)	(10.45)	(10.57)
K	N/A	N/A	N/A	(10.50)
Poles	(10.12)	(10.27)	(10.40)	(10.59)
Zeros	none	none	(10.41)	(10.58)
$H^L(s)$	(10.11)	(10.29)	(10.42)	(10.60), (10.61)

Table 10.2 Summary of properties and design equations of analog low-pass filter.

For a given set of specification parameters (band-edge frequencies and tolerances), the elliptic filter always has the smallest order of the four filter classes. Therefore, the elliptic filter is usually the preferred class for general IIR filtering applications. The other three classes are appropriate when monotone magnitude response is required in certain bands. If monotone response is required in the pass band, Chebyshev-II filter is appropriate; if monotone response is required in the stop band, Chebyshev-I filter is appropriate; if monotone response is required for all frequencies, Butterworth filter is appropriate.

The procedure `analoglp` in Program 10.1 computes the numerator and denominator polynomials of the four low-pass filter classes, as well as the poles, zeros, and constant gain. The implementation is straightforward. First, the poles and the constant gain are computed, depending on the filter class. In the case of Chebyshev-II filter, the zeros are computed as well. In the case of elliptic filter, a call is made to elliplp; see Program 10.2. This program implements the design of a low-pass elliptic filter as described in Section 10.4. Finally, the poles, zeros, and constant gain are expanded to form the two polynomials.

The procedure lpspec in Program 10.3 computes the parameters of a low-pass filter of one of the four classes according to given specifications. The inputs to the program are the four parameters ω_p, ω_s, δ_p, δ_s. The program provides the filter's order N, the frequency ω_0, the parameter ε, and, for elliptic filters, the parameter $m = k^2$. In the case of elliptic filter, a call is made to ellord, shown in Program 10.4. This program first computes the order N, using formula (10.53). Since the right side of (10.53) is not an integer in general, N is taken as the nearest larger integer. Then a search is performed over m (recall that $m = k^2$) such that (10.53) is satisfied exactly. The search is performed as follows:

1. First, m is increased in a geometric series, by 1.1 each time, until the right side of (10.53) becomes larger than N. The original m and m thus found bracket the value of m to be computed.

2. The method of *false position* is then used for finding the exact m within the brackets. In this method, a straight line is passed between the two end points and m is found such that the straight line has ordinate equal to N. With this m, the right side of (10.53) is computed again. If it is less than N, the lower end point of the bracketing interval is moved to the new m. If it is greater than N, the upper end point of the bracketing interval is moved to the new m. This iteration is terminated when the right side of (10.53) is equal to N to within 10^{-6}. The iteration is guaranteed to converge, since the right side of (10.53) is a monotone increasing function of m.

10.6   Frequency Transformations

The design of analog filters other than low pass is usually done by designing a low-pass filter of the desired class first (Butterworth, Chebyshev, or elliptic), and then transforming the resulting filter to get the desired frequency response: high pass, band pass, or band stop. Transformations of this kind are called *frequency transformations*. We first define frequency transformations in general, and then discuss special cases.

Let $f(\cdot)$ be a given rational function and let s be the Laplace transform variable. Define a transformed complex variable $\tilde{s}$ by

$$s = f(\tilde{s}).$$

This definition is *implicit*: It expresses s in terms of $\tilde{s}$, but it should be understood as defining $\tilde{s}$ in terms of s. Each of the two complex variables has a frequency variable associated with it, that is,

$$s = j\omega, \quad \tilde{s} = j\tilde{\omega}.$$

The corresponding frequency transformation is therefore

$$\omega = -jf(j\tilde{\omega}).$$

Let $H^L(s)$ denote the transfer function of a low-pass filter. Define the transformed filter $\tilde{H}^L(\tilde{s})$ by

$$\tilde{H}^L(\tilde{s}) = H^L(s)\bigg|_{s=f(\tilde{s})}. \tag{10.62}$$

The equivalent frequency-domain transformation is

$$\tilde{H}^{\mathrm{F}}(\tilde{\omega}) = H^{\mathrm{F}}(\omega)\Big|_{\omega = -jf(\tilde{\omega})}. \tag{10.63}$$

Equation (10.62) should be interpreted in the following manner: Each occurrence of the variable s in the formula for $H^{\mathrm{L}}(s)$ is replaced by the formula $f(\tilde{s})$. This yields a new function in the variable $\tilde{s}$, which is denoted by $\tilde{H}^{\mathrm{L}}(\tilde{s})$. Equation (10.63) should be interpreted in a similar manner.

Example 10.5 Let

$$H^{\mathrm{L}}(s) = \frac{1}{s+3}, \quad s = \frac{2}{\tilde{s}}.$$

Then

$$\tilde{H}^{\mathrm{L}}(\tilde{s}) = \frac{1}{(2/\tilde{s})+3} = \frac{\tilde{s}}{3\tilde{s}+2}.$$

As we see, $H^{\mathrm{L}}(s)$ is a low-pass filter, and $\tilde{H}^{\mathrm{L}}(\tilde{s})$ is a high-pass filter. $\qquad\square$

Since we have restricted $f(\cdot)$ to be a rational function, and since $H^{\mathrm{L}}(s)$ is rational, it follows that $\tilde{H}^{\mathrm{L}}(\tilde{s})$ is a rational function of $\tilde{s}$. Moreover, $f(\cdot)$ must preserve the stability of the filter. Thus, the left half s plane should be transformed to the left half $\tilde{s}$ plane, the right half s plane should be transformed to the right half $\tilde{s}$ plane, and the imaginary axis $j\omega$ should be transformed to the imaginary axis $j\tilde{\omega}$. The behavior of $\tilde{H}^{\mathrm{F}}(\tilde{\omega})$ must be according to the desired filter type in the transformed domain: high pass, band pass, or band stop. Finally, the function $f(\cdot)$ should have the minimal possible order, to minimize the complexity of the transformed filter.

The general procedure for designing an analog filter by a frequency transformation is as follows. Let us call the domain of s and ω the *design domain*, and the domain of $\tilde{s}$ and $\tilde{\omega}$ the *target domain*. It is convenient to think of the low-pass filter as normalized, that is, to think of s and ω as dimensionless numbers. The target-domain quantities $\tilde{s}$ and $\tilde{\omega}$ are measured, as usual, in radians per second. The specifications are assumed to be given in the target domain. The first step is then to convert them to specifications of a low-pass filter in the design domain, that is, to $\delta_{\mathrm{p}}, \delta_{\mathrm{s}}, \omega_{\mathrm{p}}, \omega_{\mathrm{s}}$. This conversion depends on the filter type. The next step is to design the low-pass filter $H^{\mathrm{L}}(s)$ according to these specifications, using one of the methods studied in the preceding sections. The final step is to transform the design to the target domain, using (10.62). The specific procedures for the different transformation types are described next.

10.6.1 Low-Pass to Low-Pass Transformation

Low-pass to low-pass frequency transformation is not strictly needed, because the design methods of low-pass filters we have presented are sufficiently general; however, we present it for completeness. The low-pass to low-pass transformation is

$$s = \frac{\tilde{s}}{\omega_{\mathrm{c}}}, \quad \omega = \frac{\tilde{\omega}}{\omega_{\mathrm{c}}}, \tag{10.64}$$

where ω_{c} is a positive parameter. This transformation stretches (or contracts) the frequency axis by a constant factor. For example, if $H^{\mathrm{L}}(s)$ is a low-pass filter with band-edge frequencies ω_{p} and ω_{s}, then $\tilde{H}^{\mathrm{L}}(\tilde{s}) = H^{\mathrm{L}}(\tilde{s}/\omega_{\mathrm{c}})$ is a low-pass filter with band-edge frequencies $\omega_{\mathrm{c}}\omega_{\mathrm{p}}$ and $\omega_{\mathrm{c}}\omega_{\mathrm{s}}$. The choice of ω_{c} is arbitrary; however, the following are convenient choices for the different filter classes:

1. For a Butterworth filter choose $\omega_{\mathrm{c}} = \tilde{\omega}_0$.

2. For a Chebyshev-I filter choose $\omega_c = \widetilde{\omega}_p$.

3. For a Chebyshev-II filter choose $\omega_c = \widetilde{\omega}_s$.

4. For an elliptic filter choose $\omega_c = \sqrt{\widetilde{\omega}_p \widetilde{\omega}_s}$.

All these choices cause the parameter ω_0 of the low-pass filter to be equal to 1.

Low-pass to low-pass transformation can be carried out in terms of either the numerator and denominator polynomials or the pole–zero factorization of the low-pass filter. Let

$$H^L(s) = \frac{\sum_{k=0}^{q} b_k s^{q-k}}{\sum_{k=0}^{p} a_k s^{p-k}} = \frac{b_0 \prod_{k=1}^{q}(s - u_k)}{\prod_{k=1}^{p}(s - v_k)}, \quad a_0 = 1. \tag{10.65}$$

Then the transformed filter is given by

$$\widetilde{H}^L(\tilde{s}) = \frac{\omega_c^{p-q} \sum_{k=0}^{q} b_k \omega_c^k \tilde{s}^{q-k}}{\sum_{k=0}^{p} a_k \omega_c^k \tilde{s}^{p-k}} = \frac{b_0 \omega_c^{p-q} \prod_{k=1}^{q}(\tilde{s} - u_k \omega_c)}{\prod_{k=1}^{p}(\tilde{s} - v_k \omega_c)}. \tag{10.66}$$

10.6.2 Low-Pass to High-Pass Transformation

The transformation

$$s = \frac{\omega_c}{\tilde{s}}, \quad \omega = -\frac{\omega_c}{\widetilde{\omega}}, \tag{10.67}$$

where ω_c is a positive parameter, is low pass to high pass. This is illustrated in Figure 10.13, which shows how a low-pass Butterworth filter (in the ω domain) is transformed to a high-pass Butterworth filter (in the $\widetilde{\omega}$ domain). For convenience, we have omitted the sign reversal of the frequencies. The sign reversal is immaterial, since the magnitude response is a symmetric function of the frequency.

Suppose we wish to design a high-pass filter with band edge frequencies $\widetilde{\omega}_p$ and $\widetilde{\omega}_s$. We do it by designing a low-pass filter $H^L(s)$ with $\omega_p = \omega_c/\widetilde{\omega}_p$ and $\omega_s = \omega_c/\widetilde{\omega}_s$. The tolerance parameters δ_p and δ_s used for the low-pass design are the same as the desired tolerances of the high-pass filter. Then $\widetilde{H}^L(\tilde{s}) = H^L(\omega_c/\tilde{s})$ will be a high-pass filter with the given band-edge frequencies and tolerance parameters. The convenient choices for Butterworth, Chebyshev, and elliptic filters are the same as for the low-pass to low-pass transformation.

The low-pass to high-pass transformation is a first-order rational function, so the order of $\widetilde{H}^L(\tilde{s})$ is the same as that of $H^L(s)$. To show that it preserves stability, let $\tilde{s} = \tilde{\sigma} + j\widetilde{\omega}$ be the decomposition of $\tilde{s}$ into real and imaginary parts. Then the corresponding decomposition of s is

$$s = \sigma + j\omega = \frac{\omega_c \tilde{\sigma}}{\tilde{\sigma}^2 + \widetilde{\omega}^2} - j\frac{\omega_c \widetilde{\omega}}{\tilde{\sigma}^2 + \widetilde{\omega}^2}. \tag{10.68}$$

We find that σ and $\tilde{\sigma}$ always have the same sign, which implies the stability property.

Low-pass to high-pass transformation can be carried out in terms of either the numerator and denominator polynomials or the pole–zero factorization of the low-pass filter. Let the filter $H^L(s)$ be as in (10.65). Then the transformed filter is given by

$$\widetilde{H}^L(\tilde{s}) = \frac{\tilde{s}^{p-q} \sum_{k=0}^{q} b_k \omega_c^{q-k} \tilde{s}^k}{\sum_{k=0}^{p} a_k \omega_c^{p-k} \tilde{s}^k} = \frac{b_0 \prod_{k=1}^{q}(-u_k)}{\prod_{k=1}^{p}(-v_k)} \cdot \frac{\tilde{s}^{p-q} \prod_{k=1}^{q}(\tilde{s} - \omega_c u_k^{-1})}{\prod_{k=1}^{p}(\tilde{s} - \omega_c v_k^{-1})}. \tag{10.69}$$

As we see, the number of zeros of a high-pass filter is always equal to the number of poles. If $H^L(s)$ is Butterworth or Chebyshev-I, it has no zeros. Therefore, in this case, $\widetilde{H}^L(\tilde{s})$ will have p zeros at $\tilde{s} = 0$. If $H^L(s)$ is Chebyshev-II or elliptic, it has $2\lfloor p/2 \rfloor$

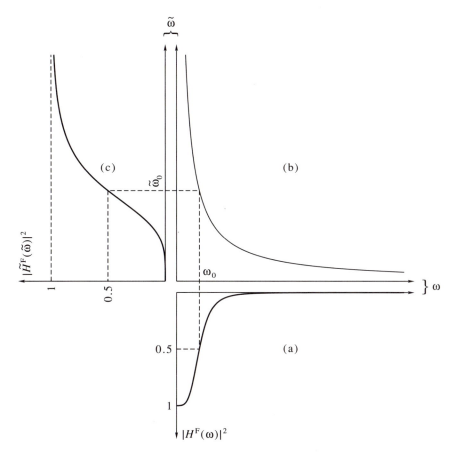

Figure 10.13 Low-pass to high-pass transformation of an analog filter: (a) frequency response of the low-pass filter (amplitude axis pointing downward); (b) $\widetilde{\omega}$ as a function of ω; (c) frequency response of the high-pass filter (amplitude axis pointing to left).

zeros on the imaginary axis. These will be transformed to zeros on the imaginary axis according to (10.69). Also, there will be an additional zero at $\widetilde{s} = 0$ if p is odd, and no additional zero if p is even.

Formula (10.69) is convenient for computer implementation of the transformation, as we shall see in Section 10.6.5. In summary, the design procedure for a high-pass analog filter is as follows:

$\boxed{\textbf{High-pass analog filter design procedure}}$

1. Choose ω_c as an arbitrary positive parameter, for example, $\omega_c = 1$.
2. Given the high-pass filter specifications $\widetilde{\omega}_p$, $\widetilde{\omega}_s$, $\widetilde{\delta}_p$, $\widetilde{\delta}_s$, let

$$\delta_p = \widetilde{\delta}_p, \quad \delta_s = \widetilde{\delta}_s, \quad \omega_p = \frac{\omega_c}{\widetilde{\omega}_p}, \quad \omega_s = \frac{\omega_c}{\widetilde{\omega}_s}. \qquad (10.70)$$

3. Design a low-pass analog filter $H^L(s)$ to meet the specifications ω_p, ω_s, δ_p, δ_s.
4. Obtain the analog high-pass filter $\widetilde{H}^L(\widetilde{s})$, using (10.69).

Example 10.6 We wish to design a high-pass filter according to the specification

$$\widetilde{\omega}_s = 0.5, \quad \widetilde{\omega}_p = 5, \quad \delta_s = 0.01, \quad \delta_p = 0.01.$$

We get, with $\omega_c = 1$,

$$\omega_p = 0.2, \quad \omega_s = 2.$$

Therefore,

$$d = 0.01432, \quad k = 0.1.$$

For Butterworth filter we get

$$N = 3, \; \omega_0 = 0.3829, \; H^L(s) = \frac{0.05614}{s^3 + 0.7658s^2 + 0.2932s + 0.05614}.$$

The low-pass to high-pass transformation gives

$$\widetilde{H}^L(\widetilde{s}) = \frac{\widetilde{s}^3}{\widetilde{s}^3 + 5.2231\widetilde{s}^2 + 13.6405\widetilde{s} + 17.8115}.$$

For Chebyshev-I filter we get

$$N = 3, \; \omega_0 = 0.2, \; \varepsilon = 0.1425, \; H^L(s) = \frac{0.01404}{s^3 + 0.4005s^2 + 0.1102s + 0.01404}.$$

The low-pass to high-pass transformation gives

$$\widetilde{H}^L(\widetilde{s}) = \frac{\widetilde{s}^3}{\widetilde{s}^3 + 7.8507\widetilde{s}^2 + 28.5325\widetilde{s} + 71.2461}.$$

For Chebyshev-II filter we get

$$N = 3, \; \omega_0 = 2, \; \varepsilon = 0.01, \; H^L(s) = \frac{0.06s^2 + 0.32}{s^3 + 1.3492s^2 + 0.9084s + 0.32}.$$

The low-pass to high-pass transformation gives

$$\widetilde{H}^L(\widetilde{s}) = \frac{\widetilde{s}^3 + 0.1875\widetilde{s}}{\widetilde{s}^3 + 2.8385\widetilde{s}^2 + 4.2160\widetilde{s} + 3.1248}.$$

For an elliptic filter we get

$$N = 3, \; \omega_0 = 0.2, \; \varepsilon = 0.1425, \; m = 0.0773,$$

$$H^L(s) = \frac{0.02116s^2 + 0.01446}{s^3 + 0.3958s^2 + 0.1084s + 0.01446}.$$

The low-pass to high-pass transformation gives

$$\widetilde{H}^L(\widetilde{s}) = \frac{\widetilde{s}^3 + 1.4631\widetilde{s}}{\widetilde{s}^3 + 7.4970\widetilde{s}^2 + 27.3713\widetilde{s} + 69.1456}.$$

$\square$

10.6.3   Low-Pass to Band-Pass Transformation

The transformation

$$s = \frac{\widetilde{s}^2 + \omega_l\omega_h}{\widetilde{s}(\omega_h - \omega_l)}, \quad \omega = \frac{\widetilde{\omega}^2 - \omega_l\omega_h}{\widetilde{\omega}(\omega_h - \omega_l)}, \tag{10.71}$$

where ω_h and ω_l are positive parameters satisfying $\omega_h > \omega_l$, is low pass to band pass, as illustrated in Figure 10.14, which shows how a low-pass Butterworth filter (in the ω domain) is transformed to a band-pass Butterworth filter (in the $\widetilde{\omega}$ domain). Since (10.71) is a second-order rational function, to each ω there correspond two values of $\widetilde{\omega}$. Therefore, a total of four values of $\widetilde{\omega}$ correspond to $\pm\omega$. Figure 10.14 shows the two positive values of $\widetilde{\omega}$ corresponding to each positive ω. Note the asymmetric form of the band-pass filter with respect to the center frequency $\sqrt{\omega_l\omega_h}$.

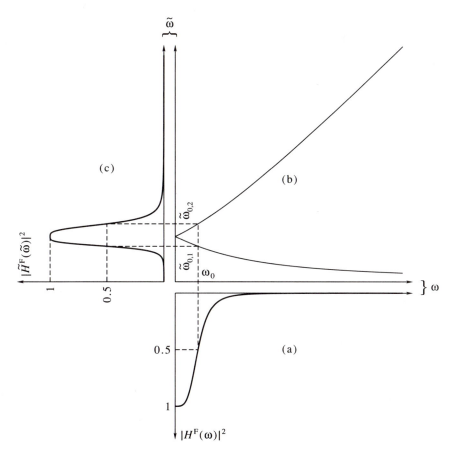

Figure 10.14 Low-pass to band-pass transformation of an analog filter: (a) frequency response of the low-pass filter (amplitude axis pointing downward); (b) $\widetilde{\omega}$ as a function of ω; (c) frequency response of the band-pass filter (amplitude axis pointing to left).

Suppose we wish to design a band-pass filter with tolerance parameters $\widetilde{\delta}_p$, $\widetilde{\delta}_{s,1}$, $\widetilde{\delta}_{s,2}$, and band-edge frequencies $\widetilde{\omega}_{s,1}$, $\widetilde{\omega}_{p,1}$, $\widetilde{\omega}_{p,2}$, $\widetilde{\omega}_{s,2}$. In general, the specifications of the two transition bands and the two stop bands are independent, therefore possibly different. On the other hand, the low-pass filter has only one transition band and one stop band. Therefore, the low-pass design must be such that all specification parameters of the band-pass filter are met or surpassed. This implies that the tolerance parameters of the low-pass filter are to be chosen as

$$\delta_p = \widetilde{\delta}_p, \quad \delta_s = \min\{\widetilde{\delta}_{s,1}, \widetilde{\delta}_{s,2}\}. \tag{10.72}$$

The transformation parameters ω_l, ω_h should be chosen in one of two ways:

1. Such that both $\widetilde{\omega}_{p,1}$ and $\widetilde{\omega}_{p,2}$ will be transformed to the same ω_p, for example, to $\omega_p = 1$. The choice

$$\omega_l = \widetilde{\omega}_{p,1}, \quad \omega_h = \widetilde{\omega}_{p,2} \tag{10.73}$$

accomplishes this, since

$$\frac{\widetilde{\omega}_{p,1}^2 - \widetilde{\omega}_{p,1}\widetilde{\omega}_{p,2}}{\widetilde{\omega}_{p,1}(\widetilde{\omega}_{p,2} - \widetilde{\omega}_{p,1})} = -1, \quad \frac{\widetilde{\omega}_{p,2}^2 - \widetilde{\omega}_{p,1}\widetilde{\omega}_{p,2}}{\widetilde{\omega}_{p,2}(\widetilde{\omega}_{p,2} - \widetilde{\omega}_{p,1})} = 1, \tag{10.74}$$

and the frequencies 1 and -1 are equivalent. With this choice, the stop-band frequencies are transformed to

$$\omega_{s,1} = \frac{\widetilde{\omega}_{s,1}^2 - \widetilde{\omega}_{p,1}\widetilde{\omega}_{p,2}}{\widetilde{\omega}_{s,1}(\widetilde{\omega}_{p,2} - \widetilde{\omega}_{p,1})}, \quad \omega_{s,2} = \frac{\widetilde{\omega}_{s,2}^2 - \widetilde{\omega}_{p,1}\widetilde{\omega}_{p,2}}{\widetilde{\omega}_{s,2}(\widetilde{\omega}_{p,2} - \widetilde{\omega}_{p,1})}. \tag{10.75}$$

A smaller value of ω_s corresponds to a narrower transition band. Therefore, the stop-band frequency of the low-pass filter must be chosen as

$$\omega_s = \min\{|\omega_{s,1}|, |\omega_{s,2}|\}, \tag{10.76}$$

to satisfy both transition-band requirements.

2. Such that both $\widetilde{\omega}_{s,1}$ and $\widetilde{\omega}_{s,2}$ will be transformed to the same ω_s, for example, to $\omega_s = 1$. The choice

$$\omega_l = \widetilde{\omega}_{s,1}, \quad \omega_h = \widetilde{\omega}_{s,2} \tag{10.77}$$

accomplishes this, since

$$\frac{\widetilde{\omega}_{s,1}^2 - \widetilde{\omega}_{s,1}\widetilde{\omega}_{s,2}}{\widetilde{\omega}_{s,1}(\widetilde{\omega}_{s,2} - \widetilde{\omega}_{s,1})} = -1, \quad \frac{\widetilde{\omega}_{s,2}^2 - \widetilde{\omega}_{s,1}\widetilde{\omega}_{s,2}}{\widetilde{\omega}_{s,2}(\widetilde{\omega}_{s,2} - \widetilde{\omega}_{s,1})} = 1, \tag{10.78}$$

and the frequencies ± 1 are equivalent. With this choice, the pass-band frequencies are transformed to

$$\omega_{p,1} = \frac{\widetilde{\omega}_{p,1}^2 - \widetilde{\omega}_{s,1}\widetilde{\omega}_{s,2}}{\widetilde{\omega}_{p,1}(\widetilde{\omega}_{s,2} - \widetilde{\omega}_{s,1})}, \quad \omega_{p,2} = \frac{\widetilde{\omega}_{p,2}^2 - \widetilde{\omega}_{s,1}\widetilde{\omega}_{s,2}}{\widetilde{\omega}_{p,2}(\widetilde{\omega}_{s,2} - \widetilde{\omega}_{s,1})}. \tag{10.79}$$

A larger value of ω_p corresponds to a narrower transition band. Therefore, the pass-band frequency of the low-pass filter must be chosen as

$$\omega_p = \max\{|\omega_{p,1}|, |\omega_{p,2}|\}, \tag{10.80}$$

to satisfy both transition-band requirements.

The low-pass to band-pass transformation is a second-order rational function, therefore the order of $\widetilde{H}^L(\widetilde{s})$ is twice that of $H^L(s)$. To show that it preserves stability, let $\widetilde{s} = \widetilde{\sigma} + j\widetilde{\omega}$ be the decomposition of $\widetilde{s}$ into real and imaginary parts. Then the corresponding decomposition of s is

$$s = \sigma + j\omega = \frac{\widetilde{\sigma}}{\omega_h - \omega_l}\left(1 + \frac{\omega_l\omega_h}{\widetilde{\sigma}^2 + \widetilde{\omega}^2}\right) + j\frac{\widetilde{\omega}}{\omega_h - \omega_l}\left(1 - \frac{\omega_l\omega_h}{\widetilde{\sigma}^2 + \widetilde{\omega}^2}\right). \tag{10.81}$$

We get that σ and $\widetilde{\sigma}$ always have the same sign, which implies the stability property.

The low-pass to band-pass transformation is not convenient to express when $H^L(s)$ is given as a ratio of two polynomials, but it is easy when $H^L(s)$ is given in a factored form. For a single zero u_k we have

$$s - u_k = \frac{\widetilde{s}^2 + \omega_l\omega_h}{\widetilde{s}(\omega_h - \omega_l)} - u_k = \frac{\widetilde{s}^2 - u_k(\omega_h - \omega_l)\widetilde{s} + \omega_l\omega_h}{\widetilde{s}(\omega_h - \omega_l)}, \tag{10.82}$$

and similarly for v_k. Let the pole–zero factorization of $H^L(s)$ be as in (10.65). Then we get, by substituting (10.82) for all poles and zeros,

$$\widetilde{H}^L(\widetilde{s}) = \frac{b_0(\omega_h - \omega_l)^{p-q}\widetilde{s}^{p-q}\prod_{k=1}^q[\widetilde{s}^2 - u_k(\omega_h - \omega_l)\widetilde{s} + \omega_l\omega_h]}{\prod_{k=1}^p[\widetilde{s}^2 - v_k(\omega_h - \omega_l)\widetilde{s} + \omega_l\omega_h]}. \tag{10.83}$$

Now (10.83) can be expanded to a ratio of two polynomials, or factored further to first-order complex factors, as needed.

The number of zeros of a band-pass filter depends on the class of the low-pass filter from which it is derived, and whether its order is even or odd. In any case, the zeros are all on the imaginary axis. See Problem 10.16 for further discussion of this point. In summary, the design procedure for a band-pass analog filter is as follows:

Band-pass analog filter design procedure

1. Given the band-pass filter specifications $\widetilde{\omega}_{p,1}, \widetilde{\omega}_{s,1}, \widetilde{\omega}_{p,2}, \widetilde{\omega}_{s,2}, \widetilde{\delta}_p, \widetilde{\delta}_{s,1}, \widetilde{\delta}_{s,2}$, choose δ_p, δ_s according to (10.72), and ω_l, ω_h according to (10.73).

2. Let $\omega_p = 1$, and compute ω_s according to (10.75), (10.76).

3. Design a low-pass analog filter $H^L(s)$ to meet the specifications $\omega_p, \omega_s, \delta_p, \delta_s$, and find its poles, zeros, and constant gain.

4. Obtain the analog band-pass filter $\widetilde{H}^L(\widetilde{s})$, using (10.83).

Example 10.7 We wish to design a band-pass filter according to the specification

$$\widetilde{\omega}_{s,1} = 0.2, \quad \widetilde{\omega}_{p,1} = 0.5, \quad \widetilde{\omega}_{p,2} = 2, \quad \widetilde{\omega}_{s,2} = 6, \quad \widetilde{\delta}_{s,1} = \widetilde{\delta}_{s,2} = 0.1, \quad \delta_p = 0.1.$$

We get, with $\omega_l = \widetilde{\omega}_{p,1}, \omega_h = \widetilde{\omega}_{p,2}$,

$$\omega_p = 1, \quad \omega_s = \min\{3.2, 3.8889\} = 3.2.$$

Therefore,

$$d = 0.048677, \quad k = 0.3125.$$

For Butterworth filter we get

$$N = 3, \quad \omega_0 = 1.2734, \quad H^L(s) = \frac{2.0647}{s^3 + 2.5467s^2 + 3.2429s + 2.0647}.$$

The low-pass to band-pass transformation gives

$$\widetilde{H}^L(\widetilde{s}) = \frac{6.9685\widetilde{s}^3}{\widetilde{s}^6 + 3.8201\widetilde{s}^5 + 10.2966\widetilde{s}^4 + 14.6087\widetilde{s}^3 + 10.2966\widetilde{s}^2 + 3.8201\widetilde{s} + 1}.$$

For Chebyshev-I filter we get

$$N = 3, \quad \omega_0 = 1, \quad \varepsilon = 0.4843, \quad H^L(s) = \frac{0.5162}{s^3 + 1.0213s^2 + 1.2716s + 0.5162}.$$

The low-pass to band-pass transformation gives

$$\widetilde{H}^L(\widetilde{s}) = \frac{1.7421\widetilde{s}^3}{\widetilde{s}^6 + 1.5320\widetilde{s}^5 + 5.8610\widetilde{s}^4 + 4.8062\widetilde{s}^3 + 5.8610\widetilde{s}^2 + 1.5320\widetilde{s} + 1}.$$

For Chebyshev-II filter we get

$$N = 3, \quad \omega_0 = 3.2, \quad \varepsilon = 0.1005, \quad H^L(s) = \frac{0.9648s^2 + 13.1732}{s^3 + 4.4972s^2 + 9.6471s + 13.1732}.$$

The low-pass to band-pass transformation gives

$$\widetilde{H}^L(\widetilde{s}) = \frac{1.4472\widetilde{s}^5 + 47.3542\widetilde{s}^3 + 1.4472\widetilde{s}}{\widetilde{s}^6 + 6.7458\widetilde{s}^5 + 24.7059\widetilde{s}^4 + 57.9513\widetilde{s}^3 + 24.7059\widetilde{s}^2 + 6.7458\widetilde{s} + 1}.$$

For an elliptic filter we get

$$N = 2, \quad \omega_0 = 1, \quad \varepsilon = 0.4843, \quad m = 0.1770, \quad H^L(s) = \frac{0.1s^2 + 1.0772}{s^2 + 1.0678s + 1.1969}.$$

The low-pass to band-pass transformation gives

$$\widetilde{H}^L(\widetilde{s}) = \frac{0.1\widetilde{s}^4 + 2.6237\widetilde{s}^2 + 0.1}{\widetilde{s}^4 + 1.6017\widetilde{s}^3 + 4.6930\widetilde{s}^2 + 1.6017\widetilde{s} + 1}.$$

$\square$

10.6.4   Low-Pass to Band-Stop Transformation

The transformation

$$s = \frac{\tilde{s}(\omega_h - \omega_l)}{\tilde{s}^2 + \omega_l \omega_h}, \quad \omega = \frac{\tilde{\omega}(\omega_h - \omega_l)}{\omega_l \omega_h - \tilde{\omega}^2}, \tag{10.84}$$

where ω_h and ω_l are positive parameters satisfying $\omega_h > \omega_l$, is low pass to band stop. This is illustrated in Figure 10.15, which shows how a low-pass Butterworth filter (in the ω domain) is transformed to a band-stop Butterworth filter (in the $\tilde{\omega}$ domain). As in the case of low-pass to band-pass transformation, to each ω there correspond two values of $\tilde{\omega}$. Therefore, a total of four values of $\tilde{\omega}$ correspond to $\pm\omega$. Figure 10.15 shows the two positive values of $\tilde{\omega}$ corresponding to each positive ω. Note the asymmetric form of the band-stop filter with respect to the center frequency $\sqrt{\omega_l \omega_h}$.

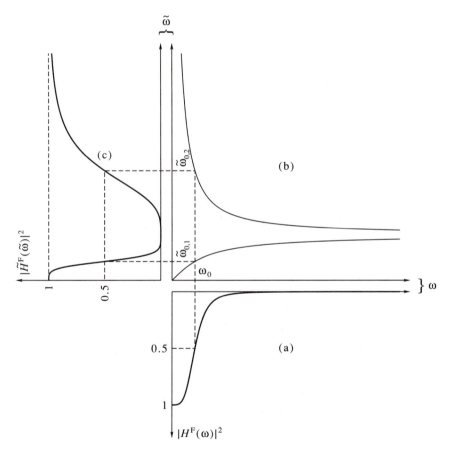

Figure 10.15 Low-pass to band-stop transformation of an analog filter: (a) frequency response of the low-pass filter (amplitude axis pointing downward); (b) $\tilde{\omega}$ as a function of ω; (c) frequency response of the band-stop filter (amplitude axis pointing to left).

Suppose we wish to design a band-stop filter with tolerance parameters $\tilde{\delta}_{p,1}$, $\tilde{\delta}_{p,2}$, $\tilde{\delta}_s$, and band-edge frequencies $\tilde{\omega}_{p,1}$, $\tilde{\omega}_{s,1}$, $\tilde{\omega}_{s,2}$, $\tilde{\omega}_{p,2}$. As in the case of band-pass filter, the low-pass design must be such that *all* specification parameters of the band-stop filter are met or surpassed. This implies that the tolerance parameters of the low-pass filter are to be chosen as

$$\delta_p = \min\{\tilde{\delta}_{p,1}, \tilde{\delta}_{p,2}\}, \quad \delta_s = \tilde{\delta}_s. \tag{10.85}$$

The transformation parameters ω_l, ω_h should be chosen in one of two ways:

1. Such that both $\tilde{\omega}_{p,1}$ and $\tilde{\omega}_{p,2}$ will be transformed to $\omega_p = 1$. The choice

$$\omega_l = \tilde{\omega}_{p,1}, \quad \omega_h = \tilde{\omega}_{p,2} \tag{10.86}$$

accomplishes this. With this choice, the stop-band frequencies are transformed to

$$\omega_{s,1} = \frac{\tilde{\omega}_{s,1}(\tilde{\omega}_{p,2} - \tilde{\omega}_{p,1})}{\tilde{\omega}_{p,1}\tilde{\omega}_{p,2} - \tilde{\omega}_{s,1}^2}, \quad \omega_{s,2} = \frac{\tilde{\omega}_{s,2}(\tilde{\omega}_{p,2} - \tilde{\omega}_{p,1})}{\tilde{\omega}_{p,1}\tilde{\omega}_{p,2} - \tilde{\omega}_{s,2}^2}. \tag{10.87}$$

The stop-band frequency of the low-pass filter must be chosen as

$$\omega_s = \min\{|\omega_{s,1}|, |\omega_{s,2}|\}, \tag{10.88}$$

to satisfy both transition-band requirements.

2. Such that both $\tilde{\omega}_{s,1}$ and $\tilde{\omega}_{s,2}$ will be transformed to $\omega_s = 1$. The choice

$$\omega_l = \tilde{\omega}_{s,1}, \quad \omega_h = \tilde{\omega}_{s,2} \tag{10.89}$$

accomplishes this. With this choice, the pass-band frequencies are transformed to

$$\omega_{p,1} = \frac{\tilde{\omega}_{p,1}(\tilde{\omega}_{s,2} - \tilde{\omega}_{s,1})}{\tilde{\omega}_{s,1}\tilde{\omega}_{s,2} - \tilde{\omega}_{p,1}^2}, \quad \omega_{p,2} = \frac{\tilde{\omega}_{p,2}(\tilde{\omega}_{s,2} - \tilde{\omega}_{s,1})}{\tilde{\omega}_{s,1}\tilde{\omega}_{s,2} - \tilde{\omega}_{p,2}^2}. \tag{10.90}$$

The pass-band frequency of the low-pass filter must be chosen as

$$\omega_p = \max\{|\omega_{p,1}|, |\omega_{p,2}|\}, \tag{10.91}$$

to satisfy both transition-band requirements.

The low-pass to band-stop transformation can be performed by transforming from low-pass to high-pass first, using (10.67), followed by a low-pass to band-pass transformation (10.71). Since each of these transformations preserves stability, so does the low-pass to band-stop transformation.

Let us write the low-pass to band-stop transformation for $H^L(s)$ in a factored form, as we did for the low-pass to band-pass transformation. For a single zero u_k we have

$$s - u_k = \frac{\tilde{s}(\omega_h - \omega_l)}{\tilde{s}^2 + \omega_l\omega_h} - u_k = \frac{(-u_k)[\tilde{s}^2 - u_k^{-1}(\omega_h - \omega_l)\tilde{s} + \omega_l\omega_h]}{\tilde{s}^2 + \omega_l\omega_h}, \tag{10.92}$$

and similarly for v_k. Let the pole–zero factorization of $H^L(s)$ be as in (10.65). Then, substituting (10.92) for all poles and zeros, we can write

$$\tilde{H}^L(\tilde{s}) = \frac{b_0 \prod_{k=1}^q (-u_k)}{\prod_{k=1}^p (-v_k)}$$
$$\cdot \frac{(\tilde{s}^2 + \omega_l\omega_h)^{p-q} \prod_{k=1}^q [\tilde{s}^2 - u_k^{-1}(\omega_h - \omega_l)\tilde{s} + \omega_l\omega_h]}{\prod_{k=1}^p [\tilde{s}^2 - v_k^{-1}(\omega_h - \omega_l)\tilde{s} + \omega_l\omega_h]}. \tag{10.93}$$

Now (10.93) can be expanded to a ratio of two polynomials, or factored further to first-order complex factors, as needed.

The number of zeros of a band-stop filter is always $2p$, and they are all on the imaginary axis. See Problem 10.17 for further discussion of this point. The low-pass to band-stop transformation can also be carried out by using the low-pass to high-pass transformation (10.69) and the low-pass to band-pass transformation (10.83) in succession. In summary, the design procedure for a band-stop analog filter is as follows:

Band-stop analog filter design procedure
---

1. Given the band-stop filter specifications $\widetilde{\omega}_{p,1}, \widetilde{\omega}_{s,1}, \widetilde{\omega}_{p,2}, \widetilde{\omega}_{s,2}, \widetilde{\delta}_p, \widetilde{\delta}_{p,1}, \widetilde{\delta}_{p,2}$, choose δ_s, δ_s according to (10.85), and ω_l, ω_h according to (10.86).

2. Let $\omega_p = 1$, and compute ω_s according to (10.87), (10.88).

3. Design a low-pass analog filter $H^L(s)$ to meet the specifications $\omega_p, \omega_s, \delta_p, \delta_s$, and find its poles, zeros, and constant gain.

4. Obtain the analog band-pass filter $\tilde{H}^L(\tilde{s})$, using (10.93).

10.6.5 MATLAB Implementation of Frequency Transformations

The procedure `analogtr` in Program 10.5 implements the frequency transformation formulas. It uses the pole–zero factorization of the low-pass filter as input and provides both the pole–zero factorization and the expanded polynomials of the transformed filter. It uses (10.66) for low-pass to low-pass transformation, (10.69) for low-pass to high-pass transformation, and (10.83) for low-pass to band-pass transformation. For low-pass to band-stop transformation it calls itself recursively twice, first performing low-pass to high-pass transformation with $\omega_c = 1$, and then performing low-pass to band-pass transformation.

10.7 Impulse Invariant Transformation

We now turn our attention to the problem of transforming an analog IIR filter into the digital domain. We assume that we are given a real, causal, stable, rational filter $H^L(s)$, designed to meet given specifications. We wish to obtain a digital IIR filter $H^z(z)$ from $H^L(s)$. The main requirements for $H^z(z)$ are as follows:

1. It should be real, causal, stable, and rational.

2. Its order should not be greater than that of $H^L(s)$. This feature is not absolutely necessary but is highly desired, since we do not wish to increase the complexity of the digital filter beyond necessity.

3. Its frequency response should be close to that of the analog filter in the frequency range of interest (the relationship between analog and digital frequencies being $\theta = \omega T$).

In addition, the transformation method should be simple, convenient to implement, and applicable to all classes and types of analog filters.

The first method we present is called the *impulse invariant method*. Let $h(t)$ be the impulse response of an analog filter, and $h[n]$ the impulse response of the desired digital filter. The impulse invariant method defines $h[n]$ as proportional to the samples of $h(t)$; in other words,

$$h[n] = Th(nT). \tag{10.94}$$

The following procedure is used to obtain the digital filter's transfer function:

1. Compute the inverse Laplace transform of $H^L(s)$ to get $h(t)$.

2. Sample the impulse response at interval T and multiply by T to obtain $h[n]$.

3. Compute the z-transform of the sequence $h[n]$.

This sequence of operations is denoted by the operator

$$H^z(z) = \{\mathcal{Z}_T H^L\}(z). \tag{10.95}$$

The name *impulse invariant* should be obvious from the definition. The constant scale factor T can be explained as follows. The DC gains of the analog and digital filters are given, respectively, by

$$H^L(0) = \int_0^\infty h(t)dt, \quad H^z(1) = \sum_{n=0}^\infty h[n] = \sum_{n=0}^\infty Th(nT). \tag{10.96}$$

The sum, including the factor T, can be regarded as a zero-order (Euler) approximation of the integral.[2] Thus, the factor T causes the two filters to have approximately equal DC gains.

Example 10.8 Consider the first-order analog filter $H^L(s) = \alpha/(s + \alpha)$. The corresponding impulse response is

$$h(t) = \alpha e^{-\alpha t} v(t),$$

so

$$h[n] = Th(nT) = \alpha T e^{-n\alpha T} v[n].$$

Therefore,

$$H^z(z) = \frac{\alpha T}{1 - e^{-\alpha T}z^{-1}}.$$

As we see, a first-order analog filter is transformed to a first-order digital filter, and stability is preserved, since $\alpha > 0$ implies $0 < e^{-\alpha T} < 1$. We shall soon see that these properties hold for general filters (not necessarily first order). The DC gain of $H^z(z)$ is

$$H^z(1) = \frac{\alpha T}{1 - e^{-\alpha T}}.$$

Since $e^{-\alpha T} \approx 1 - \alpha T$ if $\alpha T \ll 1$, we get that $H^z(1) \approx 1$ in this case. $\square$

In the next example, the impulse invariant method fails.

Example 10.9 Consider the first-order analog high-pass filter

$$H^L(s) = \frac{s}{s + \alpha} = 1 - \frac{\alpha}{s + \alpha}.$$

The corresponding impulse response is

$$h(t) = \delta(t) - \alpha e^{-\alpha t} v(t).$$

Now the presence of the delta term in $h(t)$ prevents sampling of the impulse response, so $h[n]$ is not defined. The problem is fundamental, not just technical. Suppose, for example, that we try to substitute $T^{-1}\delta[n]$ for the sampling of $\delta(t)$. This gives

$$H^z(z) = 1 - \frac{\alpha T}{1 - e^{-\alpha T}z^{-1}} = \frac{1 - \alpha T - e^{-\alpha T}z^{-1}}{1 - e^{-\alpha T}z^{-1}}.$$

The resulting digital filter will be high pass if $\alpha T \ll 1$, but not necessarily so otherwise. For example, if $\alpha T = 1$, then $H^z(z)$ becomes a low-pass filter with negative gain and an additional unit delay. $\square$

The conclusion from the preceding example is that the impulse invariant transformation is not suitable for analog filters whose transfer function is *exactly proper* (i.e., their numerator and denominator polynomials are of the same degree), since the corresponding impulse response contains a term $\delta(t)$. High-pass and band-stop filters have exactly proper transfer functions, so the impulse invariant method cannot be used for such filters. On the other hand, low-pass and band-pass filters usually have *strictly proper* transfer functions (i.e., the numerator degree is strictly less than the denominator degree). Such functions have no delta function terms in their impulse response, so the impulse invariant method can be applied to them (see, however, Problem 10.31 in this regard).

Any rational strictly proper transfer function $H^L(s)$ can be decomposed to partial fractions. If the poles of $H^L(s)$ are simple, the decomposition has the form

$$H^L(s) = \frac{b_0 s^q + b_1 s^{q-1} + \cdots + b_q}{s^p + a_1 s^{p-1} + \cdots + a_p} = \sum_{k=1}^{p} \frac{C_k}{s - \lambda_k}, \tag{10.97}$$

where p is the filter's order, $\{\lambda_k\}$ are the poles, and $\{C_k\}$ are the residues. Correspondingly,

$$h(t) = \sum_{k=1}^{p} C_k e^{\lambda_k t}, \quad h[n] = \sum_{k=1}^{p} C_k T e^{n \lambda_k T}, \tag{10.98}$$

so

$$H^z(z) = \sum_{k=1}^{p} \frac{C_k T}{1 - e^{\lambda_k T} z^{-1}}. \tag{10.99}$$

The transformed digital filter $H^z(z)$ has the following properties:

1. Its order is the same as that of the analog filter, because the common denominator on the right side has degree p.

2. Its poles are mapped according to

$$\lambda_k \longrightarrow e^{\lambda_k T}, \quad 1 \le k \le p.$$

   We have

$$\Re\{\lambda_k\} < 0 \Longrightarrow |e^{\lambda_k T}| < 1, \quad 1 \le k \le p.$$

   Consequently, a stable analog filter $H^L(s)$ is transformed to a stable digital filter $H^z(z)$.

3. The zeros of $H^z(z)$ do not bear a simple relationship to those of $H^L(s)$. When the right side of (10.99) is brought to a common denominator, the numerator will be a polynomial of degree $p - 1$ in z^{-1} in general, regardless of the degree q of the numerator of $H^L(s)$. The roots of this polynomial will not depend on $\{\lambda_k, C_k, T\}$ in an obvious manner.

These properties hold also in case the analog filter has multiple poles, but the corresponding formulas are omitted.

The procedure `impinv` in Program 10.6 implements the impulse invariant method. It first decomposes the analog transfer function into partial fractions, using the MATLAB function `residue`. It then transforms the poles and the gains as given by (10.99), and finally calls `pf2tf` (described in Section 7.4) to bring the partial fractions under a common denominator. The program issues an error message if the numerator degree of the analog filter is not smaller than the denominator degree.

The frequency response $H^f(\theta)$ of the digital filter is related to that of the analog filter $H^F(\omega)$ by the sampling theorem,

$$H^f(\theta) = \sum_{k=-\infty}^{\infty} H^F\left(\frac{\theta - 2\pi k}{T}\right) \tag{10.100}$$

(note the cancellation of $1/T$ in front of the sum, because of the inclusion of the factor T in our definition of the impulse invariant method). Since, for a rational analog filter,

$H^F(\omega)$ is never band limited, the frequency response of the digital filter *is always aliased.* If $H^L(s)$ is low pass or band pass, aliasing can be made negligible by choosing the sampling frequency $1/T$ high enough such that the fraction of energy in the range $|\omega| > \pi/T$ will be negligible. If $H^L(s)$ is high pass or band stop, the impulse invariant method cannot be used at all. We concluded this before based on the properties of the impulse response, and now we conclude it again based on the frequency response: The frequency response of these two filter classes does not decay to zero, so the right side of (10.100) does not converge.

When a digital filter needs to be designed according to given specifications, the impulse invariant method becomes problematic even in the case of low-pass or band-pass filters. Consider, for example, the design of a low-pass filter with tolerance parameters δ_p, δ_s and band-edge frequencies θ_p, θ_s. As a first attempt, we may use the given tolerance parameters for the analog filter and choose the band-edge frequencies as $\omega_p = \theta_p/T$, $\omega_s = \theta_s/T$. When transforming the analog filter designed with these parameters, using the impulse invariance method, we will usually discover that the digital filter does not meet the specifications. This happens because aliasing causes the replicas $H^F((\theta - 2\pi k)/T)$ to add up and spoil both pass-band and stop-band tolerances. A possible solution is to design the analog filter with narrower tolerances, but this requires experimentation, so the design procedure ceases to be straightforward.

Example 10.10 Recall the low-pass Chebyshev filter of the second kind designed in Example 10.3. That filter has order $N = 9$, therefore it has 9 poles and 8 zeros, making its transfer function strictly proper. As shown in Figure 10.9, its frequency response meets the stop-band specification and exceeds the pass-band specification. We transform this filter to a digital filter, using the impulse invariant method with $T = 1$. The poles of the digital filter thus obtained are

$$\alpha_{1,2} = 0.0993 \pm j0.8325, \quad \alpha_{3,4} = 0.0695 \pm j0.5584,$$
$$\alpha_{5,6} = 0.0725 \pm j0.3225, \quad \alpha_{7,8} = 0.1039 \pm j0.1386,$$
$$\alpha_9 = 0.1214.$$

The zeros of the filter are

$$\beta_{1,2} = 0.3817 \pm j2.6660, \quad \beta_{3,4} = -0.2993 \pm j0.9055,$$
$$\beta_{5,6} = -0.4315 \pm j0.4880, \quad \beta_7 = -0.2590, \quad \beta_8 = -0.0672.$$

Figure 10.16 shows the resulting frequency response. As we see, the digital filter meets neither the stop-band nor the pass-band specifications. This failure is due to the analog filter's asymptotic attenuation of only 20 dB/decade, which is not enough to prevent aliasing from spoiling the frequency response of the digital filter. □

In summary, the impulse invariant method has the advantages of preserving the order and stability of the analog filter. On the other hand, there is a distortion of the shape of the frequency response that is due to aliasing. Reducing the aliasing effect requires high sampling rates, thus limiting the usefulness of the method. Besides, the method is not applicable to all filter types. As a result of these drawbacks, this method is not in common use.

10.8 The Backward Difference Method*

As we recall, the analog-domain variable s represents differentiation. Therefore, we can try to replace s by an approximate differentiation operator in the digital domain.

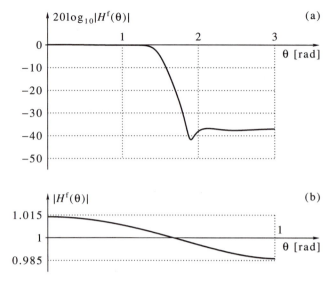

Figure 10.16 Frequency response of the Chebyshev filter in Example 10.10: (a) full range; (b) pass-band details.

A simple approximation to differentiation is given by

$$y(t) = \frac{dx(t)}{dt} \;\Longrightarrow\; y(nT) \approx \frac{x(nT) - x(nT - T)}{T},$$

so

$$Y^z(z) \approx \frac{1 - z^{-1}}{T} X^z(z).$$

This approximation suggests the s-to-z transformation

$$s \longleftarrow \frac{1 - z^{-1}}{T}. \tag{10.101}$$

The operator on the right side of (10.101) is called the *backward difference operator*, since it represents an approximate differentiation, using past and present values of the function to be differentiated. The transfer function $H^z(z)$ is then obtained as

$$H^z(z) = H^L(s) \Big|_{s = \frac{1-z^{-1}}{T}}. \tag{10.102}$$

The operation (10.102) is called *backward difference transformation*.

Example 10.11 Consider again the low-pass filter $H^L(s) = \alpha/(s + \alpha)$. We get

$$H^z(z) = \frac{\alpha T}{1 - z^{-1} + \alpha T} = \frac{\alpha T}{1 + \alpha T} \cdot \frac{1}{1 - (1 + \alpha T)^{-1} z^{-1}}. \tag{10.103}$$

As we see, the digital filter is also of first order and is stable. □

To develop a general formula for the transformed digital filter, let the filter $H^L(s)$ be given in a factored form

$$H^L(s) = \frac{b_0 \prod_{k=1}^{q} (s - u_k)}{\prod_{k=1}^{p} (s - v_k)}. \tag{10.104}$$

A single zero u_k is transformed as follows:

$$s - u_k = \frac{1}{T}(1 - z^{-1} - u_k T) = \frac{1 - u_k T}{T}\left(1 - \frac{1}{1 - u_k T} z^{-1}\right), \tag{10.105}$$

and similarly for a single pole. Therefore, the transformed filter has the factored form

$$H^z(z) = \frac{b_0 T^{p-q} \prod_{k=1}^{q}(1 - u_k T)}{\prod_{k=1}^{p}(1 - v_k T)} \cdot \frac{\prod_{k=1}^{q}[1 - (1 - u_k T)^{-1} z^{-1}]}{\prod_{k=1}^{p}[1 - (1 - v_k T)^{-1} z^{-1}]}. \tag{10.106}$$

It follows from (10.106) that the order of $H^z(z)$ is equal to that of $H^L(s)$. We now show that the transformation preserves the stability of $H^L(s)$. We have

$$z = \frac{1}{1 - sT} = \frac{1}{1 - \sigma T - j\omega T}, \tag{10.107}$$

so

$$|z| = \frac{1}{[(1 - \sigma T)^2 + (\omega T)^2]^{1/2}}. \tag{10.108}$$

We get that $\sigma < 0$ implies $|z| < 1$. If $H^L(s)$ is stable, all its poles have negative real parts, so the corresponding poles of $H^z(z)$ will have moduli less than 1. This proves that the backward difference transformation preserves stability.

Considering (10.107) further, we see that

$$\sigma = 0 \implies z = \frac{1}{1 - j\omega T} \implies z - \frac{1}{2} = \frac{1 + j\omega T}{2(1 - j\omega T)} \implies \left| z - \frac{1}{2} \right| = \frac{1}{2}. \tag{10.109}$$

This shows that the imaginary axis in the s domain is mapped to the circle of radius 0.5 centered at $z = 0.5$ in the z domain. It is *not* mapped to the circle $|z| = 1$ and, consequently, the backward difference method is not a ω-to-θ transformation. We can therefore expect that the frequency response $H^f(\theta)$ will be considerably distorted with respect to $H^F(\omega)$. The left half s plane is mapped to the disk $|z - 0.5| \leq 0.5$ in the z plane, which completely lies in the right half z plane; see Figure 10.17. An analog high-pass filter cannot be mapped to a digital high-pass filter because the poles of the digital filter cannot lie in the correct region, which is the left half of the z plane in this case. In summary, the backward difference method is crude and consequently rarely used.

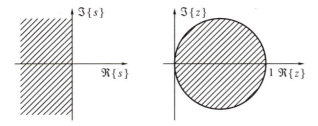

Figure 10.17 The backward difference transform from the s plane to the z plane: The imaginary axis maps to the circle, the left half plane maps to the disk.

10.9 The Bilinear Transform

10.9.1 Definition and Properties of the Bilinear Transform

The bilinear transform can be regarded as a correction of the backward difference method. It is defined by the substitution

$$s \longleftarrow \frac{2}{T} \cdot \frac{z - 1}{z + 1}. \tag{10.110}$$

Given an analog filter $H^L(s)$, the transfer function of the corresponding digital filter is

$$H^z(z) = H^L(s)\bigg|_{s = \frac{2(z-1)}{T(z+1)}}. \tag{10.111}$$

The s-to-z transformation (10.110) can be interpreted as an approximation of continuous-time integration by discrete-time trapezoidal integration; see Problem 10.32.

Example 10.12 Consider the two filters

$$H^L_1(s) = \frac{\alpha}{s+\alpha}, \quad H^L_2(s) = \frac{s}{s+\alpha}.$$

Recall that the former is low pass and the latter is high pass. We get from (10.111)

$$H^z_1(z) = \frac{\alpha}{\dfrac{2(z-1)}{T(z+1)} + \alpha} = \frac{0.5\alpha T}{1 + 0.5\alpha T} \cdot \frac{1 + z^{-1}}{1 - \dfrac{1 - 0.5\alpha T}{1 + 0.5\alpha T}z^{-1}}, \tag{10.112a}$$

$$H^z_2(z) = \frac{\dfrac{2(z-1)}{T(z+1)}}{\dfrac{2(z-1)}{T(z+1)} + \alpha} = \frac{1}{1 + 0.5\alpha T} \cdot \frac{1 - z^{-1}}{1 - \dfrac{1 - 0.5\alpha T}{1 + 0.5\alpha T}z^{-1}}. \tag{10.112b}$$

As we see, the two digital filters are first order, and they have a stable pole at $z = (1 - 0.5\alpha T)/(1 + 0.5\alpha T)$. The filter $H^z_1(z)$ is low pass (it has a zero at $z = -1$), and $H^z_2(z)$ is high pass (it has a zero at $z = 1$). □

In general, let the filter $H^L(s)$ be given in a factored form

$$H^L(s) = \frac{b_0 \prod_{k=1}^q (s - u_k)}{\prod_{k=1}^p (s - v_k)}, \tag{10.113}$$

A single zero u_k is transformed as follows:

$$s - u_k = \frac{2}{T}\left[\frac{1 - z^{-1}}{1 + z^{-1}} - 0.5Tu_k\right] = \frac{2[(1 - 0.5Tu_k) - (1 + 0.5Tu_k)z^{-1}]}{T(1 + z^{-1})}$$

$$= \frac{2(1 - 0.5Tu_k)}{T(1 + z^{-1})}\left[1 - \frac{1 + 0.5Tu_k}{1 - 0.5Tu_k}z^{-1}\right], \tag{10.114}$$

and similarly for a single pole. Therefore, the transformed filter has the factored form

$$H^z(z) = \frac{b_0(0.5T)^{p-q} \prod_{k=1}^q (1 - 0.5Tu_k)}{\prod_{k=1}^p (1 - 0.5Tv_k)}$$

$$\cdot \frac{(1 + z^{-1})^{p-q} \prod_{k=1}^q \left[1 - \dfrac{1 + 0.5Tu_k}{1 - 0.5Tu_k}z^{-1}\right]}{\prod_{k=1}^p \left[1 - \dfrac{1 + 0.5Tv_k}{1 - 0.5Tv_k}z^{-1}\right]}. \tag{10.115}$$

The procedure `bilin` in Program 10.7 implements the bilinear transformation, using the factored form (10.115). Its operation is similar to that of Program 10.5.

It is evident from (10.115) that the bilinear transform preserves the number of poles p; hence it preserves the order of the filter. The number of zeros increases from q to p when $p > q$; in this case, the additional $p - q$ zeros are at $z = -1$. Contrary to the backward difference method, the left half s plane is now mapped to the entire unit disc, rather than to a part of it. Moreover, the imaginary axis is mapped to the unit circle, see Figure 10.18. Therefore, (10.110) is a true frequency-to-frequency transformation. This is shown by the following derivation:

$$z = \frac{1 + 0.5sT}{1 - 0.5sT} = \frac{1 + 0.5\sigma T + j0.5\omega T}{1 - 0.5\sigma T - j0.5\omega T}, \tag{10.116}$$

so

$$|z| = \left[\frac{(1 + 0.5\sigma T)^2 + (0.5\omega T)^2}{(1 - 0.5\sigma T)^2 + (0.5\omega T)^2} \right]^{1/2}. \tag{10.117}$$

We see that $\sigma < 0$ if and only if $|z| < 1$, and $\sigma = 0$ if and only if $|z| = 1$. Therefore, the bilinear transform preserves stability of the transformed filter.

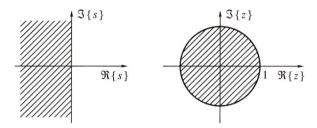

Figure 10.18 The bilinear transform from the s plane to the z plane: The imaginary axis maps to the circle, the left half plane maps to the disk.

For $\sigma = 0$ (i.e., $s = j\omega$), we get

$$z = \frac{1 + j0.5\omega T}{1 - j0.5\omega T} = e^{j\theta}, \tag{10.118}$$

where

$$\theta = 2\arctan(0.5\omega T). \tag{10.119}$$

The digital-domain frequency θ is therefore *warped* with respect to the analog frequency ω, the warping function being $2\arctan(0.5\omega T)$. At low frequencies, $\theta \approx \omega T$, which is the familiar linear relationship. The analog frequencies $\omega = \pm\infty$ are mapped to $\theta = \pm\pi$. The frequency mapping *is not aliased*; that is, the relationship between ω and θ is one-to-one. Consequently, there are no limitations on the use of the bilinear transform; it is adequate for *all* filter types.

Figure 10.19 illustrates the frequency warping introduced by the bilinear transform. The filter in this figure is a second-order, high-pass elliptic filter.

To overcome the frequency warping introduced by the bilinear transform, it is common to *prewarp* the specifications of the analog filter, so that after warping they will be located at the desired frequencies. For example, suppose we wish to design a low-pass filter with band-edge frequencies θ_p, θ_s. We transform these frequencies to corresponding analog-domain band-edge frequencies, using the inverse of (10.119), that is,

$$\omega_p = \frac{2}{T}\tan\left(\frac{\theta_p}{2}\right), \quad \omega_s = \frac{2}{T}\tan\left(\frac{\theta_s}{2}\right). \tag{10.120}$$

We then design the analog filter, using the band-edge frequencies thus obtained. After the analog filter has been transformed using the bilinear transform, the resulting digital filter will have its band-edge frequencies in the right places.

Since prewarping is performed in the beginning of the design procedure, and bilinear transformation is performed in the end, the value of T used is immaterial, as long as it is the same in both. Taking T as the sampling interval enables physical interpretation of the analog frequencies, but taking $T = 1$ or $T = 2$ may be more convenient for hand computations.

In summary, the design procedure for a digital IIR filter using the bilinear transform is as follows:

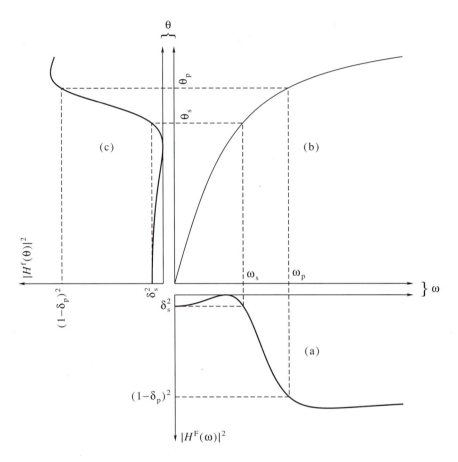

Figure 10.19 The frequency warping of the bilinear transform: (a) frequency response of the analog filter (amplitude axis pointing downward); (b) θ as a function of ω; (c) frequency response of the digital filter (amplitude axis pointing to left).

Digital IIR filter design procedure

1. Convert each specified band-edge frequency of the digital filter to a corresponding band-edge frequency of an analog filter, using (10.120). Leave the pass-band tolerance δ_p and the stop-band tolerance δ_s unchanged.

2. Design an analog filter $H^L(s)$ of the desired type, according to the transformed specifications.

3. Transform $H^L(s)$ to a digital filter $H^z(z)$, using (10.115).

Example 10.13 We wish to design a fourth-order, band-pass digital Butterworth filter with -3 dB frequencies 2000 and 3000 Hz; the sampling frequency is 8000 Hz. The standard low-pass Butterworth filter of order 2 is

$$H_0^L(s) = \frac{1}{s^2 + 1.4142s + 1}.$$

The prewarped -3 dB frequencies are

$$\omega_1 = 16,000 \tan(0.25\pi) = 16,000 \text{ rad/s},$$

$$\omega_{\text{h}} = 16,000 \tan(0.375\pi) = 38,627 \text{ rad/s}.$$

It is convenient in this case, since the frequency range of interest is in kilohertz, to specify angular frequencies in multiples of 10^4 rad/s, and time in multiples of 10^{-4} second. The low-pass to band-pass transformation is then

$$s = \frac{\tilde{s}^2 + \omega_l \omega_{\text{h}}}{(\omega_{\text{h}} - \omega_l)\tilde{s}} = \frac{\tilde{s}^2 + 6.1803}{2.2627\tilde{s}}.$$

Therefore, the band-pass Butterworth filter is

$$\tilde{H}^{\text{L}}(\tilde{s}) = \frac{1}{\left(\frac{\tilde{s}^2 + 6.1803}{2.2627\tilde{s}}\right)^2 + 1.4142\left(\frac{\tilde{s}^2 + 6.1803}{2.2627\tilde{s}}\right) + 1}$$

$$= \frac{5.1198\tilde{s}^2}{\tilde{s}^4 + 3.2\tilde{s}^3 + 17.48\tilde{s}^2 + 19.777\tilde{s} + 38.196}.$$

The corresponding digital filter is given by

$$H^z(z) = \frac{13.106\left(\frac{z-1}{z+1}\right)^2}{\left(\frac{z-1}{z+1}\right)^4 + 13.107\left(\frac{z-1}{z+1}\right)^3 + 44.749\left(\frac{z-1}{z+1}\right)^2 + 31.643\left(\frac{z-1}{z+1}\right) + 38.196}$$

$$= \frac{0.0976(1 - 2z^{-2} + z^{-4})}{1 + 1.2189z^{-1} + 1.3333z^{-2} + 0.6667z^{-3} + 0.3333z^{-4}}.$$

□

10.9.2 MATLAB Implementation of IIR Filter Design

The procedure `iirdes` in Program 10.8 combines the programs mentioned in Section 10.5 and the program for the bilinear transform to a complete digital IIR filter design program. The program accepts the desired filter class (Butterworth, Chebyshev-I, Chebyshev-II, or elliptic), the desired frequency response type (low pass, high pass, band pass, or band stop), the band-edge frequencies, the pass-band ripple, and the stop-band attenuation. The program first prewarps the digital frequencies, using sampling interval $T = 1$ (this choice is arbitrary). It then transforms the specifications to the specifications of the prototype low-pass filter. Next, the order N and the parameters ω_0, ε are computed from the specifications. The low-pass filter is designed next, transformed to the appropriate analog band, then to digital, using the bilinear transform (again with $T = 1$). The program provides both the polynomials and the pole–zero factorization of the z-domain transfer function.

10.9.3 IIR Filter Design Examples

We now illustrate IIR filter design based on the bilinear transform by several examples. We use the specification examples given in Section 8.2 and present design results that meet these specifications. We show the magnitude responses of the filters, but do not list their coefficients. You can easily obtain the coefficients, as well as the poles and zeros, with the program `iirdes`.

Example 10.14 Consider the low-pass filter whose specifications were given in Example 8.1. Butterworth, Chebyshev-I, Chebyshev-II, and elliptic filters that meet these specifications have orders $N = 27, 9, 9, 5$, respectively. Figure 10.20 shows the magnitude responses of these filters. □

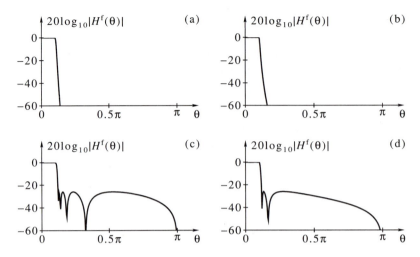

Figure 10.20 Magnitude responses of the filters in Example 10.14: (a) Butterworth; (b) Chebyshev-I; (c) Chebyshev-II; (d) elliptic.

Example 10.15 Consider the high-pass filter whose specifications were given in Example 8.2. Butterworth, Chebyshev-I, Chebyshev-II, and elliptic filters that meet these specifications have orders $N = 15, 8, 8, 5$, respectively. Figure 10.21 shows the magnitude responses of these filters.   □

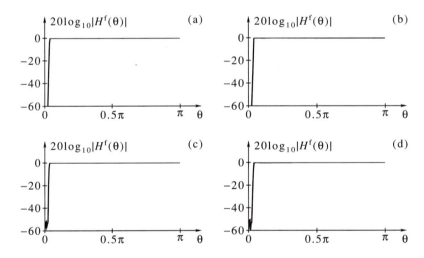

Figure 10.21 Magnitude responses of the filters in Example 10.15: (a) Butterworth; (b) Chebyshev-I; (c) Chebyshev-II; (d) elliptic.

Example 10.16 Consider the five filters whose specifications were given in Example 8.3. Here we design the band-pass filter for the second channel. Butterworth, Chebyshev-I, Chebyshev-II, and elliptic filters that meet these specifications have orders $N = 48, 22, 22, 14$, respectively. Figure 10.22 shows the magnitude responses of these filters.   □

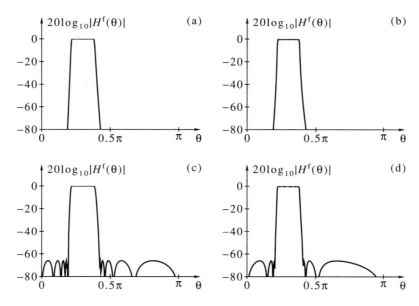

Figure 10.22 Magnitude responses of the filters in Example 10.16: (a) Butterworth; (b) Chebyshev-I; (c) Chebyshev-II; (d) elliptic.

Example 10.17 Consider the band-stop filter whose specifications were given in Example 8.4. Butterworth, Chebyshev-I, Chebyshev-II, and elliptic filters that meet these specifications have orders $N = 16, 10, 10, 8$, respectively. Figure 10.23 shows the magnitude responses of these filters. □

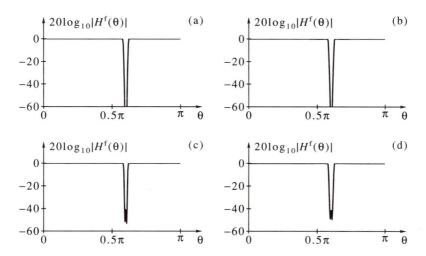

Figure 10.23 Magnitude responses of the filters in Example 10.17: (a) Butterworth; (b) Chebyshev-I; (c) Chebyshev-II; (d) elliptic.

In conclusion, the bilinear transform facilitates design of IIR filters meeting prescribed specifications without trial and error. Elliptic filters have the smallest order. Therefore, an elliptic filter should usually be the preferred choice, unless the application requires monotone response in the pass band, or the stop band, or both.

10.10   The Phase Response of Digital IIR Filters

Digital IIR filters do not have linear phase. That is, their group delay is not constant
for all frequencies. Even if we restrict ourselves to the pass band, we still find that
the group delay is not constant there. The group delay of a low-pass IIR filter typically
increases monotonically in the pass band, reaches a maximum in the transition band,
and then decreases monotonically. As a result, signals in the pass band that are fed
to such a filter are usually distorted at the output, even if the pass-band ripple of the
filter is very small. Distortion occurs, since each frequency component is delayed by a
different amount, so the relative phases of the frequency components at the output are
different from those at the input. The following example illustrates this phenomenon.

Example 10.18 Consider the following digital low-pass filter specifications:

$$\theta_p = 0.1\pi, \ \theta_s = 0.2\pi, \ \delta_p = \delta_s = 0.001.$$

We design four filters meeting these specifications, using the program iirdes. The
filters are Butterworth, Chebyshev-I, Chebyshev-II, and elliptic. We then compute the
group delay of the four filters in the frequency range $[0, 0.2\pi]$ (which includes the
pass band and the transition band), using the program grpdly. The results are shown
in Figure 10.24. As we see, the Chebyshev filter of the second kind (represented by the
dot-dashed line) has the best group delay response of the four: Its value is the smallest
at all pass-band frequencies and it is the most nearly constant in the pass band.

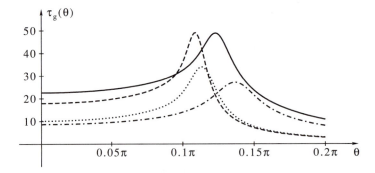

Figure 10.24 The group delay of the four filters in Example 10.18. Solid line: Butterworth,
dashed line: Chebyshev-I, dot-dashed line: Chebyshev-II, dotted line: elliptic.

To illustrate the distortion caused by the nonlinear phase, we input the signal

$$x[n] = \sum_{m=1}^{4} \frac{1}{2m-1} \sin[0.0125\pi(2m-1)n]$$

to the four filters. This signal consists of the first four odd harmonics of a square wave
whose fundamental period is 160 samples. Its highest frequency is 0.0875π, so it is
completely in the pass band of the four filters. The signal is shown in Figure 10.25.
Note that we do not depict it as vertical sticks, but as a continuous line, for better
visual appearance. The continuous line can be regarded as the continuous-time signal
corresponding to $x[n]$.

Let us denote the signal at the output of any of the four filters by $y[n]$. For meaningful comparison of $x[n]$ and $y[n]$, we must advance $y[n]$ by the group delay of
the corresponding filter at the fundamental signal frequency, that is, by $\tau_g(0.0125\pi)$.
Then, if the filter had a constant group delay and unit gain in the pass band, we would

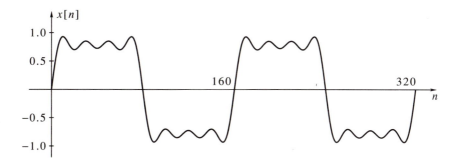

Figure 10.25 Input test signal in Example 10.18.

expect $y[n]$ to be equal to $x[n]$. In reality, we get the responses shown in Figure 10.26. The solid line in each plot shows the input $x[n]$ and the dotted line shows the advanced $y[n]$. The group delays are rounded to integer values. They are $23, 18, 9$, and 10 for the four filters, respectively. Since the differences between $y[n]$ and $x[n]$ are not clearly visible in Figure 10.26, we also show, in Figure 10.27, plots of the differences $y[n] - x[n]$. If the filter had a constant group delay, we would expect the error to be bounded by 0.001, which is the value of the pass-band ripple. As we see, the actual differences are larger by two orders of magnitude, due to the distortion caused by the nonlinear phase. The Chebyshev filter of the second kind is the best of the four, in agreement with what we saw in Figure 10.24.

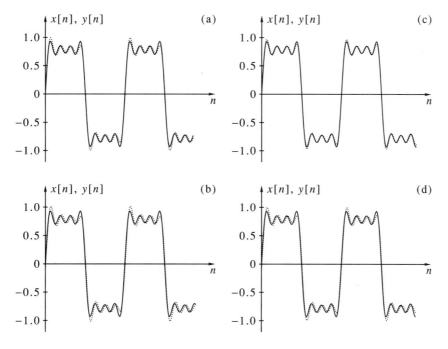

Figure 10.26 Input and output test signals in Example 10.18: (a) Butterworth; (b) Chebyshev-I; (c) Chebyshev-II; (d) elliptic. Solid line: input $x[n]$, dotted line: output $y[n]$.

In summary, nonlinear phase is a major drawback of IIR filters. However, to put this drawback in proper perspective, we reiterate that analog filters do not have linear

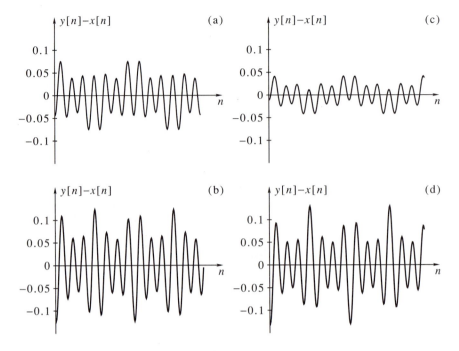

Figure 10.27 The difference between the input and output test signals in Example 10.18: (a) Butterworth; (b) Chebyshev-I; (c) Chebyshev-II; (d) elliptic.

phase either. Despite this, analog filter technology dominated electrical engineering until as recently as the early 1980s.   □

As we saw in Chapter 9, digital FIR filters can have linear phase in the pass band, thereby eliminating phase distortions completely. This is illustrated in the next example.

Example 10.19 Let us repeat Example 10.18, using an FIR filter that meets the same specifications, namely

$$\theta_p = 0.1\pi, \quad \theta_s = 0.2\pi, \quad \delta_p = \delta_s = 0.001.$$

The design is carried out using the program `firkais`. The resulting filter has order $N = 76$. We feed the signal shown in Figure 10.25 at the input of the FIR filter. Figure 10.28 shows the difference between the output signal, advanced by the group delay (which is 38 in this case), and the input signal. As we see, the difference signal is well below the pass-band tolerance and is smaller than the difference signal in Example 10.18 by three orders of magnitude. The lesson from this example is that FIR filters indeed offer a great advantage over IIR filters in eliminating distortions due to nonlinear phase response.   □

10.11 Sampled-Data Systems*

In this chapter we discussed the design of digital IIR filters. We conclude the chapter by showing how an analog system can be made to work in a digital environment. This topic is marginal to digital signal processing, but it is of great importance in digital

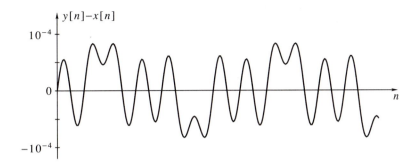

Figure 10.28 The difference between the output and the input signals in Example 10.19.

control applications. The reason is that control systems are often analog by nature and cannot be replaced by discrete-time systems. They include elements such as motors, gears, and power amplifiers. If we wish to interact with such a system digitally, we must interface to it in a proper manner and understand the effect of the interface on the system's behavior.

A *sampled-data system* is a continuous-time system whose input and output are discrete-time signals. To feed a discrete-time signal to a continuous-time system, it is necessary to convert it first to a continuous-time signal. This is usually accomplished by a zero-order hold at the input of the system. Similarly, to get a discrete-time signal from a continuous-time system, it is necessary to sample its output. Figure 10.29 shows the general structure of a sampled-data system. Such an arrangement is also known as a *hybrid system*.

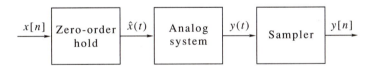

Figure 10.29 A sampled-data system.

We wish to derive the relationship between the discrete-time input signal $x[n]$ and the discrete-time output signal $y[n]$, under the condition that the continuous-time system is linear time invariant. We shall show that in this case, the two signals are related by discrete-time convolution. Thus, the sampled-data system is linear time invariant as well. We denote the impulse response of the continuous-time system by $g(t)$ and its transfer function by $G^L(s)$.

We have at the output of the ZOH,

$$\hat{x}(t) = \sum_{m=-\infty}^{\infty} x[m]h_{\text{zoh}}(t - mT). \tag{10.121}$$

Recall that the transfer function of a ZOH is

$$H_{\text{zoh}}^L(s) = \frac{1 - e^{-sT}}{s} \tag{10.122}$$

[this is obtained from (3.34) upon replacement of $j\omega$ by s]. The Laplace transform of

$\hat{x}(t)$ is therefore

$$\hat{X}^L(s) = \sum_{m=-\infty}^{\infty} x[m] H_{zoh}^L(s) e^{-msT} = \sum_{m=-\infty}^{\infty} x[m] \frac{1 - e^{-sT}}{s} e^{-msT}$$

$$= \sum_{m=-\infty}^{\infty} x[m] \frac{e^{-msT} - e^{-(m+1)sT}}{s}. \tag{10.123}$$

The Laplace transform of $y(t)$, the output of the continuous-time system, is

$$Y^L(s) = G^L(s) \hat{X}^L(s) = \sum_{m=-\infty}^{\infty} x[m] \frac{G^L(s)(e^{-msT} - e^{-(m+1)sT})}{s}$$

$$= \sum_{m=-\infty}^{\infty} x[m] F^L(s)(e^{-msT} - e^{-(m+1)sT}), \tag{10.124}$$

where we define

$$F^L(s) = \frac{G^L(s)}{s}. \tag{10.125}$$

Let $f(t)$ be the inverse Laplace transform of $F^L(s)$. Then the inverse Laplace transform of (10.124) is

$$y(t) = \sum_{m=-\infty}^{\infty} x[m]\{f(t - mT) - f(t - mT - T)\}. \tag{10.126}$$

The sampled sequence $y[n]$ is given by

$$y[n] = y(nT) = \sum_{m=-\infty}^{\infty} x[m]\{f(nT - mT) - f(nT - mT - T)\}$$

$$= \sum_{m=-\infty}^{\infty} x[m]h[n - m] = \{x * h\}[n], \tag{10.127}$$

where we define

$$h[n] = f(nT) - f(nT - T). \tag{10.128}$$

In conclusion, the input–output relationship of the block diagram in Figure 10.29 is a discrete-time convolution. Therefore, the equivalent discrete-time system is linear time invariant. Its transfer function $H^z(z)$ is computed by the following procedure:

1. Compute the inverse Laplace transform of $F^L(s)$ given in (10.124), and denote the result by $f(t)$. This is mathematically expressed by

$$f(t) = \left\{ \mathcal{L}^{-1} \left[\frac{G^L(s)}{s} \right] \right\}(t).$$

2. Sample $f(t)$ to obtain $f(nT)$.
3. Subtract from $f(nT)$ its unit-delayed version $f(nT - T)$ to get $h[n]$.
4. Compute the z-transform of $h[n]$ to obtain $H^z(z)$.

Note that we can also compute the z-transform of $f(nT)$ first, say $F^z(z)$, and then

$$H^z(z) = (1 - z^{-1})F^z(z).$$

Example 10.20 Let the continuous-time system have a transfer function

$$G^L(s) = \frac{C}{s(s + \alpha)}.$$

This system consists of an integrator and a real pole; it is a common model for a DC motor, for example. We have

$$\frac{G^L(s)}{s} = \frac{C}{s^2(s + \alpha)} = \frac{C}{\alpha s^2} - \frac{C}{\alpha^2 s} + \frac{C}{\alpha^2(s + \alpha)},$$

so

$$f(t) = \frac{Ct}{\alpha} - \frac{C}{\alpha^2} + \frac{Ce^{-\alpha t}}{\alpha^2}, \quad f(nT) = \frac{CTn}{\alpha} - \frac{C}{\alpha^2} + \frac{Ce^{-\alpha Tn}}{\alpha^2}.$$

The z-transform of this sequence is (cf. Table 7.1)

$$F^z(z) = \frac{CTz^{-1}}{\alpha(1 - z^{-1})^2} - \frac{C}{\alpha^2(1 - z^{-1})} + \frac{C}{\alpha^2(1 - e^{-\alpha T}z^{-1})}.$$

Finally,

$$H^z(z) = (1 - z^{-1})F^z(z) = \frac{CTz^{-1}}{\alpha(1 - z^{-1})} - \frac{C}{\alpha^2} + \frac{C(1 - z^{-1})}{\alpha^2(1 - e^{-\alpha T}z^{-1})}$$

$$= \frac{C[(\alpha T + e^{-\alpha T})z^{-1} - e^{-\alpha T}(1 + \alpha T)z^{-2}]}{\alpha^2(1 - z^{-1})(1 - e^{-\alpha T}z^{-1})}.$$

As we see, the equivalent discrete-time system inherits the integrator and the first-order pole from the continuous-time system, but it also has a unit delay and a zero, which the continuous-time system does not. This result—the inheritance of poles and the appearance of zeros in the discrete-time transfer function—is typical of sampled-data systems. □

10.12 Summary and Complements

10.12.1 Summary

This chapter was devoted to the design of digital infinite impulse response (IIR) filters, in particular, to design by means of analog filters. The classical analog filters have different ripple characteristics: Butterworth is monotone at all frequencies, Chebyshev-I is monotone in the stop band and equiripple in the pass band, Chebyshev-II is monotone in the pass band and equiripple in the stop band, and an elliptic filter is equiripple in all bands. Design formulas were given for these filter classes. Among the four classes, elliptic filters have the smaller order for a given set of specifications, whereas Butterworth filters have the largest order.

When an analog filter other than low pass needs to be designed, a common procedure is to design a prototype low-pass filter of the desired class, and then to transform the low-pass filter by a rational frequency transformation. Standard transformations were given for high-pass, band-pass, and band-stop filters.

After an analog filter has been designed, it must be transformed to the digital domain. The preferred method for this purpose is the bilinear transform. The bilinear transform preserves the order and stability of the analog filter. It is suitable for filters of all classes and types, and is straightforward to compute. The frequency response of the digital filter is related to that of the analog filter from which it was derived through the frequency-warping formula (10.119). At low frequencies the frequency warping is small, but at frequencies close to π it is significant. Prewarping of the discrete-time band-edge frequencies prior to the analog design guarantees that the digital filter obtained as a result of the design will meet the specifications.

Other methods for analog-to-digital filter transformation are the impulse invariant and the backward difference methods; both are inferior to the bilinear transform.

We reiterate the main advantages and disadvantages of IIR filters:

1. Advantages:
    (a) Straightforward design of standard IIR filters, thanks to the existence of well-established analog filter design techniques and simple transformation procedures.

   (b) Low complexity of implementation when compared to FIR filters, especially
       in the case of elliptic filters.

   (c) Relatively short delays, since practical IIR filters are usually minimum phase.

2. Disadvantages:

   (a) IIR filters do not have linear phase.

   (b) IIR filters are much less flexible than FIR filters in achieving nonstandard
       frequency responses.

   (c) Design techniques other than those based on analog filters are not readily
       available, and are complex to develop and implement.

   (d) Although theoretically stable, IIR filters may become unstable when their
       coefficients are truncated to a finite word length. Therefore, stability must
       be carefully verified; see Section 11.5 for further discussion.

10.12.2   Complements

1. [p. 333] The notation $T_N(x)$ for the Chebyshev polynomials is derived from the
   name *Tschebyscheff*, the German spelling of Chebyshev.

2. [p. 357] Euler approximation is based on the observation that the area under the
   function $h(t)$ between the abscissas nT and $(n + 1)T$ is approximately equal to
   the area of the rectangle having base T and height $h(nT)$, that is,

   $$\int_{nT}^{(n+1)T} h(t)dt \approx Th(nT).$$

   By dividing the entire range $[0, \infty)$ to intervals of length T each and using this
   approximation for each interval, we obtain (10.96).

10.13 MATLAB Programs

Program 10.1 Design of analog low-pass filters.

```
function [b,a,v,u,C] = analoglp(typ,N,w0,epsilon,m);
% Synopsis: [b,a,v,u,C] = analoglp(typ,N,w0,epsilon,m).
% Butterworth, Chebyshev-I or Chebyshev-II low-pass filter.
% Input parameters:
% typ: filter class: 'but', 'ch1', 'ch2', or 'ell'
% N: the filter order
% w0: the frequency parameter
% epsilon: the tolerance parameter; not needed for Butterworth
% m: parameter needed for elliptic filters.
% Output parameters:
% b, a: numerator and denominator polynomials
% v, u, C: poles, zeros, and constant gain.

if (typ == 'ell'),
   [v,u,C] = elliplp(N,w0,epsilon,m);
   a = 1; for i = 1:N, a = conv(a,[1, -v(i)]); end
   b = C; for i = 1:length(u), b = conv(b,[1, -u(i)]); end
   a = real(a); b = real(b); C = real(C); return
end
k = (0.5*pi/N)*(1:2:2*N-1); s = -sin(k); c = cos(k);
if (typ == 'but'), v  = w0*(s+j*c);
elseif (typ(1:2) == 'ch'),
   f = 1/epsilon; f = log(f+sqrt(1+f^2))/N;
   v = w0*(sinh(f)*s+j*cosh(f)*c);
end
if (typ == 'ch2'),
   v = (w0^2)./v;
   if (rem(N,2) == 0), u = j*w0./c;
   else, u = j*w0./[c(1:(N-1)/2),c((N+3)/2:N)]; end
end
a = 1; for k = 1:N, a = conv(a,[1, -v(k)]); end
if (typ == 'but' | typ == 'ch1'),
   C = prod(-v); b = C; u = [];
elseif (typ == 'ch2'),
   C = prod(-v)/prod(-u); b = C;
   for k = 1:length(u), b = conv(b,[1, -u(k)]); end;
end
if (typ == 'ch1' & rem(N,2) == 0),
   f = (1/sqrt(1+epsilon^2)); b = f*b; C = f*C; end
a = real(a); b = real(b); C = real(C);
```

Program 10.2 Design of a low-pass elliptic filter.

```
function [v,u,C] = elliplp(N,w0,epsilon,m);
% Synopsis: [v,u,C] = elliplp(N,w0,epsilon,m).
% Designs a low-pass elliptic filter.
% Input parameters:
% N: the order
% w0: the pass-band edge
% epsilon, m: filter parameters.
% Output parameters:
% v, u, C: poles, zeros, and constant gain of the filter.

flag = rem(N,2); K = ellipke(m);
if (~flag), lmax = N/2; l = (1:lmax)-0.5;
else, lmax = (N-1)/2; l = 1:lmax; end
zl = ellipj((2*K/N)*l,m); pl = 1 ./(sqrt(m)*zl);
f = prod((1-pl.^2)./(1-zl.^2));
u = w0*reshape([j*pl; -j*pl],1,2*lmax);
a = 1;
for l = 1:lmax,
   for i = 1:2, a = conv(a,[1,0,pl(l)^2]); end
end
b = 1;
for l = 1:lmax,
   for i = 1:2, b = conv(b,[1,0,zl(l)^2]); end
end
b = (f*epsilon)^2*b;
if (flag), b = -[b,0,0]; a = [0,0,a]; end
v = roots(a+b).'; v = w0*v(find(real(v) < 0));
C = prod(-v)./prod(-u);
if (~flag), C = C/sqrt(1+epsilon^2); end, C = real(C);
```

Program 10.3 The parameters of an analog low-pass filter as a function of the specification parameters.

```
function [N,w0,epsilon,m] = lpspec(typ,wp,ws,deltap,deltas);
% Synopsis: [N,w0,epsilon,k,q] = lpspec(typ,wp,ws,deltap,deltas).
% Butterworth, Chebyshev-I or Chebyshev-II low-pass filter
% parameter computation from given specifications.
% Input parameters:
% typ: the filter class:
%       'but' for Butterworth
%       'ch1' for Chebyshev-I
%       'ch2' for Chebyshev-II
%       'ell' for elliptic
% wp, ws: band-edge frequencies
% deltap, deltas: pass-band and stop-band tolerances.
% Output parameters:
% N: the filter order
% w0: the frequency parameter
% epsilon: the tolerance parameter; not supplied for Butterworth
% m: parameter supplied in case of elliptic filter.

d = sqrt((((1-deltap)^(-2)-1)/(deltas^(-2)-1))); di = 1/d;
k = wp/ws; ki =1/k;
if (typ == 'but'),
   N = ceil(log(di)/log(ki));
   w0 = wp*((1-deltap)^(-2)-1)^(-0.5/N);
   nargout = 2;
elseif (typ(1:2) == 'ch'),
   N = ceil(log(di+sqrt(di^2-1))/log(ki+sqrt(ki^2-1)));
   nargout = 3;
   if (typ(3) == '1'),
      w0 = wp; epsilon = sqrt((1-deltap)^(-2)-1);
   elseif (typ(3) == '2'),
      w0 = ws; epsilon = 1/sqrt(deltas^(-2)-1);
   end
elseif (typ == 'ell'),
   w0 = wp; epsilon = sqrt((1-deltap)^(-2)-1);
   [N,m]  = ellord(k,d);
   nargout = 4;
end
```

Program 10.4 Computation of the order of an elliptic low-pass filter.

```
function [N,m] = ellord(k,d);
% Synopsis: [N,m] = ellord(k,d).
% Finds the order and the parameter m of an elliptic filter.
% Input parameters:
% k, d: the selectivity and discrimination factors.
% Output parameters:
% N: the order
% m: the parameter for the Jacobi elliptic function.

m0 = k^2; C = ellipke(1-d^2)/ellipke(d^2);
N0  = C*ellipke(m0)/ellipke(1-m0);
if (abs(N0-round(N0))) <= 1.0e-6,
   N = round(N0); m = m0; return; end
N = ceil(N0); m = 1.1*m0;
while (C*ellipke(m)/ellipke(1-m) < N), m = 1.1*m; end
N1 = C*ellipke(m)/ellipke(1-m);
while (abs(N1-N) >= 1.0e-6),
   a = (N1-N0)/(m-m0);
   mnew = m0+(N-N0)/a;
   Nnew = C*ellipke(mnew)/ellipke(1-mnew);
   if (Nnew < N), m0 = mnew; N0 = Nnew;
   else, m = mnew; N1 = Nnew; end
end
```

Program 10.5 Frequency transformations of analog filters.

```
function [b,a,vout,uout,Cout] = analogtr(typ,vin,uin,Cin,w);
% Synopsis: [b,a,vout,uout,Cout] = analogtr(typ,vin,uin,Cin,w).
% Performs frequency transformations of analog low-pass filters.
% Input parameters:
% typ: the transformation type:
%        'l' for low-pass to low-pass
%        'h' for low-pass to high-pass
%        'p' for low-pass to band-pass
%        's' for low-pass to band-stop
% vi, ui, Cin: the poles, zeros, and constant gain of the low-pass
% w: equal to omega_c for 'l' or 'h'; a 1 by 2 matrix of
%     [omega_l, omega_h] for 'p' or 's'.
% Output parameters:
% b, a: the output polynomials
% vout, uout, Cout: the output poles, zeros, and constant gain.

p = length(vin); q = length(uin);
if (typ == 'l'),
   uout = w*uin; vout = w*vin; Cout = w^(p-q)*Cin;
elseif (typ == 'h'),
   uout = [w./uin,zeros(1,p-q)]; vout = w./vin;
   Cout = prod(-uin)*Cin/prod(-vin);
elseif (typ == 'p'),
   wl = w(1); wh = w(2); uout = []; vout = [];
   for k = 1:q,
      uout = [uout,roots([1,-uin(k)*(wh-wl),wl*wh]).']; end
   uout = [uout,zeros(1,p-q)];
   for k = 1:p,
      vout = [vout,roots([1,-vin(k)*(wh-wl),wl*wh]).']; end
   Cout = (wh-wl)^(p-q)*Cin;
elseif (typ == 's'),
   [t1,t2,t3,t4,t5] = analogtr('h',vin,uin,Cin,1);
   [t1,t2,vout,uout,Cout] = analogtr('p',t3,t4,t5,w);
end
a = 1; b = 1;
for k = 1:length(vout), a = conv(a,[1,-vout(k)]); end
for k = 1:length(uout), b= conv(b,[1,-uout(k)]); end
a = real(a); b = real(Cout*b); Cout = real(Cout);
```

Program 10.6 Impulse invariant transformation of an analog filter.

```
function [bout,aout] = impinv(bin,ain,T);
% Synopsis: [bout,aout] = impinv(bin,ain,T).
% Computes the impulse invariant transformation of an analog filter.
% Input parameters:
% bin, ain: the numerator and denominator polynomials of the
%            analog filter
% T: the sampling interval
% Output parameters:
% bout, aout: the numerator and denominator polynomials of the
%            digital filter.

if (length(bin) >= length(ain)),
   error('Analog filter in IMPINV is not strictly proper'); end
[r,p,k] = residue(bin,ain);
[bout,aout] = pf2tf([],T*r,exp(T*p));
```

Program 10.7 Bilinear transformation of an analog filter.

```
function [b,a,vout,uout,Cout] = bilin(vin,uin,Cin,T);
% Synopsis: [b,a,vout,uout,Cout] = bilin(vin,uin,Cin,T).
% Computes the bilinear transform of an analog filter.
% Input parameters:
% vi, ui, Cin: the poles, zeros, and constant gain of the
%              analog filter
% T: the sampling interval.
% Output parameters:
% b, a: the output polynomials
% vout, uout, Cout: the output poles, zeros, and constant gain.

p = length(vin); q = length(uin);
Cout = Cin*(0.5*T)^(p-q)*prod(1-0.5*T*uin)/prod(1-0.5*T*vin);
uout = [(1+0.5*T*uin)./(1-0.5*T*uin),-ones(1,p-q)];
vout = (1+0.5*T*vin)./(1-0.5*T*vin);
a = 1; b = 1;
for k = 1:length(vout), a = conv(a,[1,-vout(k)]); end
for k = 1:length(uout), b= conv(b,[1,-uout(k)]); end
a = real(a); b = real(Cout*b); Cout = real(Cout);
```

Program 10.8 Digital IIR filter design.

```
function [b,a,v,u,C] = iirdes(typ,band,theta,deltap,deltas);
% Synopsis: [b,a,v,u,C] = iirdes(typ,band,theta,deltap,deltas).
% Designs a digital IIR filter to meet given specifications.
% Input parameters:
% typ: the filter class: 'but', 'ch1', 'ch2', or 'ell'
% band: 'l' for LP, 'h' for HP, 'p' for BP, 's' for BS
% theta: an array of band-edge frequencies, in increasing
%        order; must have 2 frequencies if 'l' or 'h',
%        4 if 'p' or 's'
% deltap: pass-band ripple/s (possibly 2 for 's')
% deltas: stop-band ripple/s (possibly 2 for 'p')
% Output parameters:
% b, a: the output polynomials
% v, u, C: the output poles, zeros, and constant gain.

% Prewarp frequencies (with T = 1)
omega = 2*tan(0.5*theta);
% Transform specifications
if (band == 'l'), wp = omega(1); ws = omega(2);
elseif (band == 'h'), wp = 1/omega(2); ws = 1/omega(1);
elseif (band == 'p'),
   wl = omega(2); wh = omega(3); wp = 1;
   ws = min(abs((omega([1,4]).^2-wl*wh) ...
       ./((wh-wl)*omega([1,4])))));
elseif (band == 's'),
   wl = omega(2); wh = omega(3); ws = 1;
   wp = 1/min(abs((omega([1,4]).^2-wl*wh) ...
       ./((wh-wl)*omega([1,4])))));
end
% Get low-pass filter parameters
[N,w0,epsilon,m] = lpspec(typ,wp,ws,min(deltap),min(deltas));
% Design low-pass filter
[b,a,v1,u1,C1] = analoglp(typ,N,w0,epsilon,m);
% Transform to the required band
ww = 1; if (band == 'p' | band == 's'), ww = [wl,wh]; end
[b,a,v2,u2,C2] = analogtr(band,v1,u1,C1,ww);
% Perform bilinear transformation
[b,a,v,u,C] = bilin(v2,u2,C2,1);
```

10.14   Problems

10.1 Explain how the approximation (10.4) is derived.

10.2 This problem examines certain properties of the discrimination factor.

(a) Is the discrimination factor d defined in (10.2) typically greater than 1 or less than 1? Explain.

(b) Derive an approximation for d under the assumption that both δ_p and δ_s are much smaller than 1.

10.3 Derive (10.19). Explain the meaning of equality at the lower end of the range, and the meaning of equality at the higher end of the range.

10.4 An analog filter is required to have pass-band ripple $1 \pm \delta_p'$ and stop-band attenuation δ_s'. Show how to choose δ_p, δ_s for an analog filter $H^L(s)$, and a gain C such that the filter $CH^L(s)$ will meet the requirements.

10.5 We are given the analog filter

$$H^L(s) = \frac{1}{s^3 + 2s^2 + 2s + 1}.$$

For this filter, we are given that

$$A_p = 0.5\,\text{dB}, \quad A_s = 20\,\text{dB}.$$

Find the discrimination factor d and the selectivity factor k of this filter to four decimal places.

10.6 Let $H_1^L(s)$ and $H_2^L(s)$ be normalized Butterworth filters of orders 2 and 3, respectively. Let $H^L(s)$ be their cascade connection, that is, $H^L(s) = H_1^L(s)H_2^L(s)$.

(a) Show that $H^L(s)$ is *not* a Butterworth filter.

(b) Show that, even though $H^L(s)$ is not a Butterworth filter, it shares certain properties of a fifth order normalized Butterworth filter; specify which properties.

10.7 Prove that $T_N(x)$ is a symmetric function of x when N is even, and an antisymmetric function of x when N is odd.

10.8 Use (10.21) and show that $T_N(0) = \pm 1$ for even N, and $T_N(0) = 0$ for odd N.

10.9 Let $H^L(s)$ be a Chebyshev-I low-pass filter, with

$$N = 3, \quad \omega_p = 3, \quad \omega_s = 10, \quad \delta_s = 0.02.$$

The filter meets the specifications *exactly* in both the pass band and the stop band. Compute $H^L(s)$ and express it as a ratio of two polynomials, with coefficients accurate to 4 decimal digits.

10.10 Consider the function

$$y(x) = \frac{\log_e(x + \sqrt{x^2 - 1})}{\log_e x}.$$

(a) Prove that $y(x)$ tends to infinity as x approaches 1 from above; prove that it tends to 1 as x approaches ∞.

(b) Use MATLAB for plotting this function in the range $1.01 \leq x \leq 10$. What is the nature of this function?

(c) Use your conclusion from part b and show that, for given analog low-pass filter specifications, the order of a Chebyshev filter will always be smaller than or equal to the order of a Butterworth filter.

10.11 A second-order, low-pass analog Chebyshev-II filter is known to satisfy $H^F(1) = 0$. The filter known to have been designed with $\varepsilon = 0.1$. Compute $H^L(s)$ and express it as a ratio of two polynomials with real coefficients.

10.12 Joan, a junior engineer, is assigned the following task. She is given an analog low-pass filter and told that it is Chebyshev, but she does not know whether it is of the first or second kind. She is told that the filter was designed using the procedures described in this chapter. Joan is requested to find the transfer function of the filter, with coefficient values to four decimal digits.

Joan performs a series of tests, and finds that:

- $|H^F(0)|^2 = 1$.
- $|H^F(10)|^2 = 0.2$.
- $H^F(10.8239) = H^F(25.1313) = 0$.
- $H^F(\omega)$ is not identically zero for any frequency in the range $0 \leq \omega \leq 1000$, except for the two aforementioned frequencies.

Based on this information, help Joan to find the transfer function to the required accuracy.

10.13 We wish to sample a musical signal in order to store it on a digital tape. The sampling frequency is to be 48 kHz. It is required to design an analog low-pass antialiasing for this purpose. The specifications of the filter are

$$\omega_p = (2\pi \cdot 19)10^3, \quad \omega_s = (2\pi \cdot 24)10^3, \quad \delta_p = 0.05, \quad \delta_s = 10^{-4}.$$

As a result of the nature of music signals (whose energy typically decreases as the frequency increases), the pass-band magnitude response is required to be monotone. For each eligible filter among the ones we have studied in this chapter, compute the relevant design parameters (N, ε, ω_0, ...).

10.14 It is required to design an analog band-pass filter whose frequency response has no ripple in the pass band. The specifications are

$$\widetilde{\omega}_{s,1} = 20, \quad \widetilde{\omega}_{p,1} = 50, \quad \widetilde{\omega}_{p,2} = 20,000, \quad \widetilde{\omega}_{s,2} = 45,000.$$

$$\widetilde{A}_p = 3 \, \text{dB}, \quad \widetilde{A}_{s,1} = \widetilde{A}_{s,2} = 25 \, \text{dB}.$$

What is the minimum order of a filter that meets these specifications?

10.15 Sharon and Irwin were asked to design an analog band-pass filter according to the specifications

$$\widetilde{\delta}_p = 0.02, \quad \widetilde{\delta}_{s,1} = \widetilde{\delta}_{s,2} = 0.005,$$

$$\widetilde{\omega}_{s,1} = 5, \quad \widetilde{\omega}_{p,1} = 10, \quad \widetilde{\omega}_{p,2} = 40, \quad \widetilde{\omega}_{s,2} = 160.$$

The filter must be ripple free in all bands.

Sharon designed a band-pass Butterworth filter, as was studied in this chapter. Irwin decided to construct the band-pass filter by a cascade connection of a high-pass

Butterworth filter $H_1^L(s)$ and a low-pass Butterworth filter $H_2^L(s)$. He used band-edge frequencies $\widetilde{\omega}_{s,1}, \widetilde{\omega}_{p,1}$ for the high-pass filter and band-edge frequencies $\widetilde{\omega}_{p,2}, \widetilde{\omega}_{s,2}$ for the low-pass filter. He used the same δ_p for both filters, and the same δ_s, choosing these two parameters to satisfy the tolerances $\widetilde{\delta}_p, \widetilde{\delta}_{s,1}, \widetilde{\delta}_{s,2}$ in the overall filter.

Whose design leads to a band-pass filter of lower order?

10.16 Find the number of zeros of a band-pass filter $\widetilde{H}^L(\widetilde{s})$, derived from a low-pass filter $H^L(s)$ according to (10.83). Treat the four filter classes separately. Show that, in all cases, the zeros of the band-pass filter are on the imaginary axis.

10.17 Find the number of zeros of a band-stop filter $\widetilde{H}^L(\widetilde{s})$, derived from a low-pass filter $H^L(s)$ according to (10.93). Treat the four filter classes separately. Show that, in all cases, the zeros of the band-stop filter are on the imaginary axis.

10.18 Explain why the impulse invariant transform is not applicable to Chebyshev-II and elliptic filters of even orders (even when they are low pass or band pass).

10.19 The digital filter

$$H^z(z) = \frac{0.008502z^{-1} + 0.007272z^{-2}}{1 - 2.5060z^{-1} + 2.1470z^{-2} - 0.6252z^{-3}}$$

is known to have been obtained from a Chebyshev-I analog low-pass filter by an impulse invariant transformation with $T = 0.1$. Find the parameters ε and ω_0 of the analog filter [use MATLAB for finding the poles of $H^z(z)$].

10.20 A Chebyshev-I filter of order $N = 3$ and $\omega_0 = 1$ is known to have a pole at $s = -1 \, \text{rad/s}$.

(a) Find the other two poles of the filter and its parameter ε.

(b) The filter is transformed to the z domain using a bilinear transform with $T = 2$. Compute the transfer function of the digital filter $H^z(z)$.

10.21 A first-order analog filter $H^L(s)$ has a zero at $s = -2$, a pole at $s = -2/3$, and its DC gain is $H^L(0) = 1$. Bilinear transformation of $H^L(s)$ yields the digital filter $H^z(z) = K/(1 - \alpha z^{-1})$. Find K, α, and the sampling interval T.

10.22 We are given the digital filter

$$H^z(z) = \frac{2\theta_0^2(1 + z)^2}{(z - 1)^2 + 2\theta_0(z^2 - 1) + 2\theta_0^2(z + 1)^2},$$

where θ_0 is a given positive parameter. We are also given that $H^z(z)$ was obtained from an analog filter $H^L(s)$ by a bilinear transform with $T = 2$.

(a) What type of filter is $H^z(z)$? Low pass, high pass, band pass, or band stop? Give reasons.

(b) Compute the transfer function, as well as the poles and zeros, of the analog filter $H^L(s)$. Is this filter Butterworth, Chebyshev-I, Chebyshev-II, or elliptic?

(c) At what frequency ω_{3db} will the analog filter have an attenuation of 3 dB?

(d) At what frequency θ_{3db} will the digital filter have an attenuation of 3 dB?

10.23 We are given a digital filter $H^z(z)$ having two zeros at $z = -1$, a pole at $z = j\alpha$, and a pole at $z = -j\alpha$, where α is real, $0.6 < \alpha < 1$. The filter was obtained from an analog filter $H^L(s)$ using the bilinear transform.

(a) Draw an approximate plot of $H^f(\theta)$ in the range $0 \le \theta \le \pi$.

(b) Compute $H^L(s)$ and express it as a ratio of two polynomials, with α and T as parameters.

(c) If $\alpha = 1/\sqrt{2}$ and $T = 1$, is $H^L(s)$ Butterworth, Chebyshev-I, Chebyshev-II, or elliptic?

10.24 Figure 10.30 shows the pole–zero maps of three digital filters. For each of the three, state if it could have been obtained from Butterworth, Chebyshev-I, or Chebyshev-II analog filter, using a bilinear transform with $T = 1$. If so, write the corresponding $H^L(s)$ and state its kind. If not, explain why.

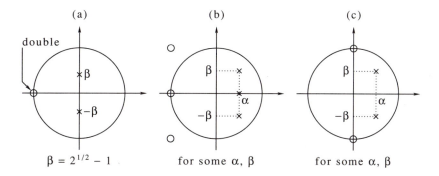

Figure 10.30 Pertaining to Problem 10.24; the circles have radii 1; ∘s indicate zeros; ×s indicate poles.

10.25 A digital IIR filter is to meet the following requirements:

- Its denominator degree p and numerator degree q should be equal.
- It must have infinite attenuation at frequency $\theta = \pi/3$.
- Its poles must be equal to those of a normalized Butterworth filter transformed to the digital domain by a bilinear transform with $T = \sqrt{2}$.
- Its DC gain must be equal to 1.
- It must have minimal order.

Find the transfer function of the filter.

10.26 As we recall, the transformation $s = \omega_0/\tilde{s}$ converts a normalized analog low-pass filter to an unnormalized high-pass filter. Let us transform s to z and $\tilde{s}$ to $\tilde{z}$, using the bilinear transform in both cases, with T as a parameter.

(a) Express $\tilde{z}$ as a function of z.

(b) Prove that the unit circle in the z plane is transformed to the unit circle in the $\tilde{z}$ plane, and find the frequency variable $\tilde{\theta}$ as a function of θ. It is convenient to define an auxiliary parameter depending on ω_0 and T.

(c) Show that the transformation from z to $\tilde{z}$ is low pass to high pass.

(d) Suppose that we are given a low-pass filter $H^z(z)$ whose pass-band cutoff frequency is θ_p, and we wish to obtain a high-pass filter in the $\tilde{z}$ domain whose pass-band cutoff frequency is $\tilde{\theta}_p$. Find ω_0 in the transformation that will achieve this.

10.27 We are given a low-pass digital Butterworth filter $H^z(z)$ whose -3 dB frequency is θ_0. We transform this filter using the substitution

$$\tilde{H}^z(\tilde{z}) = H^z(z)\bigg|_{z^{-1}=\frac{\tilde{z}^{-1}-\alpha}{1-\alpha\tilde{z}^{-1}}},$$

where $-1 < \alpha < 1$.

(a) Show that $\tilde{H}^z(\tilde{z})$ is also a low-pass filter.

(b) We want the -3 dB frequency of $\tilde{H}^z(\tilde{z})$ to be $\tilde{\theta}_0$. Find α that will accomplish this, as a function of θ_0 and $\tilde{\theta}_0$.

10.28* It is required to design a multiband digital IIR filter according to the following specifications:

- Gain 1 with $\delta_p = 0.01$ in the frequency band 0–0.2π.
- Gain 0 with $\delta_s = 0.05$ in the frequency band 0.25π–0.45π.
- Gain 1 with $\delta_p = 0.01$ in the frequency band 0.5π–0.7π.
- Gain 0 with $\delta_s = 0.05$ in the frequency band 0.75π–π.

It is proposed to build the filter as a parallel connection of two filters, $H_1^z(z)$ and $H_2^z(z)$, where the first is low pass and the second is band pass.

Design the system. Note: This problem does not have a unique solution. Any solution meeting the specifications is acceptable. However, the lower the complexity of the system, the better.

10.29* The Nth-order *Bessel polynomial* $B_N(s)$ is defined by

$$B_N(s) = \sum_{k=0}^{N} b_{N,k}s^k, \quad \text{where} \quad b_{N,k} = \frac{(2N-k)!}{2^{N-k}k!(N-k)!}.$$

The Nth-order *Bessel filter* is defined by

$$H^L(s) = \frac{1}{B_N(s)}.$$

(a) Show that Bessel polynomials satisfy the recursion

$$B_N(s) = (2N-1)B_{N-1}(s) + s^2 B_{N-2}(s), \quad B_0(s) = 1, \ B_1(s) = s + 1.$$

(b) Write a MATLAB program that computes the coefficients of the Nth-order Bessel polynomial.

(c) Compute and plot the magnitude and phase responses of a Bessel filter of orders 2 through 8. Observe in particular the phase response at low frequency. State your conclusions about the characteristics of Bessel filters.

10.30* Let $H^L(s)$ be a given analog filter. Define $g(t)$ as the response of the filter to a unit-step input, that is,

$$g(t) = \int_0^t h(\tau)d\tau.$$

Let $g[n] = g(nT)$ be the point sampling of $g(t)$. Define $H^z(z)$ as the digital filter whose response to a discrete-time, unit-step input is $g[n]$, that is,

$$g[n] = \sum_{m=0}^{n} h[m].$$

The filter $H^z(z)$ is called the *step invariant transform* of $H^L(s)$.

(a) Derive a formula and a method of computation for the step invariant transform.

(b) Compare this method with the impulse invariant transform from all points of view we have discussed.

10.31* Define the following s-to-z transformation

$$H^z(z) = H^L(s)\Big|_{s=\frac{z-1}{T}}. \qquad (10.129)$$

The operation (10.102) is called *forward difference transformation*, in analogy to (10.102). Show that this transformation does not preserve stability; that is, a stable analog filter $H^L(s)$ may be transformed to an unstable digital filter $H^z(z)$.

10.32* The purpose of this problem is to show that the bilinear transform can be interpreted as approximation of continuous-time integration by discrete-time *trapezoidal integration*.

Let $y(t)$ be the integral of $x(t)$, that is,

$$y(t) = \int_{-\infty}^{t} x(\tau)d\tau \iff Y^L(s) = \frac{X^L(s)}{s}.$$

Perform a bilinear transform on the s-domain description of the integral and obtain a z-domain transfer function relating $X^z(z)$ to $Y^z(z)$. Then transform back to time domain. Finally, interpret the resulting difference equation as approximate integration and explain the name *trapezoidal integration*. Hint: The integral relationship between $y(t)$ and $x(t)$ implies

$$y(nT) = y(nT - T) + \int_{nT-T}^{nT} x(\tau)d\tau.$$

10.33* This problem introduces the *Goertzel algorithm* [Goertzel, 1958] for computing the Fourier transform of a finite-duration signal at a single frequency point θ_0. The relation of this problem to IIR filters will become clear soon.

(a) Show that the operation

$$X^f(\theta_0) = \sum_{n=0}^{N-1} x[n]e^{-j\theta_0 n}$$

can be expressed as

$$X^f(\theta_0) = e^{-j\theta_0(N-1)}\{x * h\}[N - 1], \qquad (10.130)$$

where $h[n]$ is the impulse response of a causal IIR filter, given by

$$h[n] = e^{j\theta_0 n}, \quad n \geq 0.$$

(b) Show that the filter $h[n]$ has the transfer function

$$H^z(z) = \frac{1}{1 - e^{j\theta_0}z^{-1}}.$$

Hence show that, if $x[n]$ is a general complex sequence, (10.130) can be performed in about $4N$ real multiplications and $4N$ real additions. Note that this is no more efficient than direct computation of $X^f(\theta_0)$.

(c) Show that $H^z(z)$ can be written as

$$H^z(z) = (1 - e^{-j\theta_0}z^{-1})H_1^z(z),$$

where

$$H_1^z(z) = \frac{1}{1 - 2\cos\theta_0 z^{-1} + z^{-2}}.$$

(d) Show that (10.130) can be carried out by the following steps:

$$y[n] = \{x * h_1\}[n], \quad 0 \le n \le N - 1, \tag{10.131a}$$

$$X^f(\theta_0) = e^{-j\theta_0(N-1)}(y[N-1] - e^{-j\theta_0}y[N-2]). \tag{10.131b}$$

(e) Count the total number of real operations in (10.131). Compare it with the number of operations in direct computation of $X^f(\theta_0)$.

10.34* Consider a sampled-data system such as the one shown in Figure 10.29, except that instead of the ZOH there is a reconstructor whose impulse response is

$$h(t) = \begin{cases} 1 + \frac{t}{T}, & 0 \le t \le T, \\ 0, & \text{otherwise.} \end{cases}$$

(a) Find the transfer function $H^z(z)$ from $x[n]$ to $y[n]$.

(b) Compute $H^z(z)$ for the special case $G^L(s) = 1/s$.

10.35* Consider the sampled-data system described in Section 10.11. Suppose that the input signal $x[n]$ is delayed by Δ before being fed to the ZOH. In other words, the response of the ZOH to $x[n]$ is $h_{\text{zoh}}(t - nT - \Delta)$. Assume that $\Delta < T$. Assume also that $y[n]$ is obtained by sampling synchronously with $x[n]$ *before the delay*, that is, $y[n] = y(nT)$ as before. Show that the equivalent discrete-time system relating $y[n]$ to $x[n]$ has the impulse response

$$h[n] = u(nT) - u(nT - T),$$

where

$$u(t) = \left\{ \mathcal{L}^{-1}\left[\frac{G(s)e^{-s\Delta}}{s} \right] \right\}(t).$$

In control applications it is common to define

$$m = 1 - \frac{\Delta}{T}$$

and express $h[n]$ as a function of the parameter m. The function $H^z(z)$ is then called the *modified z-transform* of $G^L(s)$ [Jury, 1954]. Modified z-transforms are used when the input to the ZOH is delayed with respect to the system's clock, for example, because of finite computation time.

Chapter 11

Digital Filter Realization and Implementation

In the preceding two chapters we learned how to design digital filters, both IIR and FIR. The end result of the design was the transfer function $H^z(z)$ of the filter or, equivalently, the difference equation it represents. We thus far looked at a filter as a black box, whose input–output relationships are well defined, but whose internal structure is ignored. Now it is time to look more closely at possible internal structures of digital filters, and to learn how to build such filters. It is convenient to break the task of building a digital filter into two stages:

1. Construction of a block diagram of the filter. Such a block diagram is called a *realization* of the filter. Realization of a filter at a block-diagram level is essentially a flow graph of the signals in the filter. It includes operations such as delays, additions, and multiplications of signals by constant coefficients. It ignores ordering of operations, accuracy, scaling, and the like. A given filter can be realized in infinitely many ways. Different realizations differ in their properties, and some are better than others.

2. Implementation of the realization, either in hardware or in software. At this stage we must concern ourselves with problems neglected during the realization stage: order of operations; signal scaling; accuracy of signal values; accuracy of coefficients; accuracy of arithmetic operations. We must analyze the effect of such imperfections on the performance of the filter. Finally, we must build the filter—either the hardware or the program code (or both, if the filter is a specialized combination of hardware and software).

In this chapter we cover the aforementioned subjects. We begin by presenting the most common filter realizations. We describe each realization by its block diagram and by a representative MATLAB code. This will naturally lead us to state-space representations of digital filters. State-space representations are a powerful tool in linear system theory. They are useful for analyzing realizations, performing block-diagram manipulations, computing transfer functions and impulse responses, and executing a host of other applications. State-space theory is rich and we cannot hope to do it justice in a few sections. We therefore concentrate on aspects of state-space theory useful for digital filters, mainly computational ones.

The remainder of this chapter is devoted to finite word length effects in filter implementation. We concentrate on fixed-point implementations, since floating-point implementations suffer less from problems caused by finite word length. We first discuss the effect of coefficient quantization, and explore its dependence on the realization of the filter. This will lead to important guidelines on choice of realizations for different uses. Then we explore the scaling problem, and present scaling procedures for various realizations. Our next topic is noise generated by quantization of multiplication operations, also called computation noise. We present procedures for analyzing the effect of computation noise for different realizations. Finally, we briefly discuss the phenomenon of limit cycle oscillations.

11.1   Realizations of Digital Filters

11.1.1   Building Blocks of Digital Filters

Any digital system that is linear, time invariant, rational, and causal can be realized using three basic types of element:

1. A unit delay: The purpose of this element is to hold its input for a unit of time (physically equal to the sampling interval T) before it is delivered to the output. Mathematically, it performs the operation

$$y[n] = x[n-1].$$

Unit delay is depicted schematically in Figure 11.1(a). The letter "D," indicating delay, sometimes is replaced by z^{-1}, which is the delay operator in the z domain. Unit delay can be implemented in hardware by a data register, which moves its input to the output when clocked. In software, it is implemented by a storage variable, which changes its value when instructed by the program.

2. An adder: The purpose of this element is to add two or more signals appearing at the input at a specific time. Mathematically, it performs the operation

$$y[n] = x_1[n] + x_2[n] + \cdots$$

An adder is depicted schematically in Figure 11.1(b).

3. A multiplier: The purpose of this element is to multiply a signal (a varying quantity) by a constant number. Mathematically,

$$y[n] = ax[n].$$

A multiplier is depicted schematically in Figure 11.1(c). We do not use a special graphical symbol for it, but simply put the constant factor above (or beside) the signal line. A physical multiplier (in hardware or software) can multiply two signals equally easily, but such an operation is not needed in LTI filters.

Example 11.1 Consider the first-order FIR filter

$$H_1^z(z) = b_0 + b_1 z^{-1}.$$

Figure 11.2(a) shows a realization of this filter using one delay element, two multipliers, and one adder. The input to the delay element at time n is $x[n]$, and its output is then $x[n-1]$. The output of the realization is therefore

$$y[n] = b_0 x[n] + b_1 x[n-1],$$

and this is exactly the time-domain expression for $H_1^z(z)$.   □

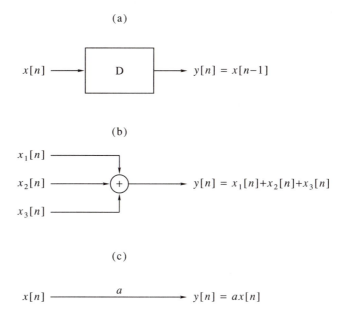

Figure 11.1 Basic building blocks for digital filter realizations: (a) unit delay; (b) adder; (c) multiplier by a constant.

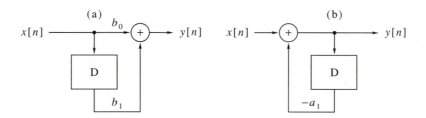

Figure 11.2 Realizations of first-order filters: (a) FIR; (b) IIR.

Example 11.2 Consider the first-order IIR filter

$$H_2^z(z) = \frac{1}{1 + a_1 z^{-1}}.$$

Figure 11.2(b) shows a realization of this filter using one delay element, one multiplier, and one adder. The input to the delay element at time n is $y[n]$, and its output is then $y[n-1]$. The output of the realization is therefore

$$y[n] = -a_1 y[n-1] + x[n],$$

and this is exactly the time-domain expression for $H_2^z(z)$. This realization is *recursive*: It builds the present value of $y[n]$ from its own past values and the input signal. Put another way, the realization uses *feedback*. The feedback enables the filter to have an infinite impulse response, although it has only one delay element. To see this, suppose that the output of the delay element has zero output at all $n < 0$, and we apply a unit-sample signal $\delta[n]$ at time $n = 0$. This causes $y[0]$ to be set to 1. At time $n = 1$, the value $y[0] = 1$ is transferred to the output of the delay element, is multiplied by $-a_1$, and is fed back to the input via the adder. The input $x[1]$ is now zero. Therefore, $y[1] = -a_1$. Continuing the same way, we see that $y[2] = a_1^2, y[3] = -a_1^3$, and in general $y[n] = (-a_1)^n$ for all $n \geq 0$. We get the impulse response of the filter at the output, as expected. $\square$

11.1.2   Direct Realizations

Let $H^z(z)$ be a rational, causal, stable transfer function. We assume, for convenience, that the orders of the numerator and denominator polynomials of the filter transfer function are equal, that is, $p = q = N$. There is no loss of generality in this assumption since, for any p and q, we can always define $N = \max\{p, q\}$ and extend either $a(z)$ or $b(z)$ to degree N by adding zero-valued coefficients. Thus $H^z(z)$ is given by

$$\frac{Y^z(z)}{X^z(z)} = H^z(z) = \frac{b(z)}{a(z)} = \frac{b_0 + b_1 z^{-1} + \cdots + b_N z^{-N}}{1 + a_1 z^{-1} + \cdots + a_N z^{-N}}. \qquad (11.1)$$

Let us introduce an auxiliary signal $u[n]$, related to the input and output signals $x[n]$ and $y[n]$ through

$$u[n] = -a_1 u[n-1] - \cdots - a_N u[n-N] + x[n], \qquad (11.2)$$

$$y[n] = b_0 u[n] + b_1 u[n-1] + \cdots + b_N u[n-N], \qquad (11.3)$$

or, in the z domain,

$$U^z(z) = \frac{1}{a(z)} X^z(z), \quad Y^z(z) = b(z) U^z(z). \qquad (11.4)$$

Figure 11.3 shows a realization of (11.2) using the three types of building block (in the figure we have used $N = 3$ as an example). By passing $u[n]$ through a chain of N delay elements, we get the signals $\{u[n-1], u[n-2], \ldots, u[n-N]\}$. Then we can form $u[n]$ as a linear combination of the delayed signals, plus the input signal $x[n]$, as is expressed in (11.2). We need N multipliers for the coefficients $\{-a_1, \ldots, -a_N\}$, and N binary adders to form the sum. The realization uses feedback: It builds the present value of the signal $u[n]$ from its own past values and the present value of the input signal $x[n]$.

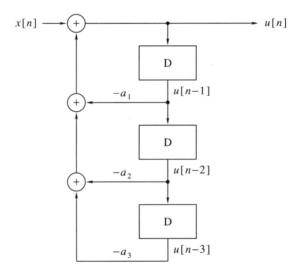

Figure 11.3 Realization of the auxiliary signal $u[n]$.

We can now use (11.3) for generating the output signal $y[n]$ from the auxiliary signal $u[n]$ and its delayed values. We do this by augmenting Figure 11.3 with $N + 1$

multipliers for the coefficients $\{b_0, \ldots, b_N\}$ and N adders. This results in the realization shown in Figure 11.4. Note that it is not necessary to increase the number of delay elements.

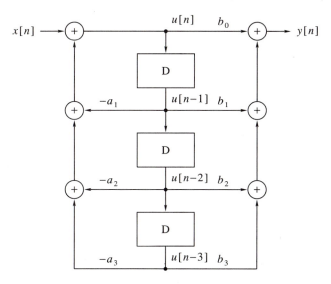

Figure 11.4 Direct realization of a digital IIR system.

The realization shown in Figure 11.4 is called a *direct realization*, or a *direct form*.[1] It has the following properties:

1. The number of delay elements N is the maximum of p and q. Assuming that the polynomials $a(z)$ and $b(z)$ have no common factor, this is the *minimum possible number of delays* in any realization of $H^z(z)$. The set of values at the outputs of the delay elements is called *the state* of the system.

2. There are $2N + 1$ multipliers and $2N$ adders. However, since some coefficients may be zero (e.g., if p and q are different), there may be a smaller number of adders and multipliers.

3. The realization is recursive, since it generates present values of its internal signals from past values of these signals. It can be shown that any realization of an IIR system must be recursive if it is required to include only a finite number of delay elements.

4. Assuming that the input signal $x[n]$ is causal, it is necessary to initialize the state components before feeding the input signal to the system. The state is usually initialized to zero. However, initialization to nonzero values is sometimes required, depending on the application.

We now explore an alternative to the direct realization. Instead of the variable $u[n]$, we introduce another auxiliary variable $v[n]$, such that

$$y[n] = -a_1 y[n-1] - \cdots - a_N y[n-N] + v[n], \tag{11.5}$$
$$v[n] = b_0 x[n] + b_1 x[n-1] + \cdots + b_N x[n-N], \tag{11.6}$$

or, in the z domain,

$$Y^z(z) = \frac{1}{a(z)} V^z(z), \quad V^z(z) = b(z) X^z(z). \tag{11.7}$$

Figure 11.5 shows a realization of (11.5) using the three types of building block (in the figure we use $N = 3$ as an example). The present value of $y[n]$ is built from its own past values and the auxiliary signal $v[n]$. To generate $y[n - N]$, we need N delay elements; these elements can be used for generating all intermediate delays, as shown in the figure. This realization effectively computes (11.5) in the z-domain form

$$Y^z(z) = (\cdots(-a_N z^{-1} - a_{N-1})z^{-1} - \cdots - a_1)z^{-1}Y^z(z) + V^z(z).$$

The state of this realization does not consist of pure delays of a single signal, but of linear combinations of different delays. As before, we need N multipliers for the coefficients $\{-a_1, \ldots, -a_N\}$, and N binary adders.

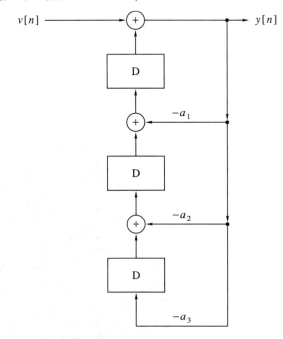

Figure 11.5 Realization of $y[n]$ from the auxiliary signal $v[n]$.

We can now use (11.6) for generating the auxiliary signal $v[n]$ from the input signal $x[n]$ and its delayed values. We do this by augmenting Figure 11.5 with $N+1$ multipliers for the coefficients $\{b_0, \ldots, b_N\}$ and N adders. This results in the realization shown in Figure 11.6. Note that it is not necessary to increase the number of delay elements, since the existing elements can take care of the necessary delays of the input signal.

The realization shown in Figure 11.6 is known as a *transposed direct realization* (or transposed direct form), for reasons explained next. The transposed direct realization shares the main properties of the direct realization. In particular, it has the same number of delays, multipliers, and binary adders. Note that, in Figure 11.6, there are $N - 1$ ternary adders and 2 binary adders, which are equivalent to $2N$ binary adders. However, the two realizations have different states. As long as the state is initialized to zero, this difference is inconsequential. However, when initialization to a nonzero state is necessary, the two realizations require different computations of the initial state.

Comparison of Figures 11.4 and 11.6 reveals that the latter can be obtained from the former by the following sequence of operations:

1. Reversal of the signal flow direction in all lines (i.e., reversal of all arrows).

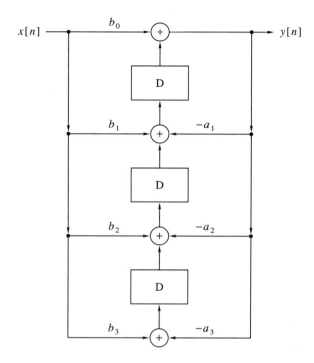

Figure 11.6 Transposed direct realization of a digital IIR system.

2. Replacing all adders by contact points, and all contact points by adders.

3. Interchanging the input and output.

This sequence of operations is called *transposition* of the realization. A known theorem in network theory states that transposition of a given realization leaves the transfer function of the realization invariant. Transposition of the transposed realization obviously gives back the original realization. Two such realizations are said to be *dual* to each other.

The procedure `direct` in Program 11.1 implements the two direct realizations of an IIR filter.[2] The MATLAB function `filter` performs the same computation more efficiently, since it is coded internally. Therefore, our program is intended for educational purposes, rather than for serious use.

11.1.3 Direct Realizations of FIR Filters

Realizations of digital FIR filters can be obtained from IIR filter realizations by specializing these realizations to the case $a(z) \equiv 1$. However, since FIR filters are usually either symmetric or antisymmetric, we can save about half the number of multiplications by exploiting symmetry. For an even-order filter we can write the filter's output as

$$y[n] = h[N/2]x[n - N/2] + \sum_{m=0}^{N/2-1} h[m](x[n - m] \pm x[n - N + m]), \tag{11.8}$$

whereas for an odd-order filter we put

$$y[n] = \sum_{m=0}^{(N-1)/2} h[m](x[n - m] \pm x[n - N + m]). \tag{11.9}$$

The plus sign is for a symmetric filter (type I or II), and the minus sign for an antisymmetric one (type III or IV). Figure 11.7 shows a realization of (11.8). This realization can be regarded as a specialization of the direct realization shown in Figure 11.4 to an FIR filter. As we see, symmetry (or antisymmetry) helps reducing the number of multiplications from $N + 1$ to $\lfloor N/2 \rfloor + 1$; the number of additions, however, is still N.

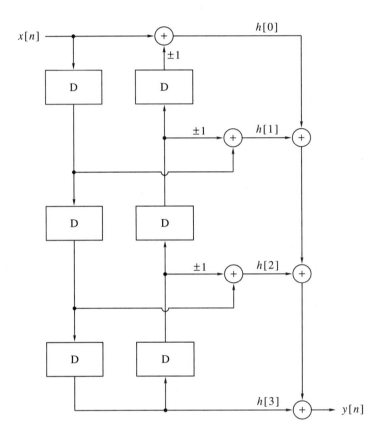

Figure 11.7 Direct realization of a symmetric or antisymmetric FIR filter.

By exploiting the rules of transposition of realizations, we get from Figure 11.7 the transposed direct realization shown in Figure 11.8.

11.1.4   Parallel Realization

Recall the partial fraction decomposition of a rational causal transfer function whose poles are simple:

$$H^z(z) = \frac{b(z)}{a(z)} = c_0 + \cdots + c_{q-p} z^{-(q-p)} + \sum_{i=1}^{p} \frac{A_i}{1 - \alpha_i z^{-1}}. \tag{11.10}$$

For most practical digital IIR filters $q \leq p$, so the c coefficients beyond c_0 are absent from the right side. Also, if α_i is complex then A_i is complex as well, and the conjugate fraction $\bar{A}_i / (1 - \bar{\alpha}_i z^{-1})$ also appears on the right side. The two can be brought together

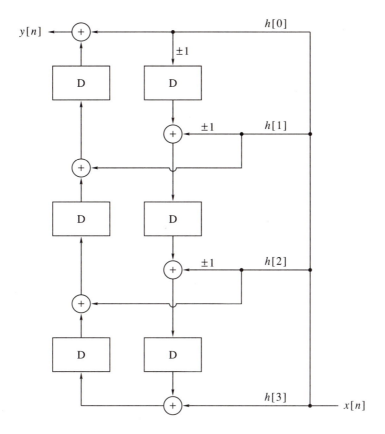

Figure 11.8 Transposed direct realization of a symmetric or antisymmetric FIR filter.

under a common denominator, yielding the real fraction

$$\frac{(A_i + \bar{A}_i) - (A_i\bar{\alpha}_i + \bar{A}_i\alpha_i)z^{-1}}{1 - (\alpha_i + \bar{\alpha}_i)z^{-1} + \alpha_i\bar{\alpha}_iz^{-2}}.$$

Let us denote the order of the filter by N instead of p, for notational consistency with the preceding section. Also, let N_1 be the number of real poles and $2N_2$ the number of complex poles, so $N = N_1 + 2N_2$. After joining complex conjugate fractions, we can bring (11.10) to the form

$$H^z(z) = c_0 + \sum_{i=1}^{N_1} \frac{f_i}{1 + e_iz^{-1}} + \sum_{i=1}^{N_2} \frac{f_{N_1+2i-1} + f_{N_1+2i}z^{-1}}{1 + e_{N_1+2i-1}z^{-1} + e_{N_1+2i}z^{-2}}, \qquad (11.11)$$

where the real numbers $\{e_i, f_i, 1 \le i \le N\}$ depend on $\{A_i, \alpha_i, 1 \le i \le N\}$.

The sum on the right side of (11.11) corresponds to a parallel connection of the individual terms. Each of the first- and second-order terms can be implemented by a direct realization. The constant term c_0 is realized by simple multiplication. Figure 11.9 illustrates the result for $N_1 = 1$, $N_2 = 1$, $N = 3$. The realization thus obtained is called *parallel realization*.

Parallel realization requires the same number of delay elements as the direct realizations. If $q = p$, it also requires the same amount of additions and multiplications. If $q < p$, the direct realizations are more economical, since in this case they require only $p + q + 1$ multiplications and additions, whereas the parallel realization still requires $2p + 1$ operations of each kind (the reason is that all the coefficients f_i will be nonzero

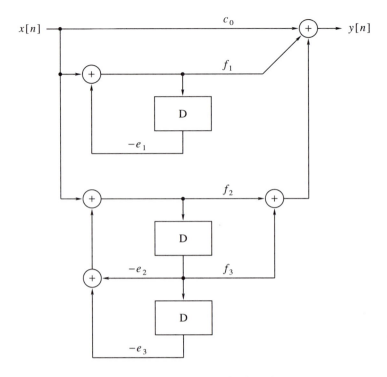

Figure 11.9 Parallel realization of a digital IIR system.

in general). As we have said, the parallel realization is limited to systems whose poles are simple. It can be extended to the case of multiple poles, but then a parallel realization is rarely used.[3] The advantages of parallel realization over direct realizations will become clear in Section 11.5, when we study the sensitivity of the frequency response of the filter to finite word length.

The procedure tf2rpf in Program 11.2 computes the parallel decomposition (11.11) of a digital IIR filter. The program first calls tf2pf (Program 7.1) to compute the complex partial fraction decomposition, then combines complex pairs to real second-order sections.

The procedure parallel in Program 11.3 implements the parallel realization of a digital IIR filter. It accepts the parameters computed by the program tf2rpf, computes the response to each second-order section separately, and adds the results, including the constant term. The program is slightly inefficient in that it treats first-order sections as second-order ones, so it is likely to perform redundant multiplications and additions of zero values. However, common IIR filters either do not have real poles (if the order is even), or they have a single real pole (if the order is odd). Therefore, the redundant operations do not amount to much. In real-time implementations, however, care should be taken to avoid them. We also reiterate that the MATLAB implementation is rather time consuming, due to the inefficiency of MATLAB in loop computations.

11.1.5 Cascade Realization

Let us assume again that $q = p = N$ for the digital filter in question. Assume for now that N is even. Recall the pole–zero factorization of the transfer function,

$$H^z(z) = b_0 \frac{\prod_{i=1}^{N}(1 - \beta_i z^{-1})}{\prod_{i=1}^{N}(1 - \alpha_i z^{-1})}. \tag{11.12}$$

Since N is even, we can always rewrite (11.12) as

$$H^z(z) = b_0 \prod_{i=1}^{N/2} \frac{(1 + h_{2i-1}z^{-1} + h_{2i}z^{-2})}{(1 + g_{2i-1}z^{-1} + g_{2i}z^{-2})}. \tag{11.13}$$

The second-order factors in the numerator and the denominator are obtained by expanding conjugate pairs of zeros or poles; hence the coefficients $\{g_i, h_i, 1 \le i \le N\}$ are real. In general, some poles or zeros may be real. Since we have assumed that N is even, their number must be even, so we can expand them in pairs as well. Thus in general, the second-order terms in (11.13) may correspond to either real or complex pairs of poles or zeros.

A product of transfer functions represents a cascade connection of the factors. Therefore, we can implement (11.13) as a cascade connection of $N/2$ sections, each of order 2. Each section can be realized by either of the two direct realizations. Figure 11.10 illustrates this connection for $N = 4$. This is called a *cascade realization*. Note that the constant gain b_0 can appear anywhere along the cascade (in Figure 11.10 it appears in the middle between the two sections). The second-order sections are also called *bi-quads*.

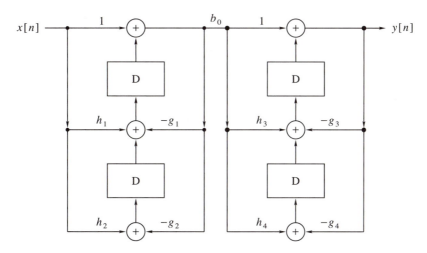

Figure 11.10 Cascade realization of a digital IIR system.

Remarks

1. Although we have assumed that N is even, the realization can be easily extended to the case of odd N. In this case there is an extra first-order term, so we must add a first-order section in cascade.

2. Although we have assumed that $p = q$, this condition is not necessary. Extra poles can be represented by sections with zero values for the h coefficients, whereas extra zeros can be represented by sections with zero values for the g coefficients.

3. The realization is minimal in terms of number of delays, additions and multiplications (with the understanding that zero-valued coefficients save the corresponding multiplications and additions).

4. The realization is nonunique, since:

    (a) There are multiple ways of pairing each second-order term in the denominator with one in the numerator. In Section 11.1.6 we discuss the pairing problem in detail.

    (b) There are multiple ways of ordering the sections in the cascade connection.

    (c) There are multiple ways of inserting the constant gain factor b_0.

5. Contrary to the parallel realization, the cascade realization is not limited to simple poles. Moreover, it does not require the condition $q \leq p$. Cascade realization is applicable to FIR filters, although its use for such filters is relatively uncommon.

11.1.6   Pairing in Cascade Realization

When cascade realization is implemented in floating point and at a high precision (such as in MATLAB), the pairing of poles and zeros to second-order sections is of little importance. However, in fixed-point implementations and short word lengths, it is advantageous to pair poles and zeros to produce a frequency response for each section that is as flat as possible (i.e., such that the ratio of the maximum to the minimum magnitude response is close to unity). We now describe a pairing procedure that approximately achieves this goal. We consider only digital filters obtained from one of the four standard filter types (Butterworth, Chebyshev-I, Chebyshev-II, elliptic) through an analog frequency transformation followed by a bilinear transform. Such filters satisfy the following three properties:

1. The number of zeros is equal to the number of poles. If the underlying analog filter has more poles than zeros, the extra zeros of the digital filter are all at $z = -1$.

2. The number of complex poles is never smaller than the number of complex zeros.

3. The number of real poles is not larger than 2. A low-pass filter has one real pole if its order is odd, and this pole may be transformed to two real poles or to a pair of complex poles by either a low-pass to band-pass or a low-pass to band-stop transformation. Except for those, all poles of the analog filter are complex, hence so are the poles of the digital filters.

The basic idea is to pair each pole with a zero as close to it as possible. By what we have learned in Section 7.6, this makes the magnitude response of the pole–zero pair as flat as possible. The pairing procedure starts at the pair of complex poles nearest to the unit circle (i.e., those with the largest absolute value) and pairs them with the nearest complex zeros. It then removes these two pairs from the list and proceeds according to the same rule. When all the complex zeros are exhausted, pairing continues with the real zeros according to the same rule. Finally, there may be left up to two real poles, and these are paired with the remaining real zeros.

The procedure `pairpz` in Program 11.4 implements this algorithm. It receives the vectors of poles and zeros, supplied by the program `iirdes` (Program 10.8) and supplies arrays of second-order numerator and denominator polynomials (a first-order pair, if any, is represented as a second-order pair with zero coefficient of z^{-2}). The routine `cplxpair` is a built-in MATLAB function that orders the poles (or zeros) in conjugate pairs, with real ones (if any) at the end. The program then selects one representative of each conjugate pair and sorts them in decreasing order of magnitude. Next the program loops over the complex poles and, for each one, finds the nearest complex zero. Every paired zero is removed from the list. The polynomials of the corresponding second-order section are computed and stored. When the complex zeros are exhausted, the remaining complex poles are paired with the real zeros using the same search procedure. Finally, the real poles are paired with the remaining real zeros.

The procedure `cascade` in Program 11.5 implements the cascade realization of a digital IIR filter. It accepts the parameters computed by the program `pairpz`. The input sequence is fed to the first section, the output is fed to the second section, and so forth. Finally, the result is multiplied by the constant gain.

The cascade realization is usually considered as the best of all those we have discussed, for reasons to be explained in Section 11.5; therefore, it is the most widely used.

11.1.7 A Coupled Cascade Realization

Direct realization of second-order sections used in cascade realization may be unsatisfactory, especially if the word length is short and the filter has poles near $z = 1$ or $z = -1$. Detailed explanation of this phenomenon is deferred until later in this chapter. We now present an alternative realization of second-order sections, which offers an improvement over a direct realization in case of a short word length. This so-called *coupled realization* is shown in Figure 11.11. The parameters α_r, α_i are the real and imaginary parts of the complex pole α of the second-order section, that is,

$$g_1 = -2\alpha_r = -2\Re\{\alpha\}, \quad g_2 = \alpha_r^2 + \alpha_i^2 = |\alpha|^2. \tag{11.14}$$

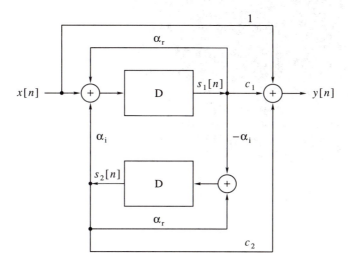

Figure 11.11 A coupled second-order section for cascade realization.

We now derive the relationship between the outputs $s_1[n]$ and $s_2[n]$ of the delay elements and the input $x[n]$. This will enable us to prove that the transfer function from $x[n]$ to $y[n]$ has the right denominator and will provide the values of the coefficients c_1, c_2 necessary to obtain the right numerator. We do the calculation in the z domain. We have from the figure,

$$zS_1^z(z) = X^z(z) + \alpha_r S_1^z(z) + \alpha_i S_2^z(z), \tag{11.15a}$$

$$zS_2^z(z) = \alpha_r S_2^z(z) - \alpha_i S_1^z(z). \tag{11.15b}$$

Or, in a matrix form,

$$\begin{bmatrix} z - \alpha_r & -\alpha_i \\ \alpha_i & z - \alpha_r \end{bmatrix} \begin{bmatrix} S_1^z(z) \\ S_2^z(z) \end{bmatrix} = \begin{bmatrix} X^z(z) \\ 0 \end{bmatrix}. \tag{11.16}$$

Therefore,

$$\begin{bmatrix} S_1^z(z) \\ S_2^z(z) \end{bmatrix} = \frac{X^z(z)}{z^2 - 2\alpha_r z + \alpha_r^2 + \alpha_i^2} \begin{bmatrix} z - \alpha_r \\ -\alpha_i \end{bmatrix} = \frac{X^z(z)}{z^2 + g_1 z + g_2} \begin{bmatrix} z - \alpha_r \\ -\alpha_i \end{bmatrix}. \tag{11.17}$$

We see that the denominator has the correct form. We can now express the output $Y^z(z)$ in terms of $X^z(z)$, $S_1^z(z)$, $S_2^z(z)$ and substitute (11.17) to get

$$Y^z(z) = X^z(z) + c_1 S_1^z(z) + c_2 S_2^z(z) = \frac{z^2 + (g_1 + c_1)z + (g_2 - c_1\alpha_r - c_2\alpha_i)}{z^2 + g_1 z + g_2} X^z(z). \tag{11.18}$$

Therefore, by choosing

$$c_1 = h_1 - g_1, \quad c_2 = \frac{g_2 - h_2 - c_1\alpha_r}{\alpha_i}, \tag{11.19}$$

we get the desired transfer function.

It follows from the preceding derivation that the section shown in Figure 11.11 can be constructed only for a pair of complex poles and does not apply to pairs of real poles. A pair of real poles must be realized by a second-order direct realization or by a cascade of two first-order sections. Each section in a coupled realization requires six multiplications and five additions, as compared with four multiplications and four additions for a direct realization. Despite the extra computations, the coupled realization is sometimes preferred to a direct realization of second-order sections, for reasons explained in Section 11.5.

11.1.8   FFT-Based Realization of FIR Filters

In Section 5.6 we showed how to use the fast Fourier transform for efficient convolution of a fixed-length sequence by a potentially long sequence. The overlap-add algorithm described there (or the overlap-save algorithm developed in Problem 5.23) can be used for FIR filter realization. The impulse response $h[n]$ is taken as the fixed-length sequence, and the input $x[n]$ as the long sequence. Table 5.2 gave the optimal FFT length as a function of the order of the filter (with N_2 in the table corresponding to $N + 1$ here). FFT-based FIR realization requires more memory than direct realization and is performed blockwise (as opposed to point by point), but is more efficient when the order of the filter is at least 18, so it is a serious candidate to consider in some applications.

11.2   State-Space Representations of Digital Filters

11.2.1   The State-Space Concept

A difference equation expresses the present output of the system in terms of past outputs, and present and past inputs. In discussing realizations of difference equations, we noted that the number of delay elements needed in any realization of the difference equation is equal to $\max\{p, q\}$. The delay elements represent the *memory* of the system, in the sense that their inputs must be stored and remembered from one time point to the next. Until now, the outputs of the delay elements were of no interest to us by themselves, only as auxiliary variables.

In this section we study a different way of representing rational LTI systems. In this representation, the outputs of the delay elements play a central role. Collected together, they are called the *state vector* of the system. Correspondingly, the representations we are going to present are called *state-space representations*. A state-space representation comprises two equations: The *state equation* expresses the time evolution of the state vector as a function of its own past and the input signal; the *output equation* expresses the output signal as a function of the state vector and the input signal.

To motivate the state-space concept, consider again the direct realization shown in Figure 11.4; introduce the notation

$$s_1[n] = u[n-1], \quad s_2[n] = u[n-2], \quad s_3[n] = u[n-3].$$

In other words, the signal $s_k[n]$ is the output of the kth delay element (starting from the top) at time n. Then we can read the following relationships directly from the figure:

$$s_1[n+1] = x[n] - a_1 s_1[n] - a_2 s_2[n] - a_3 s_3[n], \tag{11.20a}$$

$$s_2[n+1] = s_1[n], \tag{11.20b}$$

$$s_3[n+1] = s_2[n], \tag{11.20c}$$

$$y[n] = b_0 x[n] + (b_1 - b_0 a_1) s_1[n] + (b_2 - b_0 a_2) s_2[n] + (b_3 - b_0 a_3) s_3[n]. \tag{11.20d}$$

A compact way of writing (11.20) is

$$\begin{bmatrix} s_1[n+1] \\ s_2[n+1] \\ s_3[n+1] \end{bmatrix} = \begin{bmatrix} -a_1 & -a_2 & -a_3 \\ 1 & 0 & 0 \\ 0 & 1 & 0 \end{bmatrix} \begin{bmatrix} s_1[n] \\ s_2[n] \\ s_3[n] \end{bmatrix} + \begin{bmatrix} 1 \\ 0 \\ 0 \end{bmatrix} x[n], \tag{11.21}$$

$$y[n] = \begin{bmatrix} c_1 & c_2 & c_3 \end{bmatrix} \begin{bmatrix} s_1[n] \\ s_2[n] \\ s_3[n] \end{bmatrix} + b_0 x[n], \tag{11.22}$$

where

$$c_k = b_k - b_0 a_k, \quad 1 \le k \le 3. \tag{11.23}$$

These equations can be easily extended to any order N as follows. Define

$$s[n] = \begin{bmatrix} s_1[n] \\ s_2[n] \\ \vdots \\ s_N[n] \end{bmatrix}, \quad A_1 = \begin{bmatrix} -a_1 & -a_2 & \cdots & -a_{N-1} & -a_N \\ 1 & 0 & \cdots & 0 & 0 \\ \vdots & \vdots & \cdots & \vdots & \vdots \\ 0 & 0 & \cdots & 1 & 0 \end{bmatrix}, \quad B_1 = \begin{bmatrix} 1 \\ 0 \\ \vdots \\ 0 \end{bmatrix}, \tag{11.24}$$

$$c_k = b_k - b_0 a_k, \quad C_1 = \begin{bmatrix} c_1 & c_2 & \cdots & c_N \end{bmatrix}, \quad D_1 = b_0. \tag{11.25}$$

Then

$$s[n+1] = A_1 s[n] + B_1 x[n], \tag{11.26a}$$

$$y[n] = C_1 s[n] + D_1 x[n]. \tag{11.26b}$$

Equations (11.26a,b), together with the definitions (11.24), (11.25), are the state-space representation of the direct realization of the transfer function (11.1). The first is the *state equation*, and the second is the *output equation*. Together they describe the same input–output behavior as the difference equation

$$y[n] = -\sum_{k=1}^{N} a_k y[n-k] + \sum_{k=0}^{N} b_k x[n-k]. \tag{11.27}$$

Remember that, as we did for (11.1), here too we define $N = \max\{p, q\}$ and extend the sequences $\{a_k\}$, $\{b_k\}$ by zeros as necessary.

It is clear from the preceding construction that every difference equation has a corresponding state-space representation. The procedure tf2ss in Program 11.6 computes the matrices A_1, B_1, C_1, D_1 given in (11.24), (11.25). These matrices are called, respectively, the *state, input, output*, and *direct transmission* matrices. The first is $N \times N$, the second is $N \times 1$, the third is $1 \times N$, and the fourth is 1×1 (i.e., it is a scalar). A matrix such as A_1, whose first row is arbitrary, whose entries along the first subdiagonal are 1, and whose other entries are 0, is called a *top-row companion matrix*.

State-space representation is not limited to the direct realization. Let us now illustrate the state-space construction procedure for the transposed direct realization shown in Figure 11.6. As before, we denote the outputs of the delay elements by $\{s_k[n], 1 \le k \le N\}$, from top to bottom. The state vector $s[n]$ is the column vector of the $s_k[n]$. This time we define the state-space matrices as

$$A_2 = \begin{bmatrix} -a_1 & 1 & \cdots & 0 & 0 \\ \vdots & \vdots & \cdots & \vdots & \vdots \\ -a_{N-1} & 0 & \cdots & 0 & 1 \\ -a_N & 0 & \cdots & 0 & 0 \end{bmatrix}, \quad c_k = b_k - b_0 a_k, \quad B_2 = \begin{bmatrix} c_1 \\ c_2 \\ \vdots \\ c_N \end{bmatrix}. \tag{11.28}$$

$$C_2 = \begin{bmatrix} 1 & 0 & \cdots & 0 \end{bmatrix}, \quad D_2 = b_0. \tag{11.29}$$

Inspecting the connections in Figure 11.6 in relation to the definitions in (11.28), (11.29), we see that the transposed direct realization obeys the equations

$$s[n+1] = A_2 s[n] + B_2 x[n], \tag{11.30a}$$

$$y[n] = C_2 s[n] + D_2 x[n]. \tag{11.30b}$$

Equations (11.30a,b), together with the definitions (11.28), (11.29), are the state-space representation of the transposed direct realization of the transfer function (11.1). We remark that the state vector $s[n]$ in (11.30) is *different* from the one in (11.26), although we have used the same notation for convenience.

It is now clear that a given rational LTI system has more than one state-space representation, since we have demonstrated the existence of at least two such representations. We shall soon see that the number of state-space representations of a given rational LTI system is infinite. For now we note that the two representations we have derived are related as follows:

$$A_2 = A_1', \quad B_2 = C_1', \quad C_2 = B_1', \quad D_2 = D_1. \tag{11.31}$$

The two representations are said to be *dual*, or *transpose* of each other. In Section 11.1.2 we explained the sequence of operations that must be performed on a given realization to obtain its transpose. Now we see the corresponding operations on their state-space matrices: Transpose the state matrix, interchange and transpose the input and output matrices, and leave the direct transmission matrix (which is scalar in fact) unchanged.

Example 11.3 The signals $s_1[n], s_2[n]$ in Figure 11.11 are the state of the coupled realization of a second-order section. The following state equations describe this representation:

$$s[n+1] = \begin{bmatrix} \alpha_r & \alpha_i \\ -\alpha_i & \alpha_r \end{bmatrix} s[n] + \begin{bmatrix} 1 \\ 0 \end{bmatrix} x[n], \tag{11.32a}$$

$$y[n] = \begin{bmatrix} c_1 & c_2 \end{bmatrix} s[n] + x[n], \tag{11.32b}$$

where α_r, α_i are the real and imaginary parts of the complex pole α. □

11.2.2 Similarity Transformations

We now prove that a given difference equation (or a rational transfer function) has an infinite number of state-space representations. Let T be an arbitrary $N \times N$ nonsingular matrix. Define

$$\tilde{s}[n] = T^{-1} s[n], \tag{11.33}$$

and

$$\tilde{A} = T^{-1} A_1 T, \quad \tilde{B} = T^{-1} B_1, \quad \tilde{C} = C_1 T, \quad \tilde{D} = D_1. \tag{11.34}$$

Note that, whereas both s and $\tilde{s}$ depend on n, the matrix T is fixed (independent of n). Now go back to (11.26a) and premultiply it by T^{-1} to get

$$T^{-1} s[n+1] = T^{-1} A_1 s[n] + T^{-1} B_1 x[n] = T^{-1} A_1 T T^{-1} s[n] + T^{-1} B_1 x[n]. \tag{11.35}$$

Therefore, substituting (11.33) and (11.34), we get

$$\tilde{s}[n+1] = \tilde{A} \tilde{s}[n] + \tilde{B} x[n], \tag{11.36}$$

and similarly for (11.26b),

$$y[n] = C_1 T T^{-1} s[n] + D_1 x[n] = \tilde{C} \tilde{s}[n] + \tilde{D} x[n]. \tag{11.37}$$

The state-space representation (11.36), (11.37) is said to be *similar* to (11.26), and the relationships (11.33), (11.34) are called *similarity transformations*. Since there is an infinite number of $N \times N$ nonsingular matrices, there is an infinite number of state-space representations similar to (11.26). Although the state vectors of similar representations are different, the input–output relationships they describe are identical. A direct proof of this property is discussed in Problem 11.8.

11.2.3 Applications of State Space

State-space representations are of great importance in linear system theory. Here we describe only few of their many uses. We concentrate on uses relevant to topics dealt with earlier in this book. In particular, we show how state-space representations of a given system are related to its impulse response, input–output response, and transfer function.

Let

$$s[n+1] = As[n] + Bx[n], \tag{11.38a}$$

$$y[n] = Cs[n] + Dx[n], \tag{11.38b}$$

be a state-space representation of a given LTI system. Suppose the system is initially at rest, that is, $s[n] = 0$ for $n \leq 0$ and $x[n] = 0$ for $n < 0$. Now suppose we input a unit-sample signal at $n = 0$. Then the output of the system at all $n \geq 0$ is, by definition, the impulse response of the system. Let us determine the impulse response from (11.38). We have from (11.38a)

$$s[0] = 0, \; s[1] = B, \; s[2] = AB, \; s[3] = A^2B, \ldots \tag{11.39}$$

and in general

$$s[n] = \begin{cases} 0, & n \le 0, \\ A^{n-1}B, & n > 0. \end{cases} \tag{11.40}$$

Therefore, we have from (11.38b)

$$h[n] = \begin{cases} 0, & n < 0, \\ D, & n = 0, \\ CA^{n-1}B, & n > 0. \end{cases} \tag{11.41}$$

Observe how easy it is to obtain the impulse response from the state-space representation. In particular, there is no need to resort to z-transforms and partial fraction decompositions, as we did when we started from the difference equation.

Next we derive the system's response to an arbitrary (but causal) input signal. As before, we assume that the system is initially at rest. Then we get from (11.38a)

$$s[0] = 0, \; s[1] = Bx[0], \; s[2] = ABx[0] + Bx[1], \tag{11.42a}$$
$$s[3] = A^2Bx[0] + ABx[1] + Bx[2], \ldots \tag{11.42b}$$

and, in general,

$$s[n] = \begin{cases} 0, & n \le 0, \\ \sum_{k=0}^{n-1} A^{n-k-1}Bx[k], & n > 0. \end{cases} \tag{11.43}$$

Therefore, we get from (11.38b),

$$y[n] = \begin{cases} 0, & n < 0, \\ Dx[0], & n = 0, \\ \sum_{k=0}^{n-1} CA^{n-k-1}Bx[k] + Dx[n], & n > 0. \end{cases} \tag{11.44}$$

Again, we were able to compute the system's response to an arbitrary input without using z-transforms and partial fraction decompositions. Observe also from (11.41) that the relationship (11.44) is $y[n] = \{h * x\}[n]$, as expected.

The transfer function of a system described in state-space form is derived as follows. Take the z-transform of (11.38a) to obtain

$$zS^z(z) = AS^z(z) + BX^z(z), \tag{11.45}$$

hence

$$(zI_N - A)S^z(z) = BX^z(z) \implies S^z(z) = (zI_N - A)^{-1}BX^z(z). \tag{11.46}$$

Now substitute in the z-transform of (11.38b) to obtain

$$Y^z(z) = C(zI_N - A)^{-1}BX^z(z) + DX^z(z). \tag{11.47}$$

Therefore, the transfer function of the system is

$$H^z(z) = \frac{Y^z(z)}{X^z(z)} = C(zI_N - A)^{-1}B + D. \tag{11.48}$$

Direct use of (11.48) for computing the transfer function $H^z(z)$ requires the inversion of a $N \times N$ matrix containing the nonnumerical variable z. This can be done by hand if N is small, but it is tedious for large N. A computer program can be written for this purpose, but this is not straightforward either. It turns out that $H^z(z)$ can be computed without explicit inversion of $(zI_N - A)$, as we now explain.

We observe that

$$(zI_N - A)^{-1} = \frac{\mathrm{adj}(zI_N - A)}{\det(zI_N - A)}, \tag{11.49}$$

where *det* denotes the determinant of the argument, and *adj* denotes the adjugate matrix (matrix of cofactors) of the argument. Therefore,

$$H^z(z) = \frac{C\,\mathrm{adj}(zI_N - A)B}{\det(zI_N - A)} + D. \tag{11.50}$$

The elements of the adjugate matrix are polynomials in positive powers of z of degree $N - 1$ at most. Since B, C, D are constant matrices, $C\,\mathrm{adj}(zI_N - A)B$ is a polynomial of degree $N - 1$ at most.

The denominator polynomial $a(z)$ of the transfer function is the characteristic polynomial of the matrix A, so it can be computed directly from A. To compute the numerator polynomial $b(z)$ of the transfer function, observe that

$$b(z) = a(z)H^z(z), \tag{11.51}$$

or

$$b_0 + b_1 z^{-1} + \cdots + b_N z^{-N}$$
$$= (1 + a_1 z^{-1} + \cdots + a_N z^{-N})(h[0] + h[1]z^{-1} + \cdots + h[N]z^{-N} + \cdots). \tag{11.52}$$

Equating powers of z^{-k} for $0 \le k \le N$ in this equation gives

$$\begin{bmatrix} b_0 \\ b_1 \\ \vdots \\ b_N \end{bmatrix} = \begin{bmatrix} h[0] & 0 & \cdots & 0 \\ h[1] & h[0] & \cdots & 0 \\ \vdots & \vdots & \cdots & \vdots \\ h[N] & h[N-1] & \cdots & h[0] \end{bmatrix} \begin{bmatrix} 1 \\ a_1 \\ \vdots \\ a_N \end{bmatrix}. \tag{11.53}$$

The impulse response terms $\{h[n],\ 0 \le n \le N\}$ can be obtained using (11.41). Therefore, all the parameters on the right side of (11.53) can be computed, thus enabling the computation of $b(z)$.

The procedure `ss2tf` in Program 11.7 implements the computation. The characteristic polynomial is computed by the MATLAB built-in function `poly`. The terms $h[n]$ are computed using (11.41). Finally, the numerator polynomial is computed using (11.53).

11.3 General Block-Diagram Manipulation

The realizations we have seen are specific examples of discrete-time LTI networks. Since these realizations are relatively simple, we could relate them to the transfer functions they represent merely by inspection. For more complex networks, inspection may not be sufficient and more general tools are required.

A general discrete-time LTI network consists of blocks, which are LTI themselves, interconnected to make the complete system LTI. This restriction leaves us with only simple possible connections: The input to a block must be a linear combination of outputs of other blocks, with constant coefficients. In addition, some blocks may be fed from one or more external inputs. The output, or outputs, of the network are linear combinations of outputs of blocks, and possibly of the external inputs.

The preceding description can be expressed in mathematical terms as follows; see Figure 11.12.

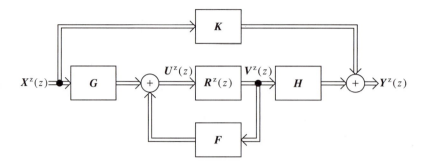

Figure 11.12 A general discrete-time LTI network.

1. Suppose we are given L LTI blocks, and let $U_l^z(z)$, $R_l^z(z)$, $V_l^z(z)$ be the input, the transfer function, and the output of the Lth block, respectively. Then we have

$$V_l^z(z) = R_l^z(z)U_l^z(z), \quad 1 \le l \le L. \tag{11.54}$$

We can write these equations collectively as

$$\mathbf{V}^z(z) = \mathbf{R}^z(z)\mathbf{U}^z(z), \tag{11.55}$$

where $\mathbf{U}^z(z)$, $\mathbf{V}_l^z(z)$ are L-dimensional vectors whose components are functions of z, and $\mathbf{R}^z(z)$ is an $L \times L$ diagonal matrix, with the functions $R_l^z(z)$ along its main diagonal.

2. Suppose we are given M external inputs whose z-transforms are $X_m^z(z)$, and P external outputs whose z-transforms are $Y_p^z(z)$. We denote the vectors built from these functions by $\mathbf{X}^z(z)$ (of dimension M) and $\mathbf{Y}^z(z)$ (of dimension P), respectively.

3. Let the block inputs be related to the block outputs and the external inputs by

$$\mathbf{U}^z(z) = \mathbf{F}\mathbf{V}^z(z) + \mathbf{G}\mathbf{X}^z(z), \tag{11.56}$$

where

(a) $\mathbf{F}$ is an $L \times L$ matrix of constants, called the *interconnection matrix*. The element $f_{i,j}$ of the matrix represents the multiplication factor of the connection from the output of the jth block to the input of the ith block. If there is no such connection, then $f_{i,j} = 0$.

(b) $\mathbf{G}$ is an $L \times M$ matrix of constants, called the *input matrix*. The element $g_{i,j}$ of the matrix represents the multiplication factor of the connection from the jth external input to the input of the ith block. If there is no such connection, then $g_{i,j} = 0$.

4. Let the external outputs be related to the block outputs and the external inputs by

$$\mathbf{Y}^z(z) = \mathbf{H}\mathbf{V}^z(z) + \mathbf{K}\mathbf{X}^z(z), \tag{11.57}$$

where

(a) $\mathbf{H}$ is a $P \times L$ matrix of constants, called the *output matrix*. The element $h_{i,j}$ of the matrix represents the multiplication factor of the connection from the output of the jth block to the ith external output. If there is no such connection, then $h_{i,j} = 0$.

(b) $\mathbf{K}$ is a $P \times M$ matrix of constants, called the *direct connection matrix*. The element $k_{i,j}$ of the matrix represents the multiplication factor of the connection from the jth external input to the ith external output. If there is no such connection, then $k_{i,j} = 0$.

5. Substitution of (11.56) in (11.55) gives

$$V^z(z) = R^z(z)FV^z(z) + R^z(z)GX^z(z), \tag{11.58}$$

so,

$$[I_L - R^z(z)F]V^z(z) = R^z(z)GX^z(z), \tag{11.59}$$

hence

$$V^z(z) = [I_L - R^z(z)F]^{-1}R^z(z)GX^z(z). \tag{11.60}$$

6. Finally, substitution of (11.60) in (11.57) gives

$$Y^z(z) = \{H[I_L - R^z(z)F]^{-1}R^z(z)G + K\}X^z(z). \tag{11.61}$$

The matrix

$$T^z(z) = H[I_L - R^z(z)F]^{-1}R^z(z)G + K \tag{11.62}$$

is the aggregate of transfer functions from the external inputs to the external outputs; that is, its element $T^z_{i,j}(z)$ relates the jth input to the ith output via

$$Y^z_i(z) = T^z_{i,j}(z)X^z_j(z), \tag{11.63}$$

provided all inputs except for the jth one are set to zero. This matrix is well defined if and only if the determinant of the matrix $[I_L - R^z(z)F]$ is nonzero. If this determinant is zero (by this we mean *identically zero*, rather than just for particular values of z), the network is said to be *singular* and its response is undefined.

Example 11.4 Consider the network shown in Figure 11.13. The matrices of this network are

$$R^z(z) = \begin{bmatrix} R^z_1(z) & 0 & 0 \\ 0 & R^z_2(z) & 0 \\ 0 & 0 & R^z_3(z) \end{bmatrix}, \quad F = \begin{bmatrix} 0 & 0 & -1 \\ 0 & 0 & 0 \\ 1 & 1 & 0 \end{bmatrix}, \quad G = \begin{bmatrix} 1 \\ 1 \\ 0 \end{bmatrix}, \tag{11.64a}$$

$$H = \begin{bmatrix} 0 & 0 & 1 \end{bmatrix}, \quad K = 0. \tag{11.64b}$$

Therefore,

$$I_3 - R^z(z)F = \begin{bmatrix} 1 & 0 & R^z_1(z) \\ 0 & 1 & 0 \\ -R^z_3(z) & -R^z_3(z) & 1 \end{bmatrix}, \tag{11.65}$$

$$[I_3 - R^z(z)F]^{-1} = \frac{1}{1 + R^z_1(z)R^z_3(z)} \begin{bmatrix} 1 & -R^z_1(z)R^z_3(z) & -R^z_1(z) \\ 0 & 1 + R^z_1(z)R^z_3(z) & 0 \\ R^z_3(z) & R^z_3(z) & 1 \end{bmatrix}, \tag{11.66}$$

and finally from (11.62),

$$T^z(z) = \frac{Y^z(z)}{X^z(z)} = \frac{[R^z_1(z) + R^z_2(z)]R^z_3(z)}{1 + R^z_1(z)R^z_3(z)}. \tag{11.67}$$

□

As we have seen, formula (11.62) is a convenient means of computing the input-output relationships of LTI networks. However, when the network has a large number of blocks with many interconnections, this formula becomes difficult to manipulate by

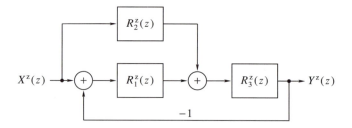

Figure 11.13 The network in Example 11.4.

hand. A computer program for computing this formula can be written, but this is not an easy task either, since the formula involves functions of z. On the other hand, state-space representations enable us to compute the matrix of transfer functions $T^z(z)$ without using (11.62) explicitly. This is done as follows.

1. We first represent each of the blocks $R_l^z(z)$ by state-space equations, say

$$s_l[n+1] = A_l s_l[n] + B_l u_l[n], \tag{11.68a}$$

$$v_l[n] = C_l s_l[n] + D_l u_l[n], \tag{11.68b}$$

so

$$R_l^z(z) = C_l(z I_{N_l} - A_l)^{-1} B_l + D_l. \tag{11.69}$$

Here N_l denotes the dimension of the state-space representation of the lth block. Any of the realizations we have studied can be used for this purpose.

2. Collecting (11.68) for all $1 \le l \le L$, we can write

$$s[n+1] = As[n] + Bu[n], \tag{11.70a}$$

$$v[n] = Cs[n] + Du[n]. \tag{11.70b}$$

Each of the matrices A, B, C, D is block diagonal, with blocks A_l, B_l, C_l, D_l on the respective main diagonals. The vector $s[n]$ is the state vector of the complete network at time n, consisting of the components of the state vectors of all blocks. The dimension of $s[n]$ is $\sum_{l=1}^{L} N_l$. The vectors $u[n]$, $v[n]$ are, respectively, the block inputs and outputs at time n.

3. Recall (11.56) and write it in the time domain as

$$u[n] = Fv[n] + Gx[n]. \tag{11.71}$$

Then substitute (11.71) in (11.70b) to get

$$u[n] = FCs[n] + FDu[n] + Gx[n], \tag{11.72}$$

so

$$u[n] = EFCs[n] + EGx[n], \quad \text{where} \quad E = (I_L - FD)^{-1}. \tag{11.73}$$

This relationship holds provided $I_L - FD$ is nonsingular. If this matrix is singular then the network is singular and its response is not defined.

4. Now substitute (11.73) in (11.70) to get

$$s[n+1] = (A + BEFC)s[n] + BEGx[n], \tag{11.74}$$

$$v[n] = (I_L + DEF)Cs[n] + DEGx[n]. \tag{11.75}$$

5. Finally substitute (11.75) in the time-domain form of (11.57) to get

$$y[n] = H(I_L + DEF)Cs[n] + (K + HDEG)x[n]. \tag{11.76}$$

Equations (11.74) and (11.76) provide a state-space representation of the network.

The procedure `network` in Program 11.8 implements the construction in MATLAB. It accepts the four connection matrices F, G, H, K and the coefficients of the transfer functions $R_i^z(z)$ as inputs. Its outputs are the state-space matrices A, B, C, D.

11.4 The Finite Word Length Problem*

In discussing filter realizations, we have so far assumed that all variables can be represented exactly in the computer, and all arithmetic operations can be performed to an infinite precision. In practice, numbers can be represented only to a finite precision, and arithmetic operations are subject to errors, since a computer word has only a finite number of bits. The operation of representing a number to a fixed precision (that is, by a fixed number of bits) is called *quantization*. Consider, for example, the digital filter

$$y[n] = -a_1 y[n-1] + b_0 x[n] + b_1 x[n-1].$$

In implementing this filter, we must deal with the following problems:

1. The input signal $x[n]$ may have been obtained by converting a continuous-time signal $x(t)$. As we saw in Section 3.5, A/D conversion gives rise to quantization errors, determined by the number of bits of the A/D.

2. The constant coefficients a_1, b_0, b_1 cannot be represented exactly, in general; the error in each of these coefficients can be up to the least significant bit of the computer word. Because of these errors, the digital filter we implement differs from the desired one: Its poles and zeros are not in the desired locations, and its frequency response is different from the desired one.

3. When we form the product $a_1 y[n-1]$, the number of bits in the result is the sum of the numbers of bits in a_1 and $y[n-1]$. It is unreasonable to keep increasing the number of bits of $y[n]$ as n increases. We must assign a fixed number of bits to $y[n]$, and quantize $a_1 y[n-1]$ to this number. Such quantization leads to an error each time we update $y[n]$ (i.e., at every time point).

4. If $x[n]$ and $y[n]$ are represented by the same number of bits, we must quantize the products $b_0 x[n]$, $b_1 x[n-1]$ every time they are computed. It is possible to avoid this error if we assign to $y[n]$ a number of bits equal to or greater than that of these products. In practice, this usually means representation of $y[n]$ in double precision.

5. The range of values of $y[n]$ that can be represented in fixed point is limited by the word length. Large input values can cause $y[n]$ to *overflow*, that is, to exceed its full scale. To avoid overflow, it may be necessary to scale down the input signal. Scaling always involves trade-off: On one hand we want to use as many significant bits as possible, but on the other hand we want to eliminate or minimize the possibility of overflow.

6. If the output signal $y[n]$ is to be fed to a D/A converter, it sometimes needs to be further quantized, to match the number of bits in the D/A. Such quantization is another source of error.

In the remaining sections of this chapter we shall study these problems, analyze their effects, and learn how they can be solved or at least mitigated.

11.5   Coefficient Quantization in Digital Filters*

When a digital filter is designed using high-level software, the coefficients of the designed filter are computed to high accuracy. MATLAB, for example, gives the coefficients to 15 decimal digits. With such accuracy, the specifications are usually met exactly (or even exceeded in some bands, because the order of the filter is typically rounded upward). When the filter is to be implemented, there is usually a need to quantize the coefficients to the word length used for the implementation (whether in software or in hardware). Coefficient quantization changes the transfer function and, consequently, the frequency response of the filter. As a result, the implemented filter may fail to meet the specifications. This was a difficult problem in the past, when computers had relatively short word lengths. Today (mid-1990s), many microprocessors designed for DSP applications have word lengths from 16 bits (about 4 decimal digits) and up to 24 bits in some (about 7 decimal digits). In the future, even longer words are likely to be in use, and floating-point arithmetic may become commonplace in DSP applications. However, there are still many cases in which finite word length is a problem to be dealt with, whether because of hardware limitations, tight specifications of the filter in question, or both. We therefore devote this section to the study of coefficient quantization effects. We first consider the effect of quantization on the poles and the zeros of the filter, and then its effect on the frequency response.

11.5.1   Quantization Effects on Poles and Zeros

Coefficient quantization causes a replacement of the exact parameters $\{a_k, b_k\}$ of the transfer function by corresponding approximate values $\{\hat{a}_k, \hat{b}_k\}$. The difference between the exact and approximate values of each parameter can be up to the least significant bit (LSB) of the computer, multiplied by the full-scale value of the parameter. For example, consider a second-order IIR filter

$$H^z(z) = \frac{b_0 + b_1 z^{-1} + b_2 z^{-2}}{1 + a_1 z^{-1} + a_2 z^{-2}}.$$

Since we are dealing only with stable filters, we know that necessarily $|a_1| < 2$ and $|a_2| < 1$ (cf. Figure 7.3). Suppose we represent each coefficient by B bits, including sign. Then, assuming we scale a_1 so that the largest representable number is ± 2, the LSB for a_1 will be $2^{-(B-2)}$, and the error $|\hat{a}_1 - a_1|$ can be up to $2^{-(B-1)}$. Similarly, assuming we scale a_2 so that the largest representable number is ± 1, the LSB of a_2 will be $2^{-(B-1)}$, and the error $|\hat{a}_2 - a_2|$ can be up to 2^{-B}.

Replacement of the exact parameters by the quantized values causes the poles and zeros of the filter to shift from their desired locations, and the frequency response to deviate from its desired shape. Usually, we are not as much interested in the deviation of the poles and zeros as in the deviation of the frequency response. However, it is instructive to explore the former first, since this will lead to important qualitative conclusions. Consider, for example, the poles of a second-order filter, given by

$$\hat{\alpha}_{1,2} = -0.5\hat{a}_1 \pm j(\hat{a}_2 - 0.25\hat{a}_1^2)^{1/2}.$$

Since $\{\hat{a}_1, \hat{a}_2\}$ can assume only a finite number of values each (equal to 2^B), the poles $\hat{\alpha}_{1,2}$ can assume only a finite number of values. Of particular interest are the possible locations of complex stable poles. These are depicted by the dots in Figure 11.14, in the case $B = 5$.

As we see from Figure 11.14, the permissible locations of complex poles of the quantized filter are not distributed uniformly inside the unit circle. In particular, small

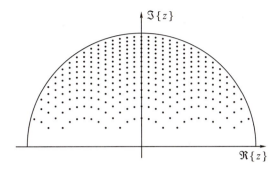

Figure 11.14 Possible locations of complex stable poles of a second-order digital filter in direct realization; number of bits: $B = 5$.

values of the imaginary part are virtually excluded. The low density of permissible pole locations in the vicinity of $z = 1$ and $z = -1$ is especially troublesome. Narrow-band, low-pass filters must have complex poles in the neighborhood of $z = 1$, whereas narrow-band, high-pass filters must have complex poles in the neighborhood of $z = -1$. We therefore conclude that high coefficient accuracy is needed to accurately place the poles of such filters.[4] Another conclusion is that high sampling rates are undesirable from the point of view of sensitivity of the filter to coefficient quantization. A high sampling rate means that the frequency response in the θ domain is pushed toward the low-frequency band; correspondingly, the poles are pushed toward $z = 1$ and, as we have seen, this increases the word length necessary for accurate representation of the coefficients.

The coupled realization of a second-order section, introduced in Section 11.1.7, can be better appreciated now. Recall that this realization is parameterized in terms of $\{\alpha_r, \alpha_i\}$, the real and imaginary parts of the complex pole. Therefore, if each of these two parameters is quantized to 2^B levels in the range $(-1, 1)$, the permissible pole locations will be distributed uniformly in the unit circle. This is illustrated in Figure 11.15 for $B = 5$. As we see, the density of permissible pole locations near $z = \pm 1$ is higher in this case.

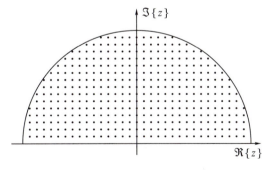

Figure 11.15 Possible locations of complex stable poles of a second-order digital filter in coupled form; number of bits: $B = 5$.

The effect of quantization of the numerator coefficients may seem, at first glance, to be the same as that of the denominator coefficients. This, however, is not the case. As we saw in Chapter 10, the zeros of an analog filter of the four common classes (Butterworth, Chebyshev-I, Chebyshev-II, elliptic) are always on the imaginary axis. Consequently, the zeros of a digital IIR filter obtained from an analog filter by a bilinear transform are always on the unit circle. Thus, the numerator of a second-order IIR filter has either real zeros at $z = \pm 1$ or a conjugate complex pair at $z = e^{\pm j\zeta}$. In the former case, the coefficients are trivial (± 1 or ± 2), so the problem of coefficient quantization does not arise. In the latter case, the numerator has the form $(1 - 2\cos\zeta z^{-1} + z^{-2})$ (up to a constant gain), so only the coefficient $2\cos\zeta$ is subject to quantization. When this coefficient undergoes a small perturbation, the zeros will shift but will stay on the unit circle. The effect of such a shift on the behavior of the filter is usually minor.

The sensitivity of the poles of a digital IIR filter to denominator coefficient quantization usually increases with the filter's order. Pole maps such as the one in Figure 11.14 are difficult to generate for orders higher than 2. An alternative way to analyze the sensitivity is to express the perturbation in a specific pole α_m as an approximate linear combination of the perturbations $\{\hat{a}_k - a_k\}$ of the coefficients, assuming that these perturbations are small and using the chain rule for partial differentiation. This analysis is carried out in Problem 11.17. The resulting formula is

$$\hat{\alpha}_m - \alpha_m \approx - \sum_{k=1}^{p} \frac{\alpha_m^{p-k}(\hat{a}_k - a_k)}{\prod_{l\neq m}(\alpha_m - \alpha_l)}. \tag{11.77}$$

As we see, poles α_l, α_m that are close to each other lead to small values of $\alpha_l - \alpha_m$, so they decrease the denominator of (11.77) and increase the error $\hat{\alpha}_m - \alpha_m$. Narrow-band filters, in particular, have their poles tightly spaced, so we can expect high sensitivity of the poles to coefficient quantization in such filters. We can already see, at least qualitatively, why parallel and cascade realizations are less sensitive to coefficient quantization than direct realizations. In parallel and cascade realizations, the second-order sections are isolated from each other, so formula (11.77) holds for each section separately. For a single section, only the distance between the two conjugate poles affects the error. Unless the conjugate poles are close to $z = \pm 1$, the sensitivity is not high. This qualitative behavior will be verified below by different means.

11.5.2 Quantization Effects on the Frequency Response

Next we examine the sensitivity of the magnitude of the filter's frequency response to coefficient quantization. We assume that $H^f(\theta)$ depends on a set of real parameters $\{p_k, 1 \le k \le K\}$. In case of a direct realization, the parameters are $\{a_i, 1 \le i \le N\}$ and $\{b_i, 0 \le i \le N\}$; in case of a parallel realization, they are c_0 and $\{e_i, f_i, 1 \le i \le N\}$; in case of a cascade realization, they are b_0 and $\{g_i, h_i, 1 \le i \le N\}$. For all IIR filter realizations, $K = 2N + 1$. In case of a direct realization of a linear phase FIR filter, the parameters are $\{h[n], 0 \le n \le \lfloor 0.5N \rfloor\}$.

Let us assume that the perturbations in the parameters resulting from quantization are small, and approximate the perturbation in the magnitude response $|H^f(\theta)|$ by a first-order Taylor series:

$$|\hat{H}^f(\theta)| - |H^f(\theta)| \approx \sum_{k=1}^{K} \frac{\partial |H^f(\theta)|}{\partial p_k}(\hat{p}_k - p_k). \tag{11.78}$$

The partial derivatives on the right side of (11.78) are given by[5]

$$\frac{\partial |H^f(\theta)|}{\partial p_k} = \Re\left\{ \frac{\partial H^f(\theta)}{\partial p_k} e^{-j\psi(\theta)} \right\}, \tag{11.79}$$

where $\psi(\theta)$ is the phase response of the filter. We note that, since

$$a(e^{j\theta}) = \sum_{k=0}^{p} a_k e^{-j\theta k}, \quad b(e^{j\theta}) = \sum_{k=0}^{q} b_k e^{-j\theta k}, \tag{11.80}$$

we have

$$\frac{\partial a(e^{j\theta})}{\partial a_k} = e^{-j\theta k}, \quad \frac{\partial a(e^{j\theta})}{\partial b_k} = 0, \tag{11.81a}$$

$$\frac{\partial b(e^{j\theta})}{\partial b_k} = e^{-j\theta k}, \quad \frac{\partial b(e^{j\theta})}{\partial a_k} = 0. \tag{11.81b}$$

The partial derivatives $\partial H^{\mathrm{f}}(\theta)/\partial p_k$ are computed as follows for the realizations we have studied:

1. For a direct realization we have

$$H^{\mathrm{f}}(\theta) = \frac{b(e^{j\theta})}{a(e^{j\theta})}, \tag{11.82}$$

therefore,

$$\frac{\partial H^{\mathrm{f}}(\theta)}{\partial b_k} = \frac{e^{-j\theta k}}{a(e^{j\theta})}, \quad \frac{\partial H^{\mathrm{f}}(\theta)}{\partial a_k} = -\frac{e^{-j\theta k} b(e^{j\theta})}{a^2(e^{j\theta})}. \tag{11.83}$$

2. The frequency response of a parallel realization has the form

$$H^{\mathrm{f}}(\theta) = c_0 + \sum_{l=1}^{L} H^{\mathrm{f}}_l(\theta) = c_0 + \sum_{l=1}^{L} \frac{b_l(e^{j\theta})}{a_l(e^{j\theta})}, \tag{11.84}$$

where the $H^{\mathrm{f}}_l(\theta)$ are of first or second order. Therefore,

$$\frac{\partial H^{\mathrm{f}}(\theta)}{\partial c_0} = 1, \tag{11.85}$$

and, if p_k is a parameter appearing in a certain $H^{\mathrm{f}}_l(\theta)$, then

$$\frac{\partial H^{\mathrm{f}}(\theta)}{\partial p_k} = \frac{\partial H^{\mathrm{f}}_l(\theta)}{\partial p_k}. \tag{11.86}$$

The right side of (11.86) can be computed as in (11.83).

3. The frequency response of a cascade realization has the form

$$H^{\mathrm{f}}(\theta) = b_0 \prod_{l=1}^{L} H^{\mathrm{f}}_l(\theta) = b_0 \prod_{l=1}^{L} \frac{b_l(e^{j\theta})}{a_l(e^{j\theta})}, \tag{11.87}$$

where the $H^{\mathrm{f}}_l(\theta)$ are of first or second order. Therefore,

$$\frac{\partial H^{\mathrm{f}}(\theta)}{\partial b_0} = \prod_{l=1}^{L} H^{\mathrm{f}}_l(\theta), \tag{11.88}$$

and, if p_k is a parameter appearing in a certain $H^{\mathrm{f}}_l(\theta)$, then

$$\frac{\partial H^{\mathrm{f}}(\theta)}{\partial p_k} = b_0 \frac{\partial H^{\mathrm{f}}_l(\theta)}{\partial p_k} \prod_{\substack{m=1 \\ m \neq l}}^{L} H^{\mathrm{f}}_m(\theta). \tag{11.89}$$

The partial derivatives on the right side of (11.89) can be computed as in (11.83).

4. For a linear phase FIR filter in direct realization we have

$$\frac{\partial H^{\mathrm{f}}(\theta)}{\partial h[n]} = \begin{cases} e^{-j\theta n} \pm e^{-j\theta(N-n)}, & n \neq 0.5N, \\ e^{-j0.5\theta N}, & n = 0.5N. \end{cases} \tag{11.90}$$

where the positive sign is for types I and II, and the negative sign is for types III and IV.

The approximate error formula (11.78) can be used for estimating the effect of coefficient quantization on the magnitude response in the following manner. Let P_k denote the full scale used for the coefficient p_k. P_k must be at least as large as $|p_k|$, and is often larger, for reasons explained below. Let q be half the quantization level relative to full scale, so $q = 2^{-B}$ (where B is the word length in bits). Then we have

$$|\hat{p}_k - p_k| \le P_k q. \tag{11.91}$$

Substitution in (11.78) leads to the approximate inequality

$$||\hat{H}^f(\theta)| - |H^f(\theta)|| \le q \sum_{k=1}^{K} \left| \frac{\partial |H^f(\theta)|}{\partial p_k} \right| P_k. \tag{11.92}$$

The function

$$S(\theta) = \sum_{k=1}^{K} \left| \frac{\partial |H^f(\theta)|}{\partial p_k} \right| P_k, \tag{11.93}$$

is the *sensitivity bound* of the filter. As we see, the sensitivity bound depends on the realization. The realization determines the coefficients p_k, their full-scale values, and the partial derivatives of the magnitude response with respect to the coefficients. In summary, the error in the magnitude response at any given frequency is approximately bounded by

$$||\hat{H}^f(\theta)| - |H^f(\theta)|| \le 2^{-B} S(\theta). \tag{11.94}$$

A common way of choosing the full-scale values P_k is as follows. For each of the numerator and denominator polynomials of a direct realization, we find the coefficient having the largest magnitude and use it as the full-scale value for *all* the coefficients of the polynomial in question. The corresponding formulas are

$$P_{\text{num}} = \max\{\text{abs}(b_i),\ 0 \le i \le N\}, \tag{11.95a}$$

$$P_{\text{den}} = \max\{\text{abs}(a_i),\ 1 \le i \le N\}. \tag{11.95b}$$

This way, the coefficient of largest magnitude in each polynomial is represented by ± 1. For parallel and cascade realizations, we scale each section separately in the same way.

Another way of choosing the full-scale values is to round the largest magnitude coefficient of each polynomial upward to the nearest integer power of 2 and take the rounded value as the full scale of all the coefficients of the polynomial. The corresponding formulas are

$$P_{\text{num}} = 2^{\lceil \log_2 (\max\{\text{abs}(b_i),\ 0 \le i \le N\}) \rceil}, \tag{11.96a}$$

$$P_{\text{den}} = 2^{\lceil \log_2 (\max\{\text{abs}(a_i),\ 1 \le i \le N\}) \rceil}. \tag{11.96b}$$

This method yields larger P_{num} and P_{den} than the ones in (11.95), but is more convenient, since the scaled coefficients are related to the true ones by simple shifts.

The sensitivity bound is useful for determining a suitable word length for implementing a given filter in a given realization. Let δ_p be the minimum tolerance of all pass bands of the filter, and δ_s the minimum tolerance of all stop bands. It is reasonable to require that the maximum deviation from the tolerance caused by quantization be no more than 10 percent of the tolerance itself (a different percentage can be chosen, depending on the application). With this requirement, (11.94) leads to

$$B \ge \log_2 \left(\frac{\max\{S(\theta) : \theta \text{ in a pass band}\}}{0.1\delta_p} \right), \tag{11.97a}$$

$$B \ge \log_2 \left(\frac{\max\{S(\theta) : \theta \text{ in a stop band}\}}{0.1\delta_s} \right). \tag{11.97b}$$

Therefore, choosing B according to the greater of the two right sides in (11.97) guarantees that the error in the magnitude response resulting from quantization will deviate only slightly (if at all) from the permitted bounds in all frequency bands.

Programs 11.9 through 11.14 compute the sensitivity bounds for the standard realizations. The procedure `sensiir` is for IIR filters in direct, parallel, and cascade realizations, whereas `sensfir` is for linear-phase FIR filters in a direct realization. Each of the two procedures returns the individual sensitivities $\partial |H^f(\theta)|/\partial p_k$ in the matrix `dHmag`, and the sensitivity bound $S(\theta)$ in the vector S. These are computed at K frequency points, in the interval specified by the input variable `theta`. The procedures `dhdirect`, `dhparal`, `dhcascad` return the partial derivatives $\partial H^f(\theta)/\partial p_k$ of the respective IIR realizations, and the full-scale values P_k of the coefficients. The full-scale values are computed by `scale2` according to (11.96).

Example 11.5 Consider the fifth-order, low-pass elliptic filter designed in Example 10.14. We compute the sensitivity bounds of this filter for the three standard realizations. Figure 11.16 shows the results. Parts a and b show the bounds of the pass band and the stop band, respectively, for the direct realization. The maximum sensitivity in the pass band is about 9×10^4, and that in the stop band is about 1100. Recall that the tolerances of this filter are $\delta_p \approx 0.011$, $\delta_s = 0.05$. The criterion (11.97) gives

$$\text{Pass-band requirement:} \quad B \geq \log_2 \left(\frac{9 \cdot 10^4}{0.0011} \right) \approx 26.3,$$

$$\text{Stop-band requirement:} \quad B \geq \log_2 \left(\frac{1100}{0.005} \right) \approx 17.7.$$

Therefore, 27 bits are sufficient for a direct realization of the filter.

Parts c and d of Figure 11.16 show the bounds of the pass band and the stop band, respectively, for the parallel realization. The maximum sensitivity at the pass band is about 15.5, and that at the stop band is about 14. The criterion (11.97) gives

$$\text{Pass-band requirement:} \quad B \geq \log_2 \left(\frac{15.5}{0.0011} \right) \approx 13.8,$$

$$\text{Stop-band requirement:} \quad B \geq \log_2 \left(\frac{14}{0.005} \right) \approx 11.5.$$

Therefore, 14 bits are sufficient for a parallel realization of the filter.

Parts e and f of Figure 11.16 show the bounds of the pass band and the stop band, respectively, for the cascade realization. The maximum sensitivity at the pass band is about 57, and that at the stop band is about 1.3. The criterion (11.97) gives

$$\text{Pass-band requirement:} \quad B \geq \log_2 \left(\frac{57}{0.0011} \right) \approx 15.7,$$

$$\text{Stop-band requirement:} \quad B \geq \log_2 \left(\frac{1.3}{0.005} \right) \approx 8.0.$$

Therefore, 16 bits are sufficient for a cascade realization of the filter.

In summary, the parallel realization is the least sensitive to coefficient quantization in this example. At present (mid-1990s), 16-bit fixed-point arithmetic, which is the most common in commercial DSP microprocessors, is suitable for both parallel and cascade realizations but is far from being sufficient for a direct realization. Even 24-bit DSP microprocessors may be insufficient for direct realizations in certain cases. □

The phenomenon seen in Example 11.5 is typical. Parallel realizations are usually best in the pass band (*best* meaning least sensitive to coefficient quantization), whereas

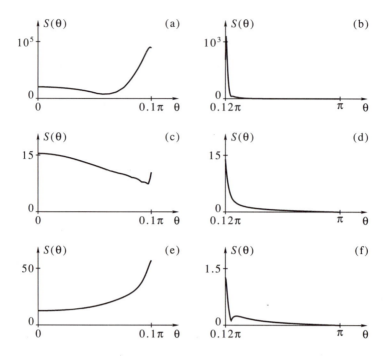

Figure 11.16 Sensitivity bounds for the filter in Example 11.5: (a) direct realization, pass band; (b) direct realization, stop band; (c) parallel realization, pass band; (d) parallel realization, stop band; (e) cascade realization, pass band; (f) cascade realization, stop band (graphs are drawn to different scales).

cascade realizations are usually best in the stop band. It is common for a practical filter to have δ_s much smaller than δ_p (the filter in Example 11.5 is an exception). Therefore, the stop-band requirement (11.97b) usually wins over the pass-band requirement (11.97a). Consequently, cascade realizations are the most commonly used. Direct realizations of IIR filters are highly sensitive to coefficient quantization and should be avoided as a rule, unless the order of the filter is low (3 at most).

Example 11.6 Consider the equiripple FIR filter of order $N = 146$, designed in Example 9.16. This filter has the same frequency bands and tolerances as the IIR filter in Example 11.5. Figure 11.17 shows the sensitivity bounds of the pass band and the stop band, respectively, for a direct realization of the filter. The maximum sensitivity is about 19 at both pass and stop bands. The tolerances of this filter are $\delta_p \approx 0.011$, $\delta_s = 0.05$. The criterion (11.97) gives

$$\text{Pass-band requirement:} \quad B \geq \log_2\left(\frac{19}{0.0011}\right) \approx 14.1,$$

$$\text{Stop-band requirement:} \quad B \geq \log_2\left(\frac{19}{0.005}\right) \approx 11.9.$$

Therefore, 15 bits are sufficient for a direct realization of this filter. □

The phenomenon seen in Example 11.6 is typical. FIR filters, even in a direct realization, are generally much less sensitive to coefficient quantization than IIR filters meeting the same specifications. Word length of 16 bits often is sufficient. Because of this property, cascade realization is seldom used for FIR filters. We reiterate that parallel realization is not applicable at all to FIR filters.

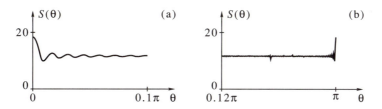

Figure 11.17 Sensitivity bounds for the filter in Example 11.6: (a) pass band; (b) stop band.

The sensitivity analysis we have presented is useful for preliminary analysis of errors due to coefficient quantization in a filter's magnitude of frequency response, and for choosing scaling and word length to represent the coefficients. After the coefficients have been quantized to the selected word length, it is good practice to compute the frequency response of the actual filter, and verify that it is acceptable. The MATLAB procedure `qfrqresp`, listed in Program 11.15, performs this computation. The program accepts the numerator and denominator coefficients of the filter, the desired realization (direct, parallel, or cascade), the number of bits, and the desired frequency points (number and range). For an FIR filter, the coefficients are entered in the numerator polynomial, and 1 is entered for the denominator polynomial. The program finds the coefficients of the desired realization and their scaling. It then quantizes the coefficients, and finally computes the frequency response by calling `frqresp` as needed.

Example 11.7 We test the filters discussed in Examples 11.5 and 11.6 with the word length found there for each case. Figure 11.18 shows the results. Only the pass-band response is shown in each case, since the stop-band response is much less sensitive to coefficient quantization for these filters. As we see, the predicted word length is indeed suitable in all cases, except for slight deviation of the response of the parallel realization at the band edge. The obvious remedy in this case is to use a word length of 16 bits, which is more practical than 14 bits anyway. We emphasize again that the response of the FIR filter, although this filter is implemented in only 15 bits, is well within the tolerance.   □

11.6 Scaling in Fixed-Point Arithmetic*

When implementing a digital filter in fixed-point arithmetic, it is necessary to scale the input and output signals, as well as certain inner signals, to avoid signal values that exceed the maximum representable number. A problem of a similar nature arises in active analog filters: There, it is required to limit the signals to voltages below the saturation levels of the operational amplifiers. However, there is an important difference between the analog and the digital cases: When an analog signal exceeds its permitted value, its magnitude is limited but its polarity is preserved. When a digital signal exceeds its value, we call it an *overflow*. An overflow in two's-complement arithmetic leads to polarity reversal: A number slightly larger than 1 changes to a number slightly larger than -1. Therefore, overflows in digital filters are potentially more harmful than in analog filters, and care is necessary to prohibit them, or to treat them properly when they occur.

The scaling problem can be stated in mathematical terms as follows. Suppose we wish to prevent the magnitude of the output signal $|y[n]|$ from exceeding a certain

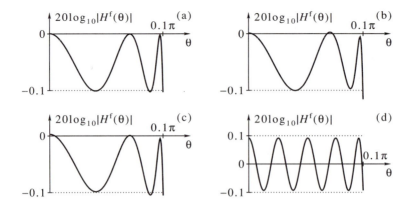

Figure 11.18 Pass-band magnitude responses of the filters in Examples 11.5, 11.6 after coefficient quantization: (a) IIR filter, direct realization, 27 bits; (b) IIR filter, parallel realization, 14 bits; (c) IIR filter, cascade realization, 16 bits; (d) FIR filter, direct realization, 15 bits.

upper limit y_{max}. Since the input and output signals are related through a convolution $y[n] = \{x * h\}[n]$, we may be able to achieve this goal by properly limiting the input signal $x[n]$. This, however, may be undesirable because limiting is a nonlinear operation that distorts the signal. A more attractive solution is to scale the impulse response by a proportionality factor c, chosen in a way that will prevent the magnitude of $c\{x * h\}[n]$ from exceeding the maximum value y_{max}. Such a proportionality factor may also be useful in the opposite case: If the dynamic range of $x[n]$ is such that $y[n]$ is much smaller in magnitude than y_{max}, we can adjust c to increase the range of $y[n]$ as needed. In either case, the transfer function $cH^z(z)$ is called a *scaled transfer function*. Our task is to find a scale factor c that will cause $y[n]$ to have as large dynamic range as possible without exceeding y_{max} in magnitude. Judicious choice of a scale factor requires knowledge of certain parameters of the input signal; accordingly, there exist several scaling methods, differing in their assumptions on the nature and properties of the input signal.

The scaling problem is obviated if floating-point arithmetic is used. However, fixed-point implementations of digital filters are very common, so familiarity with scaling methods and the effects of scaling on the properties of the filter is expedient.

11.6.1   Time-Domain Scaling

Consider an LTI filter whose impulse response and transfer function are $h[n]$ and $H^z(z)$, respectively. Assume that the input signal $x[n]$ is known to be bounded in magnitude by x_{max}, and the output signal is required to be bounded in magnitude by y_{max}. Both values are relative to the maximum value representable by the computer, usually taken as 1. The output signal is related to the input by the convolution formula

$$y[n] = \sum_{m=0}^{\infty} h[m]x[n - m]. \tag{11.98}$$

Therefore,

$$|y[n]| \le \sum_{m=0}^{\infty} |h[m]| \cdot |x[n - m]| \le x_{max} \sum_{m=0}^{\infty} |h[m]| = x_{max}\|h\|_1, \tag{11.99}$$

where

$$\|h\|_1 = \sum_{m=0}^{\infty} |h[m]|. \tag{11.100}$$

Stability of the filter $h[n]$ implies that $\|h\|_1$ is finite. As we see, the requirement $|y[n]| \le y_{max}$ will be achieved if we use a scale factor

$$c = \frac{y_{max}}{x_{max}\|h\|_1} \tag{11.101}$$

In practice, it may be convenient to round c downward to an integer power of 2, since this facilitates scaling by bit shifting.

Example 11.8 Suppose that an analog signal is sampled by a 12-bit A/D converter, then fed to a 16-bit filter whose transfer function is

$$H^z(z) = \frac{1}{1 - 0.98z^{-1}}.$$

If we place the sampled signal $x[n]$ in the lower bits of computer word (with sign extension), the corresponding x_{max} will be $1/16$. Suppose we wish to scale the output signal to $y_{max} = 1$. We have

$$\|h\|_1 = \sum_{m=0}^{\infty} 0.98^m = \frac{1}{1 - 0.98} = 50.$$

Therefore, the maximum scale factor is $c = 16/50 = 0.32$ in this case. Rounding downward to a power of 2, we get a scale factor of 0.25. Such a scaling is tantamount to shifting $x[n]$ by 2 bits to the right, thus losing 2 out of the 12 available bits. The lesson from this example is that scaling using c as in (11.101) potentially involves loss of accuracy of the input signal. □

11.6.2 Frequency-Domain Scaling

The scale factor given in (11.101) is often overconservative in its use of the dynamic range of the computer word. By this we mean that the values assumed by the output signal are usually small in magnitude compared with the full scale. Less conservative scaling is provided by frequency-domain bounds, as follows.

1. 1-norm bound: We have, by the inverse Fourier transform formula,

$$y[n] = \frac{1}{2\pi} \int_{-\pi}^{\pi} H^f(\theta)X^f(\theta)e^{j\theta n}d\theta. \tag{11.102}$$

Therefore,

$$|y[n]| \le \frac{1}{2\pi} \int_{-\pi}^{\pi} |H^f(\theta)| \cdot |X^f(\theta)|d\theta. \tag{11.103}$$

Define

$$\|X^f\|_{\infty} = \max_{\theta} |X^f(\theta)|, \quad \|H^f\|_1 = \frac{1}{2\pi} \int_{-\pi}^{\pi} |H^f(\theta)|d\theta. \tag{11.104}$$

The quantity $\|X^f\|_{\infty}$ is called the *infinity norm* of $X^f(\theta)$, and the quantity $\|H^f\|_1$ is called the *1-norm* of $H^f(\theta)$. If $\|X^f\|_{\infty}$ is finite, we can guarantee that $|y[n]| \le y_{max}$ by using the scale factor

$$c = \frac{y_{max}}{\|H^f\|_1 \|X^f\|_{\infty}}. \tag{11.105}$$

2. Infinity-norm bound: By reversing the roles of H^f and X^f in (11.105), we get the scale factor

$$c = \frac{y_{max}}{\|H^f\|_\infty \|X^f\|_1}. \tag{11.106}$$

This scale factor is applicable in case $\|X^f\|_1$, the 1-norm of the input signal, is finite. It is particularly useful when the input signal is narrow band. For example, if $x[n]$ is known to be sinusoidal, the scale factor (11.106) guarantees that the amplitude of the sinusoid at the output of the filter not exceed y_{max}.

3. 2-norm bound: Unlike the preceding bounds, the bound we introduce now is probabilistic. Define

$$\|H^f\|_2 = \left[\frac{1}{2\pi} \int_{-\pi}^{\pi} |H^f(\theta)|^2 d\theta \right]^{1/2}. \tag{11.107}$$

Note that (11.107) is identical to the square root of the noise gain of the filter [cf. (2.137)]. Let us assume that the input $x[n]$ is a wide-sense stationary signal having power spectral density $K_x^f(\theta)$. Then, using (2.134), we can bound the variance of the output signal as follows:

$$\gamma_y = \frac{1}{2\pi} \int_{-\pi}^{\pi} K_y^f(\theta) d\theta = \frac{1}{2\pi} \int_{-\pi}^{\pi} K_x^f(\theta) |H^f(\theta)|^2 d\theta \le \|K_x^f\|_\infty \|H^f\|_2^2. \tag{11.108}$$

Therefore, if we know the maximum possible value of the power spectral density of the input signal, and we require that the variance of the output signal not exceed $\gamma_{y,max}$, we get the scale factor

$$c = \left[\frac{\gamma_{y,max}}{\|H^f\|_2^2 \|K_x^f\|_\infty} \right]^{1/2}. \tag{11.109}$$

Bounding the variance of $y[n]$ *does not* guarantee boundedness of $|y[n]|$, and this is why we refer to (11.109) as probabilistic scaling. If, for example, we take $\gamma_{y,max} \approx 0.1 y_{max}^2$, the probability of $|y[n]|$ exceeding y_{max} is expected to be low.[6]

The four norms we have introduced are known to satisfy the relationships

$$\|H^f\|_1 \le \|H^f\|_2 \le \|H^f\|_\infty \le \|h\|_1. \tag{11.110}$$

This, however, does not imply that $\|H^f\|_1$ always provides the best (largest) scale factor, since the different scaling rules also depend on x_{max}, y_{max}, $\gamma_{y,max}$, $\|X^f\|_\infty$, $\|X^f\|_1$, and $\|K_x^f\|_\infty$. Therefore, judicious choice of scaling method requires knowledge about the nature of the input signal and its bounds.

11.6.3 MATLAB Implementation of Filter Norms

Exact computation of norms we have presented is not possible in general for rational filters (with the exception of $\|H^f\|_2$). However, numerical approximations can be used instead. High accuracy is not required, since the scale factor resulting from those norms is usually rounded to an integer power of 2. The procedure filnorm in Program 11.16 implements the computation of the four filter norms. The time-domain norm $\|h\|_1$ is computed by calling the function filter with a unit-sample input, and adding terms $|h[n]|$ until the relative error drops below a prescribed threshold. The norm $\|H^f\|_2$ is computed by calling nsgain (see Program 7.4) and taking the square root of the result. The norm $\|H^f\|_\infty$ is approximated by the maximum of the magnitude response, computed on a dense grid. Finally, the norm $\|H^f\|_1$ is computed by evaluating the magnitude response on a dense grid and approximating the integral by Simpson's rule.

11.6.4 Scaling of Inner Signals

Our discussion so far has concentrated on input scaling of a transfer function to avoid overflow of the output. In realizing a filter, inner (or intermediate) signals are always present, so we need to concern ourselves with potential overflow problems in such signals as well. There are four types of inner signal:

1. A signal resulting from delaying another signal. Such a signal can never overflow if the signal at the input of the delay does not.

2. A signal resulting from multiplying a signal by a constant. Assuming that both factors in the product are scaled below 1, the product is also less than 1 in magnitude, so it cannot overflow.

3. A signal resulting from adding two signals to form a partial sum, which is later used as an operand in another sum. We assume that such a signal is not used for any other purpose in the filter or outside it. Addition of two signals can potentially lead to overflow. However, two's-complement arithmetic has the following important property: If a sum of n numbers ($n > 2$) does not overflow, overflows in partial sums cancel out and do not affect the final result. For example, suppose that $x_1 + x_2 + x_3$ does not overflow, and the computation is performed in the order $(x_1 + x_2) + x_3$. Then, even if $x_1 + x_2$ overflows and the overflow is ignored, the final result will still be correct. The reason is that two's-complement is a special case of modular arithmetic [i.e., a number x is represented by $(x \bmod 2)$, assuming that the binary point is immediately to the right of the sign bit], hence standard rules of modular addition apply to it. The conclusion is that, if the filter is implemented in two's-complement arithmetic, overflows in partial sums can be ignored.

4. A signal resulting from adding two signals to form a final sum, which is needed elsewhere in the filter or outside it. Such a signal must be scaled similarly to the output signal, since its overflow is potentially harmful. This is done by computing the transfer function from the input to the signal in question and using one of the scaling methods described earlier. It is also recommended, in most applications, to detect overflows of such signals and replace the overflowed signal by the corresponding saturation value. For example, if it is detected that $x \geq 1$, x should be replaced by $1 - 2^{-(B-1)}$; if it is detected that $x < -1$, x should be replaced by -1. Most DSP microprocessors have a *saturation mode*, in which this operation is performed automatically after the addition.

Example 11.9 In the direct realization shown in Figure 11.4 there are 15 inner signals, in addition to the output signal. However, only the signals $u[n]$ and $y[n]$ are of the fourth type: The former because it is used more than once inside the filter, and the latter because it is used outside it. The transfer function from $x[n]$ to $u[n]$ is

$$\frac{U^z(z)}{X^z(z)} = \frac{1}{a(z)},$$

so one of the scaling methods explained previously should be used for it.

In the transposed direct realization shown in Figure 11.6, only the output signal $y[n]$ is of the fourth type. Any overflow generated at one of the adders along the center lane will propagate through the delays and will eventually reach $y[n]$. However, if the input is properly scaled to avoid overflow in $y[n]$, the entire realization will be insensitive to intermediate overflows. In this respect, the transposed direct realization offers an advantage over the direct realization. □

11.6.5 Scaling in Parallel and Cascade Realization

We now discuss in detail scaling procedures for parallel and cascade realizations for IIR filters. As we saw in Example 11.9, the transposed direct realization has an advantage over the direct realization, in that there is no need to scale inner signals. We therefore consider only parallel and cascade realizations in which the second-order sections are in a transposed direct realization.

Figure 11.19 illustrates a parallel connection of three second-order sections in skeletal form (delays are represented by factors z^{-1}, and summing junctions by open circles). The scaling procedure is as follows:

1. The input coefficients f_m at each section are scaled to prevent overflow of $v_k[n]$, the output signal of the section. As we have explained, this scaling depends on the assumptions made on the input signal, and on the norm used. The same scale factor should be applied to both coefficients of a specific section, but different factors are permitted at different sections. The f_m shown in Figure 11.19 are assumed to be scaled already.

2. The coefficients λ_k are computed to make the total gains of all sections (including the branch of constant gain c_0) equal to the corresponding gains in the unscaled transfer function, up to a common proportionality factor. For example, if f_1, f_2 where scaled down by 2 compared with f_3, f_4, then λ_1 must be larger than λ_2 by 2.

3. The dynamic range of $y[n]$ is checked using the scaled transfer function. If it is found that $y[n]$ can overflow, all λ_k must be decreased by a common factor, to prevent (or decrease the probability of) this overflow. If it is found that the dynamic range of $y[n]$ is small relative to full scale, all λ_k may be increased by a common factor.

4. In the preceding steps, all factors are usually taken as integer powers of 2. This way, factors greater than 1 can be implemented by left shifts, and factors smaller than 1 can be implemented by right shifts.

Example 11.10 Consider again the IIR filter discussed in Example 11.5. The parallel decomposition of the filter, with coefficients quantized to 16 bits, is

$$H^z(z) = -24984 \times 2^{-17} + \frac{8909 \times 2^{-20} - 22869 \times 2^{-20}z^{-1}}{1 - 30365 \times 2^{-14}z^{-1} + 15710 \times 2^{-14}z^{-2}}$$
$$+ \frac{-30199 \times 2^{-17} + 28813 \times 2^{-17}z^{-1}}{1 - 28072 \times 2^{-14}z^{-1} + 13038 \times 2^{-14}z^{-2}} + \frac{19504 \times 2^{-16}}{1 - 25279 \times 2^{-15}z^{-1}}.$$

Suppose we know that the input signal satisfies $\|K_x^f\|_\infty = 0.1$. We decide to use 2-norm scaling such that the variance of the signal at the output of each section, as well as at the output of the complete filter, will not exceed 0.1. This implies that the 2-norm of each section must be no larger than 1 after scaling.

The 2-norms of the three unscaled sections are computed by the program fil-norm and found to be 0.1528, 0.3805, and 0.4677. Therefore, the first section can be scaled by 4, the second by 2, and the third by 2. This is done by doubling the numerator coefficients of the second and the third sections, and quadrupling the numerator coefficients of the first.

The 2-norm of the complete filter is 0.3248, so the transfer function can be scaled by 2 to increase the dynamic range. We therefore complete the scaling procedure by taking

$$\lambda_0 = 2, \quad \lambda_1 = 0.5, \quad \lambda_2 = \lambda_3 = 1. \qquad \square$$

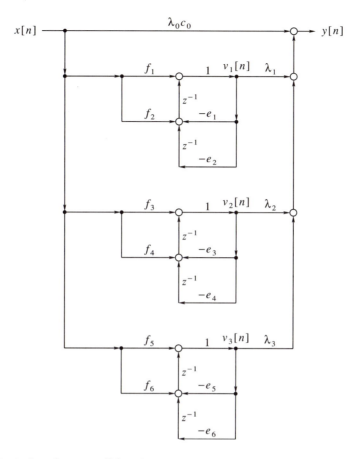

Figure 11.19 Scaling for a parallel realization (second-order sections in a transposed direct realization).

We now turn our attention to scaling of cascade realizations. Unlike a parallel realization, here we must choose the order of the sections in the cascade, since scaling depends on this order. There is no simple rule for choosing the best order in cascade realization: It depends on the nature of the input signal, the norm used for scaling, and the specific filter. Ordering rules have been proposed in the literature but are not discussed here. We use the order provided by the program `pairpz`. We reiterate that this program orders the complex poles in decreasing order of closeness to the unit circle (i.e., the poles closest to the unit circle appear first, and so on).

Figure 11.20 illustrates a cascade connection of three second-order sections. Let $H_k^z(z)$ denote the second-order sections from left to right (not including the scale factors). The scaling procedure is as follows. For each k, starting from 1 and continuing until all sections are exhausted, choose λ_k to avoid overflow in the transfer function $\prod_{i=1}^{k} \lambda_i H_i^z(z)$. At the kth stage, the factors $\{\lambda_i, \ 1 \leq i \leq k - 1\}$ have already been determined, so only λ_k is free to choose. After this procedure has been completed, the output $y[n]$ will be free of overflow, but the total gain will be $\prod_k \lambda_k$, instead of the desired value b_0. The final stage is therefore to decrease one of the λ_k such that $\prod_k \lambda_k$ will be equal to b_0 times an integer power of 2. This way, the input–output transfer function will be the desired $H^z(z)$ scaled by a power of 2, as in the case of parallel realization.

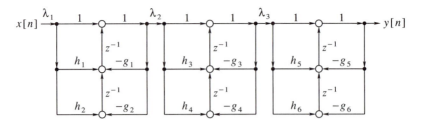

Figure 11.20 Scaling for a cascade realization (second-order sections in a transposed direct realization).

Example 11.11 We look again at the IIR filter discussed in Examples 11.5 and 11.10. The cascade decomposition of the filter, with coefficients quantized to 16 bits, is

$$H^z(z) = 29414 \times 2^{-20} \cdot \frac{1 - 28587 \times 2^{-14}z^{-1} + z^{-2}}{1 - 30365 \times 2^{-14}z^{-1} + 15710 \times 2^{-14}z^{-2}}$$
$$\cdot \frac{1 - 30493 \times 2^{-14}z^{-1} + z^{-2}}{1 - 28072 \times 2^{-14}z^{-1} + 13038 \times 2^{-14}z^{-2}} \cdot \frac{1 + z^{-1}}{1 - 25279 \times 2^{-15}z^{-1}}.$$

Assume, as in Example 11.10, that the input signal satisfies $\|K_x^f\|_\infty = 0.1$. We decide to use 2-norm scaling such that the variance of the signal at the output of each section will not exceed 0.1.

The program `filnorm` gives the following 2-norms:

$$\|H_1^f\|_2 = 1.0656, \quad \|H_1^f H_2^f\|_2 = 1.8730, \quad \|H_1^f H_2^f H_3^f\|_2 = 11.5802.$$

From the first value we get that $\lambda_1 = 0.5$, from the second that $\lambda_1\lambda_2 = 0.5$, and from the third that $\lambda_1\lambda_2\lambda_3 = 2^{-4}$. Therefore, $\lambda_2 = 1$ and $\lambda_3 = 2^{-3}$. Finally, we must decrease one of the λ_k to account for b_0. We decide to decrease λ_2 from 1 to $29414 \times 2^{-15} (\approx 0.8976)$. This gives an overall gain of 29414×2^{-19}, which is equal to $2b_0$, quantized to 16 bits. In summary,

$$\lambda_1 = 0.5, \quad \lambda_2 = 29414 \times 2^{-15}, \quad \lambda_3 = 2^{-3}. \qquad \Box$$

11.7   Quantization Noise*

11.7.1   Modeling of Quantization Noise

Addition of two fixed-point numbers represented by the same number of bits involves no loss of precision. It can, as we have seen, lead to overflow, but careful scaling usually prevents this from happening. The situation is different for multiplication. Every time we multiply two numbers of B bits each (of which 1 bit is for sign and $B - 1$ bits for magnitude) we get a result of $2B - 1$ bits (1 bit for sign and $2B - 2$ bits for magnitude). It is common, in fixed-point implementations, to immediately drop the $B - 1$ least significant bits of the product, retaining B bits (including sign) for subsequent computations. In two's-complement arithmetic, this truncation leads to a negative error, which can be up to $2^{-(B-1)}$ in magnitude. A slightly more sophisticated operation is to round the product either upward or downward so that the error does not exceed 2^{-B} in magnitude.

The error generated at any given multiplication depends on the operands. In LTI filters, one of the operands is always a constant number, whereas the other is always an instantaneous signal value. If the filter is well scaled, the instantaneous signal value will be a sizable fraction of the full scale most of the time. If, in addition, the number of bits is fairly large (say 12 or more), the error resulting from truncation (or rounding) of the product will appear random. By "appear random" we mean that it will not be random in the mathematical sense of the term, but will behave approximately as though it were random. The probability density function of the error will be approximately uniform, as shown in Figure 11.21. In case of truncation, the range of the error will be $(-2^{-(B-1)}, 0]$ [Figure 11.21(a)], whereas in case of rounding it will be $(-2^{-B}, 2^{-B}]$ [Figure 11.21(b)]. In both cases, the variance of the error will be

$$2^{B-1} \int_{-2^{-B}}^{2^{-B}} x^2 dx = \frac{2^{B-1} \times 2 \times 2^{-3B}}{3} = \frac{2^{-2B}}{3}. \tag{11.111}$$

The mean of the error will be zero in case of rounding, and -2^{-B} in case of truncation.

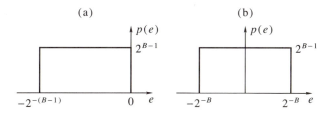

Figure 11.21 Probability density functions of quantization noise: (a) truncation; (b) rounding.

Experience shows that truncation or rounding errors at the output of a given multiplier at different times are usually uncorrelated, or approximately so. Furthermore, errors at different multipliers are usually uncorrelated as well because quantization errors, being very small fractions of multiplicands, are sufficiently random. Therefore, each error signal appears as discrete-time white noise, uncorrelated with the other error signals. This leads to the following:

> **Quantization noise model for a digital filter in fixed-point implementation:** The error at the output of every multiplier is represented as additive discrete-time white noise. The total number of noise sources is equal to the number of multipliers, and they are all uncorrelated. The variance of each noise source is $2^{-2B}/3$. The mean is zero in case of rounding, and -2^{-B} in case of truncation.

It is common to express the mean of the noise in units of least significant bit (LSB), and the variance in units of LSB^2. Since the least significant bit value is $2^{-(B-1)}$, the mean is -0.5 LSB for truncation and 0 for rounding; the variance is $(1/12)$ LSB^2 in either case. From now on, for convenience, we shall omit the negative sign of the mean in the case of truncation.

The noise generated at any point in a filter is propagated through the filter and finally appears at the output. Since the filter is linear, the output noise is added to the output signal. The additive noise at the output is undesirable, since it degrades the performance of the system. For example, if the output signal is used for detecting a sinusoidal component in noise (as discussed in Section 6.5), quantization noise decreases the SNR. If the filter is used for audio signals (e.g., speech or music), the noise may be audible and become a nuisance. It is therefore necessary to analyze the noise at the output of the filter quantitatively, and verify that it does not exceed the permitted level. If it does, the word length must be increased to decrease the mean and the variance of the quantization error at the multipliers outputs.

The effect of noise generated at a particular point in a filter on the output signal can be computed in the following manner. Let $e[n]$ be the noise at the point of interest and let μ_e, γ_e be its mean and variance, respectively. Let $G^z(z)$ be the transfer function from $e[n]$ to the filter output $y[n]$, say

$$G^z(z) = \frac{\sum_{k=0}^{N} c_k z^{-k}}{\sum_{k=0}^{N} d_k z^{-k}}. \tag{11.112}$$

The noise mean at the output is the product of μ_e and the DC gain of $G^z(z)$, that is,

$$\mu_y = G^z(1)\mu_e = \frac{\mu_e \sum_{k=0}^{N} c_k}{\sum_{k=0}^{N} d_k}. \tag{11.113}$$

The noise variance at the output is the product of γ_e and the noise gain of $G^z(z)$, that is,

$$\gamma_y = \text{NG} \, \gamma_e = \frac{\gamma_e}{2\pi} \int_{-\pi}^{\pi} |G^f(\theta)|^2 d\theta. \tag{11.114}$$

The noise gain of $G^z(z)$ can be computed from its coefficients as explained in Section 7.4.5.

Since different noise sources are uncorrelated, the noise mean at the output is the sum of the individual means, and the noise variance at the output is the sum of the individual variances. If the filter has L noise sources $\{e_l[n], 1 \le l \le L\}$, with corresponding transfer functions $\{G_l^z(z), 1 \le l \le L\}$ to the output, then the mean and the variance of the noise at the output are

$$\mu_y = \sum_{l=1}^{L} G_l^z(1)\mu_{e,l}, \quad \gamma_y = \sum_{l=1}^{L} \frac{\gamma_{e,l}}{2\pi} \int_{-\pi}^{\pi} |G_l^f(\theta)|^2 d\theta. \tag{11.115}$$

In the following we compute the output noise parameters (mean and variance) for standard filter realizations. We assume that truncation arithmetic is used. In case of rounding arithmetic, all means are zero.

11.7.2 Quantization Noise in Direct Realizations

Consider the transposed direct realization of a second-order filter. Such a realization contains five multipliers, as shown in Figure 11.22(a). We represent the error caused by each multiplication by an additive signal $e_k[n]$, $1 \le k \le 5$, at the output of the corresponding multiplier. Each of the $e_k[n]$ is white noise, and they are all uncorrelated.

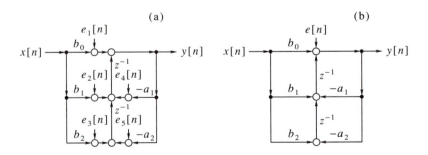

Figure 11.22 Quantization noise in a transposed direct realization: (a) separate noise sources; (b) combined noise source.

All five noise sources can be combined to a single equivalent noise source at the top adder of the realization, as shown in Figure 11.22(b). The equivalent noise is given by

$$e[n] = e_1[n] + e_2[n-1] + e_3[n-2] + e_4[n-1] + e_5[n-2]. \tag{11.116}$$

Because of the absence of correlation between the noise sources, the equivalent noise $e[n]$ has the following properties:

1. Its mean is the sum of the means, that is, $2.5\,\text{LSB}$.

2. Its variance is the sum of the variances, that is, $(5/12)\,\text{LSB}^2$.

Therefore, it is not necessary to consider the individual sources for further noise computations, only the equivalent source.

As we see from Figure 11.22(b), the transfer function from the noise to the output is

$$\frac{Y^z(z)}{E^z(z)} = \frac{1}{a(z)}. \tag{11.117}$$

Therefore,

1. The mean of the output is equal to the mean of the noise times the DC gain of the transfer function, that is,

$$\mu_y = \frac{2.5}{a(1)}\,\text{LSB} = \frac{2.5}{1 + a_1 + a_2}\,\text{LSB}. \tag{11.118}$$

2. The variance of the output is equal to the variance of the noise times the noise gain of the transfer function, that is,

$$y_y = \frac{5}{12 \times 2\pi} \int_{-\pi}^{\pi} \frac{d\theta}{|a(e^{j\theta})|^2}\,\text{LSB}^2. \tag{11.119}$$

This method of analysis applies to an IIR filter of any order in a transposed direct realization. The number of noise sources adding up to the equivalent noise is $p + q + 1$. Therefore, we need only to multiply the aforementioned formulas by $(p + q + 1)/5$.

In the case of a linear-phase FIR filter, the number of multipliers is $\lfloor 0.5N \rfloor + 1$. The denominator polynomial is $a(z) = 1$. Therefore we get in this case,[7]

$$\mu_y = \frac{\lfloor 0.5N \rfloor + 1}{2}\,\text{LSB}, \quad y_y = \frac{\lfloor 0.5N \rfloor + 1}{12}\,\text{LSB}^2. \tag{11.120}$$

Next consider the direct realization of a second-order filter. Such a realization contains five multipliers, as shown in Figure 11.23(a). As before, we represent the error caused by each multiplication by an additive signal $e_k[n]$, $1 \le k \le 5$, at the output of the corresponding multiplier. Each of the $e_k[n]$ is white noise, and they are all uncorrelated.

This time we cannot combine all five noise sources into a single equivalent source, but we can form two equivalent sources, as shown in Figure 11.23(b). The equivalent noise sources are given by

$$e_a[n] = e_1[n] + e_2[n], \quad e_b[n] = e_3[n] + e_4[n] + e_5[n]. \tag{11.121}$$

As we see from Figure 11.23(b), the transfer functions from the noise sources to the output are

$$\frac{Y^z(z)}{E_a^z(z)} = \frac{b(z)}{a(z)}, \quad \frac{Y^z(z)}{E_b^z(z)} = 1. \tag{11.122}$$

Therefore,

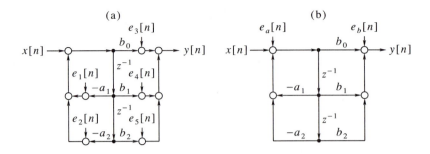

Figure 11.23 Quantization noise in a direct realization: (a) separate noise sources; (b) combined noise sources.

1. The mean of the output is

$$\mu_y = \left(\frac{b(1)}{a(1)} + \frac{3}{2}\right) \text{LSB} = \left(\frac{b_0 + b_1 + b_2}{1 + a_1 + a_2} + \frac{3}{2}\right) \text{LSB}. \tag{11.123}$$

2. The variance of the output is

$$y_y = \left(\frac{2}{12 \cdot 2\pi} \int_{-\pi}^{\pi} \frac{|b(e^{j\theta})|^2}{|a(e^{j\theta})|^2} d\theta + \frac{3}{12}\right) \text{LSB}^2. \tag{11.124}$$

This method of analysis applies to an IIR filter of any order in a direct realization with obvious modifications: Instead of the factor 2 we use p, and instead of the factor 3 we use $q + 1$.

In the case of a linear-phase FIR filter, we get

$$\mu_y = \frac{\lfloor 0.5N \rfloor + 1}{2} \text{LSB}, \quad y_y = \frac{\lfloor 0.5N \rfloor + 1}{12} \text{LSB}^2. \tag{11.125}$$

11.7.3   Quantization Noise in Parallel and Cascade Realizations

Figure 11.24 shows a scaled parallel realization, with second-order sections in transposed direct realization, and noise sources added. We observe the following:

1. There are four multipliers in each section, so the mean of the equivalent noise source $e_k[n]$ is 2 LSB and the variance is $(1/3) \text{LSB}^2$.

2. If the realization also contains a first-order section (there can be one such section at most), the mean of its equivalent noise source is 1 LSB and the variance is $(1/6) \text{LSB}^2$.

3. The scale factors λ_k are usually integer powers of 2, so they generate no extra noise (the corresponding left or right shifts are usually performed prior to truncation).

4. The amplified noise components add up at the output of the filter, as a result of the parallel connection. Also adding up at this point is the noise resulting from the constant branch c_0.

Example 11.12 Consider again the scaled parallel realization of the IIR filter discussed in Example 11.10. We wish to compute the mean and variance of the output noise to two decimal digits. This accuracy is sufficient for most practical purposes. The DC gains of the three sections (from the corresponding noise inputs) are 9.5, 12, and 4.4.

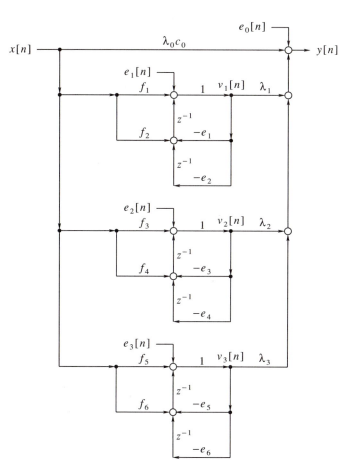

Figure 11.24 Quantization noise in a parallel realization (second-order sections in a transposed direct realization).

The noise gains are 118, 30, and 2.5. Taking into account the scale factors λ_k and adding the effect of the constant branch, we get

$$\mu_y = [2(0.5 \times 9.5 + 12) + 4.4 + 0.5]\,\text{LSB} \approx 39\,\text{LSB},$$
$$\gamma_y = [0.33(0.25 \times 118 + 30) + (2.5 \times 0.17) + 0.083]\,\text{LSB}^2 \approx 20\,\text{LSB}^2.$$

As we see, the mean output noise is about 39 least significant bits, and the standard deviation is about 4.5 least significant bits. □

Figure 11.25 shows a scaled cascade realization, with second-order sections in a transposed direct realization, and noise sources added. We observe the following:

1. There are five multipliers in each second-order section. However, one of those is identically 1, so it is not subject to quantization. Furthermore, because of the special properties of the zeros of a digital IIR filter derived from a standard analog filter, the numerator coefficients h_{2k-1}, h_{2k} are not arbitrary: h_{2k} is either 1 or −1, and h_{2k-1} is either general (if the zeros are complex) or ±2 (if the zeros are real). Therefore, the effective number of noise sources in a section is either two or three. Correspondingly, the mean of the equivalent noise source $e_k[n]$ is 1 LSB or 1.5 LSB and the variance is $(1/6)\,\text{LSB}^2$ or $(1/4)\,\text{LSB}^2$.

2. If the realization also contains a first-order section (there can be one such section at most), this section will have only one noise source, since the numerator coefficients are ± 1, and are not subject to quantization. Therefore, the mean of the equivalent noise source is 0.5 LSB and the variance is $(1/12)$ LSB^2.

3. One of the scale factors along the cascade is a general number, so it adds a noise source. The other factors are integer powers of 2, so they are not subject to quantization.

4. The transfer function from the noise source $e_k[n]$ to the output is not just $1/g_k(z)$, but $1/g_k(z)$ times the product of transfer functions of the sections further down the cascade. Therefore, the noise gain computations are more laborious than in the case of parallel realization.

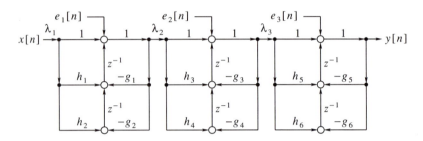

Figure 11.25 Quantization noise in a cascade realization (second-order sections in a transposed direct realization).

Example 11.13 Consider again the scaled cascade realization of the IIR filter discussed in Example 11.11. We wish to compute the mean and variance of the output noise to two decimal digits. In this example, the first two sections are second order and the third is first order. The multiplier λ_2 is subject to quantization. The DC gains from $e_1[n]$, $e_2[n]$, $e_3[n]$, and λ_2 are 29, 13, 4.4, and 3.4, respectively. The noise gains are 302, 21, 2.5, and 1.0. We get

$$\mu_y = [1.5(29 + 13) + 0.5(4.4 + 3.4)]\,\text{LSB} \approx 67\,\text{LSB},$$
$$\gamma_y = [0.25(302 + 13) + 0.083(2.5 + 1.0)]\,\text{LSB}^2 \approx 81\,\text{LSB}^2.$$

As we see, the mean output noise is about 67 LSB and the standard deviation is about 9 LSB. In this case, the noise level of the cascade realization is about twice that of the parallel realization. Different ordering of the sections will lead in general to a different (and possibly lower) noise level.

Recall, for comparison, that an FIR filter meeting the same specifications has order 146 (Example 9.16). We get from (11.120) that the mean of the output noise of the FIR filter is 37 LSB and the standard deviation is about 2.5 LSB. Therefore, the FIR filter is better than both IIR filter realizations in its output noise level.   □

11.7.4 Quantization Noise in A/D and D/A Converters

In Section 3.5.2 we introduced A/D converters, and we noted that they provide quantized signals, either truncated or rounded. The tools we developed here enable us to complete the discussion on A/D quantization and analyze this effect quantitatively. We assume, as in the case of arithmetical quantization, that the quantization level is small and the input signal is large and fast varying. Under these assumptions, A/D quantization noise can be modeled as discrete-time white noise. The mean of the noise is 0 in case of rounding, and half the least significant bit of the A/D in case of truncation. The variance is $1/12$ of the square of the least significant bit of the A/D. The mean is multiplied by the DC gain of the filter to yield the mean of the output. The variance is multiplied by the noise gain of the filter to yield the variance of the output. The noise and DC gains of the filter depend only on its total transfer function and are independent of the realization. The alignment of the least significant bit of the A/D word relative to the computer word determines the quantization level used for noise computations. Suppose, for example, that the computer word length is 16 bits and that of the A/D is 12 bits. Then the quantization level of the A/D is 2^{-15} if its output is placed at the low bits, but is 2^{-11} if placed at the high bits. However, since this placement similarly determines the dynamic range of the output signal, the *relative noise level* at the output (the signal-to-noise ratio) depends on the number of bits of the A/D, but not on their placement.

D/A converters are not, by themselves, subject to quantization. However, sometimes the word length of the D/A is shorter than that of the filter. In such cases, the signal $y[n]$ at the output of the filter must be further quantized (truncated or rounded) before it is passed to the D/A converter. We can regard this quantization as yet another source of white noise in the system. However, contrary to the preceding effects, this noise is not processed by the digital filter. Therefore, it is neither amplified nor attenuated, but appears at the output as is.

11.8 Zero-Input Limit Cycles in Digital Filters*

When a stable linear filter receives no input, and its internal state has nonzero initial conditions, its output decays to zero asymptotically. This follows because the response of each pole of the filter to initial conditions is a geometric series with parameter less than 1 in magnitude (Sections 7.5, 7.7). However, the analysis that leads to this conclusion is based on the assumption that signals in the filter are represented to infinite precision, so they obey the mathematical formulas exactly. Quantization resulting from finite word length is a nonlinear operation, so stability properties of linear systems do not necessarily hold for a filter subject to quantization. Indeed, digital filters can exhibit sustained oscillations when implemented with finite word length. Oscillations resulting from nonlinearities are called *limit cycles*.

Limit cycle phenomena are different from the noiselike behavior caused by quantization. Quantization effects are noiselike when the signal level is large and relatively fast varying, rendering the quantization error at any given time nearly independent of the errors at past times. When the signal level is low, errors caused by quantization become correlated. When the input signal is zero, randomness disappears, and the error behavior becomes completely deterministic. Oscillations in the absence of input are called *zero-input limit cycles*. Such oscillations are periodic, but not necessarily sinusoidal. They are likely to appear whenever there is feedback in the filter. Digital IIR filters always have inner feedback paths, so they are susceptible to limit cycle oscillations. On the other hand, FIR filters are feedback free, so they are immune to limit cycles. This is yet another advantage of FIR filters over IIR filters. Limit cycles may be troublesome in applications such as speech and music, because the resulting signal may be audible.

In this section we briefly discuss zero-input limit cycles in IIR filter structures. Since practical IIR implementations are almost always in parallel or cascade, we concentrate on first- and second-order sections. Mathematical analysis of limit cycles is difficult, except for first-order filters. We shall therefore omit most mathematical details, and restrict our discussion to practical techniques for analysis and remedy of limit cycle phenomena.

Example 11.14 Let us illustrate the limit-cycle phenomenon by a few simple examples.

1. Consider the first-order filter

$$y[n] = -0.625y[n-1] + x[n].$$

Suppose that we implement the filter with 4-bit rounding arithmetic, so the least significant bit is $1/8$. Let the input signal $x[n]$ be zero and $y[0] = 3/8$. Then $y[1] = -15/64$, and this will be rounded to $y[1] = -1/4$. Next, $y[2] = 5/32$, and this will be rounded to $y[2] = 1/8$. Next, $y[3] = -5/64$, and this will be rounded to $y[3] = -1/8$. Next, $y[4] = 5/64$, and this will be rounded to $y[4] = 1/8$. From this point on we will have $y[n] = 1/8$ for all even n and $y[n] = -1/8$ for all odd n. We get a constant amplitude oscillation whose frequency is $\theta_0 = \pi$ and whose amplitude is $1/8$.

2. Next consider the same filter, but implemented with 4-bit truncation arithmetic. The sequence of values of the output signal will now be

$$y[n] = 3/8, \; -1/4, \; 1/8, \; -1/8, \; 0, \; 0, \; \ldots.$$

As we see, truncation of $y[4]$ makes it zero, thus stopping the oscillations. This time there is no limit cycle.

3. Consider the filter

$$y[n] = 0.625y[n-1] + x[n],$$

with 4-bit rounding arithmetic. This time the output sequence is

$$y[n] = 3/8, \; 1/4, \; 1/8, \; 1/8, \; 1/8, \; \ldots.$$

As we see, the output reaches a constant nonzero value. This phenomenon is called *zero-frequency limit cycle*.

4. For the same filter as in part 3, but with truncation arithmetic and the same initial condition, $y[n]$ will reach zero as in part 2.

5. Consider the filter of part 3, but with truncation arithmetic and initial condition $y[0] = -3/8$. Then,

$$y[n] = -3/8, \; -1/4, \; -1/8, \; -1/8, \; -1/8, \; \ldots.$$

We get zero-frequency limit cycle again, this time with a negative constant value.

The lesson from this example is that the existence and nature of limit cycles depend on the filter in question, the quantization level, the method of quantization, and the initial conditions. □

Zero-input limit cycles in first-order filters are relatively easy to analyze. Let $y[n] = -ay[n-1] + x[n]$ be the filter in question, and assume that $x[n] = 0$. Then:

1. Zero-frequency limit cycle is possible in rounding arithmetic only if

$$\text{round}\{-2^{B-1}ay\} = 2^{B-1}y, \tag{11.126}$$

where y is the steady-state value of the output. This is equivalent to (Problem 11.25)

$$|2^{B-1}(1+a)y| \leq 0.5, \tag{11.127}$$

or

$$|y| \le \frac{2^{-B}}{1+a}. \tag{11.128}$$

This condition imposes an upper limit on the magnitude that a zero-frequency limit cycle can attain, as a function of the word length and the coefficient of the filter. The permissible magnitude range is called the *dead band* of the filter. It follows from (11.128) that limit cycle of this kind is possible only if $-1 < a \le -0.5$. Note that the dead band increases as a approaches -1.

2. Limit cycle at frequency π is possible in rounding arithmetic only if

$$\text{round}\{-2^{B-1}ay\} = -2^{B-1}y. \tag{11.129}$$

This is equivalent to

$$|2^{B-1}(1-a)y| \le 0.5, \tag{11.130}$$

or

$$|y| \le \frac{2^{-B}}{1-a}. \tag{11.131}$$

This condition again imposes an upper limit on the magnitude that a limit cycle at frequency π can attain. In this case limit cycle is possible only if $0.5 \le a < 1$. Note that the dead band increases as a approaches 1.

3. Zero-frequency limit cycle is possible in truncation arithmetic only if

$$\lfloor -2^{B-1}ay \rfloor = 2^{B-1}y. \tag{11.132}$$

This leads to the condition

$$-\frac{2^{-(B-1)}}{1+a} < y < 0. \tag{11.133}$$

This kind of limit cycle is possible only if $-1 < a < 0$.

4. Limit cycle at frequency π is possible in truncation arithmetic only if

$$\lfloor -2^{B-1}ay \rfloor = -2^{B-1}y. \tag{11.134}$$

This is equivalent to

$$0 < y < \frac{2^{-(B-1)}}{1-a}. \tag{11.135}$$

This leads to contradiction, since y cannot oscillate and remain nonnegative at all times. The conclusion is that limit cycles at frequency π are not possible in first-order filters if truncation arithmetic is used.[8]

Limit cycles in second-order filters are much more difficult to analyze. As in the case of first-order filters, limit cycles are possible at both frequencies 0 and π. However, second-order filters can also sustain limit cycles at other frequencies. The frequency and amplitude of the limit cycle (if it exists) depend on the coefficients, initial conditions, quantization method, and word length. However, in case of a second-order section they also depend on the realization. In particular, the coupled realization presented in Section 11.1.7 is less susceptible to limit cycles than a direct realization. The coupled realization is represented by the state-space equations (11.32). We now show that the coupled realization can be made completely free of zero-input limit cycles if two precautions are taken in its implementation:

1. *Magnitude truncation*, also called *rounding toward zero*, is used instead of rounding or truncation. The magnitude truncation of a number a is the number $\lfloor |a| \rfloor \text{sign}(a)$. This method of truncation is seldom implemented in hardware, so it usually requires special programming.

2. Magnitude truncation of each component of $s[n+1]$ is performed *after* both products (by α_r and α_i) are formed and added (subtracted) in double precision.

To show the coupled realization is free of zero-input limit cycles under these conditions, observe from (11.32) that

$$(s_1[n+1])^2 + (s_2[n+1])^2 = (\alpha_r s_1[n] + \alpha_i s_2[n])^2 + (-\alpha_i s_1[n] + \alpha_r s_2[n])^2$$
$$= (\alpha_r^2 + \alpha_i^2)\{(s_1[n])^2 + (s_2[n])^2\} = |\alpha|^2\{(s_1[n])^2 + (s_2[n])^2\}$$
$$< (s_1[n])^2 + (s_2[n])^2, \tag{11.136}$$

since $|\alpha|^2 < 1$. Furthermore, magnitude quantization reduces the left side still further, because it reduces each of $|s_1[n+1]|$ and $|s_2[n+1]|$ individually. The conclusion is that the sum of squares of the state-vector components strictly decreases as a function of n at least as fast as a geometric series with parameter $|\alpha|^2$. Therefore, there must come a time n for which both $|s_1[n]|$ and $|s_2[n]|$ be less than $2^{-(B-1)}$. At this point they are truncated to zero, and the filter comes to a complete rest.

Coupled realization, implemented with magnitude truncation after multiplication and addition at each stage, is therefore free of zero-input limit cycles, so it is recommended whenever such limit cycles must be prevented. However, the following drawbacks of this method should not be overlooked:

1. The realization is costly in number of operations: There are 6 multiplications and 5 additions per section (compared with 4 and 4 for direct realization), and magnitude truncation requires additional operations.

2. Magnitude truncation doubles the standard deviation of the quantization noise (compared with rounding or truncation).

Since limit cycles in second-order sections are difficult to analyze mathematically, it is helpful to develop alternative tools for exploring them. The MATLAB procedure lc2sim, listed in Program 11.17, is such a tool. This procedure tests a given second-order realization for limit cycles by simulating its zero-input response. The procedure accepts the following parameters: qtype indicates the quantization type—rounding, truncation, or magnitude truncation; when indicates whether quantization is to be performed before or after the addition; rtype enables choosing between direct and coupled realizations; apar is a vector of two parameters, either a_1, a_2 or α_r, α_i; B is the number of bits; s0 is the initial state vector; finally, n is the maximum number of points to be simulated. The program implements the realization in state space, and performs the necessary quantizations by calling the auxiliary procedure quant, listed in Program 11.18. At the end of each step, the program performs two tests. First it tests whether the state vector is identically zero. If so, the realization is necessarily free of zero-input limit cycles, since zero state at any time implies zero state at all subsequent times. Next it tests whether the state vector is the same as in the preceding step. If so, it means that zero-frequency limit cycle has been reached. If the program reaches the end of the simulation without either of these conditions being met, it declares that a limit cycle exists at nonzero frequency. The program outputs its decision in the variable flag, and also the history of the output signal $y[n]$.

The procedure lcdrive in Program 11.19 is a driver program for lc2sim. Note that it is a script file, rather than a function. Therefore, the variables must be declared in the MATLAB environment prior to running the file. The program determines the maximum duration of the simulation as a function of the magnitude of the poles. It then performs M simulations (where M is entered by the user), each time choosing the initial state at random. This is necessary because absence of a limit cycle from a specific initial state does not guarantee absence of limit cycles from other initial states. If, at any of the M simulations, the existence of limit cycle is observed, the program stops and reports occurrence. If no limit cycles are observed, the program reports that the realization appears to be free of limit cycles. The reliability of this test increases with the value assigned to M.

Example 11.15 We repeat the test of the IIR low-pass filter used in the earlier examples in this chapter. We examine each of its two sections for possible limit cycles in all 12 combinations of quantization method, quantization before or after the addition, and direct or coupled realization. Performing 100 simulations for each case, we find the following:

1. Both second-order sections are free of limit cycles if coupled realization with magnitude truncation is used, regardless of whether truncation is performed before or after the addition.

2. Both second-order sections are also free of limit cycles if a direct realization with magnitude truncation is used, provided that truncation is performed *after* the addition.

3. In all other cases, limit cycles may occur, at either zero or nonzero frequency. □

11.9 Summary and Complements

11.9.1 Summary

This chapter was devoted to three different, but related topics: digital filter realization, state-space representations of digital systems, and finite word length effects in digital filters.

Realization of a digital filter amounts to constructing a block diagram of the internal structure of the filter. Such a block diagram shows the elementary blocks of which the filter is constructed, their interconnections, and the numerical parameters associated with them. For LTI filters, only three block types are needed: delays, adders, and multipliers. Standard digital filter realizations include the direct (of which there are two forms), the parallel, and the cascade realizations. They all have the same number of delays, adders, and multipliers. However, they differ in their behavior when implemented in finite word length.

The standard realizations are special cases of the general concept of state-space representation. A state-space representation describes the time evolution of the memory variables (i.e., the state) of the filter, and the dependence of the output signal on the state and on the input signal. To a given LTI filter, there correspond an infinite number of similar state-space representations. State space is a useful tool for performing computations related to digital filters, for example, transfer function and impulse response computations. State-space representations of complex digital networks can be constructed using well-defined procedures.

Finite word length implementation affects a digital filter in several respects:

1. It causes the filter coefficients to deviate from their ideal values. Consequently, the poles, zeros, and frequency response are changed, and the filter may fail to meet the specifications.

2. The dynamic range of the various signals (input, output, and internal) becomes a problem (in fixed-point implementation), and may lead to overflows or saturations. Hence there is a need to scale the signals at various points of the filter, to maximize the dynamic range with little or no danger of overflow.

3. Quantization of multiplication operations leads to computational noise. The noise is propagated to the output and added to the output signal.

4. Digital filter structures may develop self oscillations, called limit cycles, in the absence of input.

It turns out that different realizations have different sensitivities to these effects. Direct realizations are the worst on almost all accounts. Both parallel and cascade realizations are much better, and their sensitivities are comparable. The parallel realization typically has a better pass-band behavior, whereas the cascade realization typically has a better stop-band behavior. The cascade realization is more general and more flexible than the parallel realization; therefore it is preferred in most applications of IIR filters. For FIR filters, on the other hand, direct realizations are the most common, since their sensitivity to finite word length effects, although worse than that of a cascade realization, is usually acceptable.

11.9.2   Complements

1. [p. 393] Some books use the name "direct realization II" for the structure appearing in Figure 11.4 and the name "direct realization I" for the nonminimal structure discussed in Problem 11.2; we avoid this terminology.

2. [p. 395] Programs 11.1, 11.3, and 11.5 get the input signal $x[n]$ in its entirety as a single object. In real-time applications, $x[n]$ is supplied sequentially, one value per sampling interval. In such cases it is necessary to store the state vector u, either internally as a *static* variable, or externally. For example, the MATLAB function `filter` optionally accepts the state vector as input and optionally returns its updated value. This facilitates external storage of the state.

3. [p. 398] Parallel realization of a system with multiple poles has a mixed parallel/series structure, see Kailath [1980] for details.

4. [p. 413] There are software schemes that greatly alleviate this problem. Consider, for example, the case of complex poles at $z = \rho e^{\pm j\zeta}$, close to $z = 1$. Then ρ is nearly 1 and ζ is nearly 0. Since $a_1 = -2\rho \cos \zeta$ and $a_2 = \rho^2$, a_1 is nearly -2 and a_2 is nearly 1. Write the difference equation for the auxiliary variable $u[n]$ of the direct realization as [cf. (11.2)]

$$u[n] = x[n] + u[n-1] + (u[n-1] - u[n-2])$$
$$- (a_1 + 2)u[n-1] - (a_2 - 1)u[n-2].$$

Now the modified coefficients $(a_1 + 2)$ and $(a_2 - 1)$ are both small numbers, so we can scale them up by a few bits, thus retaining a few of their lower order bits. For instance, if the modified coefficients are around 0.05, we can gain 4 significant bits this way. The products $(a_1 + 2)u[n-1]$ and $(a_2 - 1)u[n-2]$ are generated in double precision by default in most computers. These products are shifted to the right by the same number of bits used for scaling the modified coefficients, truncated to the basic word length, and added to $u[n]$. This scheme is more accurate than truncating the coefficients prior to multiplication. The addition of $u[n-1]$ and $(u[n-1] - u[n-2])$ can be done exactly, with no loss in precision. Since the filter is necessarily low pass, the signal $u[n]$ changes slowly in time, so

$|u[n-1]-u[n-2]|$ is typically small compared with the full scale of $u[n]$. If very high accuracy is required, $u[n]$ can be stored in double precision. In this case all the additions on the right side of the equation should be in double precision as well.

5. [p. 414] The differentiation that takes place here may be unfamiliar. Suppose that $x(t)$ is a complex-valued function of a real variable t. Then $y(t) = |x(t)|$ is a real-valued function of t. If $x(t)$ is differentiable, then $y(t)$ is differentiable at any point t for which $x(t) \neq 0$. The derivative of $y(t)$ can be obtained as follows. We first observe that

$$y(t) = (|x(t)|^2)^{1/2},$$

hence

$$\frac{dy(t)}{dt} = 0.5(|x(t)|^2)^{-1/2}\frac{d|x(t)|^2}{dt} = 0.5|x(t)|^{-1}\frac{d|x(t)|^2}{dt}.$$

Then we observe that

$$|x(t)|^2 = x(t)\bar{x}(t), \quad \text{hence} \quad \frac{d|x(t)|^2}{dt} = \frac{dx(t)}{dt}\bar{x}(t) + \frac{d\bar{x}(t)}{dt}x(t).$$

Therefore,

$$\frac{dy(t)}{dt} = 0.5|x(t)|^{-1}\left[\frac{dx(t)}{dt}\bar{x}(t) + \frac{d\bar{x}(t)}{dt}x(t)\right].$$

If we express $x(t)$ as $|x(t)|e^{j\psi(t)}$, where $\psi(t)$ is the phase of $x(t)$, we will get that

$$|x(t)|^{-1}x(t) = e^{j\psi(t)}, \quad |x(t)|^{-1}\bar{x}(t) = e^{-j\psi(t)}.$$

Therefore, we can express the derivative of $y(t)$ as

$$\frac{dy(t)}{dt} = 0.5\left[\frac{dx(t)}{dt}e^{-j\psi(t)} + \frac{d\bar{x}(t)}{dt}e^{j\psi(t)}\right] = \Re\left\{\frac{dx(t)}{dt}e^{-j\psi(t)}\right\}. \quad (11.137)$$

This is the formula used for obtaining (11.79).

6. [p. 422] For example, if $y[n]$ has Gaussian probability density, this probability is about 0.0016.

7. [p. 429] Quantization noise in FIR filters can be greatly reduced by performing the summation

$$y[n] = \sum_{k=0}^{N} h[k]x[n-k]$$

in double precision. In this case, the products $h[k]x[n-k]$ are not truncated at all, only the final value of $y[n]$ is truncated. This reduces the factor in (11.120) from $(\lfloor 0.5N \rfloor + 1)$ to 1. Such an implementation is recommended when the order of the filter is large and noise is a major concern (as in audio applications).

8. [p. 435] It can be shown that first-order filters can sustain limit cycles only at frequencies 0 or π.

11.10   MATLAB Programs

Program 11.1 Direct realizations of digital IIR filters.

```
function y = direct(typ,b,a,x);
% Synopsis: direct(typ,b,a,x).
% Direct realizations of rational transfer functions.
% Input parameters:
% typ: 1 for direct realization, 2 for transposed
% b, a: numerator and denominator polynomials
% x: input sequence.
% Output:
% y: output sequence.

p = length(a)-1; q = length(b)-1; pq = max(p,q);
a = a(2:p+1); u = zeros(1,pq); % u: the internal state
if (typ == 1),
    for i = 1:length(x),
        unew = x(i)-sum(u(1:p).*a);
        u = [unew,u];
        y(i) = sum(u(1:q+1).*b);
        u = u(1:pq);
    end
elseif (typ == 2),
    for i = 1:length(x),
        y(i) = b(1)*x(i)+u(1);
        u = [u(2:pq),0];
        u(1:q) = u(1:q) + b(2:q+1)*x(i);
        u(1:p) = u(1:p) - a*y(i);
    end
end
```

Program 11.2 Computation of the parallel decomposition of a digital IIR filter.

```
function [c,nsec,dsec] = tf2rpf(b,a);
% Synopsis: [c,nsec,dsec] = tf2rpf(b,a).
% Real partial fraction decomposition of b(z)/a(z). The polynomials
% are in negative powers of z. The poles are assumed distinct.
% Input parameters:
% a, b: the input polynomials
% Output parameters:
% c: the free polynomial; empty if deg(b) < deg(a)

nsec = []; dsec = []; [c,A,alpha] = tf2pf(b,a);
while (length(alpha) > 0),
  if (imag(alpha(1)) ~= 0),
    dsec = [dsec; [1,-2*real(alpha(1)),abs(alpha(1))^2]];
    nsec = [nsec; [2*real(A(1)),-2*real(A(1)*conj(alpha(1)))]];
    alpha(1:2) = []; A(1:2) = [];
  else,
    dsec = [dsec; [1,-alpha(1),0]]; nsec = [nsec; [real(A(1)),0]];
    alpha(1) = []; A(1) = [];
  end
end
```

Program 11.3 Parallel realization of a digital IIR filter.

```
function y = parallel(c,nsec,dsec,x);
% Synopsis: y = parallel(c,nsec,dsec,x).
% Parallel realization of an IIR digital filter.
% Input parameters:
% c: the free term of the filter.
% nsec, dsec: numerators and denominators of second-order sections
% x: the input sequence.
% Output:
% y: the output sequence.

[n,m] = size(dsec); dsec = dsec(:,2:3);
u = zeros(n,2); % u: the internal state
for i = 1:length(x),
  y(i) = c*x(i);
  for k = 1:n,
    unew = x(i)-sum(u(k,:).*dsec(k,:)); u(k,:) = [unew,u(k,1)];
    y(i) = y(i) + sum(u(k,:).*nsec(k,:));
  end
end
```

Program 11.4 Pairing of poles and zeros to real second-order sections.

```
function [nsec,dsec] = pairpz(v,u);
% Synopsis: [nsec,dsec] = pairpz(v,u).
% Pole-zero pairing for cascade realization.
% Input parameters:
% v, u: the vectors of poles and zeros, respectively.
% Output parameters:
% nsec: matrix of numerator coefficients of second-order sections
% dsec: matrix of denom. coefficients of second-order sections.

if (length(v) ~= length(u)),
  error('Different numbers of poles and zeros in PAIRPZ'); end
u = reshape(u,1,length(u)); v = reshape(v,1,length(v));
v = cplxpair(v); u = cplxpair(u);
vc = v(find(imag(v) > 0)); uc = u(find(imag(u) > 0));
vr = v(find(imag(v) == 0)); ur = u(find(imag(u) == 0));
[temp,ind] = sort(abs(vc)); vc = vc(fliplr(ind));
[temp,ind] = sort(abs(vr)); vr = vr(fliplr(ind));
nsec = []; dsec = [];
for n = 1:length(vc),
   dsec = [dsec; [1,-2*real(vc(n)),abs(vc(n))^2]];
   if (length(uc) > 0),
      [temp,ind] = min(abs(vc(n)-uc)); ind = ind(1);
      nsec = [nsec; [1,-2*real(uc(ind)),abs(uc(ind))^2]];
      uc(ind) = [];
   else,
      [temp,ind] = min(abs(vc(n)-ur)); ind = ind(1);
      tempsec = [1,-ur(ind)]; ur(ind) = [];
      [temp,ind] = min(abs(vc(n)-ur)); ind = ind(1);
      tempsec = conv(tempsec,[1,-ur(ind)]); ur(ind) = [];
      nsec = [nsec; tempsec];
   end
end
if (length(vr) == 0), return
elseif (length(vr) == 1),
   dsec = [dsec; [1,-vr,0]]; nsec = [nsec; [1,-ur,0]];
elseif (length(vr) == 2),
   dsec = [dsec; [1,-vr(1)-vr(2),vr(1)*vr(2)]];
   nsec = [nsec; [1,-ur(1)-ur(2),ur(1)*ur(2)]];
else
   error('Something wrong in PAIRPZ, more than 2 real zeros');
end
```

Program 11.5 Cascade realization of a digital IIR filter.

```
function y = cascade(C,nsec,dsec,x);
% Synopsis: y = cascade(C,nsec,dsec,x).
% Cascade realization of an IIR digital filter.
% Input parameters:
% C: the constant gain of the filter.
% nsec, dsec: numerators and denominators of second-order sections
% x: the input sequence.
% Output:
% y: the output sequence.

[n,m] = size(dsec);
u = zeros(n,2); % u: the internal state
dsec = dsec(:,2:3); nsec = nsec(:,2:3);
for i = 1:length(x),
   for k = 1:n,
      unew = x(i)-sum(u(k,:).*dsec(k,:));
      x(i) = unew + sum(u(k,:).*nsec(k,:));
      u(k,:) = [unew,u(k,1)];
   end
   y(i) = C*x(i);
end
```

Program 11.6 Computation of the state-space matrices corresponding to a given transfer function.

```
function [A,B,C,D] = tf2ss(b,a);
% Synopsis: [A,B,C,D] = tf2ss(b,a).
% Converts a transfer function to direct state-space realization.
% Inputs:
% b, a: the numerator and denominator polynomials.
% Outputs:
% A, B, C, D: the state-space matrices

p = length(a)-1; q = length(b)-1; N = max(p,q);
if (N > p), a = [a,zeros(1,N-p)]; end
if (N > q), b = [b,zeros(1,N-q)]; end
A = [-a(2:N+1); [eye(N-1), zeros(N-1,1)]];
B = [1; zeros(N-1,1)];
C = b(2:N+1) - b(1)*a(2:N+1);
D = b(1);
```

Program 11.7 Computation of the transfer function corresponding to given state-space matrices.

```
function [b,a] = ss2tf(A,B,C,D);
% Synopsis: [b,a] = tf2ss(A,B,C,D).
% Converts a state-space realization to a transfer function.
% Inputs:
% A, B, C, D: the state-space matrices
% Outputs:
% b, a: the numerator and denominator polynomials.

a = poly(A); N = length(a)-1; h = zeros(1,N+1); h(1) = D; tmp = B;
for i = 1:N, h(i+1) = C*tmp; tmp = A*tmp; end
b = a*toeplitz([h(1);zeros(N,1)],h);
```

Program 11.8 Construction of a state-space representation of a digital network.

```
function [A,B,C,D] = network(F,G,H,K,Rnum,Rden);
% Synopsis: [A,B,C,D] = network(F,G,H,K,Rnum,Rden).
% Builds a state-space representation of a digital network.
% Input parameters:
% F, G, H, K: network connection matrices
% Rnum: rows contain numerator coefficients of blocks
% Rden: rows contain denominator coefficients of blocks
% Output parameters:
% A, B, C, D: state-space matrices.

[L,Nnum] = size(Rnum); [L,Nden] = size(Rden);
A = []; B = []; C = []; D = []; N = 0;
for l = 1:L,
   rnum = Rnum(l,:); rden = Rden(l,:);
   while (rnum(length(rnum)) == 0),
      rnum = rnum(1:length(rnum)-1); end
   while (rden(length(rden)) == 0),
      rden = rden(1:length(rden)-1); end
   [At,Bt,Ct,Dt] = tf2ss(rnum,rden); Nt = length(Bt);
   A = [A,zeros(N,Nt); zeros(Nt,N),At];
   B = [B,zeros(N,1); zeros(Nt,l-1),Bt];
   C = [C,zeros(l-1,Nt); zeros(1,N),Ct];
   D = [D,zeros(l-1,1); zeros(1,l-1),Dt];
   N = N + Nt;
end
E = eye(L)-F*D;
if (rank(E) < L), error('Network is singular'), end
E = inv(E); A = A + B*E*F*C; B = B*E*G;
C = H*(eye(L) + D*E*F)*C; D = K + H*D*E*G;
```

Program 11.9 Sensitivity bound for the magnitude response of an IIR filter to coefficient quantization.

```
function [dHmag,S] = sensiir(typ,b,a,K,theta);
% Synopsis: [dHmag,S] = sensiir(typ,b,a,K,theta).
% Computes the sensitivity bound for the magnitude response of
% an IIR filter to coefficient quantization.
% Input parameters:
% typ: 'd' for direct realization
%      'p' for parallel realization
%      'c' for cascade realization
% b, a: numerator and denominator polynomials
% K: number of frequency points
% theta: frequency interval (2-element vector).
% Output parameters:
% dHmag: the partial derivative matrix, M by K, where M is the
%        number of coefficients in the realization
% S: the sensitivity bound, 1 by K.

Hangle = exp(-j*angle(frqresp(b,a,K,theta)));
if (typ == 'd'),
   [dH,sc] = dhdirect(b,a,K,theta);
elseif (typ == 'p'),
   [c,nsec,dsec] = tf2rpf(b,a);
   [dH,sc] = dhparal(nsec,dsec,c,K,theta);
elseif (typ == 'c'),
   c = b(1); v = roots(a); u = roots(b);
   [nsec,dsec] = pairpz(v,u);
   [dH,sc] = dhcascad(nsec,dsec,c,K,theta);
end
[M,junk] = size(dH);
dHmag = real(dH.*(ones(M,1)*Hangle));
S = sum(abs((sc*ones(1,K)).*dHmag));
```

Program 11.10 Partial derivatives of the frequency response of an IIR filter in direct realization with respect to the coefficients.

```
function [dH,sc] = dhdirect(b,a,K,theta);
% Synopsis: [dH,sc] = dhdirect(b,a,K,theta).
% Computes the derivatives of the magnitude response of an
% IIR filter in direct realization with respect to the
% parameters, and a scaling vector for the parameters.
% Input parameters:
% b, a: the numerator and denominator polynomials
% K: number of frequency points
% theta: frequency interval (2-element vector).
% Output parameters:
% dH: matrix of partial derivatives of |H(theta)|
% sc: a scaling vector.

dHn = []; dHd = []; scn = []; scd = [];
H = frqresp(b,a,K,theta);
for k = 0:length(b)-1,
   dHn = [dHn; frqresp([zeros(1,k),1],a,K,theta)];
end
for k = 1:length(a)-1,
   dHd = [dHd; -frqresp([zeros(1,k),1],a,K,theta).*H]; end
   scn = scale2(b)*ones(length(b),1);
   scd = scale2(a)*ones(length(a)-1,1);
   dH = [dHn; dHd]; sc = [scn; scd];
end
```

Program 11.11 Partial derivatives of the frequency response of an IIR filter in parallel realization with respect to the coefficients.

```
function [dH,sc] = dhparal(nsec,dsec,c,K,theta);
% Synopsis: [dH,sc] = dhparal(nsec,dsec,c,K,theta).
% Computes the derivatives of the magnitude response of an
% IIR filter in parallel realization with respect to the
% parameters, and a scaling vector for the parameters.
% Input parameters:
% nsec, dsec, c: parameters of the parallel realization
% K: number of frequency points
% theta: frequency interval (2-element vector).
% Output parameters:
% dH: matrix of partial derivatives of |H(theta)|
% sc: a scaling vector.

dHn = []; dHd = []; scn = []; scd = [];
[M,junk] = size(nsec);
for k = 1:M,
   if (dsec(k,3) == 0),
      [dHt,sct] = dhdirect(nsec(k,1),dsec(k,1:2),K,theta);
      dHn = [dHn; dHt(1,:)];   dHd = [dHd; dHt(2,:)];
      scn = [scn; sct(1)]; scd = [scd; sct(2)];
   else,
      [dHt,sct] = dhdirect(nsec(k,:),dsec(k,:),K,theta);
      dHn = [dHn; dHt(1:2,:)]; dHd = [dHd; dHt(3:4,:)];
      scn = [scn; sct(1)*ones(2,1)];
      scd = [scd; sct(2)*ones(2,1)];
   end
end
dH = [dHn; dHd; ones(1,K)]; sc = [scn; scd; scale2(c)];
```

Program 11.12 Partial derivatives of the frequency response of an IIR filter in cascade realization with respect to the coefficients.

```
function [dH,sc] = dhcascad(nsec,dsec,c,K,theta);
% Synopsis: [dH,sc] = cascad(nsec,dsec,c,K,theta).
% Computes the derivatives of the magnitude response of an
% IIR filter in cascade realization with respect to the
% parameters, and a scaling vector for the parameters.
% Input parameters:
% nsec, dsec, c: parameters of the cascade realization
% K: number of frequency points
% theta: frequency interval (2-element vector).
% Output parameters:
% dH: matrix of partial derivatives of |H(theta)|
% sc: a scaling vector.

dHn = []; dHd = []; scn = []; scd = [];
cntd = 0; cntn = 0;
[M,junk] = size(nsec); H = ones(1,K);
for k = 1:M,
   if (nsec(k,3) ~= 0 & abs(nsec(k,2)) ~= 2),
      Ht = frqresp(nsec(k,:),dsec(k,:),K,theta);
      [dHt,sct] = dhdirect(nsec(k,:),dsec(k,:),K,theta);
      H = Ht.*H;
      dHn = [dHn; dHt(2,:)./Ht]; cntn = cntn+1;
      dHd = [dHd; dHt(4:5,:)./(ones(2,1)*Ht)];
      cntd = cntd+2;
      scn = [scn; sct(2,1)];
      scd = [scd; sct(4:5,1)];
   end
end
dHn = c*(ones(cntn,1)*H).*dHn; dHd = c*(ones(cntd,1)*H).*dHd;
dH = [dHn; dHd; H]; sc = [scn; scd; scale2(c)];
```

Program 11.13 Full scale of a vector of coefficients in fixed-point filter implementation.

```
function s = scale2(a);
% Synopsis: s = scale2(a).
% Finds a power-of-2 full scale for the vector a.

s =  exp(log(2)*ceil(log(max(abs(a)))./log(2)));
```

Program 11.14 Sensitivity bound for the magnitude response of a linear-phase FIR filter to coefficient quantization.

```
function [dHmag,S] = sensfir(h,K,theta);
% Synopsis: [dHmag,S] = sensfir(h,K,theta).
% Computes the sensitivity bound for the magnitude response of
% a linear-phase FIR filter to coefficient quantization.
% Input parameters:
% h: vector of coefficients
% K: number of frequency points
% theta: frequency interval (2-element vector).
% Output parameters:
% dHmag: the partial derivative matrix, M by K, where M is the
%        number of coefficients in the realization
% S: the sensitivity bound, 1 by K.

Hangle = exp(-j*angle(frqresp(h,1,K,theta)));
N = length(h) - 1; dH = [];
if (sign(h(1))==sign(h(N+1))), pm = 1; else, pm = -1; end
for k = 0:floor((N-1)/2),
   dH = [dH; frqresp( ...
   [zeros(1,k),1,zeros(1,N-1-2*k),pm,zeros(1,k)],1,K,theta)];
end
if (rem(N,2) == 0),
   dH = [dH; frqresp([zeros(1,N/2),1,zeros(1,N/2)],1,K,theta)];
end
sc = scale2(h);
[M,junk] = size(dH);
dHmag = real(dH.*(ones(M,1)*Hangle));
S = sc*sum(abs(dHmag));
```

Program 11.15 Frequency response of a filter subject to coefficient quantization.

```
function H = qfrqresp(typ,B,b,a,K,theta);
% Synopsis: H = qfrqresp(typ,B,b,a,K,theta).
% Computes the frequency response of a filter subject
% to coefficient quantization.
% Input parameters:
% typ: 'd' for direct, 'p' for parallel, 'c' for cascade
% b, a: numerator and denominator polynomials
% K: number of frequency points
% theta: frequency interval (2-element vector).
% Output parameters:
% H: the frequency response.

if (typ == 'd'),
   scn = (2^(B-1))/scale2(b); b = (1/scn)*round(scn*b);
   scd = (2^(B-1))/scale2(a); a = (1/scd)*round(scd*a);
   H = frqresp(b,a,K,theta);
elseif (typ == 'p'),
   [c,nsec,dsec] = tf2rpf(b,a);
   sc = (2^(B-1))/scale2(c); c = (1/sc)*round(sc*c);
   [M,junk] = size(nsec); H = c;
   for k = 1:M,
      nt = nsec(k,:); dt = dsec(k,:);
      if (dt(3) == 0), dt = dt(1:2); nt = nt(1); end
      scn = (2^(B-1))/scale2(nt); nt = (1/scn)*round(scn*nt);
      scd = (2^(B-1))/scale2(dt); dt = (1/scd)*round(scd*dt);
      H = H + frqresp(nt,dt,K,theta);
   end
elseif (typ == 'c'),
   c = b(1); v = roots(a); u = roots(b);
   [nsec,dsec] = pairpz(v,u);
   sc = (2^(B-1))/scale2(c); c = (1/sc)*round(sc*c);
   [M,junk] = size(nsec); H = c;
   for k = 1:M,
      nt = nsec(k,:); dt = dsec(k,:);
      if (dt(3) == 0), dt = dt(1:2); nt = nt(1:2); end
      scn = (2^(B-1))/scale2(nt); nt = (1/scn)*round(scn*nt);
      scd = (2^(B-1))/scale2(dt); dt = (1/scd)*round(scd*dt);
      H = H.*frqresp(nt,dt,K,theta);
   end
end
```

Program 11.16 The four norms of a rational filter.

```
function [h1,H1,H2,Hinf] = filnorm(b,a);
% Synopsis: [h1,H1,H2,Hinf] = filnorm(b,a).
% Computes the four norms of a rational filter.
% Input parameters:
% b, a: the numerator and denominator polynomials.
% Output parameters:
% h1: sum of absolute values of the impulse response
% H1: integral of absolute value of frequency response
% H2: integral of magnitude-square of frequency response
% Hinf: maximum magnitude response.

[h,Z] = filter(b,a,[1,zeros(1,99)]);
h1 = sum(abs(h)); n = 100;  h1p = 0;
while((h1-h1p)/h1 > 0.00001),
   [h,Z] = filter(b,a,zeros(1,n),Z);
   h1p = h1; h1 = h1 + sum(abs(h)); n = 2*n;
end

H2 = sqrt(nsgain(b,a));

N = 2 .^ ceil(log(max(length(a),length(b))-1)/log(2));
N = max(16*N,512)+1; temp = abs(frqresp(b,a,N));
Hinf = max(temp);
temp = [1,kron(ones(1,(N-1)/2-1),[4,2]),4,1].*temp;
H1 = sum(temp)/(3*(N-1));
```

Program 11.17 Zero-input limit cycle simulation for a second-order filter.

```
function [flag,y] = lc2sim(qtype,when,rtype,apar,B,s0,n);
% Synopsis: [flag,y] = lc2sim(qtype,when,rtype,apar,B,s0,n).
% Zero-input limit cycle simulation for a second-order filter.
% Input parameters:
% qtype: 't': truncate, 'r': round, 'm': magnitude truncate
% when:  'b': quantize before summation, 'a': after
% rtype: 'd': direct realization, 'c': coupled realization
% apar:  [a1, a2] for direct, [alphar, alphai] for coupled
% B:     number of bits, s0:    initial state
% n:     maximum number of time points to simulate.
% Output parameters:
% flag:  0: no LC, 1: DC LC, 2: other LC
% y:     the output signal.

s = [quant(s0(1),qtype,B), quant(s0(2),qtype,B)]; sp = s;
apar(2) = quant(apar(2),'r',B);
if (abs(apar(1)) >= 1), apar(1) = 2*quant(apar(1)/2,'r',B);
else, apar(1) = quant(apar(1),'r',B); end
y = zeros(1,n); flag = 2;
for i = 1:n,
if (rtype == 'd'),
temp1 = -apar(1)*s(1); temp2 = -apar(2)*s(2); s(2) = s(1);
if (when == 'b'),
   s(1) = quant(temp1,qtype,B) + quant(temp2,qtype,B);
else, s(1) = quant(temp1+temp2,qtype,B); end; y(i) = s(1);
else,
temp1 = apar(1)*s(1); temp2 = apar(2)*s(2);
temp3 = -apar(2)*s(1); temp4 = apar(1)*s(2);
if (when == 'b'),
   s(1) = quant(temp1,qtype,B) + quant(temp2,qtype,B);
   s(2) = quant(temp3,qtype,B) + quant(temp4,qtype,B);
else,
   s(1) = quant(temp1+temp2,qtype,B);
   s(2) = quant(temp3+temp4,qtype,B);
end; y(i) = s(1);
end
if (s(1) == 0 & s(2) == 0),
   flag = 0; y = y(1:i); break; end
if (s(1) == sp(1) & s(2) == sp(2)),
   flag = 1; y = y(1:i); break; end
sp = s; end
```

Program 11.18 Quantization by rounding, truncation, or magnitude truncation.

```
function aq = quant(a,qtype,B);
% Synopsis: aq = quant(a,qtype,B).
% Quantizes a number.
% Input parameters:
% a: the input number, assumed a fraction.
% qtype: 't': truncation, 'r': rounding,
%        'm': magnitude truncation
% B:     number of bits

fs = 2^(B-1);
aq = a*fs;
if (qtype == 't'),
   aq = floor(aq)/fs;
elseif (qtype == 'r'),
   aq = round(aq)/fs;
elseif (qtype == 'm'),
   aq = (sign(aq)*floor(abs(aq)))/fs;
else
   error('Unrecognized qtype in QUANT')
end
```

Program 11.19 A driver program for lc2sim.

```
disp('Make sure qtype, when, rtype, a1, a2, B, M are defined');
r = roots([1,a1,a2]);
if (max(abs(r)) >= 1),
   disp('Input filter is unstable'); return, end
if (imag(r(1)) == 0),
   disp('Poles are real'); return, end
if (rtype == 'c'), apar = [real(r(1)),imag(r(1))];
else, apar = [a1,a2]; end
n = ceil(-2*B*log(2)/log(abs(r(1))));
for i = 1:M,
   s0 = rand(1,2)-0.5*ones(1,2);
   flag = lc2sim(qtype,when,rtype,apar,B,s0,n);
   if (flag == 1),
      disp('DC limit cycle exists!'); return
   elseif (flag == 2),
      disp('Non-DC limit cycle exists!'); return
   end
end
disp('Apparently limit cycle free!');
```

11.11   Problems

11.1 Figure 11.26 shows a realization of a digital filter.

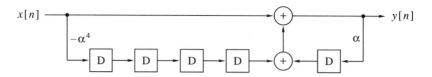

Figure 11.26 Pertaining to Problem 11.1.

(a) Find the impulse response of the filter.

(b) Draw a realization of the filter having the minimum possible number of delays.

(c) For what values of α will the filter have a linear phase (exact or generalized)?

(d) Compute the response of the filter to the input

$$x[n] = \begin{cases} 0, & n < 0, \\ 1, & n \geq 0. \end{cases}$$

11.2 Construct a realization of an Nth-order IIR filter using $2N$ delay elements, N for the input signal $x[n]$, and N for the output signal $y[n]$.

11.3 It is required to build a digital filter having the impulse response

$$h[n] = \begin{cases} \alpha^m, & n = mM, \ m \geq 0, \\ 0, & \text{otherwise}, \end{cases}$$

where M is a fixed positive integer, $M > 1$, and $|\alpha| < 1$. Construct a realization of the filter having a minimum number of delays. How many operations per time point are needed in this realization for a general input signal $x[n]$?

11.4 The filter

$$H^z(z) = 1 + z^{-1} + \cdots + z^{-(N-1)}$$

is called a *moving-average* filter, or a *boxcar* filter. It is an FIR filter whose coefficients are all 1. Direct realization of this filter requires $N-1$ additions and no multiplications per time point. Derive a realization of this filter that requires only two additions and no multiplications per time point, but N delay elements. Draw a block diagram of this realization. Hint: The realization looks like IIR, but it is FIR.

11.5 Consider the digital system shown in Figure 11.27. Assume that the input of the top delay is fed with the constant positive number A at time $n = 0$ (which also causes $y[0]$ to be equal to A), and that of the bottom delay is fed with zero. At later times the system receives no input.

(a) Derive a closed-form mathematical expression for the output signal $y[n]$. Hint: If you work in the z domain, consult Table 7.1.

(b) Suggest a possible application of this system and specify the main advantage of generating $y[n]$ this way.

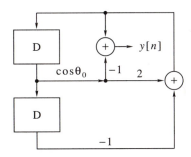

Figure 11.27 Pertaining to Problem 11.5.

11.6 It is required to design a tunable band-pass filter whose pass band is from $\theta_0 - \Delta\theta$ to $\theta_0 + \Delta\theta$, where $\Delta\theta$ is fixed and θ_0 is adjustable externally. The filter is to be type-I FIR, designed by the windowing method.

Show how to implement the filter using $N + 1$ multipliers, N adders, and $0.5N$ trigonometric function generators. The inputs to the filter are the input signal $x[n]$ and the frequency θ_0. Hint: Use the trigonometric identity

$$\sin\alpha - \sin\beta = 2\sin[0.5(\alpha - \beta)]\cos[0.5(\alpha + \beta)].$$

11.7 Let $\{A, B, C, D\}$ be the matrices of a state-space realization, and $H^z(z)$ the corresponding transfer function. Prove that the transposed realization, whose matrices are $\{A', C', B', D\}$, has the same transfer function.

11.8 Let $\{A, B, C, D\}$ be the matrices of a state-space realization, and $\{\tilde{A}, \tilde{B}, \tilde{C}, \tilde{D}\}$ those of a similar realization [as defined in (11.33) and (11.34)]. Use either (11.41) or (11.48) and show that both realizations have the same transfer function.

11.9 This problem discusses state-space representation of a parallel realization.

(a) Let the two LTI systems $H_1^z(z)$, $H_2^z(z)$ be connected in parallel. Let $\{A_i, B_i, C_i, D_i,\ i = 1, 2\}$ be state-space representations of the two systems. Construct a state-space representation for the parallel connection.

(b) Suppose we wish to construct a state-state representation (11.74), (11.76) for the parallel realization of an IIR filter. We take $\{A_l, B_l, C_l, D_l, 1 \le l \le N_1 + N_2\}$ as the state-space matrices of the individual first- or second-order sections. What are the matrices F, G, H, K needed in the construction discussed in Section 11.3?

11.10 This problem discusses state-space representation of a cascade realization.

(a) Repeat Problem 11.9(a) for two systems $H_1^z(z)$, $H_2^z(z)$ connected in cascade.

(b) Suppose we wish to construct a state-state representation for the cascade realization of an IIR filter. We take $\{A_l, B_l, C_l, D_l, 1 \le l \le \lceil 0.5N \rceil\}$ as the state-space matrices of the individual second-order sections. What are the matrices F, G, H, K needed in the construction discussed in Section 11.3?

11.11 Write the state-space matrices of two different state-space representations of an FIR filter

$$H^z(z) = h[0] + h[1]z^{-1} + \cdots + h[N]z^{-N}.$$

11.12 We are given the state-space matrices

$$A = \begin{bmatrix} 3 & -0.25 \\ 5 & 0 \end{bmatrix}, \quad B = \begin{bmatrix} 1 \\ 1 \end{bmatrix}, \quad C = \begin{bmatrix} 2 & -1 \end{bmatrix}, \quad D = 0.$$

(a) Draw a block diagram of the state-space representation corresponding to these matrices.

(b) Write down the difference equation of the system.

(c) Compute the impulse response of the system.

(d) Suggest a realization having the same transfer function as the that of the given system, but simpler than the one you drew in part a.

11.13 The purpose of this problem is to derive a state-space version of the bilinear transform. The derivation is based on the trapezoidal integration interpretation of the bilinear transform, introduced in Problem 10.32.

(a) Suppose we have an analog filter expressed in the state-space form

$$\frac{ds(t)}{dt} = As(t) + Bx(t),$$

$$y(t) = Cs(t) + Dx(t).$$

Use the trapezoidal integration derived in Problem 10.32 for the approximation

$$s(nT + T) = s(nT) + 0.5T[As(nT + T) + Bx(nT + T) + As(nT) + Bx(nT)],$$

$$y(nT) = Cs(nT) + Dx(nT).$$

(b) Define

$$\tilde{A} = (I_N - 0.5TA)^{-1}(I_N + 0.5TA), \quad \tilde{B} = 0.5T(I_N - 0.5TA)^{-1}B.$$

Also define the discrete-time, state-space vector as

$$\tilde{s}(nT) = s(nT) - \tilde{B}x(nT).$$

Show that $x(nT)$, $y(nT)$, $\tilde{s}(nT)$ have the discrete-time, state-space representation

$$\tilde{s}(nT + T) = \tilde{A}\tilde{s}(nT) + (I_N + \tilde{A})\tilde{B}x(nT),$$

$$y(nT) = C\tilde{s}(nT) + (D + C\tilde{B})x(nT).$$

(c) Write a MATLAB program `bilinss` that implements the bilinear transformation in state-space form. The inputs to the program are the numerator and denominator polynomials of the analog filter. The program should convert those polynomials to continuous-time, state-space form, using `tf2ss`; compute the discrete-time, state-space matrices, as found in part b; convert to a z-domain transfer function, using `ss2tf`.

11.14 In this problem we introduce *lattice* realizations of minimum-phase FIR filters, by way of a second-order example.

Let

$$a(z) = 1 + a_1 z^{-1} + a_2 z^{-2},$$

and define

$$\rho_1 = -\frac{a_1}{1 + a_2}, \quad \rho_2 = -a_2.$$

(a) Show that the roots of $a(z)$ are inside the unit circle if and only if $|\rho_1| < 1$ and $|\rho_2| < 1$. Hint: Recall Example 7.2.

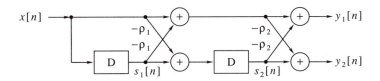

Figure 11.28 A second-order FIR lattice.

(b) Write down the state equations of the realization shown in Figure 11.28. Take the state vector as shown in the figure.

(c) Use the state equations you have found in part b and show that the transfer function from $x[n]$ to $y_1[n]$ is

$$\frac{Y_1^z(z)}{X^z(z)} = a(z).$$

(d) Find the transfer function from $x[n]$ to $y_2[n]$, and relate it to $a(z)$.

The structure in Figure 11.28 is called an *FIR lattice*. In Section 13.4.4 we will extend this structure to a minimum-phase FIR filter of any order.

11.15 In this problem we introduce a lattice realization of an all-pole IIR filter (i.e., an IIR filter that has no zeros), again by way of a second-order example. Let $a(z), \rho_1, \rho_2$ be as in Problem 11.14.

(a) Write down the state equations of the realization shown in Figure 11.29. Take the state vector as shown in the figure. Certain elements of the matrix A require special care; account for all interconnections. Obtaining the correct result in part b will serve as a verification of the correctness of your solution to this part.

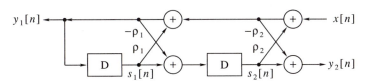

Figure 11.29 A second-order all-pole IIR lattice.

(b) Use the state equations you have found in part b and show that the transfer function from $x[n]$ to $y_1[n]$ is

$$\frac{Y_1^z(z)}{X^z(z)} = \frac{1}{a(z)}.$$

(c) Find the transfer function from $x[n]$ to $y_2[n]$. What special property does this transfer function have? Hint: Recall Problem 7.29.

The structure in Figure 11.29 is called an *all-pole IIR lattice*. In Section 13.4.4 we will extend this structure to an all-pole IIR filter of any order.

11.16 This problem explores a realization of a second-order section of an IIR filter based on a lattice structure. It can serve as an alternative to the coupled realization introduced in Section 11.1.7. The realization is shown in Figure 11.30. It is obtained from the all-pole IIR lattice of Figure 11.29 by adding a chain of multipliers/adders to form a new output $y[n]$.

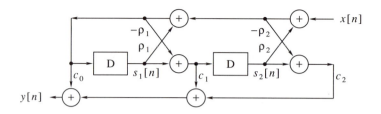

Figure 11.30 A second-order pole–zero IIR lattice.

(a) Find the transfer function from $x[n]$ to $y[n]$ in Figure 11.30.

(b) Suppose we wish to realize the second-order transfer function

$$H^z(z) = \frac{b_0 + b_1 z^{-1} + b_2 z^{-2}}{1 + a_1 z^{-1} + a_2 z^{-2}}.$$

We already know from Problem 11.15 how to choose ρ_1, ρ_2 to get the required denominator. Compute the coefficients c_0, c_1, c_2 to get the required numerator.

The structure in Figure 11.30 is called a *pole–zero IIR lattice*. It can be extended to filters $b(z)/a(z)$ of any order, see Oppenheim and Schafer [1989] for details. Here, however, we are interested in it only as a candidate for a second-order section in a cascade realization.

11.17* The purpose of this problem is to derive the sensitivity of a root of a polynomial to small perturbations in the coefficients of the polynomial. Let

$$a(z) = \sum_{k=0}^{p} a_k z^{-k} = \prod_{i=1}^{p}(1 - \alpha_i z^{-1}).$$

Suppose that all roots α_k are distinct.

(a) Find $\partial a(z)/\partial a_k$ from the sum representation of $a(z)$, and $\partial a(z)/\partial \alpha_l$ from the product representation.

(b) Use the result of part a and show that

$$\frac{\partial a(z)}{\partial \alpha_l}\bigg|_{z=\alpha_m} = \begin{cases} -\alpha_m^{-1} \prod_{i \neq m}(1 - \alpha_i \alpha_m^{-1}), & l = m, \\ 0, & l \neq m. \end{cases}$$

(c) Substitute $z = \alpha_m$ in the chain rule

$$\frac{\partial a(z)}{\partial a_k} = \sum_{l=1}^{p} \frac{\partial a(z)}{\partial \alpha_l}\frac{\partial \alpha_l}{\partial a_k}$$

and use part b for deriving an expression for $\partial \alpha_m/\partial a_k$.

(d) Assuming that the differences $\hat{a}_k - a_k$ are small, show that the deviation in the pole α_m is given by the approximate formula (11.77).

11.18* Let $H^z(z) = b(z)/a(z)$.

(a) If p_k is a coefficient appearing only in $a(z)$, show that

$$\frac{\partial |H^f(\theta)|}{\partial p_k} = -|H^f(\theta)|\Re\left\{ \frac{1}{a(e^{j\theta})} \cdot \frac{\partial a(e^{j\theta})}{\partial p_k} \right\}.$$

(b) If p_k is a coefficient appearing only in $b(z)$, show that

$$\frac{\partial |H^f(\theta)|}{\partial p_k} = |H^f(\theta)|\Re\left\{ \frac{1}{b(e^{j\theta})} \cdot \frac{\partial b(e^{j\theta})}{\partial p_k} \right\}.$$

11.19* Recall that the coupled realization of a second-order section, as presented in Section 11.1.7, as a denominator polynomial

$$a(z) = 1 - 2\alpha_r z^{-1} + (\alpha_r^2 + \alpha_i^2)z^{-2},$$

where α_r, α_i are the real and imaginary parts of the complex pole. The purpose of this problem is to compare the sensitivity of $|H^f(\theta)|$ to the parameters α_r, α_i with the sensitivity to a_1, a_2, in a direct realization of a second-order section.

(a) Compute $\partial a(e^{j\theta})/\partial\alpha_r$ and $\partial a(e^{j\theta})/\partial\alpha_i$.

(b) By Problem 11.18, it is sufficient to compare the two realizations by examining the corresponding values of $\Re\{[1/a(e^{j\theta})][\partial a(e^{j\theta})/\partial p_k]\}$. Derive these expressions for $p_k = a_1$, $p_k = a_2$ in a direct realization, and for $p_k = \alpha_r$, $p_k = \alpha_i$ in the coupled realization.

(c) Compute the four expressions in part b for $\theta = 0$. Assuming that the poles of the second-order section are in the vicinity of $z = 1$, draw your conclusions about the relative sensitivity of the two realizations to coefficient quantization at low frequencies.

11.20* Explain why coefficient quantization in a linear-phase FIR filter preserves the linear-phase property.

11.21* Discuss potential finite word length effects in the realization you have obtained in Problem 11.4.

11.22* Write a MATLAB procedure `qtf` that converts a transfer function to either parallel or cascade realization, then scales and quantizes the coefficients to a desired number of bits. The calling syntax of the function should be

$$[c,nsec,dsec,sc,sn,sd] = qtf(b,a,typ,B);$$

The input parameters are as follows:

- `b, a`: the numerator and denominator polynomial coefficients,
- `typ`: 'c' for cascade, 'p' for parallel,
- `B`: number of bits.

The output parameters are as follows:

- `c`: the constant coefficient,
- `nsec`: matrix of numerators of second-order sections,
- `dsec`: matrix of denominators of second-order sections,
- `sc`: scale factor for c,
- `sn`: scale factors for numerators,
- `sd`: scale factors for denominators.

The procedure should use the procedures `tf2rpf`, `scale2`, and `pairpz`, described in this chapter.

11.23* Use MATLAB for computing the possible pole locations of a second-order pole-zero lattice filter, assuming that the parameters ρ_1, ρ_2 are quantized to $B = 5$ bits. Note that these parameters can assume values only in the range $(-1, 1)$. Draw a diagram in the style of Figures 11.14 and 11.15, and interpret the result.

11.24* Consider the digital system described in Problem 11.5 and shown in Figure 11.27. Assume the same initial conditions and input as in Problem 11.5.

(a) Discuss possible problems resulting from (i) quantization of $\cos\theta_0$ to a finite word length and (ii) quantization of the multiplier's output.

(b) Simulate the system in MATLAB, using 10-bit fixed-point arithmetic with truncation. Use the function quant for this purpose. Take $A = 0.875$. For θ_0 examine two cases: one such that $\cos\theta_0 = 0.9375$, and one such that $\cos\theta_0 = 0.9375 + 2^{-10}$.

(c) Let the simulated system run for $0 \le n \le 5000$ and store the output signal $y[n]$. Compute the theoretical waveform $y[n]$ as found in part a. Plot the error between the theoretical waveform and the one obtained from the simulation. Repeat for the two values of θ_0 specified in part b. Report your results and conclusions.

11.25* Derive (11.127) from (11.126). Hint: In general,

$$\text{round}\{x\} = m \quad \text{if and only if} \quad m - 0.5 \le x \le m + 0.5.$$

11.26* Modify the scheme suggested in item 4 of Section 11.9.2 to the case of a second-order filter whose complex poles are near $z = -1$.

Chapter 12

Multirate Signal Processing

In our study of discrete-time signals and systems, we have assumed that all signals in a given system have the same sampling rate. We have interpreted the time index n as an indicator of the physical time nT, where T is the sampling interval. A *multirate system* is characterized by the property that signals at different sampling rates are present. An example that you may be familiar with, although perhaps not aware of its meaning, is the audio compact-disc player. Today's CD players often carry the label "8X oversampling" (or a different number). Thus, the digital signal read from the CD, whose sampling rate is 44.1 kHz, is converted to a signal whose sampling rate is 8 times higher, that is, 352.8 kHz. We shall get back to this example in due course and explain the reason for this conversion.

Multirate systems have gained popularity since the early 1980s, and their uses have increased steadily since. Such systems are used for audio and video processing, communication systems, general digital filtering, transform analysis, and more. In certain applications, multirate systems are used out of necessity; in others, out of convenience. One compelling reason for considering multirate implementation for a given digital signal processing task is computational efficiency. A second reason is improved performance.

The two basic operations in a multirate system are decreasing and increasing the sampling rate of a signal. The former is called decimation, or down-sampling. The latter is called expansion, or up-sampling. A more general operation is sampling-rate conversion, which involves both decimation and expansion. These are the first topics discussed in this chapter: first in the time domain and then in transform domains.

Proper sampling-rate conversion always requires filtering. Linear filters used for sampling-rate conversion can be implemented efficiently. It turns out that sampling-rate conversion is sometimes advantageous in digital filtering even when the input and the output of the filter are needed at the same rate. This happens when the filter in question has a small bandwidth compared with the input sampling rate. Linear filtering in multirate systems is the next topic discussed in this chapter.

The second major topic of the chapter is *filter banks*. A filter bank is an aggregate of filters designed to work together and perform a common task. A typical filter bank has either a single input and many outputs, or many inputs and a single output. In the former case it is called an analysis bank; in the latter, a synthesis bank. Analysis banks are used for applications such as splitting a signal into several frequency bands. Synthesis banks are most commonly used for combining signals previously split by an analysis bank. The simplest form of a filter bank is the two-channel bank, which splits a signal into two, or combines two signals into one. We present several types of two-channel filter bank and discuss their properties and applications. Finally, we extend our discussion to more general filter banks. The subject of filter banks is rich, and our treatment of it serves only as a brief introduction.

12.1 Decimation and Expansion

Decimation can be regarded as the discrete-time counterpart of sampling. Whereas in sampling we start with a continuous-time signal $x(t)$ and convert it to a sequence of samples $x[n]$, in decimation we start with a discrete-time signal $x[n]$ and convert it to another discrete-time signal $y[n]$, which consists of *subsamples* of $x[n]$.   Thus, the formal definition of M-fold *decimation*, or *down-sampling*, is

$$y[n] = x[nM], \quad n \in \mathbb{Z}. \tag{12.1}$$

Figure 12.1 illustrates 3-fold decimation. Note that the samples corresponding to $n = \ldots, -2, 1, 4, \ldots$ and $n = \ldots, -1, 2, 5, \ldots$ are lost in the decimation. In general, the samples of $x[n]$ corresponding to $n \neq kM$ are lost in M-fold decimation. This is similar to point sampling of a continuous-time signal $x(t)$, in which values corresponding to $t \neq nT$ are lost.

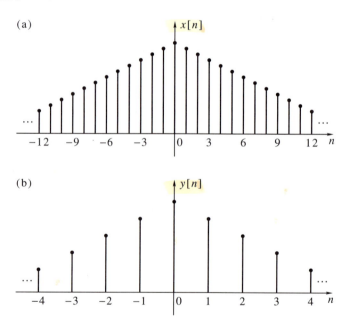

Figure 12.1 Decimation of a discrete-time signal by a factor of 3: (a) the original signal; (b) the decimated signal.

Figure 12.1 shows the samples of the decimated signal $y[n]$ spaced three times wider than the samples of $x[n]$. This is not a coincidence. In real time, the decimated signal indeed appears at a rate slower than that of the original signal by a factor M. If the sampling interval of $x[n]$ is T, then that of $y[n]$ is MT.

Expansion is another operation on a discrete-time signal that yields a discrete-time signal. It is related to reconstruction of a discrete-time signal, an operation that yields a continuous-time signal. However, the relation is less obvious than in the case of sampling and decimation. The mathematical definition of L-fold *expansion*, or *up-sampling*, is

$$y[n] = \begin{cases} x[n/L], & \text{if } n/L \text{ is an integer,} \\ 0, & \text{if } n/L \text{ is noninteger.} \end{cases} \qquad (12.2)$$

Figure 12.2 illustrates 3-fold expansion. Expansion is an information-preserving operation: All samples of $x[n]$ are present in the expanded signal $y[n]$. The samples of the expanded signal appear at a rate faster than that of the original signal by a factor L. If the sampling interval of $x[n]$ is T, then that of $y[n]$ is T/L.

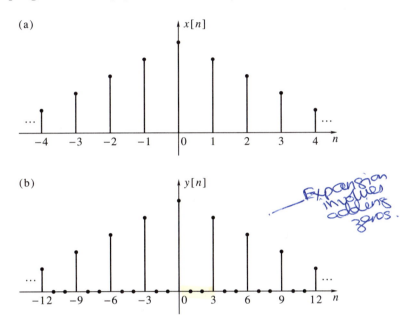

Figure 12.2 Expansion of a discrete-time signal by a factor of 3: (a) the original signal; (b) the expanded signal.

Expansion is also called *interpolation* in the digital signal processing literature. However, this term does not well describe the operation depicted in Figure 12.2, since the inserted values are identically zero and are not related to the signal values $x[n]$, which they purport to interpolate. This is in contrast with reconstruction by a zero-order hold, in which the reconstructed signal between sampling points has the value of the most recent sample. The term *expansion* better describes this operation; we reserve the term *interpolation* for another operation, to be described later.

We shall use the following notations for decimation and expansion:

$$\text{Decimation:} \qquad y[n] = x_{(\downarrow M)}[n],$$
$$\text{Expansion:} \qquad y[n] = x_{(\uparrow L)}[n].$$

Figure 12.3 shows common block-diagram descriptions of an M-fold decimator and an L-fold expander.

A MATLAB expression for decimation is

```
y = x(1:M:length(x));
```

There are a number of ways to implement expansion in MATLAB; for example,

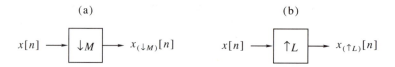

(a)   (b)

$x[n] \longrightarrow \boxed{\downarrow M} \longrightarrow x_{(\downarrow M)}[n]$ $x[n] \longrightarrow \boxed{\uparrow L} \longrightarrow x_{(\uparrow L)}[n]$

Figure 12.3 Block-diagram notations for (a) M-fold decimation and (b) L-fold expansion.

```
y = reshape([x; zeros(L-1,length(x))],1,L*length(x));
```

Decimation and expansion are both linear operations (Problem 12.1). However, neither is time invariant. To show this, consider the following counterexample. Let $x[n]$ be the sequence

$$x[n] = \begin{cases} 1, & n \text{ even,} \\ 0, & n \text{ odd,} \end{cases}$$

and let $v[n] = x[n-1]$, that is, the right shift of $x[n]$ by 1. Then the 2-fold decimation of $x[n]$ is the sequence $x_{(\downarrow 2)}[n] = 1$, whereas the 2-fold decimation of $v[n]$ is the sequence $v_{(\downarrow 2)}[n] = 0$. The latter is obviously not a shift by 1 of the former. Also, the 2-fold expansion of $x[n]$ is

$$x_{(\uparrow 2)}[n] = \begin{cases} 1, & n \text{ is divisible by 4,} \\ 0, & n \text{ otherwise,} \end{cases}$$

whereas the 2-fold expansion of $v[n]$ is

$$v_{(\uparrow 2)}[n] = \begin{cases} 1, & n - 2 \text{ is divisible by 4,} \\ 0, & n \text{ otherwise.} \end{cases}$$

Here the latter sequence is the right shift by 2 of the former, instead of right shift by 1.

Decimation and expansion do not commute in general. For example, if we expand $x[n]$ by a factor L and then decimate the result by the same factor, we get $x[n]$ back. However, if we decimate first and then expand, we get the sequence

$$y[n] = \begin{cases} x[n], & \text{if } n/L \text{ is an integer,} \\ 0, & \text{if } n/L \text{ is noninteger.} \end{cases}$$

In one special case, as the next theorem shows, expansion and decimation do commute.

Theorem 12.1 If M and L are relatively prime, then decimation by M and expansion by L are commutative operations, that is,

$$\{x_{(\downarrow M)}\}_{(\uparrow L)}[n] = \{x_{(\uparrow L)}\}_{(\downarrow M)}[n]. \tag{12.3}$$

Proof We have, by the definitions of decimation and expansion,

$$\{x_{(\downarrow M)}\}_{(\uparrow L)}[n] = \begin{cases} x[nM/L], & nM \text{ is divisible by } L, \\ 0, & \text{otherwise,} \end{cases} \tag{12.4}$$

$$\{x_{(\uparrow L)}\}_{(\downarrow M)}[n] = \begin{cases} x[nM/L], & n \text{ is divisible by } L, \\ 0, & \text{otherwise.} \end{cases} \tag{12.5}$$

When M and L are coprime, nM is divisible by L if and only if n itself is divisible by L, hence equality of the two expressions follows. $\qquad \square$

Example 12.1 Let $M = 3$ and $L = 2$. Then both sequences in (12.3) are

$$x[0], 0, x[3], 0, x[6], 0, \ldots .$$

On the other hand, let $M = 6$ and $L = 4$. Then,

$$\{x_{(\downarrow 6)}\}_{(\uparrow 4)}[n] = x[0], 0, 0, 0, x[6], 0, 0, 0, x[12], 0, \ldots,$$

but

$$\{x_{(\uparrow 4)}\}_{(\downarrow 6)}[n] = x[0], 0, x[3], 0, x[6], 0, \ldots .$$

12.2 Transforms of Decimated and Expanded Sequences

The analysis and understanding of decimation and expansion are greatly facilitated by their transforms. We shall use the following notations for the z-transforms and Fourier transforms of a decimated sequence:

$$\{\mathcal{Z}x_{(\downarrow M)}\}(z) = X^z_{(\downarrow M)}(z) = \{X^z(z)\}_{(\downarrow M)}, \tag{12.6a}$$

$$\{\mathcal{F}x_{(\downarrow M)}\}(\theta) = X^f_{(\downarrow M)}(\theta) = \{X^f(\theta)\}_{(\downarrow M)}. \tag{12.6b}$$

For the transforms of expanded sequences we shall use

$$\{\mathcal{Z}x_{(\uparrow L)}\}(z) = X^z_{(\uparrow L)}(z) = \{X^z(z)\}_{(\uparrow L)}, \tag{12.7a}$$

$$\{\mathcal{F}x_{(\uparrow L)}\}(\theta) = X^f_{(\uparrow L)}(\theta) = \{X^f(\theta)\}_{(\uparrow L)}. \tag{12.7b}$$

In each case, the third notation is inaccurate, but sometimes it is more convenient than the other two, especially when we want to apply decimation or expansion to a product of transforms.

As we have seen, decimation and expansion are linear, but not time-invariant operations. Therefore, we do not expect them to be described by transfer functions. In other words, we do not expect to get expressions such as

$$X^z_{(\downarrow M)}(z) = H^z(z)X(z), \quad \text{or} \quad X^z_{(\uparrow L)}(z) = H^z(z)X^z(z).$$

However, these two transforms can be expressed in terms of $X(z)$, as we now show.

Consider decimation first. Let $\{c_M[n]\}$ be the *comb* signal

$$c_M[n] = \sum_{k=-\infty}^{\infty} \delta[n - kM] \tag{12.8}$$

(see Figure 12.4). The comb signal is the digital counterpart of the impulse train (cf. Figure 2.4). By (4.6), we can express $c_M[n]$ as

$$c_M[n] = \frac{1}{M} \sum_{m=0}^{M-1} W_M^{mn}, \quad \text{where} \quad W_M = e^{j2\pi/M}. \tag{12.9}$$

Note the similarity of this equality to Poisson's formula (2.48).

Using the comb signal, we can express the decimated signal as

$$x_{(\downarrow M)}[n] = x[nM]c_M[nM], \tag{12.10}$$

and its z-transform as

$$X^z_{(\downarrow M)}(z) = \sum_{n=-\infty}^{\infty} x[nM]c_M[nM]z^{-n} = \sum_{n=-\infty}^{\infty} x[n]c_M[n]z^{-n/M}. \tag{12.11}$$

The last equality follows because $c_M[n]$ is identically zero for n that is not an integer multiple of M, so the sum on the right of (12.11) includes only indices that are integer

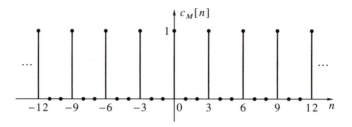

Figure 12.4 The comb signal $c_M[n]$ (shown for $M = 3$).

multiples of M. Now, substitution of (12.9) in (12.11) gives

$$X^z_{(\downarrow M)}(z) = \frac{1}{M} \sum_{n=-\infty}^{\infty} \sum_{m=0}^{M-1} x[n] W_M^{mn} z^{-n/M} = \frac{1}{M} \sum_{m=0}^{M-1} \left[\sum_{n=-\infty}^{\infty} x[n](z^{1/M} W_M^{-m})^{-n} \right]$$

$$= \frac{1}{M} \sum_{m=0}^{M-1} X^z(z^{1/M} W_M^{-m}). \tag{12.12}$$

Example 12.2 Let $x[n] = \alpha^n$, $n \geq 0$. Then $x_{(\downarrow M)}[n] = \alpha^{Mn}$, $n \geq 0$. We have

$$X^z(z) = \frac{z}{z - \alpha}, \quad X^z_{(\downarrow M)}(z) = \frac{z}{z - \alpha^M}.$$

Therefore, we get from (12.12) the nontrivial identity

$$\frac{z}{z - \alpha^M} = \frac{1}{M} \sum_{m=0}^{M-1} \frac{z^{1/M} W_M^{-m}}{z^{1/M} W_M^{-m} - \alpha}. \tag{12.13}$$

□

The meaning of formula (12.12) is perhaps not obvious, but it will become clear when we pass from the z-transform to the Fourier transform of $x_{(\downarrow M)}[n]$. We do this by substituting $z = e^{j\theta}$ in (12.12), obtaining

$$X^f_{(\downarrow M)}(\theta) = \frac{1}{M} \sum_{m=0}^{M-1} X^f \left(\frac{\theta - 2\pi m}{M} \right). \tag{12.14}$$

This result bears an obvious resemblance to the sampling theorem [see (3.9)]. We can therefore think of (12.14) as the discrete-time sampling theorem. The derivation of (12.12), from which we obtained (12.14), is similar to the proof of Theorem 3.1.

The implications of (12.14) on aliasing caused by decimation are similar to those in the case of sampling of a continuous-time signal. In general, if $X^f(\theta)$ occupies the entire bandwidth from $-\pi$ to π, then $X^f_{(\downarrow M)}(\theta)$ will be aliased due to the superposition of the M shifted and frequency-scaled transforms in (12.14). This is depicted in Figure 12.5, which illustrates the aliasing phenomenon for $M = 3$. Part a of the figure shows the Fourier transform of the original signal. Part b shows the M shifted and frequency-scaled replicas (three in this case). Part c shows the superposition of the three replicas, which yields the transform of the decimated signal. As we see in part c, the Fourier transform of the decimated signal has a shape different from that of the original signal.

If the signal $x[n]$ is band limited to $\theta \in [-\pi/M, \pi/M]$, the decimated signal $x_{(\downarrow M)}[n]$ is alias free, as depicted in Figure 12.6. As we see, the Fourier transform of the decimated signal occupies the entire bandwidth, because of the M-fold expansion of the frequency scale, but is not aliased. We can regard the condition that the Fourier

We need to bandlimit to $[-\pi/m, \pi/m]$ as in decimation, The bandwidth of spectra expands by M.

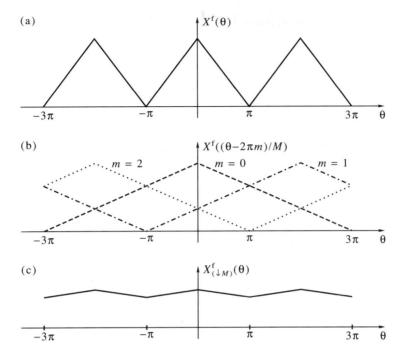

Figure 12.5 Aliasing caused by decimation: (a) Fourier transform of the original signal; (b) shifted and frequency-scaled replicas of part a; (c) Fourier transform of the decimated signal.

transform is nonzero only for $\theta \in [-\pi/M, \pi/M]$ as Nyquist's no-aliasing condition for decimation of discrete-time signals.

The preceding is not the only case in which the decimated signal is not aliased. Another important case is that of a complex band-pass signal whose Fourier transform is nonzero only in the range $\theta \in [(2k - 1)\pi/M, (2k + 1)\pi/M]$ for some integer k. Figure 12.7 illustrates such a case, for $M = 3$ and $k = 1$. The Fourier transform has an asymmetrical shape in this example, to emphasize that it belongs to a complex signal, rather than a real signal. As we see, the Fourier transform of the decimated signal occupies the entire bandwidth, and it is not apparent from looking at it that the original signal was band pass. Nonetheless, there is no aliasing, so no information about the original signal was lost in the decimation process.

Computation of the transforms of expanded signals is much easier than that of decimated signals. The z-transform of an expanded signal is given by

$$X_{(\uparrow L)}^z(z) = \sum_{n=-\infty}^{\infty} x_{(\uparrow L)}[nL]z^{-nL} = \sum_{n=-\infty}^{\infty} x[n]z^{-nL} = X^z(z^L). \tag{12.15}$$

and the Fourier transform by

$$X_{(\uparrow L)}^f(\theta) = X^f(L\theta). \tag{12.16}$$

Figure 12.8 illustrates the expansion operation in the frequency domain. Part a shows the Fourier transform of the original signal and part b shows the transform of the expanded signal. As we see, the shape of the Fourier transform is contracted by a factor L in the frequency axis and is repeated L times in the range $[-\pi, \pi]$. Other than the contraction, the shape of the transform is preserved, confirming our conclusion that expansion does not lead to aliasing.

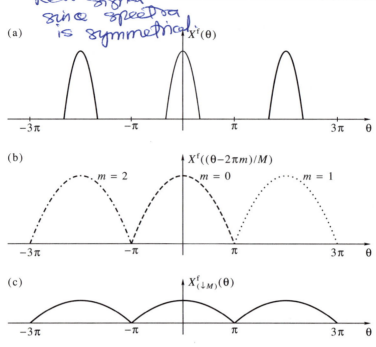

Real signal since spectra is symmetrical

Figure 12.6 Alias-free decimation of a band-limited signal: (a) Fourier transform of the original signal; (b) shifted and frequency-scaled replicas of part a; (c) Fourier transform of the decimated signal.

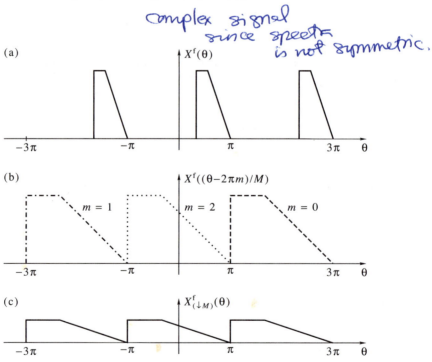

Complex signal since spectra is not symmetric.

Figure 12.7 Alias-free decimation of a complex band-pass signal: (a) Fourier transform of the original signal; (b) shifted and frequency-scaled replicas of part a; (c) Fourier transform of the decimated signal.

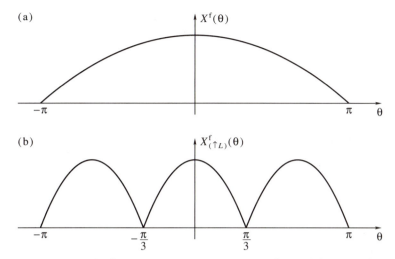

Figure 12.8 Expansion in the frequency domain: Fourier transform of the original signal (a) and the expanded signal (b).

We have seen how the spectra of decimated and expanded signals are related to those of the corresponding original signals. When showing the spectra as a function of the variable θ, they appear stretched by a factor M in the case of decimation, and compressed by a factor L in the case of expansion. However, we recall that the variable θ is related to the physical frequency ω by the formula $\omega = \theta/T$, where T is the sampling interval. We also recall that the sampling interval of a decimated signal is M times larger than that of the original signal. Similarly, the sampling interval of an expanded signal is L times smaller than that of the original signal. The conclusion is that the frequency range of the spectra, expressed in terms of ω, is not affected by either decimation or expansion. For example, suppose that the signal $x[n]$ was obtained from a continuous-time signal having bandwidth of $\pm 100\,\text{Hz}$ by sampling at $T = 0.001$ second. Then $x[n]$ occupies the range $\pm 0.2\pi$ in the θ domain. Now suppose that $y[n]$ is obtained from $x[n]$ by 5-fold decimation. Then the spectrum of $y[n]$ is alias free and it occupies the range $\pm\pi$ in the θ domain. The sampling rate of $y[n]$ is 0.005 second, so its physical bandwidth is $\pm 100\,\text{Hz}$, same as that of the original continuous-time signal. If we now expand $y[n]$ by a factor of 5 to get the signal $z[n]$, the spectrum of $z[n]$ will exhibit five replicas of the basic shape, and one of those will occupy the range $\pm 0.2\pi$ in the θ domain. Since the sampling interval of $z[n]$ is 0.001 second, the corresponding physical frequency is again $\pm 100\,\text{Hz}$.

12.3 Linear Filtering with Decimation and Expansion

12.3.1 Decimation

Since decimation, like sampling, leads to potential aliasing, it is desirable to precede the decimator with an antialiasing filter. Unlike sampling, here the input signal is already in discrete time, so we use a digital antialiasing filter. The antialiasing filter, also called the *decimation filter*, should approximate an ideal low-pass filter with cutoff frequency π/M. This is illustrated in Figure 12.9, for $M = 4$. In this figure, the input signal $x[n]$ has bandwidth slightly over $\pm\pi/4$. The decimation filter $H^{\text{f}}(\theta)$ eliminates the spectral components outside the frequency range $[-\pi/4, \pi/4]$, resulting in the signal $v[n]$. This signal is decimated by 4, yielding the output signal $y[n]$, whose spectrum occupies the entire bandwidth. Note that $Y^f(\theta)$ is faithful to the part of $X^f(\theta)$ in the range $[-\pi/4, \pi/4]$, but the information outside this range is lost.

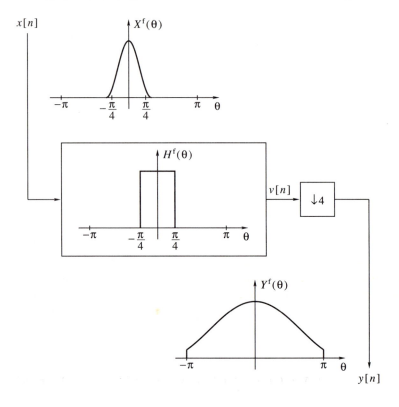

Figure 12.9 Passing a signal through an antialiasing filter prior to decimation.

Linear filtering followed by M-fold decimation is described mathematically by

$$y[n] = \sum_i h[i]x[nM - i] = \sum_i x[i]h[nM - i]. \tag{12.17}$$

Since we need only one out of every M outputs of the filter, we have a potential for M-fold computational saving. When the filter $h[n]$ is IIR, it is difficult to take advantage of this potential. However, when $h[n]$ is FIR, computational saving can be realized easily. For an Nth-order FIR filter, the necessary number of multiply–add operations is about N/M per input data point. We defer a discussion on programming of decimation with FIR filtering until the next section.

The following point concerning operation count in FIR decimation filters is worth noting. It is often reasonable to design the filter such that its transition bandwidth $\Delta\theta$ is a fixed fraction of the cutoff frequency, say $\Delta\theta = \mu\pi/M$, where μ depends on the specific application. As we know from FIR filter design theory, the order N of the filter is approximately inversely proportional to the transition bandwidth, the proportionality factor depending on the required stop-band attenuation and pass-band ripple [see (9.56c) or (9.79)]. Thus, N is approximately proportional to M. Therefore, the number of operations per input data point, N/M, is approximately independent of the decimation ratio M and is determined only by the relative steepness of the transition region, the stop-band attenuation, and the pass-band ripple.

Example 12.3 With $\Delta\theta = \mu\pi/M$, we get from (9.56c) and (9.79),

$$\frac{N}{M} \approx \frac{-20\log_{10}(\min\{\delta_p, \delta_s\}) - 7.95}{2.285\mu\pi} \quad \text{(Kaiser window design)}, \tag{12.18a}$$

$$\frac{N}{M} \approx \frac{-20\log_{10}\sqrt{\delta_p\delta_s} - 13}{2.32\mu\pi} \quad \text{(equiripple design)}. \tag{12.18b}$$

For example, if $\delta_p = 0.01$, $\delta_s = 0.001$, and $\mu = 0.1$, the required number of operations per input data point is about 73 in a Kaiser window filter, or about 51 in an equiripple filter. This example illustrates that the computational cost of decimation can be quite heavy in practice. □

Example 12.4 In Section 3.5 we studied physical A/D converters. There we saw that the two major parameters by which an A/D converter is judged are the number of bits and the speed. In applications such as audio or biomedical signals, speed requirements are easy to meet, since the signal bandwidth is low (e.g., up to 20 kHz for audio signals). On the other hand, a large number of bits may be required, resulting in complex and expensive hardware. We now show how an A/D with a relatively high speed and a relatively low number of bits can be made to give a higher number of bits at a reduced speed.

Suppose that the bandwidth of the input signal is f_m and, instead of sampling at the Nyquist rate $2f_m$ we sample at a higher rate f_{sam}, such that $M = 0.5f_{sam}/f_m$ is an integer. As we saw in Section 11.7.4, the variance of the quantization noise of the A/D is $2^{-2(B-1)}/12$, where B is the number of bits. The noise is white, so its power spectral density is constant over the frequency range $[-0.5f_{sam}, 0.5f_{sam}]$. Now, since the signal bandwidth is f_m, we can pass the output signal of the A/D through a low-pass filter with cutoff frequency f_m. Assuming that the filter is ideal, it would reduce the noise variance by a factor M without harming the signal. After doing so, we can decimate the signal by M. The discrete-time signal thus obtained will be sampled at the Nyquist rate, but the variance of the noise will only be $M^{-1} \cdot 2^{-2(B-1)}/12$, and the noise will still be white at the new sampling rate. Thus, we have effectively gained $0.5\log_2 M$ bits. Each doubling of the sampling rate provides an extra half-bit, or each quadrupling of the rate provides an extra bit. In reality the effective gain is less than half a bit, since the decimation filter is not ideal, so it does not completely eliminate the noise in the stop band. In Section 14.7 we shall use this idea again; see page 586. □

12.3.2 Expansion

Expansion does not cause aliasing, but it yields undesired replicas of the signal's spectrum in the range $[-\pi, \pi]$, as we saw in Figure 12.8. These replicas can be eliminated by a digital low-pass filter following the expander, as shown in Figure 12.10 (for $L = 3$). In this figure, the input signal $x[n]$ occupies the entire bandwidth. The expanded signal $v[n]$ has three replicas in the frequency domain, each occupying a bandwidth of $2\pi/3$. The filter approximates an ideal low-pass filter with cutoff frequency $\pi/3$ (or π/L in general), and retains only the base-band replica. The spectrum of the output signal $y[n]$ is band limited to $\pm\pi/3$ (or $\pm\pi/L$ in general).

The combination of expansion followed by low-pass filtering is called *interpolation*, and the low-pass filter is called an *interpolation filter*, or an *anti-imaging filter*. Whereas decimation may or may not be preceded by a decimation filter, expansion is almost always followed by an interpolation filter, because without such a filter the expanded signal has little use.

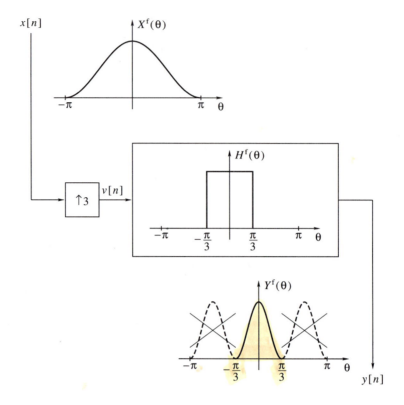

Figure 12.10 Passing an expanded signal through an interpolation filter. The dashed lines in $Y^f(\theta)$ indicate parts of the spectrum that are eliminated by the filter.

Sort of
time $y[n] = \sum x(\tau) \, h(t-\tau).$

L-fold expansion followed by linear filtering is described mathematically by

$$y[n] = \sum_i x[i]h[n - Li]. \qquad (12.19)$$

↑ as rest of inserts are zeros.

As in the case of decimation, we can reduce the amount of computation by a factor L, compared with regular convolution, in case the filter $h[n]$ is FIR. The required number of multiply–add operations is about N/L per output data point. We shall discuss the programming of interpolation with FIR filtering in the next section.

The comments we have made regarding the operation count apply to expansion as well. If the transition bandwidth is set to a fixed fraction of the cutoff frequency, then the number of operations per output data point is approximately independent of the interpolation ratio and is determined only by the relative steepness of the transition region and the required attenuation and ripple tolerances.

Let us now discuss interpolation in more detail. Consider an interpolation filter $h[n]$ that is an ideal low-pass filter with cutoff frequency π/L. Such a filter is IIR and not causal, so it cannot be implemented, but we can still explore its theoretical behavior. The impulse response of the filter is

$$h[n] = \frac{1}{2\pi} \int_{-\pi/L}^{\pi/L} Le^{j\theta n} d\theta = \text{sinc}\left(\frac{n}{L}\right). \qquad (12.20)$$

Substitution in (12.19) gives

$$y[n] = \sum_{i=-\infty}^{\infty} x[i]\text{sinc}\left(\frac{n - iL}{L}\right). \qquad (12.21)$$

This result is similar to Shannon's interpolation formula (3.23). Indeed, ideal interpolation of a discrete-time signal can be viewed as ideal reconstruction of that signal (i.e., its conversion to a band-limited, continuous-time signal), followed by resampling at a rate L times higher than the original rate. If we apply this operation to (3.23), we will get precisely (12.21). Practical interpolation approximates the ideal sinc filter by a causal filter, usually chosen to be FIR.

Example 12.5 Music signals on compact discs are sampled at 44.1 kHz. When the signal is converted to analog, faithful reconstruction is required up to 20 kHz, with only little distortion. As we saw in Section 3.4, this is extremely difficult to do with analog hardware, because of the sinc-shaped frequency response of the zero-order hold and the small margin available for extra filtering (from 20 to 22.05 kHz). A common technique for overcoming this problem is called *oversampling,* and it essentially consists of the following steps:

1. The digital signal is expanded by a certain factor, typically 8, followed by an interpolation filter. The sampling rate of the resulting signal is now 352.8 kHz, but its bandwidth is still only 22.05 kHz.

2. The interpolated signal is input to a zero-order hold. The frequency response of the ZOH is sinc shaped, and its bandwidth (to the first zero crossing) is 352.8 kHz.

3. The output of the ZOH is low-pass filtered to a bandwidth of 20 kHz by an analog filter. Such a filter is relatively easy to design and implement, since we have a large margin (between 20 kHz and 352.8 kHz) over which the frequency response can decrease gradually. The bandwidth of the digital signal is limited to 22.05 kHz, so the analog filter will little affect it.

Figure 12.11 illustrates this procedure. Part a shows a 20 kHz sinusoidal signal sampled at 44.1 kHz, denoted by $x[n]$. The five samples represent a little over two periods of the signal. Such a discrete-time signal would be extremely hard to reconstruct faithfully by means of a zero-order hold followed by an analog filter. Part b shows an 8-fold interpolation of $x[n]$, denoted by $y[n]$. Part c shows the signal $\hat{y}(t)$, reconstructed from $y[n]$ by a zero-order hold. Finally, part d shows the output signal $z(t)$, obtained by passing $\hat{y}(t)$ through a low-pass filter with cutoff frequency $f_p = 20$ kHz. For simplicity, we have not shown the delays introduced by the interpolation filter and the analog low-pass filter. The oversampling technique is implemented nowadays in all CD players and in most digital processing systems of music signals.   □

12.3.3 Sampling-Rate Conversion

A common use of multirate signal processing is for sampling-rate conversion. Suppose we are given a digital signal $x[n]$ sampled at interval T_1, and we wish to obtain from it a signal $y[n]$ sampled at interval T_2. Ideally, $y[n]$ should be spectrally identical to $x[n]$. The techniques of decimation and interpolation enable this operation, provided the ratio T_1/T_2 is a rational number, say L/M. We distinguish between two possibilities:

1. $T_1 > T_2$, meaning that the sampling rate should be increased. This is always possible without aliasing.

2. $T_1 < T_2$, meaning that the sampling rate should be decreased. This is possible without aliasing only if $x[n]$ is band limited to a frequency range not higher than $\pm \pi T_1/T_2$. If $x[n]$ does not fulfill this condition, a part of its frequency contents must be eliminated to avoid aliasing.

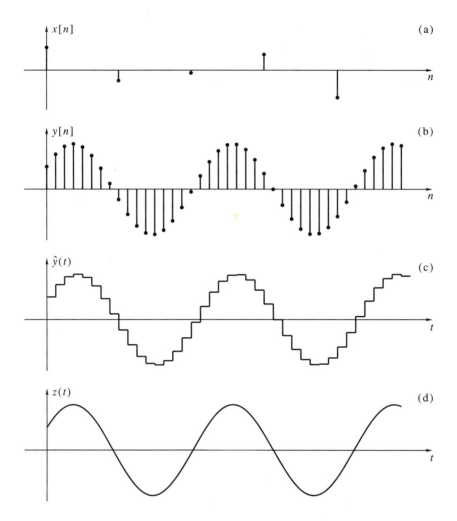

Figure 12.11 Demonstration of oversampling in compact-disc music signal reconstruction: (a) a 20 kHz sinusoidal signal sampled at 44.1 kHz; (b) 8-fold interpolation of the sampled signal; (c) zero-order hold reconstruction of the interpolated signal; (d) analog low-pass filtering of the reconstructed signal.

Sampling-rate conversion can be accomplished by L-fold expansion, followed by low-pass filtering and M-fold decimation. Figure 12.12 depicts this sequence of operations. The low-pass filter performs both interpolation of the expanded signal and antialiasing. If the sampling rate is to be increased, then $L > M$. The low-pass filter should then have a cutoff frequency π/L. If the sampling rate is to be decreased, then $L < M$. The low-pass filter should then have a cutoff frequency π/M. In this case, the filter will eliminate a part of the signal's frequency contents if its original bandwidth is higher than $\pi L/M$. Thus, the sampling-rate conversion filter should always have a cutoff frequency $\pi/\max\{L, M\}$.

Sampling-rate conversion is described mathematically by

$$y[n] = \sum_i x[i]h[Mn - Li]. \tag{12.22}$$

The number of multiply–add operations can be estimated as follows. Assume we have P input points, so the number of output points is PL/M, and for each output point we

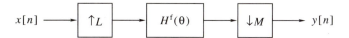

Figure 12.12 Sampling-rate conversion by expansion, filtering, and decimation.

need about N/L operations. Thus, the number of operations is about N/M per input point, or N/L per output point. The length of the filter is proportional to $\max\{L, M\}$. We get that if $L > M$ (sampling rate increase), the number of operations per output point is approximately constant, whereas if $L < M$ (sampling rate decrease), the number of operations per input point is approximately constant.

12.4 Polyphase Filters

As we saw in Section 12.3, linear filtering with decimation, interpolation, and sampling-rate conversion allows potential computational savings. However, it is not always obvious how to take advantage of this potential. *Polyphase filters* is a name given to certain realizations of multirate filtering operations, which facilitate computational savings in both software and hardware implementations. As we shall see, polyphase implementations are useful for FIR filters, but not for IIR.

12.4.1 The Multirate Identities

In preparation for our discussion of polyphase filters, we now derive two useful identities, known as the *multirate identities*.

The decimation identity Suppose we decimate a discrete-time signal $x[n]$ by a factor M and then pass it through a filter $H^z(z)$, as shown in Figure 12.13(a). The decimation identity asserts that this operation is equivalent to expanding the impulse response of $H^z(z)$ by a factor M, then feeding $x[n]$ to the new filter, and finally decimating by a factor M, as shown in Figure 12.13(b).

(a) (b)

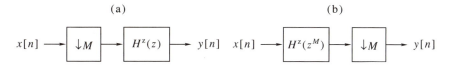

Figure 12.13 The decimation identity.

The decimation identity is easy to understand in the time domain. Both operations are expressed by the time-domain formula

$$y[n] = \sum_k h[k]x[(n-k)M].$$ (12.23)

A z-domain proof is given as follows. The output in Figure 12.13(b) has a z-transform

$$\{X^z(z)H^z(z^M)\}_{(\downarrow M)} = \frac{1}{M}\sum_{m=0}^{M-1} X^z(z^{1/M}W_M^{-m})H^z((z^{1/M}W_M^{-m})^M)$$

$$= \frac{1}{M}\sum_{m=0}^{M-1} X^z(z^{1/M}W_M^{-m})H^z(z) = H^z(z)\{X^z(z)\}_{(\downarrow M)}.$$ (12.24)

The right side is precisely the z-domain description of Figure 12.13(a).

The expansion identity Suppose we pass a discrete-time signal $x[n]$ through a filter $H^z(z)$ and then expand it by a factor L, as shown in Figure 12.14(a). The expansion identity asserts that this operation is equivalent to expanding both the signal and the impulse response of $H^z(z)$ by a factor L, then feeding the expanded signal to the new filter, as shown in Figure 12.14(b).

(a)   (b)

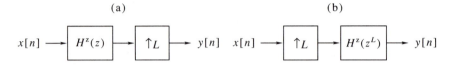

Figure 12.14 The expansion identity.

The expansion identity is easy to understand in the z domain. Both operations are expressed by the z-domain formula

$$Y^z(z) = H^z(z^L)X^z(z^L). \tag{12.25}$$

A time-domain proof is given as follows. The output in Figure 12.14(b) is given by

$$y[n] = \sum_k h_{(\uparrow L)}[k]x_{(\uparrow L)}[n-k] = \begin{cases} \sum_i h[i]x\left[\dfrac{n}{L} - i\right], & n \text{ is divisible by } L, \\ 0, & \text{otherwise.} \end{cases} \tag{12.26}$$

The right side is precisely the time-domain description of Figure 12.14(a).

12.4.2   Polyphase Representation of Decimation

Consider again the operation (12.17), which represents filtering followed by decimation, and express i as

$$i = i'M + m, \quad 0 \leq m \leq M - 1.$$

Then we get

$$y[n] = \sum_{i=0}^{N} h[i]x[nM - i] = \sum_{m=0}^{M-1} \sum_{i'} h[i'M + m]x[nM - i'M - m]$$

$$= \sum_{m=0}^{M-1} \sum_{i'} h[i'M + m]x[(n - i')M - m]. \tag{12.27}$$

For each $0 \leq m \leq M - 1$, define the sequence $p_m[i']$ as

$$p_m[i'] = h[i'M + m], \tag{12.28}$$

and the sequence $u_m[n]$ as

$$u_m[n] = x[nM - m]. \tag{12.29}$$

Substitution of these definitions in (12.27) then gives

$$y[n] = \sum_{m=0}^{M-1} \{p_m * u_m\}[n]. \tag{12.30}$$

Let us interpret this result:

1. For each m, the sequence $p_m[i']$ is obtained by advancing (i.e., left shifting) the sequence $h[n]$ by m, then decimating by M.

2. For each m, the sequence $u_m[n]$ is obtained by delaying (i.e., right shifting) the sequence $x[n]$ by m, then decimating by M.

3. The output $y[n]$ is obtained by performing M convolutions of the sequences $u_m[n]$ with the corresponding FIR filters $p_m[i']$, then adding the results.

The M FIR filters $p_m[i']$ are called the *polyphase* components of $h[n]$. Their corresponding transfer functions are

$$P_m^z(z) = \sum_{i'} p_m[i']z^{-i'} = \sum_{i'} h[i'M + m]z^{-i'}. \tag{12.31}$$

The transfer function $H^z(z)$ can be expressed in terms of the polyphase components as

$$H^z(z) = \sum_{m=0}^{M-1} z^{-m} P_m^z(z^M). \tag{12.32}$$

Therefore, the z-domain description of filtering followed by decimation is

$$Y^z(z) = \sum_{m=0}^{M-1} \{P_m^z(z^M)z^{-m}X^z(z)\}_{(\downarrow M)}. \tag{12.33}$$

By the decimation identity, this can be written as

$$Y^z(z) = \sum_{m=0}^{M-1} P_m^z(z)\{z^{-m}X^z(z)\}_{(\downarrow M)}. \tag{12.34}$$

However, by our definition of the sequences $u_m[n]$ we have

$$U_m^z(z) = \{z^{-m}X^z(z)\}_{(\downarrow M)}. \tag{12.35}$$

Therefore,

$$Y^z(z) = \sum_{m=0}^{M-1} P_m^z(z)U_m^z(z), \tag{12.36}$$

which is precisely the z-domain description of (12.30).

Figure 12.15 illustrates how the polyphase components are obtained from a given FIR sequence. Figure 12.16 illustrates the operation (12.30), or its z-domain equivalent (12.36). The block diagram in this figure is called the *polyphase decomposition* of the filtering and decimation operation. The procedure ppdec in Program 12.1 implements filtering and decimation by the polyphase decomposition.

An alternative way of implementing the polyphase decomposition is to replace the delay chain by a *commutator*, that is, an M-position switch that routes the input data to the polyphase filters in succession. This implementation is shown in Figure 12.17. Note the order in which the commutator arm is switched: At time $n = 0$ it is at the top position, at time $n = 1$ it is at the bottom position, at time $n = 2$ at the prior-to-bottom position, and so on. In other words, the rotation is counterclockwise, starting at the top position. In software implementation, the commutator is conceptual, rather than physical, and switching is done by appropriate program statements.

We finally remark the notations $p_m[n]$ and $P_m^z(z)$ for the polyphase components are ambiguous, since they depend not only on $h[n]$ and on m, but also on the decimation factor M. To avoid cumbersome notation, we shall continue to omit the dependence on M, but keep in mind that M has to be specified for the polyphase components to be uniquely defined.

12.4.3 Polyphase Representation of Expansion

Consider again the operation (12.19), and express n as

$$n = n'L + L - 1 - l, \quad 0 \le l \le L - 1.$$

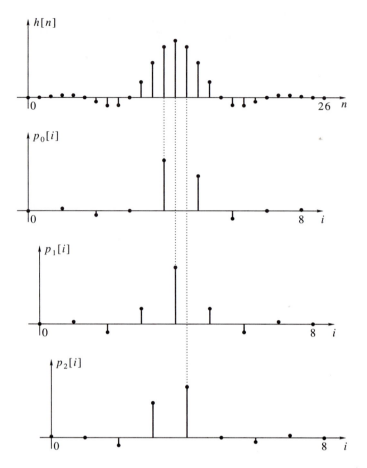

Figure 12.15 The polyphase components of an FIR filter, illustrated for $N = 26$ and $M = 3$.

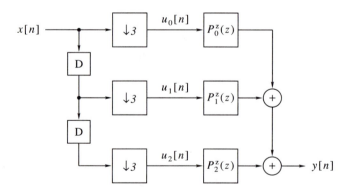

Figure 12.16 Polyphase decomposition of filtering and M-fold decimation (shown for $M = 3$).

Then we get

$$y[n'L + L - 1 - l] = \sum_i x[i]h[(n' - i)L + L - 1 - l]. \tag{12.37}$$

For each $0 \le l \le L - 1$, define the sequence $q_l[n']$ as

$$q_l[n'] = h[n'L + L - 1 - l], \tag{12.38}$$

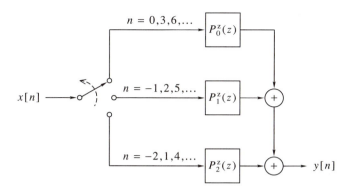

Figure 12.17 Polyphase decomposition of filtering and M-fold decimation with a commutator instead of delays and decimation (shown for $M = 3$).

and the sequence $v_l[n']$ as

$$v_l[n'] = y[n'L + L - 1 - l].$$ (12.39)

Substitution of these definitions in (12.37) then gives

$$v_l[n'] = \{x * q_l\}[n'].$$ (12.40)

We also observe that $y[n]$ can be expressed in terms of the $v_l[n']$ as

$$y[n] = \sum_{l=0}^{L-1} \{v_l\}_{(\uparrow L)}[n - L + 1 + l].$$ (12.41)

This formula is verified as follows. By definition of L-fold expansion, the value $\{v_l\}_{(\uparrow L)}[n - L + 1 + l]$ is equal to $v_l[(n - L + 1 + l)/L]$ when $(n - L + 1 + l)/L$ is integer, and is equal to 0 otherwise. By (12.39), this is equal to $y[n]$. For each n, exactly one of the $\{(n - L + 1 + l)/L, \ 0 \le l \le L - 1\}$ is integer, so only one term in the sum (12.41) is nonzero.

As we see, $q_l[n']$ can be regarded as the polyphase components of $h[n]$ corresponding to a decimation factor L, indexed in reversed order. The sequence of operations for computing $y[n]$ is therefore as follows:

1. Convolve the input sequence $x[n]$ with each of the polyphase components $q_l[n']$ to obtain the sequences $v_l[n']$.

2. Expand each sequence $v_l[n']$ by a factor L.

3. Delay the expanded sequences $\{v_l\}_{(\uparrow L)}[n]$ by $L - l - 1$ time units and add them. Note that the addition is fictitious, since at each time point only one of the sequences has nonzero value.

The transfer function $H^z(z)$ can be expressed in terms of the polyphase components as

$$H^z(z) = \sum_{l=0}^{L-1} z^{-(L-1-l)} Q_l^z(z^L).$$ (12.42)

Therefore, the z-domain description of expansion followed by filtering is

$$Y^z(z) = \sum_{l=0}^{L-1} z^{-(L-1-l)} Q_l^z(z^L) X_{(\uparrow L)}^z(z).$$ (12.43)

By the expansion identity, this can be written as

$$Y^z(z) = \sum_{l=0}^{L-1} z^{-(L-1-l)} \{Q_l^z(z)X^z(z)\}_{(\uparrow L)}. \tag{12.44}$$

However, by our definition of the sequences $v_l[n]$ we have

$$V_l^z(z) = Q_l^z(z)X^z(z). \tag{12.45}$$

Therefore,

$$Y^z(z) = \sum_{l=0}^{L-1} z^{-(L-1-l)} \{V_l^z(z)\}_{(\uparrow L)}, \tag{12.46}$$

which is precisely the z-domain description of (12.41). Figure 12.18 illustrates this sequence of operations. The block diagram in this figure is called the *polyphase decomposition* of the expansion and filtering operation. The procedure `ppint` in Program 12.2 implements expansion and filtering by the polyphase decomposition.

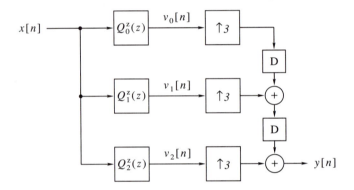

Figure 12.18 Polyphase decomposition of L-fold expansion and filtering (shown for $L = 3$).

An alternative way of implementing the polyphase decomposition is to replace the delay chain by a commutator, as we did in the case of decimation. This implementation is shown in Figure 12.19.

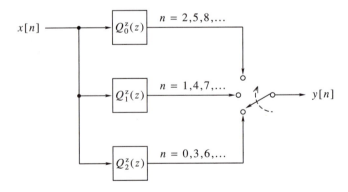

Figure 12.19 Polyphase decomposition of L-fold expansion and filtering with a commutator instead of expansion and delays (shown for $L = 3$).

12.4.4   Polyphase Representation of Sampling-Rate Conversion

Sampling-rate conversion can be done either by efficient interpolation followed by brute-force decimation or by brute-force interpolation followed by efficient decimation. Both approaches are wasteful, however, since neither achieves the minimum possible rate of computation. To get the maximum possible computational efficiency, we need to attack the problem from the root.

Consider again the operation (12.22) and express n and i as

$$n = n'L + l, \quad 0 \le l \le L - 1,$$

$$i = i'M + m, \quad 0 \le m \le M - 1.$$

Then we get

$$y[n'L + l] = \sum_{m=0}^{M-1} \sum_{i'} x[i'M + m]h[(n' - i')ML + lM - mL]. \tag{12.47}$$

It is convenient, in this case, to break the sum over m into two parts. Define $m_0 = \lfloor lM/L \rfloor$, so that

$$y[n'L + l] = \sum_{m=0}^{m_0} \sum_{i'} x[i'M + m]h[(n' - i')ML + lM - mL]$$

$$+ \sum_{m=m_0+1}^{M-1} \sum_{i'} x[i'M + m]h[(n' - 1 - i')ML + lM + (M - m)L]. \tag{12.48}$$

For each $0 \le l \le L - 1$ and $0 \le m \le M - 1$, define the following sequences:

$$r_{l,m}[i'] = \begin{cases} h[i'ML + lM - mL], & m \le m_0, \\ h[i'ML + lM + (M - m)L], & m_0 + 1 \le m \le M - 1. \end{cases} \tag{12.49}$$

$$u_m[i'] = x[i'M + m], \tag{12.50}$$

$$v_l[n'] = y[n'L + l]. \tag{12.51}$$

Substitution of these definitions in (12.48) then gives

$$v_l[n'] = \sum_{m=0}^{m_0} \{u_m * r_{l,m}\}[n'] + \sum_{m=m_0+1}^{M-1} \{u_m * r_{l,m}\}[n' - 1]. \tag{12.52}$$

Finally we have, as in (12.41),

$$y[n] = \sum_{l=0}^{L-1} \{v_l\}_{(\uparrow L)}[n - l]. \tag{12.53}$$

The filters $r_{l,m}[i']$ are the polyphase components for sampling-rate conversion. There is a total of ML components. Note the slight irregularity in their definition, because of the need to distinguish between m below and above m_0. The sequences $u_m[i']$ are shifts and M-fold decimations of the input sequence, whereas $v_l[n']$ are shifts and L-fold decimations of the output sequence.

The procedure ppsrc in Program 12.3 is a MATLAB implementation of sampling-rate conversion by polyphase decomposition. As a special case, it can perform pure decimation (use $L = 1$), or pure interpolation (use $M = 1$). The program is perhaps not as transparent as the programs for polyphase decimation and interpolation, but it follows equations (12.49) through (12.53) closely.

12.5   Multistage Schemes*

When decimation by a large factor is required, it is often advantageous to divide the process into several stages. Such a division is possible if the decimation factor M is a composite number. Suppose that $M = M_1 M_2$; then we can perform M-fold decimation as illustrated in Figure 12.20. We first filter the input signal $x[n]$ by a low-pass filter $H_1^z(z)$, then decimate by M_1, filter the decimated signal $u[n]$ by a low-pass filter $H_2^z(z)$, and finally decimate by M_2. The output signal $y[n]$ is decimated by M with respect to $x[n]$. The filters $H_1^z(z)$, $H_2^z(z)$ should be designed such that aliasing will be below a prescribed level and overall pass-band and stop-band tolerances will be met. This two-stage scheme often leads to computational savings, as we shall illustrate.

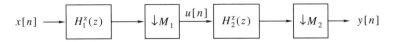

Figure 12.20 Multistage decimation.

Let us discuss the specifications of the two low-pass filters in two-stage decimation. We assume that the stop-band lower frequency of the overall system is $\theta_s = \pi/M = \pi/M_1 M_2$, and the pass-band upper frequency is $\theta_p < \theta_s$. We also assume that tolerances δ_p, δ_s are given for the overall system. Figure 12.21(a) shows the desired frequency response of a low-pass filter in single-stage decimation. In this figure we take $M_1 = 3$, $M_2 = 2$, $M = 6$.

Let $\theta_{p,1}$, $\theta_{s,1}$ be the band-edge frequencies of the first filter. Since we are going to decimate by a factor M_1, it appears as though we need $\theta_{s,1} = \pi/M_1$ to avoid aliasing in the first decimation. Further thinking reveals that we can do with a higher stop-band frequency. After M_1-fold decimation, the first replica on the right of the zeroth replica is going to be centered around $\theta = 2\pi/M_1$, and stretched to the left down to $\theta = 2\pi/M_1 - \theta_{s,1}$. We can therefore take

$$\frac{2\pi}{M_1} - \theta_{s,1} = \frac{\pi}{M}, \quad \text{or} \quad \theta_{s,1} = \frac{2\pi}{M_1} - \frac{\pi}{M} = \frac{(2M_2 - 1)\pi}{M},$$

and still avoid aliasing in the frequency range up to π/M. This is illustrated in Figure 12.21(b). There will be aliasing at frequencies higher than π/M, but this aliasing is of no concern, since it is going to be eliminated at the next stage.

The pass-band upper frequency $\theta_{p,1}$ should be identical to θ_p of the overall system. It can be higher, but there is no advantage in this. The pass-band tolerance should be smaller than δ_p, since pass-band ripple will be present at the next stage as well. It is reasonable to require pass-band tolerance $0.5\delta_p$ at each of the two stages. The stop-band tolerance should be δ_s, the same as for the overall system.

Figure 12.21(c) shows the desired frequency response of the second filter. Now the sampling rate has already been reduced M_1-fold, so the band-edge frequencies of the second filter should be $\theta_{p,2} = M_1 \theta_p$ and $\theta_{s,2} = M_1 \theta_s = \pi/M_2$ in the new θ domain. The pass-band and stop-band tolerances should be $0.5\delta_p$ and δ_s, respectively. The following example illustrates multistage decimation.

Example 12.6 We are given a signal sampled at 96 kHz. The useful information in the signal is in the frequency band 0 to 3 kHz. It is decided to decimate the signal by $M = 12$. The required tolerances are $\delta_p = 0.01$, $\delta_s = 0.001$. A single-stage decimation filter must have $\theta_p = \pi/16$, $\theta_s = \pi/12$. Suppose we use equiripple design for the decimation filter. According to (9.79), the necessary order is $N = 244$. Since the

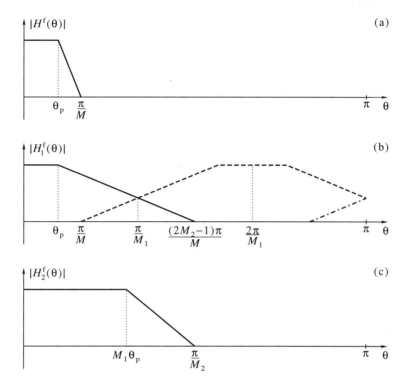

Figure 12.21 Frequency responses of filters for multistage decimation: (a) response of a single-state decimation filter; (b) response of the first-stage filter, dashed and dot-dashed lines show replicas generated by decimation; (c) response of the second-stage filter.

decimation ratio is 12, the number of operations per input data point is about 20.3. Now consider two-stage decimation, with $M_1 = 3$, $M_2 = 4$. The specifications of the first filter are

$$\theta_{p,1} = \pi/16, \quad \theta_{s,1} = 7\pi/12, \quad \delta_{p,1} = 0.005, \quad \delta_{s,1} = 0.001.$$

The corresponding order of an equiripple filter is $N_1 = 11$. Since the decimation ratio is 3, the number of operations per input data point is about 3.7. The specifications of the second filter are

$$\theta_{p,2} = 3\pi/16, \quad \theta_{s,2} = \pi/4, \quad \delta_{p,2} = 0.005, \quad \delta_{s,2} = 0.001.$$

The corresponding order of an equiripple filter is $N_2 = 88$. Since the decimation ratio is now 12, the number of operations per input data point is about 7.3. In summary, two-stage implementation requires about 11 operations per input point, or about half the number in single-stage implementation. You are asked to repeat these calculations when the roles of M_1 and M_2 are reversed (Problem 12.13). □

Multistage schemes can be used for interpolation equally well. Suppose that the interpolation factor is composite, say $L = L_1 L_2$. Then we can perform L-fold interpolation as illustrated in Figure 12.22. We first expand the input signal $x[n]$ by L_1 and pass it through a low-pass filter $H_1^z(z)$, then expand by L_2, and finally pass through a low-pass filter $H_2^z(z)$. The output signal $y[n]$ is interpolated by L with respect to $x[n]$. The filters $H_1^z(z)$, $H_2^z(z)$ should be designed such that images will be suppressed below a prescribed level and overall pass-band and stop-band tolerances will be met. This two-stage scheme often leads to computational savings, as we shall illustrate.

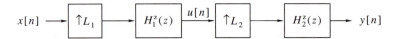

Figure 12.22 Multistage interpolation.

Let us discuss the specifications of the two low-pass filters in two-stage interpolation. Figure 12.23(a) shows the Fourier transform of the input signal. We assume that the frequency response decays to zero at the high frequency range, to allow for finite transition band of the interpolation filter. If this is not the case, the interpolation filter will inevitably attenuate a part of the high-frequency components. Figure 12.23(b) shows the Fourier transform of the signal expanded by L_1, before and after the filter. The frequency response of $H_1^z(z)$ in this figure is assumed to be equal to $U^f(\theta)$ in magnitude. Clearly $\theta_{s,1}$ must be equal to π/L_1. Figure 12.23(c) shows the Fourier transform of the signal expanded by L_2, before and after the filter, and the magnitude response of $H_2^z(z)$. As we see, $\theta_{s,2}$ can be as large as $(2L_1 - 1)\pi/L$ and still suppress all images. The pass-band tolerances $\delta_{p,1}, \delta_{p,2}$ should be taken as $0.5\delta_p$ each, as in the case of decimation. The stop-band tolerances $\delta_{s,1}, \delta_{s,2}$ should be taken according to the desired level of image suppression.

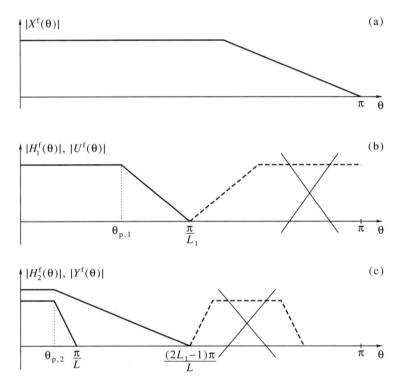

Figure 12.23 Frequency responses of signals and filters for multistage interpolation: (a) response of the input signal; (b) response of the first-stage filter, dashed line shows image suppressed by the filter; (c) response of the second-stage filter and of the output signal.

It is clear from the preceding description that multistage interpolation is dual to multistage decimation. For example, if we reverse the procedure described in Example 12.6 (with $L_1 = 4$, $L_2 = 3$), we will find that the filters have the same orders, and the same computational saving can be achieved.

Multistage schemes are also useful for sampling-rate conversion systems. As an example, consider the problem of transferring digitally recorded music from digital tape to compact disc. Digital audiotapes use sampling rate 48 kHz and compact discs use 44.1 kHz. The required ratio L/M is therefore 147/160. Single-stage sampling-rate conversion would require expanding the input signal to over 7 MHz. Factoring L and M into primes, we find that $147 = 7 \times 7 \times 3$, and $160 = 2^5 \times 5$. We can therefore build a three-stage system: one stage with $L_1/M_2 = 7/10$, a second with $L_2/M_2 = 7/8$, and a third with $L_3/M_3 = 3/2$. Such a system would be much more efficient than a single-stage system and would obviate the need for signals in the megahertz range.

12.6 Filter Banks*

12.6.1 Subband Processing

An important application of multirate signal processing is *subband processing.* Subband processing is based on splitting the frequency range into M segments (subbands), which together encompass the entire range. Each subband is processed independently, as called for by the specific application. If necessary, the subbands are recombined, after processing, to form an output signal whose bandwidth occupies the entire frequency range.

Example 12.7 Consider a multiple-user communication system in which several (say M) users share a single carrier channel, whose bandwidth is sufficient to accommodate all M users. One method of sharing the channel between users is by *frequency division multiplexing.* In this method, a subband of width $1/M$ of the entire channel is allocated to each user. The receiver of such a system has to split the received signal into subbands, then examine the contents of each separately. Traditional systems of this kind used analog band-pass filtering for this purpose. However, modern systems rely more and more on digital filtering. In this application, there is usually no need to reconstruct the full bandwidth signal again after processing. □

The division of a signal into subbands is usually accomplished by a *filter bank.* As the name implies, a filter bank is a collection of band-pass filters, all processing the same input signal. Figure 12.24 illustrates the structure of a filter bank. In this figure, the transfer functions of the band-splitting filters are $\{G_m^z(z), 0 \le m \le M - 1\}$. These filters are also known as the *analysis bank.* The outputs of the analysis filters, the signals $u_m[n]$, are fed into the subband processing system, resulting in the processed subband signals $v_m[n]$. If there is a need to reconstruct the signal after processing, this is usually accomplished by a second bank of filters, called the *synthesis bank.* In Figure 12.24, the transfer functions of the synthesis filters are $\{H_m^z(z), 0 \le m \le M-1\}$.

The output $y[n]$ of the filter bank will practically always be different from the input signal $x[n]$, mainly because of the block performing the subband processing (e.g., the compression method described in Example 12.8; see page 493). However, suppose we eliminate the subband processing block and short-circuit each signal $u_m[n]$ to the corresponding $v_m[n]$. It is highly desirable to design the filter bank such that, in this case, $y[n]$ will be equal to $x[n]$, except for constant delay. A filter bank for which (in the absence of subband processing) $y[n]$ differs from $x[n]$ only by a constant delay and a constant scale factor, is said to have the *perfect reconstruction* property. Whether a

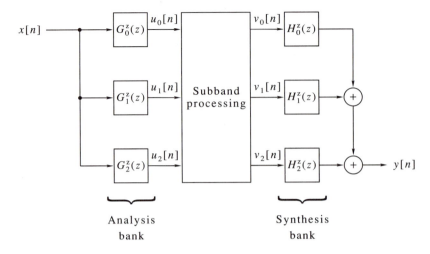

Figure 12.24 Subband processing with a filter bank.

given filter bank has this property depends on the design of the analysis and synthesis filters. For now, we say only that perfect reconstruction is highly desirable, but not at all easy to achieve.

Figure 12.25 shows a typical division of the frequency range into subbands. In this figure, all subbands have the same width. However, this is by no means a necessity. In certain applications it is of advantage to use subbands of different widths. When all subbands have the same width, the filter bank is called *uniform*. The following points are worth noting:

1. The zeroth subband usually occupies the range $[-\pi/M, \pi/M]$. The filter $g_0[n]$, which is responsible for this subband, has real coefficients.

2. All the other subbands are *nonsymmetric* as a function of the frequency θ, implying that their coefficients are complex.

3. The organization of the subbands for odd M slightly differs from that for even M. In the case of even M, band number $M/2$ is split evenly between positive and negative frequencies.

4. It is common for the subbands to overlap slightly, as shown in Figure 12.25. This overlap is necessary to prevent gaps in the spectrum, since physical filters have finite transition bands.

12.6.2 Decimated Filter Banks

The signals $u_m[n]$ shown in Figure 12.24 are band limited. If the filter bank is uniform, each has a bandwidth of about $2\pi/M$. As we saw in Section 12.2, such a signal can be decimated by a factor up to M without being aliased. This is true whether the signal is low pass (cf. Figure 12.6) or band pass (cf. Figure 12.7). In either case, the decimated signal occupies the entire frequency band (if decimated M-fold). The filter bank shown in Figure 12.24 is wasteful, since it does not use this property of the subband signals.

Figure 12.26 shows a modified filter bank, in which each subband signal is decimated by a factor K. The filters in this bank are complex in general (except when $K = 2$). The decimation factor can be either equal to M or smaller (but not larger, since this will lead to aliasing in general). This is called a *decimated filter bank*. If $K = M$, it is

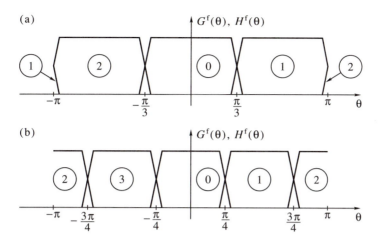

Figure 12.25 Division of the frequency range into bands in subband processing: (a) odd number (shown for $M = 3$); (b) even number (shown for $M = 4$).

called a *maximally decimated filter bank.* Note that we are using this terminology only for uniform filter banks. Before recombination, the processed subband signals should be expanded by the same factor K (unless sampling-rate conversion is required), then interpolated by the narrow-band filters $h_m[n]$, and finally combined to form $y[n]$.

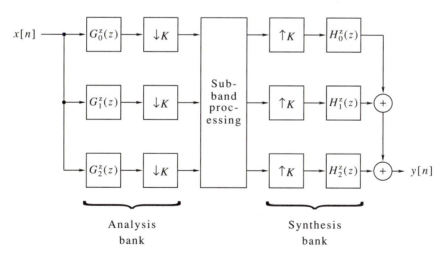

Figure 12.26 A decimated filter bank.

Maximally decimated filter banks achieve the maximum possible computational efficiency, because the subband signals have the lowest possible rate. However, the decimation and interpolation filters $g_m[n]$ and $h_m[n]$ are typically required to meet tight specifications. For this and other reasons, it is sometimes preferable to decimate by a factor smaller than the highest possible. We continue to discuss only maximally decimated filter banks.

12.7 Two-Channel Filter Banks*

The simplest filter bank separates the input signal into two bands, therefore it is called a *two-channel filter bank*. Figure 12.27 shows a maximally decimated, two-channel filter bank. The left side shows the analysis part, and the right side—the synthesis part. The analysis filter $G_0^z(z)$ is low pass, and its pass band should be approximately $[0, 0.5\pi]$. The analysis filter $G_1^z(z)$ is high pass, and its pass band should be approximately $[0.5\pi, \pi]$. The same holds for the synthesis filters $H_0^z(z)$ and $H_1^z(z)$, respectively. The outputs of the analysis filters, $u_0[n]$ and $u_1[n]$, are decimated by 2. The figure does not show the subband processing part, but assumes that the decimated signals are immediately expanded by 2, to yield the signals $v_0[n]$ and $v_1[n]$. These are passed through the synthesis filters and combined to form the output signal $y[n]$. The filters in a two-channel filter bank have real coefficients. They can be either FIR or IIR; here we limit ourselves to FIR filters.[1]

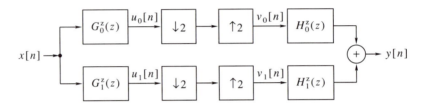

Figure 12.27 A maximally decimated, two-channel filter bank.

12.7.1 Properties of Two-Channel Filter Banks

Before we discuss the design of a two-channel filter bank, we analyze its properties and, in particular, explain what errors are likely to arise and how they can be eliminated or minimized. We do this by computing the z-transform of the output signal as a function of the z-transform of the input and the transfer functions of the four filters. The intermediate signals $u_0[n]$ and $u_1[n]$ have the z-transforms

$$U_0^z(z) = G_0^z(z)X^z(z), \quad U_1^z(z) = G_1^z(z)X^z(z). \tag{12.54}$$

Using Problem 12.4, we can express the z-transforms of $v_0[n]$ and $v_1[n]$ as

$$V_0^z(z) = 0.5U_0^z(z) + 0.5U_0^z(-z), \tag{12.55a}$$

$$V_1^z(z) = 0.5U_1^z(z) + 0.5U_1^z(-z). \tag{12.55b}$$

Substituting (12.54) in (12.55b) gives

$$V_0^z(z) = 0.5G_0^z(z)X^z(z) + 0.5G_0^z(-z)X^z(-z), \tag{12.56a}$$

$$V_1^z(z) = 0.5G_1^z(z)X^z(z) + 0.5G_1^z(-z)X^z(-z). \tag{12.56b}$$

Therefore,

$$Y^z(z) = H_0^z(z)V_0^z(z) + H_1^z(z)V_1^z(z) = 0.5[G_0^z(z)H_0^z(z) + G_1^z(z)H_1^z(z)]X^z(z)$$
$$+ 0.5[G_0^z(-z)H_0^z(z) + G_1^z(-z)H_1^z(z)]X^z(-z). \tag{12.57}$$

Formula (12.57) expresses the output signal as a sum of two components. The first is the response to the input signal $X^z(z)$, and the second is the response to the alias component $X^z(-z)$ (see Problem 12.5). To eliminate the alias component from the output signal, it is necessary and sufficient to have

$$G_0^z(-z)H_0^z(z) + G_1^z(-z)H_1^z(z) = 0. \tag{12.58}$$

There are many ways of satisfying this condition. A convenient one is to take the synthesis filters as

$$H_0^z(z) = 2G_1^z(-z), \quad H_1^z(z) = -2G_0^z(-z). \tag{12.59}$$

This choice not only eliminates the alias component, but also satisfies the basic requirement from the synthesis filters, namely that $H_0^z(z)$ be low pass and $H_1^z(z)$ be high pass (Problem 12.15).

Note that elimination of aliasing in the output signal does not imply its elimination in any of the channels separately. The decimated signals $u_0[2n]$ and $u_1[2n]$ are aliased in general. This aliasing is exactly canceled by the summing operation at the output of the filter bank, provided (12.58) is satisfied. It would not be desirable to eliminate the aliasing in $u_0[2n]$ and $u_1[2n]$ individually, because this would inevitably lead to considerable distortions in the frequency range around 0.5π, due to the finite transition bands of the analysis filters.

When condition (12.59) is used, the output formula (12.57) becomes

$$Y^z(z) = [G_0^z(z)G_1^z(-z) - G_0^z(-z)G_1^z(z)]X^z(z). \tag{12.60}$$

In this case, the filter bank is linear time invariant and its transfer function is

$$F^z(z) = G_0^z(z)G_1^z(-z) - G_0^z(-z)G_1^z(z). \tag{12.61}$$

An ideal filter bank would satisfy $F^z(z) = 1$, therefore $y[n] = x[n]$. Such a filter is not realizable, however, since all four filters must be causal. Next best is to have

$$F^z(z) = cz^{-l} \tag{12.62}$$

for a constant c and a nonnegative integer l. In this case $y[n] = cx[n-l]$, so $y[n]$ is a distortion-free reconstruction of $x[n]$. The filter bank that satisfies (12.62) is said to have the *perfect reconstruction* property.

A filter bank that does not satisfy (12.62) leads to distortions. In general,

$$F^z(z) = A(\theta)e^{j\phi(\theta)}.$$

If $A(\theta) \not\equiv c$, the filter bank has *amplitude distortion*. If $\phi(\theta)$ is not linear, it has *phase distortion*. A filter bank that does not have the perfect reconstruction property can have either kind of distortion, or both.

12.7.2 Quadrature Mirror Filter Banks

An analysis filter pair satisfying

$$G_1^z(z) = G_0^z(-z), \quad \text{or} \quad G_1^f(\theta) = G_0^f(\theta - \pi) \tag{12.63}$$

is called a *quadrature mirror filter bank*, or QMF bank for short [Esteban and Galand, 1977]. Note that (12.63) can also be expressed in the form

$$G_1^f(0.5\pi + \theta) = \bar{G}_0^f(0.5\pi - \theta). \tag{12.64}$$

We see that $G_1^f(\cdot)$ is the mirror image of $G_0^f(\cdot)$ around the point $\theta = 0.5\pi$: The magnitude is symmetric, and the phase is antisymmetric with respect to this point. This is the reason for the name *quadrature mirror filter*. We also see that $G_1^f(\theta)$ is high pass if $G_0^f(\theta)$ is low pass.

The transfer function of a QMF bank is obtained by substituting (12.64) in (12.61). This gives

$$F^z(z) = [G_0^z(z)]^2 - [G_0^z(-z)]^2. \tag{12.65}$$

Problem 12.16 explores the limitations on FIR filter banks resulting from the quadrature mirror restriction. As shown there, such a bank can have perfect reconstruction only if $G_0^z(z)$ contains two nonzero coefficients at most. An FIR filter having only two nonzero coefficients is too simple to be of any practical use. Therefore, the use of FIR QMF banks behooves us to forfeit the perfect reconstruction property.

A QMF bank whose prototype filter $G_0^z(z)$ has linear phase is free of phase distortion. To show this, let

$$G_0^f(\theta) = A(\theta)e^{-j0.5\theta N}, \quad G_0^f(\theta - \pi) = A(\theta - \pi)e^{-j0.5(\theta - \pi)N}, \qquad (12.66)$$

where N is the order of $G_0^z(z)$ and $A(\theta)$ is its amplitude function. We get from (12.65),

$$F^f(\theta) = [A^2(\theta) - (-1)^N A^2(\theta - \pi)]e^{-j\theta N}. \qquad (12.67)$$

Since $A(\theta)$ is either symmetric or antisymmetric,

$$|F^f(0.5\pi)| = |A^2(0.5\pi) - (-1)^N A^2(0.5\pi)|. \qquad (12.68)$$

Therefore, if we choose an even-order N, we will get $F^f(0.5\pi) = 0$. This implies that N must be odd, for otherwise we get an unacceptable amplitude distortion at $\theta = 0.5\pi$. Assuming that N is odd, we get from (12.67),

$$F^f(\theta) = [A^2(\theta) + A^2(\theta - \pi)]e^{-j\theta N} = [|G_0^f(\theta)|^2 + |G_0^f(\theta - \pi)|^2]e^{-j\theta N}. \qquad (12.69)$$

The amplitude distortion of a linear-phase QMF bank is

$$|F^f(\theta)| - 1 = |G_0^f(\theta)|^2 + |G_0^f(\theta - \pi)|^2 - 1. \qquad (12.70)$$

In practice, the amplitude distortion can be made very small by a careful design of $G_0^z(z)$. Distortions of 0.01 and smaller can be achieved, using numerical optimization techniques. Such techniques have been derived by Johnston [1980] and Jain and Crochiere [1983]; we do not discuss them here.

12.7.3   Perfect Reconstruction Filter Banks

To get perfect reconstruction FIR filter banks, we must abandon the quadrature mirror property (12.63). Smith and Barnwell [1984], and Mintzer [1985] developed FIR filter banks having the perfect reconstruction property. They replaced (12.63) by the condition

$$G_1^z(z) = (-z)^{-N} G_0^z(-z^{-1}). \qquad (12.71)$$

Filters satisfying (12.71) are called *conjugate quadrature filters*, or CQF. The coefficients of $G_1^z(z)$ are related to those of $G_0^z(z)$ by (Problem 12.21)

$$g_1[n] = (-1)^n g_0[N - n]. \qquad (12.72)$$

Substituting (12.71) in (12.61), we get

$$F^z(z) = z^{-N} G_0^z(z) G_0^z(z^{-1}) - (-z)^{-N} G_0^z(-z) G_0^z(-z^{-1}). \qquad (12.73)$$

Assuming further that N is chosen to be odd, we get

$$F^z(z) = z^{-N}[G_0^z(z) G_0^z(z^{-1}) + G_0^z(-z) G_0^z(-z^{-1})]. \qquad (12.74)$$

A sufficient condition for (12.74) to be of the form (12.62) is that

$$G_0^z(z) G_0^z(z^{-1}) + G_0^z(-z) G_0^z(-z^{-1}) = 1. \qquad (12.75)$$

Comparing this with (8.73), we see that (12.75) means that $G_0^z(z) G_0^z(z^{-1})$ is a zero-phase, half-band filter. We also see that

$$G_0^f(\theta)\bar{G}_0^f(\theta) + G_0^f(\theta - \pi)\bar{G}_0^f(\theta - \pi) = |G_0^f(\theta)|^2 + |G_0^f(\theta - \pi)|^2 = 1. \qquad (12.76)$$

A filter $G_0^z(z)$ satisfying (12.76) is said to be *power symmetric*. Thus, a filter $G_0^z(z)$ is power symmetric if $G_0^z(z)G_0^z(z^{-1})$ is a zero-phase half band. The design procedure of a CQF bank therefore concentrates on finding an odd-order, power-symmetric filter whose pass band is approximately $[0, 0.5\pi]$, and such that required pass-band and stop-band tolerances are met. Once such a filter has been found, the filter bank design is completed by using (12.71) and (12.59).

The design of a filter $G_0^z(z)$ that meets the aforementioned requirements is not straightforward. The first step is to design a zero-phase, half-band filter $R^z(z)$. Then this filter must be factored as

$$R^z(z) = G_0^z(z)G_0^z(z^{-1}). \tag{12.77}$$

Such a factorization is called *spectral factorization*, and is possible if and only if

$$R^f(\theta) \geq 0, \quad \text{for all} \quad -\pi \leq \theta \leq \pi. \tag{12.78}$$

Necessity of this condition is obvious, since $R^f(\theta) = |G_0^f(\theta)|^2$. Sufficiency will follow from the procedure described next. We are therefore faced with two tasks: (1) to design a zero-phase, half-band filter that meets (12.78), and (2) to carry out the spectral factorization (12.77). We now describe methods for performing these two tasks.

Nonnegative half-band filter design We describe two methods for designing a half-band filter that satisfies (12.78), one based on windows and one based on the Parks–McClellan algorithm. In the first method, we start by choosing a window $w[n]$ of even length $2M + 2$. We convolve this window with itself to get a window $\{w * w\}[n]$ of length $4M + 3$, and normalize the new window so that its middle coefficient will be 1. The new window has a nonnegative kernel function $[W^f(\theta)]^2$. We then apply this window to the ideal impulse response of a zero-phase, half-band filter, which is

$$h_d[n] = 0.5\mathrm{sinc}(0.5n). \tag{12.79}$$

The resulting impulse response is half band of order $4M + 2$, since $h_d[n]$ is half band (see Problem 8.11). Furthermore, the convolution of $H_d(\theta)$ with $[W^f(\theta)]^2$ is nonnegative for all θ, being the integral of two nonnegative functions. This method is simple to implement but requires experimentation with M and the window type, for meeting the tolerances. Also, as with all window-based methods, it does not yield a filter of the smallest possible order.

The second method is based on the idea developed in Problem 9.35 [Vaidyanathan and Nguyen, 1987]. There it is shown how to design an equiripple half-band filter, starting from a one-band filter designed by the Parks–McClellan algorithm. Suppose the half-band filter thus designed, $R_1^z(z)$, has an order $4M + 2$ and tolerances δ_p, δ_s. We assume that the coefficients of $R_1^z(z)$ are in the range $-(2M + 1) \leq n \leq 2M + 1$, so this filter is zero phase. Since the filter is equiripple, $R_1^f(\theta)$ is bounded from below by $-\delta_s$. Therefore, if we take

$$R^z(z) = \frac{0.5}{0.5 + \delta_s}[R_1^z(z) + \delta_s], \tag{12.80}$$

then we will have $R^f(\theta) \geq 0$, as required. The coefficients of $R^z(z)$ are obtained from those of $R_1^z(z)$ by

$$r[n] = \frac{0.5}{0.5 + \delta_s}(r_1[n] + \delta_s\delta[n]). \tag{12.81}$$

The factor $0.5/(0.5 + \delta_s)$ is necessary, because a zero-phase, half-band filter must satisfy $R^f(0.5\pi) = 0.5$ (see Problem 8.11). Since $R_1^f(0.5\pi) = 0.5$, this is exactly the factor that meets this requirement.

The frequency response $R^f(\theta)$ obtained by the aforementioned procedure will be exactly zero at the points of local minima in the stop band of $R_1^f(\theta)$. The spectral factorization procedure described next is numerically sensitive to these zeros. The accuracy of spectral factorization is greatly improved if $R^f(\theta)$ is strictly greater than zero at all frequencies, as in the case of window design. This can be achieved by a slight change in the procedure: We replace δ_s by δ_s', slightly larger than δ_s (for example, $\delta_s' = \mu\delta_s$, where μ is slightly larger than 1). You are asked to compute the pass-band and stop-band tolerances of $R^z(z)$ as a function of δ_p, δ_s, and δ_s' (Problem 12.22).

Spectral factorization Having obtained a filter $R^z(z)$ of order $4M + 2$, we are looking for a filter $G_0^z(z)$ of order $2M + 1$ that will satisfy (12.77). We observe that, if β is a zero of $G_0^z(z)$, then β^{-1} is a zero of $G_0^z(z^{-1})$, hence both β and β^{-1} are zeros of $R^z(z)$. Since we have designed $R^z(z)$ such that $R^f(\theta) > 0$ for all θ, $R^z(z)$ has no zeros on the unit circle.[2] Therefore, the $4M + 2$ zeros of $R^z(z)$ can be divided into two sets: the set $\{\beta_k, \ 1 \le k \le 2M + 1\}$ of zeros inside the unit circle, and the set of their reciprocals $\{\beta_k^{-1}, \ 1 \le k \le 2M + 1\}$. We can now construct $G_0^z(z)$ as

$$G_0^z(z) = C \prod_{k=1}^{2M+1} (1 - \tilde{\beta}_k z^{-1}), \tag{12.82}$$

where each of the $\tilde{\beta}_k$ can be taken as either β_k or β_k^{-1}. We thus have $2^{(2M+1)}$ different choices for $G_0^z(z)$, each satisfying (12.77). In particular, choosing $\tilde{\beta}_k = \beta_k$ for all k makes $G_0^z(z)$ a minimum-phase filter. In general, no choice can make $G_0^z(z)$ a linear-phase filter.

The constant C in (12.82) is computed as follows. We know that

$$r[0] = \sum_{n=0}^{2M+1} (g_0[n])^2 = 0.5, \tag{12.83}$$

because of the half-band property of $R^z(z)$. We therefore compute the coefficients of $G_0^z(z)$ by expanding (12.82) first with $C = 1$. We then take C^2 as half the reciprocal of the sum of squares of the coefficients, and multiply the vector of coefficients by C. The new coefficients now satisfy (12.83).

The procedure cqfw in Program 12.4 designs a CQF bank using the window-based method and the spectral factorization method we have described. It accepts an even-length window as input and provides the four filters (analysis and synthesis) as outputs. The program first convolves the window with itself, normalizes the result, and calls firdes to design a half-band filter with this window. The filter thus obtained is factored to its roots, and the set of roots inside the unit circle is selected. These are expanded to form the polynomial $G_0^z(z)$, which is then normalized to satisfy (12.83). Finally, $G_1^z(z)$ is obtained according to (12.71), and $H_0^z(z)$, $H_1^z(z)$ according to (12.59).

12.7.4 Tree-Structured Filter Banks

Splitting of a signal into two bands has limited usefulness. However, splitting into 2^L channels is possible for any integer L using a tree-structured filter bank, as shown in Figure 12.28 (using $L = 2$ as an example). Each of the two outputs of the first level is passed to a two-channel bank at a second level. This process can be continued according to the desired number of subbands. The filter pairs shown in Figure 12.28 are all equal. This, however, is not necessary; different pairs can be used at different levels.

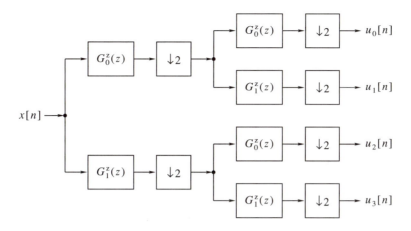

Figure 12.28 A tree-structured analysis filter bank.

The synthesis tree-structured filter bank corresponding to the bank in Figure 12.28 is shown in Figure 12.29. If the outputs of the analysis bank are fed to the synthesis bank, the complete system will have perfect reconstruction provided each two-channel bank has perfect reconstruction.

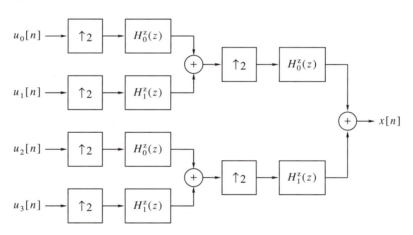

Figure 12.29 A tree-structured synthesis filter bank.

A full-blown tree, such as the those shown in Figures 12.28 and 12.29, is not always needed. A tree can be pruned by eliminating parts of certain levels. For example, the two bottom filters at the second level in the tree shown in Figure 12.28 can be eliminated. Then the filter bank will have only three outputs, two decimated by 4 and one decimated by 2. The synthesis bank must be constructed in a dual manner, to preserve the perfect reconstruction property.

Example 12.8 Compression of speech signals is highly desirable in all applications requiring transmission or storage of speech. Examples include commercial telephony, voice communication by radio, and storage of speech for multimedia applications. When speech is compressed and later reconstructed, the resulting quality usually undergoes a certain degradation. A common measure of speech quality is the mean opinion score (MOS). The MOS of a compressed speech is obtained as follows. An ensemble of people is recruited and listens to a sample of the reconstructed speech. Each person is asked to give a score between 1 to 5, where:

- 5 is for an excellent quality with imperceptible degradation;
- 4 is for a good quality with perceptible, but not annoying, degradation;
- 3 is for a fair quality with perceptible and slightly annoying degradation;
- 2 is for a poor quality with annoying, but not objectionable, degradation;
- 1 is for a bad quality with objectionable degradation.

The individual scores are averaged to give the MOS. For example, MOS of 4.5 and above is usually regarded as *toll quality*, meaning that it can be used for commercial telephony. MOS of 3 to 4 is regarded as *communications quality*, and is acceptable for many specialized applications. MOS lower than 3 is regarded as *synthetic quality*.

Speech signals are typically converted to a digital form by sampling at a rate 8 kHz or higher (up to about 11 kHz). A speech signal sampled at 8 kHz and quantized to 8 bits per sample has MOS of about 4.5. The corresponding rate is 64,000 bits per second. The *compression ratio* is defined as the number of bits per second before compression divided by that after compression.

One of the first speech compression techniques put into use was *subband coding*. In subband coding, the frequency range of the sampled speech is split into a number subbands by an analysis filter bank. The spectrum of a speech signal is decidedly nonuniform over the frequency range, so its subbands have different energies. This makes it possible to quantize each subband with a different number of bits. Reconstruction of the compressed speech consists of decoding each subband separately, then combining them to a full bandwidth signal using a synthesis filter bank.

The following scheme, due to Crochiere [1981], is typical of subband coding. The signal is sampled at 8 kHz, and split into four subbands, as shown in Figure 12.28. Each of these bands thus has a bandwidth of 1 kHz. Next, the band 0 to 1 kHz is split further into two bands. We therefore get a pruned tree with five bands: 0–0.5, 0.5–1, 1–2, 2–3, and 3–4 kHz. Three quantization schemes have been proposed:

1. Quantization to 5 bits per sample in the first two bands, 4 bits per sample in the third and fourth bands, and 3 bits per sample in the fifth band. The bit rate thus obtained is

$$(2 \times 5 \times 1000) + (2 \times 4 \times 2000) + 3 \times 2000 = 32,000.$$

This scheme achieves MOS of 4.3 [Daumer, 1982].

2. Quantization to 5 bits per sample in the first two bands, 4 bits per sample in the third band, 3 bits per sample in the fourth band, and 0 bits per sample in the fifth band (i.e., no use of this band). The bit rate thus obtained is

$$(2 \times 5 \times 1000) + 4 \times 2000 + 3 \times 2000 = 24,000.$$

This scheme achieves MOS of 3.9 [Daumer, 1982].

3. Quantization to 4 bits per sample in the first two bands, 2 bits per sample in the third and fourth bands, and 0 bits per sample in the fifth band. The bit rate thus obtained is

$$(2 \times 4 \times 1000) + (2 \times 2 \times 2000) = 16,000.$$

This scheme achieves MOS of 3.1 [Daumer, 1982]

Subband coding has been superseded by more efficient techniques and is not considered a state-of-art speech compression technique any more. However, it has recently gained popularity in compression of high-fidelity audio. A typical raw bit rate for high-fidelity audio is about 0.7 megabit per second per channel. The MPEG[3] standard for audio compression defines 32 subbands and allows compression down to 128 kilobits per second per channel. This represents a compression ratio of about 5.5 and gives almost CD-like music quality.   □

12.7.5 Octave-Band Filter Banks

An *octave-band filter bank* is a special kind of a pruned-tree filter bank, constructed according to the following rule: At each level, the high-pass output is pruned and the low-pass output is forwarded to the next level. Figure 12.30 shows a three-level octave-band filter bank. The left half forms the analysis bank, and the right half the synthesis bank. In this figure we have used a self-explanatory concise depiction of filtering–decimation and expansion–filtering.

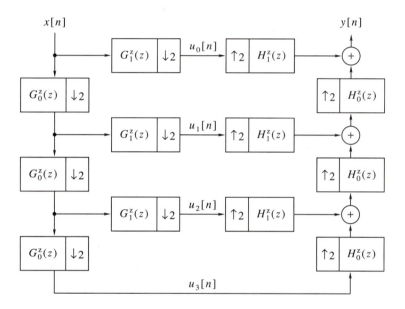

Figure 12.30 An octave-band filter bank.

The bandwidths of the output signals of the analysis bank are in octaves. For example, in Figure 12.30, the bandwidth of $u_0[n]$ is $[0.5\pi, \pi]$, that of $u_1[n]$ is $[0.25\pi, 0.5\pi]$, that of $u_2[n]$ is $[0.125\pi, 0.25\pi]$, and that of $u_3[n]$ is $[0, 0.125\pi]$. An L-level bank divides the full bandwidth into L octaves and yields $L + 1$ decimated signals. The last two signals have the same bandwidth, namely $2^{-L}\pi$. As in the case of a general pruned-tree filter bank, perfect reconstruction holds for the complete bank if it holds for the two-channel bank.

Octave-band filter banks are closely related to *wavelets*. The subject of wavelets has received enormous popularity in recent years, but is not discussed in this book; see Vetterli and Kovacevic [1995].

12.8 Uniform DFT Filter Banks*

12.8.1 Filter Bank Interpretation of the DFT

Uniform DFT filter banks offer a simple, yet powerful way of implementing a maximally decimated filter bank. As a motivation for this topic, consider a simple example. Suppose we pass the signal $x[n]$ through a chain of $M - 1$ delays and denote the output of the mth delay by

$$x_m[n] = x[n - m]. \tag{12.84}$$

Now compute, at each time point n, the M-point conjugate DFT of the vector

$$\{x_0[n], x_1[n], \ldots, x_{M-1}[n]\}$$

(the reason for conjugation will be explained soon). Denote by $u_m[n]$ the mth component of the output vector at time n. Then

$$u_m[n] = \sum_{i=0}^{M-1} x_i[n] W_M^{im} = \sum_{i=0}^{M-1} x[n - i] W_M^{im}. \tag{12.85}$$

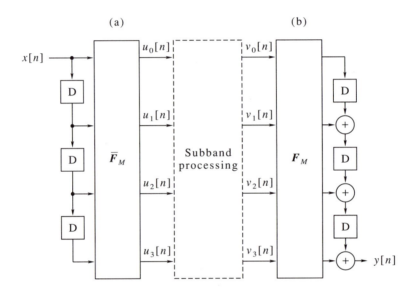

Figure 12.31 A simple uniform DFT filter bank ($M = 4$): (a) analysis bank; (b) synthesis bank.

The preceding operations are shown schematically in Figure 12.31(a), for $M = 4$. Recall that F_M is the DFT matrix [as defined in (4.19)] and $\bar{F}_M$ is its conjugate. As we see, equation (12.85) represents a linear convolution between the (potentially infinite) sequence $x[n]$ and the length-M sequence

$$g_m[n] = W_M^{nm}, \quad 0 \le n \le M - 1. \tag{12.86}$$

Therefore, this sequence can be regarded as the impulse response of a length-M FIR filter, and (12.85) can be regarded as the result of passing $x[n]$ through this filter. Since there are M such filters, we get a filter bank. Let us compute the transfer functions and frequency responses of these filters. We have

$$G_m^z(z) = \sum_{n=0}^{M-1} W_M^{nm} z^{-n} = \frac{1 - (zW_M^{-m})^{-M}}{1 - (zW_M^{-m})^{-1}}, \tag{12.87}$$

and

$$G_m^f(\theta) = \frac{1 - e^{-jM(\theta - 2\pi m/M)}}{1 - e^{-j(\theta - 2\pi m/M)}} = D\left(\theta - \frac{2\pi m}{M}, M\right) e^{-j0.5(\theta - 2\pi m/M)(M-1)}, \qquad (12.88)$$

where $D(\cdot, \cdot)$ is the Dirichlet kernel, defined in (6.7). Consider in particular the filter corresponding to $m = 0$. The transfer function and the frequency response of this filter are given by

$$G_0^z(z) = 1 + z^{-1} + \cdots + z^{-(M-1)} = \frac{1 - z^{-M}}{1 - z^{-1}}, \qquad (12.89)$$

$$G_0^f(\theta) = D(\theta, M)e^{-j0.5\theta(M-1)}. \qquad (12.90)$$

Then we see from (12.87) and (12.89) that

$$G_m^z(z) = G_0^z(zW_M^{-m}). \qquad (12.91)$$

Therefore, all M filters in the bank are obtained from the single prototype filter $G_0^z(z)$ through the relationship (12.91). Similarly,

$$G_m^f(\theta) = G_0^f(\theta - 2\pi m/M), \qquad (12.92)$$

so the frequency responses of the M filters are obtained from that of $G_0(\theta)$ by right shifts of $2\pi m/M$ in the frequency domain.

We summarize the properties of the DFT filter bank shown in Figure 12.31(a) as follows:

1. All filters in the bank have length M, equal to the size of the bank (which is also equal to the size of the DFT).

2. All filters are obtained from a single prototype filter $G_0^z(z)$ by the operation (12.91), which represents shifting in the frequency domain. The shifting is to the right because we used conjugate DFT in (12.85). If we were to use direct DFT, $G_m^f(\theta)$ would be related to $G_0^f(\theta)$ by shifting to the left.

3. The frequency response of the filter $G_0^z(z)$ has the shape of a Dirichlet kernel. It is therefore low pass, but its stop-band attenuation is only 13.5 dB and its pass-band response is mediocre. It must therefore be considered as an unsatisfactory filter for most applications.

4. The filters $G_m^z(z)$, $m \geq 1$ are band pass, having the same pass-band and stop-band properties as $G_0^z(z)$. Note that their coefficients are complex in general.

5. The entire filter bank can be implemented with $M - 1$ delays, which is the number of delays needed for a single filter.

Figure 12.31(b) shows how the filter bank outputs can be processed to get back the original signal $x[n]$. In the absence of subband processing, $v_m[n] = u_m[n]$ for all n and m. We first pass the vector

$$\{v_0[n], v_1[n], \ldots, v_{M-1}[n]\}$$

through a direct DFT matrix F_M. By the properties of the DFT, this yields the vector

$$\{Mx[n], Mx[n-1], \ldots, Mx[n-M+1]\}.$$

Observing how the chain of delays and adders is arranged in Figure 12.31(b), we conclude that the output $y[n]$ is equal to $M^2 x[n - M + 1]$. Therefore, the uniform DFT filter bank has the perfect reconstruction property: Its output is equal to the input up to a constant delay and a constant scale factor.

The uniform DFT filter bank shown in Figure 12.31 is not decimated. All signals have the same rate as the input signal $x[n]$. Further examination of this figure reveals that it is highly redundant. We will not lose any information if we decimate the delay outputs by M before feeding them to the matrix $\bar{F}_M$. Then, at time n, the input vector to this matrix will be

$$\{x[nM], x[nM - 1], \ldots, x[nM - M + 1]\}.$$

Consecutive vectors are now made of consecutive nonoverlapping segments of length M of the input signal. Now the DFT is performed once every M points, instead of at each time point. The synthesis bank is treated in a similar manner: the DFT output vector is expanded M-fold and passed through a delay chain. Now the output $y[n]$ is equal to $Mx[n - M + 1]$, so the perfect reconstruction property is preserved. Figure 12.32 illustrates the analysis and synthesis DFT filter banks in their decimated forms. In this figure we have omitted the subband processing block for simplicity.

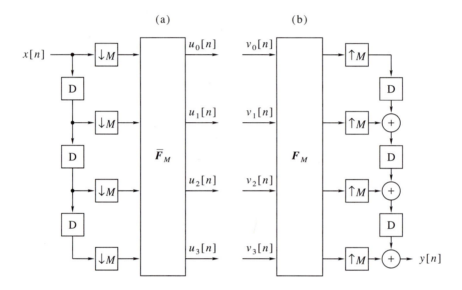

Figure 12.32 A maximally decimated version of the uniform DFT filter bank shown in Figure 12.31: (a) analysis bank; (b) synthesis bank.

The maximally decimated uniform DFT filter bank shown in Figure 12.32 can also be implemented with commutators at the input and at the output instead of the delay chains, as in Figures 12.17 and 12.19.

12.8.2 Windowed DFT Filter Banks

The DFT filter bank we have described suffers from a major drawback, which severely limits its practicality: the inferior frequency response characteristics of the filters $G_m^z(z)$. A simple modification partially corrects this problem: Multiply the input vector to the analysis DFT by a length-M window $\{w[m], \ 0 \leq m \leq M - 1\}$. Then (12.85) changes to

$$u_m[n] = \sum_{i=0}^{M-1} x[n - i]w[i]W_M^{im}, \tag{12.93}$$

so the impulse responses of the filters become

$$g_m[n] = w[n]W_M^{nm}, \quad 0 \leq n \leq M - 1. \tag{12.94}$$

The frequency response of the prototype filter becomes

$$G_0^f(\theta) = \sum_{n=0}^{M-1} w[n]e^{-j\theta n}, \tag{12.95}$$

which is the kernel function of the window. Thus, we can control the stop-band attenuation of the filter by an appropriate choice of window, for example, by using a Kaiser
window with α chosen according to the desired stop-band attenuation. The pass-band
behavior cannot be controlled in this manner, however, and may remain unsatisfactory
for certain applications.

The synthesis filter bank can be modified to achieve perfect reconstruction by multiplying the outputs of the synthesis DFT by the sequence $\{1/w[m], \ 0 \le m \le M - 1\}$.
This is possible if and only if none of the window coefficients is zero.

12.8.3 A Uniform DFT Filter Bank of Arbitrary Order

The reason for the unsatisfactory characteristics of the uniform DFT filter banks presented above is the limit on the order of the individual filters, that is, the restriction
that it must be equal to $M - 1$. We now show how this restriction can be removed, thus
enabling the filters to meet any desired pass-band and stop-band specifications.

The defining property of a uniform DFT filter bank with an arbitrary prototype
filter $G_0^z(z)$ is

$$G_m^z(z) = G_0^z(zW_M^{-m}), \quad 1 \le m \le M - 1. \tag{12.96}$$

In other words, all band-pass filters are obtained from the prototype filter by shifting
the frequency response along the θ axis by an integer multiple of $2\pi/M$.

Suppose we have designed $G_0^z(z)$ to meet certain transition bandwidth, pass-band
ripple, and stop-band attenuation. Let

$$G_0^z(z) = \sum_{i=0}^{M-1} z^{-i} P_i^z(z^M) \tag{12.97}$$

be the polyphase decomposition of $G_0^z(z)$. Then $G_m^z(z)$ is given by

$$G_m^z(z) = G_0^z(zW_M^{-m}) = \sum_{i=0}^{M-1} (zW_M^{-m})^{-i} P_i^z(z^M W_M^{-mM}) = \sum_{i=0}^{M-1} W_M^{im} z^{-i} P_i^z(z^M). \tag{12.98}$$

Let $x[n]$ be the input sequence and $u_m[n]$ the output of the mth filter. Then

$$U_m^z(z) = \sum_{i=0}^{M-1} W_M^{im} P_i^z(z^M) z^{-i} X^z(z). \tag{12.99}$$

By the decimation identity, we can write the z-transform of the decimated sequence
$\{u_m\}_{(\downarrow M)}[n]$ as

$$\{U_m^z\}_{(\downarrow M)}(z) = \sum_{i=0}^{M-1} W_M^{im} P_i^z(z) \{z^{-i} X^z(z)\}_{(\downarrow M)}. \tag{12.100}$$

Figure 12.33 shows an implementation of (12.100). As we see, the sequence of
operations is as follows:

1. Delay and decimate the input signal $x[n]$ (a commutator can be used for this).
2. Pass each of the decimated sequences through the appropriate polyphase component of $G_0^z(z)$.
3. Perform M-point conjugate DFT on the output vector of the polyphase filters at
   a rate $1/M$ of the input signal rate.

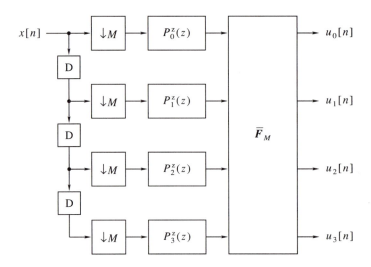

Figure 12.33 A maximally decimated uniform DFT analysis filter bank (shown for $M = 4$).

The structure in Figure 12.33 is called a *maximally decimated uniform DFT analysis filter bank*. The DFT filter bank shown in Figure 12.32(a) is a special case, where all the polyphase components are $P_m^z(z) = 1$.

The filter bank shown in Figure 12.33 can be implemented efficiently. Denote the order of the prototype filter by N. Each polyphase filter performs approximately N/M complex operations every M samples of the input signal. Together they perform N/M complex operations per sample of the input signal. Then, about $0.5M \log_2 M$ complex operations are needed for the DFT every M samples of the input signal, or $0.5 \log_2 M$ per sample. The total is about $N/M + 0.5 \log_2 M$ complex operations per sample of the input signal. At this cost we get M filtering–decimation operations, each of order N. Furthermore, we have the freedom of choosing any prototype filter according to the given specifications.

The procedure `udftanal` in Program 12.5 gives a MATLAB implementation of a maximally decimated uniform DFT analysis filter bank. It is similar to `ppdec`, except that the filtered decimated signals are not combined, but passed as a vector to the IFFT routine. Note also that the MATLAB function `ifft` operates on each column of the matrix u individually, and that it gives an additional scale factor $1/M$.

Figure 12.34 shows a maximally decimated uniform DFT synthesis filter bank. The filters $Q_m^z(z)$ are the polyphase components of the prototype synthesis filter $H_0^z(z)$, indexed in reversed order, according to the convention used for expansion [cf. (12.38)]. The sequence of operations carried out by the synthesis bank is as follows:

1. Perform M-point DFT on the input vector.

2. Pass each of the DFT outputs through the appropriate polyphase component.

3. Expand, delay, and sum the polyphase filters' outputs (a commutator can be used for this).

The procedure `udftsynt` in Program 12.6 gives a MATLAB implementation of a maximally decimated uniform DFT synthesis filter bank.

We recall that the simple uniform DFT filter bank shown in Figure 12.32 has the perfect reconstruction property. This property is not shared by a general uniform DFT filter bank. To see the source of the problem, assume we connect the filter banks in Figures 12.33 and 12.34 in tandem, so $v_m[n] = u_m[n]$. Then the conjugate DFT and the

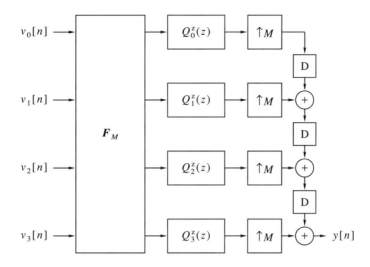

Figure 12.34 A maximally decimated uniform DFT synthesis filter bank (shown for $M = 4$).

DFT cancel each other, except for a scale factor M. The output $Y^z(z)$ can be expressed in terms of the input $X^z(z)$ as follows:

$$Y^z(z) = M \sum_{i=0}^{M-1} \{\{z^{-i}X^z(z)\}_{(\downarrow M)} P_i^z(z)Q_i^z(z)\}_{(\uparrow M)} z^{-(M-1-i)}. \tag{12.101}$$

Using (12.12) for the operation ($\downarrow M$), we get

$$Y^z(z) = \sum_{i=0}^{M-1} \left\{ \left[\sum_{m=0}^{M-1} (z^{1/M}W_M^{-m})^{-i} X^z(z^{1/M}W_M^{-m}) \right] P_i^z(z)Q_i^z(z) \right\}_{(\uparrow M)} z^{-(M-1-i)}. \tag{12.102}$$

Using (12.15) for the operation ($\uparrow M$), we get

$$Y^z(z) = \sum_{i=0}^{M-1}\sum_{m=0}^{M-1} (zW_M^{-m})^{-i} X^z(zW_M^{-m}) P_i^z(z^M)Q_i^z(z^M) z^{-(M-1-i)}$$

$$= z^{-(M-1)} \sum_{m=0}^{M-1} X^z(zW_M^{-m}) \sum_{i=0}^{M-1} W_M^{mi} P_i^z(z^M)Q_i^z(z^M). \tag{12.103}$$

Hence we arrive at the following conclusion: A maximally decimated uniform DFT analysis-synthesis filter bank has the perfect reconstruction property if and only if the polyphase components of the prototype filters satisfy

$$\frac{1}{M} \sum_{i=0}^{M-1} W_M^{mi} P_i^z(z^M)Q_i^z(z^M) = cz^{-l}\delta[m] = \begin{cases} cz^{-l}, & m = 0, \\ 0, & m \neq 0, \end{cases} \tag{12.104}$$

for a nonzero constant c and an integer delay l. The left side of (12.104) is the inverse DFT of the sequence $\{P_i^z(z^M)Q_i^z(z^M),\ 0 \le i \le M - 1\}$. Therefore, (12.104) holds if and only if

$$P_i^z(z^M)Q_i^z(z^M) = cz^{-l}, \quad 0 \le i \le M - 1. \tag{12.105}$$

This condition has been derived by Portnoff [1980].

The simple uniform DFT filter bank has the perfect reconstruction property, since

$$P_i^z(z) = Q_i^z(z) = 1, \quad 0 \le i \le M - 1.$$

Similarly, the windowed uniform DFT filter bank has the perfect reconstruction

property, since
$$P_i^z(z) = w[i], \ Q_i^z(z) = 1/w[i], \quad 0 \le i \le M - 1.$$

Nontrivial solutions to (12.105), in which the prototype filters $G_0^z(z)$, $H_0^z(z)$ are FIR of order larger than M, have been found by Farkash and Raz, but we shall not discuss them here. Nontrivial solutions also exist if we allow IIR filters, but these are not of practical interest for this application.

We remark, in conclusion, that the lack of perfect reconstruction does not mean that the filter bank is not useful. DFT filter banks have many applications, especially when analysis alone is required, without synthesis.

12.9   Summary and Complements

12.9.1   Summary

This chapter has served as an introduction to the area of multirate signal processing, an area of increasing importance and popularity.

The basic multirate operations are decimation and expansion. Decimation is similar to sampling, in that it aliases the spectrum in general, unless the signal has sufficiently low bandwidth prior to decimation. Expansion does not lead to aliasing, but it generates images in the frequency domain. Therefore, decimation and expansion are almost always accompanied by filtering. A decimation filter acts like an antialiasing filter: It precedes the decimator and its purpose is to limit the signal bandwidth to $\pm\pi/M$, where M is the decimation ratio. An expansion filter acts to interpolate the expanded signal, so it is usually called an interpolation filter. It succeeds the expander and its purpose is to limit the signal bandwidth to $\pm\pi/L$, where L is the expansion ratio.

Sampling-rate conversion is a combined operation of expansion, filtering, and decimation. A sampling-rate converter changes the sampling rate of a signal by a rational factor L/M.

Filters used for decimation, interpolation, and sampling-rate conversion are usually FIR. Decimation and interpolation allow for considerable savings in the number of operations. Polyphase filter structures are particularly convenient for this purpose.

Filter banks are used for either separating a signal to several frequency bands (analysis banks) or for combining signals at different frequency bands to one signal (synthesis banks). To increase computational efficiency and to reduce the data rate, filter banks are usually decimated; that is, the signal rate at each band is made proportional to the bandwidth. A useful property that an analysis–synthesis filter bank pair may have is perfect reconstruction. Perfect reconstruction means that a signal passing through an analysis bank and through a synthesis bank is unchanged, except for a delay and a constant scale factor.

The simplest filter banks have two channels. The simplest of those are quadrature mirror filter banks. QMF banks do not have perfect reconstruction, but they can be made nearly so if designed properly. Conjugate quadrature filter banks, on the other hand, do have perfect reconstruction. These are, however, more difficult to design. Also, the individual filters do not have linear phase.

A filter bank having more than two channels can be built from two-channel filter banks, by connecting them in a tree structure. A tree-structured filter bank can be full or pruned. A special case of a pruned tree is the octave-band filter bank. In such a bank, the bandwidths of the signals occupy successive octave ranges.

When there is a need for a filter bank having more than two channels, a tree-structured, two-channel bank is not necessarily the most efficient. Of the many schemes for M-channel filter banks developed in recent years, we presented only the uniform DFT filter bank. A uniform DFT analysis filter bank uses a single prototype filter, whose polyphase components are implemented separately, and their outputs undergo length-M DFT. The synthesis filter inverts the DFT operation and reconstructs the signal by a prototype interpolation filter implemented by its polyphase components. A uniform DFT filter bank is easy to design and implement, but it does not possess the perfect reconstruction property in general (except for simple cases, whose usefulness is limited).

The first book completely devoted to multirate systems and filter banks is by Crochiere and Rabiner [1983]. Recent books on this subject are by Vaidyanathan [1993], Fliege [1994], and Vetterli and Kovacevic [1995].

12.9.2 Complements

1. [p. 488] Two-channel IIR filter banks are discussed in Vaidyanathan [1993, Sec. 5.3].

2. [p. 492] Spectral factorization by root computation is also possible if $R^z(z)$ has zeros on the unit circle. Each such zero must have multiplicity 2. One out of every two such zeros is included in the set of zeros of $G_0^z(z)$. However, root-finding programs are sensitive to double zeros. Even worse, a small error in δ_s (when equiripple design is used) may cause the condition $R^f(\theta) \geq 0$ to be violated. In this case, the double zero splits into two simple zeros on the unit circle, rendering the zero selection procedure impossible.

3. [p. 495] MPEG is an acronym for Motion Picture Experts Group, a standard adopted by the International Standards Organization (ISO) for compression and coding of motion picture video and audio. The definitive documents for this standard are ISO/IEC JTC1 CD 11172, *Coding of Moving Pictures and Associated Audio for Digital Storage Media up to 1.5 Mbits/s*, 1992, and ISO/IEC JTC1 CD 13818, *Generic Coding of Moving Pictures and Associated Audio*, 1994.

12.10   MATLAB Programs

Program 12.1 Filtering and decimation by polyphase decomposition.

```
function y = ppdec(x,h,M);
% Synopsis: y = ppdec(x,h,M).
% Convolution and M-fold decimation, by polyphase decomposition.
% Input parameters:
% x: the input sequence
% h: the FIR filter coefficients
% M: the decimation factor.
% Output parameters:
% y: the output sequence.

lh = length(h); lp = floor((lh-1)/M) + 1;
p = reshape([reshape(h,1,lh),zeros(1,lp*M-lh)],M,lp);
lx = length(x); ly = floor((lx+lh-2)/M) + 1;
lu = floor((lx+M-2)/M) + 1; % length of decimated sequences
u = [zeros(1,M-1),reshape(x,1,lx),zeros(1,M*lu-lx-M+1)];
u = flipud(reshape(u,M,lu)); % the decimated sequences
y = zeros(1,lu+lp-1);
for m = 1:M, y = y + conv(u(m,:),p(m,:)); end
y = y(1:ly);
```

Program 12.2 Expansion and filtering by polyphase decomposition.

```
function y = ppint(x,h,L);
% Synopsis: y = ppint(x,h,L).
% L-fold expansion and convolution, by polyphase decomposition.
% Input parameters:
% x: the input sequence
% h: the FIR filter coefficients
% L: the expansion factor.
% Output parameters:
% y: the output sequence.

lh = length(h); lq = floor((lh-1)/L) + 1;
q = flipud(reshape([reshape(h,1,lh),zeros(1,lq*L-lh)],L,lq));
lx = length(x); ly = lx*L+lh-1;
lv =lx + lq; % length of interpolated sequences
v = zeros(L,lv);
for l = 1:L, v(l,1:lv-1) = conv(x,q(l,:)); end
y = reshape(flipud(v),1,L*lv);
y = y(1:ly);
```

Program 12.3 Sampling-rate conversion by polyphase decomposition.

```
function y = ppsrc(x,h,L,M);
% Synopsis:  y = ppsrc(x,h,L,M).
% Sampling-rate conversion by polyphase filters.
% Input parameters:
% x: the input sequence
% h: the conversion filter
% L, M: the interpolation and decimation factors.
% Output parameters:
% y: the output sequence

ML = M*L; lh = length(h); lx = length(x);
ly = floor((L*lx+lh-2)/M)+1; % length of the result
K = floor((lh-1)/ML)+1; % max length of polyphase components
r = zeros(ML,K); % storage for polyphase components
for l = 0:L-1, % build polyphase components
for m = 0:M-1,
   temp = h(rem(l*M+(M-m)*L,ML)+1:ML:lh);
   if (length(temp) > 0),
   r(M*l+m+1,1:length(temp)) = temp; end
end, end
x = [reshape(x,1,lx),zeros(1,M)]; % needed for the 1 delay
lx = lx + M;
lu = floor((lx-1)/M) + 1; % length of the sequences u_m
x = [x, zeros(1,M*lu-lx)];
x = reshape(x,M,lu); % now the rows of x are the u_m
y = zeros(L,K+lu-1);
for l = 0:L-1, % loop on sequences v_l
for m = 0:M-1, % loop on sequences u_m
   if (m <= floor(l*M/L)),
      temp = x(m+1,:);
   else
      temp = [0,x(m+1,1:lu-1)];
   end
   y(l+1,:) = y(l+1,:) + conv(r(M*l+m+1,:),temp);
end, end
y = reshape(y,1,L*(K+lu-1));
y = y(1:ly);
```

Program 12.4 Conjugate quadrature filter design by windowing.

```
function [g0,g1,h0,h1] = cqfw(w);
% Synopsis: [g0,g1,h0,h1] = cqfw(w).
% Designs a Smith-Barnwell CQF filter bank by windowing.
% Input parameters:
% w: the window; must have even length, which will also be
%    the length of the filters.
% Output parameters:
% g0, g1: the analysis filters
% h0, h1: the synthesis filters

N = length(w)-1; % order of the output filters
w = conv(w,w); w = (1/w(N+1))*w;
r = firdes(length(w)-1,[0,0.5*pi,1],w);
rr = roots(r);
g0 = real(poly(rr(find(abs(rr) < 1))));
g0 = reshape(g0/sqrt(2*sum(g0.^2)),1,N+1);
h1 = (-1).^(0:N).*g0; g1 = fliplr(h1);
h0 = 2*fliplr(g0); h1 = 2*h1;
```

Program 12.5 A maximally decimated uniform DFT analysis filter bank, implemented by polyphase filters.

```
function u = udftanal(x,g,M);
% Synopsis: u = udftanal(x,g,M).
% Maximally decimated uniform DFT analysis filter bank.
% Input parameters:
% x: the input sequence
% g: the FIR filter coefficients
% M: the decimation factor.
% Output parameters:
% u: a matrix whose rows are the output sequences.

lg = length(g); lp = floor((lg-1)/M) + 1;
p = reshape([reshape(g,1,lg),zeros(1,lp*M-lg)],M,lp);
lx = length(x); lu = floor((lx+M-2)/M) + 1;
x = [zeros(1,M-1),reshape(x,1,lx),zeros(1,M*lu-lx-M+1)];
x = flipud(reshape(x,M,lu)); % the decimated sequences
u = [];
for m = 1:M, u = [u; conv(x(m,:),p(m,:))]; end
u = ifft(u);
```

Program 12.6 A maximally decimated uniform DFT synthesis filter bank, implemented by polyphase filters.

```
function y = udftsynt(v,h,M);
% Synopsis: y = udftsynt(v,h,M).
% Maximally decimated uniform DFT synthesis filter bank.
% Input parameters:
% v: a matrix whose rows are the input sequences
% h: the FIR filter coefficients
% M: the expansion factor.
% Output parameters:
% y: the output sequence

lh = length(h); lq = floor((lh-1)/M) + 1;
q = flipud(reshape([reshape(h,1,lh),zeros(1,lq*M-lh)],M,lq));
v = fft(v);
y = [];
for m = 1:M, y = [conv(v(m,:),q(m,:)); y]; end
y = y(:).';
```

12.11　Problems

12.1 Prove that both decimation and expansion are linear operations.

12.2 Let $x[n] = \cos(\theta_0 n + \phi_0)$. Express $x_{(\downarrow M)}[n]$ in the time and frequency domains. Discuss possible aliasing of the decimated signal as a function of the parameters θ_0 and M.

12.3 Repeat Problem 12.2 for $x_{(\uparrow L)}[n]$.

12.4 A signal $x[n]$ is decimated by M, and the result is expanded by M to yield a signal $y[n]$. Express the z-transform of $y[n]$ in terms of the z-transform of $x[n]$.

12.5 The z-transform of a signal obtained by M-fold decimation followed by M-fold expansion contains M components, as you have obtained in Problem 12.4. Express each of these components in the time domain as a function of $x[n]$, and interpret the result. Discuss the special case $M = 2$. The last $M - 1$ terms in the sum are called the *alias components* of $x[n]$.

12.6 We wish to perform linear interpolation of a signal $x[n]$ by a factor L. Linear interpolation is defined by

$$y[nL + i] = \frac{L - i}{L} x[n] + \frac{i}{L} x[n + 1], \quad -\infty < n < \infty, \ 0 \le i \le L - 1.$$

Show that linear interpolation can be performed by the standard expansion/filtering scheme (12.19), and find the impulse response of the interpolation filter.

12.7 Consider the moving-average filter introduced in Problem 11.4. Suppose the output is to be decimated by N.

(a) Derive an expression for the Fourier transform of the decimated output as a function of the Fourier transform of the input.

(b) Suggest an efficient implementation of the filter–decimator.

12.8 This problem examines the possibility of representing IIR filters by polyphase components.

(a) Find the polyphase components $P_m^z(z)$ of the first-order IIR filter

$$H^z(z) = \frac{1}{1 - \alpha z^{-1}}.$$

What is the order of each polyphase component?

(b) Use part a for finding the polyphase components $P_m^z(z)$ for the second-order IIR filter (where A and α are complex)

$$H^z(z) = \frac{A}{1 - \alpha z^{-1}} + \frac{\bar{A}}{1 - \bar{\alpha} z^{-1}}.$$

What is the order of each component?

(c) Use parts a and b for finding the polyphase components of any IIR filter whose poles are simple from its partial fraction decomposition (11.11). What is the order of each component?

(d) Suppose we implement IIR filtering followed by decimation as in Figure 12.16, using the polyphase components derived in part c. Would that lead to computational saving? If so, by what factor?

12.9 This problem shows how to implement a *fractional delay* using interpolation and decimation. The ideal frequency response of fractional delay is

$$H^f(\theta) = e^{-j\theta\tau},$$

where τ is noninteger. We assume that τ is rational, that is, $\tau = l/M$, where l, M are coprime positive integers. The idea is to interpolate the input signal $x[n]$ by a factor M, then delay the interpolated signal by l time units, and finally decimate the result by a factor M.

(a) Explain why this method works in principle, why it only approximates the ideal frequency response, and what errors are introduced in the process.

(b) Show that only one polyphase component of the interpolation filter is actually used in this procedure.

12.10 A discrete-time signal $x[n]$ is reconstructed using a causal reconstructor $h(t)$ [cf. (3.31)]:

$$\hat{x}(t) = \sum_{m=-\infty}^{\lfloor t/T \rfloor} x[m]h(t - mT).$$

The reconstructed signal is sampled at interval T/L, where L is an integer, to yield a discrete time signal $y[n] = \hat{x}(nT/L)$. Write an expression for $y[n]$ in terms of $x[n]$, and relate the result to the material in Sections 10.7 and sec:decexplinfil.

12.11 Let $h[n]$ be the impulse response of a linear-phase FIR filter. Are the polyphase components of the filter linear phase in general? What about the special case $M = 2$? Distinguish between even and odd N.

12.12* Suppose we need to design a low-pass digital FIR filter with $\theta_s \ll \pi$. The tolerance parameters are δ_p, δ_s. The output rate must be equal to the input rate. The standard approach is to design the filter by one of the methods studied in Chapter 9. Figure 12.35 shows a multirate alternative to the standard approach. The signal is low-pass filtered by $H_1^z(z)$, whose specification parameters are $\theta_p, \theta_s, \tilde{\delta}_{p,1}, \tilde{\delta}_{s,1}$. Then the output is decimated by a factor $M = \lfloor \pi/\theta_s \rfloor$. The decimated signal $u[n]$ is immediately expanded by a factor M to yield the signal $v[n]$. Finally, $v[n]$ is low-pass filtered by $H_2^z(z)$, whose specification parameters are $\theta_p, \theta_s, \tilde{\delta}_{p,2}, \tilde{\delta}_{s,2}$.

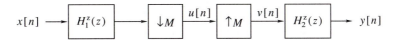

Figure 12.35 Pertaining to Problem 12.12.

The main idea behind the multirate scheme is that the two filters will require only N_1/M and N_2/M operations per point, respectively (where N_1, N_2 are the respective orders). Therefore, we may hope to achieve a lower operation count than in common unirate filtering. The possibility of achieving this is not guaranteed, however, because each of the two filters must meet tighter tolerances than the unirate filter. In addition, decimation leads to aliasing, so $\tilde{\delta}_{s,1}$ must be set independently of the other parameters to keep the aliasing level below a prescribed limit.

The aim of this problem is to determine the tolerance parameters of the two low-pass filters, to estimate their orders, and to determine the conditions under which the multirate scheme is more efficient than a unirate filter. We assume that equiripple design is used for all filters, so the order formula (9.79) is applicable.

(a) Show that the total pass-band ripple of the output signal $y[n]$ can be up to $\tilde{\delta}_{p,1} + \tilde{\delta}_{p,2}$.

(b) Show that the total stop-band attenuation of the output signal is equal to $\tilde{\delta}_{s,2}$.

(c) Find $\tilde{\delta}_{p,1}, \tilde{\delta}_{p,2}$ that will minimize $N_1 + N_2$ subject to the constraint

$$\tilde{\delta}_{p,1} + \tilde{\delta}_{p,2} = \delta_p,$$

where δ_p is fixed. Hint: Use (9.79) and note that $\tilde{\delta}_{s,1}, \tilde{\delta}_{s,2}, M$, and $|\theta_p - \theta_s|$ are fixed.

(d) For $\tilde{\delta}_{p,1}, \tilde{\delta}_{p,2}$ found in part c, find a condition on M as a function of $\delta_p, \delta_{s,1}, \delta_{s,2}$ such that

$$\frac{N_1 + N_2}{M} < N,$$

where N is the order of a unirate filter that meets the specifications. This condition can be used for deciding when multirate low-pass filtering is preferable to unirate filtering.

(e) Consider a low-pass filter required to meet the following specifications:

$$\theta_p = 0.02\pi, \ \theta_s = 0.04\pi, \ \delta_p = 0.05, \ \delta_s = 0.001.$$

Assume that $\tilde{\delta}_{s,1} = 0.001$. Compute the number of operations per data point in unirate filtering and in multirate filtering.

12.13* Repeat Example 12.6 for $M_1 = 4$, $M_2 = 3$. Compare the efficiency of this scheme to the one presented in Example 12.6.

12.14* Design a three-stage up-sampling system for compact disc, as described in Example 12.5. The input and output signal rates are 44.1 and 352.8 kHz. Each stage should up-sample the signal by 2. Assume that the input signal contains no energy at frequencies higher than 20 kHz. The required tolerances are $\delta_p = 0.001$, $\delta_s = 0.0001$. As an answer, give the specifications and the orders of the three filters (there is no need to give the coefficients).

12.15* Show that $H_0^z(z)$ and $H_1^z(z)$ defined in (12.59) are low pass and high pass, respectively.

12.16* Consider a QMF filter bank constructed from FIR filters $G_0^z(z), G_1^z(z)$, satisfying (12.63). Show that the only case in which the transfer function $F^z(z)$ can be of the form (12.65) is when $G_0^z(z)$ has two nonzero coefficients at most. Hint: Express $F^z(z)$ as a product of two FIR transfer functions. The conclusion from this result is that QMF FIR filter banks constructed according to the condition (12.63) cannot have the perfect reconstruction property, except when $G_0^z(z)$ has a particularly simple form, which is of little use. Therefore, to preserve both FIR and perfect reconstruction properties, the restriction (12.63) must be removed.

12.17* Let the analysis filter $G_0^z(z)$ of a QMF bank be expressed in terms of its two polyphase components $P_{0,0}^z(z), P_{0,1}^z(z)$.

(a) Express each of the three other filters in the bank, $G_1^z(z), H_0^z(z), H_1^z(z)$, in terms of $P_{0,0}^z(z)$ and $P_{0,1}^z(z)$.

(b) Construct a realization of the complete bank (analysis and synthesis) in terms of the polyphase components, using the polyphase identities.

12.18* A *two-channel time division multiplexer* (TDM) is a device that accepts two signals $x_0[n]$, $x_1[n]$ and outputs the signal

$$y[n] = \begin{cases} x_0[m], & n = 2m, \\ x_1[m], & n = 2m + 1. \end{cases}$$

(a) Show how to implement a two-channel TDM using two expanders and a delay element.

(b) Show how to split the signal $y[n]$ back into its components using two decimators and a delay element.

12.19* A *two-channel frequency division multiplexer* (FDM) is a device that accepts two signals $x_0[n]$, $x_1[n]$ and outputs a signal $y[n]$ such that $y[n]$ has twice the rate (hence twice the bandwidth) of each of the $x_i[n]$, and the spectrum of $y[n]$ in the band $[0, 0.5\pi]$ replicates the full spectrum of $x_0[n]$, whereas the spectrum of $y[n]$ in the band $[0.5\pi, \pi]$ replicates the full spectrum of $x_1[n]$.

(a) Show how to implement a two-channel FDM using two expanders and two filters (assume that the filters are ideal).

(b) Show how to split the signal $y[n]$ back into its components using two decimators and two filters (assume that the filters are ideal).

12.20* The scheme shown in Figure 12.36 is called a *two-channel transmultiplexer*.

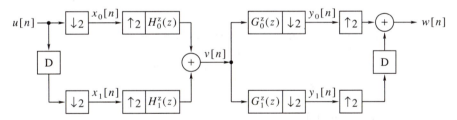

Figure 12.36 A two-channel transmultiplexer (pertaining to Problem 12.20).

(a) Explain what a transmultiplexer performs. Hint: Interpret the relationships between $\{x_0[n], x_1[n]\}$ and $u[n]$, between $v[n]$ and $\{x_0[n], x_1[n]\}$, between $\{y_0[n], y_1[n]\}$ and $w[n]$, and between $w[n]$ and $\{y_0[n], y_1[n]\}$. Use your solutions to Problems 12.18 and 12.19.

(b) Derive the z-transform relationships between the pair $\{Y_0^z(z^2), Y_1^z(z^2)\}$ and the pair $\{X_0^z(z^2), X_1^z(z^2)\}$.

(c) Let the synthesis filters be related to the analysis filters according to

$$H_0^z(z) = 2z^{-1} G_1^z(-z), \quad H_1^z(z) = -2z^{-1} G_0^z(-z)$$

[this is the same as (12.59) except for the additional delay]. Show that this makes $Y_0^z(z^2)$ independent of $X_1^z(z^2)$ and $Y_1^z(z^2)$ independent of $X_0^z(z^2)$. Such a dependence is called *cross-talk*. The aforementioned choice of synthesis filters eliminates cross-talk, in the same way that aliasing is eliminated in a two-channel filter bank.

(d) Show that taking $G_0^z(z)$ and $G_1^z(z)$ to be conjugate quadrature filters leads to perfect reconstruction from $X_0^z(z^2)$ to $Y_0^z(z^2)$ and from $X_1^z(z^2)$ to $Y_1^z(z^2)$.

12.21* Derive (12.72) from (12.71).

12.22* Refer to the design method of nonnegative half-band filters (12.80), but with $\delta_s' = \mu \delta_s$ instead of δ_s. Express the pass-band and stop-band tolerances $\widetilde{\delta}_p$, $\widetilde{\delta}_s$ of $R^z(z)$ in terms of δ_p, δ_s, and μ. Then invert these relationships and find δ_p, δ_s necessary to obtain prescribed values of $\widetilde{\delta}_p$, $\widetilde{\delta}_s$ (assuming μ is fixed).

12.23* Let each of the filters in an L-level, two-channel, tree-structured analysis filter bank have order N. Estimate the total number of operations per input data point.

12.24* Let each of the filters in an L-octave analysis filter bank (such as the one shown in Figure 12.30) have order N. Estimate the total number of operations per input data point.

12.25* Write a MATLAB program that implements an octave-band analysis filter bank. The inputs to the program are the two analysis filters (assumed to be FIR), the number of octaves L, and the input signal. The program outputs are the $L + 1$ output signals.

$$MSE = \emptyset \quad \text{optimize}$$
$$\text{wrt } a.$$

$$\frac{\partial}{\partial x}\left[E\left[(x - \hat{x})J \right]^2 \right] = \emptyset$$

leads to
auto correlation
if YuleWalker.

$$Ra = -r$$

$$e(n) = (d(n) - \Sigma \omega(v))^2$$

$$\frac{\partial e(n)^2}{\partial \omega} = \phi$$

Chapter 13

$$E[d(n)] = E[e(n)e^*(n)]$$

Analysis and Modeling of Random Signals*

The random signals encountered in this book until now were chiefly white noise sequences. White noise sequences appeared in Section 6.5, in the context of frequency measurement, and in Section 11.7, in the context of quantization noise. We devote this chapter to a more general treatment of random signals, not necessarily white noise sequences. Randomness is inherent in many physical phenomena. Communication, radar, biomedicine, acoustics, imaging—are just a few examples in which random signals are prevalent. It is not exaggeration to say that knowledge of DSP cannot be regarded as satisfactory if it does not include methods of analysis of random signals. The methods presented in this chapter are elementary. We begin by discussing simple methods for estimating the power spectral density of a WSS random signal: the periodogram and some of its variations. Most of the chapter, however, is devoted to parametric modeling of random signals. We present general rational models for WSS random signals, then specialize to all-pole, or autoregressive models, and finally discuss joint signal modeling by FIR filters.

13.1 Spectral Analysis of Random Signals

In Chapter 6 we discussed the measurement of sinusoidal signal parameters, both without and with additive noise. Sinusoidal signals have a well-defined shape. If we know the amplitude, the frequency, and the initial phase of a sinusoidal signal, we can exactly compute its value at any desired time. Let us look, for example, at a 10-second recording of power-line voltage. Depending on where you live, it will have a frequency of 50 or 60 Hz, and its amplitude will be between $\sqrt{2} \cdot 110$ and $\sqrt{2} \cdot 240$ volts, so we will see either 500 or 600 periods. If we look at another 10-second recording, taken later, we will again see the same number of periods, with the same amplitude. Only the initial phase may be different, depending on the starting instant of the recording.

Not all signals encountered in real life are sinusoidal. In particular, we often must deal with signals that are random to a certain extent. Even the power-line signal, if examined carefully, will be seen to exhibit fluctuations of amplitude and frequency. As another example, suppose that we are interested in measuring the time variation of the height of ocean waves. We pick a spot and observe the height of the water at that spot as a function of time. Figure 13.1 shows a possible waveform of such a measurement and the magnitude of its Fourier transform (in dB). Like the power-line voltage, this waveform is oscillatory. However, its frequency and amplitude are not constant, and the overall shape bears little resemblance to a sinusoid. Moreover, if we repeat the experiment at a later time, we will record a waveform whose overall shape may resemble the one shown in Figure 13.1, but the details will most likely be different. Ocean waves are an example of a random signal.

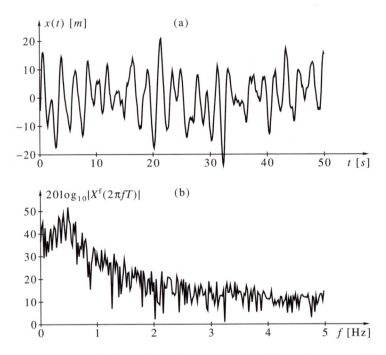

Figure 13.1 Ocean wave heights: (a) as a function of time; (b) as a function of frequency.

The method of spectral analysis we described in Chapter 6, consisting of windowing followed by examining the magnitude of the windowed DFT, is not satisfactory when applied to random signals. When the signal is random, so is its Fourier transform. Thus, the shape of the DFT will vary from experiment to experiment, limiting the information we can extract from it. Furthermore, it can be shown mathematically that increasing the length of the sampled sequence does not help; the longer DFTs show more details of the frequency response, but most of those details are random and vary from experiment to experiment. The randomness of the DFT is evident in Figure 13.1(b).

A possible solution, in case of a stationary random signal, is to arrange the samples in (relatively short) segments of equal length, compute the DFT of each segment, and average the resulting DFTs. Averaging reduces the randomness and provides a relatively smooth spectrum. The average spectrum displays the macroscopic characteristics of the signal and suppresses the random details. The more segments we use for averaging, the better the smoothing and the stronger the randomness suppression. However, averaging is effective only if we know beforehand that our signal is stationary during the entire time interval spanning the union of all segments. For example, it makes no sense to average ocean wave height measurements if some are taken when the sea is calm and some when it is stormy.

Once we understand how averaging works in principle, the question arises: Exactly what entities should we use for averaging? Averaging the DFTs themselves is not the right answer. The DFT is a complex sequence, and the relative phases of the DFTs of different segments are likely to be random. When we average complex numbers having random phases, we are most likely to end up with a number close to zero, which is not a meaningful result. To preserve the spectral information, we average the *square magnitudes* of the DFTs. The term *periodogram* is a synonym for the square magnitude of the DFT.[1] Suppose we have a total of NL data points of the random signal $x[n]$. We divide these points into L consecutive segments of length N each. Then, the *averaged periodogram* is defined as

$$\hat{K}_x^f(\theta) = \frac{1}{L} \sum_{l=0}^{L-1} \left\{ \frac{1}{N} \left| \sum_{n=0}^{N-1} x[n+lN]e^{-j\theta n} \right|^2 \right\}. \tag{13.1}$$

The justification for the averaging procedure in (13.1) and for the notation $\hat{K}_x^f(\theta)$ comes from the Wiener–Khintchine theorem, which we studied in Section 2.9. As we recall from (2.131), the expectation of each of the square magnitudes of the transforms

$$\frac{1}{N} \left| \sum_{n=0}^{N-1} x[n+lN]e^{-j\theta n} \right|^2$$

converges to the power spectral density $K_x^f(\theta)$ as N tends to infinity. The right side of (13.1) is the empirical mean of the individual square magnitudes of the transforms; therefore we may hope that if both L and N are allowed to approach infinity, the empirical mean will converge to the theoretical mean and we will get[2]

$$\lim_{\substack{N \to \infty \\ L \to \infty}} \hat{K}_x^f(\theta) = K_x^f(\theta) \tag{13.2}$$

In practice, the simple averaged periodogram defined in (13.1) is rarely used. Instead, we use a *windowed averaged periodogram*, defined by

$$\hat{K}_x^f(\theta) = \frac{1}{L} \sum_{l=0}^{L-1} \left\{ \frac{1}{N} \left| \sum_{n=0}^{N-1} w[n]x[n+lN]e^{-j\theta n} \right|^2 \right\}, \tag{13.3}$$

where $w[n]$ is a window chosen by the user. It can be shown that, if the limit (13.2) holds for the simple averaged periodogram, it also holds for the windowed averaged periodogram. In reality, neither N nor L is infinite. Practice shows that, for finite N and L, windowing helps smoothing the randomness of $\hat{K}_x^f(\theta)$ and gives better results. The shape of the periodogram depends not only on the window, but also on the choice of N and L. These parameters must be chosen to make their product equal to the total number of given data points. Thus, increasing L necessarily decreases N and vice versa. In general, increasing N provides more details in $\hat{K}_x^f(\theta)$, but also increases its randomness. Increasing L makes $\hat{K}_x^f(\theta)$ smoother, with fewer details and less randomness.

Figure 13.2 depicts the periodogram obtained by averaging 100 DFTs of simulated ocean wave measurements, of the kind shown in Figure 13.1. Each segment is 50 seconds long and contains 500 samples. The Hann window was used for all segments. The frequency scale was converted to a physical frequency in hertz. The information in the periodogram should be interpreted as follows: Its value at each frequency indicates the energy density at that frequency. Thus, in Figure 13.1, low-frequency waves are relatively strong, the strongest ones having a frequency of about 0.5 Hz. The energy density decays gradually as the frequency increases. We also notice the slight randomness of the graph, because averaging is not perfect when the number of data points is finite.

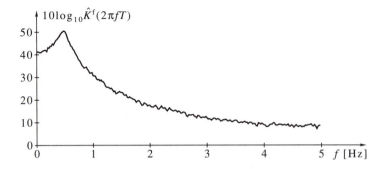

Figure 13.2 Averaged windowed periodogram of ocean wave heights.

When dividing the total number of data into segments, it is common to let the segments overlap, as shown in Figure 13.3; overlap of 50 percent is typical.

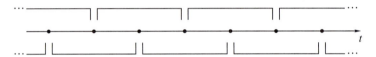

Figure 13.3 Segment overlapping for periodogram averaging.

The rationale for segment overlap is as follows: Since we window each segment, the relative weight given to the points near the ends of the segment is less than the weight given to the points near the center. When 50 percent overlap is used, points near the end of one segment will be near the center of a neighboring segment, as is evident from Figure 13.3. Therefore, all data points will be equally represented on the average. The method of averaging periodograms with windowing and overlapping was developed by P. D. Welch [1970] and is known as the *Welch periodogram*. The Welch periodogram is a common and useful method for spectral analysis of stationary random signals.

The mathematical expression of the Welch periodogram is as follows. Assume that the segment length N is even and that the total number of segments to be averaged is L. Then, with 50 percent overlap, we have

$$\hat{K}_x^f(\theta) = \frac{1}{L} \sum_{l=0}^{L-1} \left\{ \frac{1}{N} \left| \sum_{n=0}^{N-1} w[n] x[n + 0.5lN] e^{-j\theta n} \right|^2 \right\}. \tag{13.4}$$

The frequencies θ can be taken as integer multiples of $2\pi/N$ (usual DFT), integer multiples of $2\pi/M$ for $M > N$ (zero-padded DFT), or dense points in a small frequency interval (chirp Fourier transform).

Figure 13.4 summarizes the stages of practical spectral analysis of random signals. The continuous-time signal from the physical source is filtered to avoid aliasing and then sampled at or above the Nyquist rate. It is then divided into segments, with optional overlap. Each segment is windowed, then undergoes FFT. Variations on simple FFT include zero-padded FFT to increase display resolution and chirp Fourier transform to display only part of the frequency range. Absolute value and squaring is then performed on the output of the FFT, followed by optional averaging. As we have explained, averaging should be restricted to signals that are known to be random and stationary. In any case, the number of segments to be averaged should be chosen such that the total time interval for all segments does not exceed the stationarity time of the signal. Finally, the result is converted to dB and sent to a graphical display or other uses.

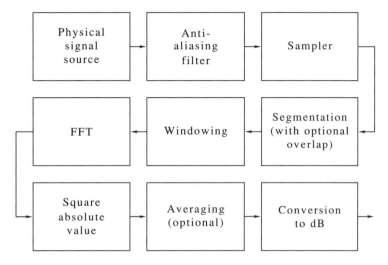

Figure 13.4 A complete short-time spectral analysis scheme.

The procedure `stsa` in Program 13.1 implements spectral analysis in MATLAB. It provides for optional overlapping and averaging, as well as optional zero padding or chirp Fourier transform. The window has to be supplied externally. The rows of the output matrix are the DFTs of the segments (or averages thereof, depending on the input parameters). The program does not convert the results to dB, so this must be done externally if desired.

Example 13.1 In Example 3.13 we introduced the digital communication signaling method BPSK. In this example we introduce the signaling method *offset quadrature phase-shift keying*, or OQPSK, and examine its frequency domain characteristics.

Suppose we are given a sequence of bits $b[n]$, appearing every T seconds. We split the sequences into two subsequences, one consisting of the even-index bits and the other of the odd-index bits, that is,

$$b_e[n] = b[2n], \quad b_o[n] = b[2n + 1].$$

Each of the subsequences has a rate $1/2T$ and they are offset by T seconds (hence the word *offset* in the name). We generate an NRZ signal for each sequence, as in Example 3.13. We use the sequence $b_e[n]$ to modulate the phase of a cosine waveform and the sequence $b_o[n]$ to modulate the phase of a sine waveform. Finally, we subtract the two waveforms. This sequence of operations is described by the formula

$$x(t) = p_r(t) \cos(\omega_0 t) - p_i(t) \sin(\omega_0 t), \qquad (13.5)$$

where

$$p_r(t) = \begin{cases} 1, & b_e[n] = 0, \\ -1, & b_e[n] = 1, \end{cases} \quad 2nT \le t < 2(n + 1)T, \qquad (13.6a)$$

$$p_i(t) = \begin{cases} 1, & b_0[n] = 0, \\ -1, & b_0[n] = 1, \end{cases} \quad (2n+1)T \le t < (2n+3)T. \quad (13.6b)$$

The signals $p_r(t)$ and $p_i(t)$ are depicted in Figure 13.5(a, b).

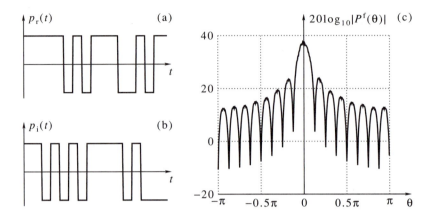

Figure 13.5 An offset quadrature phase-shift keying signal: (a) real part; (b) imaginary part; (c) spectrum.

An alternative way to express (13.5) is as

$$x(t) = \Re\{p(t)e^{j\omega_0 t}\}, \quad \text{where} \quad p(t) = p_r(t) + jp_i(t). \quad (13.7)$$

The signal $p(t)$ is called the *complex envelope* of the modulated signal. The spectrum of the complex envelope provides all the information about the frequency-domain characteristics of the modulated signal, since multiplication by $e^{j\omega_0 t}$ shifts the spectrum by ω_0, and the operation $\Re$ generates the conjugate image at $-\omega_0$.

Figure 13.5(c) illustrates the spectrum of the complex envelope of an OQPSK signal (*spectrum* herewith refers to the estimated PSD, computed by averaged periodogram). It was generated from 4096 random bits using the program stsa discussed earlier. The sampling rate is 8 times the bit duration. The segment length is $N = 512$, corresponding to 64 bits. There is no overlap between adjacent segments, and all segments (64 in total) are averaged to provide one spectral plot. As we see, OQPSK has relatively high side lobes. This feature results from the discontinuities of the modulating waveform and is typical of many digital communication signaling methods. High side lobes are undesirable, since they potentially interfere with neighboring communication channels. Communication regulations in most countries (FCC regulations in the United States) forbid high side lobes such as the ones shown in Figure 13.5(c).

A common way to reduce the spectral side lobes is to filter the signal $p(t)$ before feeding it to the RF modulator, that is, to form the signal

$$q(t) = \{p * h\}(t). \quad (13.8)$$

The filtered complex envelope $q(t)$ is then used for modulating the carrier as in (13.7). The low-pass filter $h(t)$ used for this purpose typically has bandwidth equal to that of the main lobe of $P^F(\omega)$ or slightly higher. Figure 13.6(a, b) illustrates the filtered signal and Figure 13.6(c), generated in the same way as Figure 13.5(c), shows its spectrum. The bandwidth of the filtered signal is approximately $1/2T$.

The complex envelope $p(t)$ has the property that $|p(t)| = \sqrt{2}$, so unfiltered OQPSK signal is a *constant envelope* signal. Constant envelope signals are highly desirable, since they are convenient to amplify by analog circuitry. The filtering operation has an

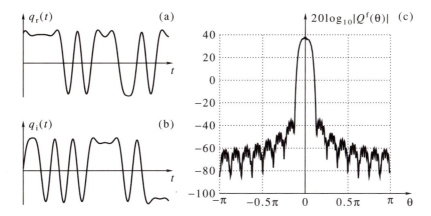

Figure 13.6 A filtered offset quadrature phase-shift keying signal: (a) real part; (b) imaginary part; (c) spectrum.

undesirable side effect, however. The constant envelope property is lost in the filtering process; the absolute value of $q(t)$ varies by a factor of about 2 between minimum and maximum.

Many communication systems, especially ones designed to be inexpensive and power efficient, hard limit the envelope signal in the process of generating the final transmitted signal. Hard limiting is represented mathematically (at least approximately) by the operation

$$r(t) = Ae^{j \angle q(t)}, \tag{13.9}$$

where A is the minimum value of $|q(t)|$ (or slightly smaller), and $\angle q(t)$ is the phase of $q(t)$. The signal $r(t)$ contains the same phase information as $q(t)$ (it is the phase that carries the information about the bits!), but has a constant envelope and is therefore easier to handle by the electronic circuitry. However, hard limiting introduces discontinuities to the derivative of the signal, so it inevitably increases the side-lobe level again. This phenomenon is called *spectrum regrowth* and is an undesirable side effect of hard limiting.

Figure 13.7(a, b) illustrates the hard-limited signal, and Figure 13.7(c) shows its spectrum. The spectral side lobes are now only slightly higher than those before hard limiting and in any case much lower than those of the unfiltered OQPSK signal. We conclude that an OQPSK signal has modest spectral regrowth as a result of hard limiting. This is one of the most attractive features of this signaling method. By comparison, filtering a BPSK signal followed by envelope hard limiting causes severe spectrum regrowth, which almost nullifies the filtering effect. □

13.2 Spectral Analysis by a Smoothed Periodogram

The Welch periodogram, presented in the preceding section, has limited use if the data length is relatively short, because it may be difficult to perform effective segmentation in such a case. We now introduce another spectrum estimation method for random signals—the *smoothed periodogram* method of Blackman and Tukey [1958]. The smoothed periodogram method performs smoothing of the square magnitude of the DFT in the frequency domain *without* segmentation and averaging. It is therefore useful in cases where the data sequence is short.

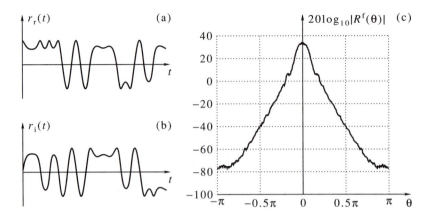

Figure 13.7 A filtered and hard-limited offset quadrature phase-shift keying signal: (a) real part; (b) imaginary part; (c) spectrum.

The smoothed periodogram of a zero-mean sequence $x[n]$ of length N is defined as

$$\hat{K}_x^f(\theta) = \frac{1}{2\pi} \int_{-\pi}^{\pi} N^{-1}|X^f(\theta - \lambda)|^2 W^f(\lambda) d\lambda, \qquad (13.10)$$

where $W^f(\lambda)$ is the kernel function of an odd-length symmetric window $\{w[m], -M \leq m \leq M\}$. Equation (13.10) is a frequency-domain convolution between the periodogram of the sequence and the kernel function; it therefore acts to smooth the randomness of $|X^f(\theta)|^2$. Spectral lines, resulting from periodicities in $x[n]$, will be smeared as well. Therefore, we face the usual window trade-off: side-lobe level versus main-lobe width.

The convolution operation (13.10) is not convenient to perform in the frequency domain. The corresponding time-domain expression is a multiplication of the window $w[m]$ by the inverse Fourier transform of $N^{-1}|X^f(\theta)|^2$. This, in turn, can be found as follows:

$$N^{-1}|X^f(\theta)|^2 = N^{-1} \left| \sum_{n=0}^{N-1} x[n]e^{-j\theta n} \right|^2 = N^{-1} \left(\sum_{n=0}^{N-1} x[n]e^{-j\theta n} \right) \overline{\left(\sum_{l=0}^{N-1} x[l]e^{-j\theta l} \right)}$$

$$= N^{-1} \sum_{n=0}^{N-1} \sum_{l=0}^{N-1} x[n]x[l]e^{-j\theta(n-l)} = \sum_{m=-(N-1)}^{N-1} \hat{k}_x[m]e^{-j\theta m}, \qquad (13.11)$$

where

$$\hat{k}_x[m] \triangleq N^{-1} \sum_{i=0}^{N-1-|m|} x[i]x[i + |m|], \quad -(N-1) \leq m \leq N-1. \qquad (13.12)$$

In passing to the rightmost form of (13.11), we made the substitutions

$$m = n - l, \quad i = l \text{ if } m \geq 0, \quad i = n \text{ if } m < 0.$$

Therefore, the inverse Fourier transform of $N^{-1}|X^f(\theta)|^2$ is the finite-length sequence $\{\hat{k}_x[m], -(N-1) \leq m \leq N-1\}$. It follows that (13.10) can be expressed as

$$\hat{K}_x^f(\theta) = \sum_{m=-M}^{M} \hat{k}_x[m]w[m]e^{-j\theta m}, \qquad (13.13)$$

assuming that $M \leq N - 1$.

The quantity $\hat{\kappa}_x[m]$ can be interpreted as an estimate of the covariance $\kappa_x[m]$, defined in (2.119). Comparing (13.13) with (2.122), we see that the smoothed periodogram $\hat{K}_x^f(\theta)$ can be interpreted as an estimate of $K_x^f(\theta)$, obtained by (1) replacing the true covariance sequence $\kappa_x[m]$ by the estimated covariance sequence $\hat{\kappa}_x[m]$ and (2) truncation and windowing of the estimated covariance sequence.

If the mean of the random signal $x[n]$ is not zero, we define its *estimated mean* as

$$\hat{\mu}_x = N^{-1} \sum_{n=0}^{N-1} x[n] \tag{13.14}$$

and the *estimated covariance* as

$$\hat{\kappa}_x[m] \triangleq N^{-1} \sum_{i=0}^{N-1-|m|} (x[i] - \hat{\mu}_x)(x[i + |m|] - \hat{\mu}_x). \tag{13.15}$$

Again, the smoothed periodogram is given by (13.13).

Under certain conditions on the random signal $x[n]$, $\hat{K}_x^f(\theta)$ can be made to converge to $K_x^f(\theta)$ as N tends to infinity, by a proper choice of the window $w[m]$ and its length M as a function of N. However, in the present application, N is relatively small, so there is no guarantee that $\hat{\kappa}_x[m]$ will be a good approximation of $\kappa_x[m]$ or that $\hat{K}_x^f(\theta)$ will be a good approximation of $K_x^f(\theta)$.

We reiterate the computational procedure for the smoothed periodogram of a sequence $x[n]$. First we compute $\hat{\kappa}_x[m]$, according to (13.12) if $x[n]$ is zero mean, or to (13.15) if it is not. We then multiply $\hat{\kappa}_x[m]$ by the window $w[m]$, and finally compute the DFT of the result. Basic, zero-padded, or chirp DFT can be used. The procedure smooper in Program 13.2 implements this computation. The program first checks that the window length is odd. It then subtracts from $x[n]$ its mean and generates the sequence $\hat{\kappa}_x[m]$ by convolving the input sequence with its reversed copy. The product of $\hat{\kappa}_x[m]$ and $w[m]$ is formed next, taking care to center the latter with respect to the former. If the length $2M + 1$ of $w[m]$ is shorter than $2N - 1$, the product $\hat{\kappa}_x[m]w[m]$ will contain $N - M - 1$ zeros at each end. The FFT of the product is then computed. The length of the FFT is $2N - 1$, out of which the first N elements are retained, corresponding to the positive frequencies. The FFT introduces linear phase, because it treats the sequence as causal, although this sequence is symmetric with respect to $m = 0$. By taking the absolute value of the FFT we remove the linear phase. This program does not provide zero-padded or chirp DFT.

In practical use of the smoothed periodogram, the window length is always smaller than $2N - 1$, but not much smaller, to prevent excessive smearing of periodic components that may be present. Typical values of the ratio M/N are between 0.2 and 0.5. Weak windows, such as Hann or Hamming, are more commonly used than strong windows, such as Blackman or Kaiser.

Example 13.2 The data we examine in this example—annual average sunspot numbers—are not directly related to electrical engineering, but they are important to the history of random signal analysis, since they have been used on numerous occasions to test methods of random signal analysis. Sunspot statistics are important in astronomy and radio communications.[3] Sunspot numbers vary irregularly, but approximately follow a periodic pattern, with a period of about 11 years. Figure 13.8(a) shows the annual average sunspot numbers in the years 1749–1924. These numbers are known as *Wolfer's sunspot data*; they are available in the file sunspot.dat (see page vi for information on how to download this file). The quasi-periodicity of the sequence is evident from the figure.

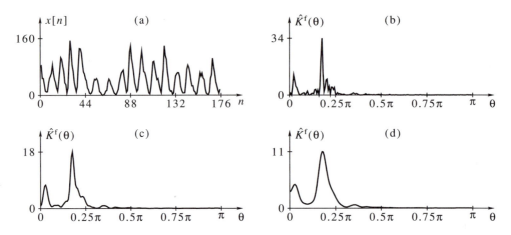

Figure 13.8 Wolfer's sunspot data: (a) annual sunspot numbers, 1749-1924; (b) periodogram; (c) smoothed periodogram, $M = 88$; (d) smoothed periodogram, $M = 44$ (parts b, c, d are drawn to different scales; maximum values shown are in thousands).

The periodogram (i.e., square magnitude of the Fourier transform) of the sunspot numbers is shown in Figure 13.8(b). The randomness of the data leads to randomness of the periodogram, but a dominant peak is apparent, at a frequency slightly below $(1/11)\,\text{year}^{-1}$. The periodogram was first used for this purpose by Schuster [1906a]. We show two smoothed periodograms of the sequences [Figure 13.8(c, d)], both using a Hamming window, with $M = 88, 44$, respectively (i.e., $1/2$ and $1/4$ of the data length). As we see, a shorter window leads to a smoother periodogram. In both cases, the dominant peak has a frequency of about $(1/11)\,\text{year}^{-1}$.

The three periodograms exhibit, in addition to the main 11-year periodicity, a second periodicity with a period of about 70 years. Since the number of data is 176, there are only slightly over two full periods of this apparent periodicity. It is therefore not entirely clear whether the additional periodicity represents a true physical phenomenon or whether it is an artifact. □

13.3   Rational Parametric Models of Random Signals

The spectrum estimation methods presented in the preceding two sections do not rely on specific assumptions on the signal $x[n]$, except that it is wide-sense stationary. Such methods are said to be *nonparametric*. A second class of spectrum estimation methods relies on a *parametric model* of the signal in question. For example, the sinusoids-in-noise model (6.74), studied in Section 6.5.3, is a parametric signal model. The parameters of this model are the amplitudes, frequencies, and initial phases $\{A_k, \theta_k, \phi_k, 1 \le k \le M\}$ of the sinusoidal components.

A general parametric model for a discrete-time, zero-mean, WSS random signal $x[n]$ is its expression as the output of a difference equation excited by discrete-time white noise, that is, as

$$x[n] = -a_1 x[n-1] - \cdots - a_p x[n-p] + b_0 v[n] + b_1 v[n-1] + \cdots + b_q v[n-q],$$
$$(13.16)$$

where $v[n]$ is zero-mean white noise with variance y_v. The linear system represented by (13.16) is assumed to be causal and stable. The random variable $v[n]$ is assumed to be uncorrelated with past values of $x[n]$, that is,

$$E(v[n]x[n-m]) = 0 \quad \text{for all } n \text{ and all } m > 0. \tag{13.17}$$

The white noise $v[n]$ is called the *innovation* of $x[n]$.

The model (13.16) is called *rational*, or *pole–zero*, for obvious reasons. The orders p, q and the parameters $\{a_1, \ldots, a_p, b_0, \ldots, b_q, \gamma_v\}$ characterize the model. Two important special cases of this model are as follows:

1. When $p = 0$ we get

$$x[n] = b_0 v[n] + b_1 v[n-1] + \cdots + b_q v[n-q]. \tag{13.18}$$

This is called an *all-zero*, or a *moving-average* model, or MA.

2. When $q = 0$ we get

$$x[n] = -a_1 x[n-1] - \cdots - a_p x[n-p] + v[n]. \tag{13.19}$$

This is called an *all-pole*, or an *autoregressive* model, or AR.[4]

The general pole–zero model (13.16) is also called an *autoregressive moving-average* model, or ARMA.

As we recall from (2.135), the power spectral density of $x[n]$ satisfying the ARMA model (13.16) is given by

$$K_x^f(\theta) = \gamma_v \left| \frac{b(e^{j\theta})}{a(e^{j\theta})} \right|^2, \tag{13.20}$$

where $a(z)$ and $b(z)$ are, respectively, the denominator and numerator polynomials of the transfer function corresponding to the difference equation (13.16). In the special case of a moving-average model, $a(z) = 1$, whereas in the special case of an autoregressive model, $b(z) = 1$. Because of the stability requirement on (13.16), the poles [the roots of $a(z)$] are necessarily inside the unit circle. It can be shown that, because of the requirement (13.17), the zeros [the roots of $b(z)$] must also be inside the unit circle. In other words, the rational transfer function $b(z)/a(z)$ is minimum phase, as defined in Section 8.4.7.

The *ARMA modeling problem* for a discrete-time, zero-mean WSS random signal can be stated as follows: Given a finite-length segment of the signal, say $\{x[n], 0 \le n \le N-1\}$, find a set of parameters in the difference equation (13.16) that will describe the given signal, subject to the stability and minimum-phase requirements on $a(z)$ and $b(z)$, with $v[n]$ being white noise. The MA and AR modeling problems are likewise defined. For a WSS signal whose mean μ_x is nonzero, the model in question (ARMA, AR, or MA) applies to (or is sought for) $x[n] - \mu_x$.

The following important point should be borne in mind: When modeling a physical random signal by means of a model such as (13.16), (13.18), or (13.19), usually we do not mean that the signal necessarily obeys the model exactly. More often, the model provides, or aims at providing, only an approximate description of the signal. The "goodness" of such an approximation can be measured by various criteria. A common criterion is the accuracy to which the true PSD of the signal is approximated by the rational formula (13.20). On the other hand, when *deriving properties* of a model, we always assume that the model holds exactly for the signal in question. Thus, an ARMA signal is a signal obeying (13.16) exactly, an MA signal is one obeying (13.18) exactly, and an AR signal is one obeying (13.19) exactly. ARMA, MA, and AR signals are almost never encountered as physical signals, but serve as approximate models for such signals.

Both ARMA and MA modeling are relatively complicated subjects, which are not treated further in this book. AR modeling, on the other hand, is much simpler; we discuss AR modeling in detail in the next section.

13.4 Autoregressive Signals

The autoregressive signal model (13.19) has a unique place among parametric models for discrete-time random signals. The theory of AR signals is rich and elegant, but also simple and easy to utilize. In this section we present basic methods for AR signal modeling, algorithms related to these methods, and their implementation. An extended discussion of autoregressive signals can be found in Porat [1994, Chapters 2, 5, 6, and 8], including proofs of results and facts given in this section without proofs. To avoid repetition, we will not refer to Porat [1994] explicitly each time we state a fact without a proof. Extended discussions of the material in this section can also be found in Marple [1987], Kay [1988], Therrien [1992], and Hayes [1996].

In autoregressive modeling of a random signal, we have to distinguish between two cases: (1) when we are given the covariance sequence $\kappa_x[m]$ of the signal; (2) when we are given a set of measured values of the signal, but not its covariances. We will first treat the former case and then proceed to the latter.

13.4.1 The Yule–Walker Equations

The Yule-Walker equations establish the relationships between the covariance sequence $\kappa_x[m]$ of $x[n]$ and the parameters $\{y_v, a_1, \ldots, a_p\}$. Yule [1927] first introduced the AR model and applied it to Wolfer's sunspot data; see Example 13.2. Walker [1931] extended Yule's work. The Yule-Walker equations are derived as follows. Multiply (13.19) by $x[n - m]$, where $m > 0$, and take the expectation of both sides. Recall that $\kappa_x[m] = E(x[n]x[n - m])$, cf. (2.119), and use (13.17) to get

$$\kappa_x[m] = -a_1\kappa_x[m - 1] - \cdots - a_p\kappa_x[m - p], \quad m > 0. \tag{13.21}$$

In particular, by collecting the equalities (13.21) for $1 \le m \le p$ and expressing them in a matrix–vector notation, we get

$$\begin{bmatrix} \kappa_x[0] & \kappa_x[-1] & \ldots & \kappa_x[-p + 1] \\ \kappa_x[1] & \kappa_x[0] & \ldots & \kappa_x[-p + 2] \\ \vdots & \vdots & \ldots & \vdots \\ \kappa_x[p - 1] & \kappa_x[p - 2] & \ldots & \kappa_x[0] \end{bmatrix} \begin{bmatrix} a_1 \\ a_2 \\ \vdots \\ a_p \end{bmatrix} = - \begin{bmatrix} \kappa_x[1] \\ \kappa_x[2] \\ \vdots \\ \kappa_x[p] \end{bmatrix}. \tag{13.22}$$

When we repeat the operation leading to (13.21) for $m = 0$, we get

$$\kappa_x[0] = -a_1\kappa_x[-1] - \cdots - a_p\kappa_x[-p] + E(v[n]x[n]). \tag{13.23}$$

The cross-correlation $E(v[n]x[n])$ is nonzero. To find it, multiply (13.19) by $v[n]$ and use (13.17) to get

$$E(v[n]x[n]) = E(v[n])^2 = y_v. \tag{13.24}$$

Substitution of (13.24) in (13.23) gives

$$y_v = \kappa_x[0] + a_1\kappa_x[-1] + \cdots + a_p\kappa_x[-p]. \tag{13.25}$$

Recalling that the sequence $\kappa_x[m]$ is symmetric, we can rewrite these equations as

$$
\begin{bmatrix}
\kappa_x[0] & \kappa_x[1] & \cdots & \kappa_x[p-1] \\
\kappa_x[1] & \kappa_x[0] & \cdots & \kappa_x[p-2] \\
\vdots & \vdots & \cdots & \vdots \\
\kappa_x[p-1] & \kappa_x[p-2] & \cdots & \kappa_x[0]
\end{bmatrix}
\begin{bmatrix}
a_1 \\ a_2 \\ \vdots \\ a_p
\end{bmatrix}
= -
\begin{bmatrix}
\kappa_x[1] \\ \kappa_x[2] \\ \vdots \\ \kappa_x[p]
\end{bmatrix},
\tag{13.26a}
$$

$$
y_v = \kappa_x[0] + a_1 \kappa_x[1] + \cdots + a_p \kappa_x[p].
\tag{13.26b}
$$

Equations (13.26) are the Yule–Walker equations. The first enables the computation of the parameters $\{a_1, \ldots, a_p\}$ as a function of the covariance sequence values $\{\kappa_x[0], \ldots, \kappa_x[p]\}$. The second enables the computation of the innovation variance y_v. The procedure yw in Program 13.3 solves the Yule–Walker equations.

Example 13.3 Let us solve the Yule–Walker equations for a first-order AR model. We get from (13.26a)

$$
a_1 = -\frac{\kappa_x[1]}{\kappa_x[0]}
\tag{13.27}
$$

and

$$
y_v = \kappa_x[0] + a_1 \kappa_x[1] = \kappa_x[0] - \frac{\kappa_x^2[1]}{\kappa_x[0]}.
\tag{13.28}
$$

□

13.4.2 Linear Prediction with Minimum Mean-Square Error

The preceding derivation of the Yule–Walker equations relies only on the AR model (13.19) and on the property (13.17) of the innovation $v[n]$. We now examine the AR model from a different point of view—that of linear prediction—and introduce an optimality criterion that also leads to the Yule–Walker equations.

Suppose we wish to predict the random variable $x[n]$ from the past p values of the signal $\{x[n-k],\ 1 \le k \le p\}$. For example, suppose we wish to predict today's market value of a certain stock from its values in the last p days. Let us restrict ourselves to a *linear prediction* strategy, that is, to an expression of the form

$$
\hat{x}[n] = -a_1 x[n-1] - \cdots - a_p x[n-p],
\tag{13.29}
$$

where $\hat{x}[n]$ denotes the predicted value of $x[n]$ and $\{-a_1, \ldots, -a_p\}$ are the coefficients of the linear combination. Let $v[n]$ be the error between the true $x[n]$ and its predicted value $\hat{x}[n]$. Then we get from (13.29)

$$
v[n] = x[n] - \hat{x}[n] = \sum_{k=0}^{p} a_k x[n-k], \quad \text{where} \quad a_0 \triangleq 1.
\tag{13.30}
$$

We propose the following criterion for the predictor $\hat{x}[n]$: Choose $\{a_1, \ldots, a_p\}$ to minimize $E(v[n])^2$, the mean-square prediction error. The predictor thus obtained is called *linear MMSE predictor*, where MMSE stands for *minimum mean-square error*. The number p of past values used for prediction is called the *order* of the predictor. Each order has a set of linear MMSE predictor coefficients associated with it.

The problem of finding the values of $\{a_1, \ldots, a_p\}$ that minimize $E(v[n])^2$ is similar to the one solved in Section 9.5, and is likewise solved by differentiation and equating the partial derivatives to zero. We get

$$\frac{\partial E(v[n])^2}{\partial a_l} = 2E\left(\frac{\partial v[n]}{\partial a_l}v[n]\right) = 2E\left(x[n-l]\sum_{k=0}^{p}a_kx[n-k]\right)$$

$$= 2\sum_{k=0}^{p}a_kE(x[n-l]x[n-k]) = 2\sum_{k=0}^{p}a_k\kappa_x[l-k] = 0, \quad 1 \le l \le p. \qquad (13.31)$$

As we see, the resulting set of equations is identical to (13.22). Furthermore, the minimum value of $E(v[n])^2$ is given by

$$E(v[n])^2 = E\left(\sum_{k=0}^{p}a_kx[n-k]\right)\left(\sum_{l=0}^{p}a_lx[n-l]\right)$$

$$= \sum_{k=0}^{p}\sum_{l=0}^{p}a_ka_lE(x[n-k]x[n-l]) = \sum_{k=0}^{p}\sum_{l=0}^{p}a_ka_l\kappa_x[l-k]$$

$$= \sum_{k=0}^{p}a_k\kappa_x[-k] + \sum_{l=1}^{p}a_l\left(\sum_{k=0}^{p}a_k\kappa_x[l-k]\right). \qquad (13.32)$$

However, by (13.31), the second term on the right side of (13.32) is zero. Therefore,

$$E(v[n])^2 = \gamma_v = \sum_{k=0}^{p}a_k\kappa_x[-k]. \qquad (13.33)$$

In summary, (13.31) and (13.33) are identical to (13.22) and (13.25), respectively. The conclusion we have reached is thus:

> The linear MMSE predictor of order p is characterized by coefficients $\{a_1,\dots,a_p\}$ and error variance γ_v satisfying the Yule-Walker equations. Therefore, if the signal $x[n]$ is AR of order p, the linear predictor coefficients coincide with the model parameters, and the prediction error $v[n]$ coincides with the innovation.

We emphasize the following point: The linear MMSE predictor is defined for *any* discrete-time WSS signal, whether autoregressive or not, and is *always* obtained by solving the Yule-Walker equations. When $x[n]$ is a true AR signal of order p, and only then, the error $v[n]$ is white noise and satisfies (13.17).

The solution of the Yule-Walker equations can be used to construct a polynomial

$$a(z) = 1 + a_1z^{-1} + \cdots + a_pz^{-p}. \qquad (13.34)$$

The transfer function from the error signal $v[n]$ to $x[n]$ is

$$\frac{X^z(z)}{V^z(z)} = \frac{1}{a(z)}. \qquad (13.35)$$

An important property of $a(z)$ is that it is *always stable*, whether $x[n]$ is a true AR signal of order p or not. We will not prove this property here.

13.4.3 The Levinson–Durbin Algorithm

A numerical solution of p equations in p unknowns, as needed in solving the Yule-Walker equations, requires about p^3 multiply-add operations. As we have said, solutions of different orders provide linear MMSE predictors of different orders. Often we are interested in solving the Yule-Walker equations of all orders from 1 up to a certain maximum order p. This enables us to compare different predictors and decide which one is the best for the specific application. Such a solution requires $\sum_{i=1}^{p}i^3 \approx p^4/4$

multiply-add operations. The Levinson–Durbin algorithm, which we derive in this section, solves the Yule-Walker equations of all orders $1 \le i \le p$ in approximately p^2 multiply-add operations. This is considerably less than p^3 (or $p^4/4$) if p is large. For example, if $p = 10$, the number of operations is reduced by a factor of 10 if only the tenth-order solution is required, and by a factor of 25 if the solutions of all orders up to 10 are required. Beside reducing the computational requirements, the Levinson–Durbin algorithm leads to some interesting results and applications in stability theory and digital filter realizations. We shall discuss these applications after presenting the basic theory of the Levinson–Durbin algorithm. Levinson [1947] first presented the algorithm (in a more general setting than described here), and Durbin [1960] derived the version presented here.

Since we are now dealing with AR models of different orders, let us rewrite the definition of an ith-order AR model as [cf. (13.19)]

$$x[n] = -a_{i,1}x[n-1] - \cdots - a_{i,i}x[n-i] + v_i[n], \tag{13.36}$$

where $\{a_{i,l}, \ 1 \le l \le i\}$ are the coefficients of the ith-order model, and $v_i[n]$ is the ith-order prediction error. We will also use the following definitions:

1. K_i: the symmetric Toeplitz matrix shown in (13.26a), but of dimensions $i \times i$ (i.e., containing values of the covariance sequence up to $\kappa_x[i-1]$).

2. $a_i = [a_{i,1}, \ldots, a_{i,i}]'$: the solution of the ith-order Yule-Walker equation (13.26a).

3. $\kappa_i = [\kappa_x[1], \ldots, \kappa_x[i]]'$: the right side of the ith-order Yule-Walker equation (without the minus sign).

4. $\tilde{a}_i$: the vector a_i in reversed order.

5. $\tilde{\kappa}_i$: the vector κ_i in reversed order.

6. $a^*_{i+1} = [a_{i+1,1}, \ldots, a_{i+1,i}]'$: the vector consisting of the first i components of the solution of the $(i+1)$st-order Yule-Walker equation.

7. $a_{i+1,i+1}$: the last component of the solution of the $(i+1)$st-order Yule-Walker equation.

8. s_i: the variance $E(v_i[n])^2$ of the ith-order prediction error, given by (13.26b).

The ith-order Yule-Walker equations can be expressed as

$$K_i a_i = -\kappa_i, \quad K_i \tilde{a}_i = -\tilde{\kappa}_i, \quad s_i = \kappa_x[0] + \kappa'_i a_i = \kappa_x[0] + \tilde{\kappa}'_i \tilde{a}_i. \tag{13.37}$$

The first is the Yule-Walker equation for the ith-order predictor. The second is a direct consequence of the first: It follows from the fact that reversing the order of the rows and columns of the matrix K_i leaves this matrix unchanged. The third is a rewriting of (13.26b).

The Yule-Walker equation for the $(i+1)$st-order predictor is

$$\begin{bmatrix} K_i & \tilde{\kappa}_i \\ \tilde{\kappa}'_i & \kappa_x[0] \end{bmatrix} \begin{bmatrix} a^*_{i+1} \\ a_{i+1,i+1} \end{bmatrix} = -\begin{bmatrix} \kappa_i \\ \kappa_x[i+1] \end{bmatrix}. \tag{13.38}$$

Let us assume that we have solved the Yule-Walker equations of order i, and we wish to solve (13.38) for a^*_{i+1} and $a_{i+1,i+1}$. We can expand this equation to

$$K_i a^*_{i+1} + \tilde{\kappa}_i a_{i+1,i+1} = -\kappa_i, \tag{13.39a}$$

$$\tilde{\kappa}'_i a^*_{i+1} + \kappa_x[0]a_{i+1,i+1} = -\kappa_x[i+1]. \tag{13.39b}$$

Solving (13.39a) for a^*_{i+1} and using the first two equalities in (13.37) gives

$$a^*_{i+1} = -K_i^{-1}\kappa_i - a_{i+1,i+1}K_i^{-1}\tilde{\kappa}_i = a_i + a_{i+1,i+1}\tilde{a}_i. \tag{13.40}$$

Substituting (13.40) in (13.39b) and solving for $a_{i+1,i+1}$ gives

$$a_{i+1,i+1} = -\frac{\kappa_x[i+1] + \tilde{\boldsymbol{\kappa}}_i' \boldsymbol{a}_i}{\kappa_x[0] + \tilde{\boldsymbol{\kappa}}_i' \tilde{\boldsymbol{a}}_i} = -\frac{\kappa_x[i+1] + \tilde{\boldsymbol{\kappa}}_i' \boldsymbol{a}_i}{s_i}. \tag{13.41}$$

The parameter s_i is updated as follows:

$$\begin{aligned} s_{i+1} &= \kappa_x[0] + \boldsymbol{\kappa}_{i+1}' \boldsymbol{a}_{i+1} = \kappa_x[0] + \boldsymbol{\kappa}_i' \boldsymbol{a}_{i+1}^* + \kappa_x[i+1]a_{i+1,i+1} \\ &= \kappa_x[0] + \boldsymbol{\kappa}_i' \boldsymbol{a}_i + a_{i+1,i+1}\boldsymbol{\kappa}_i' \tilde{\boldsymbol{a}}_i + \kappa_x[i+1]a_{i+1,i+1} \\ &= s_i + a_{i+1,i+1}\{\kappa_x[i+1] + \boldsymbol{\kappa}_i' \tilde{\boldsymbol{a}}_i\} = s_i(1 - a_{i+1,i+1}^2). \end{aligned} \tag{13.42}$$

The parameter

$$\rho_{i+1} = -a_{i+1,i+1} = \frac{\kappa_x[i+1] + \tilde{\boldsymbol{\kappa}}_i' \boldsymbol{a}_i}{s_i} = \frac{\kappa_x[i+1] + \sum_{l=1}^{i} a_{i,l}\kappa_x[i+1-l]}{s_i} \tag{13.43}$$

is called the *partial correlation coefficient*, or the *reflection coefficient*, of order $i+1$, and is defined for all $i \geq 0$. Equations (13.41), (13.40), (13.42), in this order of evaluation, comprise the ith step of the Levinson-Durbin algorithm. The parameter s_0 is equal to $\kappa_x[0]$, since $\hat{x}[n] = 0$ and $v_0[n] = x[n]$ for a zero-order predictor. For convenience, we rewrite the complete Levinson-Durbin algorithm, including initialization.

1. Define

$$s_0 = \kappa_x[0]. \tag{13.44}$$

2. For $i = 0, 1, \ldots, p-1$ do the following:

$$\rho_{i+1} = \frac{\kappa_x[i+1] + \sum_{l=1}^{i} a_{i,l}\kappa_x[i+1-l]}{s_i}, \tag{13.45a}$$

$$s_{i+1} = s_i(1 - \rho_{i+1}^2), \tag{13.45b}$$

$$a_{i+1,l} = a_{i,l} - \rho_{i+1}a_{i,i+1-l}, \quad 1 \leq l \leq i, \tag{13.45c}$$

$$a_{i+1,i+1} = -\rho_{i+1}. \tag{13.45d}$$

The procedure `levdur` in Program 13.4 implements the Levinson-Durbin algorithm.

The number of operations needed for the ith step of the algorithm is determined as follows: In (13.45a) we have i additions and i multiplications and likewise in (13.45c). In (13.45a) we also have one division, and in (13.45b) we have two multiplications and one addition. In total, ith step requires $2i + 2$ multiplications, $2i + 1$ additions, and one division. The complete algorithm thus requires $p^2 + p$ multiplications, p^2 additions, and p divisions. As we said in the beginning of this section, the computational efficiency of the Levinson-Durbin algorithm is the main reason for its importance to AR modeling.

The parameters s_i and ρ_{i+1} deserve special attention. The former is, by definition, the variance of the prediction error of the ith-order AR model. The product $s_i\rho_{i+1}$ is the covariance between $y[n-i-1]$ and the prediction error of the ith-order AR model at time n; see Problem 13.9. By its definition, s_i is nonnegative for all i. It therefore follows from (13.45b) that $|\rho_{i+1}| \leq 1$ for all i. Furthermore, s_i is monotonically decreasing in i, that is, $s_{i+1} \leq s_i$ for all i. This is not surprising, since increasing the model order implies better ability to predict the signal's present value from its past, hence a smaller variance of the prediction error. Two extreme cases are as worth noting:

1. If $x[n]$ is a true AR signal of order p, then $s_p = \gamma_v$, and also $s_i = \gamma_v$ for all $i > p$. Therefore, it follows from (13.45b) that in this case $\rho_{i+1} = 0$ for all $i \geq p$. In other words, for a true AR signal of order p, the Levinson-Durbin algorithm stops naturally after p steps: The parameters s_i and $a_{i,l}$ cease to change, and the reflection coefficients become identically zero for $i \geq p$. On the other hand, if

$x[n]$ is not a true AR signal of any order, the algorithm does not stop naturally, but s_i continues to decrease and the $a_{i,l}$ continue to change as i increases.

2. If $\rho_{i+1} = \pm 1$ for some i, then s_{i+1}, the prediction error of order $i+1$, becomes zero. Therefore, in this case, $x[n]$ becomes exactly predictable from its $i+1$ past values. Random signals for which this happens are called *deterministic*. Problem 13.12 provides an example of a signal that is deterministic according to this definition. Most physical random signals are not deterministic; that is, they are not exactly predictable from their past, no matter how large the order of the predictor.

13.4.4 Lattice Filters

The FIR filter $a(z)$, used to obtain the innovation $v[n]$ from $x[n]$, can be realized by either of the two direct realizations described in Section 11.1 (note that this filter is neither symmetric nor antisymmetric, so the realizations described in Section 11.1.3 are not applicable). Likewise, the IIR filter $1/a(z)$, used to obtain $x[n]$ from $v[n]$, can be realized by any of the realizations described in Section 11.1: direct, parallel, or cascade. In this section we use the Levinson–Durbin algorithm to derive new realizations of these filters, called *lattice realizations*.

Let us define

$$v_i[n] = \sum_{l=0}^{i} a_{i,l}x[n - l], \quad \tilde{v}_i[n] = \sum_{l=0}^{i} a_{i,i-l}x[n - l], \quad a_{i,0} = 1,$$

$$\tag{13.46}$$

$$a_i(z) = \sum_{l=0}^{i} a_{i,l}z^{-l}, \quad \tilde{a}_i(z) = \sum_{l=0}^{i} a_{i,i-l}z^{-l}.$$

$$\tag{13.47}$$

The signal $v_i[n]$ is the result of passing $x[n]$ through the FIR filter obtained from the ith-order linear MMSE predictor. Similarly, the signal $\tilde{v}_i[n]$ is the result of passing $x[n]$ through the reversed-order filter. We can use (13.45c,d) to express the $(i + 1)$st-order signals $v_{i+1}[n]$ and $\tilde{v}_{i+1}[n]$ in terms of the ith-order signals as follows:

$$v_{i+1}[n] = \sum_{l=0}^{i+1} a_{i+1,l}x[n - l] = x[n] + \sum_{l=1}^{i}(a_{i,l} - \rho_{i+1}a_{i,i+1-l})x[n - l] - \rho_{i+1}x[n - i - 1]$$

$$= v_i[n] - \rho_{i+1}\tilde{v}_i[n - 1],$$

$$\tag{13.48}$$

$$\tilde{v}_{i+1}[n] = \sum_{l=0}^{i+1} a_{i+1,i+1-l}x[n - l]$$

$$= -\rho_{i+1}x[n] + \sum_{l=1}^{i}(a_{i,i+1-l} - \rho_{i+1}a_{i,l})x[n - l] + x[n - i - 1]$$

$$= \tilde{v}_i[n - 1] - \rho_{i+1}v_i[n].$$

$$\tag{13.49}$$

Figure 13.9(a) illustrates the dependence of the $(i + 1)$st-order signals on the ith-order ones. The signal $\tilde{v}_i[n]$ is delayed by one time unit and, together with $v_i[n]$, fed to a butterfly-like section that performs the operations expressed in (13.48) and (13.49). Figure 13.9(b) depicts this operation as a black box. The matrix $R_{i+1}(z)$ is the 2×2 z-domain matrix

$$R_{i+1}(z) = \begin{bmatrix} 1 & -\rho_{i+1}z^{-1} \\ -\rho_{i+1} & z^{-1} \end{bmatrix}.$$

$$\tag{13.50}$$

The inputs to the section $R_1(z)$, $v_0[n]$ and $\tilde{v}_0[n]$, are both identically equal to $x[n]$. When fed to this section, they yield $v_1[n]$ and $\tilde{v}_1[n]$. When these are fed to the

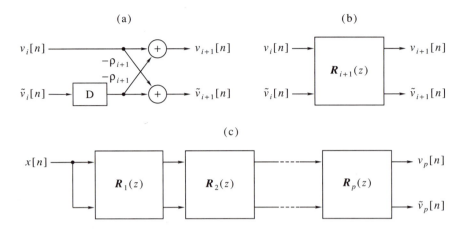

Figure 13.9 An FIR lattice filter: (a) a typical section; (b) schematic drawing of a typical section; (c) the complete lattice.

section $R_2(z)$, they yield $v_2[n]$ and $\tilde{v}_2[n]$, and so on. By connecting p such sections in cascade, as shown in Figure 13.9(c), we get $v_p[n]$ and $\tilde{v}_p[n]$. Therefore, the transfer function from $x[n]$ to $v_p[n]$ in the realization shown in Figure 13.9(c) is precisely $a_p(z)$. Likewise, the transfer function from $x[n]$ to $\tilde{v}_p[n]$ is $\tilde{a}_p(z)$, the reversed-order filter. The realization shown in Figure 13.9(c) is called an *FIR lattice*, owing to its visual appearance. We already encountered a special case of this realization in Problem 11.14.

The FIR lattice shown in Figure 13.9(c) is obviously of order p, since it contains p delay elements; it is FIR because it contains no feedback paths. It has the following properties:

1. The lattice structure is *nested*, in the sense that all intermediate-order FIR filters $a_i(z)$ are embedded in the pth-order filter. When we feed the signal $x[n]$ at the input, we get the ith-order error signal $v_i[n]$ at the output of the ith section for all $1 \le i \le p$. We can therefore change the desired filter's order simply by picking the desired $v_i[n]$, and there is no need to modify the filter's structure or its coefficients.

2. The internal gains ρ_{i+1} are less than 1 in magnitude, so the filter can be conveniently implemented in fixed-point arithmetic, and there is no need to scale the gains.

3. The lattice structure is known to have low sensitivity to quantization noise (we will not elaborate on this topic here).

4. The realization requires $2p$ multiplications and additions, which is twice the necessary minimum. Therefore, it is wasteful compared with the direct realizations. This drawback may make the lattice realization problematic in time-critical applications.

To derive a lattice realization for $1/a_p(z)$, we use (13.48) to express $v_i[n]$ in terms of $v_{i+1}[n]$ and $\tilde{v}_i[n-1]$:

$$v_i[n] = v_{i+1}[n] + \rho_{i+1}\tilde{v}_i[n-1]. \tag{13.51}$$

Figure 13.10(a) illustrates the operations expressed in (13.49) and (13.51). Figure 13.10(b) depicts these operations as a black box. The matrix $S_{i+1}(z)$ expresses the z-domain dependence of $v_i[n]$ and $\tilde{v}_{i+1}[n]$ on $v_{i+1}[n]$ and $\tilde{v}_i[n]$. To derive this matrix, substitute (13.51) in (13.49) to get

$$\tilde{v}_{i+1}[n] = (1 - \rho_{i+1}^2)\tilde{v}_i[n-1] - \rho_{i+1}v_{i+1}[n]. \qquad (13.52)$$

Therefore,

$$S_{i+1}(z) = \begin{bmatrix} 1 & \rho_{i+1}z^{-1} \\ -\rho_{i+1} & (1 - \rho_{i+1}^2)z^{-1} \end{bmatrix} \qquad (13.53)$$

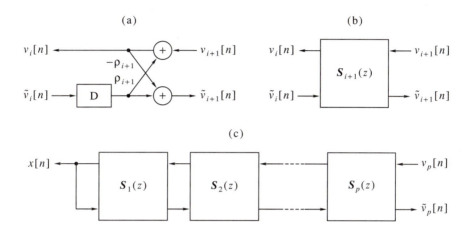

Figure 13.10 An IIR lattice filter: (a) a typical section; (b) schematic drawing of a typical section; (c) the complete lattice.

The upper output of the section $S_1(z)$, $v_0[n]$, is short-circuited to the input $\tilde{v}_0[n]$, since both are identically equal to $x[n]$. The p sections are connected as shown in Figure 13.10(c). When $v_p[n]$ is fed at the upper right input, we get $x[n]$ at the upper left output and $\tilde{v}_p[n]$ at the lower right output. Therefore, the transfer function from $v_p[n]$ to $x[n]$ in the realization shown in Figure 13.10(c) is precisely $1/a_p(z)$. Also, the transfer function from $v_p[n]$ to $\tilde{v}_p[n]$ is $\tilde{a}_p(z)/a_p(z)$.

The realization shown in Figure 13.10(c) is called an *IIR lattice*. We already encountered a special case of this realization in Problem 11.15. The realization is of order p, since it contains p delay elements; it is IIR because it contains feedback paths. It has properties similar to those of the FIR lattice. However, the nesting property does not hold for the IIR filter as for the FIR filter. To realize the intermediate-order all-pole filter $1/a_i(z)$, we must disable the higher order sections by resetting the coefficients $\{\rho_{i+1}, \ldots, \rho_p\}$ to zero. Stability of the IIR lattice is guaranteed as long as all ρ_{i+1} are less than 1 in magnitude.

Both FIR and IIR lattice realizations can be constructed for any stable polynomial $a(z)$ and without regard to any relation to a random signal. For this, we need to compute the set of reflection coefficients $\{\rho_{i+1}, 0 \le i \le p-1\}$ associated with $a(z)$. These coefficients can be computed by carrying out the Levinson–Durbin in reversed order. The reversed-order Levinson–Durbin algorithm is derived as follows: It is obvious from the construction of the FIR lattice that

$$\begin{bmatrix} a_{i+1}(z) \\ \tilde{a}_{i+1}(z) \end{bmatrix} = R_{i+1}(z) \begin{bmatrix} a_i(z) \\ \tilde{a}_i(z) \end{bmatrix}, \qquad (13.54)$$

so

$$\begin{bmatrix} a_i(z) \\ \tilde{a}_i(z) \end{bmatrix} = [\mathbf{R}_{i+1}(z)]^{-1} \begin{bmatrix} a_{i+1}(z) \\ \tilde{a}_{i+1}(z) \end{bmatrix} = \frac{1}{(1 - \rho_{i+1}^2)z^{-1}} \begin{bmatrix} z^{-1} & \rho_{i+1}z^{-1} \\ \rho_{i+1} & 1 \end{bmatrix} \begin{bmatrix} a_{i+1}(z) \\ \tilde{a}_{i+1}(z) \end{bmatrix}.$$

$$(13.55)$$

In particular,

$$a_i(z) = (1 - \rho_{i+1}^2)^{-1}[a_{i+1}(z) + \rho_{i+1}\tilde{a}_{i+1}(z)], \qquad (13.56)$$

where $\rho_{i+1} = -a_{i+1,i+1}$. The reversed-order Levinson–Durbin algorithm is initialized by setting $i + 1 = p$ and $a_p(z) = a(z)$, the given stable polynomial. The iteration proceeds downward, until $i = 0$ is reached. Then we have all the coefficients ρ_{i+1}, needed to construct the (FIR or IIR) lattice filter. Equation (13.56) is seen to be identical to (7.61); see Section 7.4.4. In fact, the reversed-order Levinson–Durbin algorithm is identical to the Schur–Cohn stability test. Stability of $a(z)$ guarantees that $|\rho_{i+1}| < 1$ for all $0 \le i \le p - 1$.

13.4.5   The Schur Algorithm

The Levinson–Durbin algorithm can be used to obtain the sequence of reflection coefficients from the sequence of covariances, computing the intermediate-order model coefficients along the way. The *Schur algorithm* [Schur, 1917, 1918] enables the computation of the reflection coefficients from the covariance sequence without explicitly computing the AR coefficients.

To derive the Schur algorithm, let us input the sequence $\kappa_x[n]$ to the FIR lattice shown in Figure 13.9(c) (note that this input sequence is not causal). Denote the upper and lower output sequences of the $(i+1)$st section by $c_{i+1}[n]$ and $\tilde{c}_{i+1}[n]$, respectively. Then,

$$c_{i+1}[n] = \kappa_x[n] + \sum_{l=1}^{i+1} a_{i+1,l}\kappa_x[n - l], \qquad (13.57a)$$

$$\tilde{c}_{i+1}[n] = \sum_{l=0}^{i} a_{i+1,i+1-l}\kappa_x[n - l] + \kappa_x[n - i - 1]. \qquad (13.57b)$$

It follows from the $(i + 1)$st-order Yule–Walker equation that

$$c_{i+1}[n] = 0, \quad 1 \le n \le i + 1, \qquad (13.58a)$$

$$\tilde{c}_{i+1}[n] = 0, \quad 0 \le n \le i. \qquad (13.58b)$$

Furthermore, it follows from (13.37) that

$$\tilde{c}_{i+1}[i + 1] = \sum_{l=0}^{i} a_{i+1,i+1-l}\kappa_x[i + 1 - l] + \kappa_x[0] = s_{i+1}. \qquad (13.59)$$

Observing the lattice structure and letting $\kappa_x[n]$ be the input sequence, we see that the sequences $c_i[n]$ and $\tilde{c}_i[n]$ obey the relationships

$$c_{i+1}[n] = c_i[n] - \rho_{i+1}\tilde{c}_i[n - 1], \qquad (13.60a)$$

$$\tilde{c}_{i+1}[n] = \tilde{c}_i[n - 1] - \rho_{i+1}c_i[n]. \qquad (13.60b)$$

Since, by (13.58a), $c_{i+1}[i + 1] = 0$, we get from (13.60a)

$$\rho_{i+1} = \frac{c_i[i + 1]}{\tilde{c}_i[i]}. \qquad (13.61)$$

The initial conditions of the Schur algorithm are

$$c_0[n] = \tilde{c}_0[n] = \kappa_x[n], \quad \text{for all } n. \qquad (13.62)$$

The algorithm then iterates (13.61) and (13.60) for all $0 \leq i \leq p-1$. Note that, because of (13.58), (13.60a) needs to be computed only for $i + 2 \leq n \leq p$ and (13.60b) needs to be computed only for $i + 1 \leq n \leq p$. At the end of the iteration, s_p can be obtained from (13.59). For convenience, we rewrite the complete Schur algorithm, including initialization.

1. Define

$$c_0[n] = \tilde{c}_0[n] = \kappa_x[n], \quad 0 \leq n \leq p. \tag{13.63}$$

2. For $i = 0, 1, \ldots, p - 1$ do the following:

$$\rho_{i+1} = \frac{c_i[i+1]}{\tilde{c}_i[i]}, \tag{13.64a}$$

$$c_{i+1}[n] = c_i[n] - \rho_{i+1}\tilde{c}_i[n-1], \quad i + 2 \leq n \leq p, \tag{13.64b}$$

$$\tilde{c}_{i+1}[n] = \tilde{c}_i[n-1] - \rho_{i+1}c_i[n], \quad i + 1 \leq n \leq p. \tag{13.64c}$$

3. Set

$$s_p = \tilde{c}_p[p]. \tag{13.65}$$

We leave it to the reader to develop a MATLAB procedure for the Schur algorithm (Problem 13.15).

The Schur algorithm has the same computational complexity as the Levinson-Durbin algorithm. However, it avoids the operation (13.45a), which requires i additions to complete. In contrast, the operations in (13.64) require only one addition each. This facilitates parallel computation of these operations, hence the Schur algorithm can be advantageously implemented on a parallel computer; see Hayes [1996, Sec. 5.2.6] for further discussion of this subject.

13.4.6 AR Modeling from Measured Data

The autoregressive model obtained by solving the Yule-Walker equations requires knowledge of the covariance sequence $\kappa_x[m]$. In reality, this sequence is seldom known, at least not exactly. Typically, we are given only a finite set of measurements, say $\{x[n], \ 0 \leq n \leq N - 1\}$. If we wish to use these measurements for AR modeling of $x[n]$, an obvious approach is to compute a set of estimated covariances first, say $\{\hat{\kappa}_x[m], \ 0 \leq m \leq p\}$, then use the estimated covariances in the Yule-Walker equations, in lieu of the true covariances $\{\kappa_x[m], \ 0 \leq m \leq p\}$. The estimated covariances can be computed using the formula

$$\hat{\kappa}_x[m] \triangleq N^{-1} \sum_{i=0}^{N-1-|m|} x[i]x[i + |m|], \quad 0 \leq m \leq p \tag{13.66}$$

[cf. (13.12) and recall we are assuming that the signal has zero mean]. The procedure kappahat in Program 13.5 computes the sequence $\hat{\kappa}_x[m]$. The corresponding AR model coefficients will be denoted by $\{\hat{a}_1, \ldots, \hat{a}_p\}$. We can also input the estimated covariances to the Levinson-Durbin algorithm and obtain AR models of increasing orders, with coefficients $\{\hat{a}_{i,l}, \ 1 \leq i \leq p, \ 1 \leq l \leq i\}$.

The polynomial $a_i(z)$ obtained from the solution of the ith-order Yule-Walker equation is always stable, because $\kappa_x[m]$ is the true covariance sequence of a WSS signal (whether autoregressive or not). Since the estimated covariances $\hat{\kappa}_x[m]$ differ from the true covariances $\kappa_x[m]$, it is not obvious whether the polynomial $\hat{a}_i(z)$ obtained by solving the Yule-Walker equation with $\hat{\kappa}_x[m]$ is necessarily stable. The following result is of considerable importance:

Theorem 13.1 If the estimated covariances $\hat{\kappa}_x[m]$ are computed using (13.66) and substituted in the ith-order Yule–Walker equations, then the polynomial

$$\hat{a}_i(z) \triangleq 1 + \hat{a}_{i,1}z^{-1} + \cdots + \hat{a}_{i,i}z^{-i} \qquad (13.67)$$

obtained from the solution of these equations is stable for any i in the range $1 \le i \le N-1$. $\qquad\qquad\square$

A formal proof of this theorem is difficult and will not be given here. However, we give a heuristic argument that is almost as good as a proof, as follows: A fundamental property of the covariance sequence $\kappa_x[m]$, besides being real and symmetric, is that its Fourier transform $K_x^f(\theta)$ is nonnegative; see (2.125). Now, the sequence $\{\hat{\kappa}_x[m], -(N-1) \le m \le N-1\}$ is real and symmetric, and according to (13.11) it also has a nonnegative Fourier transform, namely $N^{-1}|X^f(\theta)|^2$. Therefore, this sequence is (or at least *can be*) a bona fide covariance sequence of *some* WSS signal, albeit not necessarily the signal $x[n]$ itself. Therefore, $\hat{a}_i(z)$ obtained from this sequence is also a linear MMSE predictor for *some* WSS signal, hence is stable.

If we wish to model the given signal by an IIR lattice filter, we need estimates of the reflection coefficients, say $\hat{\rho}_{i+1}$. These can be obtained by the Schur algorithm and using the estimated covariances. Since $\{\hat{\kappa}_x[m], -(N-1) \le m \le N-1\}$ is a valid covariance sequence, the estimated reflection coefficients are guaranteed to be less than 1 in magnitude.

The estimated polynomial $\hat{a}_i(z)$, together with the estimate $\hat{s}_i$ of the input-signal variance, given by

$$\hat{s}_i = \sum_{k=0}^{i} \hat{a}_{i,k}\hat{\kappa}_x[k]. \qquad (13.68)$$

can be used to construct an estimate of the PSD of $x[n]$ [cf. (13.20)]:

$$\hat{K}_x^f(\theta) = \hat{s}_i \left| \frac{1}{\hat{a}_i(e^{j\theta})} \right|^2. \qquad (13.69)$$

This is called an *autoregressive spectrum estimate* of $x[n]$. Note that $\hat{K}_x^f(\theta)$ is not unique, since it depends on the order i of the linear MMSE predictor. Therefore, we get in fact a family of AR spectrum estimates, one for each i.

Example 13.4 We continue to examine Wolfer's sunspot data we met in Example 13.2, this time fitting an AR model to these data. We feed the data (after subtracting the mean value and computing the estimated covariances) to the Levinson–Durbin algorithm and run the algorithm up to order $p = 50$. Figure 13.11(a) shows the parameters s_i for $2 \le i \le 50$. We recall that s_i measures the variance of the ith-order prediction error. As we see, between a second-order model and a 50th-order one, there is only a moderate decrease of the error variance, from 250 to 170.

Figure 13.11(b) shows the PSD computed from (13.69), using a second-order AR model. This was the model used by Yule in his 1927 paper. As we see, the estimated PSD is a very smooth, hence probably is a very crude depiction of the true PSD. Nonetheless, it displays a peak at frequency of about $(1/11)\,\text{year}^{-1}$, as obtained from the smoothed periodograms in Example 13.2.

Figure 13.11(c) shows the PSD computed from (13.69), using a 50th-order AR model. Such a model was beyond anyone's ability to compute in Yule's day but can be easily computed today. As we see, the estimated PSD shows many details not shown in Figure 13.11(b). Some of these details may be random artifacts, but some may be real.

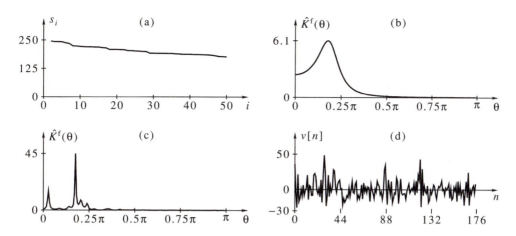

Figure 13.11 AR modeling of Wolfer's sunspot data: (a) the prediction-error variance as a function of the model order; (b) AR spectrum of order 2; (c) AR spectrum of order 50; (d) the prediction error for a 50th-order model (parts b, c are drawn to different scales; maximum values shown are in thousands).

In particular, the 70-year period found in the smoothed periodograms is seen here also.

Figure 13.11(d) shows the prediction error $v[n]$ corresponding to the 50th-order model. Since $s_{50} \approx 170$, the standard deviation of $v[n]$ is about 13. As seen from the figure, the peak error reaches about 50 occasionally. The practical implication of these numbers is: Based on the AR model, we may not be able to predict next year's sunspot number to an accuracy better than ± 13 on the average, and the error may be up to 50 in extreme cases. $\qquad\qquad\qquad\qquad\qquad\qquad\qquad\qquad\qquad\qquad\qquad\qquad$ $\square$

13.4.7 AR Modeling by Least Squares

An alternative to the Yule–Walker method for estimating AR parameters from given measurements of the signal is provided by the *least-squares* method. Let us assume again that the values $\{x[n],\ 0 \le n \le N-1\}$ are given. Then we can form the prediction-error equation for a pth-order model

$$v[n] = x[n] - \hat{x}[n] = \sum_{k=0}^{p} a_k x[n-k], \quad \text{where} \quad a_0 \overset{\triangle}{=} 1, \ p \le n \le N - 1. \quad (13.70)$$

Note that the specified range of n makes use of all available data. Now define the average square error

$$V = \frac{1}{N-p} \sum_{n=p}^{N-1} v^2[n]. \quad (13.71)$$

The *least-squares* AR model of order p is the set of parameters $\{a_k,\ 1 \le k \le p\}$ for which V is minimum. The least-squares criterion is similar to the MMSE criterion, presented in Section 13.4.2. However, instead of minimizing the *expected* square error, which relies on the unknown covariance sequence $\kappa_x[m]$, it seeks to minimize the *actual* square error, which depends only on the given data.

To solve the problem of minimizing V, we proceed as in Section 13.4.2: We differentiate with respect to the unknown parameters and equate the partial derivatives to zero. We get

$$
\frac{\partial V}{\partial a_l} = \frac{2}{N-p} \sum_{n=p}^{N-1} \frac{\partial v[n]}{\partial a_l} v[n] = \frac{2}{N-p} \sum_{n=p}^{N-1} x[n-l] \left(\sum_{k=0}^{p} a_k x[n-k] \right)
$$

$$
= \frac{2}{N-p} \sum_{k=0}^{p} a_k \left(\sum_{n=p}^{N-1} x[n-l]x[n-k] \right) = 0, \quad 1 \le l \le p. \tag{13.72}
$$

Define

$$
\xi_{l,k} = \frac{1}{N-p} \sum_{n=p}^{N-1} x[n-l]x[n-k]. \tag{13.73}
$$

Let Ξ_p be the $p \times p$ matrix constructed from $\{\xi_{l,k},\ 1 \le l,k \le p\}$ and let ξ_p be the p-dimensional vector constructed from $\{\xi_{l,0},\ 1 \le l \le p\}$. Then we can rewrite (13.72) as

$$
\Xi_p a_p + \xi_p = 0 \implies a_p = -\Xi_p^{-1} \xi_p. \tag{13.74}
$$

The vector of parameters a_p obtained by solving (13.74) is the pth-order, least-squares estimate for the measured data $\{x[n],\ 0 \le n \le N-1\}$.

The minimum value of V is given by

$$
V_{\min} = \frac{1}{N-p} \sum_{n=p}^{N-1} \left(\sum_{k=0}^{p} a_k x[n-k] \right) \left(\sum_{l=0}^{p} a_l x[n-l] \right) = \sum_{k=0}^{p} \sum_{l=0}^{p} a_k a_l \xi_{l,k}
$$

$$
= \sum_{k=0}^{p} a_k \xi_{0,k} + \sum_{l=1}^{p} a_l \left(\sum_{k=0}^{p} a_k \xi_{l,k} \right). \tag{13.75}
$$

However, by (13.74), the second term on the right side of (13.75) is zero. Therefore,

$$
V_{\min} = \sum_{k=0}^{p} a_k \xi_{0,k}. \tag{13.76}
$$

The value of $V_{\min}$ can serve as an estimate of γ_v.

The least-squares solution usually provides a more accurate AR model than the Yule–Walker solution when N, the number of data, is relatively small. Therefore, it is often preferred to the Yule–Walker solution in such cases. However, the least-squares solution suffers from two major drawbacks:

1. Contrary to the polynomial obtained from the solution of the Yule–Walker equation, the polynomial obtained from the least-squares solution (13.74) is not guaranteed to be stable, although the occurrence of instability is rare. Stability of the polynomial can be verified, if needed, using the Schur–Cohn test.

2. The matrix Ξ_p is not Toeplitz, so the solution of (13.74) normally requires p^3 operations, rather than p^2 (as required by the Levinson–Durbin algorithm). Efficient algorithms have been derived for solving (13.74) in a number of operations proportional to p^2 [Friedlander et al., 1979; Porat et al., 1982]. However, these algorithms are somewhat complicated and not easy to program, therefore they are not in common use. Since the computational complexity of the least-squares method is higher than that of the Levinson–Durbin method, the former is seldom used in time-critical applications.

We finally remark that the Yule–Walker (or Levinson–Durbin) solution to the AR modeling problem is also called the *autocorrelation method*, and the least-squares solution is also called the *covariance method*. These names are common in speech modeling applications; in Section 14.2 we shall study such an application.

13.5 Joint Signal Modeling

The autoregressive model, discussed in the preceding section, is a way of modeling a single signal. In numerous signal processing applications, we are interested in modeling the relationship between a pair of signals, say $x[n]$, $y[n]$. The signals may be the input and output of a system whose nature and parameters are unknown. For example, $x[n]$ may represent the input of a communication channel and $y[n]$ the output of the channel, and we may wish to model the dynamic behavior of the channel. More generally, the signals may have been generated by some common mechanism. For example, $x[n]$ and $y[n]$ may represent, respectively, a neurological signal (so-called *electromyogram*, or EMG) and a brain signal (so-called *electroencephalogram*, or EEG) measured in response to some stimulus.

Two discrete-time WSS random signals $x[n]$, $y[n]$ are said to be *jointly wide-sense stationary* if each of them is WSS and their cross-covariance $E\{(y[n+m] - \mu_y)(x[n] - \mu_x)\}$ depends only on the difference m between their respective time indices. The corresponding *cross-covariance sequence* is then defined as [cf. (2.119)]

$$\kappa_{yx}[m] = E\{(y[n+m] - \mu_y)(x[n] - \mu_x)\} = E(y[n+m]x[n]) - \mu_y\mu_x. \quad (13.77)$$

In particular, when the signals are zero mean, we have $\kappa_{yx}[m] = E(y[n+m]x[n])$. Note that, in contrast with the covariance sequence $\kappa_x[m]$, the cross-covariance sequence is not symmetric in m.

In this section we study a simple model for a pair of jointly WSS signals: that of a linear, causal, FIR filter. We assign to $x[n]$ the role of input and to $y[n]$ the role of output, although these roles are not necessarily implied by physical considerations. Our model is thus

$$y[n] = b_0 x[n] + b_1 x[n-1] + \cdots + b_q x[n-q]. \quad (13.78)$$

The parameters of the model are the order q and the coefficients $\{b_0, b_1, \ldots, b_q\}$. We note that (13.78) is not a moving-average model, despite being similar to (13.18), because we do not assume here that $x[n]$ is white noise. As in the case of an autoregressive model, we initially assume that the covariances and cross-covariances of the two signals are known, and later consider the case where they are unknown. Based on knowledge of these sequences, we wish to compute the parameters of the model.

Let us proceed in the same way as we derived the Yule–Walker equation: Multiply (13.78) by $x[n-m]$, $m \geq 0$, and take expected values of both sides to obtain

$$\kappa_{yx}[m] = b_0 \kappa_x[m] + b_1 \kappa_x[m-1] + \cdots + b_q \kappa_x[m-q]. \quad (13.79)$$

Now collect $q+1$ such equations, for $0 \leq m \leq q$, and arrange them in a matrix–vector form:

$$\begin{bmatrix} \kappa_x[0] & \kappa_x[1] & \cdots & \kappa_x[q] \\ \kappa_x[1] & \kappa_x[0] & \cdots & \kappa_x[q-1] \\ \vdots & \vdots & \cdots & \vdots \\ \kappa_x[q] & \kappa_x[q-1] & \cdots & \kappa_x[0] \end{bmatrix} \begin{bmatrix} b_0 \\ b_1 \\ \vdots \\ b_q \end{bmatrix} = \begin{bmatrix} \kappa_{yx}[0] \\ \kappa_{yx}[1] \\ \vdots \\ \kappa_{yx}[q] \end{bmatrix}. \quad (13.80)$$

Equation (13.80) is known as the *Wiener equation*, or sometimes the *Wiener–Hopf equation*, and its solution—the Wiener solution to the modeling problem (13.78).[5] The Wiener solution can also be derived using a minimum mean-square error approach as in Section 13.4.2; see Problem 13.16. The procedure wiener in Program 13.6 solves the Wiener equation.

The Wiener equation can be solved iteratively, in a number of operations proportional to q^2, like the Levinson–Durbin algorithm. In fact, Levinson's work of 1947 concerned the solution of the Wiener equation (13.80) rather than the Yule–Walker equations (13.26). We now derive this solution; we call the resulting algorithm the *joint Levinson algorithm*. We introduce the following notations, in addition to the ones defined in Section 13.4.3 for the derivation of the Levinson–Durbin algorithm.

1. $\boldsymbol{b}_{i-1} = [b_{i-1,0}, \ldots, b_{i-1,i-1}]'$: the solution of the $(i-1)$st-order Wiener equation.

2. $\boldsymbol{b}_i^* = [b_{i,0}, \ldots, b_{i,i-1}]'$: the vector consisting of the first i components of the solution of the ith-order Wiener equation.

3. $b_{i,i}$: the last component of the solution of the ith-order Wiener equation.

4. $\boldsymbol{\lambda}_{i-1} = [\kappa_{yx}[0], \ldots, \kappa_{yx}[i-1]]'$: the right side of the $(i-1)$st-order Wiener equation.

The ith-order Wiener equation can be written as

$$\begin{bmatrix} \boldsymbol{K}_i & \tilde{\boldsymbol{\kappa}}_i \\ \tilde{\boldsymbol{\kappa}}_i' & \kappa_x[0] \end{bmatrix} \begin{bmatrix} \boldsymbol{b}_i^* \\ b_{i,i} \end{bmatrix} = \begin{bmatrix} \boldsymbol{\lambda}_{i-1} \\ \kappa_{yx}[i] \end{bmatrix}. \tag{13.81}$$

Let us assume that we have solved the Wiener equation of order i, and we wish to solve (13.81) for $\boldsymbol{b}_i^*$ and $b_{i,i}$. We can expand this equation to

$$\boldsymbol{K}_i \boldsymbol{b}_i^* + \tilde{\boldsymbol{\kappa}}_i b_{i,i} = \boldsymbol{\lambda}_{i-1}, \tag{13.82a}$$

$$\tilde{\boldsymbol{\kappa}}_i' \boldsymbol{b}_i^* + \kappa_x[0] b_{i,i} = \kappa_{yx}[i]. \tag{13.82b}$$

Solving (13.82a) for $\boldsymbol{b}_i^*$ and using the second equality in (13.37) (the solution of the reversed ith-order Yule–Walker equation) gives

$$\boldsymbol{b}_i^* = \boldsymbol{K}_i^{-1} \boldsymbol{\lambda}_{i-1} - \boldsymbol{K}_i^{-1} \tilde{\boldsymbol{\kappa}}_i b_{i,i} = \boldsymbol{b}_{i-1} + b_{i,i} \tilde{\boldsymbol{a}}_i. \tag{13.83}$$

Substituting (13.83) in (13.82b) and solving for $b_{i,i}$ gives

$$b_{i,i} = \frac{\kappa_{yx}[i] - \tilde{\boldsymbol{\kappa}}_i' \boldsymbol{b}_{i-1}}{\kappa_x[0] + \tilde{\boldsymbol{\kappa}}_i' \tilde{\boldsymbol{a}}_i} = \frac{\kappa_{yx}[i] - \tilde{\boldsymbol{\kappa}}_i' \boldsymbol{b}_{i-1}}{s_i}. \tag{13.84}$$

A convenient way for implementing the joint Levinson algorithm is to add it on the top of the Levinson–Durbin algorithm, because the latter computes the parameters $\tilde{\boldsymbol{a}}_i$ and s_i needed in (13.83), (13.84). We add the computation of $b_{0,0}$ to the initialization step, using

$$b_{0,0} = \frac{\kappa_{yx}[0]}{\kappa_x[0]}. \tag{13.85}$$

We then rewrite (13.83), (13.84) for $i+1$ and add them at the ith step, right after (13.45d):

$$b_{i+1,i+1} = \frac{\kappa_{yx}[i+1] - \sum_{l=0}^{i} b_{i,l} \kappa_x[i+1-l]}{s_{i+1}}, \tag{13.86a}$$

$$b_{i+1,l} = b_{i,l} + b_{i+1,i+1} a_{i+1,i+1-l}, \quad 0 \le l \le i. \tag{13.86b}$$

The joint Levinson algorithm should be iterated for $0 \le i \le q-1$, where q is the desired filter's order. The procedure `jlev` in Program 13.7 implements the joint Levinson algorithm.

When the covariances $\kappa_x[m]$ and $\kappa_{yx}[m]$ are not available, only measured sequences $x[n]$ and $y[n]$, we can use these sequences to construct estimates of the covariances. The sequence $\hat{\kappa}_x[m]$ can be constructed as in (13.66) and the sequence $\hat{\kappa}_{yx}[m]$ by

$$\hat{\kappa}_{yx}[m] \triangleq N^{-1} \sum_{i=0}^{N-1-m} x[i] y[i+m], \quad 0 \le m \le q. \tag{13.87}$$

The corresponding model coefficients will be denoted by $\{\hat{b}_0, \hat{b}_1, \ldots, \hat{b}_q\}$. A least-squares model can also be computed in lieu of the Wiener solution, as done in Section 13.4.7 for the autoregressive model. The details of the least-squares approach are left as a problem; see Problem 13.17.

Example 13.5 Consider the problem of transferring digital information through a base-band channel, that is, a channel capable of transmitting unmodulated signals. Such channels are rare in practice; however, real modulated channels have equivalent base-band models, so the application we present here is applicable to modulated channels equally well. The signaling method we discuss in this example is *pulse amplitude modulation* (PAM). In this method, digital information is represented by symbols that can take a finite number of real values. For example, possible symbol values are ± 1 in binary PAM, $\{\pm 1, \pm 3\}$ in quaternary PAM, etc. To each symbol we assign a pulse whose waveform is fixed and whose amplitude is proportional to the symbol value. The pulses are then transmitted in a sequence. Denoting the symbol sequence by $u[m]$, the symbol interval by T, and the pulse waveform by $p(t)$, the transmitted waveform is

$$s(t) = \sum_{m=-\infty}^{\infty} u[m]p(t - mT). \tag{13.88}$$

In the simplest case, the waveform $p(t)$ is a rectangle of duration T, but other waveforms are also used.

Let us assume that the channel is linear time invariant, and denote its impulse response by $h(t)$. Then the received signal is (disregarding noise and other effects)

$$x(t) = \{h * s\}(t) = \sum_{m=-\infty}^{\infty} u[m] \int_{-\infty}^{\infty} h(\tau)p(t - \tau - mT)d\tau. \tag{13.89}$$

Sampling the received signal at interval T gives the discrete-time signal

$$x[n] = x(nT) = \sum_{m=-\infty}^{\infty} u[m] \int_{-\infty}^{\infty} h(\tau)p(nT - \tau - mT)d\tau$$

$$= \sum_{m=-\infty}^{\infty} u[m]g[n - m], \tag{13.90}$$

where

$$g[n] = \int_{-\infty}^{\infty} h(\tau)p(nT - \tau)d\tau, \quad n \in \mathbb{Z}. \tag{13.91}$$

The sequence $g[n]$ is the equivalent discrete-time impulse response of the channel. If this sequence were equal to a unit sample $\delta[n]$, we would get $x[n] = u[n]$, so we would be able to recover the transmitted symbol sequence exactly. In reality, $g[n]$ is almost never equal to $\delta[n]$. Therefore, $x[n]$ is affected by symbol values at other times, not just by $u[n]$. This phenomenon is called *intersymbol interference* (ISI), because symbols belonging to times $m \neq n$ interfere with the symbol at time n and hinder our ability to detect it faithfully. ISI is a major limiting factor of the performance of narrow-band communication channels.

An *equalizer* is a linear time invariant system designed to invert, or cancel, the convolution operation (13.90). The transfer function of an ideal equalizer should be equal to the inverse of $G^z(z)$, the transfer function of the equivalent discrete-time channel. An ideal equalizer is not realizable, except in rare cases, for the following reasons:

1. The impulse response $h(t)$ of the channel is usually unknown (or at least not exactly known), so its discrete-time equivalent $g[n]$ is not available.

2. Even if $g[n]$ is known, $G^z(z)$ is not necessarily rational, so its inverse is not necessarily rational either.

3. $G^z(z)$ does not necessarily have a causal and stable inverse $1/G^z(z)$. This is true even when $G^z(z)$ itself is rational, causal, and stable. If you have solved Problem 8.12, you know the reason: $G^z(z)$ may be nonminimum phase; that is, it may contain zeros outside the unit circle. In such a case, $G^z(z)$ may possess a stable inverse, but this inverse will not be causal. Therefore, to determine $u[n]$, an ideal equalizer must use future values of $x[n]$ (potentially infinite in number).

A common alternative to an ideal, unrealizable, equalizer is an approximate FIR equalizer. Such an equalizer is defined by the relationship

$$y[n] = b_0 x[n] + b_1 x[n-1] + \cdots + b_q x[n-q]. \tag{13.92}$$

The order q is usually chosen to be even, say $q = 2N$, and the equalizer's coefficients are chosen to satisfy

$$\{b * g\}[n] \approx \delta[n - N]. \tag{13.93}$$

The approximation here means that the coefficient $\{b * g\}[N]$ should be close to unity and all other coefficients should be close to zero. This equalizer "looks into the future" in the sense that its output at time n, $y[n]$, is approximately equal to $u[n - N]$, so the reconstructed symbol value corresponding to time $n-N$ depends on the received signal values $x[n - N + 1]$ through $x[n]$, which are in its future (it also depends on the past and present signal values $x[n - 2N]$ through $x[n - N]$).

Practical equalizers need to estimate the coefficients $\{b_0, b_1, \ldots, b_q\}$, since these coefficients depend on the unknown impulse response of the channel. In a *minimum mean-square error equalizer*, the coefficients are obtained by solving the Wiener equation (13.80). A common way to estimate the covariances $\kappa_x[m]$ and $\kappa_{yx}[m]$ is as follows: During initial operation, a fixed sequence of symbols $u[n]$, known to the receiver, is transmitted. The equalizer sets $y[n] = u[n - N]$; it then collects the received sequence $x[n]$ corresponding to the known transmitted sequence, and uses the two sequences to construct $\hat{\kappa}_x[m]$ and $\hat{\kappa}_{yx}[m]$ as in (13.66), (13.87). The estimated covariances are substituted in (13.80) and this equation is solved to yield $\{\hat{b}_0, \hat{b}_1, \ldots, \hat{b}_q\}$. An alternative procedure is to use the two sequences in a *least-squares equalizer*, as discussed in Problem 13.17. The fixed sequence $u[n]$ is called a *training sequence*, because it serves to train the equalizer. Figure 13.12 shows a block diagram of the equalizer. The left position of the switch corresponds to a training mode.

After the training sequence has been exhausted, the equalizer is normally switched to a *decision-directed mode*, indicated by the right position of the switch in Figure 13.12. In this mode, the equalizer's output $\hat{y}[n] = \{b * x\}[n]$ is fed to a decision circuit, whose task is to detect the transmitted symbols. The detected symbol corresponding to time $n - N$, denoted by $\hat{u}[n - N]$, is obtained by rounding $\hat{y}[n]$ to the nearest legal symbol value (e.g., one of $\pm 1, \pm 3$ for a quaternary PAM). The sequence of detected symbols is used to construct the equalizer's reference sequence $y[n]$ according to the rule $y[n] = \hat{u}[n - N]$. Then, the equalizer continues to update the coefficients $\{\hat{b}_0, \hat{b}_1, \ldots, \hat{b}_q\}$. In reality, *adaptive equalization* is normally employed, but we do not discuss this subject here; see Gitlin et al. [1992, Chapter 8]. □

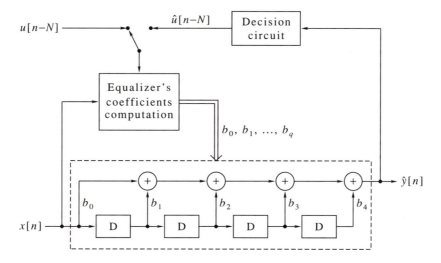

Figure 13.12 Channel equalizer, illustrated for $q = 4$, $N = 2$; switch in left position: training mode; switch in right position: decision-directed mode.

13.6 Summary and Complements

13.6.1 Summary

In this section we discussed spectrum estimation methods and parametric modeling methods for WSS random signals. Simple short-time spectral analysis is of limited use in the case of random signals, because of the random appearance of the Fourier transform of such signals. Randomness can be smoothed by averaging the square magnitudes of the DFTs of successive segments. The amount of smoothing depends on the number of segments being averaged. This number, in turn, is limited by the length of time during which the signal can be assumed stationary. Among the various methods of averaging, the most popular is the one by Welch. An alternative to the Welch periodogram is the smoothed periodogram, useful mainly when the data length is too short for effective averaging.

Rational parametric models for WSS random signals were introduced next, in particular autoregressive models. The parameters of an AR model are related to the covariance sequence of the signal by the Yule–Walker equations. The Levinson–Durbin algorithm facilitates efficient solution of the Yule–Walker equations. This algorithm also leads to lattice realizations of the FIR filter $a(z)$ and the IIR filter $1/a(z)$ corresponding to the AR model. When the covariance sequence of the signal is not known, it can be estimated from actual measurements of the signal, and used in the Levinson–Durbin algorithm in place of the true covariances. An alternative approach is to obtain the model parameters by minimizing the sum of squares of the measured prediction error.

Rational models are also useful in joint modeling of two random signals. We examined the special case of an FIR model, and derived the Wiener solution for this model. The joint Levinson algorithm facilitates an efficient solution to the Wiener equation.

The parametric techniques presented in this chapter naturally lead to the field of *adaptive signal processing*. This field is concerned with real-time, time-varying modeling of signals and systems. For a comprehensive, up-to-date exposition of adaptive signal processing, see Haykin [1996].

13.6.2   Complements

1. [p. 515] The term *periodogram* was introduced by Schuster in his landmark paper [1906b]; it is derived from *a diagram of periods*, since its most common use is for finding periodic components in a signal.

2. [p. 515] The limit (13.2) holds for random signals that are *ergodic in the mean square*. We do not define such signals here, nor do we deal with their mathematical theory, but we implicitly assume that all random signals mentioned in this book are ergodic in the mean square.

3. [p. 523] Expressing a signal as a linear combination of other signals plus noise (or error) is called *regression* in statistics. Equation (13.19) is an expression of $x[n]$ as a linear combination of *its own past values* and the noise $v[n]$, therefore it is called *autoregression*.

4. [p. 521] Radio communication at frequencies 3 to 30 megahertz (so-called "short waves") is made mainly through the ionosphere: The waves are transmitted from the ground upward and are reflected by the ionosphere back to the ground. Communication ranges of thousands of kilometers are made possible this way and are used mainly by radio amateurs all over the world. High sunspot activity increases the range of frequencies that can be transmitted via the ionosphere and improves the quality of communication. Therefore, radio amateurs need to be aware of the current sunspot cycle status.

5. [p. 537] The name *Wiener equation* for (13.80), though widespread, diminishes the contribution of Norbert Wiener, which was incomparably deeper than the straightforward solution (13.80) to the simple modeling problem (13.78). Wiener solved the problem of computing the *causal* minimum mean-square error filter in full generality (originally in continuous time). His work, performed as part of the Second World War effort, was published in a classified report in 1942 and later reprinted [Wiener, 1949]. The name also does injustice to A. N. Kolmogorov, who was the first to develop a mean-square prediction theory for discrete-time WSS signals [Kolmogorov, 1941].

13.7 MATLAB Programs

Program 13.1 Short-time spectral analysis.

```
function X = stsa(x,N,K,L,w,opt,M,theta0,dtheta);
% Synopsis: X = stsa(x,N,K,L,w,opt,M,theta0,dtheta).
% Short-time spectral analysis.
% Input parameters:
% x: the input vector
% N: segment length
% K: number of overlapping points in adjacent segments
% L: number of consecutive DFTs to average
% w: the window (a row vector of length N)
% opt: an optional parameter for nonstandard DFT:
%       'zp' for zero padding
%       'chirpf' for chirp Fourier transform
% M: length of DFT if zero padding or chirp was selected
% theta0, dtheta: parameters for chirp FT.
% Output:
% X: a matrix whose rows are the DFTs of the segments
%    (or averaged segments).

lx = length(x); nsec = ceil((lx-N)/(N-K)) + 1;
x = [reshape(x,1,lx), zeros(1,N+(nsec-1)*(N-K)-lx)];
nout = N; if (nargin > 5), nout = M; else, opt = 'n'; end
X = zeros(nsec,nout);
for n = 1:nsec,
   temp = w.*x((n-1)*(N-K)+1:(n-1)*(N-K)+N);
   if (opt(1) == 'z'), temp = [temp, zeros(1,M-N)]; end
   if (opt(1) == 'c'), temp = chirpf(temp,theta0,dtheta,M);
   else, temp = fftshift(fft(temp)); end
   X(n,:) = abs(temp).^2;
end
if (L > 1),
   nsecL = floor(nsec/L);
   for n = 1:nsecL, X(n,:) = mean(X((n-1)*L+1:n*L,:)); end
   if (nsec == nsecL*L+1),
      X(nsecL+1,:) = X(nsecL*L+1,:); X = X(1:nsecL+1,:);
   elseif (nsec > nsecL*L),
      X(nsecL+1,:) = mean(X(nsecL*L+1:nsec,:));
      X = X(1:nsecL+1,:);
   else, X = X(1:nsecL,:); end
end
```

Program 13.2 A smoothed periodogram.

```
function s = smooper(x,w);
% Synopsis: s = smooper(x,w).
% Computes the smoothed periodogram of the data vector x.
% Input parameters:
% x: the data vector
% w: the window; must have odd length.
% Output:
% s: the smoothed periodogram, of length equal to that of x.

if (rem(length(w),2) == 0),
   error('Window in SMOOPER must have an odd length');
end
x = reshape(x,1,length(x));
x = x - mean(x);
kappa = (1/length(x))*conv(x,fliplr(x));
n = 0.5*(length(kappa)-length(w));
s = fft([zeros(1,n),w,zeros(1,n)].*kappa);
s = abs(s(1:length(x)));
```

Program 13.3 Solution of the Yule–Walker equations.

```
function [a,gammav] = yw(kappa);
% Synopsis: [a,gammav] = yw(kappa).
% Solves the Yule-Walker equations.
% Input:
% kappa: the covariance sequence value from 0 to p.
% Output parameters:
% a: the AR polynomial, with leading entry 1.
% gammav: the innovation variance

p = length(kappa)-1;
kappa = reshape(kappa,p+1,1);
a = [1; -toeplitz(kappa(1:p,1))\kappa(2:p+1,1)]';
gammav = a*kappa;
```

Program 13.4 The Levinson-Durbin algorithm.

```
function [a,rho,s] = levdur(kappa);
% Synopsis: [a,rho,s] = levdur(kappa).
% The Levinson-Durbin algorithm.
% Input:
% kappa: the covariance sequence values from 0 to p.
% Output parameters:
% a: the AR polynomial, with leading entry 1
% rho: the set of p reflection coefficients
% s: the innovation variance.

p = length(kappa)-1;
kappa = reshape(kappa,p+1,1);
a = 1; s = kappa(1); rho = [];
for i = 1:p,
   rhoi = (a*kappa(i+1:-1:2))/s; rho = [rho,rhoi];
   s = s*(1-rhoi^2);
   a = [a,0]; a = a - rhoi*fliplr(a);
end
```

Program 13.5 Computation of the estimated covariance sequence.

```
function kappa = kappahat(x,p);
% Synopsis: kappa = kappahat(x,p).
% Generate estimated covariance values of a data sequence.
% Input parameters:
% x: the data vector
% p: maximum order of covariance.
% Output parameters:
% kappa: the vector of kappahat from 0 through p.

x = x - mean(x); N = length(x);
kappa = sum(x.*x);
for i = 1:p,
   kappa = [kappa, sum(x(1:N-i).*x(i+1:N))];
end
kappa = (1/N)*kappa;
```

Program 13.6 Solution of the Wiener equation.

```
function b = wiener(kappax,kappayx);
% Synopsis: b = wiener(kappax,kappayx).
% Solves the Wiener equation.
% Input parameters:
% kappax: the covariance sequence of x from 0 to q
% kappayx: the joint covariance sequence of y and x from 0 to q.
% Output:
% b: the Wiener filter.

q = length(kappax)-1;
kappax = reshape(kappax,q+1,1);
kappayx = reshape(kappayx,q+1,1);
b = (toeplitz(kappax)\kappayx)';
```

Program 13.7 The joint Levinson algorithm.

```
function b = jlev(kappax,kappayx);
% Synopsis: b = jlev(kappax,kappayx).
% The joint Levinson algorithm.
% Input parameters:
% kappax: the covariance sequence of x from 0 to q
% kappayx: the joint covariance sequence of y and x from 0 to q.
% Output:
% b: the Wiener filter.

q = length(kappax)-1;
kappax = reshape(kappax,q+1,1);
kappayx = reshape(kappayx,q+1,1);
a = 1; s = kappax(1); b = kappayx(1)/kappax(1);
for i = 1:q,
   rho = (a*kappax(i+1:-1:2))/s;
   s = s*(1-rho^2);
   a = [a,0]; a = a - rho*fliplr(a);
   bii = (kappayx(i+1) - b*kappax(i+1:-1:2))/s;
   b = [b+bii*fliplr(a(2:i+1)), bii];
end
```

13.8 Problems

13.1 Write the Welch periodogram formula (13.4) in case the overlap is not 50 percent, but a given number of points K (where $K < N$).

13.2 Extend the idea in Problem 6.17 to the case where $x(t)$ is a complex OQPSK signal, as defined in Example 13.1. Suggest an operation on $y(nT)$ which will enable estimation of ω_0 using DFT, determine the sampling interval, and find the loss in output SNR.

13.3 For a signal $x[n]$ of length N, define

$$S^d[k] = N^{-1}|X^d[k]|^2, \quad 0 \le k \le N - 1.$$

  (a) Find the relationship between $s[n]$, the inverse DFT of $S^d[k]$, and the sequence $\hat{\kappa}_x[m]$ defined in (13.12). Hint: Use (13.11).

  (b) Obtain $x_a[n]$ from $x[n]$ by zero padding to length $M \ge 2N - 1$, and let

$$S_a^d[k] = N^{-1}|X_a^d[k]|^2, \quad 0 \le k \le M - 1.$$

  Find the relationship between $s_a[n]$, the inverse DFT of $S_a^d[k]$, and the sequence $\hat{\kappa}_x[m]$.

  (c) Suggest a replacement for kappahat that uses the result of part b.

13.4 Solve the Yule–Walker equation (13.26a) for $p = 2$ and obtain an explicit solution for a_1, a_2.

13.5 Write the Yule–Walker equations (13.26a,b) explicitly for $p = 1$. Now assume that, in these equations, a_1 and γ_v are known and solve explicitly for $\kappa_x[0], \kappa_x[1]$.

13.6 Repeat Problem 13.5 for $p = 2$ and solve for $\kappa_x[0], \kappa_x[1], \kappa_x[2]$ as a function of γ_v, a_1, a_2. Remark: This is quite tedious, but the final expressions are not overly complicated.

13.7 Generalize Problems 13.5 and 13.6 to an arbitrary order p as follows: Assume that γ_v and $\{a_1, \ldots, a_p\}$ are known, and write down a set of linear expressions for the unknown variables $\{\kappa_x[0], \ldots, \kappa_x[p]\}$. Do not attempt to solve this system of equations explicitly, since this is quite complicated. Hint: You may find it useful to take $0.5\kappa_x[0]$ as an unknown variable, rather than $\kappa_x[0]$.

13.8 Write a MATLAB procedure that solves the set of equations you have obtained in Problem 13.7. Test your procedure by computing $\{\kappa_x[0], \ldots, \kappa_x[p]\}$ for a given pth-order polynomial $a(z)$ (ensure that this polynomial is stable!) and a given positive constant γ_v. Feed the result to the procedure yw, and verify that you get back the polynomial $a(z)$ and the constant γ_v.

13.9 Show that the numerator of the reflection coefficient ρ_{i+1}, defined in (13.43), is the covariance between $x[n - i - 1]$ and $v_i[n]$, the prediction error of the ith-order AR model at time n.

13.10 Show that the variance of $\tilde{v}_i[n]$, defined in (13.46), is equal to the variance of $v_i[n]$.

13.11 Show that ρ_{i+1}, defined in (13.43), is the correlation coefficient between $v_i[n]$ and $\tilde{v}_i[n-1]$.

13.12 Let

$$x[n] = \sin(\theta_0 n + \phi_0),$$

where ϕ_0 is a random variable whose distribution is uniform in the range $[-\pi, \pi)$, and θ_0 is fixed.

(a) Prove that the mean of $x[n]$ is zero.

(b) Prove that the covariance sequence of $x[n]$ is

$$\kappa_x[m] = 0.5 \cos(\theta_0 m),$$

hence $x[n]$ is a WSS signal. Hint for parts a, b: The distributions of both $\sin \phi_0$ and $\cos \phi_0$ are symmetric about 0; therefore their means are zero.

(c) Prove that $x[n]$ satisfies the identity

$$x[n] - 2\cos\theta_0 x[n-1] + x[n-2] = 0.$$

(d) Explain why part c implies that $x[n]$ is deterministic in the sense defined in Section 13.4.3.

(e) Derive formulas for ρ_1 and ρ_2 obtained by the Levinson–Durbin algorithm when fed with the covariance sequence $\kappa_x[m]$ given in part b. What is $|\rho_2|$ in this case? What can you conclude about s_2? Reconcile this result with the one in part d.

13.13 Explain how to use the IIR lattice for realizing an all-pass filter. Hint: Recall Problem 7.29.

13.14 Generate the following signal in MATLAB:

$$x[n] = \sin(2\pi 0.17n) + 0.5\sin(2\pi 0.18n), \quad 0 \le n \le 127.$$

(a) Is it possible to measure the frequencies of the two sinusoids using the windowed DFT technique studied in Chapter 6? Answer based on theoretical arguments and, if necessary, experiment with MATLAB.

(b) Use the procedures kappahat and yw to generate a 50th-order AR model to $x[n]$. Compute and plot the magnitude response of the AR model, using frqresp (use 1000 points for the plot). Is it possible to measure the frequencies from the AR magnitude response?

(c) Add noise to $x[n]$, that is, form

$$y[n] = x[n] + v[n],$$

where $v[n]$ is generated using s*randn(1,128). Choose s as 0.1, then 0.2, and finally 0.4. Conclude whether it is possible to measure the frequencies from the AR magnitude response at each of the three noise levels.

(d) State your conclusions from this experiment.

13.15 Write a MATLAB procedure for the Schur algorithm; test it by comparing its output with that of the procedure levdur.

13.16 In the model (13.78), let $e[n]$ denote the error between the two sides. Find the parameters $\{b_0, b_1, \ldots, b_q\}$ that minimize $E(e^2[n])$. Show that this leads to the Wiener equation (13.80).

13.17 Develop a least-squares solution to the modeling problem (13.78). Assume that the sequences $\{x[n], y[n],\ 0 \le n \le N - 1\}$ are given. Define $e[n]$ as the difference between the two sides of (13.78). Let

$$V = \frac{1}{N - q} \sum_{n=q}^{N-1} e^2[n]. \tag{13.94}$$

Finally, find the values of $\{b_0, b_1, \ldots, b_q\}$ that minimize V, similarly to the derivation in Section 13.4.7.

Chapter 14

Digital Signal Processing Applications*

Throughout this book, we have encountered applications of digital signal processing in examples and problems. This chapter is exclusively devoted to applications. We assume that you have mastered most of the book by now, and are ready to explore DSP as it is used in real life. It is impossible, in a single chapter, to do justice to all but a small sample of the DSP world. Our small sample comprises seven topics.

First, we present signal compression; we have already touched upon this topic in Example 12.8, in the context of filter banks and subband coding. Here we present a compression method considerably more important than subband coding: the discrete cosine transform. The DCT has been standardized in recent years for image and motion picture compression. Our treatment of DCT-based signal compression is limited to temporal signals, however, and we will not deal with image compression.

The second topic is speech modeling and compression by a technique called linear predictive coding (LPC). LPC is not a new technique: its principles have been known since the mid-1960s, and it has been applied to speech since the mid-1970s. However, recent years have seen substantial developments in this area. In particular, cellular telephone services now use LPC as a standard. Here we present, as an example, the speech compression and coding technique used in the Pan-European Digital Mobile Radio System, better known by the French acronym GSM (Groupe Special Mobile).

The third topic concerns modeling and processing of musical signals. Compared with other natural signals, musical signals are characterized by an orderly structure; this makes them convenient for modeling and interpretation. On the other hand, high fidelity is an extremely important factor in our enjoyment of music; this makes synthesis of musical signals difficult and challenging.

The fourth topic is probably the one getting the most attention in today's technical world: communication. Until recently, electronic circuitry in wireless digital communication systems has been chiefly analog. However, DSP is rapidly taking over, and future communication systems will undoubtedly be based on digital processing. As an example of this vast field, we present a digital receiver for frequency-shift keying (FSK) signals.

The fifth topic presents an example from the biomedical world: electrocardiogram (ECG) analysis. We have chosen ECG since, of all electrical signals measured from the human body, it is probably the easiest to analyze, at least at a basic level. Our example concerns the use of ECG for measurement of heart rate variations.

The last two topics are concerned with technology. In the first, we discuss micro-processors for DSP applications. We have chosen, as an example, a particular DSP chip of the mid-1990s vintage: the Motorola DSP56301. We use this example for illustrating common trends and techniques in DSP hardware today. Among all topics covered in this book, this will probably be the fastest to become obsolete, since digital technology progresses at an enormously fast rate. Finally, we present a modern A/D converter technology: sigma–delta A/D. Sigma–delta A/D converters provide a fine example of the advantages gained by combining very large-scale integration (VLSI) technology and digital signal processing principles.

14.1   Signal Compression Using the DCT

Any information stored digitally is inherently finite, say of N bits. *Compression* is the operation of representing the information by N_c bits, where $N_c < N$. Compression is useful for economical reasons: it saves storage space, transmission time, or transmission bandwidth. The ratio N/N_c is called the *compression ratio*. The greater the compression ratio, the better the compression.

There are two basic types of compression: lossless and lossy. Lossless compression is defined by the property that the original information can be retrieved *exactly* from the compressed information. Mathematically, lossless compression is an invertible operation. For example, compression of text must be lossless, since otherwise the text cannot be exactly retrieved. The highest possible lossless compression ratio of a given information is related to the *entropy* of the source of information, a term perhaps known to those who have studied information theory (but which we shall not attempt to define here). Typical ratios achievable with lossless compression are 2 to 3. The best-known compression methods are the *Huffman code* [Huffman, 1952] and the *Ziv–Lempel algorithms* [Ziv and Lempel, 1977, 1978]. We shall not discuss lossless compression further here; see Cover and Thomas [1991] for a detailed exposition of this subject.

When data are subjected to lossy compression, the original information cannot be retrieved exactly from the compressed information. Mathematically, lossy compression is a noninvertible operation. The advantage of lossy compression over lossless one is that much higher compression ratios can be achieved. However, lossy compression is limited to applications in which we can tolerate the loss. For example, speech signals can be compressed at high ratios (10 and above), and the quality of the reconstructed speech will be only slightly inferior to the original speech. Most people can tolerate some distortion of a speech signal without impairment of their ability to comprehend its contents. Therefore, compression is highly useful for transmitting speech over telephone lines or via wireless channels. Even higher compression ratios can be obtained for images. Compression is very useful for storing images, since image storage is highly space consuming. Still higher compression ratios can be achieved for motion pictures, a fact that is useful for storing motion pictures on digital media (such as video discs) and in video transmission (video conferencing, digital TV).

There are many methods for lossy signal compression. Here we describe the principle of operation of compression by orthonormal transforms. Consider a signal $x[n]$ of length N; let $\boldsymbol{O}_N$ be an $N \times N$ orthonormal matrix, and $X^o[k]$ the result of transforming $x[n]$ by $\boldsymbol{O}_N$, that is,

$$\boldsymbol{X}_N^o = \boldsymbol{O}_N \boldsymbol{x}_N. \tag{14.1}$$

The transform is selected with an attempt to decorrelate the components of the transformed signal X_N^o. Successful decorrelation often results in nonuniform distribution of the magnitudes of the components of X_N^o. Typically, there are a few components whose magnitudes are large, and a large number of components whose magnitudes are small. We now explain how this nonuniform distribution is used for compression.

Assume that we represent each component of x_N by B bits, so the complete vector requires NB bits. In general, the vector X_N^o is represented by NB bits or more. (Extra bits may be needed to guarantee that there will be no overflows in the additions.) The next step in the compression procedure takes advantage of the nonuniform distribution of the magnitudes of the $X_N^o[k]$ through *quantization*: Each component $X_N^o[k]$ is represented by a number of bits proportional to its magnitude. As a result, the total number of bits needed for representing the complete vector X_N^o will often be smaller than NB. In an extreme example of quantization, every $X_N^o[k]$ smaller in magnitude than a certain threshold is completely neglected, that is, replaced by zero. We denote the compressed version of X_N^o, resulting from the quantization procedure, by $\hat{X}_N^o$.

Because the transform is orthonormal, the signal can be recovered exactly from its (uncompressed) transform by

$$x_N = O_N' X_N^o. \tag{14.2}$$

Correspondingly, an approximate (lossy) reconstruction of the signal from its compressed version is obtained by the operation

$$\hat{x}_N = O_N' \hat{X}_N^o. \tag{14.3}$$

In summary, signal compression by an orthonormal transform is carried out by the following operations:

1. Transforming the signal using (14.1).

2. Modifying the transform values by quantization. This should be done according to a rule designed to guarantee that distortion due to quantization will not exceed a certain percent of the total signal energy.

The modified transformed values are used for storage or transmission of the information. Compression is efficient if the total number of bits in these values is much smaller than NB. Note that, because of orthonormality, the total energy of the quantization error in the transform domain is equal to that in the signal domain. Signal reconstruction is carried out by the operation (14.3). The result again consists of NB bits, but these are not identical to the bits of the original signal.

In choosing an orthonormal transform for signal compression, there are two major considerations:

1. The compression ratios that can be achieved for the class of signals in question, under the constraint that distortion due to compression (which affects the *quality* of the compressed signal) be within acceptable limits.

2. The amount of computation needed to carry out the compression and reconstruction operations.

For a general orthonormal transform, each of the operations (14.1) and (14.3) requires a number of operations proportional to N^2. This is prohibitively large in many applications. However, the transforms we have studied all have fast versions, requiring a number of operations on the order of $N \log_2 N$. The fast Fourier transform, the discrete cosine and sine transforms, and the Hartley transform are all candidates for signal compression.

Of the aforementioned transforms, the DCT has been shown to be particularly effective for first-order autoregressive signals. Recall from (13.19) that a first-order AR signal obeys the first-order difference equation

$$x[n] = a_1 x[n-1] + v[n], \tag{14.4}$$

where a_1 is a constant parameter, smaller than 1 in magnitude, and $v[n]$ is white noise. In particular, when a_1 is nearly 1, the signal varies slowly over time. Natural images often have characteristics similar to first-order autoregressive signals, so the DCT is an effective compression technique for them. Image processing is beyond the scope of this book, so we just mention two simple facts, which can be understood without delving deeply into this subject:

1. An image is a two-dimensional array of real numbers. The numbers represent the intensity levels of the points (*pixels*) in the image.

2. A two-dimensional transform such as the DFT or DCT can be performed by transforming each row of the image individually as a one-dimensional signal, then transforming each column of the result individually; similarly for an inverse transform.

We now illustrate signal compression by DCT. We choose DCT-II, which is the one in standard use for image compression applications; see Section 4.9.2 for the definition and properties of DCT-II. We use a simple quantization rule for compression, as follows. The components of the transform are sorted in order of decreasing magnitudes. Denote the sorted components by $\{X_{\text{sort}}^{c2}[k],\ 0 \le k \le N - 1\}$. Since the transform is orthonormal, it satisfies Parseval's energy conservation property, that is,

$$E \triangleq \sum_{n=0}^{N-1} (x[n])^2 = \sum_{k=0}^{N-1} (X^{c2}[k])^2. \tag{14.5}$$

Let β be a number smaller than 1, representing the fraction of energy we wish to preserve in the compressed signal (e.g., $\beta = 0.99$). The first M components (after sorting) are retained to their full precision, where M is the minimum value for which

$$\sum_{k=0}^{M-1} (X_{\text{sort}}^{c2}[k])^2 \ge \beta E. \tag{14.6}$$

The values of $\{X_{\text{sort}}^{c2}[k],\ k \ge M\}$ are set to zero. The compressed information consists of $\{X_{\text{sort}}^{c2}[k],\ 0 \le k \le M - 1\}$ and their indices in the original transform vector.

Example 14.1 The signal shown in Figure 14.1(a) is first-order AR, with $a_1 = 0.99$. Its DCT-II components are shown in Figure 14.1(b), in their natural order. As we see, the signal has random appearance. Its time variation is generally slow, but it also has high-frequency components. The DCT-II components have large magnitudes for small values of k (corresponding to low frequencies), and small magnitudes for large values of k. We compressed the signal using a threshold parameter $\beta = 0.99$. In this case, the number of coefficients after compression is $M = 50$. Figure 14.2(a) shows the result of reconstructing the compressed signal as a solid line. For reference, we show the original signal as a dotted line. As we see, the reconstructed signal follows the outline of the original signal very well, but not the high-frequency oscillations. Figure 14.2(b) emphasizes this effect by showing the error signal $\hat{x}[n] - x[n]$ scaled by 5. The error signal has most of its energy in the high frequencies, a phenomenon typical of this type of compression. We can regard the compression–reconstruction procedure as a special kind of low-pass filtering of the input signal.

For comparison, we compress the same signal using the same procedure, but with the DFT matrix. Since the DFT matrix is complex, only the first $N/2$ coefficients are used for compression. These and their complex conjugates are then used for reconstruction. For a threshold $\beta = 0.99$, 34 complex coefficients, equivalent to 68 real coefficients, are needed. The conclusion is that DCT-II has a higher compression ratio than DFT in this case.   □

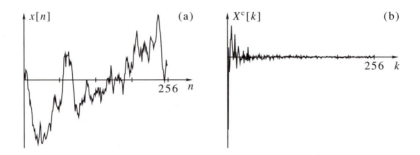

Figure 14.1 The signal (a) and its DCT (b) in Example 14.1.

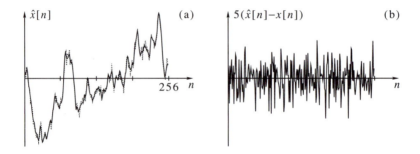

Figure 14.2 The reconstructed signal (a) and the error signal (b) in Example 14.1 (the error signal is scaled by 5).

In summary, the discrete cosine transform is a proven effective compression tool. It has become the method of choice in internationally accepted image compression standards, such as JPEG (from the *Joint Photographic Experts Group*) for still image compression, and MPEG (from the *Motion Picture Expert Group*) for motion pictures compression. See Pennebaker and Mitchell [1993] for further reading.

14.2   Speech Signal Processing

In Chapter 12 we described speech compression by subband coding; see Example 12.8, page 493. In this section we examine an application of digital signal processing to modeling and compression of speech signals. We describe a modeling method known as *linear predictive coding*, and its application to the speech compression standard GSM. At the time this book is written, GSM has become an almost worldwide standard for cellular telephone communications (although not in the United States).

14.2.1 Speech Modeling

Human voice is generated mainly by the vocal cords. The vibration of the vocal cords excites the *vocal tract*, which includes the larynx (the part of the respiratory tube containing the vocal cords), the pharynx (the part between the larynx and the mouth), the mouth cavity, and the nasal cavity. Sound waves are emitted from the lips and, to a lesser degree, from the nose.

The basic units of speech are called *phonemes*. The phonemes in the English language are listed in Rabiner and Juang [1993, page 25]. They are broadly classified into the following groups: (1) vowels, (2) diphthongs, (3) semivowels, (4) consonants. The consonants are further divided into nasals, voiced stops, unvoiced stops, voiced fricatives, unvoiced fricatives, whisper, and affricates. Most phonemes are *voiced*; that is, they are generated by the vocal cords. A few are *unvoiced*; that is, the vocal cords are not involved in their generation.

In this section we discuss modeling of voiced sound, which is largely the most important part of speech. The signal generated by the vocal cords has roughly the shape of a quasi-periodic impulse train. By *quasi-periodic* we mean that (1) the intervals between successive impulses are not exactly constant, but vary slightly and (2) the amplitudes of different impulses are not exactly constant. The mathematical description of a discrete-time, quasi-periodic impulse train is

$$u[n] = \sum_{m=-\infty}^{\infty} c_m \delta[n - n_m] = \begin{cases} c_m, & n = n_m, \\ 0, & \text{otherwise.} \end{cases} \tag{14.7}$$

The impulses occur at instants n_m and their amplitudes are c_m. The impulses are relatively sparse; that is, the differences $n_m - n_{m-1}$ are much larger than 1. These differences are nearly, but not exactly, constant. The reciprocal of the average interval between successive impulses is called the *pitch frequency*, or the pitch for short. The pitch is characteristic of the person. The pitch of adult males is typically 80–120 Hz, and that of adult females is about 150–300 Hz. Children have a higher pitch. When a person sings, the pitch varies according to the melody. Pitch varies to a certain extent even during normal speaking. For example, a question is articulated by raising the pitch toward the end.

The vocal tract is commonly modeled by a time-varying linear system. During a single phoneme, the parameters of the system are approximately constant; however, they vary widely from phoneme to phoneme, and this is why the phonemes sound different. The frequency response of the linear system representing a particular phoneme typically exhibits several peaks. The frequencies of these peaks are called *formants*. The assumption that the linear system is time invariant during the phoneme is rather crude: in reality, it varies during the phoneme. However, for relatively short time intervals (up to about 30 milliseconds) it is common to assume that the vocal tract system is LTI.

The common LTI model for the vocal tract is as a *rational, all-pole* transfer function, that is,

$$H^z(z) = \frac{1}{a(z)}, \quad \text{where} \quad a(z) = 1 + a_1 z^{-1} + \cdots + a_p z^{-p}. \tag{14.8}$$

Typical values of p are 8 to 12. In summary, a common short-time model for voiced speech is

$$y[n] = -a_1 y[n-1] - \cdots - a_p y[n-p] + u[n], \tag{14.9}$$

with $u[n]$ as in (14.7).

Example 14.2 The file female-w.mat contains a speech fragment 0.3 second long, of a female saying "why" (see page vi for information on how to download this file). The sampling frequency is 11,025 Hz. You can hear it by loading the file into the MATLAB environment and issuing the command sound(s,11025) (assuming that your computer is equipped with a sound system). Figure 14.3 shows the speech waveform. Part a shows the full 0.3 second and part b zooms in on 30 milliseconds in the middle. As we see, the signal indeed looks like a response to a sequence of nearly periodic impulses.   □

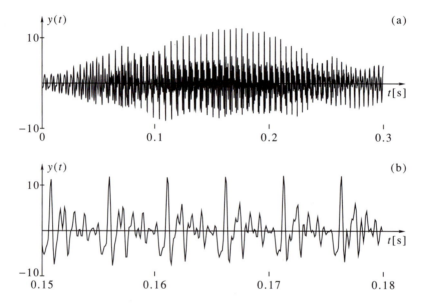

Figure 14.3 The speech signal in Example 14.2 ("why" pronounced by a female): (a) full waveform; (b) 30-millisecond segment.

The voiced speech model (14.7), (14.9) is known as *linear predictive coding* (LPC). Recalling what we have studied in Section 13.3, we see that (14.9) is essentially an autoregressive model for $y[n]$. However, in the present context, the excitation signal $u[n]$ is not white noise and $y[n]$ is not a random WSS signal. Still, $y[n]$ can be thought of comprising a predicted value—the linear combination of past values on the right side of (14.9)—and an error term $u[n]$. If $u[n] = 0$, the prediction is exact. A nonzero value of $u[n]$ can be interpreted as a correction term to the prediction, to make it exact. The term LPC has become standard in the speech literature, so we will prefer it to the term "AR model" in this application.

The LPC model is useful for applications such as speech generation by computer (synthetic speech) and speech compression. Speech compression by LPC modeling consists of the following conceptual stages:

1. Division of the sampled speech into short intervals, typically 10–30 milliseconds long. These intervals, called *frames*, can be overlapping or nonoverlapping.

2. Computation of a set of LPC parameters $\{a_k, 1 \le k \le p\}$ in the model (14.9), describing the vocal tract system for the phoneme being pronounced during the

present frame. This is done either by the Yule-Walker method, as described in Sections 13.4.1 and 13.4.6 (the autocorrelation method), or by least-squares modeling, as described in Section 13.4.7 (the covariance method). In the former case, the data $y[n]$ in the frame are sometimes windowed. Although data windowing can hardly be justified from a theoretical viewpoint,[1] experience shows that it tends to improve the quality of the model.

3. Computation of the excitation signal $u[n]$.

4. Modeling of the excitation signal, in particular its pitch and amplitude during the frame. Such modeling may lead to a secondary excitation signal, which generates $u[n]$ via another LTI system.

5. Coding and compression of the variables found by this procedure: the LPC parameters, the excitation signal parameters, and the secondary excitation signal (if present). The compressed speech consists of the coded variables in every frame; the rate of the compressed speech is the number of bits in the coded variables divided by the duration of the frame.

Reconstruction of a speech signal from its compressed representation is basically a reversal of the aforementioned process. The variables in each frame are decoded, the secondary excitation signal is regenerated, used for building the primary excitation signal $u[n]$, which is used in turn for reconstructing the speech signal $y[n]$.

Example 14.2 (continued) The signal contained in the file **female-w.mat** and plotted in Figure 14.3 is divided into frames of 160 samples each (corresponding to about 14.5 milliseconds). For each frame, the estimated covariance sequence $\hat{\kappa}_y[m]$ is computed as in (13.66), and the Yule-Walker equation (13.26a) is solved to give a corresponding vector of $\{a_k\}$. This way, 20 vectors are generated (since there are 20 frames). We use an LPC model of order $p = 10$. Figure 14.4 shows the magnitude response of the modeling filter $1/a(z)$ for two frames: the 11th and the 16th. The frequency range in the plot is up to 4 kHz, since the speech signal was band limited to 3.8 kHz prior to sampling. In the first graph we see two formants, at frequencies about 1 and 2.5 kHz. In the second we see three formants, at frequencies about 0.6, 1, and 2.8 kHz.

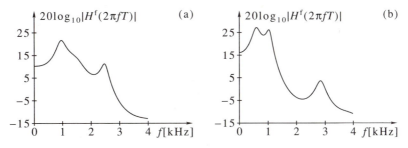

Figure 14.4 Magnitude responses of the modeling filter in Example 14.2: (a) the 11th frame; (b) the 16th frame.

The excitation signal $u[n]$ is generated by passing $y[n]$ through the FIR filter $a(z)$, as expressed in (14.9). At each frame, the coefficients of $a(z)$ are set to their respective values obtained from the solution of the Yule-Walker equation (13.26a). The signal thus obtained is shown in Figure 14.5. Part a shows the full time interval, and part b zooms in on 30 milliseconds in the middle. As we see, the signal looks approximately like a nearly periodic impulse train.

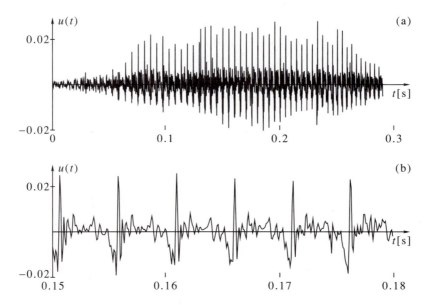

Figure 14.5 The excitation signal in Example 14.2: (a) full waveform; (b) 30-millisecond segment.

Examination of Figure 14.5 reveals a change in pitch during the syllable "why." The pitch frequency is the lowest in the beginning and increases gradually. We find, by detecting the epochs in the $u[n]$ waveform and computing the time intervals between them, that the pitch frequency is 170 Hz in the beginning and increases linearly to 260 Hz at the end.   □

14.2.2 Modeling of the Excitation Signal

A simple way of modeling the excitation signal $u[n]$ is as a perfectly periodic impulse train during the frame. The pitch frequency and the amplitude of the impulses need to be estimated from $u[n]$. This modeling ignores pitch and amplitude variations, and assigns zero values to $u[n]$ between the impulses. It was the common way of speech compression and generation in older systems (mainly during the 1980s). Speech reconstructed from a perfectly periodic impulse train excitation sounds metallic. Therefore, LPC modeling was considered a low-quality compression technique during that time. In recent years, more sophisticated approaches have been developed for modeling of the excitation signal. These methods have considerably improved the quality of LPC modeling, and this technique is used today for many commercial applications. Here we describe a modeling method of the excitation signal based on linear prediction.

A perfectly periodic impulse train starting at $n = 0$, extending to infinity, and having unity amplitude and period M has the z-transform

$$U^z(z) = \sum_{n=0}^{\infty} z^{-Mn} = \frac{1}{1 - z^{-M}}. \tag{14.10}$$

Ideally, $u[n]$ can be generated as the response of the linear time-invariant system $G^z(z) = 1/(1 - z^{-M})$ to a single impulse at $n = 0$, that is,

$$u[n] = u[n - M] + \delta[n]. \tag{14.11}$$

The model (14.11) is not adequate, however, for representing the actual excitation signal. First, this model is not stable, since its poles are *on* the unit circle. Second, in reality $u[n]$ is not perfectly periodic, the amplitude of the impulses is not constant, and $u[n]$ assumes nonzero values between the impulses.

A model that is conceptually derived from (14.11) but better represents the true excitation signal is

$$u[n] = \eta u[n - M] + e[n], \qquad (14.12)$$

where $0 < \eta < 1$. We can interpret (14.12) as a linear prediction of $u[n]$ from $u[n - M]$, where $e[n]$ serves as a correction term. We call $e[n]$ the *secondary excitation signal*. In contrast with the LPC model (14.9), which predicts $y[n]$ from its immediate past values, here we predict $u[n]$ from a value M points in the past. Therefore, it is common to refer to (14.12) as *long-term prediction*.

The long-term prediction model (14.12) is good if the energy of $e[n]$ is small. Therefore, we can compute η by requiring that $\sum_n (e[n])^2$ be minimum. We have

$$\sum_n (e[n])^2 = \sum_n (u[n] - \eta u[n - M])^2. \qquad (14.13)$$

Differentiating with respect to η, equating the derivative to zero, and solving for η gives

$$\eta = \frac{\sum_n u[n]u[n - M]}{\sum_n (u[n - M])^2}. \qquad (14.14)$$

In practice, we do not know the period M exactly. Furthermore, M is not even well defined, since the period is not exactly constant, even during short intervals. The solution is to compute η from (14.14) for all $M_{min} \le M \le M_{max}$, where the upper and lower limits on M are determined from the lower and upper limits on the pitch frequency, respectively. It can be shown that the value of η closest to 1 yields the minimum value of $\sum_n (e[n])^2$. All integer values of M in the selected range must be examined, since η is relatively sensitive to small deviations of M from its optimal value. The computation should be carried out for every frame, where the frame can be the same as the LPC frame or shorter. However, large values of M typically involve values of $u[n]$ across frame boundaries.

Example 14.2 (continued) We illustrate the result of modeling the excitation signal $u[n]$ obtained from the speech signal in the file `female-w.mat`. We use the same frame length as for the LPC modeling, that is, 160 samples. The values of η and M found in the 20 frames are

$$M = 67, 65, 63, 63, 62, 62, 60, 59, 58, 57, 57, 56, 55, 53, 52, 51, 49, 47, 45, 43;$$
$$\eta = 0.37, 0.68, 0.70, 0.72, 0.67, 0.73, 0.80, 0.94, 0.79, 0.75, 0.90, 0.92, 0.81,$$
$$0.82, 0.94, 0.89, 0.56, 0.70, 0.83, 0.85.$$

As we see, the optimal M decreases gradually, thereby tracking the increasing pitch frequency. The parameter η is close to 1 most of the time, but occasionally decreases.

Figure 14.6 shows the secondary excitation signal $e[n]$ obtained by the filtering operation

$$e[n] = u[n] - \eta u[n - M], \qquad (14.15)$$

with η and M varying from frame to frame. As we see, the secondary excitation has average amplitude lower than that of the primary excitation, and it looks much more random. □

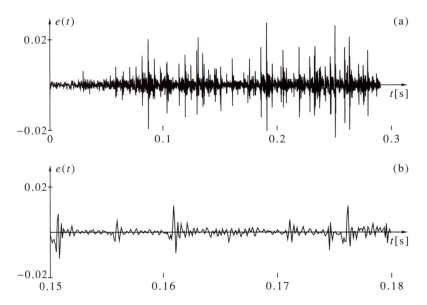

Figure 14.6 The secondary excitation signal in Example 14.2: (a) full waveform; (b) 30-millisecond segment.

14.2.3 Reconstruction of Modeled Speech

Reconstruction, or synthesis, of a speech signal is conceptually straightforward. Assuming that we are given the secondary waveform $e[n]$, the long-term prediction parameters η and M, and the LPC parameters $\{a_k\}$, the signal $y[n]$ can be reconstructed as shown in Figure 14.7. The filter $1/a(z)$ is guaranteed to be stable, because the roots of $a(z)$ are inside the unit circle. The filter $1/(1 - \eta z^{-M})$ is stable, because $0 < \eta < 1$.

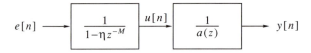

Figure 14.7 Speech reconstruction by LPC model and long-term prediction.

In implementing Figure 14.7, the following point should be kept in mind: Both reconstruction filters are IIR, and they are not time invariant, because their parameters are updated periodically (every frame). When the parameters are updated, it is necessary to preserve continuity of the state. It would be a gross mistake to initialize the state of the filters to zero at the beginning of every frame. Preserving the state of the LPC synthesis filter is straightforward. Preserving the state of the long-term synthesis filter is tricky, since the dimension of the state vector of this filter is M, and M varies from frame to frame. The solution is to keep a state vector whose dimension is the maximum value of M, say $M_{\max}$. The order of the corresponding IIR filter is then $M_{\max}$, but its coefficients are extremely sparse: Only one is nonzero at any given time. The location of the nonzero coefficient varies from frame to frame.

When a speech signal is reconstructed as in Figure 14.7, it will ideally reproduce the original signal. This happens when there is no compression. Compression distorts the parameters and the secondary excitation signal, hence the reconstructed speech; this is discussed next.

14.2.4 Coding and Compression

The term *coding* means a certain transformation of the model variables. The term *compression*, as we already know, means reduction of the bit rate. All the building blocks of the model are candidates for coding and compression. LPC modeling has been the basis for numerous compression schemes. The quality of speech reconstructed from its compressed model depends heavily on a successful choice of coding and compression schemes, and on the bit rate. Experience shows that the quality of LPC modeling is most sensitive to the way the excitation signal $u[n]$ is compressed and is less sensitive to compression of the LPC parameters. LPC parameters typically require 1.5 to 2.5 kilobits per second after compression and coding. Research on LPC in recent years has focused on the compression of the excitation signal.

Example 14.3 We give here a fairly detailed description of GSM, the compression standard used in the Pan-European Digital Mobile Radio System. It is based on LPC modeling followed by modeling of the excitation signal, as described previously. A major advantage of this standard is that the quality of the reconstructed speech depends little on the language or accent. This is necessary in Europe, where many languages and accents are used. By comparison, standards for North America are typically optimized to English, and the reconstructed speech suffers degradation when other languages or foreign accents are used.

The standard we describe is GSM 06.10, and is documented by the European body [GSM, 1991]. The input signal is sampled at 8 kHz, 13 bits, or 104 kilobits per second. The rate of the compressed speech is 13 kilobits per second, so the compression ratio is 8. The main features of the standard follow.

1. Offset compensation. The input signal is passed through the filter

$$G_1^z(z) = \frac{1 - z^{-1}}{1 - \alpha z^{-1}}, \quad \text{where} \quad \alpha = 32,735 \times 2^{-15} \approx 0.999. \tag{14.16}$$

   The purpose of this filter is to remove any DC component that may be present, without affecting the signal in any other way.

2. Pre-emphasis. The DC-free signal is passed through the filter

$$G_2^z(z) = 1 - \beta z^{-1}, \quad \text{where} \quad \beta = 28,180 \times 2^{-15} \approx 0.86. \tag{14.17}$$

   The purpose of this filter is to emphasize high frequencies, partly compensating for their low energy in the input signal. High-frequency pre-emphasis improves the quality of the LPC model. At the last stage of reconstruction de-emphasis of high frequencies is performed (the reverse of pre-emphasis).

3. The signal at the output of $G_2^z(z)$ is divided into frames of 160 samples each, or 20 milliseconds. The operations described next are performed for every frame.

4. The parameters $\{\kappa_y[k],\ 0 \le k \le 8\}$ are computed, as defined in (13.66). Thus, the order of the LPC model is $p = 8$.

5. Instead of computing the parameters a_k of the LPC model, the reflection coefficients $\{\rho_k,\ 1 \le k \le p\}$ are computed. This is performed by means of the Schur algorithm (13.64).

6. The parameters ρ_k are transformed to an equivalent set of parameters, called the *log-area ratio* (LAR), and defined as

$$\text{LAR}[k] = \log_{10} \left(\frac{1 + \rho_k}{1 - \rho_k} \right). \tag{14.18}$$

This transformation helps to improve the quality of the parameters after quantization. In practice, an approximation to the log function is used.

7. The LAR parameters are quantized, using $7 - \lceil k/2 \rceil$ bits for LAR[k]. Thus, the numbers of bits are 6, 6, 5, 5, 4, 4, 3, 3, or a total of 36 bits for the eight parameters. We skip the details of the quantization scheme.

8. The parameters ρ_k are reconstructed from the quantized LAR parameters, and used for filtering the speech signal and for computing the excitation signal $u[n]$. Filtering is performed using the IIR lattice discussed in Section 13.4.4 and shown in Figure 13.10.

9. The values of $u[n]$ in each frame are further divided into 4 subframes of length 40 each (or 5 milliseconds). All subsequent operations are performed on each subframe.

10. Long-term prediction is performed on $u[n]$ to determine the parameters M and η. In the GSM standard, the former is called *lag*, and the latter *gain*. The range of M is between 40 and 120, corresponding to pitch frequencies in the range 67 to 200 Hz. If the pitch is higher, a fraction of the pitch (or a multiple of the period) is detected. The algorithm is conceptually as described in Section 14.2.2, but differs in the details. The main difference is that instead of correlating $u[n]$ with its own past values, it is correlated with past values of the *reconstructed* signal. This leads to minimization of the energy of the error between the actual excitation and the predicted reconstructed excitation, which helps to improve the quality of the compressed signal.

11. The parameter M is coded in 7 bits and the parameter η is quantized to 2 bits. This requires 9 bits in each subframe.

12. The secondary excitation signal $e[n]$ is computed as in (14.15). In the GSM standard, this signal is called *regular pulse excitation* (RPE). Next it is necessary to compress this signal. Compression of $e[n]$ involves the following steps:

    (a) The signal is convolved with a tenth-order, low-pass FIR filter $G_3^z(z)$ whose coefficients, multiplied by 2^{13}, are

    $$g[0] = g[10] = -134, \ g[1] = g[9] = -374, \ g[2] = g[8] = 0,$$
    $$g[3] = g[7] = 2054, \ g[4] = g[6] = 5741, \ g[5] = 8192.$$

    Since the length of $e[n]$ is 40, the result of the convolution has length 50, but only the 40 middle terms are retained. We denote the 40 values thus obtained by $x[n]$. The purpose of the low-pass filter is to limit the bandwidth of the secondary excitation signal to about $\pi/3$.

    (b) The objective is now to decimate $x[n]$ by 3. This is done by considering the four decimated sequences

    $$x_l[i] = x[l + 3i], \quad 0 \le l \le 3, \ 0 \le i \le 12.$$

    The value of l for which $\sum_{i=0}^{12}(x_l[i])^2$ is the largest is selected, and the corresponding subsequence $x_l[i]$ (of length 13) is retained. All other points of $x[n]$ are discarded.

(c) The parameter

$$x_{max} = \max\{|x_l[i]|,\ 0 \le i \le 12\}$$

is computed.

(d) The integer l is coded in 2 bits, the parameter x_{max} is quantized to 6 bits, and each of $x_l[i]$ is quantized to 3 bits. Nonlinear quantization is used, the details of which we omit.

In summary, compression and coding of the secondary excitation signal results in 56 bits per subframe: 7 for M, 2 for η, 2 for l, 6 for x_{max}, and 39 for the 13 values of $x_l[i]$. The total number of bits per frame is therefore 224. Adding to this the 36 bits used for the LPC modeling stage, we get 260 bits per frame, or 13 kilobits per second.

13. The encoder also contains most of what is needed in the decoder (at the receiving end). It reconstructs a distorted version of $e[n]$ by expanding $x_l[i]$, starting at the time point l and using simple zero filling for the missing points. It then reconstructs a distorted version of the primary excitation using the parameters M and η. This signal is used in the next subframe for the new values of M and η.

14. The decoder includes, in addition to the part that reconstructs the primary excitation, the LPC reconstruction filter and a de-emphasis filter whose transfer function is the inverse of $G_2^z(z)$.

In summary, the GSM standard is relatively simple, yet efficient and robust. It achieves mean opinion score (see Example 12.8) of about 3.5. Higher compression ratios can be achieved at this quality, but more complex algorithms are needed, hence more expensive hardware. □

14.3 Musical Signals

Audio signals generated by musical instruments typically have a harmonic structure. A simple mathematical model of a musical signal is

$$x(t) = a(t) \sum_{m=1}^{\infty} c_m \cos(2\pi m f_0 t + \phi_m), \tag{14.19}$$

where:

1. f_0 is the fundamental frequency, or the *pitch*.

2. c_m is the amplitude and ϕ_m is the phase of the mth harmonic.

3. $a(t)$ is the *envelope*. It is a low-frequency signal, which modulates the amplitude of the sound.

The frequency f_0 is determined by the note being played. In Western music, the frequencies of musical instruments obey a geometric series formula

$$f_i = f_{ref}q^i, \quad q = 2^{1/12}, \tag{14.20}$$

where f_{ref} is a reference frequency. The standard reference frequency is 440 Hz, and the corresponding note is called A (also known as *la*). The ratio $2^{1/12}$ is called a *semitone*. The notes B, C, D, E, F, G (also known as *ti, do, re, mi, fa, sol*, respectively) are 2, 3, 5, 7, 8, and 10 semitones above A. These are called the *natural notes*. Between A and B we have the note called A-sharp or B-flat, depending on the scale being used. Similarly, we have sharp and flat notes between C and D, D and E, F and G, and G and A. The note 12 semitones above A is again called A, and is said to be an *octave* higher. The ratio between two notes an octave apart is exactly 2. The seven natural notes A through G and their octaves are played by the white keys of the piano. The sharp and flat notes are played by the black keys. There are twelve semitones in an octave.

The *range* of a musical instrument is the set of notes it can play. This corresponds to the range of indices i in (14.20). For example, the range of a piano is $-48 \leq i \leq 39$. The corresponding frequencies are 27.5 Hz and 4186 Hz (the lowest note is A and the highest is C).

Western music is largely based on *diatonic scales*. A diatonic scale consists of 7 notes per octave. The diatonic C major scale consists of the natural notes C, D, E, F, G, A, B and their octaves. There are 12 major keys, each starting at a different semitone. The frequency ratios of the 7 notes of all major keys are identical. There are three kinds of minor key: natural, harmonic, and melodic. There are 12 keys of each of these kinds. The natural A minor scale consists of the natural notes A, B, C, D, E, F, G. In the harmonic A minor scale, the G is replaced by G-sharp. In the melodic A minor scale, the F and G is replaced by F-sharp and G-sharp, respectively, when the scale is ascending. These two notes revert to natural when the scale is descending.

The harmonic structure of a musical instrument is the series of amplitudes of the various harmonics relative to the fundamental frequency (measured in dB). The harmonic structure depends on the type of the instrument, the individual instrument, the note played, and the way it is played. Different instruments have their typical harmonic structures. It is the differences in harmonic structure, as well as the envelope, that make different instruments sound differently. The relative phases of the various harmonics are of little importance, since the human ear is almost insensitive to phase.

The envelope $a(t)$ is characteristic of the instrument, and also depends on the note and the way the note is played. For example, notes in bow instruments (such as violin or cello) and wind instruments (such as flute or horn) can be sustained for long times. Notes played on plucked string instruments (such as guitar) and keyboard instruments (such as piano) are limited in the time they can be sustained.

The model above does not represent all acoustic effects produced by musical instruments. For example, *vibrato* is a form of periodic frequency modulation around the nominal frequency of the note; *glissando* is a rapid succession of notes that sounds almost as a continuously varying pitch frequency.

Many instruments are capable of playing chords. A *chord* is a group of notes played either simultaneously or in a quick succession. For example, the C major chord consists of C, E, and G, and possibly a few of their octaves. The A minor chord consists of A, C, and E, and possibly a few of their octaves. When a chord is played, the signal can be described to a good approximation as superposition of the signals (14.19) of the individual notes.

Example 14.4 The files `cello.bin`, `guitar.bin`, `flute.bin`, and `frhorn.bin` contain about 1 second sounds of cello, classical guitar, flute, and French horn, respectively (see page vi for information on how to download these files). The note played is A in each case. The envelope waveforms of these four instruments are shown in Figure 14.8. As we see, the cello has a gradual rise of the amplitude, followed by a gradual decay. The guitar is characterized by a steep rise when the string is plucked, followed by a steep decay after it is released. The flute has a characteristic low-frequency amplitude modulation. The French horn has a steady amplitude during the time the note is played.

Figure 14.9 shows a 10-millisecond waveform segment of each instrument. As a comparison of the waveforms suggests, the pitch of the cello is 220 Hz, that of the guitar and the French horn is 440 Hz, and that of the flute is 880 Hz.

Figure 14.10 shows the magnitude DFT of a 186-millisecond segment (8192 samples) of each instrument. We plot only the frequency range 0 to 11 kHz, since the energy at higher frequencies is negligible.   □

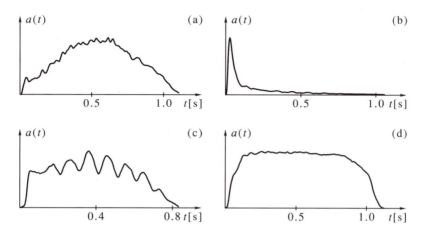

Figure 14.8 Envelope waveforms of musical instruments: (a) cello; (b) classical guitar; (c) flute; (d) French horn.

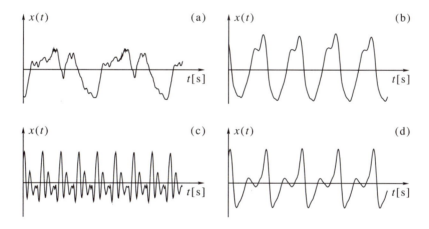

Figure 14.9 Waveforms of musical instruments, note played is A, 10-millisecond segments: (a) cello; (b) classical guitar; (c) flute; (d) French horn.

The model (14.19) can be applied for the synthesis of musical signals. To synthesize a note of a particular instrument, we need to know the characteristic envelope signal of the instrument and its characteristic harmonic structure. Chords are synthesized by superposition of individual notes. This method of synthesizing musical signals is called *additive synthesis*. It is relatively simple to implement, and you are encouraged to program it in MATLAB and test it, using the information in the waveforms of the four instruments. The musical quality of signals synthesized this way is not high, however, and present-day synthesis methods of higher quality have rendered the additive synthesis method all but obsolete.

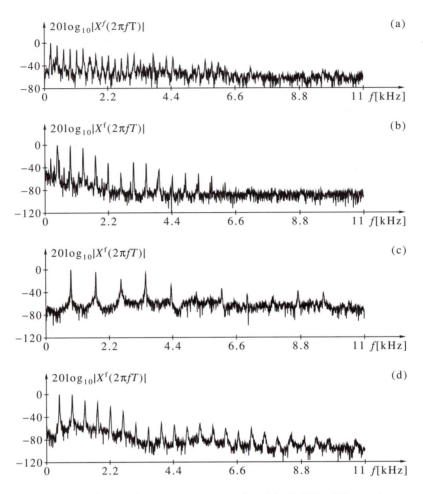

Figure 14.10 Spectra of musical instruments, note played is A, 186-millisecond segments: (a) cello; (b) classical guitar; (c) flute; (d) French horn.

14.4   An Application of DSP in Digital Communication

Digital communication has been in use for many years. However, until recently, the circuitry used for generation, transmission, and reception of digital communication signals was chiefly analog. Recent years have seen a growing trend of replacing analog functions needed in digital communications by digital algorithms, implemented on DSP microprocessors. Today this still applies mainly to base-band signals (either before modulation or after demodulation), because modulated signals are still too fast varying to be handled by digital means in most applications. In this book, we have already described several digital communication techniques in various examples and exercises. Here we describe, as an example of DSP application in communication systems, a reasonably complete system for receiving digital communication signals.

Since we have already met BPSK and OQPSK, we choose another common signaling method this time: frequency-shift keying (FSK). To make our example more interesting, we choose four-level FSK (rather than the simpler binary FSK). In four-level FSK we group the bits in pairs, and denote each pair by a *symbol*. For example, let us assign the symbols $-3, -1, 1, 3$ to the bit pairs $00, 01, 11, 10$, respectively. Then we allocate a frequency to each symbol, so we need four different frequencies. Suppose we wish to transmit at a rate f_0 symbols per second, which is the same as $2f_0$ bits per second. Then, the time interval allocated to each symbol is $T_0 = 1/f_0$. During each interval of length T_0, we transmit a signal at the frequency corresponding to the symbol in that interval. Thus, the frequency of the transmitted signal changes according to the transmitted symbols. The task of the receiver is to find, at each interval, which frequency was transmitted and to assign to it the corresponding symbol.

14.4.1 The Transmitted Signal

The *base-band signal* is piecewise constant, representing each symbol by a constant level. In the typical base-band signal shown in Figure 14.11, we have assigned levels $-3, -1, 1, 3$ to the different symbols. The mathematical description of the base-band signal is

$$s(t) = \sum_{m=-\infty}^{\infty} u[m]\text{rect}\left(\frac{t - mT_0}{T_0}\right), \tag{14.21}$$

where $u[m]$ is the sequence of symbols, and rect$(\cdot)$ is the rectangular function (2.32).

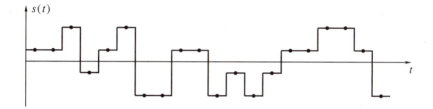

Figure 14.11 Modulating waveform of a four-level FSK signal.

Frequency modulation is defined by the mathematical operation

$$x(t) = \cos[2\pi f_c t + \phi(t)], \quad \text{where} \quad \phi(t) = 2\pi K_f \int_0^t s(\tau)d\tau, \tag{14.22}$$

f_c is the carrier frequency, and K_f is a constant. The *instantaneous frequency* of $x(t)$ is the time derivative of its instantaneous phase, divided by 2π, that is,

$$f(t) = f_c + K_f s(t). \tag{14.23}$$

The *peak frequency deviation* of an FSK signal is defined as the maximum value of $K_f|s(t)|$. In our case, since the maximum value of $|s(t)|$ is 3, the peak frequency deviation is $3K_f$. An important parameter of an FSK signal is the ratio between the peak frequency deviation and the symbol rate f_0. Denoting this ratio by β, we get that $K_f = f_0\beta/3$. Therefore,

$$f(t) = f_c + (\beta/3)f_0 s(t). \tag{14.24}$$

The parameter β controls the bandwidth of the modulated signal: the larger is β, the larger the bandwidth. The choice $\beta = 1.5$ has a special meaning. Recall that the separation between adjacent levels of $s(t)$ is 2. Therefore, the separation between the corresponding frequencies, with $\beta = 1.5$, is exactly f_0. Thus, during the symbol interval T_0, there is a difference of exactly one cycle between the numbers of cycles corresponding to adjacent symbols. We shall use this value of β in our system. Figure 14.12 shows the spectrum of the FSK signal $x(t)$ in this case. This figure was obtained by sampling the signal at a rate $16f_0$ and averaging 128 DFT magnitudes of 512 points each. We have plotted the spectrum around the carrier frequency f_c for convenience. Distinctive in this figure are the four spectral lines at frequencies $f_c \pm 0.5f_0$ and $f_c \pm 1.5f_0$. Otherwise, the spectrum is flat up to $\pm 1.5f_0$, with a sharp decay and relatively low side lobes at higher deviations from the carrier frequency. The bandwidth of $x(t)$ is about $\pm 2f_0$ around the carrier frequency.

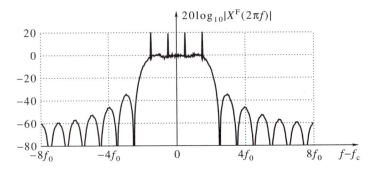

Figure 14.12 Spectrum of a four-level FSK signal.

14.4.2 The Received Signal

The received signal is delayed with respect to the transmitted signal, due to the finite propagation time, and is shifted by an unknown phase; it is given by

$$r(t) = \cos[2\pi f_c(t - t_0) + \phi(t - t_0) + \phi_0'] = \cos[2\pi f_c t + \phi(t - t_0) + \phi_0], \quad (14.25)$$

where t_0 is an unknown delay and $\phi_0 = \phi_0' - 2\pi f_c t_0$ is an unknown phase. The delay is constant if both transmitter and receiver are stationary, but it can change over time if there is relative motion between them. For example, if the receiver is located on a car moving at 30 m/s then, since the speed of light is 3×10^8 m/s, t_0 will change at a rate up to 0.1 microsecond per second. If the transmitter is located on a satellite moving at 7000 m/s, the rate of change of t_0 can be over 20 microseconds per second. The received signal also includes a noise component. However, for the time being we ignore the noise; we shall consider its effect when we examine the performance of the receiver.

If we were to build an analog receiver, we would construct a circuit for extracting the instantaneous frequency from $r(t)$. Such a circuit is called an *FM discriminator*. Since we aim at a digital receiver, we shall not use an analog FM discriminator. However, we still need to convert the signal to a sufficiently low frequency, to process it digitally. This is done by simple DSB demodulation, as explained in Problem 3.43. The result is a *low-IF* signal. We explained the concepts of IF and superheterodyne in Example 3.12. In the present case we call it low-IF, because it is close to the base band.

Assuming that we are going to use a sampling rate Lf_0 for the A/D converter (where L is an integer), it is convenient to choose the IF frequency as $0.25Lf_0$. The frequency of the local oscillator should then be $f_c - 0.25Lf_0$. However, physical oscillators are never completely accurate; an error of a few parts per million is typical. Since typical values of f_c are in the hundreds-of-megahertz range, a relative error of 10^{-6} in the local oscillator amounts to a few hundred hertz. This is not negligible with respect to f_0, which is typically in kilohertz. Therefore, we shall later have to face the problem of estimating the center frequency of the received signal relative to the receiver reference frequency. Let us therefore express the demodulated signal as

$$y(t) = \cos[2\pi(0.25Lf_0 + \Delta f)t + \phi(t - t_0) + \phi_0], \tag{14.26}$$

where Δf is the *carrier offset*, resulting from the frequency error. The demodulated signal is passed through an analog antialiasing filter, and from it to the A/D. The discrete-time low-IF signal is thus

$$y[n] = y(n/Lf_0) = \cos[2\pi(0.25Lf_0 + \Delta f)(n/Lf_0) + \phi(n/Lf_0 - t_0) + \phi_0]$$
$$= \cos[2\pi(0.25 + \Delta f/Lf_0)n + \phi(n/Lf_0 - t_0) + \phi_0]. \tag{14.27}$$

14.4.3 Choosing the Sampling Rate

We now choose the sampling rate of the digital receiver. We have already constrained the rate to be an integer multiple of the frequency f_0, so it remains to determine the factor L. To do this, we must assume an upper limit on $|\Delta f|$, say Δf_{max}.

Recall that the bandwidth of the FSK signal is $\pm 2f_0$ around its center frequency, so after sampling it becomes $\pm 2/L$ (in units of $\theta/2\pi$). Since the center frequency is $0.25 + \Delta f/Lf_0$, the signal occupies the band

$$\left[0.25 + \frac{\Delta f}{Lf_0} - \frac{2}{L}, \ 0.25 + \frac{\Delta f}{Lf_0} + \frac{2}{L}\right].$$

The following constraints must be imposed:

1. Since the signal $y[n]$ is real, its spectrum is two sided. Therefore, when $\Delta f = -\Delta f_{max}$, the lower band-edge frequency must be greater than zero, for otherwise the positive and negative frequency bands will overlap and distort the spectral contents. This leads to the constraint

$$0.25 - \frac{\Delta f_{max}}{Lf_0} - \frac{2}{L} > 0. \tag{14.28}$$

2. When $\Delta f = \Delta f_{max}$, the higher band-edge frequency must be less than 0.5, for otherwise we will get aliasing. This leads to the constraint

$$0.25 + \frac{\Delta f_{max}}{Lf_0} + \frac{2}{L} < 0.5. \tag{14.29}$$

When (14.28) and (14.29) are solved for L, they both yield

$$L > 4\left(2 + \frac{\Delta f_{max}}{f_0}\right). \tag{14.30}$$

We already see one advantage of choosing a nominal IF frequency of $0.25Lf_0$: This way, both constraints lead to the same lower limit on L. Later we shall see another advantage of this choice.

To be specific, let us assume that $\Delta f_{max} = f_0$. Then we get from (14.30) that $L > 12$. A convenient number is $L = 16$, and this will be our choice. With this value of L, the signal occupies the band $[1/16, 5/16]$ when $\Delta f = -\Delta f_{max}$, or the band $[3/16, 7/16]$ when $\Delta f = \Delta f_{max}$.

14.4.4 Quadrature Signal Generation

The information about the symbols is in the time derivative of the phase function of $y(t)$. Symbol detection is complicated by the presence of the three unknown parameters: the phase ϕ_0, the delay t_0, and the carrier offset Δf. A common way of making the detector insensitive to unknown phase is to generate the *quadrature signal*

$$y_q[n] = \sin[2\pi(0.25 + \Delta f/Lf_0)n + \phi(n/Lf_0 - t_0) + \phi_0]. \tag{14.31}$$

The signal $y[n]$ itself is called the *in-phase* signal, and the pair $\{y[n], y_q[n]\}$ is called the I-Q signals. We explain later how the availability of I-Q signals overcomes the problem of unknown phase.

The signal $y_q[n]$ can be generated by passing $y[n]$ through a Hilbert transformer. Therefore, the first component of our digital receiver will be a digital Hilbert transformer. To design a digital Hilbert transformer, we need to define the parity of the order (even or odd), the pass-band width, and the pass-band ripple. Note that a Hilbert transformer does not have stop bands, but it does have two transition bands, around $\theta = 0$ and $\theta = \pi$.

Since we are going to use $\{y[n], y_q[n]\}$ as a pair, and since $y_q[n]$ is going to be delayed by the group delay of the Hilbert transformer, we must delay the in-phase signal $y[n]$ by the same amount. Choosing an even order makes the group delay integer, so $y[n]$ must be delayed by an integer number of samples, and this is a simple operation. On the other hand, choosing an odd order makes the group delay fractional, and then we must delay $y[n]$ by an interpolated-delay filter. Therefore, we use a Hilbert transformer of an even order.

To determine the width of the pass band, recall the analysis in Section 14.4.3. There we concluded that the lowest frequency of $y[n]$ when $\Delta f = -\Delta f_{max}$ is 1/16 of the sampling rate, and the highest frequency when $\Delta f = \Delta f_{max}$ is 7/16 of the sampling rate. To accommodate any value of Δf within the specified range, we choose the pass band as $[\pi/8, 7\pi/8]$.

A typical pass-band ripple δ_p for our application is between 0.01 and 0.001. The former is marginal, whereas the latter is conservative. Designing a Hilbert transformer by the Parks–McClellan algorithm, we find that an order $N = 30$ is the minimum for obtaining $\delta_p = 0.01$, whereas $N = 38$ gives $\delta_p = 0.001$; here we choose $N = 38$.

14.4.5 Complex Demodulation

Having obtained the in-phase and quadrature components, we can describe them as a complex signal, as we did in Section 9.2.5:

$$y_a[n] = y[n] + jy_q[n] = \exp\{j[2\pi(0.25 + \Delta f/Lf_0)n + \phi(n/Lf_0 - t_0) + \phi_0]\}. \tag{14.32}$$

The signal $y_a[n]$ is the *analytic signal* of $y[n]$. We emphasize again the need to delay $y[n]$ by half the order of the Hilbert transformer (that is, by 19 in this case): Failing to do so will prevent the system from working properly.

We recall that the spectrum of the analytic signal occupies only positive frequencies. It is now convenient to shift the spectrum to the base band. Ideally we would like to shift by $2\pi(0.25 + \Delta f/Lf_0)$, and then the spectrum would be centered around zero. However, since Δf is unknown, we shift only by 0.5π. To perform this shift, we must multiply $y_a[n]$ by $e^{-j0.5\pi n}$, that is, to compute

$$z[n] = y_a[n]e^{-j0.5\pi n} = \exp\{j[2\pi(\Delta f/Lf_0)n + \phi(n/Lf_0 - t_0) + \phi_0]\}. \tag{14.33}$$

The operation (14.33) is called *complex demodulation*. We have $e^{-j0.5\pi n} = (-j)^n$, so

$$z[n] = z_r[n] + jz_i[n] = (y[n] + jy_q[n])(-j)^n.$$

This can be written as

$$z_r[n] = \begin{cases} y[n], & n \bmod 4 = 0, \\ y_q[n], & n \bmod 4 = 1, \\ -y[n], & n \bmod 4 = 2, \\ -y_q[n], & n \bmod 4 = 3, \end{cases} \qquad z_i[n] = \begin{cases} y_q[n], & n \bmod 4 = 0, \\ -y[n], & n \bmod 4 = 1, \\ -y_q[n], & n \bmod 4 = 2, \\ y[n], & n \bmod 4 = 3. \end{cases} \qquad (14.34)$$

As we see, complex demodulation requires no computations, only rearrangement of the data and sign reversals. This simplification occurs because we have chosen the low-IF frequency as 0.25 of the sampling rate.

14.4.6 Symbol Detection: Preliminary Discussion

The parts that we have discussed so far, the Hilbert transformer and the complex demodulator, are just the front end of the receiver. The main task, extracting the symbols from $z[n]$, is still ahead of us. This task is considerably complicated by the unknown time delay t_0 and the unknown carrier offset Δf. Before devising methods for tackling these difficulties, it is helpful to discuss how we would have performed symbol detection if these two parameters were zero. In such a case, the signal $z[n]$ during the mth symbol would be

$$z[n] = \exp\left\{j\left[2\pi\frac{0.5u[m](n-Lm)}{L} + \phi_0\right]\right\}, \quad Lm \le n \le Lm + L - 1, \qquad (14.35)$$

where $u[m]$ is the mth symbol. Since $u[m]$ can assume only four values, we can perform *matched filtering* to the four possible waveforms. By this we mean the following: We define four complex filters of length L each, and denote their impulse responses by $g_k[n]$, where $k = -3, -1, 1, 3$. The impulse responses are

$$g_k[n] = \exp\left[-j2\pi\frac{0.5k(L-1-n)}{L}\right], \quad 0 \le n \le L - 1. \qquad (14.36)$$

We convolve $z[n]$ with each of the four filters and sample the outputs at time $n = Lm + L - 1$ to get

$$\{z * g_k\}[Lm + L - 1] = e^{j\phi_0}\sum_{l=0}^{L-1}\exp\left[j2\pi\frac{0.5(u[m]-k)(L-1-l)}{L}\right]. \qquad (14.37)$$

Since $u[m] - k$ is an integer multiple of 2 for all possible values of k and $u[m]$, we get

$$\{z * g_k\}[Lm + L - 1] = \begin{cases} Le^{j\phi_0}, & k = u[m], \\ 0, & \text{otherwise.} \end{cases} \qquad (14.38)$$

Therefore, by observing the absolute values of $\{z * g_k\}[Lm + L - 1]$ for the four filters, we can determine the symbol $u[m]$: It is the value of k for which the result is not zero. In practice, due to noise, none of the outputs will be zero, so we choose k for which the absolute value is the largest of the four. By taking the absolute value, we eliminate the unknown phase ϕ_0.

If you have solved Problem 8.21, you know that matched filtering is optimal, in the sense of maximizing the signal-to-noise ratio at the filter output when white noise is added to the input signal. Therefore, the matched filtering scheme is the best for symbol detection, provided we have zero (or perfectly known) carrier offset and delay.

If the unknown delay and carrier offset are nonzero, (14.38) will not hold any more. All four outputs will be nonzero in general, and we may well get that the largest absolute value is not at the right k. Reliable detection in the presence of carrier and timing

offsets is a major problem in digital communication. We must devise a way of estimating the unknown frequency and delay parameters before we can use the matched
filters.

14.4.7   FM to AM Conversion

One way of estimating the carrier and timing offsets is to extract the derivative of the
phase function $\phi(n/Lf_0 - t_0)$ from $z[n]$ and use the derivative for estimating these
two parameters. Extraction of the phase derivative is called *FM-to-AM conversion*, or
FM discrimination. Carrier and timing estimation by FM discrimination is not necessarily the best approach (from accuracy point of view), but it is simple and convenient.
Besides, there is a lot to learn from seeing how such a scheme works, and our aim here
is to teach. Therefore, we now build a digital FM discriminator.

If a continuous-time complex signal $z(t)$ were available, we could have extracted
the phase derivative as follows. Since

$$z(t) = \exp\{j[2\pi\Delta f t + \phi(t - t_0) + \phi_0]\}, \tag{14.39}$$

we have

$$\frac{dz(t)}{dt} = j\left[2\pi\Delta f + \frac{d\phi(t - t_0)}{dt}\right] z(t). \tag{14.40}$$

Therefore,

$$2\pi\Delta f + \frac{d\phi(t - t_0)}{dt} = -j\frac{dz(t)}{dt}\frac{\bar{z}(t)}{|z(t)|^2}. \tag{14.41}$$

Division by $|z(t)|^2$ is necessary here, since $z(t)$ has nonunity amplitude in general.
Until now we ignored the signal amplitude, because it was immaterial; now we must
account for it.

To approximate (14.41) using the discrete-time signal, we must approximate the
differentiation. In Section 9.2.4 we learned how to build digital differentiators. Unfortunately, those differentiators are of little use here. The reason is that $d\phi(t - t_0)/dt$
is discontinuous at points where the symbol changes value; therefore, a differentiator
with a long memory (i.e., a filter with a long impulse response) will lead to significant
distortions. A simple alternative is suggested by noting that

$$\frac{z[n]\bar{z}[n - 1]}{|z[n]\bar{z}[n - 1]|} = \exp\left\{j\left[\frac{2\pi\Delta f}{Lf_0} + \phi\left(\frac{n}{Lf_0} - t_0\right) - \phi\left(\frac{n - 1}{Lf_0} - t_0\right)\right]\right\}, \tag{14.42}$$

therefore,

$$\arcsin\left(\frac{\Im\{z[n]\bar{z}[n - 1]\}}{|z[n]\bar{z}[n - 1]|}\right) = \frac{2\pi\Delta f}{Lf_0} + \phi\left(\frac{n}{Lf_0} - t_0\right) - \phi\left(\frac{n - 1}{Lf_0} - t_0\right). \tag{14.43}$$

We have from (14.22),

$$\phi\left(\frac{n}{Lf_0} - t_0\right) - \phi\left(\frac{n - 1}{Lf_0} - t_0\right) = \pi f_0 \int_{(n-1)/Lf_0-t_0}^{n/Lf_0-t_0} s(\tau)d\tau = \frac{\pi u[m]}{L} \tag{14.44}$$

whenever the interval $[(n - 1)/Lf_0 - t_0, n/Lf_0 - t_0]$ falls within the mth symbol. If there
is a symbol change in this interval, there is a discontinuity and the left side of (14.44)
assumes a different value. We finally get from (14.43) and (14.44),

$$u[m] + \frac{2\Delta f}{f_0} = \frac{L}{\pi}\arcsin\left(\frac{\Im\{z[n]\bar{z}[n - 1]\}}{|z[n]\bar{z}[n - 1]|}\right), \tag{14.45}$$

except at the discontinuity points.

We denote the discrete-time signal defined by the right side of (14.45) as $\hat{s}[n]$, because it represents an estimate of $s(t - t_0)$, except for the additive term $2\Delta f / f_0$. The signal $\hat{s}[n]$ defines our digital FM discriminator.

Figure 14.13 shows the waveform of $\hat{s}[n]$ at the output of the FM discriminator, corresponding to the waveform $s(t)$ shown in Figure 14.11. Here we compensated for the delay t_0 to enable comparison between the waveforms $\hat{s}[n]$ and $s(t)$. Part a shows the result when there is no noise. As we see, even in this case $\hat{s}[n]$ is not identical to $s(t)$, for two reasons: the nonideality of the Hilbert transformer and the transients caused by the approximate differentiator (14.45) at the discontinuity points. Part b shows the result when there is white noise at SNR of 12 dB at the output of the A/D. As we see, reliable detection of the transmitted symbols is not likely to be easy in this case.

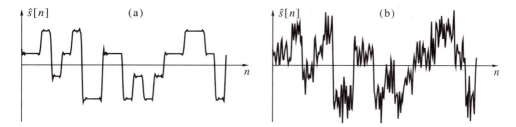

Figure 14.13 Output signal of the FM discriminator: (a) infinite SNR; (b) SNR = 12 dB.

14.4.8 Timing Recovery

Timing recovery means identifying the transition instants between symbols. Since the signal is sampled L times per symbol, our goal is to identify the transition instants up to the nearest sample. This will leave us with timing error that can be up to $\pm 0.5 T_0 / L$. It is possible to reduce the timing error below this value, using interpolation, but this complicates the system and we shall not attempt it here.

If the signal $\hat{s}[n]$ were clean, as in Figure 14.13(a), we could obtain the transition instants from the peaks of the magnitude-difference signal $|\hat{s}[n] - \hat{s}[n-1]|$ [which is approximately proportional to the magnitude of the derivative of $s(t)$]. However, when $\hat{s}[n]$ is noisy, as in Figure14.13(b), we must filter $\hat{s}[n]$ before we can use it for timing recovery.

A simple and effective timing recovery circuit is shown in the upper part of Figure 14.14. This circuit also contains a part related to symbol detection (lower part of the figure). We now explain the timing recovery part, then proceed to the symbol detection part.

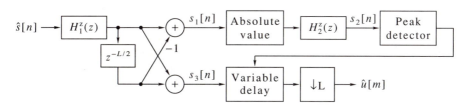

Figure 14.14 Timing recovery and matched filtering.

1. The transfer function $H_1^z(z)$ is

$$H_1^z(z) = 1 + z^{-1} + \cdots + z^{-(0.5L-1)}. \tag{14.46}$$

Let $s_1[n]$ denote the output of the subtractor in the upper line of Figure 14.14. Then, as we see from the figure, the transfer function from $\hat{s}[n]$ to $s_1[n]$ is

$$H^z(z) = H_1^z(z)[1 - z^{-0.5L}]$$
$$= 1 + z^{-1} + \cdots + z^{-(0.5L-1)} - z^{-0.5L} - z^{-(0.5L+1)} - \cdots - z^{-(L-1)}. \tag{14.47}$$

Therefore, $s_1[n]$ is related to $\hat{s}[n]$ by

$$s_1[n] = \sum_{k=0}^{0.5L-1} \hat{s}[n-k] - \sum_{k=0.5L}^{L-1} \hat{s}[n-k]. \tag{14.48}$$

Suppose that a symbol change has occurred in the interval between $n_0 T_0$ and $(n_0 + 1)T_0$, and the symbol values were $u[m-1]$ and $u[m]$ before and after the change, respectively. Then, assuming that $\hat{s}[n]$ is identical to $s[n]$ (except for the unknown delay), we get from (14.48) that

$$s_1[n_0] = 0, \ s_1[n_0 + 0.5L] = 0.5L(u[m] - u[m-1]), \ s_1[n_0 + L] = 0. \tag{14.49}$$

Also, $s_1[n]$ changes linearly between n_0 and $n_0 + 0.5L$, and between $n_0 + 0.5L + 1$ and $n_0 + L$. Figure 14.15 illustrates the operation of the filter $H^z(z)$ on the sequence $\hat{s}[n]$. We conclude that $s_1[n]$ exhibits a local extremum each time there is a change in the symbol value. The extremum occurs $0.5L$ time points after the change; it is positive if $u[m] > u[m-1]$ and negative otherwise.

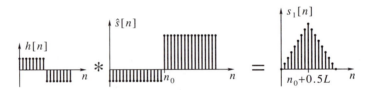

Figure 14.15 Operation of the early–late gate filter.

The filter $H^z(z)$ is a discrete-time version of an *early–late gate synchronizer*. This filter estimates the center of gravity of the transition area. In case of an ideal signal, the center of gravity coincides with the transition instant. In reality, $\hat{s}[n]$ is different from $s[n]$ due to noise and the nonideal operation of the FM discriminator. Then each of the two halves of the filter (the early and the late gates) attenuates the noise in the corresponding symbol by averaging over $0.5L$ samples. The filter $H_1^z(z)$ can be implemented efficiently by noting that

$$H_1^z(z) = \frac{1 - z^{-0.5L}}{1 - z^{-1}}. \tag{14.50}$$

The difference equation corresponding to (14.50) requires only two additions and no multiplications per time point; see Problem 11.4.

2. The transition instants can be found, in principle, by finding the local maxima of $|s_1[n]|$. Ideally, these maxima will be exactly L samples apart. However, when two adjacent symbols are equal, there is no local maximum. Also, noise may cause the points of local maximum to move randomly from their true locations, a phenomenon known as *timing jitter*. It is therefore desirable to filter the timing jitter by averaging the estimated transition instants over time. This operation is accomplished by the filter $H_2^z(z)$.

The desired impulse response of $H_2^z(z)$ is a train of impulses spaced L samples apart, that is,

$$h_2[n] = \sum_{k=0}^{\infty} \delta[n - kL]. \tag{14.51}$$

When such an impulse train is convolved with $|s_1[n]|$, it will exhibit peaks at the peaks of $|s_1[n]|$, but it will also perform averaging when $|s_1[n]|$ is subject to timing jitter. The problem with $h_2[n]$ is that its memory is too strong, since it averages an ever-increasing number of peaks. In practice, we want $h_2[n]$ to forget the past gradually, since the delay t_0 may vary slowly because of changes in the distance between the transmitter and the receiver (if either or both are in motion). The sequence

$$h_2[n] = \sum_{k=0}^{\infty} \alpha^k \delta[n - kL], \quad 0 < \alpha < 1 \tag{14.52}$$

is a decaying impulse train. When this sequence is convolved with $|s_1[n]|$, it attenuates past peaks exponentially. The closer α to 1, the longer is the memory of the filter. The transfer function corresponding to (14.52) is

$$H_2^z(z) = \frac{1}{1 - \alpha z^{-L}}. \tag{14.53}$$

This filter requires one multiplication and one addition per sample.

3. The output of $H_2^z(z)$, denoted by $s_2[n]$, has its local peaks spaced L samples apart most of the time. This signal is passed to a peak detector, which is responsible for finding these peaks. The peak detector operates as follows. Initially it finds the largest value among the last L consecutive points of $s_2[n]$ and marks the time of this peak. It then idles for $L - M - 1$ samples, where M is a small number, typically 1 or 2. Then it examines $2M + 1$ consecutive values of $s_2[n]$. In most cases, it finds the maximum at the $(M + 1)$st (i.e., the middle) point, which is L samples after the preceding peak. The same cycle then repeats itself continuously. Occasionally, the maximum will move a sample or two forward or backward. If this happens due to noise, it will usually correct itself later. If the true delay has changed by a physical motion, the peak detector will start yielding the new transition instants. The output of the peak detector—the sequence of estimated transition instants— is passed to the matched filter, to be discussed next.

Figure 14.16 shows typical waveforms of the signals in the timing recovery circuit. Part a shows the signal $s_1[n]$ when there is no noise, whereas part b shows this signal when there is noise at SNR of 6 dB. Part c shows the signal $s_2[n]$ when there is no noise, whereas part d shows this signal when there is noise at SNR of 6 dB. The signal $s_2[n]$ is shifted to the left by $0.5L$ samples, so its peaks indicate the true transition instants. We shall discuss parts e and f in Section 14.4.9.

14.4.9 Matched Filtering

The signal $s[n]$ has the property that its waveform is the same for all four symbol values; only its amplitude depends on the symbol. Therefore, unlike the four matched filters discussed in Section 14.4.6, we only need one matched filter for $\hat{s}[n]$. The symbol is then detected based on the amplitude of the matched filter output.

Since the level of $s[n]$ is constant during each symbol, the matched filter is a rectangular window of length L, that is,

$$G^z(z) = 1 + z^{-1} + \cdots + z^{-(L-1)}. \tag{14.54}$$

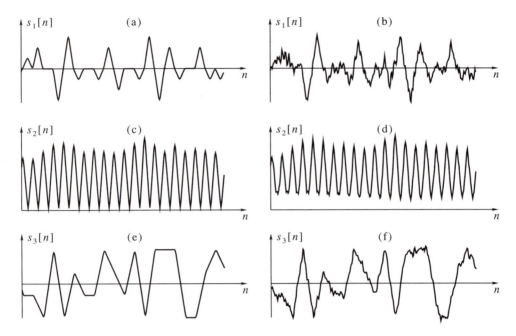

Figure 14.16 Waveforms in the timing recovery and matched filter circuits: (a) the signal $s_1[n]$ at infinite SNR; (b) the signal $s_1[n]$ at SNR = 6 dB; (c) the signal $s_2[n]$ at infinite SNR; (d) the signal $s_2[n]$ at SNR = 6 dB; (e) the signal $s_3[n]$ at infinite SNR; (f) the signal $s_3[n]$ at SNR = 6 dB.

Recalling the definition (14.46) of $H_1^z(z)$, we see that

$$G^z(z) = H_1^z(z)[1 + z^{-0.5L}]. \tag{14.55}$$

The response of $G^z(z)$ to $\hat{s}[n]$, to be denoted by $s_3[n]$, is generated at the output of the bottom adder shown in Figure 14.14.

The proper time to sample the output of the matched filter is at the end of every symbol. This happens $0.5L$ samples after every peak of $s_2[n]$. The purpose of the variable delay block is to perform this delay, relative to the current estimate of the transition instant. The delayed signal $s_3[n]$ is then sampled once every L time points, and the resulting signal $\hat{u}[m]$ is an estimate of the mth symbol (up to a constant scale factor L). Figure 14.16(e) shows the waveform of $s_3[n]$ when there is no noise, whereas Figure 14.16(f) shows this signal when there is noise at SNR of 6 dB.

14.4.10 Carrier Recovery and Symbol Detection

The estimated symbol $\hat{u}[m]$ is, except for noise, equal to $u[m] + 2\Delta f/f_0$, as seen from (14.45). Therefore, the frequency error Δf appears as a DC term. If we remove this DC term, we will be able to detect the symbol $u[m]$ as the number closest to $\hat{u}[m]$ in the set $\{-3, -1, 1, 3\}$. The carrier recovery and symbol detection algorithm we now describe has two modes of operation: an *initialization* mode and a *decision-directed* mode.

1. **Initialization mode:** We collect a set of M consecutive samples of $\hat{u}[m]$, where M is fairly large (say 100 or more). If the transmitted symbols are sufficiently random, the average of the true values of $u[m]$ must be nearly zero. We therefore

estimate Δf by

$$\widehat{\Delta f} = \frac{f_0}{2M} \sum_{m=1}^{M} \hat{u}[m]. \tag{14.56}$$

We now form the corrected sequence

$$u_1[m] = \hat{u}[m] - \frac{2\widehat{\Delta f}}{f_0} \tag{14.57}$$

and use it for detecting the symbols according to the quantization rule

$$u_2[m] = 2\,\text{round}\{0.5u_1[m] + 1.5\} - 3 \tag{14.58a}$$

$$u_3[m] = \max\{\min\{u_2[m], 3\}, -3\}. \tag{14.58b}$$

We see that $u_3[m]$ is the number closest to $u_1[m]$ in the set $\{-3, -1, 1, 3\}$, so it serves as the estimated value of $u[m]$. We now refine the estimated frequency error by replacing (14.56) with

$$\widehat{\Delta f} = \frac{f_0}{2M} \sum_{m=1}^{M} (\hat{u}[m] - u_3[m]). \tag{14.59}$$

2. **Decision-directed mode:** Each time a new value of $\hat{u}[m]$ is generated, we have new information about Δf. This information can be used for updating the estimated value of Δf. If we do this for every m, we get a sequence $\widehat{\Delta f}[m]$ of estimates. To understand the principle of operation of the decision-directed mode, we need to carry the following mathematical derivation. Let

$$\hat{u}[m] = u[m] + \frac{2\Delta f}{f_0} + v[m], \tag{14.60}$$

where $v[m]$ is the additive noise. If we generate $u_1[m]$ using (14.57b) with $\widehat{\Delta f}[m-1]$, the most recently available estimate of Δf, we get

$$u_1[m] = \hat{u}[m] - \frac{2\widehat{\Delta f}[m-1]}{f_0} = u[m] + \frac{2(\Delta f - \widehat{\Delta f}[m-1])}{f_0} + v[m]. \tag{14.61}$$

If we now generate $u_3[m]$ using (14.58), we can assume that $u_3[m] = u[m]$ most of the time, so we can subtract $u_3[m]$ from $u_1[m]$ to obtain

$$u_1[m] - u_3[m] = \frac{2(\Delta f - \widehat{\Delta f}[m-1])}{f_0} + v[m]. \tag{14.62}$$

Therefore, $u_1[m] - u_3[m]$ is proportional to the error in the frequency estimate, up to the additive noise. We can use this error signal for updating $\widehat{\Delta f}[m]$ as

$$\frac{2\widehat{\Delta f}[m]}{f_0} = \frac{2\widehat{\Delta f}[m-1]}{f_0} + \eta(u_1[m] - u_3[m]), \tag{14.63}$$

where η is a positive small number. Substitution of (14.62) in (14.63) gives

$$\frac{2\widehat{\Delta f}[m]}{f_0} = (1 - \eta)\frac{2\widehat{\Delta f}[m-1]}{f_0} + \eta\frac{2\Delta f}{f_0} + \eta v[n]. \tag{14.64}$$

As we see, the transfer functions from $2\Delta f/f_0$ and $v[n]$ to $2\widehat{\Delta f}[m]/f_0$ are both $\eta/[1-(1-\eta)z^{-1}]$. The DC gain of this transfer function is 1, so $\widehat{\Delta f}[m]$ approaches Δf in steady state. The noise gain is (see Problem 7.37)

$$\text{NG} = \frac{\eta^2}{1 - (1 - \eta)^2} = \frac{\eta}{2 - \eta} \approx 0.5\eta. \tag{14.65}$$

Therefore, the noise is considerably attenuated if η is small.

Figure 14.17 shows the decision-directed system for carrier recovery. The state variable $\widehat{\Delta f}[m]$ is initialized to the value obtained by (14.59) during the initialization mode. The quantizer is as given in (14.58). Note that the system operates at the symbol rate, which is L times lower than the sampling rate of the receiver.

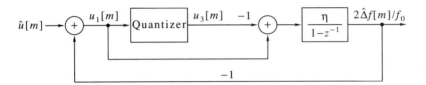

Figure 14.17 Decision-directed carrier recovery.

14.4.11 Improved Carrier Recovery and Symbol Detection

The decision-directed carrier recovery circuit shown in Figure 14.17 provides both the carrier offset and the symbols, so it appears as though the receiver is complete. However, the reliability of symbol detection of this circuit is dubious when the SNR is low. The main reason for this is the FM discriminator, which is, as we have seen, sensitive to both noise and discontinuities of the instantaneous frequency. If, however, the timing and carrier offset estimates provided by the circuits described previously are good enough, symbol detection can be greatly improved. Moreover, the carrier offset estimate can be improved as well. This is done by invoking the matched filters $g_k[n]$ introduced in Section 14.4.6. We make the following assumptions:

1. The timing recovery circuit described in Section 14.4.8 is accurate to within half the sampling interval.

2. The error $\delta f \triangleq \Delta f - \widehat{\Delta f}$, with $\widehat{\Delta f}$ obtained by the circuit described in Section 14.4.10, is a small fraction of the symbol rate f_0.

Let us demodulate the signal $z[n]$ given in (14.33) as follows:

$$z_1[n] \triangleq z[n] \exp\left\{-j2\pi \frac{\widehat{\Delta f} n}{Lf_0}\right\} = \exp\left\{j\left[2\pi \frac{\delta f n}{Lf_0} + \phi\left(\frac{n}{Lf_0}\right) + \phi_0\right]\right\}. \quad (14.66)$$

Note that t_0 does not appear in (14.66), since we assume that timing offset has been compensated. During the mth symbol, $z_1[n]$ is given by [cf. (14.35)]

$$z_1[n] = \exp\left\{j\left[2\pi \frac{0.5u[m](n - Lm)}{L} + 2\pi \frac{\delta f n}{Lf_0} + \phi_0\right]\right\}, \quad Lm \le n \le Lm + L - 1. \quad (14.67)$$

Let us convolve $z_1[n]$ with each of the four matched filters $g_k[n]$, defined in (14.36), and sample the outputs at time $n = Lm + L - 1$. We then get, similarly to (14.37),

$$\{z_1 * g_k\}[Lm + L - 1]$$
$$= \sum_{l=0}^{L-1} \exp\left\{j2\pi\left[\frac{0.5(k - u[m])(L - 1 - l)}{L} + \frac{\delta f(Lm + L - 1 - l)}{Lf_0}\right] + j\phi_0\right\}$$
$$= \exp\left\{j2\pi\left[\frac{\delta f m}{f_0} + \frac{0.25(k - u[m])(L - 1)}{L} + \frac{0.5\delta f(L - 1)}{Lf_0}\right] + j\phi_0\right\}$$
$$\cdot D\left(2\pi\left[\frac{0.5(k - u[m])}{L} + \frac{\delta f}{Lf_0}\right], L\right). \quad (14.68)$$

For $k = u[m]$, the magnitude of the right side of (14.68) is $D(\delta f / L f_0, L)$, which is nearly L if $|\delta f / f_0| \ll 1$. For $k \neq u[m]$, the argument of the Dirichlet kernel is in a side lobe, so the magnitude will be considerably less than L. Usually, therefore, unless the SNR is low, the magnitude of the matched filter corresponding to $k = u[m]$ will be the largest of the four. The conclusion is that for small timing and carrier errors, the matched filters $g_k[n]$ can be used for symbol detection, with only slight performance degradation with respect to the case of zero timing and carrier errors.

Considering (14.68) further, we see that the output of the matched filter of largest magnitude contains information about δf, which can be used for estimating this parameter. We have

$$z_2[m] = \{z_1 * g_{u[m]}\}[Lm + L - 1]$$

$$= D\left(\frac{2\pi \delta f}{L f_0}, L\right) \exp\left\{j 2\pi \left[\frac{\delta f m}{f_0} + \frac{0.5 \delta f (L - 1)}{L f_0}\right] + j\phi_0\right\}. \qquad (14.69)$$

The sequence $z_2[m]$ is obtained by taking, at each m, the complex output of the matched filter whose magnitude is the largest of the four. As we see, this sequence is a complex exponential in the discrete-time variable m, with frequency $\theta_0 = 2\pi \delta f / f_0$. Therefore, we can estimate δf from the DFT of N consecutive values of this sequence, as we learned in Section 6.5. The number N is not necessarily large; $N = 32$ or 64 is often sufficient. The estimate of δf is added to $\widehat{\Delta f}$ used for forming the signal $z_1[n]$. It is desirable, in most applications, to repeat the estimation of δf periodically, since the carrier offset may vary due to component aging and environmental conditions such as temperature and vibrations.

14.4.12 Summary

In this section we described a digital receiver for four-level FSK signals. The receiver consists of front end, FM discriminator, timing recovery circuit, carrier recovery circuit, and symbol detection circuit. The front end includes a Hilbert transformer and a complex demodulator. The FM discriminator extracts the real modulating signal from the complex frequency-modulated signal. The timing recovery circuit determines the symbol transition instants. The carrier recovery circuit estimates the carrier offset and compensates for it. Finally, the symbol detection circuit decides which symbol was transmitted at each interval of T_0 seconds. Symbol detection can be performed using the real signal, but it is better to perform matched filtering on the complex frequency-modulated signal for this purpose. The outputs of the matched filters can also be used for improving carrier offset estimation.

The main computational load of the system is in the Hilbert transformer. As we have seen, the Hilbert transformer requires about 38 real operations per sample (the order of the filter). The four matched filters together require 4 complex operations per sample, since the length of each filter is L and their outputs are decimated by L. The FM discriminator requires only few operations, one being an arcsine operation. The other parts require only few computations, thanks to the simplicity of the filters they use.

Similar techniques to those described here can be used for other types of digital communication signals; see Frerking [1994] for a detailed exposition of digital techniques in communication systems.

14.5 Electrocardiogram Analysis

A typical ECG signal, the electrical signal measured from the heart, is shown in Figure 14.18. An ECG signal looks roughly like an impulse train whose frequency is the heart rate.

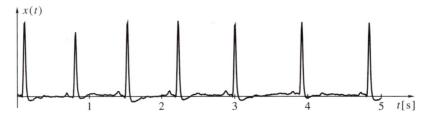

Figure 14.18 Typical electrocardiogram.

Close examination of a single ECG pulse reveals a characteristic pattern, as shown in Figure 14.19. The highest positive wave is called R. Shortly before and after the R wave there are negative waves, called Q and S, respectively. Before the Q wave there is a positive wave called the P, and after the S wave there is another positive wave called T. Both P and T are typically much lower than R. The Q, R, and S waves together are called the QRS complex.

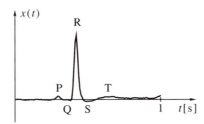

Figure 14.19 A single ECG wave.

The properties of the QRS complex—its rate of occurrence and the times, heights, and widths of its components—provide a wealth of information to the cardiologist on various pathological conditions of the heart. ECG instruments have been used by cardiologists for many years. In common instruments, the ECG signal is plotted on a chart recorder, and its evaluation is done manually. In modern instruments, processing of ECG signals is done digitally. Typical sampling frequencies of ECG signals are from 100 to 250 Hz.

In this section we discuss a particular application of ECG: measurement of the heart rate. The heart rate (the pulse) is not constant, even for a healthy person in a relaxed condition. One type of heart rate variation is related to control of the respiratory system and has a period of about 4–5 seconds, the normal breathing interval. Other variations are related to the control effects of the autonomic nervous system; these have periods of about 10–50 seconds. Various heart rate irregularities are developed by cardiac pathologies.

The ECG signal we use for our analysis is available in the file ecg.bin (see page vi for information on how to download this file). This signal has been sampled at a frequency of 128 Hz and contains 2 minutes of ECG of a healthy person in a relaxed condition.

Observation of the ECG signal reveals that it is slightly noisy, and the noise frequency is nearly half the sampling rate. This noise is most likely due to power mains interference, which is 50 Hz in this case. For detailed analysis of the QRS complex, this noise should be filtered. However, for heart rate determination, which relies on the R waves alone, the noise can be ignored to first approximation.

To determine the heart rate, we need to find the intervals between successive R points. We use the procedure locmax (Program 6.4) for this. To avoid finding the noise peaks and the P and T peaks, we limit the signal from below to a large fraction of its maximum value. We then sort the indices of the local peaks in increasing order and find their differences. The following MATLAB fragment implements this procedure:

```
[y,ind] = locmax(max(ecg,0.4*max(ecg)));
dt = (1/128)*diff(sort(ind));
```

The division by 128 converts the heart intervals to seconds.

Figure 14.20(a) shows the sequence of heart intervals $T[n]$. The mean interval is 0.887 second, corresponding to a mean rate of 1.13 beats per second. The instantaneous interval varies in the range 0.7–1 second, so the instantaneous rate varies in the range 1–1.4 beats per second.

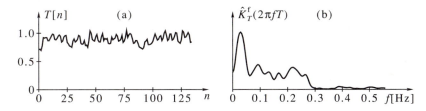

Figure 14.20 The sequence of heart intervals (a) and its estimated spectrum (b).

Next we wish to find periodicities in the heart rate. Since the sequence of heart intervals looks rather random, finding periodicities by observing the magnitude DFT does not appear promising. The Welch periodogram, presented in Section 13.1, is not adequate for this purpose either; the data length ($N = 135$) is too short for effective segmentation. We therefore use the smoothed periodogram, presented in Section 13.2, for analyzing the spectrum of the heart interval sequence $T[n]$. Before we show the results, we need to discuss the physical interpretation of the frequency variable in this application. The sequence $T[n]$, as shown in Figure 14.20(a), is in discrete time. However, we can interpret the sampling interval of this sequence as the average heart interval, which is 0.887 second in this case. Therefore, the physical frequency f is related to the variable θ of the periodogram by $f = \theta/(2\pi \times 0.887)$.

Figure 14.20(b) shows the smoothed periodogram of the sequence $T[n]$. This periodogram was computed with Hamming window of length 101. We can see three peaks, indicating the existence of three periodic components. The two lower ones have frequencies 0.03 and 0.093 Hz. These are associated with the control effects of the autonomic nervous system. The higher frequency is 0.22 Hz, and is associated with the control of the respiratory system.

14.6 Microprocessors for DSP Applications

Texas Instruments, in 1982, introduced the first microprocessor specifically designed for DSP applications—the TMS32010. Since then, several manufacturers have developed DSP microprocessors (also called DSP chips) of their own. The leading DSP chip manufacturers, at the time this book is written, are (in alphabetical order):

1. **Analog Devices:** the ADSP-21xx family of 16-bit, fixed-point chips and the ADSP-21xxx family of 32-bit, floating-point chips (each x refers to a decimal digit in the designation of a member of the family).

2. **AT&T:** the ADSP16xx family of 16-bit, fixed-point chips and the ADSP32xx family of 32-bit, floating-point chips.

3. **Motorola:** the DSP56xxx family of 24-bit, fixed-point chips and the DSP96xxx family of 32-bit, floating-point chips.

4. **NEC:** the μPD77xxx family of 16-bit and 24-bit, fixed-point chips.

5. **Texas Instruments:** the TMS320Cxx families of 16-bit, fixed-point and 32-bit, floating-point chips.

Besides those, there are numerous smaller manufacturers of both general-purpose and special-purpose DSP chips. We shall not attempt to do justice to all here.

In this section, we first discuss general concepts related to DSP chips. We then describe a single fixed-point chip of mid-1990s vintage: the Motorola DSP56301. We have chosen this particular chip arbitrarily. By the time you read this book, the chip will most probably be obsolete, thanks to the rapid progress of chip technology. However, we believe that the basic concepts and principles will last longer.

14.6.1   General Concepts

To appreciate the benefits offered by DSP microprocessors, we consider, as an example, the FIR filtering operation

$$y[n] = \sum_{k=0}^{N} h[k]x[n-k]. \tag{14.70}$$

Let us dissect this computation, as performed by a simple single-instruction, single-data (SISD) processor. Assume we have stored the $N+1$ coefficients in a storage vector $\boldsymbol{h}$, and the $N+1$ most recent values of the input signal in a storage vector $\boldsymbol{x}$. Just prior to the execution of the computations corresponding to time n, the vector $\boldsymbol{x}$ holds $\{x[n], x[n-1], \ldots, x[n-N]\}$. At time n, we need to perform the following operations:

1. Set a temporary variable y to 0. When the operation is complete, y will hold the computed value of $y[n]$.

2. Set a loop counter k to 0.

3. Repeat the following operations until $k = N$:

   (a) Load $h[k]$ from the kth location of $\boldsymbol{h}$ to the CPU.

   (b) Load $x[n-k]$ from the kth location of $\boldsymbol{x}$ to the CPU.

   (c) Multiply $h[k]$ by $x[n-k]$ and optionally round the result to single precision.

   (d) Load y, add to it the product, and store back in y.

   (e) Increase the loop counter k by 1, check if $k > N$ and exit if true.

4. Output y to its destination (e.g., to a D/A converter).

5. Shift the elements of $\boldsymbol{x}$ one place to the right, deleting $x[n-N]$ and making room for $x[n+1]$ in the first position.

6. Input $x[n+1]$ from its source (e.g., from an A/D converter) and store in the first position of $\boldsymbol{x}$.

In addition, each instruction needs to be read from the program memory into the control unit of the CPU.

As we see, a single FIR update operation can take many CPU cycles if implemented on a SISD computer. Let us now explore a few possibilities for expediting this procedure, at the expense of additional hardware.

1. Suppose we have two memory areas that can be accessed simultaneously. Then we can keep h in one area, x in the second, and load $h[k]$ and $x[n - k]$ simultaneously.

2. We can keep the temporary variable y in a CPU register, thus eliminating its loading from memory and storing back at each count of k. Such a register is called an *accumulator*. Furthermore, we can let the accumulator have a double length compared with that of $h[k]$ and $x[n-k]$. Then the product need not be rounded, but can be added directly to the current value of y. The combination of multiplier, double-length accumulator, and double-length adder is called multiplier-accumulator, or MAC for short.

3. We can use *hardware loop control*, which causes the sequence of operations in part 2 to repeat itself automatically $N + 1$ times, without explicit program control.

4. We can use a *circular buffer* for the vector x. The principle of operation of a circular buffer is illustrated in Figure 14.21. A pointer indicates the location of the most recent data point $x[n]$. Older data points are stored clockwise from the pointer. After $y[n]$ is computed, $x[n - N]$ is replaced with $x[n + 1]$ and the pointer moves counterclockwise by one position to point at $x[n + 1]$. As we see, all storage variables but one do not change their positions in the buffer, and only one variable is replaced.

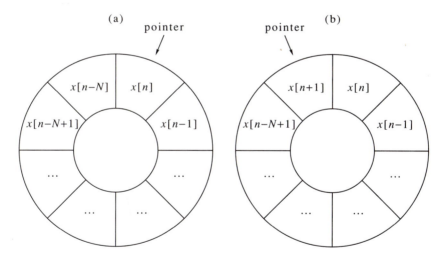

Figure 14.21 A circular buffer: (a) before the nth time point; (b) after the nth time point.

In practice, a circular buffer is implemented by using modular arithmetic for the memory address; that is, the pointer advances by 1 modulo $N + 1$ each time k is increased. At the end of the loop, we decrease the pointer by 1 modulo $N + 1$. It is convenient to store the filter coefficients $h[k]$ in a circular buffer as well. However, the pointer of this buffer is not decreased by 1 at the end of the loop.

A processor that takes advantage of all these possibilities can, in the limit, perform all operations needed at each count of k in a single machine cycle. To achieve this limit, parallel processing and pipelining are necessary. The former refers to performing independent operations simultaneously (such as loading the two multiplicands from memory). The latter refers to performing operations belonging to different cycles at the same time. For example, addition of a product $h[k]x[n-k]$ to the accumulator can be performed while the multiplier already computes $h[k+1]x[n-k-1]$.

14.6.2 The Motorola DSP56301

The Motorola DSP56301 is a member of the DSP56300 family of 24-bit microprocessors [Motorola, 1995]. Its CPU is built around a multiplier, capable of multiplying 24-bit numbers and producing a 48-bit result, and two 56-bit accumulators. The CPU gets its operands from two independent memory areas, denoted by X and Y. A 48-bit X register holds data transferred to and from the X memory, and a 48-bit Y register holds data transferred to and from the Y memory. The inputs to the multiplier can come only from the X and Y registers. The output of the multiplier can be added to (or subtracted from) either of the two accumulators, denoted by A and B. The outputs of A and B can be moved to either X or Y memory areas.

Figure 14.22 shows the structure of the DSP56301 registers. Each of X and Y consists of two parts: X1, X0, Y1, and Y0. These parts can be used as independent 24-bit registers, or combined to form 48-bit registers. The DSP56301 arithmetic is *fractional*: The numbers represented by a 24-bit word are in the range -1 to $1 - 2^{-23}$, and the numbers represented by a 48-bit word are in the range -1 to $1 - 2^{-47}$. Usually, X0 and X1 are used for 24-bit numbers each, as are Y0 and Y1. The main reason for combining them is evident when we wish to multiply two double-precision numbers, or a double-precision number by a single-precision number. Such multiplications cannot be performed by the multiplier in a single step, but in a sequence of steps.

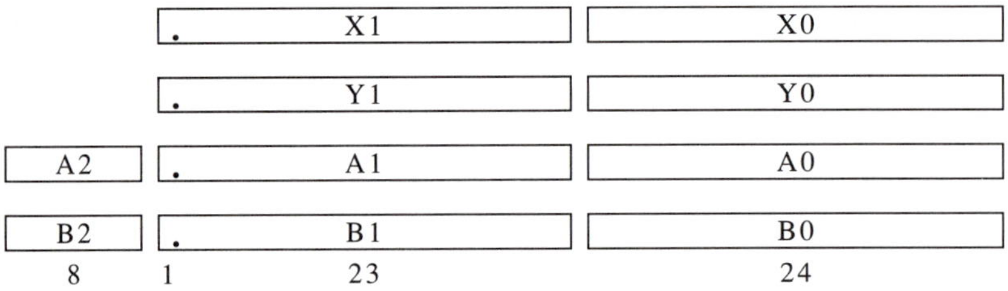

Figure 14.22 The registers of the Motorola DSP56301 microprocessor.

The A and B accumulators contain 56 bits each. The 8-bit parts A2 and B2 are called extension registers. They extend the range of the two accumulators to about ± 256. In other words, these two accumulators have both an integer and a fractional part, as shown in Figure 14.22. It follows that when the multiplicands are in single precision, no quantization noise is generated, since the product is accumulated in double precision. Also, overflow up to ± 256 does not lead to loss of information, since the overflow bits are properly stored in the extension register. This greatly alleviates the scaling problem in both FIR and IIR filters.

The MAC performs a multiply–add operation in two machine cycles. However, since it is pipelined, a new multiply–add can be initiated at every cycle, thus yielding an effective computation rate of one multiply–add per cycle.

The DSP56301 microprocessor contains several other features that enhance digital signal processing applications:

1. The accumulators can be switched to a saturation mode, as explained in Section 11.6.4. In saturation mode, the extension accumulators A2, B2 are not used, and A, B are limited to fractional values.

2. Regardless of whether the accumulators are in saturation mode, data are moved around in a *saturation transfer* mode. Thus, when the number in an accumulator is larger than 1 in magnitude, the number passed back to the X and Y registers or to memory is saturated with the proper sign.

3. The result of a MAC operation can optionally be multiplied by 2 or divided by 2. This is useful for implementing block floating-point FFT, as explained in Section 5.3.4. It is also useful for implementing second-order sections of IIR filters since, as we saw in Section 11.6, the denominator coefficients g_i of the sections are usually scaled by 0.5.

4. There is hardware loop control, enabling automatic repetition of either a single instruction or a block of instructions a desired number of times (with no latency).

5. The DSP56301 has two rounding modes: two's-complement rounding and convergent rounding. The two differ only in the way the number 0.5 is rounded.[2]

6. There are special instructions to facilitate double-precision multiplication, as well as division (which, however, require more than one machine cycle).

7. The availability of two accumulators is convenient for implementing complex arithmetic, for example, in FFT.

8. The CPU can be switched to a 16-bit mode, in which single-precision numbers have 16 bits and double-precision numbers have 32 bits. The extension accumulators A2, B2 continue to have 8 bits and fulfill the same tasks as in 24-bit mode.

9. There are eight 24-bit address registers, each having three fields. The field Rk (where $0 \leq k \leq 7$) holds the memory address, the field Nk holds the offset with respect to that address, and the field Mk contains information related to the mode of offset calculation. There are three modes of offset calculation: linear, modular, and reverse carry. The first simply adds the offset to the memory address; it is useful for accessing arrays of data. The second adds the offset to the memory address, modulo a given positive number; it is useful for implementing a circular buffer, as explained previously. The third implements bit-reversed offset; it is useful for loading or storing data in radix-2 FFT, as we recall from Section 5.3.

10. The DSP56301 has five on-chip memories: 2K X random-access memory (RAM), 2K Y RAM, 3K program RAM, 1K instruction cache, and 192 words bootstrap read-only memory (ROM) (K is 1024 words; in this case, each word is 24 bits long).

11. The DSP56301 has host interface to industry standard buses, enabling connections to other computers, as well as synchronous and serial interfaces to various peripherals.

We now illustrate how the FIR convolution operation (14.70) is implemented on the DSP56301. The following assembler code fragment performs this calculation for a single time point n.

```
movep   y:input,y:(r4)                                      ; line 1
clr     a               x:(r0)+,x0   y:(r4)+,y0             ; line 2
rep     #N                                                  ; line 3
mac     x0,y0,a         x:(r0)+,x0   y:(r4)+,y0             ; line 4
macr    x0,y0,a                      (r4)-                   ; line 5
movep   a,y:output                                          ; line 6
```

Here is an explanation of this code fragment:

1. The order of the filter N is given by the constant value #N.

2. The memory area X is assumed to hold the coefficients $h[k]$ in an increasing order, in a circular buffer of length $N + 1$. The address register R0 holds the address of $h[0]$. The modifier field M0 holds the number N. This causes R0 to be incremented modulo $N + 1$ when needed (the number in the modifier field is 1 less than the modulus).

3. The memory area Y is assumed to hold the signal samples $h[n-k]$ in a decreasing order, in a circular buffer of length $N + 1$. The address register R4 holds the address of the most recent data point. The modifier field M4 holds the number N. This causes R4 to be incremented modulo $N + 1$ when needed.

4. In line 1, the sample $x[n]$ is loaded from an input port mapped to the memory address y:input (e.g., from an A/D converter) and stored in the Y memory area, in the address specified by the contents of R4.

5. In line 2, the accumulator A is cleared. At the same time, the registers X0 and Y0 are loaded with $h[0]$ and $x[n]$, respectively. When loading is complete, R0 and R4 are incremented. Therefore, R0 now contains the address of $h[1]$ and R1 contains the address of $x[n-1]$.

6. Line 3 instructs the CPU to perform the next instruction (in line 4) N times.

7. In line 4, the product $h[k]x[n-k]$ is calculated for all $0 \le k \le N-1$ and added to the contents of A. Each time, the next coefficient and data sample are loaded to X0 and Y0, and the address registers R0, R4 are incremented.

8. In line 5, the product $h[N]x[n-N]$ is calculated, added to the contents of A, and the result is rounded. Now A1 contains the number $y[n]$. We note that both R0 and R4 have been incremented a total of $N + 1$ times. Therefore, they now point again at $h[0]$ and $x[n]$, respectively. By decrementing R4, we cause it to point at $x[n-N]$. This is the address to be overwritten by $x[n+1]$ at the next time point.

9. In line 6, the accumulator contents, $y[n]$, is sent to an output port (e.g., a D/A converter) mapped to the memory address y:output.

14.7 Sigma–Delta A/D Converters

In Example 12.4 we demonstrated the possibility of trading speed and accuracy in A/D converters. However, the technique presented there can gain only half a bit accuracy for each doubling of the sampling rate. In this section we describe a state-of-the-art technique for A/D converter implementation that further exploits the speed–accuracy trade-off. This technique, called *sigma–delta A/D*, provides a fine example of the advantages gained by combining VLSI technology and digital signal processing principles. As we shall see, sigma–delta A/D converters require internal A/D and D/A converters of only few bits; all other bits are gained by increasing the sampling rate. In the extreme case, only 1-bit internal A/D and D/A are required. One-bit converters are extremely simple to implement: 1-bit A/D is just a sign detector, and 1-bit D/A is just a short circuit (or a constant gain).

Looking back at Example 12.4, we realize that only half a bit is gained for each doubling of the sampling rate because of the whiteness of the quantization noise. The key to improving the system is therefore to distribute the noise energy nonuniformly over the frequency band, to push most of the energy to higher frequencies. Noise at higher frequencies will be eliminated by the decimation filter, thus reducing the quantization noise level at the output of the filter. Figure 14.23(a) shows a block diagram of such a system. This is called a first-order sigma–delta A/D converter.

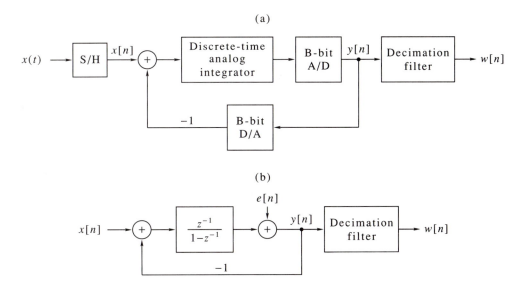

Figure 14.23 First-order sigma–delta A/D converter: (a) physical block diagram; (b) equivalent block diagram.

A first-order sigma–delta A/D converter operates as follows. The analog input signal is first sampled at a high rate f_{sam}. The output of the S/H is an analog (i.e., continuous-amplitude) discrete-time signal $x[n]$. An analog discrete-time error signal is generated by subtracting from $x[n]$ a reconstructed version of the output signal. The error signal is fed to a discrete-time analog integrator. Such an integrator is commonly implemented using switched-capacitor VLSI technology.[3] The actual A/D converter, assumed to have B bits, is placed at the output of the integrator. This is where the quantization noise $e[n]$ is generated. The A/D output $y[n]$ is a digital signal at a rate f_{sam}. This signal is forwarded to the decimation filter. It is also fed back to the input via a D/A converter that also has B bits. Since the number of bits is the same in the A/D and the D/A, together they form a unity system (neglecting secondary effects). Thus, the output of the integrator is fed back to its input at an inverted polarity. The name *sigma–delta* is because of the integration and subtraction performed by the circuit.

To analyze the operation of the first-order sigma–delta converter, we represent it by an equivalent block diagram, as shown in Figure 14.23(b). The transfer function of the integrator is $z^{-1}/(1 - z^{-1})$. We therefore get, by straightforward computation,

$$Y^z(z) = z^{-1}X^z(z) + (1 - z^{-1})E^z(z). \tag{14.71}$$

As we see, the transfer function from $x[n]$ to $y[n]$ is pure delay, so the input signal is not distorted. The noise $e[n]$, on the other hand, undergoes differencing, which is approximately differentiation. Therefore, the noise at $y[n]$ has little energy at low frequencies, as desired.

Let us assume that the signal bandwidth is f_m and that the decimation filter is an ideal low pass with cutoff frequency $\theta_0 = 2\pi f_m / f_{sam}$. Then the variance of the noise at the output of the decimation filter is given by

$$y_w = \frac{2^{-2(B-1)}}{12} \cdot \frac{1}{2\pi} \int_{-\theta_0}^{\theta_0} |1 - e^{-j\theta}|^2 d\theta = \frac{2^{-2(B-1)}}{12} \cdot \frac{1}{\pi} \int_{-\theta_0}^{\theta_0} (1 - \cos\theta) d\theta$$

$$= \frac{2^{-2(B-1)}}{12} \cdot \frac{2}{\pi} (\theta_0 - \sin\theta_0). \tag{14.72}$$

If $\theta_0 \ll \pi$ (as it will usually be in practical systems) we can use the approximation

$$\sin\theta_0 \approx \theta_0 - \frac{\theta_0^3}{6};$$

then we will get

$$y_w \approx \frac{2^{-2(B-1)}}{12} \cdot \frac{2}{\pi} \cdot \frac{\theta_0^3}{6} = \frac{2^{-2(B-1)}}{12} \cdot \frac{\pi^2}{3} \cdot \left(\frac{2f_m}{f_{sam}}\right)^3. \tag{14.73}$$

Because the cubic dependence of y_w on $2f_m/f_{sam}$, the noise variance decreases by 9 dB with each doubling of the sampling rate. This is equivalent to an additional 1.5 bit for each doubling of the sampling rate. However, the factor $\pi^2/3$ detracts about 1 bit (explain why!). Also, since the decimation filter is not ideal, the noise in the stop band is not completely eliminated, so in practice the equivalent number of bits is smaller.

As an example, consider the sigma–delta A/D converter for speech applications, described in Leung et al. [1988]. This A/D converter uses 1-bit A/D at the output of the integrator, and a sampling frequency of 4 MHz. The signal bandwidth is 4 kHz. Therefore $f_{sam}/2f_m = 500$, so the equivalent number of bits is

$$1 + 1.5\log_2 500 - 1 \approx 13.$$

Being able to produce 13 bits from a 1-bit A/D may sound incredible, but it is true.

The sigma–delta idea can be exploited yet further. The block diagram shown in Figure 14.24 represents a second-order sigma–delta A/D converter. Each block $z^{-1}/(1 - z^{-1})$ is a switched-capacitor integrator. The output $y[n]$ is related to the input $x[n]$ and the noise $e[n]$ through

$$Y^z(z) = z^{-1}X^z(z) + (1 - z^{-1})^2 E^z(z). \tag{14.74}$$

The variance of the noise at the output of the decimation filter is

$$y_w \approx \frac{2^{-2(B-1)}}{12} \cdot \frac{\pi^4}{5} \cdot \left(\frac{2f_m}{f_{sam}}\right)^5. \tag{14.75}$$

Therefore, a second-order sigma–delta A/D converter gives 2.5 bits for each doubling of the sampling rate. From this we must subtract about 4.5 bits because of the factor $\pi^4/5$.

As an example, consider the sigma–delta A/D converter for high-fidelity audio applications, described in Sarhang-Nejad and Temes [1993]. This A/D converter uses 4-bit A/D and a sampling frequency of 5.25 MHz. The signal bandwidth is 20.5 kHz. Therefore $f_{sam}/2f_m = 128$, so the equivalent number of bits is

$$4 + 2.5\log_2 128 - 4.5 \approx 17.$$

In practice, this A/D converter gives only 16 bits.

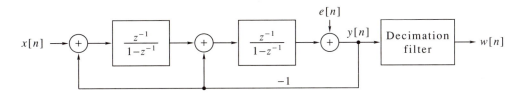

Figure 14.24 Second-order sigma–delta A/D converter.

14.8 Summary and Complements

14.8.1 Summary

We devoted this chapter to applications of digital signal processing and DSP technology. We presented applications from speech, music, communication, biomedicine, and signal compression. We then discussed features of current DSP microprocessors, and state-of-the-art A/D converter technology.

If you have mastered the contents of this book, you are ready for pursuing many advanced topics in digital signal processing, whether related to the aforementioned applications or not. Here is a selected list of such topics.

1. *Image processing* is a natural and highly important extension of signal processing. Images are two-dimensional signals: Instead of varying over time, they vary over the x and y coordinates of the image. Video (or motion picture) is a three-dimensional signal: It varies over the x and y coordinates of each frame, and the frames vary over time. Image processing has many aspects similar to conventional signal processing—sampling, frequency-domain analysis, z-domain analysis, filtering—and many unique aspects.

2. *Statistical signal processing* is concerned with the analysis and treatment of random signals: modeling, estimation, adaptive filtering, detection, pattern recognition.

3. *Speech processing* is concerned with speech signals and includes operations such as compression, enhancement, echo cancellation, speaker separation, recognition, speech-to-text and text-to-speech conversion.

4. *Biomedical signal processing* is concerned with signals generated by the human body, with the auditory and visual systems, with medical imaging, with artificial organs, and more.

5. *Array signal processing* is concerned with the utilization of sensor arrays for localization and reception of multiple signals. Array signal processing has long been used for military applications (for both electromagnetic and underwater acoustic signals), but has been extended to commercial applications in recent years.

6. *DSP technology* is concerned with general-purpose and application-specific architectures, parallel processing, VLSI implementations, converters, and more.

I shall let a pen worthier than mine write the final word:

> *'Tis pleasant, sure, to see one's name in print;*
> *a book's a book, although there's nothing in't.*
> Lord Byron (1788–1824)

14.8.2   Complements

1. [p. 557] If $\{y[n],\ 0 \le n \le N - 1\}$ is a WSS signal and $w[n]$ is a window, then $\{y[n]w[n],\ 0 \le n \le N - 1\}$ cannot be WSS, because

$$E(y[n + m]w[n + m]y[n]w[n]) = w[n + m]w[n]\kappa_y[m],$$

and this depends on n as well as on m.

2. [p. 585] Suppose we want to round a binary number from double precision to single precision. Let x_0 be the least significant part, x_1 the most significant part, and y the LSB of x_1. Then, if $x_0 < 0.5$, x_1 is rounded downward (i.e., it is not changed); if $x_0 > 0.5$, x_1 is rounded upward (i.e., increased by 1 LSB). The question is how to handle the case $x_0 = 0.5$. In two's-complement rounding, x_1 is always rounded upward. In convergent rounding, x_1 is rounded downward (not changed) if $y = 0$, or upward (increased by 1 LSB) if $y = 1$.

3. [p. 587] A switched-capacitor integrator is based on the property that the voltage across a capacitor is proportional to the integral of the current passing through it. The current is controlled by the input voltage such that each sampling interval, a charge proportional to the input voltage is added to the capacitor. This makes the switched capacitor a discrete-time, continuous-amplitude integrator.

Bibliography

Page numbers at which a reference is mentioned appear in brackets after the reference.

Antoniou, A., *Digital Filters: Analysis, Design, and Applications,* 2nd ed., McGraw–Hill, New York, 1993. [10, 343]

Barker, R. H., "The Pulse Transfer Function and Its Application to Sampling Servomechanisms," *Proc. IEE*, vol. 99, part IV, pp. 302–317, 1952. [230]

Blackman, R. B. and Tukey, J. W., *The Measurement of Power Spectra,* Dover Publications, New York, 1958. [173, 519]

Burrus, C. S., "Efficient Fourier Transform and Convolution Algorithms," in *Advanced Topics in Signal Processing,* J. S. Lim and A. V. Oppenheim, eds., Prentice Hall, Englewood Cliffs, NJ, 1988. [154]

Churchill, R. V. and Brown, J. W., *Introduction to Complex Variables and Applications,* 4th ed., McGraw–Hill, New York, 1984. [230]

Clements, M. A. and Pease, J. W., "On Causal Linear Phase IIR Digital Filters," *IEEE Trans. Acoust., Speech, Signal Process.,* vol. ASSP-37, pp. 479–484, April 1989. [267]

Cooley, J. W. and Tukey, J. W., "An Algorithm for the Machine Computation of Complex Fourier Series," *Math. Comput.,* 19, pp. 297–301, April 1965. [133]

Crochiere, R. E., "Sub-Band Coding," *Bell System Tech. J.,* pp. 1633–1654, September 1981. [494]

Crochiere, R. E. and Rabiner, L. R., *Multirate Digital Signal Processing,* Prentice Hall, Englewood Cliffs, NJ, 1983. [503]

Cover, T. M. and Thomas, J. A., *Elements of Information Theory,* John Wiley, New York, 1991. [551]

Daumer, W. R., "Subjective Evaluation of Several Efficient Speech Coders," *IEEE Trans. Commun.,* pp. 662–665, April 1982. [494]

Dolph, C. L., "A Current Distribution for Broadside Arrays Which Optimizes the Relationship Between Beam Width and Side-Lobe Level," *Proc. IRE,* vol. 34, 6, pp. 335–348, June 1946. [175]

Dupré, L., *BUGS in Writing,* Addison–Wesley, Reading, MA, 1995. [xii]

Durbin, J., "The Fitting of Time-Series Models," *Rev. Inst. Int. Statist.,* 28, pp. 233–243, 1960. [527]

Esteban, D. and Galand, C., "Application of Quadrature Mirror Filters to Split Band Voice Coding Schemes," *Proc. IEEE Int. Conf. Acoust., Speech, Signal Process.,* pp. 191–195, May 1977. [489]

Farkash, S. and Raz, S., "The Discrete Gabor Expansion—Existence and Uniqueness", *Signal Processing*, to appear. [502]

Fliege, N. J., *Multirate Digital Signal Processing*, John Wiley, New York, 1994. [503]

Frerking, M. E., *Digital Signal Processing in Communication Systems*, Van Nostrand Reinhold, New York, 1994. [579]

Friedlander, B., Morf, M., Kailath, T., and Ljung, L., "New Inversion Formulas for Matrices Classified in Terms of Their Distance from Toeplitz Matrices," *Linear Algebra Appl.*, 27, pp. 31-60, 1979. [536]

Gabel, R. A. and Roberts, R. A., *Signals and Linear Systems*, 3rd ed., John Wiley, New York, 1987. [33]

Gardner, W. A., *Introduction to Random Processes*, Macmillan, New York, 1986. [33]

Gitlin, R. D., Hayes, J. F., and Weinstein, S. B., *Data Communication Principles*, Plenum Press, New York, 1992. [540]

Goertzel, G., "An Algorithm for Evaluation of Finite Trigonometric Series", *Am. Math. Mon.*, vol. 65, pp. 34-45, January 1958. [387]

Good, I. J., "The Interaction Algorithm and Practical Fourier Analysis," *J. R. Statist. Soc., Ser. B*, 20, pp. 361-375, 1958; addendum, 22, pp. 372-375, 1960. [154]

GSM, *ETSI-GSM Technical Specification, GSM 06.10, Version 3.2.0*, UDC: 621.396.21, European Telecommunications Standards Institute, 1991. [561]

Haddad, R. A. and Parsons, T. W., *Digital Signal Processing; Theory, Applications, and Hardware*, Computer Science Press, New York, 1991. [10]

Harris, F. J., "On the Use of Windows for Harmonic Analysis with the Discrete Fourier Transform," *Proc. IEEE*, 66, pp. 51-83, January 1978. [174, 200]

Hayes, M. H., *Statistical Digital Signal Processing and Modeling*, John Wiley, New York, 1996. [524, 533]

Haykin, S., *Communication Systems*, 3rd ed., John Wiley, 1994. [80]

Haykin, S., *Adaptive Filter Theory*, 3rd ed., Prentice Hall, Englewood Cliffs, NJ, 1996. [541]

Haykin, S. and Van Veen, B., *Signals and Systems*, John Wiley, New York, 1997. [33]

Huffman, D. A., "A Method for the Construction of Minimum Redundancy Codes," *Proc. IRE*, 40, 1098-1101, 1952. [551]

Hurewicz, W., Chapter 5 in *Theory of Servomechanisms*, H. M. James, N. B. Nichols, and R. S. Phillips, eds., MIT Radiation Laboratory Series, Vol. 25, McGraw-Hill, New York, 1947. [230]

Jackson, L. B., *Digital Filters and Signal Processing*, 3rd ed., Kluwer, Boston, 1996. [10]

Jain, V. K. and Crochiere, R. E., "A Novel Approach to the Design of Analysis/Synthesis Filter Banks," *Proc. IEEE Int. Conf. Acoust., Speech, Signal Process.*, pp. 228-231, April 1983. [490]

Johnston, J. D., "A Filter Family Designed for Use in Quadrature Mirror Filter Banks," *Proc. IEEE Int. Conf. Acoust., Speech, Signal Process.*, pp. 291-294, April 1980. [490]

Jury, E. I., "Synthesis and Critical Study of Sampled-Data Control Systems," *AIEE Trans.*, vol. 75, pp. 141-151, 1954. [230]

Kailath, T., *Linear Systems*, Prentice Hall, Englewood Cliffs, NJ, 1980. [34, 438]

Kay, S. M., *Modern Spectral Estimation, Theory and Applications,* Prentice Hall, Englewood Cliffs, NJ, 1988. [524]

Kolmogorov, A. N., "Stationary Sequences in Hilbert Space" (in Russian), *Bull. Math. Univ. Moscow,* 2(6), pp. 1–40, 1941; ["Interpolation und Extrapolation von stationären zufälligen Folgen," *Bull. Acad. Sci. USSR Ser. Math.,* 5, pp. 3–14, 1941]. [542]

Kuc, R., *Introduction to Digital Signal Processing,* McGraw–Hill, New York, 1988. [10]

Kuo, F. F. and Kaiser, J. F., *System Analysis by Digital Computer,* Chapter 7, John Wiley, New York, 1966. [175]

Kwakernaak, H. and Sivan, R., *Modern Signals and Systems,* Prentice Hall, Englewood Cliffs, NJ, 1991. [33]

Leung, B., Neff, R., Gray, P., and Broderson, R., "Area-Efficient Multichannel Oversampled PCM Voice-Band Coder," *IEEE J. Solid State Circuits,* pp. 1351–1357, December 1988. [588]

Levinson, N., "The Wiener RMS (Root Mean Square) Error Criterion in Filter Design and Prediction," *J. Math. Phys.,* 25, pp. 261–278, 1947. [527]

Linvill, W. K., "Sampled-Data Control Systems Studied Through Comparison of Sampling and Amplitude Modulation," *AIEE Trans.,* vol. 70, part II, pp. 1778–1788, 1951. [230]

MacColl, L. A., *Fundamental Theory of Servomechanisms,* Van Nostrand, New York, 1945. [230]

Markushevich, A. I., *Theory of Functions of a Complex Variable,* Chelsea Publishing Company, New York, 1977. [80]

Marple, S. L., *Digital Spectral Analysis with Applications,* Prentice Hall, Englewood Cliffs, NJ, 1987. [524]

Mintzer, F., "Filters for Distortion-Free Two-Band Multirate Filter Banks," *IEEE Trans. Acoust., Speech, Signal Process.,* vol. ASSP-33, pp. 626–630, June 1985. [490]

Motorola, *DSP56300 24-Bit Digital Signal Processor Family Manual,* Motorola, Inc., Austin, TX, 1995. [584]

Nyquist, H., "Certain Topics in Telegraph Transmission Theory," *AIEE Trans.,* pp. 617–644, 1928. [80]

Oppenheim, A. V. and Schafer, R. W., *Digital Signal Processing,* Prentice Hall, Englewood Cliffs, NJ, 1975. [10, 57]

Oppenheim, A. V. and Schafer, R. W., *Discrete-Time Signal Processing,* Prentice Hall, Englewood Cliffs, NJ, 1989. [10, 458]

Oppenheim, A. V. and Willsky, A. S., with Young, I. T., *Signals and Systems,* Prentice Hall, Englewood Cliffs, NJ, 1983. [33]

Papoulis, A., *Probability, Random Variables, and Stochastic Processes,* 3rd ed., McGraw–Hill, New York, 1991. [33]

Parks, T. W. and Burrus, C. S., *Digital Filter Design,* John Wiley, New York, 1987. [10]

Parks, T. W. and McClellan, J. H., "A Program for the Design of Nonrecursive Digital Filters with Linear Phase," *IEEE Trans. Circuit Theory,* vol. CT-19, pp. 189–194, March 1972(a). [306]

Parks, T. W. and McClellan, J. H., "Chebyshev Approximation for the Design of Linear Phase Finite Impulse Response Digital Filters," *IEEE Trans. Audio Electroacoust.,* vol. AU-20, pp. 195–199, August 1972(b). [306]

Pennebaker, W. B. and Mitchell, J. L., *JPEG: Still Image Data Compression Standard,* Van Nostrand Reinhold, New York, 1993. [554]

Porat, B., *Digital Processing of Random Signals: Theory and Methods,* Prentice Hall, Englewood Cliffs, NJ, 1994. [524]

Porat, B., Friedlander, B., and Morf, M., "Square-Root Covariance Ladder Algorithms," *IEEE Trans. Autom. Control,* AC-27, pp. 813–829, 1982. [536]

Portnoff, M. R., "Time-Frequency Representation of Digital Signals and Systems Based on Short-Time Fourier Analysis," *IEEE Trans. Acoust., Speech, Signal Process.,* vol. ASSP-28, pp. 55–69, January 1980. [501]

Proakis, J. G. and Manolakis, D. G., *Introduction to Digital Signal Processing,* 2nd ed., Macmillan, New York, 1992. [10]

Rabiner, L. and Juang, B., *Fundamentals of Speech Recognition,* Prentice Hall, Englewood Cliffs, NJ, 1993. [555]

Ragazzini, J. R. and Zadeh, L. A., "The Analysis of Sampled-Data Systems," *AIEE Trans.,* vol. 71, part II, pp. 225–234, November 1952. [230]

Rao, K. R. and Yip, P., *Discrete Cosine Transform,* Academic Press, San Diego, CA, 1990. [123]

Remez, E. Ya., "General Computational Methods of Chebyshev Approximations," Atomic Energy Translation 4491, Kiev, USSR, 1957. [306]

Rihaczek, A. W., *Principles of High-Resolution Radar,* Peninsula Publishing, Los Altos, CA, 1985. [155]

Roberts, R. A., and Mullis, C. T., *Digital Signal Processing,* Addison–Wesley, Reading, MA, 1987. [10]

Rudin, W., *Principles of Mathematical Analysis,* McGraw–Hill, New York, 1964. [35]

Sarhang-Nejad, M. and Temes, G., "A High-Resolution Multibit Sigma-Delta ADC with Digital Correction and Relaxed Amplifier Requirements," *IEEE J. Solid State Circuits,* pp. 648–660, June 1993. [588]

Schur, I., "On Power Series Which are Bounded in the Interior of the Unit Circle," *J. Reine Angew. Math.* (in German), vol. 147, pp. 205–232, 1917; vol. 148, pp. 122–125, 1918 (translated in: I. Gohberg, ed., *I. Schur Methods in Operator Theory and Signal Processing*, Birkhäuser, Boston, MA, 1986). [532]

Schuster, A., "On the Periodicities of Sunspots," *Philos. Trans. R. Soc. Ser. A,* vol. 206, pp. 69–100, 1906a. [522]

Schuster, A., "The Periodogram and Its Optical Analogy," *Proc. R. Soc. London, Ser. A,* vol. 77, pp. 136–140, 1906b. [542]

Shannon, C. E., "Communication in the Presence of Noise," *Proc. IRE,* 37, pp. 10–21, 1949. [80]

Smith, M. J. T. and Barnwell III, T. P., "A Procedure for Designing Exact Reconstruction Filter Banks for Tree Structured Subband Coders," *Proc. IEEE Int. Conf. Acoust., Speech, Signal Process.,* pp. 27.1.1–27.1.4, San Diego, CA, March 1984. [490]

Strum, R. D. and Kirk, D. E., *First Principles of Discrete Systems and Digital Signal Processing,* Addison–Wesley, Reading, MA, 1989. [10]

Therrien, C. W., *Discrete Random Signals and Statistical Signal Processing,* Prentice Hall, Englewood Cliffs, NJ, 1992. [524]

Thomas, L. H., "Using a Computer to Solve Problems in Physics," *Applications of Digital Computers,* Ginn & Co., Boston, MA, 1963. [154]

Tsypkin, Ya. Z., "Theory of Discontinuous Control," *Autom. Telemach.,* No. 3, 1949; No. 4, 1949; No. 5, 1950. [230]

Vaidyanathan, P. P., *Multirate Systems and Filter Banks,* Prentice Hall, Englewood Cliffs, NJ, 1993. [503]

Vaidyanathan, P. P. and Nguyen, T. Q., "A Trick for the Design of FIR Half-Band Filters," *IEEE Trans. Circuits Syst.,* vol. CS-34, pp. 297–300, March 1987. [326, 491]

Van Trees, H. L., *Detection, Estimation, and Modulation Theory,* Part I, John Wiley, New York, 1968. [196]

Vetterli, M. and Kovacevic, J., *Wavelets and Subband Coding,* Prentice Hall, Englewood Cliffs, NJ, 1995. [495, 503]

Walker, G., "On Periodicity in Series of Related Terms," *Proc. R. Soc. London, Ser. A,* 131, pp. 518–532, 1931. [524]

Welch, P. D., "The Use of the Fast Fourier Transform for the Estimation of Power Spectra," *IEEE Trans. Audio Electroacoust.,* vol. AU-15, pp. 70–73, June 1970. [516]

Whittaker, E. T., "On the Functions Which Are Represented by the Expansions of the Interpolation Theory," *Proc. R. Soc. Edinburgh,* 35, pp. 181–194, 1915. [80]

Wiener, N., *Extrapolation, Interpolation, and Smoothing of Stationary Time Series with Engineering Applications,* MIT Press, Cambridge, MA, 1949. [542]

Winograd, S., "On Computing the Discrete Fourier Transform," *Math. Comput.,* 32, pp. 175–199, January 1978. [154]

Yule, G. U., "On a Method for Investigating Periodicities in Disturbed Series with Special Reference to Wolfer's Sunspot Numbers," *Philos. Trans. R. Soc. London, Ser. A,* 226, pp. 267–298, 1927. [524]

Ziv, J. and Lempel, A., "A Universal Algorithm for Sequential Data Compression," *IEEE Trans. Inform. Theory,* vol. IT-23, pp. 337–343, 1977. [551]

Ziv, J. and Lempel, A., "Compression of Individual Sequences by Variable Rate Coding," *IEEE Trans. Inform. Theory,* vol. IT-24, pp. 530–536, 1978. [551]

Index